Second Edition
PHYSICS
AND THE PHYSICAL PERSPECTIVE

HENRY O. HOOPER
University of Maine at Orono

PETER GWYNNE

HARPER & ROW, PUBLISHERS
SAN FRANCISCO

Cambridge London
Hagerstown Mexico City
New York São Paulo
Philadelphia Sydney

1817

Sponsoring Editor: Malvina Wasserman
Project Editor: Eva Marie Strock
Production Coordinator: Marian Hartsough
Designer: Nancy B. Benedict
Illustrator: J & R Technical Services
Cover Designer: Nancy B. Benedict
Cover Laser Photograph: Laser Images, Inc.
Cover Separator: Focus 4
Compositor: York Graphic Services
Printer and Binder: Kingsport Press

PHYSICS AND THE PHYSICAL PERSPECTIVE, Second Edition
Copyright © 1980 by Harper & Row, Publishers, Inc.

All rights reserved. Printed in the United States of America. No part of this book may be used or reproduced in any manner whatsoever without written permission except in the case of brief quotations embodied in critical articles and reviews. For information address Harper & Row, Publishers, Inc., 10 East 53rd St., New York NY 10022.

Library of Congress Cataloging in Publication Data

Hooper, Henry O.
    Physics and the physical perspective.

    Includes index.
    1. Physics.   I. Gwynne, Peter, 1941–   joint author.   II. Title.
QC23.H759   1980   530   79-24708
ISBN 0-06-042912-7

Contents

Preface   xi

Useful Information   xiv

The Nature of Physics   1
1.1 From Archeology to Zoology   3
1.2 Recent and Modern   3
    Special Topic: *The Urge to Unify*   6

Motion   7
2.1 Length   9
2.2 Time   13
2.3 Vectors   15
2.4 Speed and Velocity (A Scalar and a Vector)   22
2.5 Relative Velocities   23
2.6 Average Speed and Instantaneous Speed (Two Scalars)   27
2.7 Acceleration   29
2.8 Projectile Motion (Vectors in Two Dimensions)   38
    Special Topic: *Higher, Faster, Farther*   52

Forces 55

- 3.1 Newton's Laws of Motion 57
- 3.2 The Second Law 60
- 3.3 Mass and Weight 61
- 3.4 Units of Force 63
- 3.5 The Second Law in Action 65
- 3.6 Newton's Third Law 66
- 3.7 Applying Newton's Laws 69
- 3.8 Friction 77
- Special Topic: *Return to a Small Planet* 88

Universal Gravitation 89

- 4.1 Kepler's Laws of Planetary Motion 92
- 4.2 Newton's Law of Universal Gravitation 94
- 4.3 Moving Round Curves 99
- 4.4 Gravitation and the Solar System 102
- 4.5 Artificial Satellites 107
- 4.6 When the Tide Rushes In 108
- 4.7 That Floating Feeling 111
- 4.8 Fleeing the Center 114
- 4.9 Earth's Gravitational Field 114
- Special Topic: *An Open and Shut Universe* 121

Energy 123

- 5.1 The Real Meaning of Work 125
- 5.2 Kinetic Energy: Energy of Movement 128
- 5.3 Potential Energy: Energy of Position 130
- 5.4 Conservation of Energy 133
- 5.5 Swinging the Changes 135
- 5.6 Gravitational Potential Energy 139
- 5.7 Power 143
- Special Topic: *Producing Energy for the People* 149

6 Impulse and Momentum   153

- 6.1 Momentum   154
- 6.2 Impulse   155
- 6.3 Conservation of Momentum   158
- 6.4 Coming Together   161
- Special Topic: *The Impulsive Athlete*   171

7 Systems of Many Particles   173

- 7.1 The Two Centers   175
- 7.2 Tau is for Torque   177
- 7.3 Getting at the Center of Gravity   181
- 7.4 Moving Around   187
- 7.5 Rotational Dynamics   191
- 7.6 The Gyroscope   197
- Special Topic: *A Revolutionary Means of Staying on the Straight and Narrow*   207

8 Matter under Stress   209

- 8.1 Solids, Liquids, and Gases   211
- 8.2 Density and Relative Density   213
- 8.3 Under Pressure   216
- 8.4 Stresses and Strains   219
- 8.5 Surface Tension   226
- Special Topic: *Bone As a Building Material*   232

9 Fluids under Pressure   234

- 9.1 The Pressure of the Atmosphere   237
- 9.2 Archimedes and His Principle   243
- 9.3 Fluids in Equilibrium   249
- 9.4 The Bernoulli Effect   252
- 9.5 Viscosity: A Real Drag   256
- Special Topic: *The Circulatory System*   262

10 Thermal Energy and Thermal Properties of Matter 264

- 10.1 Taking the Temperature 266
- 10.2 Expanding on Expansion 270
- 10.3 Moving in All Directions 273
- 10.4 A Question of Calories 277
- 10.5 Hidden Heat 281
- 10.6 The Mechanical Equivalent 283
- 10.7 Heat, Temperature, Thermal Energy, and Kinetic Theory 285
- 10.8 The Transfer of Heat 286
  - Special Topic: *Holding Down Those Heating Bills* 294

11 Ideal Gases, and Thermodynamics 295

- 11.1 Chemical Interlude 300
- 11.2 The Gas Law 301
- 11.3 The Ideal Gas 305
- 11.4 Maxwell's Speed Distribution 309
- 11.5 Vapor Pressure—The Key to Good Breath 312
- 11.6 Relative Humidity—The Comfort Barrier 314
- 11.7 Laws of Thermodynamics 316
  - Special Topic: *The Cruel Air, the Cruel Sea* 328

12 Making Waves 330

- 12.1 Why Waves Wave 333
- 12.2 Simple Harmonic Motion 336
- 12.3 Wave Speed 341
- 12.4 Periodic Waves 346
- 12.5 Reflection 349
- 12.6 Standing Waves and Resonance 352
- 12.7 And the Beats Go On 362
- 12.8 Sounds—Physical Versus Subjective 363
- 12.9 The Doppler Effect 367
  - Special Topic: *Hear, Hear: The Inside Story of the Ear* 379

13 Electrostatics 381

- 13.1 Getting a Charge Out of Electricity 382
- 13.2 Conductors and Insulators 385
- 13.3 Introducing the Electroscope 387
- 13.4 Coulomb's Law 389
- 13.5 Electric Fields 393
- 13.6 Potential Difference and Potential Energy 398
- 13.7 Potential Difference and Potential Gradient 405
- 13.8 Equipotentials and Electric Field Lines 407
  - Special Topic: *How Millikan Measured the Ultimate Electric Charge* 415

14 Current Electricity 417

- 14.1 The Reason for Resistance 421
- 14.2 Energy and Electricity 427
- 14.3 Ohm's Law in Action 430
- 14.4 Kirchoff's Laws 435
- 14.5 Measuring Methods 439
  - Special Topic: *How the Nervous System Works* 450

15 Moving Charges and Magnetic Fields 451

- 15.1 The Earth's Magnetic Field 455
- 15.2 Electric Charges in Magnetic Fields 455
- 15.3 The Magnetic Field and Electric Currents 466
- 15.4 Current Loops and Magnets 473
- 15.5 Meters and Motors 476
  - Special Topic: *Earth's Changing Magnetic Field* 486

16 Induced Electric Currents 487

- 16.1 Induction in the Beginning 490
- 16.2 The Contrariness of Nature 492

16.3 The AC Generator 497
16.4 Stepping Up and Stepping Down 502
Special Topic: *The Oscilloscope* 512

17 Alternating Current and Electromagnetic Waves 516

17.1 Changing the Character of Circuits 517
17.2 AC Versus DC Power 525
17.3 Capacitive and Inductive Circuits 527
17.4 Electrical Oscillations 530
17.5 AC to DC 533
17.6 The Transistor and Beyond 536
17.7 Maxwell's Equations 538
17.8 Electromagnetic Waves 540
Special Topic: *Is Anybody Out There?* 548

18 The Nature of Light 552

18.1 Huygens Sees the Light 554
18.2 The Interference of Light 561
18.3 Other Forms of Interference 566
18.4 Diffraction: Interference at the Edges 572
18.5 Polarization 577
18.6 Holography: Photographs Without Lenses 580
Special Topic: *The Search for c* 587

19 Geometrical Optics 589

19.1 Through a Glass Darkly 590
19.2 Lenses 596
19.3 Picturing the Image 599
19.4 Calculating the Image 602
19.5 The Lens-Makers' Equation 607
19.6 Lenses in Combination 609
19.7 Aberrations of Lenses 615
19.8 Reflections on Mirrors 616
Special Topic: *The Eye: The Window of the Brain* 626

20 Relativity 628

20.1 Relative Motion 630
20.2 The Special Theory of Relativity 635
20.3 Space and Time 637
20.4 The Transformations of Lorentz 643
20.5 Time Dilation 647
20.6 Velocity Transformations 652
20.7 Mass and Energy 655
20.8 The General Theory of Relativity 659
Special Topic: *The General Theory of Relativity: Into the Fourth Dimension* 662

21 Approaching the Atom 664

21.1 The Radiation Riddle 665
21.2 The Photoelectric Effect 669
21.3 The Discovery of the Nucleus 676
21.4 The Hydrogen Spectrum 681
21.5 The Atom According to Bohr 683
21.6 Confirming Bohr's Theory 686
Special Topic: *The Solution Without a Problem* 697

22 Moving Into the Atom 700

22.1 The Compton Effect 701
22.2 The Schizophrenic Nature of Light and Matter 705
22.3 The Proof of the Duality 708
22.4 The Quantized Atom 711
22.5 The Nature of Matter Waves 714
22.6 The Uncertainty Principle 716
22.7 Quantum Numbers 721
22.8 Pauli and the Periodic Table 726
Special Topic: *The Men Who Made a Scientific Revolution* 732

x Contents

Inside the Nucleus 734

23.1 Now, the Neutron 736
23.2 Nuclear Forces 739
23.3 Nuclear Binding Energy 740
23.4 The Stability of Nuclei 743
23.5 Radioactive Decay 744
23.6 The Decay of Radioisotopes 749
23.7 Nuclear Reactions 753
23.8 Beyond the Nucleus 761
   Special Topic: *Selected Species from the Subnuclear Zoo* 765

Radiation

24.1 The Generation of an X-Ray Beam 769
24.2 Radiation Detectors 774
24.3 How Radiation Affects Matter 779
24.4 Particle Radiation 783
24.5 Radiation Dosimetry 785
24.6 Radiation in Medicine 788
   Special Topic: *The Radiation Around Us* 792

Epilogue 794

Appendixes 796

A Mathematics 796
B Conversions 814
C Physical Constants 816
D Tables of Useful Data 817
E Natural Trigonometric Functions 818
F Table of Elements, Abundant Isotopes 820

Answers to Odd-Numbered Problems 823

Index 828

Preface

Physics and the Physical Perspective, second edition, is a textbook primarily for students who require a familiarity with but not an exhaustive knowledge of the science of physics: those who need the basics of physics for careers in medicine, dentistry, the biological sciences, agriculture, architecture, geology, oceanography, law, business, or as a general understanding of the physical world. Elementary physics texts and courses are often classified by the level of mathematics employed; this is a *noncalculus* text in which high school algebra is used extensively and trigonometry is used as necessary to provide insights and better understanding of the principles of physics. Because many students tend to have problems with the mathematics in this course, reviews of mathematical concepts are included where they are first used. In addition, Appendix A presents a detailed overview of mathematical concepts. Throughout the book, applications of a non-engineering nature should help students see the relevance of physics to their career goals.

Physics and the Physical Perspective, second edition, brings a new approach to teaching the subject at this level. The experience of a physics professor, who has taught this course for more than fifteen years at three very different universities, and the skills of a science writer trained in physical science have been combined to produce a text that is lively, readable, and informative for students while maintaining the intellectual integrity of the course. In addition, this edition benefitted greatly from the intense review of Mario Iona, professor of physics at the University of Denver and author of the "Would You Believe?" column concerned with textbook accuracy for *The Physics Teacher.* Mario contributed valuable suggestions for the text's revision. In addition, he reviewed substantial portions of the second edition text material during the first proof stages. Iona's critical comments increased greatly the clarity and exactness of the text.

Suggestions from first edition users were incorporated in expanding one or two sections of many chapters and clarifying a number of statements and problems. In addition, the comments from this extensive survey were used in restructuring the second edition. For example, based upon user comments, Chapter 2, Motion, was shortened by delaying a discussion of circular motion until Chapter 4, Universal Gravitation, where the circular motion and centripetal force are necessary applications to planetary motion. Power is introduced in Chapter 5, Energy, and is discussed again in several later chapters. A discussion of simple machines is introduced as a Special Topic in Chapter 5.

In Chapter 9, Fluids under Pressure, the presentation of topics was altered from the first edition so that viscous flow is now the last topic introduced. Thermodynamics was moved from Chapter 10, Thermal Energy and Thermal Properties of Matter, to the end of Chapter 11, Ideal Gases, and Thermodynamics, enabling simple examples of thermodynamics to be based upon the ideal gas law. The discussion of entropy was expanded. A Special Topic on home insulation, which introduces the R-factor and degree days, was added to Chapter 10. In Chapter 12 the discussion of simple harmonic motion (Section 12.2) was expanded to make it more complete.

Chapter 17 was completely reorganized and rewritten to include new material on ac circuits and ac power, along with material from the first edition's Chapter 25 on capacitance, inductance, diodes, and transistors. Chapter 17 ends with an introduction to electromagnetic waves via an elementary discussion of Maxwell's equations. The discussion of polarization formerly in Chapter 17 is now in Chapter 18, Section 18.5, after a discussion of the wave properties of light.

The Instructors Guide includes several suggestions for adapting the text to various academic-year calendars. Brief comments on the highlights of each chapter are presented to help instructors use the text in their courses. Suggestions for assignments, including lecture demonstrations and laboratory experiments, are also included, along with sample exam questions (some multiple choice) and answers to the even-numbered problems.

We use several features to brighten up what otherwise would be dry textual material. One element is the *narrative style* throughout the book, making each chapter a cohesive and complete story in itself. Each *chapter* is *outlined* on its opening page; unlike the first edition, the chapter contents now are contained within one page for concise reference.

At least two *Short Subjects* are interspersed throughout most chapters. These subjects are essentially "asides" to the text presentation. Although the subjects are not required reading—students can receive all the information they need from each chapter by concentrating solely on the basic text—they provide material that both interests the students and breaks up the text, making the going just a little easier. By showing how physics abuts the real world, the descriptive Short Subjects complement the text, ranging from 200 to 1000 words and including such topics as the history of the metric system, flywheels, holography, and biographies of Newton, Einstein, and Boyle.

There is a descriptive *Special Topic* at the end of each chapter. The Special Topics, from 1000 to 1500 words in length, treat, in greater depth, a subject of interest that is related to points discussed within chapters. Timely subjects the Special Topics cover include "The Urge to Unify," a discussion of a search for a unified field theory, "Producing Energy for the People," "Bone As a Building Material," and "Holding Down Those Heating Bills."

Within each chapter there are several other elements that serve both pedagogical and motivational purposes; these elements include definitions in **boldface,** important statements highlighted in color and placed in the margin, and important marginal notes and reminders. Each chapter contains many worked-out illustrative examples that clearly indicate techniques for setting up and solving typical problems. End-of-chapter material includes, besides a Special Topic, a summary, questions, and numerous problems. The extensive sets of problems are grouped by subject matter; within each group problems are graded from least to more difficult as the number of the problem increases. The appendixes are useful tables of physical constants, conversion factors, trigonometric tables, and, as mentioned, an extensive math review. Answers to all odd-numbered problems are at the end of the text. The answers were reviewed thoroughly by both authors, Mario Iona, and independent physicists.

As in the first edition, we use the mks system of units throughout the text, to attempt to minimize any confusion about units. The SI abbreviations are introduced at their first occurrence and thereafter used in the book. However, because the cgs and British systems are still used by many people, including scientists and engineers, we introduce them early in the book and use them occasionally in examples and problems.

Henry O. Hooper
Peter Gwynne

Useful Information

60 mi/h = 88 ft/s
1 mi = 1.61 km
1 kg weighs 9.8 N at earth's surface
1-lb weight has mass of 0.453 6 kg
2.2 lb of weight has mass of 1 kg
1 cal = 4.184 J, 1 kcal = 4184 J
1 eV = 1.602×10^{-19} J
1 atm = 76.0 cmHg = 1.01×10^5 N/m² = 14.7 lb/in.²
931.5 MeV = 1 u $\times c^2$ (speed of light)
1 u = 1.66×10^{-27} kg

2π rad = 1 r = 360°
Area of circle = πR^2
Circumference of circle = $2\pi R$
Volume of sphere = $\frac{4}{3}\pi R^3$
$\pi = 3.14 = \frac{22}{7}$; $\sqrt{2} = 1.41$; $\sqrt{3} = 1.73$
sin 0° = cos 90° = 0
sin 90° = cos 0° = 1
sin 30° = cos 60° = $\frac{1}{2}$
sin 60° = cos 30 = 0.866 = $\sqrt{3}/2$
sin 45° = cos 45° = 0.707 = $1/\sqrt{2}$

Physical Constants
Speed of light $c = 2.9979 \times 10^8$ m/s
Planck's constant $h = 6.625 \times 10^{-34}$ J·s
Charge on electron $e = 1.602 \times 10^{-19}$ C
Rest mass of electron $m_e = 9.11 \times 10^{-31}$ kg
Rest mass of proton $m_p = 1.67 \times 10^{-27}$ kg
Electric constant $k = 8.98 \times 10^9$ N·m²/C²
Universal gravitation constant $G = 6.673 \times 10^{-11}$ N·m²/kg²
Acceleration due to gravity $g = 32.2$ ft/s² = 9.8 m/s² at earth's surface

1 The Nature of Physics

1.1 FROM ARCHEOLOGY TO ZOOLOGY

1.2 RECENT AND MODERN

SHORT SUBJECT: Powers of 10

Summary

SPECIAL TOPIC: The Urge to Unify

What is physics and why is it so important? Scientists have long tended to justify their pursuit of knowledge in an expression similar to that used by the British climber George Mallory to explain why he tried to conquer Mount Everest: "Because it's there." Certainly few physicists, or other scientists for that matter, start out on their avocation with anything but a desire to add to humanity's fund of knowledge, however abstruse their contribution may appear to the average person in the street, to scientists in other fields, and even to their colleagues in the next office. But the past half-decade has seen the emergence of articulate critics who argue that science in general, and often physics in particular, has caused more harm than good by creating nuclear bombs, nuclear power plants, aerosol sprays, vehicles that emit pollutants, and other devices that, in the critics' view, are harming our environment. These critics all too often confuse science—the pursuit of new knowledge—with technology—the application of that knowledge. The message is clear: in this age of skepticism, since scientific research is largely funded by taxpayers through government grants, scientists must justify their existence in more concrete terms than the mysterious interest their fields of research hold for them personally.

For physicists, an important part of this type of justification lies in the basic nature of their subject. Physics is truly a fundamental science, encompassing a range of subject matter from atoms to galaxies and even beyond, into the miniature world of subatomic particles and the unimaginably large arena of the nature of the universe. The first people who rationally could be called scientists were the priests and sages of ancient Babylon, Egypt, and China, who plotted the movements of the stars and planets through the heavens by using the evidence of their naked eyes. The man who was first to use what we regard today as the modern scientific method of moving from observation to theory to prediction to confirmation was also a physicist—a former medical student named Galileo Galilei, whose unstinting support around the turn of the seventeenth century of the theory that the earth orbits around the sun forced an intellectual struggle with the Roman Catholic Church which echoed throughout the civilized world.

Less than 50 years (yr) after Galileo died, another physicist, Isaac Newton, reached what has been called "the greatest generalization achieved by the human mind" in his elegant yet simple laws of motion and the law of universal gravitation: the law that describes how the world goes round. Closer to our time, no scientist has ever achieved as much public adulation as physicist Albert Einstein, whose theory of

Figure 1.1 Galileo: source of an intellectual struggle. (*Culver*)

relativity was a mental tour de force comparable only with Newton's laws.

The study of physics also has a compelling quality of scientific frontiership. Physicists exploring the structure of the atom in the university laboratories of Cambridge, Göttingen, and other centers of learning in the 1920s did so in the unspoken belief that they were on the cutting edge of scientific endeavor. Today, physicists at the giant particle accelerators of Batavia (Illinois), Geneva (Switzerland), and elsewhere are probing into the very depths of the atom with equal confidence that their efforts will ultimately reveal the nature of matter.

1.1 FROM ARCHEOLOGY TO ZOOLOGY

But science cannot live by philosophical significance alone. In their quest for the nature of matter and energy and the relationship between them, physicists are aware of the needs of other sciences. One rough and ready rule of the scientific bureaucracy is that the line for scarce research grants should be headed by those branches of science which offer most in the way of spin-offs to other fields of research. On this criterion physics stands supreme. The techniques and concepts of physics have been adapted to every other science imaginable, from archeology to molecular biology, from oceanography to zoology. The remarkable advances of recent years that have brought molecular biologists ever closer to understanding the nature of life had their genesis in knowledge of the structure of such complex organic molecules as DNA—knowledge obtained through use of the x-ray analysis techniques developed in physics laboratories. New insights into the lives of our ancestors and the animals that coexisted with them in prehistoric times have been made possible by carbon dating, another technique borrowed from physicists. Experts in energy and raw materials now seek to discover new fields containing oil, natural gas, and mineral ores by using satellite photographs whose production depends on the work of physicists studying rocketry and atomic spectroscopy. Nuclear power traces its lineage directly back to the theoretical physics of the 1930s. And such medical technologies as ultrasonic probing and cobalt therapy have moved rapidly from physics theories to physics experiments to applications in health care.

1.2 RECENT AND MODERN

The techniques and concepts of physics, then, permeate the whole realm of science and technology. Yet much of present-day physics, at least that which makes its way into textbooks, seems to have an ancient, if not positively outdated, air about it. After all, the law of universal gravitation was discovered scarcely half a century after the Mayflower deposited its human cargo at Plymouth Point. The experiments that led to the large-scale generation of electrical energy were performed more than 30 yr before the American Civil War. And the bulk of fundamental work on the nature of the atom was carried out by scientists for whom a trip to their colleagues across the Atlantic was an exploit that required a total of at least 10 days travel time. For all its mental

Powers of 10

Since physicists deal with a huge range of numbers—speeds from 0 to 186 000 miles per second (mi/s), distances from trillionths of meters to trillions of meters—they need a simple shorthand to express such diversity. The simple system is the powers of 10 notation, sometimes called scientific notation, whose essence is to express all multiples of 10 as powers of that number. For example,

$$100 = 10 \times 10 = 10^2$$
$$1000 = 10 \times 10 \times 10 = 10^3$$

Number		Power of 10		Prefix	Symbol
1 000 000 000 000	=	10^{12}	one trillion	tera	T
1 000 000 000	=	10^{9}	one billion	giga	G
1 000 000	=	10^{6}	one million	mega	M
1 000	=	10^{3}	one thousand	kilo	k
100	=	10^{2}	one hundred	hecto	h
10	=	10^{1}	ten	deka	da
1	=	10^{0}	one		
0.1	=	10^{-1}	one-tenth	deci	d
0.01	=	10^{-2}	one-hundredth	centi	c
0.001	=	10^{-3}	one-thousandth	milli	m
0.000 001	=	10^{-6}	one-millionth	micro	μ
0.000 000 001	=	10^{-9}	one-billionth	nano	n
0.000 000 000 001	=	10^{-12}	one-trillionth	pico	p

All submultiples of 10, such as one-tenth and one-hundredth, can be expressed as negative powers of 10. For example,

$$\frac{1}{10} = 10^{-1}$$

$$\frac{1}{100} = \frac{1}{10 \times 10} = \frac{1}{10^2} = 10^{-2}$$

$$\frac{1}{1000} = \frac{1}{10 \times 10 \times 10} = \frac{1}{10^3} = 10^{-3}$$

The multiples of 10 have common names and, in the metric system, the prefixes and symbols shown in the table at the top of this page. These prefixes or symbols are placed in front of a unit name, such as meter, ampere, or second, which thus becomes kilometer (km), microampere (μA), or nanosecond (ns), respectively.

When performing calculations using powers of 10 notation we must recall one or two simple rules for handling exponents. When we multiply two power expressions with the same base, we add the exponents algebraically, so that

$$10^x \times 10^y = 10^{x+y}$$

If $x = 2$ and $y = 3$, then

$$10^2 \times 10^3 = 10^5$$

If $x = 2$ and $y = -2$, then

$$10^2 \times 10^{-2} = 10^0 = 1$$

When we divide power expressions with the same base, we subtract the exponent

$$\frac{10^m}{10^n} = 10^{m-n}$$

If $m = 4$ and $n = 2$, then

$$\frac{10^4}{10^2} = 10^{4-2} = 10^2 = 100$$

Here are four illustrative examples employing powers of 10:

(a) $(1.6 \times 10^{-19})(3 \times 10^{10}) =$
$(1.6 \times 3)(10^{-19+10}) = 4.8 \times 10^{-9}$

(b) $(6.02 \times 10^{23})(2000)(1.6 \times 10^{-19})$
$= (6.02 \times 2 \times 1.6)(10^{23+3-19})$
$= 19.26 \times 10^7 = 1.93 \times 10^8$

(c) $\frac{(0.0022)(4000)}{(0.07)(4198)}$

$= \frac{(2.2 \times 10^{-3})(4.0 \times 10^3)}{(7 \times 10^{-2})(4.198 \times 10^3)}$

$= \frac{2.2 \times 4.0}{7 \times 4.198}(10^{-3+3+2-3})$

$= 0.2995 \times 10^{-1} = 3.0 \times 10^{-2}$

(d) $\frac{(3 \times 10^{10})(6.02 \times 10^{23})}{(1.6 \times 10^{-19})(6.625 \times 10^{-34})}$

$= \frac{(3 \times 6.02)(10^{10+23+19+34})}{(1.6 \times 6.625)}$

$= 1.70 \times 10^{86}$

Figure 1.2 Newton, Einstein, Gell Mann: modern scientists all. (*Culver; Wide World; Floyd Clark, Cal Tech*)

stimulation, this type of physics has, at first glance, little connection with radio telescopes, particle accelerators, lasers, transistors, integrated circuits, and all the other devices and endeavors that are part of modern physics.

The fallacy here is in the use of the word "modern." In terms of reasoning and methodology, Newton's derivation of his law of universal gravitation was quite as modern as Einstein's formulation of his theory of relativity more than two centuries later and Murray Gell Mann's conceptualization of "the eightfold way," an approach to the fundamental nature of matter that is slowly bringing order out of the chaos of the subnuclear world more than a decade after its publication. Gell Mann's and Einstein's theories were more recent than Newton's, but certainly not more modern in the strict scientific sense of the word. When applied under appropriate circumstances, Newton's theory is no less valid today than it was when he formulated it. The theory proved its value, if proof were needed, in the exploits of the Apollo astronauts who flew to the moon not by luck but on a course dictated by Newton's laws and interpreted by the giant space agency computers.

Thus, the physics that appears in the next 23 chapters is entirely modern, although not necessarily recent. Just about every topic is relevant to and important in at least one other subject outside the realm of physics. Every step along the way opens up new possibilities for the next step in understanding the nature of our world.

SUMMARY

1. Physics is the science that deals with matter and energy. Its subject matter ranges from subnuclear particles to cosmology, and physics provides answers to some of the most philosophically significant questions asked by people.

2. The concepts, techniques, and instruments of physics have wide application in other sciences, both physical and biological.

Special Topic
The Urge to Unify

The primary purpose of physics, as of most of the rest of science, is unification. Physicists seek to group similar phenomena under a single umbrella of principles or sets of equations. This is a timeless theme; for example, the fifth century B.C. Greek philosopher Thales regarded water as the fundamental essence of all matter. His contemporary, Pythagoras, thought that all the contents of the universe could be understood through the medium of a few integral numbers. Today, physicists understand the extraordinary diversity of creation in terms of four forces: gravity, electromagnetism, and the weak and strong forces that hold atomic nuclei together. Many specialists think that the four forces are merely different manifestations of the same truly fundamental principle. The effort to develop a "unified field theory" that would account for all four forces in this way is perhaps the most ambitious and profound undertaking in all of science.

The groundwork has been laid over the centuries. In the early 1600s Galileo emerged as the first effective unifier, as we shall see in Chap. 2. He brought together the physics of motion on the earth and elsewhere. Later in the same century Isaac Newton carried the process a strong step forward. He realized, as we shall see in Chap. 4, that the force responsible for holding the moon in orbit around the earth is exactly the same as that force which pulls a ripe apple from the branch of a tree. About 100 yr ago James Clerk Maxwell devised the set of four equations, featured in Chap. 17, that link electricity and magnetism. And in his special theory of relativity, published in 1905 and highlighted in Chap. 20, Albert Einstein removed the distinction between space and time, joining the two dimensions in a single entity that has ever since given physics students fits and science fiction buffs hours of amusement.

Even Einstein could not unify forces, however, for all his efforts during the 25 yr before his death in 1955. He was, in fact, tackling only part of the problem, although it was the most difficult part: he tried to combine just the forces of gravity and electromagnetism. Meanwhile, the weak and strong nuclear forces had been discovered, and Japanese physicist Hideki Yukawa was unsuccessful in his attempt to unify them. The situation seemed to fit a quip of Austrian theorist Wolfgang Pauli: "What God hath put asunder, no man shall ever join."

Some physicists continued to believe, however, that God had not created four completely independent forces; rather, He had linked them at a very profound and hard-to-detect level. In 1956, Julian Schwinger suggested that the electromagnetic and weak nuclear forces might be combined in a so-called gauge theory. This theory posits that the electromagnetic and weak nuclear forces between objects arise when specific elementary particles race back and force between the objects. It was 15 yr later, after inspired theoretical work by Sheldon Glashow, Steven Weinberg, Abdus Salam, John Ward, and Gerard 't Hooft, among others, that a realistic theory linking these two forces emerged. This theory made predictions about reactions between subnuclear particles—experiments upheld these predictions. By 1979, much of the world of physics accepted that these two of the four forces of nature were, at root, a single force. All that remained was to find the intermediate particle involved in the weak force. Experimenters are confident that the entity, the intermediate boson, will be spotted in one of a generation of new particle accelerators now under construction.

The theorists are now working hard on the next stage: bringing the strong nuclear force into the scheme, the scheme they optimistically term "grand unification." This work, which involves deep understanding of elementary particle physics, is progressing well. The most difficult problem of all will be incorporating gravity into the system. Gravity is the most familiar natural force to the average person but the most complex for the physicist. Nevertheless, a few theorists are working on a concept called **supergravity** that they hope might eventually lead to a total unification and the completion of a 2500-yr dream. The theorists' work, and that of today's physicists generally, is built on the foundations of physics and physicists past, as the following 23 chapters will reveal.

2 Motion

2.1 **LENGTH**

   SHORT SUBJECT: A Man for All Sciences
   SHORT SUBJECT: From Ancient to Modern

2.2 **TIME**

   SHORT SUBJECT: The How, Why, and Where of Metric Measure

2.3 **VECTORS**

   SHORT SUBJECT: Vector Addition and Subtraction
   SHORT SUBJECT: Trigonometry Review

2.4 **SPEED AND VELOCITY (A SCALAR AND A VECTOR)**

2.5 **RELATIVE VELOCITIES**

2.6 **AVERAGE SPEED AND INSTANTANEOUS SPEED (TWO SCALARS)**

2.7 **ACCELERATION**

   SHORT SUBJECT: The Slope—A Mathematical Definition and an Application to Instantaneous Speed
   SHORT SUBJECT: Equations of Motion for Constant Acceleration

2.8 **PROJECTILE MOTION (VECTORS IN TWO DIMENSIONS)**
   Summary
   Questions
   Problems

   SPECIAL TOPIC: Higher, Faster, Farther

A fighter plane screams at supersonic speed over a scarred war zone, twisting and weaving to avoid a heat-seeking missile on its tail. A javelin hurled by a world-class athlete soars more than 90 meters (m) before burying itself in the turf. A hair-thin beam of subnuclear particles whirls around the circumference of a giant particle accelerator before exiting, striking a thin slab of metal, and producing a stream of new particles. An emperor penguin moves clumsily on two feet toward the steel-blue waters of the Ross Sea, then plunges in and starts to swim with effortless grace. The planets drift in their stately paths around the sun, spinning inexorably through the eons. All these scenes are examples of motion; for all their dissimilarities, all take place according to the same universal set of physical laws. Because much of modern physics is concerned with motion in one form or another, an understanding of these laws is essential to an overall grasp of the subject.

Sages have been fascinated by movement since people began to analyze their world. The Greek philosopher Aristotle, for example, divided movements into two classes. Natural motions, according to his schema, were those that took place with no obvious stimulus: These were vertical motions, such as the descent of a rock to its natural resting place, the earth; or the ascent of smoke into the atmosphere, where gases were supposed to reside; and the daily circular motions of the sun, stars, and planets about the earth. Violent motions, by contrast, resulted from the application of force to an object, such as the effect of a bowstring on an arrow or a boat's oars on the water.

Philosophers also turned their eyes to the heavens and tried to work out the rationale of the movements of the stars and planets relative to the earth. Their theorizing produced almost as many systems as there were philosophers, but the longest lasting view was that of the Alexandrian astronomer, Ptolemy, who lived in the second century A.D. Taking as his basic starting point the self-evident (to him) truth that the earth was fixed at the center of the universe, Ptolemy devised an incredible series of convoluted circles in which the planets were supposed to travel around the earth. This cumbersome system survived until well into the eighteenth century, although in 1543 an obscure Polish scholar named Nicholas Copernicus published his theory that all the planets, including the earth, actually traveled around the sun. So publicity shy was Copernicus that his theory all but went unnoticed during his lifetime and for 25 yr after his death in 1543. It was left to Galileo Galilei, a somewhat abrasive Italian scientist, to spread the word on the heliocentric theory, thus defying the hierarchy of the Church, which regarded the theory as heretical.

The heliocentric idea was actually advanced by Aristarchus, among others, nearly 2000 yr before Copernicus. However, it was largely ignored.

The Church's attitude was not entirely unreasonable because the heliocentric theory was in conflict with Aristotelian physics; thus the acceptance of the theory necessitated the development of a new physics. A prime ingredient in the new approach was Galileo's principle of inertia. But before detailing the present understanding of motion that was initiated by Galileo, we need definitions of the basic factors and concepts involved in kinematics—the aspects of motion, excluding mass and force.

2.1 LENGTH

One basic notion in physics is that of an interval of space, what we call a distance. To determine a distance from one position in space to another, we make a count by some means, such as pacing off with equal steps or laying off with a ruler. Length is simply a measured distance, and the units in which we measure it are completely arbitrary. The British system of inches, feet, and yards, used almost everywhere in the United States outside scientific laboratories, is perhaps more arbitrary than most, being based on the folklore system developed in the Dark Ages. Scientists predominantly use the metric system, which is considerably more rational. The meter was originally chosen as one-ten-millionth of the length of the arc from the earth's equator to one of its poles; although the geodetic survey on which this definition was based is now known to have been in error, the meter has remained at its originally stated length, which is defined as the distance between two scratches on a platinum-iridium bar kept at Sèvres, France. Today, however, the meter is defined in terms of the length of light waves emitted by particular atoms of krypton gas and in principle can be reproduced in any laboratory in the world.

Unlike the British system, the different units of length in the metric scale are related simply by multiples of 10. A centimeter, for example, is one-hundredth of a meter; a millimeter is one-thousandth of a meter; and a kilometer is 1000 m (see Table 2.1).

Two derived physical units are related to length. **Area is length multiplied by length** and in the metric system is expressed in square meters; **volume is area multiplied by length** and has units of cubic meters.

Table 2.1 Units of Length

British Units	Metric Units	Conversion Relations
12 inches (in.) = 1 foot (ft)	10 millimeters (mm) = 1 centimeter (cm)	1 in. = 2.54 cm
3 ft = 1 yard (yd)	10 cm = 1 decimeter (dm), a rarely used unit	1 m = 39.37 in.
5280 ft = 1 mile (mi)	10 dm = 1 meter (m)	1 km = 0.621 mi
	1000 m = 1 kilometer (km)	1 mile = 1.61 km

A Man for All Sciences

Obtuse, acerbic, caustic, opinionated, contentious, argumentative, obstinate. These adjectives and others equally disparaging were applied to Galileo Galilei by his academic contemporaries in the universities and the Church. This characterization was partly justified because Galileo, by which name he is universally known, did not like to be criticized. But the Italian scientist's exploits in creating the science of mechanics and revolutionizing astronomy made Galileo a veritable giant among the pygmies who carped at him.

Galileo originally intended to be a doctor, on the advice of his mathematician father, who argued that there was much more money in healing ailments than in proving theorems. But once the young student happened to hear a lecture on geometry, he was sold on mathematics and managed to persuade his father to let him study the subject. Galileo's first major discovery came in 1581, when the 17-yr-old was still a medical student at the University of Pisa. Using his pulse as an automatic timer, and hence demonstrating the value of his medical training, Galileo discovered that the time a pendulum took to swing back and forth was independent of the amplitude of the swing.

A few years later Galileo threw himself into the subject of falling objects. Using the basic experiments outlined in this chapter, he laid the foundations for the science of kinematics—foundations on which students and professors still build today.

Galileo. (*Culver*)

Then in the early 1600s, after he had moved to Padua, Galileo began to survey the heavens by using the newly invented telescope and to wage an ardent battle to establish the Copernican theory that the earth revolves around the sun. The Roman Catholic hierarchy of the time was suspicious of a scheme that envisioned the heavens as less than perfect residences of the angels and relegated the earth from its dominant position at the center of the universe; in 1616 the Church authorities forced the scientist into silence.

Sixteen years later, sensing a detente in the Church, Galileo published the classic *Dialogue on the Two Chief World Systems,* a tract that took the form of an argument between upholders of the earth-centered and sun-centered theories in front of a neutral observer. The Pope of the time, Urban VIII, was led to believe that the character who argued the earth-centered view—and who had the worse of the argument—was a satire on himself, and so summoned Galileo to appear before the Inquisition. The scientist, nearly blind because of his telescopic investigations of the sun, was forced to recant his views, although legend says that after the renunciation he muttered the words "eppur si muove"—"but still it [the earth] moves." Galileo died in 1642, but the scientific revolution he wrought lived on and flourished.

From Ancient to Modern

The corner supermarket advertises a special on 1-liter (L) bottles of Coke. A healthy newborn baby weighs in at 3.2 kg. Wild applause greets the announcement over the Yankee Stadium's loudspeaker system that a tape-measured home run traveled 121 m. Strange as they might sound, such incidents could be standard fare in America by the 1990s, as they are today in most of the rest of the world. Although the metric system may seem quite unfamiliar to Americans brought up with the units of pounds, feet, and pints, it is in fact far more rational than the traditional system of weights and measures, and so now pressures are growing daily to convert the everyday measurements used in the United States to the metric system.

The so-called British system of weights and measures, which is now on its way out in Britain, has its roots in ancient history. Written records from early Babylonian and Egyptian sources and the Old Testament indicate that the ancient civilizations routinely used such limbs as the forearms, hands, and fingers to measure lengths; the periods of the sun, moon, stars, and planets to mark the passage of time; and numbers of plant seeds to determine volumes and weights. The term carat, for example, which jewelers still use to measure gems, derives from the carob seed.

Inevitably, each civilization measured according to its own traditions; in fact, the Tower of Babel could be regarded as a commentary on the diversity of measurements as well as languages. The Romans rationalized the situation to some extent, dividing the *pes,* or foot, into 12 equal *unciae,* from which the latter-day terms inch and ounce are derived. But it was really the Anglo-Saxons who defined the units that measure out our lives. The inch, for example, was standardized as the length between the knuckle and tip of the thumb. The foot, which was originally defined as four palm widths, or 16 fingers, was eventually settled on as the length of 36 barleycorns. The yard came from the Saxon word *gird,* meaning the circumference of a person's waist. It was originally defined as the length of the sash that Saxon kings wore around their waists. Later, however, King Edgar deemed that the yard should be the distance between the tip of his nose and his thumb, and this standard was legalized by King Henry I in the twelfth century. A fathom was set at the width of a Viking's reach. An acre was the amount of land that could be plowed in a day by a yoke of oxen. Another agricultural unit was the furrow-long, or what is now called the furlong, established in early sixteenth century England as 220 yards (yd). Late in the same century, Queen Elizabeth I decided that the traditional Roman mile of 5000 ft should be replaced by one of 8 furlongs, or 5280 ft.

The standardized, comprehensive Anglo-Saxon system became the basis of all world trade for the next three centuries, despite occasional murmurings in Europe and the United States in favor of a more logical, scientifically based system. The main impetus for what we now know as the metric system came from revolutionary France at the end of the eighteenth century. In 1790, the National Assembly asked the French Academy of Sciences to "deduce an invariable standard for all the measures and all the weights." The meter (from the Greek word for a measure, *metron*) was the fundamental unit of the system, defined as one ten-millionth of the distance from the North Pole to the equator along the earth's meridian that runs through Dunkirk, France, and Barcelona, Spain. This unit was the basis of length, area, and volume measurement, and the standard of mass—roughly defined as the amount of matter in an object—was related to it, being defined as the mass of a liter, or cubic decimeter, of water.

The metric system was hardly popular in its early years; not until 1840 was its use made mandatory in France. Other countries slowly followed the lead of France, however, and the explosion of scientific knowledge at the turn of the century benefitted from the use of a simple decimal system. During this century, the metric system has spread into everyday usage throughout the world—apart from the United States and a handful of small nations.

EXAMPLE 2.1

Express in meters and kilometers the sum of 1723 mm, 664.9 cm, 13.47 dm, and 1.362 m.

Solution

$$
\begin{aligned}
1723 \text{ mm} &= 1.723 \text{ m} \\
664.9 \text{ cm} &= 6.649 \text{ m} \\
13.47 \text{ dm} &= 1.347 \text{ m} \\
1.362 \text{ m} &= \underline{1.362 \text{ m}} \\
& 11.081 \text{ m}
\end{aligned}
$$

Expressed in km, 1 km = 1000 m, so 11.081 m = 0.011 081 km = 1.1081×10^{-2} km.

EXAMPLE 2.2

How tall is a 6-foot (ft) man in centimeters and meters?

Solution

$$6 \text{ ft} \times \frac{12 \text{ in.}}{1 \text{ ft}} \times \frac{2.54 \text{ cm}}{1 \text{ in.}} = 183 \text{ cm}$$

$$183 \text{ cm} \times \frac{1 \text{ m}}{100 \text{ cm}} = 1.83 \text{ m}$$

Notice that we have written down the conversion factors with their units and canceled units in a manner similar to that used with numbers. Although this may seem unnecessary, it is a good practice because it often prevents careless errors. The unit remaining after cancellation of the units should be the one we desire; if it is not, then we have made an error.

EXAMPLE 2.3

Compute the area of the floor of a rectangular living room 15 ft wide and 30 ft long. Give the answers in square feet (ft²) and square meters (m²).

Solution

The area of a rectangle is the length times its width; thus, in square feet,

$$\text{Area} = 15 \text{ ft} \times 30 \text{ ft} = 450 \text{ ft}^2$$

We could convert the 450 ft² to square meters directly if we knew how many square feet were equivalent to 1 m². However, for practice in handling conversions we can use a conversion factor from Table 2.1. One conversion many people remember is that 1 in. = 2.54 cm. Hence,

$$1 \text{ ft} = 1 \text{ ft} \times (12 \text{ in./ft}) \times (2.54 \text{ cm/in.}) = 30.48 \text{ cm} = 0.3048 \text{ m}$$

Hence

$$15 \text{ ft} = 15 \text{ ft} \times 0.3048 \text{ m/ft} = 4.572 \text{ m}$$

and

$$30 \text{ ft} = 30 \text{ ft} \times 0.3048 \text{ m/ft} = 9.144 \text{ m}$$

The area of the living room in square meters can be computed as:

$$\text{Area} = 4.572 \text{ m} \times 9.144 \text{ m} = 41.81 \text{ m}^2$$

Keeping a consistent number of significant figures, this area becomes 41.8 m^2.

When converting we cannot—or at least should not—gain significant figures because that would imply greater precision than the figures have. Thus, to be consistent, we express 15 ft as 4.6 m, not 4.572 m. We cannot improve on the accuracy of our original numbers by just converting numbers, a point that we should remember when we use pocket calculators to help us with our computations. This is very important when we are dealing with experimental measurements, where the number of significant figures in a measurement may reflect the accuracy with which the measurement was made.

EXAMPLE 2.4

Express the speed of 60 miles per hour (mi/h) in feet/second (ft/s) and kilometers/hour (km/h).

Solution

To convert 60 mi/h into ft/s, we use data from Table 2.1, 5280 ft = 1 mi:

$$1 \text{ h} = 1 \text{ h} \times 60 \text{ s/minute (min)} \times 60 \text{ min/h} = 3600 \text{ s}$$

$$60 \text{ mi/h} = 60 \text{ mi/h} \times 5280 \text{ ft/mi} \times \frac{1}{3600 \text{ s/h}} = 88 \text{ ft/s}$$

To convert miles per hour to kilometers per hour, we again resort to data from Table 2.1, 1 mi = 1.61 km:

$$60 \text{ mi/h} = 60 \text{ mi/h} \times 1.61 \text{ km/mi} = 96.6 \text{ or } 97 \text{ km/h}$$

2.2 TIME

If we wish to keep any record of motion, we must be able to measure the intervals of time that elapse between separate parts of the motion. These intervals can occur in such varied circumstances as the swift passage of light from the sun to a sunbather on earth, the arcing movement of a curve ball from a pitcher's hand to the catcher's glove, and the slow progress of a train from one station to another. Measuring these intervals basically involves counting the number of times a

The How, Why, and Where of Metric Measure

Valuable as they are in rationalizing measurements worldwide, the standard weights and measures would have no scientific use unless they could be reproduced in practical situations. The existence of a standard meter, for example, would be pointless unless meter rules could be regularly checked for accuracy against that standard or an accurate copy of it. The definition of the second of time would be worthless if clocks could not be set in terms of the standard second. To make such checks of everyday measurements simpler, scientists in recent years have defined a growing number of the fundamental measures in terms of atomic quantities that are readily reproducible anywhere in the world—or the universe.

It was not always so. The original definition of the meter, as noted, was the length of one ten-millionth of a quadrant of the earth. In 1889, after fresh measurements showed that the geodetic survey which produced this standard was in error, the meter was engraved more solidly on the hearts and minds of physicists. An international body known as the General Conference on Weights and Measures defined the meter as the distance between two lines engraved on a metal bar at a temperature of 0°Celsius (C). The bar, a rustproof platinum-iridium alloy, was kept at the International Bureau of Weights and Measures in Sèvres, France, just outside Paris, and copies of it were sent to standards laboratories throughout the world.

Standardization against a bar of metal was a complicated procedure, however, and in 1960 the General Conference on Weights and Measures redefined the meter in atomic terms. According to its new definition, the fundamental unit of length in the metric system is the length of 1 650 763.73 of the light waves emitted by particular atoms of the gas krypton under strictly specified conditions. Measuring such atomic emissions is a relatively simple matter for any sophisticated laboratory.

The times are also changing in regards to the basic definition of the second. In 1960 the General Conference on Weights and Measures agreed to define the second as 1/31 556 925.9747 of the duration of the year 1900. Specification of a particular year was necessary because the actual duration of the year varies slightly, owing to irregularities in the earth's motion. But just 7 yr after defining the second astronomically, the General Conference had a change of heart. The experts redefined it in atomic terms as the duration of 9 192 631 770 periods of the radiation absorbed by a particular atomic form of the metal cesium. These radiations are measured by using a standard instrument known as the atomic clock, and thus the standard second is as readily reproducible as the standard meter.

At present only one of the fundamental units is not defined in terms of physical phenomena generally reproducable—the standard of mass, which can be regarded as a measure of the amount of matter in any object. The metric unit of mass is the kilogram, which is defined as the mass of a platinum-iridium cylinder kept at the International Bureau of Weights and Measures in Sèvres. But with improvements in physical measurements of atomic quantities, even this standard seems destined to be replaced eventually by a simply reproducible definition based on the masses of atoms.

regularly occurring event, such as the ticking of a clock, actually happens during the time interval in question. One such regularly occurring event that can be used to measure relatively long periods of time is the rotation of the earth on its axis. For shorter intervals of time we might use the unvarying swing of a pendulum. And an extremely brief period can be measured by counting the number of light waves emitted by a particular kind of atom during the period.

Fortunately, all systems of measuring time are related to the second, which was originally defined as 1/86 400 of a day (the average period between passages of the sun across the meridian). Because phenomena on the atomic scale are more constant, can be measured more precisely than astronomical ones, and can be duplicated everywhere, the second is now based on the number of periods of the radiation absorbed by one isotope of the element cesium under strictly controlled conditions.

2.3 VECTORS

If we talk about temperature or volume, we are referring to quantities that are completely independent of direction. But if we refer to length, we often imply a direction as well as a simple number.

Length and time are measurable physical quantities. Before we study their relationship in motion, we require a grasp of a more complicated concept: that of directionality. If we talk about temperature, say, the 95°F high of a sweltering summer day, or volume, the 1-cubic meter (m^3) capacity of a large refrigerator, we are referring to quantities that are completely independent of direction. But if we refer to length—traveling 10 km to visit a friend in the next town—we often imply a direction as well as a simple number. Similarly, the volume of a swimming pool is totally unrelated to any direction, whereas the actual change in position a swimmer achieves in the same pool is entirely dependent on the direction in which the swimmer moves throughout the activity. By swimming in a circle, the water enthusiast will sooner or later end up at the starting point; but if the swimmer travels exactly the same length in a straight line, he will finish up exactly that length away from his starting point. And if the swimmer takes a zigzag path in the water, he can end up, according to the direction of the zigs and zags, at any point between the starting point and the straight-line distance.

In the word of physics, **quantities that are independent of direction, among them temperatures, volumes, masses, and energies, are known as scalars. Quantities such as directed distances, forces, torques, weights, and momenta that do depend on directions are called vector quantities. A directed distance is referred to by the special term** *displacement.*

Scalars can only be added to or subtracted from other scalars; likewise, vectors can be added to or subtracted from other vectors only.

The inherent difference between scalars and vectors is most obvious when we perform simple addition and subtraction with quantities of the different types. Obviously, the scalars can only be added to or subtracted from other scalars, and vectors can be similarly added to or subtracted from other vectors only. But whereas these operations with scalars are just arithmetic exercises, addition and subtraction of vectors is complicated by their possession of directionality. In fact, the direction of vectors is just as important as their magnitude when we add or subtract them. When we deal with scalars, 2 plus 2 always equals 4, but in the case of a vector of magnitude 2 and another vector of magnitude 2, they can add up to any vector length between 0 and 4, depending on the relative directions of the two vectors involved.

This situation can be understood by referring to a couple of tug-of-war teams that are evenly matched. If they are pulling against each other on a rope [Fig. 2.1(a)], the net force on the rope is zero; if they both combine to pull in the same direction, the net force on the wall is twice the force exerted by each team individually [Fig. 2.1(b)]; and, if

Figure 2.1 Tug of war. Two teams pulling on a rope in different ways.

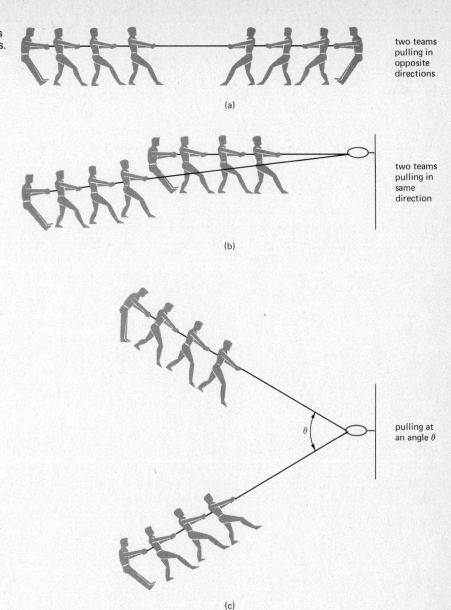

they insert the rope through a ring and then pull in slightly different directions, the net force they exert on the wall is somewhere between these two quantities [Fig. 2.1(c)]. In the latter case, the exact magnitude of the net force on the wall depends on the angle between the two halves of the rope.

To consider vector addition and subtraction in formal mathematical terms, let us return to the concept of distance. The vector quantity that describes length traveled in a specific direction, **the displacement, is**

Vector Addition and Subtraction

A. Addition of vectors directed along the same line

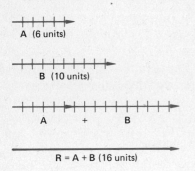

B. Subtraction of vectors directed along the same line

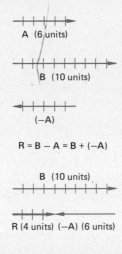

C. Vector addition of two vectors not along the same line

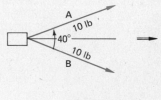

simply the directed distance between the starting and finishing positions of a moving object during the time interval of interest. Consider, for example, a yacht that travels exactly 20 nautical miles in a northeasterly direction from its original anchorage, anchors while its crew eats lunch, and then makes another 12 nautical miles, again in a northeasterly direction. The total displacement of the vessel during the day is the arithmetic sum of the two displacements, that is, 32 nautical miles in a northeasterly direction—because both displacements are in exactly the same direction. (In other words, the two vectors are along the same straight line.)

Alternately, let us consider the case if the crew sails the yacht 20 nautical miles to the northeast before lunch and then reverses direction and sails 12 nautical miles southwest, that is, back along its original path. Here we subtract the distance of the second journey from that of the first, to obtain the net displacement of 8 nautical miles in a northeasterly direction, even though the total path length or distance traveled is still 32 nautical miles.

But what if the craft sails 20 nautical miles northeast and then puts in 12 nautical miles due south? Here the two vectors are not in a

Figure 2.2 Vector diagram for the trip of a sailboat sailing 20 nautical miles northeast and then 12 nautical miles due south. The resultant **R** would be a trip of 14.3 mi at 8.6° north of east.

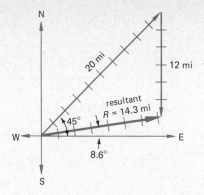

straight line. Thus, the net displacement sailed by the yacht cannot be computed by simple addition or subtraction of the magnitudes, and the direction between the yacht's starting and finishing points is not related in any obvious way to the two directions in which it actually sailed.

When adding or subtracting vectors that are not in a straight line, we must devise a means of accounting for both the magnitude and directionality of the individual vectors. We can do this pictorially, using graphic arrows to represent the vectors. The shaft of each arrow is placed parallel to the direction of the vector it represents, with the arrowhead pointing in the corresponding direction. The length of each arrowshaft is made proportional to the magnitude of the vector it represents. For example, if we were dealing with vectors like those that have magnitudes of 12 and 20 nautical miles, we might select a convenient scale of 4 nautical miles, represented by a length of 1 cm on our vector diagram. Hence 12 nautical miles would be represented by an arrow 3 cm long.

By representing vectors this way **we can sum them graphically by placing the head of one arrow at the tail of the next,** as shown in Fig. 2.2 for the yacht that sails 20 nautical miles northeast and then 12 nautical miles due south. We then obtain the net displacement traveled by the boat by joining the tail of the arrow that represents the first leg of the journey to the head of the arrow representing the final leg. In this case the yacht's overall progress is 14.3 nautical miles in a direction 8.6° north of east. Figure 2.3 shows similar graphical analysis in cases where the second leg of 12 nautical miles is sailed in directions other than due south.

This same method of adding vectors can be applied to cases of more then two vector quantities. Consider, for example, a person on a shopping trip who travels from home to a variety of stores before stopping at a restaurant for lunch. If we wish to work out the overall displacement between home and the lunch stop, we represent each stage of the journey—between one store and the next—by a vector arrow and sum all the vectors in the same way as in the yacht example, (Fig. 2.4).

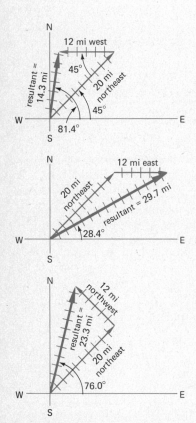

Figure 2.3 Vector diagrams for three more sailboat trips. The boats sail 20 nautical miles northeast and then 12 nautical miles in different directions. The scale is different from that of Fig. 2.2.

Trigonometry Review

Because right triangle trigonometry is fundamental to physics, this section outlines its functions and relations. For the diagram of a right triangle, the functions of the angle θ are given as:

$$\sin\theta = \frac{\text{opposite side}}{\text{hypotenuse}} = \frac{b}{c}$$

$$\cos\theta = \frac{\text{adjacent side}}{\text{hypotenuse}} = \frac{a}{c}$$

$$\tan\theta = \frac{\text{opposite side}}{\text{adjacent side}} = \frac{b}{a}$$

From these definitions it is clear that the angle ϕ, which is simply $90° - \theta$, has functions such that $\sin\theta = \cos\phi$ and $\cos\theta = \sin\phi$.

In most cases we use the functions of the angles listed in App. E to determine sides of a right triangle such as components of a vector. We can rewrite the definitions of the trig functions for sine and cosine to define the two rectilinear components (b and a) of a vector represented by the hypotenuse c:

$$b = c\sin\theta \quad \text{and} \quad a = c\cos\theta$$

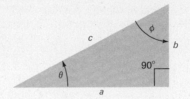

Also recall that the sides a, b, and c are related in magnitude by the Pythagorean theorem:

$$c^2 = a^2 + b^2$$

This relationship is useful in determining the magnitude of a vector (c) if its components a and b are known.

Another method of adding and subtracting vectors is more formal in a mathematical sense. This involves selecting a pair of graphic axes, usually at right angles to each other, and splitting each individual vector into two components, one along one axis, the other along the other axis. **A component is the projection of a vector on a particular axis.**

For a yacht that sails 20 nautical miles northwest and then 12 nautical miles due south, for example, the simplest axes to choose are

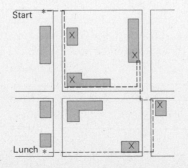

Figure 2.4 Representation of a shopping trip with a vector diagram to find the resultant displacement from the start to the lunch stop. Each stop on the way is indicated by an X. Several vectors in the diagram are actually the sum of two vectors, one horizontal and one vertical.

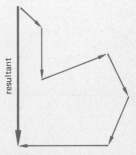

those in the south-north and west-east directions. These are referred to as the Cartesian y and x axes, respectively (Fig. 2.5).

Plotting the vectors that represent the two separate legs of the vessel's voyage on the axes enables us to split each leg into an x and a y component. Each component is the length of the vector along that particular axis. For the northwesterly leg of the yacht's journey, the x and y components are equal in size, both being 14.14 nautical miles. However, the x component is negative and the y component is positive, according to the usual sign convention. For the southerly leg, the calculation is simpler: the x component is plainly zero and the y component is 12 nautical miles, that is, 12 nautical miles in the direction opposite to that in which the positive axis points.

Having obtained the x and y components of each leg, we add the x and y values separately to obtain the x and y components of the yacht's total displacement. Because the two axes form two sides of a right-angle triangle, we can then obtain the net displacement of the yacht by simple

Figure 2.5 Yacht trip. Vector addition using x and y components.

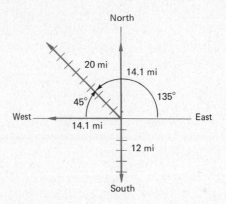

Vector	x Component	y Component
20 mi	$20 \cos 135° = -14.14$ mi	$20 \sin 135° = +14.14$ mi
12 mi	0	-12 mi
Resultant (sum)	-14.14 mi	$+2.14$ mi

(a) Resolution of vectors into their x and y components. The components of the 20-mi vector could also be computed using the supplement of the 135° angle, or 45°.

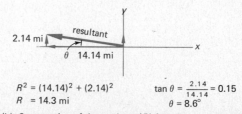

$R^2 = (14.14)^2 + (2.14)^2$ $\tan \theta = \dfrac{2.14}{14.14} = 0.15$
$R = 14.3$ mi $\theta = 8.6°$

(b) Construction of the resultant (R) from its x and y components. The acute angle θ was calculated rather than its supplement.

trigonometry. The displacement is the hypotenuse of a vector triangle whose sides are the x and y components of the total displacement (see our fifth Short Subject: Trigonometry Review).

EXAMPLE 2.5

A ship steams 8 km east and then 6 km northeast. Represent these two displacements by arrows, and determine the distance and direction of the ship from the starting point by both graphic and trigonometric methods.

Solution

First, select a convenient scale, one neither too large nor too small. A scale that uses centimeters is usually more convenient than one that employs inches because of the more common and more easily handled decimal subdivisions. In the diagram (Fig. 2.6), the scale is 1 cm = 2 km. Choosing A as the starting point, lay off the line AB horizontally to the right (east) and 4 cm (8 km) in length. Then represent the second displacement by drawing the arrow BC from B in a direction 45° north of east and 3 cm (6 km) long. The arrow drawn from A to C represents the resultant displacement. Measure the length AC; it is 6.5 cm, which is 13 km on the scale chosen. Measuring the angle BAC with a protractor gives 19° for its value. Thus, the resultant displacement is 13 km in a direction 19° north of east.

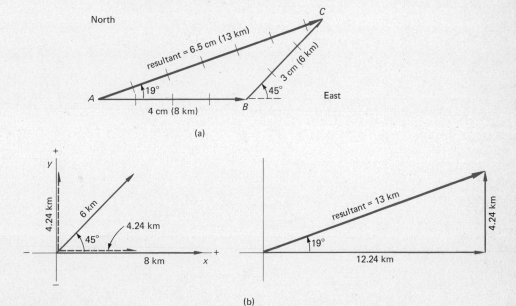

Figure 2.6 (a) Combination of two vectors by the graphic method. **AC** is the resultant of vectors **AB** and **BC**. Scale: 1 cm = 2 km. (b) Addition of two vectors using trigonometry.

To solve this vector addition problem using trigonometry, the two vectors shown in Fig. 2.6(b) on an *xy* coordinate system must be resolved into their *x* and *y* components. These are summarized in the following table:

Vector	x Component	y Component
6 km	+6 cos 45° = +4.24 km	+6 sin 45° = +4.24 km
8 km	+8 km	0
Resultant	+12.24 km	+4.24 km

The *x* and *y* components of the resultant are just the respective sums of all the *x* components and all the *y* components. To obtain the magnitude of the resultant, we use the Pythagorean theorem

$$(\text{Resultant})^2 = (12.24 \text{ km})^2 + (4.24 \text{ km})^2$$

or

$$\text{Resultant} = \sqrt{149.8 + 18.0} = 12.95 \text{ km or } 13 \text{ km}$$

The angle θ that the resultant makes with the positive *x* axis is obtained from the tangent (tan) of the angle:

$$\tan \theta = \frac{\text{opposite side}}{\text{adjacent side}} = \frac{4.24}{12.24} = 0.346$$

The angle whose tangent is 0.346 is 19° as determined from a trigonometry table.

2.4 SPEED AND VELOCITY (A SCALAR AND A VECTOR)

Problems that involve displacement simply scratch the surface of the physics of motion. To understand the basics of motion (kinematics) better, we must introduce the concepts that relate displacement and distance to time in studies of motion. These concepts, one a scalar, the other a vector, are speed and velocity.

The simpler of the two, because it is the scalar, is speed. **Speed is the ratio of the distance traveled by any object,** irrespective of its direction, **to the time it takes to travel that distance.** Actually, we should use the term average speed since we are determining the average value of the speed over the time interval we are considering. Mathematically, this is expressed as

$$\bar{v} = \frac{s}{t} \tag{2.1}$$

where $\bar{v}$ is the average speed of an object between two points, s is the distance traveled between those points, and t is the time the object takes to travel that distance. Note that it is common to denote an average quantity by placing a bar over the symbol.

Velocity, the vector, is related to speed in the same way that displacement is related to distance. Velocity is the ratio of the displacement—which takes account of direction as well as distance—to time interval. Thus, a car driver looking at her speedometer obtains only a measure of her speed alone, but an airplane navigator referring to the plane's instruments (including the compass) calculates the vehicle's velocity—a necessity because winds can alter the direction in which planes travel.

The units in which speed is most conveniently measured depend on what sort of objects are doing the traveling. If we are talking about the movement of trains or planes, we generally use miles per hour or kilometers per hour. When we refer to subnuclear particles, which move extraordinarily fast, it is more convenient to express their speeds in terms of centimeters per second, kilometers per second, or even fractions of the speed of light. And if by chance we decide to discuss the progress of turtles, we might find it best to express their speed in meters per day.

A car driver looking at her speedometer obtains only a measure of her speed alone, but an airplane navigator referring to the plane's instruments calculates the vehicle's velocity.

EXAMPLE 2.6

Light travels from the sun to the earth in about 8.33 min. The speed of light is approximately 186 000 mi/s, or 300 000 km/s. Compute the distance of the earth from the sun.

Solution

We use the fact that the speed equals the distance traveled divided by the time. Hence, the distance from earth to sun equals the speed of light times the time for light to travel from the sun to the earth:

$$s = \bar{v}t = (186\,000 \text{ mi/s}) \times 8.33 \text{ min} \times 60 \text{ s/min} = 93 \times 10^6 \text{ mi}$$

or

$$s = (300\,000 \text{ km/s}) \times 8.33 \text{ min} \times 60 \text{ s/min} = 150 \times 10^6 \text{ km}$$

2.5 RELATIVE VELOCITIES

The value of vector addition and subtraction in solving problems of movement is immediately evident in the concept of relative motion. When we describe the movement of any object in the universe, be it a star or an atom, an elephant or a flea, we must specify the benchmark —the object relative to which we measure the velocity of another object—that we use to observe that motion. A flea's velocity, for example, might be measured in relation to the animal on which it lives. An airplane's velocity is measured relative to the ground beneath it. A planet's velocity is normally specified relative to the star around which it revolves. In every case, the benchmark we use is also moving. **All motion is relative.** The animal on which the flea moves is moving on the surface of the earth; the earth is moving relative to the sun; and the sun is moving relative to the center of its galaxy. In fact, there is no object or location in the whole universe that is "entirely still." The concept that all motion is relative is of vital importance in Einstein's special theory of relativity and post-Einsteinian developments that are even now at the frontier of physics research.

All motion is relative.

Getting back to earth, it is clear that since all motion is relative, we must specify the object or reference frame with respect to which we make our measurements. Physicists normally choose the earth for their frame of reference. However, we are often interested in the relative velocity of two similar objects, such as two airplanes whose velocities are each known relative to the earth. Consider, for example, a passenger plane whose normal air speed is 600 mi/h and that makes a round trip between New York and Paris in windy conditions. This velocity is measured relative to the atmosphere through which the plane moves. On the outward flight, let us imagine that the winds blow steadily at 100 mi/h in the same direction as the plane is flying. By adding the two vectors, whose direction is along the same straight line, we obtain a velocity of 700 mi/h relative to the ground in the direction of New York to Paris.

If the same winds are blowing in the same direction on the return journey, that is, opposite to the direction of the plane's motion, the velocity of the plane relative to the earth is obtained by vector addition—obtained by subtraction of the magnitudes. The result is 500 mi/h along the straight line between Paris and New York.

To take this reasoning a step further, imagine that the jet plane traveling at a velocity of 500 mi/h relative to the earth (on its return journey from Paris to New York) passes a turboprop airliner traveling in exactly the same direction at a speed of 200 mi/h relative to the earth. (The velocity of the turboprop, of course, is 200 mi/h *in the direction from Paris to New York*.) To passengers and crew in the turboprop, the jet plane appears to be approaching at a speed of 500 − 200, that is, 300 mi/h. In other words, the velocity of the jet plane relative to the turboprop is 300 mi/h in the same direction as the turboprop is traveling.

As seen from the jet, the turboprop appears to be flying backward! It is an odd sensation; people in the jet plane see the turboprop coming toward them, tail first, as the speedier jet overtakes the other airliner. In this case the velocity of the turboprop relative to the jet is 300 mi/h in the direction opposite to that of the jet's motion.

EXAMPLE 2.7

You are driving a car along a straight expressway at 55 mi/h relative to the ground and approach another car moving in the same direction at 45 mi/h relative to the expressway. What is the speed of one car relative to the other, and what is the velocity of the slower car relative to you?

Solution

The speed of one car relative to the other is clearly the "common sense" difference between the two speeds: 10 mi/h. However, we should be more complete, and to compute the relative velocity of the slower car with respect to your car, we must be very careful. Consider the direction both cars move relative to the road to be positive. The relative velocity of the slower car with respect to your car is the vector difference between the two velocities measured with respect to the road. Since the vectors lie along a straight line, we take the difference $+45$ mi/h $- 55$ mi/h $= -10$ mi/h. Hence, the slower car moves in the opposite direction to you as measured relative to your faster moving car. This is true both before and after you pass the car.

These examples of relative velocity are simple since they deal with motions in a straight line. The concept can also be applied to forms of motion that are not aligned. Figure 2.7, for example, shows two men trying to paddle a canoe across a river that is running a strong current. At full stretch they can paddle at a speed of 6 mi/h relative to completely still water. To discover their speed relative to the riverbank when the water is moving at a speed of 3 mi/h relative to the same bank, we construct the vector diagram in Fig. 2.7. The vector sum of the two velocities gives us not only the speed of the intrepid canoeists

Figure 2.7 Two canoeists are able to paddle at a speed of 6 mi/h relative to still water. They attempt to cross a river where the water moves at 3 mi/h relative to the bank. The resultant velocity of the canoe relative to the river bank is given by a vector sum.

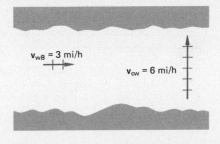

v_{cB} = velocity of the canoe relative to the bank
v_{cw} = velocity of the canoe relative to the water
v_{wB} = velocity of the water relative to the bank

$\tan \theta = \dfrac{3 \text{ mi/h}}{6 \text{ mi/h}} = \dfrac{1}{2}$

$\theta = 26.6°$

$v_{cB} = \sqrt{v_{wB}^2 + v_{cw}^2}$

$v_{cB} = \sqrt{(3 \text{ mi/h})^2 + (6 \text{ mi/h})^2}$

$v_{cB} = \sqrt{(9 + 36)(\text{mi/h})^2}$

$v_{cB} = 6.71$ mi/h

relative to the bank but also the direction in which the combination of their effort and the river's current actually takes the canoe—in other words, the canoe's velocity relative to the riverbank.

EXAMPLE 2.8

An ocean liner is traveling at 18 km/h. A passenger on deck walks toward the rear of the ship at a rate of 4.0 km/h. After walking 30 m in a straight line, she turns and walks at 4.0 km/h toward the rail of the boat. What is her velocity relative to the water while she is walking to the (*a*) rear, (*b*) rail?

Solution

(*a*) Although she walks toward the rear of the ship, she actually will be moving forward, relative to the water, if we take the forward, or positive, direction to be that in which the ship moves relative to the water. The ship moves at a velocity of 18 km/h, whereas the woman moves at a velocity of -4 km/h relative to the ship. Hence the woman's velocity relative to the water is $+18$ km/h -4 km/h $= 14$ km/h in the direction the ship is moving. (*b*) Now, when the woman walks at right angles to the direction of the ship, we can determine her resultant velocity with respect to the water by drawing a vector diagram such as that shown in Fig. 2.8. We represent the forward velocity of the woman, that is, the forward velocity of the ship, 18 km/h, as a vertical vector. We add to that the velocity of the woman with respect to the ship, namely, 4 km/h to the right. The resultant is a vector from the tail of the vector representing the 18-km/h component of her velocity to the head of the vector representing the velocity of the woman across the ship. The resultant makes an angle θ with the forward direction. The angle is calculated from its tangent:

$$\tan \theta = \frac{4 \text{ km/h}}{18 \text{ km/h}} = 0.22 \quad \text{or} \quad \theta = 12.5°$$

The magnitude of the resultant R is calculated by using the Pythagorean theorem:

$$R^2 = (18 \text{ km/h})^2 + (4 \text{ km/h})^2$$

$$R = \sqrt{324 + 16} = 18.4 \text{ km/h}$$

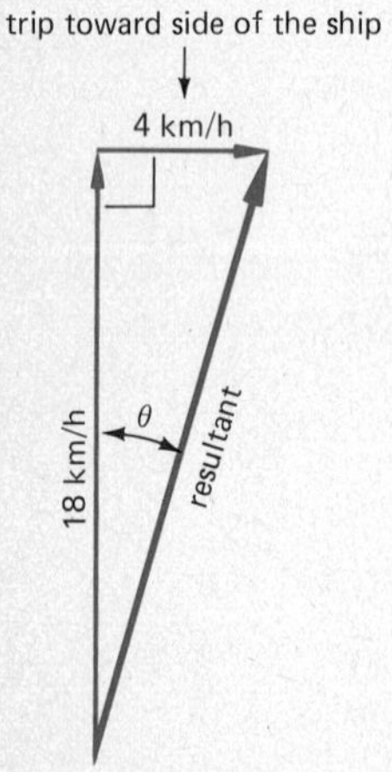

Figure 2.8 Vector diagram of a walk on the deck of a ship.

This result could also be obtained graphically.

EXAMPLE 2.9

An airplane is flying toward a destination 200 mi due east of its starting point, and the wind is northwest, that is, blowing from the northwest, at 30 mi/h. The pilot wishes to make the trip in 40 min. Where should he head the plane? At what air speed should he fly?

Figure 2.9 Vector diagram of an airplane flying in a cross wind.

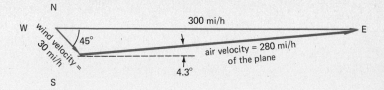

Solution

To solve this problem it is best to construct a vector diagram (Fig. 2.9) using the velocities. The problem requires that the airplane cover the 200 mi in 40 min, or 0.67 ($\frac{2}{3}$) of an hour. The desired ground speed of the plane will, therefore, equal 200 mi/0.67 h, or 300 mi/h. The ground velocity is a vector due east. A northwest wind comes from the northwest and moves toward the southeast. The airplane's air speed must be the remaining side of a vector triangle in which the ground speed vector is the resultant. This problem is solved graphically in Fig. 2.9.

2.6 AVERAGE SPEED AND INSTANTANEOUS SPEED (TWO SCALARS)

The timetable given here is a rough interpretation of Amtrak's actual schedule. The authors assume no responsibility for passenger delays.

According to the Amtrak timetable, express trains that carry passengers between Boston and New York, with a half-dozen intermediate stops, take 4.5 h for the 240-mi journey. Using Eq. (2.1), we see that the average speed of trains between the two cities is 240 mi/4.5 h = 53.3 mi/h.

At various points on the journey, however, the speedometer in the engineer's cab may register a speed of anything between 0 and 80 mi/h. The average speed, 53.3 mi/h, is the speed of the train over the whole journey, taking into account all delays and speedups on the way, and the reading on the speedometer is the instantaneous speed at the time the engineer glances at the instrument's dial.

The relation between average and instantaneous speed is best depicted diagrammatically. Figure 2.10 is a graphic representation of the journey of the Yankee Clipper between Boston and New York City. We have plotted time along the x axis and distance along the y axis. At the points representing New Haven and Providence on the graph, the train spends a certain amount of time without making any progress; here, it has scheduled stops of 10 min. At New London, by contrast, the stopping time is very short, just enough to allow passengers to get on and off the train. Between stations, the distance increases rapidly in relation to time.

Remembering that the train's speed is the ratio of the distance it travels to the time it takes to go that far, we can easily calculate the train's average speed over the whole journey by drawing a line between the starting and finishing points on the graph. The slope of this line represents precisely the total distance traveled by the train divided by the time it took, that is, its average speed between Boston and New York.

Figure 2.10 Distance versus time for the Boston-to-New York train.

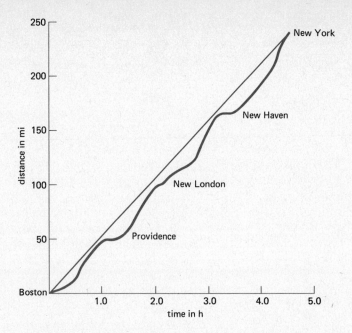

By drawing similar lines between points on the curve and calculating their slope, we can work out the average speed of the train between any two points on its journey—between two of the stations, for example, or just between any two locations along the track. Mathematically, this is expressed as follows:

$$\bar{v} = \frac{s_2 - s_1}{t_2 - t_1} = \frac{\Delta s}{\Delta t} \quad (2.2)$$

where $\bar{v}$ is the train's average speed between two points, for example, mileage markers, along the track s_1 and s_2, which the train passes at times t_1 and t_2 after it sets out on its trip from New York. The Greek letter delta (Δ) means "change"; Δs and Δt represent a change in distance and a time interval, respectively.

Plainly, we can decrease the distances and time intervals between any two points on the journey to very small values. As the points get closer, the line joining them eventually will become simply a tangent to the curve of time against distance. The slope of this tangent is the value of the instantaneous speed of the train and is effectively the speed registered on the train's speedometer. (See our sixth Short Subject: The Slope.)

Let us now return briefly to the concept of average speed. The graph in Fig. 2.11 indicates the speeds recorded on a car's speedometer in miles per hour during the early minutes of the car's journey along a straight freeway, starting at a traffic light on the entrance road to the freeway as the light turns green. As the graph shows, the speed of the

Figure 2.11 Graph of speed versus time.

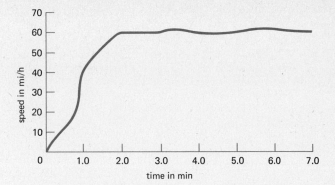

car changes according to a definite pattern until it settles down to an almost steady 60 mi/h. The starting speed is 0 mi/h. In the first minute of the trip, the vehicle's speed rises from that figure to 40 mi/h. The average speed during this minute—calculated approximately by adding the speed at the end of the minute to that at the beginning and dividing by 2, rather than from the distance the car has traveled—works out at $(40 + 0)/2$, or, 20 mi/h. During the second minute, the speed of the car moves up from 40 mi/h, to 60 mi/h, giving an average over this minute of $(40 + 60)/2$, or 50 mi/h. And during the third minute, the average speed is the approximately steady speed during that time, 60 mi/h.

These averages, however, do not provide any indication of just how the speed of the car changed during each of the first 3 min. Most importantly, they do not tell us anything about the changes in the instantaneous velocity of the car. To examine the changes in speed it is convenient to introduce a new concept, acceleration. Recall that the instantaneous speed is only the magnitude of a more important quantity: the instantaneous velocity. Even for motion along a straight line, direction is important because motion of an object can be either forward or backward; that is, there are two options. We need a concept bearing the same relationship to velocity that velocity does to displacement—the concept is acceleration (a vector quantity).

2.7 ACCELERATION

Acceleration is a familiar word to almost everyone, and in physics it is an extremely important quantity. We must therefore define it carefully and exactly. When we "take off" in our car from a stoplight at the instant the light turns green, we say that we accelerate. In this case we have changed our speed in a positive manner, increasing it with time. When we put on our brakes in a car to stop the machine we "decelerate," a special type of acceleration referring to a decrease in speed with time. **This time rate of change of speed—or of velocity in cases where we are also concerned with changes in the direction of travel—is known as** *acceleration.* Whereas velocity is the ratio of the directed distance

The Slope—A Mathematical Definition and an Application to Instantaneous Speed

The study of physics involves the examination of how one variable changes with respect to others. In most cases we simplify our study by examining how the single variable changes in response to just one other variable while holding all the others constant. We can often visualize how variables depend on each other by plotting one variable against another on a graph. Usually one plots the dependent variable on the vertical axis. To represent the rate at which one variable changes with respect to the other, we shall make frequent use of the term *slope*.

As an example, Fig. A(a) shows the displacement of an object as a function of time. For this particular situation the displacement changes with time at a uniform rate over the total time described by this graph. The displacement at time $t = 0$ is taken as s_0 measured with respect to some landmark or point of reference. The rate at which the displacement changes is constant for this particular motion. We can see this by taking the ratio $\Delta s/\Delta t$ at different times, which is the average speed. If we consider the interval from $t = 0$ s to $t = 1$ s, then $\Delta s = 3$ m $- 2$ m $= 1$ m, and $\Delta t = 1$ s $- 0$ s $= 1$ s. Hence, the average speed over the first second is

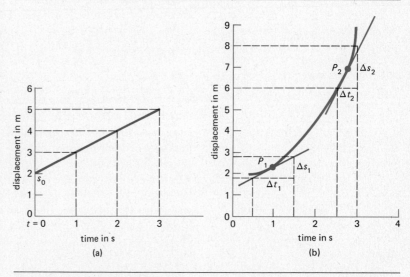

Figure A Computation of speed using the slope of a line. (a) A straight line or uniform speed. (b) Slope taken at two points along a path where the speed changes.

$\bar{v} = \Delta s/\Delta t = 1$ m/s. Over the second second, $\bar{v} = \Delta s/\Delta t = (4$ m $- 3$ m$)/(2$ s $- 1$ s$) = 1$ m/s. Clearly, the value of $\Delta s/\Delta t$ is constant and the same at any instant of time. The quantity $\Delta s/\Delta t$ **taken for this straight-line function is called the slope.** It is generally a useful quantity, as in this case, where it is the speed.

We can extend our definition of slope to functions where the dependence of one variable upon the other is not constant or uniform. Consider the graph of displacement versus time shown in the accompanying figure. Here, the displacement increases for a given time interval as a function of time. That is, as time increases, the graph of displacement versus time

traveled (the displacement) to the time taken for the journey, **acceleration is the ratio of the change in velocity to the time over which this change occurs.**

It must be stressed that acceleration is a vector involving the changes in a vector quantity, the velocity, a fact that leads to some complication, rises more steeply toward vertical. The slope, defined as $\Delta s/\Delta t$, will be different depending on at what times we try to measure this quantity. More precisely, the slope is the time derivative of the displacement, a quantity that forms the basis of calculus. For our use we shall define **the slope at any point on the curve as the slope of a line drawn tangent to the curve at the point of interest.** Two such lines are drawn on Fig. A(b) to illustrate how we find the slope graphically at points P_1 and P_2. For example, at point P_1 the slope will equal

$$\frac{\Delta s_1}{\Delta t_1} = \frac{2.8 \text{ m} - 1.8 \text{ m}}{1.5 \text{ s} - 0.5 \text{ s}} = 1 \text{ m/s}$$

and at point P_2 the slope will equal

$$\frac{\Delta s_2}{\Delta t_2} = \frac{8 \text{ m} - 6 \text{ m}}{3 \text{ s} - 2.5 \text{ s}}$$

$$= \frac{2 \text{ m}}{0.5 \text{ s}} = 4 \text{ m/s}$$

At point P_2 the rate of change of displacement with time is four times greater than at point P_1. Since we have defined $\Delta s/\Delta t$ as speed, this means the speed at time 2.75 s is 4 m/s, and at $t = 1$ s the speed is only 1 m/s. Furthermore, these speeds calculated from the slopes of the tangent lines to these two points represent instantaneous speeds and not average values. The importance of the slope is that it allows us to determine the instantaneous speed.

To summarize this discussion of slope, we could represent the vertical axis as the y axis and the horizontal axis as the x axis and then define the slope as the quantity $\Delta y/\Delta x$ determined for the straight line tangent to a curve at a specific point on the curve. Often students find it helpful to remember slope as "rise over run" for a straight line.

EXAMPLE

Using the data in Fig. 2.11, determine graphically the slope of the curve at time $t = 1$ s and again at time $t = 2.5$ s. What is the particular value of the slope of this v-versus-t curve called?

Solution

We can answer the question just posed by noting that the slope is $\Delta y/\Delta x$ in general. Hence, for a v-versus-t graph, $\Delta y/\Delta x$ corresponds to $\Delta v/\Delta t$, which is acceleration. We now evaluate the instantaneous acceleration after 1 s and 2.5 s by taking the slope of the curve at these two times. At time $t = 1$ s draw, as carefully as you can, the tangent to the curve. Then identify two points on the straight line and drop lines perpendicular to the v and t axes. Then determine $\Delta v/\Delta t$. If we pick these two points as those where the tangent crosses the horizontal line at $v = 60$ mi/h and $v = 20$ mi/h, then the corresponding times are approximately $t = 1.6$ min and 0.6 min. The slope will, therefore, have the value

Instantaneous acceleration

$$= \text{slope} = \frac{60 \text{ mi/h} - 20 \text{ mi/h}}{1.6 \text{ min} - 0.6 \text{ min}}$$

$$= 40 \text{ mi}/(\text{h} \cdot \text{min})$$

In other words, at time $t = 1$ s the speed is changing at a rate of 40 mi/h/min.

To compute the slope at $t = 2.5$ s, we again draw a line tangent to the v-versus-t curve. Clearly, such a line would be horizontal, and the slope would equal zero since there is no change in the parameter plotted $\Delta y = \Delta v = 0$ on the vertical scale.

as we shall see. Acceleration is generally measured in meters per second per second (abbreviated m/s²), feet per second per second (abbreviated ft/s²), or even in mixed units such as miles per hour per second.

Because velocity has both magnitude and direction, any vehicle or other object can accelerate in either of two ways. The magnitude of the

A model electric train moving around a circular track at constant speed is also accelerating because the direction in which it moves is constantly changing.

velocity, the speed, can change, or the direction in which the object or vehicle is moving can change, or both can change simultaneously. Clearly, an object that speeds up or slows down even if it still maintains its original direction is undergoing acceleration. However, a model electric train moving around a circular track at constant speed is also accelerating because the direction in which it moves is constantly changing. For a more general motion, such as that of a car on the highway or a train on its tracks, both the magnitude of the velocity, the speed, and the direction of motion are changing constantly. To simplify our discussion somewhat, we can for now consider only rectilinear motion, motion along a straight line. This enables us to consider acceleration as changes only in the magnitude of the velocity, that is, changes in speed. Later we shall examine how changes in direction of the velocity are described by an acceleration vector.

Like speed and velocity, acceleration can be measured in both average and instantaneous terms. Average acceleration is given by the simple formula written in vector symbols:

$$\overline{\mathbf{a}} = \frac{\mathbf{v} - \mathbf{v}_0}{t - t_0} = \frac{\Delta \mathbf{v}}{\Delta t} \tag{2.3}$$

where the velocity changes from $\mathbf{v}_0$ to $\mathbf{v}$ during an interval of time Δt. For rectilinear motion, $\Delta \mathbf{v}/\Delta t$ represents the change in speed computed algebraically. When this time interval becomes so small that it is difficult or "effectively impossible" to measure, the same formula gives the instantaneous acceleration rather than just the average. In graphic terms, the instantaneous acceleration at any time is the slope of the tangent to the curve that relates speed to time at that particular time.

In a large number of cases involving motion, the acceleration is constant. When this occurs, we can extend this equation, and those equations relating speed and displacement, by simple algebra to provide a series of ready-made formulas that relate acceleration, speed, distance, and time. These formulas, derived on page 35, are the algebraic keys for unlocking a number of simple motion problems.

One of the most common examples of constant, or uniform, acceleration is the type of movement known as **free fall.** This refers to the fall of any object—a penny, an elephant, or an airplane—under the influence of gravity alone. This condition precludes the force of air resistance. Ideally, therefore, we require a vacuum to illustrate free fall motion; the fall of a hammer and a feather dropped by Apollo 15 astronaut David Scott on the airless lunar surface (see Fig. 2.12) is an excellent example of free fall. This type of motion can also be induced artificially on earth, inside a vacuum chamber. And if, as is often the case in scientific experiments, we accept a certain amount of imprecision as unavoidable, we can regard the fall to earth of a small solid spherical object like a ball bearing as a practical example of free fall. In this particular example the effect of air resistance in holding up the ball

Figure 2.12 Astronaut David Scott demonstrates that all objects fall toward the moon with the same acceleration. (*NASA*)

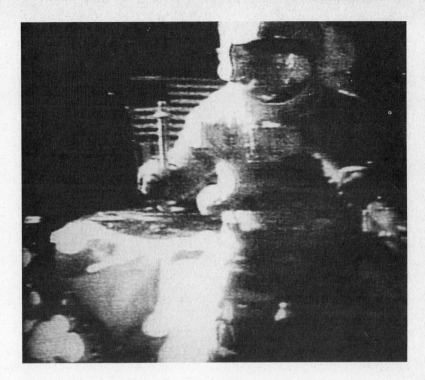

The incident in which Galileo reputedly demonstrated the principle to the skeptical intellectuals by toppling cannonballs of various weights off the leaning tower of Pisa cannot be substantiated.

bearing is thousands of times less than the effect of earth's gravity in pulling it down.

Galileo first discovered the basic physics of free fall motion in an even less precise series of experiments. He rolled metal balls down a plane, a wooden track inclined to the horizontal, forcing gravity to compete against not only air resistance but also friction. But because gravity was easily the predominant force acting on the balls, Galileo was able to see the forest for the trees; that is, Galileo was able to interpret the somewhat rough measurements he made in terms of a universal physical law.

Galileo measured the time the balls took to pass equally spaced marks on the track when the track was inclined at a variety of angles to the horizontal. From these measurements Galileo worked out the average velocities and accelerations of the balls during their roll down the track. For each inclination of the track, the balls had constant acceleration. Furthermore, the actual value of acceleration increased with increasing inclination. Extending his readings to a ball rolling down a vertical track, Galileo could calculate the acceleration of the ball undergoing free fall. Modern measurements yield a value for this free fall acceleration of 32.2 ft/s^2, 9.8 m/s^2, or 980 cm/s^2. This is actually the free fall acceleration of the ball.

Free fall acceleration is exactly the same for any object, be it a tiny ball bearing or a huge cannonball, a feather or an elephant.

Galileo's best-known contribution to the science of kinematics was his discovery that this free fall acceleration is exactly the same for any object, be it a tiny ball bearing or a huge cannonball, a feather or an elephant. Given an absence of air resistance, each object will fall to earth with exactly the same acceleration.

The free fall acceleration, that is, the acceleration caused by the earth's gravity, is identified by the symbol g. This acceleration is one of the most often utilized factors in any calculation involving movement on or around the earth.

Of course, the value of g is different on different planets. Space scientists must use a value for acceleration due to gravity in their computations of a lunar landing different from the one they use in a splashdown situation on earth. In fact, the value of g also changes to some extent on earth, varying basically with altitude but also depending on latitude, topography, and subsurface structure. This variation has been observed by ultrasensitive equipment.

The g value can also be observed and measured with reasonable accuracy by using a camera and a regularly blinking light known as a strobe light. A billiard ball is dropped across the camera's field of view in a dark room; its progress in falling is captured on film at regular intervals when the blinking strobelight illuminates it, as shown in Fig. 2.13. Table 2.2 is a typical set of such measurements from which the value of g is easily derived, using the basic equations on page 35.

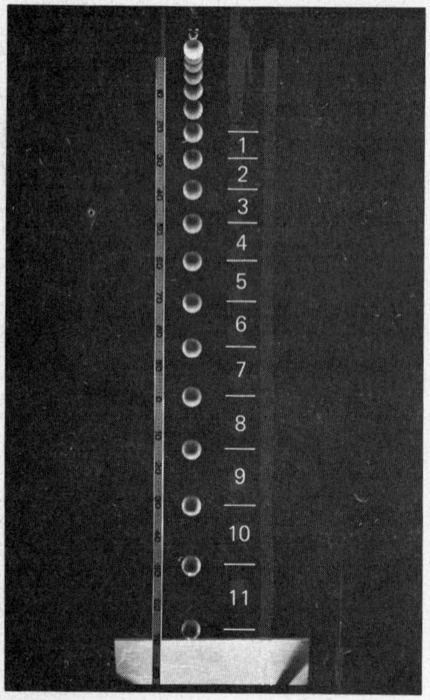

Figure 2.13 A flash photograph of a falling billiard ball. The position scale is in centimeters, and the time interval between successive positions of the ball is $\frac{1}{30}$ s. This motion is analyzed in Table 2.2. (*From PSSC Physics, D. C. Heath & Company, Lexington MA, 1965*).

Table 2.2 Calculated Values of the Acceleration of the Falling Ball Whose Position at Each 1/30 s Is Shown in Fig. 2.13

Interval Number	Displacement Δx(cm)	Average Speed $\Delta x/\Delta t = v$(cm/s)	Change in Average Speed Δv(cm/s)	Acceleration $\Delta v/\Delta t$(m/s^2)
1	7.70	231		
2	8.75	263	32	9.6
3	9.80	294	31	9.3
4	10.85	326	32	9.6
5	11.99	360	34	10.2
6	13.09	393	33	9.9
7	14.18	425	32	9.6
8	15.22	457	32	9.6
9	16.31	489	32	9.6
10	17.45	524	35	10.5
11	18.52	556	32	9.6
			Average	9.75

Distance measurements are estimated to a fraction of a millimeter. The acceleration calculated for each interval remains constant within the limits of the measurement.

Equations of Motion for Constant Acceleration

When the acceleration is constant, the displacements, velocities, and accelerations of moving objects are related by a series of simple equations. Consider the initial values of time, displacement, and velocity as $t = 0$, $s = 0$, and $v = v_0$. First let us determine the position as a function of time. Begin with the definition of the average velocity

$$\bar{v} = \frac{s - s_0}{t - t_0}$$

which for our set of initial conditions reduces to

$$s = \bar{v}t \tag{2.4}$$

The average velocity over time t will just equal one-half the sum of the original velocity and the velocity at time t, or

$$\bar{v} = \frac{v + v_0}{2} \tag{2.5}$$

For this special case of constant acceleration, the average velocity is equal to the instantaneous velocity just halfway between the velocity at the beginning and the end of the time interval. Substituting this value for the average velocity into Eq. (2.4) yields

$$s = \bar{v}t = \left(\frac{v + v_0}{2}\right)t \tag{2.6}$$

If we now use the definition of acceleration [Eq. (2.3)], we can obtain a value for the instantaneous velocity v, which can be substituted into Eq. (2.6):

$$a = \frac{v - v_0}{t - t_0} \tag{2.3}$$

Since $t_0 = 0$,

$$v = v_0 + at \tag{2.7}$$

Substituting into Eq. (2.6),

$$s = \left(\frac{v_0 + at + v_0}{2}\right)t$$

$$= v_0 t + \tfrac{1}{2}at^2 \tag{2.8}$$

Equation (2.8) allows one to calculate directly the distance traveled after time t in terms of the initial speed v_0 and the *constant* acceleration.

With a little additional algebra, Eq. (2.6) can be combined with Eq. (2.7) to yield an equation in which time does not explicitly appear. For example,

$$s = \bar{v}t = \left(\frac{v + v_0}{2}\right)t \tag{2.6}$$

Using Eq. (2.7), $v = v_0 + at$, we can solve for t:

$$t = \frac{v - v_0}{a}$$

Substituting this into Eq. (2.6) yields

$$s = \left(\frac{v + v_0}{2}\right)t$$

$$= \left(\frac{v + v_0}{2}\right)\left(\frac{v - v_0}{a}\right)t$$

$$= \frac{v^2 - v_0^2}{2a}$$

$$v^2 = v_0^2 + 2as \tag{2.9}$$

We can summarize these results for uniform (that is, constant) acceleration in three equations, which we shall use again and again throughout this text:

$$v = v_0 + at \tag{2.7}$$
$$s = v_0 t + \tfrac{1}{2}at^2 \tag{2.8}$$
$$v^2 = v_0^2 + 2as \tag{2.9}$$

EXAMPLE 2.10

While an automobile is in second gear, its speed along a straight road increases from 10 km/h (about 6 mi/h) to 35 km/h (about 21 mi/h) in 3 s. Compute the average acceleration.

Solution

Since by definition the average acceleration is the time rate of change of velocity, note that the velocity along a straight line changes by 25 km/h in 3 s; so it changed at an average rate (25 km/h)/3 each second or (8.33 km/h)/s. Calculating the average acceleration in a formal manner, we obtain

$$\bar{a} = \frac{v - v_0}{t} = \frac{35 \text{ km/h} - 10 \text{ km/h}}{3 \text{ s}} = \frac{25 \text{ km/h}}{3 \text{ s}} = \frac{8.33 \text{ km/h}}{\text{s}}$$

The result is read as 8.33 km/h/s which means that, on the average, the speed has increased by 8.33 km/h during each second. *Note:* 8.33 km/h = 5.17 mi/h.

EXAMPLE 2.11 An automobile maintains a constant acceleration in the direction of motion of 8 m/s². If its initial speed was 20 m/s, what will its speed be after 6 s?

Solution Using the definition of acceleration [(Eq. 2.3)], or Eq. (2.7),

$$a = \frac{v - v_0}{t - t_0}$$

Since $t_0 = 0$, then $v = v_0 + at = 20 \text{ m/s} + (8 \text{ m/s}^2)(6 \text{ s}) = 68 \text{ m/s}$.

EXAMPLE 2.12 A moving object increases its speed uniformly from 2 to 4 m/s in 2 min. What is its average speed? How far did it travel in the 2 min?

Solution The average speed can be found directly from the usual technique for an "average" as given in Eq. (2.5):

$$\bar{v} = \frac{v + v_0}{2} = \frac{4 \text{ m/s} + 2 \text{ m/s}}{2} = 3 \text{ m/s}$$

The distance traveled is just obtained from Eq. (2.2). From the definition of average velocity $\bar{v} = \Delta s / \Delta t$,

$$\Delta s = \bar{v} \Delta t = (3 \text{ m/s})(2 \text{ min} \times 60 \text{ s/min}) = 360 \text{ m}$$

EXAMPLE 2.13 How far in feet does a body fall during the first second after it is dropped?

Solution The beginning velocity is 0. Since the acceleration due to gravity is 32 ft/s², the velocity at the end of 1 s is about 32 ft/s. Hence the average velocity for the first second of fall, $\bar{v}_1$, is

$$\bar{v}_1 = \frac{v_0 + v_1}{2} = \frac{0 + 32 \text{ ft/s}}{2} = 16 \text{ ft/s}$$

$$s_1 = \bar{v}t = 16 \text{ ft/s} \times 1 \text{ s} = 16 \text{ ft}$$

or

$$s_1 = v_0 t + \tfrac{1}{2}at^2$$
$$= 0 + \tfrac{1}{2} \times 32 \text{ ft/s}^2 \times (1 \text{ s})^2 = 16 \text{ ft}$$

EXAMPLE 2.14

An object is dropped from a stationary helicopter hovering at a height of 1000 ft above the earth's surface. Neglecting the effect of friction, compute in English units the (a) velocity of the object 3 s after release, (b) distance fallen during the first 3 s, (c) distance fallen during the third second.

Solution

(a) Because the acceleration is 32 ft/s², the body is falling faster each second by 32 ft/s. Thus, starting from rest, the velocity at the end of 1 s will be 32 ft/s, at the end of 2 s, 32 ft/s + 32 ft/s = 64 ft/s, and at the end of 3 s, 32 ft/s + 32 ft/s + 32 ft/s = 96 ft/s, or

$$v = v_0 + at$$
$$v_3 = 0 + 32 \text{ ft/s}^2 \times 3 \text{ s} = 96 \text{ ft/s}$$

(b) The average velocity for the 3-s period from $t = 0$ to $t = 3$ s is

$$\bar{v}_3 = \frac{v_0 + v_3}{2} = \frac{0 + 96 \text{ ft/s}}{2} = 48 \text{ ft/s}$$

so $s_3 = \bar{v}_3 t = 48 \text{ ft/s} \times 3 \text{ s} = 144 \text{ ft}$ or,

$$s = v_0 t + \tfrac{1}{2}at^2$$
$$s_3 = 0 + \tfrac{1}{2} \times 32 \text{ ft/s}^2 \times (3 \text{ s})^2 = 144 \text{ ft}$$

(c) Because the velocity at the beginning of the third second is 64 ft/s and at the end of the third (3d) second is 96 ft/s,

$$\bar{v}_{3d} = \frac{64 \text{ ft/s} + 96 \text{ ft/s}}{2} = \frac{160 \text{ ft/s}}{2} = 80 \text{ ft/s}$$

$$s_{3d} = \bar{v}_{3d} t = 80 \text{ ft/s} \times 1 \text{ s} = 80 \text{ ft}$$

This answer of 80 ft could also be obtained by computing separately the distances fallen in 3 s (144 ft) and in 2 s:

$$s_2 = \bar{v}_2 t = \frac{0 + 64 \text{ ft/s}}{2} \times 2 \text{ s} = 64 \text{ ft}$$

Subtracting the latter from the former,

$$s_{3d} = s_3 - s_2 = 144 \text{ ft} - 64 \text{ ft} = 80 \text{ ft}$$

EXAMPLE 2.15

A person standing on a bridge high over the Colorado River throws a stone vertically downward with an initial velocity of 500 cm/s. Neglecting air friction, compute the (*a*) speed of the stone after 4 s, (*b*) distance through which the stone falls during the first 4 s, and (*c*) distance fallen during the fourth second.

Solution

Since all quantities are **measured downward,** they can all be taken as **positive** (note that then $g = +9.8$ m/s²).

(*a*) $v = v_0 + at$. After 4 s,

$$v_4 = v_0 + gt = 5 \text{ m/s} + 9.8 \text{ m/s}^2 \times 4 \text{ s} = 44.2 \text{ m/s}$$

(*b*) The distance fallen during the first 4 s is given by

$$s = v_0 t + \tfrac{1}{2} at^2$$

$$s_4 = v_0 t + \tfrac{1}{2} gt^2 = 5 \text{ m/s} \times 4 \text{ s} + (\tfrac{1}{2})(9.8 \text{ m/s}^2) \times (4 \text{ s})^2 = 98.4 \text{ m}$$

(*c*) The distance traveled during the fourth second can be computed by finding the distance traveled during the first 3 s and subtracting the result from that obtained in (*b*):

$$s_3 = v_0 t + \tfrac{1}{2} gt^2 = 5 \text{ m/s} \times 3 \text{ s} + (\tfrac{1}{2})(9.8 \text{ m/s}^2)(3 \text{ s})^2 = 59.1 \text{ m}$$

Distance traveled during the fourth second = $s_4 - s_3 = 98.4$ m $-$ 59.1 m = 39.3 m.

2.8 PROJECTILE MOTION (VECTORS IN TWO DIMENSIONS)

A fruitful area of application of the principle of free fall motion is the movement of projectiles (Fig. 2.14). A projectile might be a mortar shell shot out of a gun, a baseball hurled toward the catcher by a right fielder trying to cut off a run at the plate, a snowflake blown out of the rear end of a snow blower on a freezing winter morning, an arrow in flight, or even a bomb released from an airplane.

Neglecting the effects of air resistance, as physicists frequently do to simplify the problem, it is clear that the change in vertical motion of a projectile is caused by gravity alone. The projectile moves in the horizontal direction with a constant velocity—the horizontal component of the velocity imparted to it by the blow that set it in motion. In other words, if there were no air, the only acceleration of a projectile from the moment it leaves the gun barrel, the fielder's hand, the snow blower, the bow, or the airplane would be that due to the earth's gravity, which acts in a vertical direction downward; there is no

If there were no air, the only acceleration of a projectile would be that due to the earth's gravity.

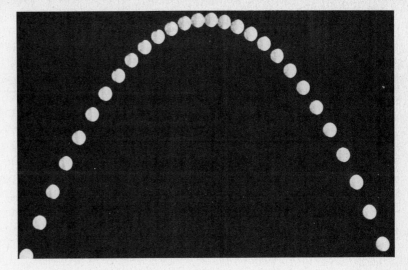

Figure 2.14 This photograph was obtained in a timed strobe that results in a picture taken at regular intervals, in this case every $\frac{1}{30}$ s. Note that the horizontal distance traveled during each time interval is essentially the same. The result is a constant velocity in the horizontal direction. This is not true for the accelerated motion in the vertical direction. (*Kodansha Co. Ltd., Tokyo*)

horizontal acceleration whatever on the projectile, even though it is moving partly in a horizontal direction. We shall justify and explain this in Chap. 3 when we examine Newton's laws.

Because projectiles move in both horizontal and vertical directions simultaneously, their movements can best be understood by using vectors. The simplest way to analyze the motion is to set up a grid in which the x axis represents the horizontal direction and the y axis represents the vertical direction.

The diagram in Fig. 2.15 shows the motion of a projectile viewed broadside and plotted on just such a grid. After being shot into the air at the point at which both x and y have the value zero, the projectile arcs upward and then downward in a perfectly symmetrical parabolic path before hitting the ground, that is, returning to the line along which y has the value zero. **The horizontal distance traveled by the projectile along the x axis is called the *range*.**

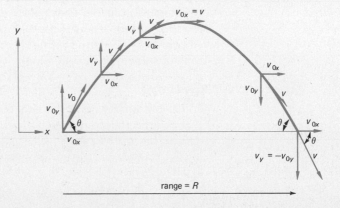

Figure 2.15 Projectile motion. At time $t = 0$, velocity $\mathbf{v} = \mathbf{v}_o$, with components $v_{ox} = v_o \cos \theta$ and $v_{oy} = v_o \sin \theta$.

The velocity of the projectile is continually changing during its flight, as shown in Fig. 2.15. This alteration is actually less complicated than it first appears. We can split up the velocity vector into its horizontal (x) and vertical (y) components, as shown in Fig. 2.15. The horizontal component of the velocity remains constant throughout the flight of the projectile, and the vertical component has a maximum value upward at the start of the projectile's flight, falls to zero at the peak of the projectile's soaring curve, and then rapidly builds up in the downward direction as the projectile drops to the ground. By the time it reaches the ground, the projectile is traveling downward with the same vertical speed that it possessed in the upward direction at the start of its flight—assuming that it hits the ground on a level with its starting point. The motion is symmetrical about a vertical line through the highest point reached by the projectile.

The vertical component of a projectile's velocity can actually be separated from the total velocity visually if one stands in a line with the projectile's starting and finishing points. From afar, the projectile seems simply to rise up in the air and then fall back directly. A typical example of this sort of observation, which illustrates the relative nature of motion, occurs in baseball when the batter hits a pop fly toward second base. A spectator in the deep center field bleachers will see the ball rise up and then down and realize only that it also moved horizontally when he sees the second baseman, shortstop, or center fielder make the catch.

Dealing with projectile motion somewhat more analytically, we can separate the x and y components of the velocity at any point during the flight. If the projectile is fired with an initial speed of v_0 at an angle θ to the horizontal, simple trigonometry shows that the horizontal component of its velocity (which remains the same throughout the flight) is given by the equation

$$v_{0x} = v_0 \cos \theta$$

In the special case in which θ is 90°, the horizontal component of the velocity is zero; this is, of course, free fall motion in which the projectile is hurled vertically up in the air.

Remembering that distance traveled at a constant speed is equal to the speed multiplied by the time of travel, we can obtain the horizontal component x of the projectile's displacement at any time t after the motion started by the equation

$$x = v_{0x}t = (v_0 \cos \theta)t$$

The most important horizontal distance we are likely to need is that between the projectile's starting and finishing points. This is the range, denoted by the symbol R. Its value can be computed as long as we know the total time the projectile spends in the air. If this total time is symbolized as T, then

$$R = (v_0 \cos \theta) T$$

The projectile's motion in the vertical direction is an entirely different problem from that in the horizontal direction because the vertical velocity is not constant. Instead, the projectile is accelerating vertically throughout the flight under the influence of earth's gravity. The value of this acceleration is g, and because the earth is pulling the projectile downward, in the opposite direction to that of the y axis in the graphic grid, we give this a negative sign. In other words, the constant vertical acceleration of the projectile is $-g$.

Trigonometry provides us with the original vertical velocity of the projectile:

$$v_{0y} = v_0 \sin \theta$$

And using the simple equations of motion derived on page 35, we can quickly obtain the formulas for the projectile's height (y) and the vertical component of its velocity (v_y) at any time t after it is blasted into the air. The equations are

$$y = v_{0y}t + \tfrac{1}{2}at^2 = v_0(\sin \theta)t - \tfrac{1}{2}gt^2$$

$$v_y = v_{0y} + at = v_{0y} - gt = v_0 \sin \theta - gt$$

It is rather easy, using these equations and a few observations concerning the motion of the projectile, to compute the time T during which the projectile is in flight. To compute the total time of flight in Fig. 2.15, we use the observation that the projectile is at the very top of its arc halfway through the flight. At this time, which from symmetry is clearly one-half the total time in flight, the projectile's vertical velocity is zero. Actually, at this highest point in the trajectory the projectile must stop moving in the vertical direction for an instant as it "turns around" and begins to experience an increase in its speed in a downward direction. We can set $v_y = 0$ at this time, $t = T/2$ (one-half the time of flight T), in our equation for the vertical speed and obtain the time of flight:

$$v_y = v_{0y} + at = v_{0y} - gt$$

$$0 = v_{0y} - g\frac{T}{2}$$

In other words, the time to reach the highest point is

$$\frac{T}{2} = \frac{v_{0y}}{g}$$

and the total time in flight becomes

$$T = \frac{2v_{0y}}{g} = \frac{2v_0 \sin \theta}{g}$$

This value for the total time in air for this symmetrical trajectory can now be used in our expression for the range to provide an expression for the range in terms of the initial velocity of the projectile and the acceleration due to gravity:

$$R = (v_0 \cos\theta)T = (v_0 \cos\theta)\frac{(2v_0 \sin\theta)}{g}$$

$$R = \frac{2v_0^2}{g}\cos\theta \sin\theta = \frac{v_0^2}{g}\sin 2\theta$$

These simple equations can be used, along with a basic knowledge of vectors, to solve problems involving projectiles of every variety, from golf balls to cannon shells.

EXAMPLE 2.16

A golf ball leaves a tee at an angle of 53° with a velocity of 30 m/s. Find (a) its velocity after 4 s, (b) the maximum height the ball reaches, (c) the range of the ball. Refer to Fig. 2.16.

Solution

This problem involves several vectors because at times the ball moves up and at times it moves down. The acceleration **a** is downward, so $\mathbf{a} = \mathbf{g} = -9.8$ m/s², and **v** will be taken positive when upward. Of course, the displacement in the vertical direction will always be positive as long as the ball is above the horizontal position from which it was hit.

Golfers often end up with negative displacements in the y direction because tees are frequently at higher altitudes than the fairways.

(a) To find the velocity after 4 s we must first obtain the components of the initial velocity at $t = 0$:

$$v_{0x} = +v_0 \cos\theta = 30 \text{ m/s} \times \cos 53° = 18.05 \text{ m/s}$$
$$v_{0y} = +v_0 \sin\theta = 30 \text{ m/s} \times \sin 53° = 23.96 \text{ m/s}$$

$v_x = v_{0x} =$ constant; hence after 4 s the horizontal component of the velocity will still be 18.05 m/s. To find the y component of the velocity, we employ

$$v_y = v_{0y} + at$$
$$= +v_{0y} - gt = +23.96 \text{ m/s} - 9.80 \text{ m/s}^2 \times 4 \text{ s}$$
$$= (23.96 - 39.20) \text{ m/s} = -15.24 \text{ m/s}$$

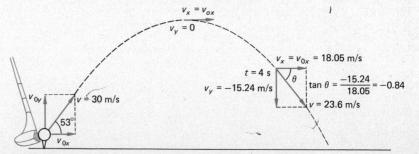

Figure 2.16 Projectile motion of a golf ball.

The negative sign means that the ball has passed its highest point and is dropping. To compute the velocity, we must compute both the magnitude and direction as specified by θ on the diagram.

To compute the instantaneous speed, we use the Pythagorean theorem:

$$v^2 = v_x^2 + v_y^2$$
$$v = \sqrt{v_x^2 + v_y^2} = \sqrt{(18.05 \text{ m/s})^2 + (15.24 \text{ m/s})^2}$$
$$= \sqrt{558.1 \text{ (m/s)}^2} = 23.62 \text{ m/s}$$

To find the direction, we use the tangent of the angle θ:

$$\tan \theta = \frac{-15.24}{18.05} = -0.84 \quad \text{or} \quad \theta = -40.2°$$

Note that since the y component of the velocity is negative, this angle ($-40.2°$) measured with respect to the horizontal is below the horizontal (see Fig. 2.16).

(b) To find the maximum height reached by the ball, we use the fact that $v_y = 0$ when the ball reaches the highest point. Also recall that the time (t) to reach this highest point is one-half the total time the ball is in the air; that is, $t = T/2$

$$v_y = v_{0y} + at$$
$$v_y = v_{0y} - g\frac{T}{2}$$

Since $v_y = 0$, then

$$0 = +23.96 \text{ m/s} - 9.8 \text{ m/s}^2 \frac{T}{2}$$

$$\frac{T}{2} = \frac{23.96 \text{ m/s}}{9.8 \text{ m/s}^2} = 2.44 \text{ s}$$

$$y = (v_0 \sin \theta)t - \tfrac{1}{2}gt^2$$

$$y_{\max} = (v_0 \sin \theta)\frac{T}{2} - \frac{1}{2}g\left(\frac{T}{2}\right)^2$$

$$= (23.96 \text{ m/s})(2.44 \text{ s}) - \tfrac{1}{2}(9.8 \text{ m/s}^2)(2.44 \text{ s})^2 = 29.29 \text{ m}$$

(c) To find the range of the ball, we can use the symmetry of the projectile motion and note that the time to reach the maximum height, $t = 2.44$ s, is just one-half the time T the ball is in the air. Since the ball travels at a constant speed in the horizontal direction, we have for the range R

$$R = v_{0x}T = 18.05 \text{ m/s} \times 2(2.44 \text{ s}) = 88.1 \text{ m}$$

As an alternative route, we could use our previously developed expression for the range in terms of the initial parameters:

$$R = v_{0x}T = \frac{(v_0 \cos\theta)(2v_0 \sin\theta)}{g} = \frac{2v_0^2 \cos\theta \sin\theta}{g}$$

$$R = \frac{2(30 \text{ m/s})^2(\cos 53°)(\sin 53°)}{9.8 \text{ m/s}^2} = 88.3 \text{ m}$$

This agrees with the previous result within rounding-off errors.

EXAMPLE 2.17

A baseball is thrown vertically upward from the roof of a tall building with an initial speed of 20 m/s. Compute the (*a*) time required to reach its maximum height, (*b*) maximum height, (*c*) position and velocity after 1.0 s, (*d*) ball's position and velocity after 5 s.

Solution

To consider this problem, refer to Fig. 2.17. We shall arbitrarily consider vectors downward as negative and vectors upward as positive.
(*a*) The given data are:

$s = 0$ at the top of the building
$v_0 = +20$ m/s (upward)
$a = -g = -9.8$ m/s² downward

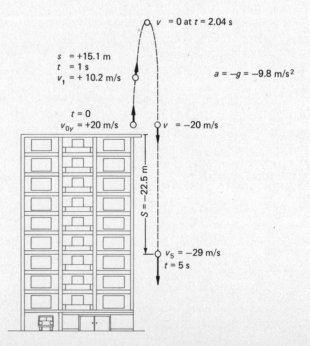

Figure 2.17 A ball thrown vertically upward from the top of a building.

To find the time required to reach the maximum height, we use the fact that the instantaneous velocity of the ball is zero at its highest point. To find the time to reach the maximum height, we use

$$v = v_0 + at$$

$$t = \frac{v - v_0}{a} = \frac{0 - v_0}{-g} = \frac{-20 \text{ m/s}}{-9.8 \text{ m/s}^2} = 2.04 \text{ s}$$

(b) The maximum height occurs when $t = 2.04$ s, so

$$s = v_0 t + \tfrac{1}{2} a t^2$$
$$s = v_0 t - \tfrac{1}{2} g t^2$$
$$s_{\max} = (20 \text{ m/s})(2.04 \text{ s}) - \tfrac{1}{2}(9.8 \text{ m/s}^2)(2.04 \text{ s})^2 = 20.4 \text{ m}$$

The $s_{\max}$ value could also be found using

$$s = \bar{v} t = \frac{(v + v_0)}{2} t = \frac{v_0}{2} t = \frac{20 \text{ m/s}}{2}(2.04 \text{ s}) = 20.4 \text{ m}$$

(c) To find the position and velocity after 1.0 s:

$$s = v_0 t + \tfrac{1}{2} a t^2$$
$$s = v_0 t - \tfrac{1}{2} g t^2$$
$$s = (20 \text{ m/s})(1 \text{ s}) - \tfrac{1}{2}(9.8 \text{ m/s}^2)(1 \text{ s})^2 = 15.1 \text{ m}$$
$$v = v_0 + at = v_0 - gt$$
$$v_1 = 20 \text{ m/s} - 9.8 \text{ m/s}^2 (1 \text{ s}) = (20 - 9.8) \text{ m/s}$$
$$= 10.2 \text{ m/s, upward}$$

(d) To find the position and velocity after 5 s:

$$s = v_0 t - \tfrac{1}{2} g t^2$$
$$s_5 = (20 \text{ m/s})(5 \text{ s}) - \tfrac{1}{2}(9.8 \text{ m/s}^2)(5 \text{ s})^2$$
$$s_5 = (100 - 122.5) \text{ m} = -22.5 \text{ m}$$

The negative sign means that the ball is below the level of the roof of the building.

The velocity of the ball after 5 s is given by

$$v = v_0 + at$$
$$v_5 = 20 \text{ m/s} - (9.8 \text{ m/s}^2 \times 5 \text{ s})$$
$$v_5 = (20 - 49) \text{ m/s} = -29 \text{ m/s}$$

The negative sign indicates that the velocity is downward.

EXAMPLE 2.18

A diver in Acapulco dives off the top edge of a vertical cliff 200 m above the ocean at a speed of 3 m/s in a horizontal direction (see Fig. 2.18) (a) How long will it be before he hits the water? (b) What is the

Figure 2.18 A Mexican diver diving from a cliff into the sea in projectile motion.

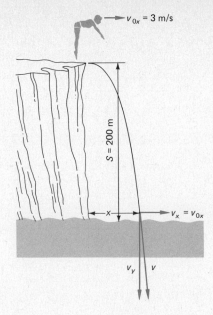

horizontal distance from the foot of the cliff to the point where the diver hits the water? (c) What are the horizontal and vertical components of the diver's velocity just prior to striking the ocean?

Solution

(a) The time required to reach the water is determined only by the free fall, vertical motion. Initially, the velocity in the y direction is zero. The diver will drop 200 m in time t. Note that in this problem we again shall consider all upward vectors as positive and all downward vectors as negative:

$$y = v_{0y}t + \tfrac{1}{2}at^2 = -\tfrac{1}{2}gt^2 \quad \text{since } a = -g$$

To compute the time of drop, we use the fact that the downward displacement is negative:

$$t^2 = \frac{y}{-\tfrac{1}{2}g} = -\frac{-200 \text{ m}}{-\tfrac{1}{2}(9.8 \text{ m/s}^2)} = +40.82 \text{ s}^2$$

$$t = 6.39 \text{ s}$$

(b) The horizontal distance covered is determined from knowledge of the initial speed in the horizontal direction, which remains constant, and that the diver is in the air for 6.39 s:

$$x = v_{0x}t = 3 \text{ m/s} \times 6.39 \text{ s} = 19.2 \text{ m}$$

(c) To find the horizontal and vertical components of the velocity just before the diver strikes the water, we separate the motion into horizontal motion, for which the speed remains constant, and the vertical motion, which is just free fall:

$$v_x = v_{0x} = \text{constant} = +3 \text{ m/s}$$
$$v_y = v_{0y} - gt = -gt = -9.8 \text{ m/s}^2 \times 6.39 \text{ s}$$
$$= -62.6 \text{ m/s, downward}$$

We could carry this problem further to compute the vector resultant of v_x and v_y, which is the velocity of the diver just as he strikes the water. However, in this case the diver almost hits the water in a vertical direction since his horizontal component of velocity is so small, about one-twentieth the vertical component.

SUMMARY

1. Length is an interval of distance. Area is length multiplied by length; volume is area multiplied by length.
2. Scalar quantities are those that are independent of direction. Vectors have a magnitude, and they depend upon direction. The shortest distance between two points, including the distance's direction, is called displacement.
3. Vectors not in a straight line are added and subtracted using simple trigonometry.
4. Average speed is a scalar quantity, the ratio of the distance an object travels to the time of travel. Average velocity is a vector, the ratio of an object's displacement to the time interval during which the displacement occurs.
5. Relative velocities of objects are obtained using vector addition and subtraction.
6. The average speed is the slope of a line connecting the starting and finishing points on a graph of distance against time. The instantaneous speed at any point is the tangent to the curve of distance against time at that point. Instantaneous velocity is the instantaneous speed directed along the tangent to the path.
7. Acceleration, a vector quantity, is the time rate at which velocity changes.
8. An important example of constant acceleration is free fall, the motion of any object under the influence of gravity alone. In the absence of air resistance, all objects fall to the earth with exactly the same acceleration.
9. Projectiles are real-life examples of free fall motion; they involve extra motion with a constant velocity in the horizontal direction.

QUESTIONS

1. What is the difference between velocity and speed? Distance and displacement? Constant velocity and constant acceleration?
2. Jane and Dick swing on identical swings. Describe the motion of Jane's swing as seen by Dick if both swings start together with identical pushes (provided by Sally).
3. Make a vector diagram of your displacement in your travels from class during a typical day. If you end up back in your room or at your car, what is your displacement? Estimate the distance you covered.
4. On your next trip in a car as a passenger, record the speedometer and odometer readings at various times and determine your acceleration at startups and stops. Compare your average and instantaneous velocities. Perhaps you can make a speed-versus-time graph.
5. Think of a way to clock cars, and with some friends set up your own speed trap on campus. Determine the average speed of several cars.
6. If you are in an elevator that undergoes free fall and you drop your pencil and books, what happens?
7. Name an object comparable in size with the following lengths: (a) 10^7 m, (b) 10^3 m, (c) 10 m, (d) 0.1 m, (e) 10^{-3} m, (f) 10^{-6} m, (g) 10^{-12} m.
8. Several vectors, not all in the same plane, add to produce a resultant of magnitude zero. What is the minimum number that will accomplish this condition?
9. Is it possible to have motion in a straight line with the velocity and acceleration always in opposite directions?

PROBLEMS

Unit Conversions

1. A basketball player is 6 ft 6 in. tall. What is his height in meters? In centimeters?
2. Portland, Maine, is approximately 100 mi from Boston, Massachusetts. How far is this distance in kilometers?
3. Perform the following conversions: (*a*) 17.3 cm to m, (*b*) 50 mi/h to ft/s, (*c*) 50 mi/h to km/h, (*d*) 186 000 mi/s to m/s.
4. What is the conversion factor between cubic feet and cubic meters?
5. A living room in a house is 13 ft 4 in. wide and 15 ft long. How many square feet of carpet will be required to cover the floor? Express your answer in square inches and square meters.

Vectors

6. A vector $\mathbf{F} = 30$ units to the northeast. What are the magnitudes and directions of the vectors $2\mathbf{F}$, $-\mathbf{F}$, $2\mathbf{F} + 3\mathbf{F}$, $2\mathbf{F} - 3\mathbf{F}$?
7. Charlie Brown leaves his house and follows the following route: From his house he travels three blocks east, three blocks north, three blocks east, five blocks south, and eight blocks west. Assuming all blocks are 120 m long, (*a*) what distance does he cover, (*b*) what is his displacement, and (*c*) in what direction must he walk if he were to head *directly* toward home?
8. A woman runs north 2.4 km and then southwest a distance of 1.6 km. How far and in what direction will she have to run to return to her starting point?
9. A man travels 16 km eastward and then 24 km northward. What is the distance he traveled and the resultant displacement?
10. A girl pulls a sled with a force of 50 N at an angle of 30° with the horizontal. Find the horizontal and vertical components of her pull.
11. An airplane leaves a runway and climbs at a 20° angle with respect to the horizontal at a constant speed of 320 km/h. What are the horizontal and vertical components of its velocity?
12. Eastport, Maine, is about 300 mi northeast of Boston, Massachusetts. How far east of Boston is Eastport, Maine? How far north is it?
13. A man is trying to fell a tree in a certain direction to avoid destroying his rose bushes. He does not trust his accuracy in notching the tree alone, so he decides to tie a rope to a tall branch 20 m above the ground. Standing 15 m from the base of the tree, with all his strength he pulls with a force of 220 lb. With what force is he pulling horizontally, and what is the vertical component of his pull on the tree?
14. Two forces (vectors) at right angles have a resultant of 20 N. If one force is 16 N, what is the other?
15. Two boys are pulling a large object with a rope. They loop the rope about the object, and each pulls on a different end, with two pieces of the rope making an angle of 80°. If one pulls with 80 lb and the other with 100 lb, what is the magnitude and direction of the resultant of these two vectors?
16. Draw vector diagrams to find the magnitude and direction of the sum of the vectors (resultant) for each situation pictured at the bottom of this page.
17. Find the resultant of the vectors shown analytically in Prob. 16 by taking the rectilinear components.

Speed

18. A student drives from Orono to Houlton, Maine, 120 mi in 3 h. What is her average speed?
19. A fast movie camera can take pictures at a rate of 2400 frames per minute. How many seconds elapse between successive pictures?
20. Express in meters the distance light travels in 1 yr (1 light-year).
21. The blood speed can be measured by tagging water molecules. If the blood travels a distance of 2.5 cm in an average of 0.70 s, what is the average blood speed?
22. A high school sprinter runs the 60-m dash in 7.2 s. What was his average speed during the race?
23. A driver on a turnpike is distracted by a sight on the edge of the road. If her glance is only 0.5 s and she is traveling at 55 mi/h (25 m/s), how far does she travel blindly?
24. A car traveling along a straight highway maintains a speed of 60 mi/h for 1 h and then 30 mi/h for

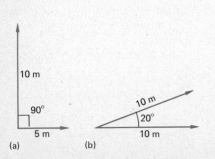
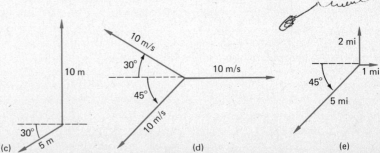

2 h. What was the average velocity over the 3-h interval?

25. A marksman fires a bullet at 720 m/s and hits a target that is 200 m from him. If he hears the sound of the bullet hitting the target 0.88 s after firing, what is the speed of sound in air?

Relative Motion and Vectors

26. A small motor boat takes 30 s to travel 90 m upstream and 15 s to travel the same distance downstream. Calculate the speed of the river's current and the speed of the boat relative to the water both upstream and downstream.

27. A boat is coming into a dock at a speed of 10 km/h with respect to the water. At the same time the tide is running in the same direction at a velocity of 5 km/h. (a) How fast is the dock approaching the boat? (b) What is the velocity of the boat with respect to the dock? (c) Of the dock with respect to the boat? (Use a few words in your answer.)

28. A defenseman on an ice hockey team is skating backward at a speed of 4 m/s relative to the ice while a forward from the other team skates toward him at a speed of 5 m/s relative to the ice. The forward hits the puck, sending it directly toward the defenseman at 30 m/s relative to the ice. (a) What is the velocity of the forward as measured by the defenseman? (b) What is the velocity of the puck relative to the forward? Relative to the defenseman? (c) How long will it take for the puck to travel from the forward to the defenseman if they are 15 m apart when the puck is hit?

29. An airplane maintains a heading due north at 600 km/h airspeed. It is flying through winds that move directly west at 200 km/h. In what direction is the plane moving with respect to the ground? What is the airplane's speed with respect to the ground?

30. An airplane is headed due east at 480 km/h relative to the ground. The wind is blowing on the airplane in a southwesterly direction at 100 km/h. Where is the plane with respect to its starting point after 0.5 h?

31. Charlie Brown throws his baseball while riding in the back seat of a convertible that is traveling at 30 m/s. If he throws the baseball horizontally at a speed of 15 m/s relative to the car in a direction perpendicular to the direction of the motion of the car, what is the velocity, speed, and direction of the ball relative to the ground?

32. In Prob. 31, if the ball hits the ground 0.5 s after leaving the car and the car continues moving in a straight line, how far apart will the ball and car be when the ball hits the ground?

33. A balloon is rising with a vertical component of velocity of 2 m/s while the wind is blowing horizontally at a speed of 5 m/s relative to the ground. In which direction does the balloon move?

34. A motor boat whose speed through still water is 12 m/s is attempting to cross a river 400 m wide in which the water flows at 6 m/s. The skipper desires to reach the point directly across the bank. In which direction should she head her boat, and how long will it take to cross the river?

35. A swimmer who can maintain a speed of 2 km/h relative to still water wishes to swim across a river 0.5 km wide. The water moves downstream at a speed of 9 km/h relative to the river bank. If the swimmer crosses the river always stroking toward the opposite shore perpendicular to the river bank, (a) in what direction will the swimmer move relative to the bank? (b) How long will it take the swimmer to cross the river? (c) How far downstream relative to the point directly across the river from where she began will the swimmer finally land?

Acceleration and Motion Along a Straight Line

36. The position of a bike rider moving along a straight road is given by the following:

t(s)	0	1	2	3	4	5
s(m)	0	2.3	8.4	17.1	27.2	37.5

t(s)	6	7	8	9	10
s(m)	46.8	53.9	57.6	56.7	50

(a) What is the average speed of the rider over the first 5 s of the trip? Over the first 8 s? Over the entire 10 s? (b) What is the average value of speed during the interval of time $t = 3$ s to 6 s? During $t = 6$ s to 10 s? (c) Plot these data on a graph and draw a smooth curve so that you can estimate the speed at $t = 4.5$ s and 1.5 s. (d) Draw the tangent to the curve at $t = 1$ s, 4 s, and 9 s. Determine the instantaneous speed at these times, and compare them with speeds determined earlier. (e) What is the average acceleration over interval $t = 1$ to 4 s?

37. A girl rides her bicycle on a trip about town, and her position as a function of time is shown in Fig. 37 (p. 50). (a) What is the instantaneous speed at $t = 0.2$ h, $t = 1.0$ h, and $t = 1.6$ h? (b) When is she traveling the fastest? (c) What is the average speed during the first 0.6 h? During the first 1.2 h?

38. From the graph in Fig. 38 (p. 50) of velocity versus time for a car on a straight road, determine the acceleration at $t = 20$ s, $t = 48$ s, $t = 60$ s.

39. Compare the accelerations of a car and a bicycle for the case of the car increasing its speed from 45 to 50 mi/h in the same time that the cyclist goes from rest to 5 mi/h.

40. A rocket starting from rest with constant acceleration acquires a vertical velocity of 160 m/s in 12 s. Find the (a) acceleration and (b) distance above ground after 12 s.

41. Starting from rest, an automobile

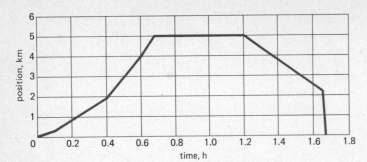

Figure 37

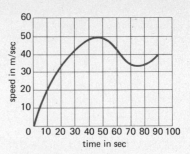

Figure 38

with constant acceleration attains a final speed of 5 m/s in 8 s. Find the (a) acceleration, (b) distance traveled in 8 s, (c) speed after 2 s.

42. Starting from rest, an automobile reaches a final speed along a straight road 10 m long in a time of 8 s. Find the average acceleration, distance traveled, and speed after 2 s.

43. A skier on a downhill schuss is accelerating at a constant acceleration of 5 m/s². She starts from rest. (a) How fast is she moving after 5 s? (b) How far has she moved in these 5 s? (c) When she reaches a speed of 30 m/s, how far has she traveled?

44. A skater is gliding over the ice with a speed of 3 m/s. Another skater pushes her from behind in a straight line, producing a steady acceleration of 1.3 m/s² until her speed is 18 m/s. How long did the push last?

45. A truck accelerates at a rate of 4 m/s². How much time does it take and how far does it travel while accelerating from 5 to 13 m/s?

46. The driver of a car traveling at 20 m/s (45 mi/h) slams on his brakes, stopping the car in 3 s. Find the (a) acceleration, (b) distance traveled during the 3 s, (c) speed at the end of the first second.

47. A driver in a car moving at 20 m/s slams on her brakes and stops in 4 s. Assume the acceleration is uniform, and find the (a) acceleration, (b) distance traveled during the 4 s, and (c) speed 3 s after she applied the brakes.

48. A driver sees a deer on the edge of a straight roadway. He slows down from 20 to 10 m/s in a distance of 90 m. Find (a) his acceleration and (b) the time during which he traveled the 90 m.

49. A subway train increases its speed from 24 km/h to 80 km/h by a constant acceleration over 0.8 km. What was its acceleration?

50. An automobile starts from rest on a straight road and for 6 s accelerates at 4 m/s² and then travels for 8 more seconds at a constant velocity. Find the (a) total distance traveled and (b) final velocity.

51. If an ant makes a trip of 200 m, how long will it take him if he first travels 2 cm/s for 50 m, then decelerates in 1 m until his speed is 1 cm/s and continues traveling at this speed?

52. A subway train starts from rest at a station and has an acceleration of 4 ft/s² for 10 s. It then runs at a constant speed for 30 s and decelerates at 8 ft/s² until it stops at the next station. Find the total distance covered.

53. Two sprinters in the 60-m dash are timed at 7.2 and 7.4 s. How far apart are the two runners when the first one crosses the finish line?

54. A trackee runs the mile in 4.22 min; another runs it in 4.37 min. (a) Calculate the average speed of each runner. (b) How far apart were they when the winner crossed the finish line?

55. A lady sprints down the street at 6 m/s after a bus that has just pulled away from the curb with an acceleration of 0.5 m/s². She is 20 m from the bus. Can she catch the bus? If so, how far from where the bus left the curb?

56. A fox and a rabbit start from rest at the same instant, with the fox initially a distance B behind the rabbit. Both begin to move at the same time in the same direction, with the rabbit moving with an acceleration of 1.3 m/s² and the fox moving with an acceleration of 2 m/s². The fox catches the rabbit after the rabbit has moved 260 m. (a) How long does it take for the fox to overtake the rabbit? (b) How far was the fox behind the rabbit initially (B)? (c) What were the speeds of the fox and the rabbit when they were abreast?

57. An automobile and a truck start from rest at the same instant, with the automobile initially at some distance behind the truck. The truck has a constant acceleration of 4 ft/s² and the automobile an acceleration of 6 ft/s². The automobile overtakes the truck after the truck has moved 150 ft. (a) How long does it take the auto to overtake the truck? (b) How far was the auto behind the truck initially? (c) What is the velocity of each when they are abreast?

58. A Ford car is at rest at a stop light, and just as the light changes, the Ford begins a constant acceleration of 10 mi/(h · s). At the same time, a VW moving at a constant speed of 30 mi/h passes the Ford. (a) If they move along a straight road, how long will it take before the Ford overtakes the VW? (b) How far will they have traveled when this happens?

Free Fall Motion

59. A motorcycle crashes into a tree at 40 mi/h. From what height would the motorcycle have to be dropped to hit the ground at the same speed?

60. (a) What is the instantaneous velocity of a falling ball, dropped from rest, at the end of 3 s? (b) What is the instantaneous velocity at the end of 4 s? (c) What is the average velocity over the fourth second? (d) What is the distance traveled during the fourth second?

61. A clever farmer wishes to determine the depth of his well, so he drops a stone into the well and hears a splash 5 s later. (a) How deep was the well? Disregard the time required for the sound to reach the top of the well. To see if neglecting the time for the sound of the splash of the rock to reach the top of the well is unrealistic, (b) compute the speed of the stone just as it hits the water after dropping 5 s, and then (c) compute the average speed of the stone. (d) Compare that average speed with the speed of sound: 330 m/s. What do you conclude?

62. With what minimum upward speed must a ball be thrown so that a boy on a 5-m-high balcony can catch it easily?

63. A ball is thrown vertically upward with a speed of 14 m/s. (a) How high does it rise? (b) How long does it take to reach this highest point? (c) How long is it in the air? (d) What is the velocity of the ball just before it hits the ground?

64. A ball is thrown upward at a velocity of 60 m/s. Calculate the velocity (magnitude and direction) at the end of 2 s, 3 s.

65. A small ball is thrown vertically upward with an initial speed of 20 m/s from the top of a 25-m-high building. (a) How high does the ball go? (b) How long does it take before the ball passes the top of the building on its way down? (c) With what speed is the ball moving just before it hits the ground?

66. Ty Cobb hits a pop-up straight up at 30 m/s. How far was it from his bat and in what direction was it traveling after 1 s? 4 s?

67. A snowball is thrown vertically downward from the edge of a cliff, leaving the thrower's hand with a speed of 10 m/s. Find the velocity after 2 s, assuming no air resistance.

68. Wyle Coyote climbs to the top of Evan's Notch, a height of 1250 ft. He is planning to drop a 1-ton rock upon the roadrunner. (a) How long will it take the boulder to make the trip? (b) If the roadrunner is 300 ft away and traveling at a speed of 10 ft/s, when should Wyle push the rock?

69. A ball is thrown vertically upward and just reaches the top of a building 15 m high. At the very same instant the ball is released, a second ball is dropped from rest at the top of the building. At what height do the two balls pass each other?

70. A package is dropped 40 m off the ground from a balloon that is rising steadily at 20 m/s. What is the velocity and distance of the package from the ground after (a) 1 s, (b) 2 s, and (c) 3 s?

71. A balloon is descending at 6 m/s when a sandbag is released. If the balloon continues to descend, but at 4 m/s, where is the sandbag relative to the balloon 2 s later, and how fast is the sandbag moving relative to the earth and relative to the balloon?

Projectile Motion

72. Jim Rice makes contact with a fast ball 4 ft from the ground, sending it toward the left field fence, 360 ft away. The ball is hit so it takes off at a 45° angle such that its range will be 400 ft. The left field fence is 30 ft high. Will he have another home run?

73. If an Olympic athlete could throw the shot 18.31 m in Helsinki, where $g = 981.91$ cm/s^2, how much further can he throw it on the moon where g is one-sixth that at Helsinki?

74. At the instant a ball is thrown horizontally, an identical ball is dropped from a position adjacent to the original ball. Which ball strikes the floor first?

75. A girl standing on a 6-ft-high bridge pitches a penny into the wishing pool below. The penny leaves her hand, going up at an angle of 45° with the horizontal. If the penny's initial velocity is 5 ft/s, (a) what is the velocity just as it hits the water? (b) Where does the penny hit the water? Assume the girl releases the penny 3 ft above the bottom of the bridge.

76. A tennis ball rolls off a 1.22-m-high table at a speed of 0.2 m/s. (a) How long is the ball in the air? (b) What is the horizontal component of the displacement from the table's edge if the ball was initially rolling at right angles to the table's edge?

77. A quarterback releases a pass with an initial horizontal speed v_0 from a height of 8 ft above the ground. He wishes to hit his receiver down field a distance of 15 yd. What is the minimum value of the initial horizontal speed v_0 that will allow the receiver to make a shoestring catch just before the ball hits the playing field?

78. An airplane flying at 100 m/s loses a wheel at 2100 m. (a) How long before the wheel strikes the earth? (b) How far does it travel horizontally? (c) Find the horizontal and vertical components of the wheel's velocity when it strikes.

Special Topic
Higher, Faster, Farther

Wyomia Tyus takes the tape. (*Wide World*)

Speed, length, and time are such fundamental concepts that we tend to take them for granted. But our everyday experience with them is really extremely limited. The range of speeds, distances, and lengths of time that we encounter in our daily lives is far removed from the range investigated by physicists.

The span of three-score years and ten that the Bible gives as human life expectancy is minute in comparison with all the eons of geologic time. Human beings are newborn babes in the universe, having emerged as recognizably human in just the past few million years. This time contrasts with the evolutionary time span of the horseshoe crab, which has lived on earth for at least the past 200 million years, and of life itself on earth, which emerged in the form of blue-green algae about 3 billion years ago.

Astronomical time is even more vast. Cosmologists believe that our solar system of sun and planets was formed about 4.5 billion years ago, and the universe itself appears to be at least 15 billion years old, give or take a few billion years. One can gain some feeling for the hugeness of these figures by seeing how long it takes to count up to a thousand.

Ultrashort periods of time are just as difficult to envision. To the uninitiated eye, the result of turning on a light switch appears to be an instantaneous brightening of the light bulb. But in fact, the brightening takes a significant portion of a second. First, the switch must operate—a process that can take as much as a few millionths of a second. Then, the signal must travel from the switch to the light bulb, taking another few billionths to millionths of a second in the process. Next, the electrical element in the bulb must warm up and glow; this process takes a few hundredths of a second. Finally, the light takes a few more billionths of a second to get to the eye of the person who flicked on the switch. Some of these factors are strictly limited by physical laws; the electrical signal and the light cannot travel any faster, for example. But improved technology can speed up the switching process. In fact, laser-controlled switches have been developed that take no more than just a few picoseconds (trillionths of a second) to switch from on to off and vice versa.

Our understanding of speed is as subjective as that of time. We have a feel only for those speeds in the narrow range of our experience. An experienced driver can often tell without looking at the speedometer when the car is staying inside the speed limit or is exceeding it, but the same person in a supersonic plane can only marvel at the sensation of speed.

The greatest running speed ever recorded by a human being was that of sprinter Robert Lee Hughes, who was clocked at 27.89 mi/h (12.37 m/s) between the 60- and 75-yd markers of a 100-yd race in St. Louis in 1963. That speed made Hughes truly the world's fastest human. For the record, the world's fastest woman is Wyomia Tyus, who traveled at 23.78 mi/h (10.63 m/s) during a sprint in Kiev, Russia, in 1965.

The fastest human to travel on land with the assistance of an engine was Gary Gabelich, of the United States. In 1970, at the Bonneville Salt Flats in Utah, Gabelich drove a vehicle powered by liquid natural gas over a measured kilometer at an amazing 631.368 mi/h (282.25 m/s) (faster than the cruising speed of most passenger planes). The fastest air speed ever achieved is more than three times as fast as that, namely 2070.12 mi/h (925.43 m/s), reached by Col. Robert L. Stephens and Lt. Col. Daniel Andre of the United States Air Force while piloting a Lockheed YF-12A in 1965.

Even this figure seems insignificant when compared with the speed record for outer space. In May 1969 Apollo 10 astronauts Tom Stafford, John Young, and Eugene Cernan achieved a blistering 24 791 mi/h (11 083 m/s) as they approached the top of the earth's atmosphere on their way back from the vicinity of the moon. The fastest speed by an unpiloted spacecraft is even more impressive. As it swooped within 81 000 mi of the giant planet Jupiter in December 1973, the Pioneer 10 spacecraft was moving at more than 82 000 mi/h (37 000 m/s).

The ultimate speed, according to theory as well as observation, is the speed of light. Its value is 186 282.4 mi/s, or $(2.997\ 924\ 58 \pm 0.000\ 000\ 01) \times 10^8$ m/s. Albert Einstein's special theory of relativity postulates that no object can accelerate to a speed greater than this, although some theoretical physicists believe that particles can exist which spend all their existence traveling faster than the speed of light. The existence of these particles, which are called tachyons, after the Greek word for speed, remains speculative, but physicists recognize that many subatomic particles travel in particle accelerators at speeds close to that of light. Other fast movers are the galaxies close to the edge of the universe, some of which are evidently moving away from our solar system at rates approaching three-quarters of the speed of light.

These all but unimaginable speeds are matched by the near-incomprehensible distances of cosmic objects. In fact, conventional units of distance are inadequate to define the distances between the stars and the galaxies. Thus, astronomers measure these interstellar spaces using units based on geometry or time. The most common astronomical unit is the parsec, which is defined as the distance at which the

Artist's concept of Pioneer spacecraft. (*NASA*)

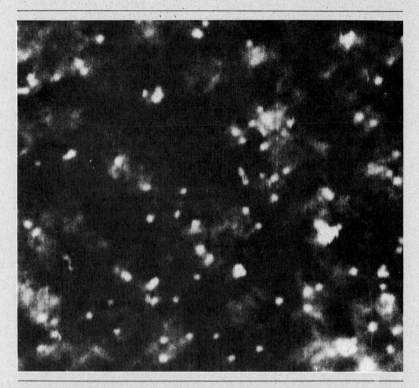

Micrograph images of individual uranium atoms magnified 7.5 million times taken with a newly developed electron microscope. (*University of Chicago*)

radius of the earth's orbit about the sun subtends an angle of just 1 second—a 360th of a degree. One parsec is approximately 3.262 times a more familiar unit, the light-year. A light-year is the distance that a pulse of light travels in a full year. A light-year is measured as about $9.460\ 528\ 436 \times 10^{12}$ km, or roughly about $5.878\ 499\ 833 \times 10^{12}$ mi. Our sun, at a distance of roughly 93 000 000 mi, is round the cosmic corner, just 8 light-minutes from earth. But the next nearest star, Proxima Centauri, is 4.2 light-years away, and the farthest galaxies yet observed are astonishingly billions of light-years distant. The light from those galaxies that astronomers view in their telescopes has taken these billions of years to travel from the outermost parts of the universe. Thus, astronomers see these galaxies as they were in those incredibly distant times ago.

At the other end of the scale is the equally incomprehensible magnitude of the ultrasmall. The smallest size that one can resolve with the naked eye is about 10^{-4} m (0.1 mm), and the best optical microscopes can differentiate objects with diameters of no more than half the wavelength of the light used to observe them, about 3×10^{-7} m. Such devices as the electron microscope and the scanning ion microscope can actually provide pictures of objects as small as individual atoms and molecules, whose diameters are typically approximately 10^{-10} m.

Forces

3.1 NEWTON'S LAWS OF MOTION

3.2 THE SECOND LAW

3.3 MASS AND WEIGHT

3.4 UNITS OF FORCE
SHORT SUBJECT: Weighing and Massing

3.5 THE SECOND LAW IN ACTION

3.6 NEWTON'S THIRD LAW

3.7 APPLYING NEWTON'S LAWS
Translational Equilibrium

3.8 FRICTION

SHORT SUBJECT: Getting a Grip

Summary
Questions
Problems

SPECIAL TOPIC: Return to a Small Planet

Chapter 2 opened with five examples of motion. To gain a fuller understanding of these examples, we must now delve into the causes of the motion in each case.

The fighter plane, for example, is driven forward through the air by the gases thrust backward through the nozzles of its jets. The javelin is set on its arching path by the muscular power of the athlete's arm. The subatomic particles are accelerated by electric fields and magnetic fields, which draw them through the atom smasher (particle accelerator) at higher and higher velocities. The emperor penguin makes its ungainly way across the ice by the muscle power of its feet and then uses its powerful wings to drive its body through the water. And the planets move around their sun, confined in their orbits, by the force of gravitational attraction. The common factor that produces the change in motion in every case is the application of a **force** to the object in question.

Forces not only produce changes in motion called acceleration; they can also distort the object on which they act. Often they can cause both acceleration and distortion, as any boxer whose head has been jolted backward by an opponent's swift right cross can testify (Fig. 3.1). For

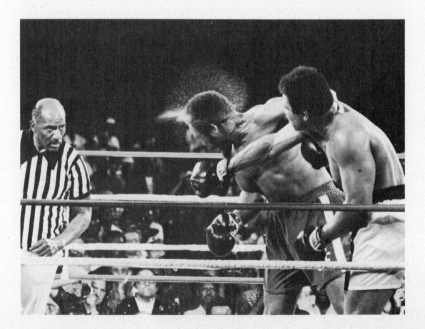

Figure 3.1 Force in action! It causes distortion and motion as George Foreman takes a pounding from Muhammad Ali. (*Wide World*)

simplicity, however, we shall assume in this chapter that the objects affected by forces are rigid enough to withstand distortion and that the only effect of the forces on them is, therefore, a change in their motion. In later chapters we shall examine cases in which forces produce distortions rather than changes in motion.

3.1 NEWTON'S LAWS OF MOTION

The relationship between force and acceleration is summarized in three basic laws of motion propounded in 1687 by the gifted British physicist Sir Isaac Newton. The laws are basic insofar as they are generalizations of experience rather than mathematical derivations. The few brief sentences that outline the laws contain conceptual ideas that require an abundance of examples to be understood fully.

The first law, as quoted from Newton's epochal scientific work *The Principia,* is as follows: "*Every body perseveres in its state of rest or of uniform motion in a straight line unless it is compelled to change that state by forces impressed thereon.*" The key word in the first law is "perseveres." An object in the absence of an unbalanced (net) force acting upon it will remain at rest or continue in uniform motion in a straight line. The law seems to combine two different sets of circumstances: bodies at rest and objects moving with a constant velocity (uniform speed in a straight line). What both situations have in common is zero acceleration. The state of rest is no more than a state of constant velocity with the particular magnitude of zero. The law recognizes that objects resist changes in motion. An object will not accelerate without the application of a net, that is, an unbalanced, external force.

Implicit in the first law is a concept that Galileo had earlier termed **inertia.** But although Galileo introduced the idea, Newton was the first to realize its significance. **Inertia is the tendency of an object,** whether an atom, a 10-ton truck, or a planet, **to resist a change in its motion,** and hence continue in its state of rest or uniform motion in a straight line. It is harder, for example, to start moving from rest an object the size of a planet, slow it down once it is moving, or turn it from a straight-line motion than it is to do the same with a 10-ton truck because the planet has greater inertia. More realistically, it is harder to start up and slow down a 10-ton truck than it is to perform the same task with a Volkswagen because the truck has more inertia than the Volkswagen. Which would you prefer to have to push to a gas station?

We experience inertia in many everyday situations. The lurch that affects standing passengers in a bus or subway car as the vehicle starts forward is a typical example. The vehicle itself, being a solid object, accelerates forward under the force of the motor. This force is not applied immediately to the standing passengers, however, because they are linked only loosely to the vehicle, through the soles of their feet. Thus, as the vehicle starts to move forward, the passengers tend to stay in their state of rest. The back of the vehicle, in effect, moves toward them, causing them to lurch toward the rear. To prevent themselves from falling, the passengers must find a force to accelerate them along

Inertia is the tendency of an object to resist a change in its motion.

A turning bus experiences a "change" in its motion—an acceleration—even if it turns with uniform speed.

with the bus. Typically, they do this by balancing themselves, applying a force of their own, as they grasp the back of a seat or a strap. Similarly, when a vehicle stops suddenly, the tendency of the passengers to continue in their state of motion in a straight line causes them to start falling toward the front of the bus; the standees must overcome their motion by linking themselves with the force that is decelerating the vehicle, again by grabbing a seat back or a strap. When the bus goes around a turn, standing passengers will continue to move in a straight line rather than turn with the bus because of their inertia *until* they grab something attached firmly to the bus that will pull them along with the turning bus.

A more desperate situation occurs when an automobile crashes head-on into a large tree. Under the impact of the force exerted by the tree on the car, the car stops almost instantaneously. The inertia of the car's driver and passengers, however, carries them forward in the direction they were traveling until some force acts upon them to halt their movement. All too often that force is exerted by the windshield, the dashboard, or the steering wheel, sometimes with fatal results. Somehow, a way must be found to minimize the effect of the decelerating forces on occupants of a crashing vehicle because the force on the persons cannot be eliminated altogether. For the safety-conscious, a seat belt and shoulder harness are ways of applying the restraining force to overcome the driver's or passenger's motion without drastic results.

An air bag provides an even greater opportunity to reduce injuries in an automobile accident because it distributes the decelerating force more evenly over the potential victim's body and extends the time during which the force acts.

Plainly, then, safety standards for automobiles have their basis in Newton's first law of motion; the pity is that the law was not applied to car safety until almost 100 yr after the invention of the car.

We might also note in passing that Newton's first law applies as strongly to the tree with which the car collides as to the car itself. The tree started at rest, and remained at rest despite the collision because the inertia of the tree and the earth in which it is solidly implanted is much greater than that of the car.

Safety standards for automobiles have their basis in Newton's first law of motion; the pity is that the law was not applied to car safety until almost 100 yr after the invention of the car.

Many military passenger airplanes were designed with rear-facing seats so that in the event of a crash, the back of the seat would provide a means of transmitting the large short-term decelerating force to the whole body. Forward-facing passengers experience the force in concentrated form just at the point where the seat belt rests. Astronauts reenter the earth's atmosphere facing backward on their way home from space. Of course, when the astronauts are launched, they face forward so that the large accelerating forces can be transmitted in a uniform manner to their entire bodies. The accelerating or decelerating forces affect the organs and the circulatory systems for they also have inertia. Experiments on rocket sleds have illustrated clearly that one's body may be damaged internally as well as externally if acceleration or deceleration is too rapid [see Fig. 3.2(a)].

Other examples of inertia in action are less dramatic. When a chilly householder shovels snow from the driveway to a snowbank, both the shovel and the snow are set in motion. Then the householder stops the movement of the shovel, allowing the snow to continue to move on to the snowbank. As another demonstration, a loose ax head may be "forced" more tightly on to the handle by swiftly striking the end of the handle on the ground. The handle stops while the inertia of the ax head carries it further onto the handle [see Fig. 3.2(b)].

A more dramatic demonstration of inertia is presented by the magician's standard trick, removing a tablecloth from under a table full of china dishes without disturbing any of the place settings. The secret of the trick is to remove the cloth with a quick jerk. Because of its

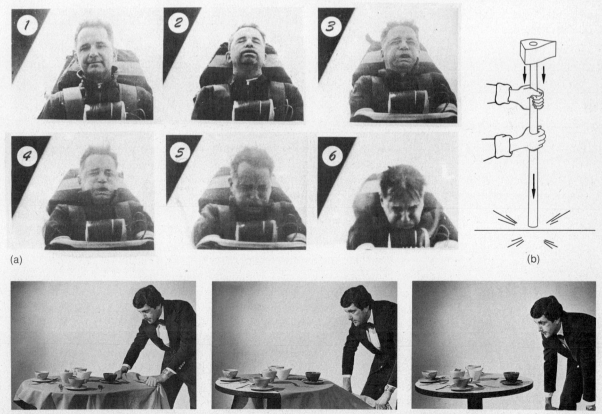

Figure 3.2 Newton's first law in real life. (a) What happens to the volunteer in the rocket sled when his vehicle slams to a halt? (*U.S. Air Force*) (b) How can you tighten an ax head on the ax handle? (c) How can the magician remove the tablecloth without removing or upsetting the dishes? (*Arthur Manfredi*)

inertia, the china remains at rest unless a force is exerted on it. Removal of the tablecloth as illustrated in Fig. 3.2(c) provides such a force, in the friction between the cloth and the dishes. But if the cloth is whipped away fast enough, that force acts on the place settings for such a short time that it has almost no effect; the plates, cups, and saucers remain in place. If the tablecloth is pulled away very slowly, the force has much longer to act; in this case, the cloth drags all the pieces along with it, in their upright positions. And if the cloth is yanked away quickly, but not quite quickly enough, the force gives the tableware a jolt and succeeds only in spilling all the pieces on the table. Tablecloths and tables full of china are obviously best left to magicians, but students can perform the same experiment with a minimal risk if they substitute a sheet of paper for the tablecloth and a glass of water or a can of beer for the best china.

3.2 THE SECOND LAW

Newton's first law of motion tells us how an object behaves in the absence of a force and so, by implication, what a force does to the motion of any object. The second law defines the effects of forces. The forces themselves can be defined as either pushes or pulls. More generally, a force is an agent of change. An interpretation of the second law can be stated, using the terminology introduced in the previous section, as: **The time rate of change in motion is proportional to the net force impressed and is made in the direction of the straight line in which the force is impressed.**

The phrase "time rate of change in motion" can simply be expressed as acceleration. Mathematically, therefore, we can express Newton's second law of motion as:

$$\mathbf{F}_{net} \propto \mathbf{a} \qquad \text{or} \qquad \mathbf{F}_{net} \propto \frac{\Delta \mathbf{v}}{\Delta t} \tag{3.1}$$

where $\mathbf{F}_{net}$ is the net force acting on any particular object and $\mathbf{a}$ is the acceleration that this net force produces. **The net force is, of course, the vector sum of all the individual forces acting on the object.**

Several examples in Fig. 3.3 illustrate the meaning of the net force. Forces are found to be vectors, and the direction of each individual force applied to an object is important. In Fig. 3.3 we assume that all the forces acting on an object are shown, and it is obviously clear that in each case the net force depends upon the direction of the force as well as upon its magnitude. In the last diagram in Fig. 3.3, two equal forces (vectors) act in a symmetrical manner. The vertical components of the two equal forces (10 lb sin 30°) are themselves equal but are oppositely directed, so that the resultant, or net, force acts in a horizontal direction. The two equal forces have been added graphically in the figure.

The proportionality factor that relates force and the resulting acceleration is known as the inertial mass and is represented by the symbol

Figure 3.3 Applied force and net force, considering motion only in the horizontal plane.

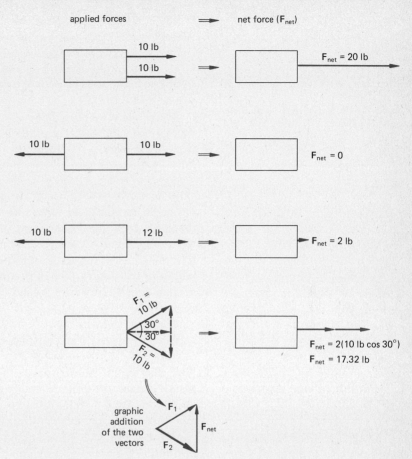

m. Thus, we can rewrite the proportionality relation in equation form:

$$\mathbf{F}_{net} = m\mathbf{a} \tag{3.2}$$

The mass, or rather the inertial mass, as defined by this equation is a very basic property of matter and deserves its own section.

3.3 MASS AND WEIGHT

Newton's first law led us to the concept of inertia, a property of every body. Newton's second law proceeds a step further and provides us with a quantitative measure of inertia, namely, **mass**. The greater the mass of an object, the greater its resistance to a change in motion—the greater its inertia. In terms of the second law, we see that the greater the mass of an object for a given applied force, the smaller the change in motion—the smaller its acceleration. **Mass, then, is a measure of inertia and is related to the quantity of matter in a body.** Since "quantity of matter" in a body is related to volume, density, and weight, the expression is confusing. Hence, it is more exact and meaningful to define mass as the measure of the inertia, or sluggishness, that a body exhibits in response to a net force applied to it.

The equation $\mathbf{F}_{net} = m\mathbf{a}$ shows us that the more massive the object, the greater the force required to give it a particular acceleration. If the Ford assembly line breaks down and starts to put Pinto engines into LTDs, for example, we can be sure that the strange hybrids will start off from rest more slowly than normal Pintos. Each car has the same engine, but each LTD has a far greater mass.

At this point we must study one of the subtler, but most important, distinctions in physics: the difference between mass and weight. In the case of the relative performances of Pintos and LTDs equipped with Pinto engines, the most obvious cause, from the standpoint of Newton's second law, of the hybrid's slower acceleration would appear to be its extra mass. We have a specific definition for mass; now for a definition of **weight. Weight is the force with which an object is attracted toward the earth. It is, simply, the force of gravity acting on an object.** If an object accelerates because of the force of gravity alone, it will undergo free fall, and near the earth's surface it will have an acceleration that we labeled g in Chap. 2. We can relate the weight force to the free fall acceleration by Newton's second law,

$F = ma$
Weight = mass × acceleration due to gravity: $W = mg$ \hfill (3.3)

Weight and mass are proportional to one another.

Note that weight and mass are proportional to one another. Near the earth's surface, the proportionality factor is g, a constant for all objects, as we saw from Galileo's work. We shall see in Chap. 4 that the acceleration due to gravity g varies slightly over the earth, decreasing as one moves from sea level to a mountain top. Furthermore, as you travel into space, the gravity acting on you and the resulting acceleration of gravity both change. However, your mass remains constant. Mass is a scalar quantity, fundamental to the object under consideration. Your mass will not change if you take a trip to the moon; however, the attractive force holding you to the moon's surface, your weight on the moon, will be reduced to about one-sixth your weight on earth. Of course your "free fall acceleration" on the moon would be one-sixth that on the earth's surface. The high mobility of astronauts on the moon arises from the reduction of their "weight" on the moon. Their space suits with life-support system weigh 330 lb on earth, and individual astronauts typically weigh about 150 lb on earth; thus they each carry a total load of about 480 lb weight on earth. But on the moon, this is a weight of $\frac{1}{6} \times 480$ lb = 80 lb. With the same muscles that they possess on earth, the astronauts can certainly jump about on the moon with ease.

As you travel into space, the gravity acting on you and the resulting acceleration of gravity both change.

In summary, one's mass, a scalar quantity, remains constant, but one's weight varies considerably as one travels about the universe. Mass is an extremely important concept because it is an invariant quantity that is conserved under almost all circumstances.

3.4 UNITS OF FORCE

The confusion between mass and weight is not clarified by our common usage of the British engineering system, whereas scientists everywhere and the general public in most other countries use the metric system. In the British engineering system of units, weight is correctly expressed in pounds; the pound is *defined* as the basic unit of force. The unit of mass is derived from this, and from the basic unit of acceleration, in feet per second per second. The British unit (rarely used) of mass is actually known as the slug, which is derived from the word "sluggish," which is quite fitting, for the more massive an object is, the more sluggish it is when it is to be moved. According to the second-law formula, 1 slug is the mass that is accelerated to 1 ft/s^2 by a force of 1 lb.

In the meter, kilogram, second (mks) metric system, the fundamental unit of mass is the kilogram (Table 3.1). This is the mass of the platinum-iridium cylinder kept alongside the (now outdated) standard meter in the International Bureau of Weights and Measures near Paris. Because mass is always constant, the value of the kilogram mass of the cylinder never varies with temperature or position, as do some of the other standards. If we wished to have a standard kilogram mass of our own, for calibration purposes we would compare the weight of our mass with the weight of the standard kept in Paris. Using an equal-arm balance, we would compare "weights" at the same place (same value of g), which would be equivalent to comparing masses.

In the small-scale centimeter, gram, second (cgs) units of the metric system, the standard unit of mass is the gram, which is one-thousandth of a kilogram.

The unit for mass together with the units for acceleration can be employed in the second-law equation to define the basic unit of force. In the larger-scale system, **the unit of force is known as the newton (N) and is defined as the force required to give a 1-kg mass an acceleration of 1 m/s^2.**

$$1 \text{ N} = 1 \text{ kg} \times 1 \text{ m/s}^2 = 1 \text{ kg} \cdot \text{m/s}^2$$

According to the smaller-scale units, the basic unit of force is the dyne (dyn). **1 dyn is the amount of force necessary to give a mass of 1 g an acceleration of 1 cm/s^2.** Remembering that 1 kg is 10^3 g and 1 m is 10^2 cm, we can compute arithmetically that 1 N equals 10^5 dyn.

Table 3.1 Metric Units of Mass

10 milligrams (mg)	= 1 centigram (cg)
10 cg	= 1 decigram (dg)
10 dg	= 1 gram (g)
1000 g	= 1 kilogram (kg)
1000 kg	= 1 metric ton

Weighing and Massing

For physicists, mass and weight are interrelated, and differentiating the two concepts is one of the most difficult tasks for most students of science. One of the simpler means of showing the difference between mass and weight depends on ready access to a spacecraft and two types of weighing devices: a spring scale and an equal-arm balance.

First, in our laboratory, imagine that we measure rice in a sack, by adding grains of rice to the sack while it is on one arm of an equal-arm balance. A standard 1-kg mass rests on the other arm of the balance. When the arms exactly balance, we know that the mass of the sack of rice is exactly 1 kg. Because the force of gravity on the earth is exactly the same on the standard kilogram and the sack of rice, we know that the rice weighs exactly the same as the kilogram mass. We then weigh the sack of rice on the spring balance, which gives a reading of 9.8 N. This is the upward force exerted by the spring on the mass and hence the downward force of gravity on the 1-kg mass. (Weights have the units of force, and a kilogram mass weighs the product of 1 kg and the 9.8 m/s^2 acceleration caused by earth's gravity.)

Then, we transport the complete setup to the moon, taking care not to spill any of the rice, and repeat the two operations. Once more, the standard kilogram mass and the sack of rice exactly balance on the equal-arm balance, indicating that the mass of the rice is unchanged as a result of the journey. But the weight of the sack of rice, as recorded by the spring scale, is just 1.6 N, indicating that the moon's gravity is just one-sixth that of earth's. More ambitious space travel would reveal a general rule: that the mass of any object remains constant throughout the universe as its weight varies with the strength of the local gravity.

If we wish to show that mass does not change as an object is moved about in the universe, we might apply to the object a known force, such as a horizontal push on a frictionless surface, and then measure the acceleration of the object. The ratio of the force to the measured acceleration will be constant everywhere. This inertial mass, defined by an experiment in which gravity plays no role, does not have any connection with the gravitational mass. However, gravitational and inertial mass are found by experiment to be the same. In fact, Einstein in his general theory of relativity stated this fact in his principle of equivalence. This principle is at the foundation of the general theory of relativity and states: **There is no way of distinguishing between gravitational effects and acceleration through space.** It is a consequence of the peculiar feature of gravitational interactions that acceleration does not depend upon any property of a body that is being accelerated. We shall examine the gravitational force in more detail in Chap. 4.

Plainly, the metric system weight of anything, whether a diet-conscious person, a sack of potatoes, or a planet, should really be measured in dynes or newtons, the units of force. However, commonly people mistakenly refer to weights of a few grams, or so many kilograms. Of course, they mean that "the object weighs as much as X kilograms do."

Two simple examples illustrate the difference between mass and weight in the two different systems of units. Using the formula $m = W/g$, we can obtain the mass of a man who weighs 180 lb. Here,

$$m = \frac{180 \text{ lb}}{32 \text{ ft/s}^2} = 5.6 \text{ slugs}$$

Similarly, the weight of the standard cylinder of metal that is the benchmark for the kilogram of mass is given by the equation $W = mg$. Since the acceleration due to gravity in metric units is 9.8 m/s², we find that

$$W = 1 \text{ kg}(9.8 \text{ m/s}^2) = 9.8 \text{ N}$$

One final note about the confusion between mass and weight: The conversion for 1 kg between the metric and British engineering units is normally given as 2.2 lb equals 1 kg. This is plainly improper because mass and weight (force) are different quantities. The kilogram is a unit of mass, and in the British engineering system the 2.2 lb is a weight (force). The correct conversion is between 9.8 N and 2.2 lb, both being the weight of a 1-kg mass at sea level on the earth.

3.5 THE SECOND LAW IN ACTION

Examples 3.1 to 3.3 will demonstrate the application of Newton's second law.

EXAMPLE 3.1

Compute the net uniform force necessary to give an object whose mass is 4 kg a velocity of 10 m/s in a straight line 5 s after starting from rest.

Solution

We are given the object's mass, which is constant. To obtain the force, we require its acceleration also. Since the problem states that a *uniform* net force is applied for 5 s, we know that the acceleration is constant; it can thus be computed from the definition of acceleration. The acceleration is the rate of change of velocity, and since the velocity changes by 10 m/s in 5 s, it must change by one-fifth as much in 1 s or by 2 m/s each second, or

$$a = \frac{\Delta v}{\Delta t} = \frac{10 \text{ m/s} - 0}{5 \text{ s} - 0} = 2 \text{ m/s}^2$$

Hence, $F_{net} = ma = 4 \text{ kg} \times 2 \text{ m/s}^2 = 8 \text{ N}$

EXAMPLE 3.2

Compute the net constant force necessary to slow a 2-ton (4000 lb, or 1.78×10^4 N) car down from 60 mi/h (96.6 km/h) to 30 mi/h in 10 s.

Solution

Here again, we must set up the factors for the simple second-law equation by calculating the car's acceleration in appropriate units and converting its weight to mass in appropriate units. First, the speeds must be converted to feet per second as follows:

The conversion of 60 mi/h to 88 ft/s is an extremely useful one to remember for problems that involve this type of transition.

$$60 \text{ mi/h} = 60 \, \frac{\text{mi}}{\text{h}} \times 5280 \, \frac{\text{ft}}{\text{mi}} \times \frac{1}{60 \times 60 \, \frac{\text{s}}{\text{h}}} = 88 \text{ ft/s}$$

with all the units canceled as one would cancel symbols in an algebraic equation.

Similarly, 30 mi/h = 44 ft/s.

To compute acceleration, we note that

$$a = \frac{\Delta v}{\Delta t} = \frac{v_{final} - v_{initial}}{t} = \frac{44 \text{ ft/s} - 88 \text{ ft/s}}{10 \text{ s}} = -4.4 \text{ ft/s}^2$$

The negative sign means that the acceleration is negative, that is, a deceleration. Determining the mass of the car from its weight, we have

$$m = \frac{W}{g} = \frac{2(2000) \text{ lb}}{32 \text{ ft/s}^2} = 125 \text{ slugs}$$

Thus, substituting the values into Newton's second-law equation,

$$F_{net} = ma = (125 \text{ slugs}) \times (-4.4 \text{ ft/s}^2) = -550 \text{ lb}$$

Since the direction of the velocity was taken as positive, the negative answer is consistent with a retarding force; it means that the force is opposite in direction to the velocity.

EXAMPLE 3.3

A car with a mass of 1000 kg accelerates from rest to a speed of 30 m/s over a distance of 120 m. Assuming that a constant net force was applied to the car, what was its magnitude?

Solution

To calculate the uniform force, we must first compute the uniform acceleration it causes. The average speed is 15 m/s over the 120 m. Hence, the 120 m will be covered in 120 m/(15 m/s), or 8 s. The change in speed in that time is 30 m/s, so the acceleration will be (30 m/s)/8 s, or 3.75 m/s². We could also obtain the same result using Eq. (2.9), which applies in the case of constant acceleration:

$$v^2 = v_0^2 + 2as$$

Thus,

$$a = \frac{v^2 - v_0^2}{2s} = \frac{(30 \text{ m/s})^2 - 0}{2(120 \text{ m})} = 3.75 \text{ m/s}^2$$

Hence, substituting into the second-law equation,

$$F_{net} = ma = 10^3 \text{ kg} \times 3.75 \text{ m/s}^2 = 3.75 \times 10^3 \text{ N}$$

3.6 NEWTON'S THIRD LAW

When a carpenter hits the head of a nail with a hammer, the hammer exerts a force on the nail that drives it into a piece of wood. At the same time, the nail exerts an equal force on the hammer that stops the hammer's motion in midswing.

Forces always occur in pairs.

In the engines of a jet airplane, expanding hot gases produced by the explosion of a mixture of alcohol and oxygen or low-grade gasoline and oxygen whistle, at very high speeds, out of small nozzles facing the rear of the plane. In response, the plane itself surges forward through the air.

The basic truth illustrated by these two common events is that forces always occur in pairs. The force of the hammer driving the nail is exactly matched by the force of the nail on the hammer; the force that sends the hot gas surging backward through the nozzles of the jet engines is matched by an equal force that propels the plane forward. The concept is stated in Newton's third law of motion, which can be paraphrased as follows: **When any object exerts a force on another object, the second object also exerts an equal and opposite force on the first.** Alternatively, we might say that for every action, there is an equal and opposite reaction.

Everyday life is full of other examples of this particular law: the recoil of a rifle when it fires a bullet, the crazy zigzag movement of a balloon that has been blown up and then released, and the movement of a canoe through the water as the canoeist pushes backward on the water with the oars.

In many cases the effect of the action or reaction is too small to be visible. Every time we step on the ground, for example, we exert a downward force on the earth, and the earth exerts an equal upward force on us. The earth shows no apparent effect of all the footsteps it experiences because of its huge mass. However, if we could somehow get on to a very small asteroid and then jump off it, we would move upward while the small asteroid would be displaced downward. A more realistic illustration is stepping out of a rowboat on to a dock. If the

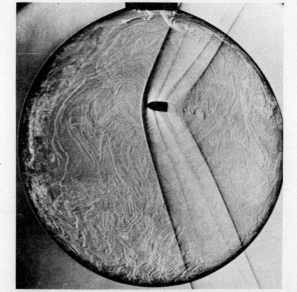

Figure 3.4 .22 calibre bullet passing about midway through a soap bubble. (*United Press International*)

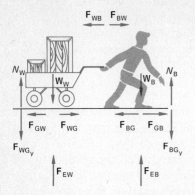

Figure 3.5 Small boy pulling a heavily loaded wagon. We shall assume that the boy pulls the wagon with only a horizontal force, although he is actually pulling upward and sideways, as indicated by the angle of his pulling arm.

boat is not firmly tied up, it will move away from the dock because you pushed it in that direction as it pushed you toward the dock.

Two important points must be remembered when we apply Newton's third law: The forces of action and reaction act **on different objects;** it is the vector sum of all the forces acting **on** a particular object that determines whether the object accelerates. These points are illustrated by the following paradox: A boy is pulling on a heavily loaded wagon. He exerts a force on the wagon, but according to the third law, the wagon pulls back on him with an equal and opposite force. Thus, it would seem that the boy cannot possibly move the wagon, no matter how strong he is.

The horizontal forces in Fig. 3.5 are as follows: $\mathbf{F}_{WB}$ = the force the wagon exerts *on* the boy; $\mathbf{F}_{BW}$ = the force the boy exerts *on* the wagon; $\mathbf{F}_{GW}$ = the force the ground exerts *on* the wagon; $\mathbf{F}_{WG}$ = the force the wagon exerts *on* the ground; $\mathbf{F}_{BG}$ = the force of the boy pushing on the ground; and $\mathbf{F}_{GB}$ = the force of the ground pushing back on the boy. Note that we have indicated all the forces in the vertical direction. There are four equal pairs of vertical forces. Since there is no acceleration of the wagon in the vertical direction, the net force on the wagon in the vertical direction must be zero, by Newton's second law. This is likewise true for the boy if we assume he moves only in the horizontal direction. We also neglect any frictional forces that might exist between the wagon wheel and its axis.

According to Newton's third law of motion, the forces $\mathbf{F}_{WB}$ and $\mathbf{F}_{BW}$ are equal but opposite, as are the forces $\mathbf{F}_{GW}$ and $\mathbf{F}_{WG}$ and the forces $\mathbf{F}_{BG}$ and $\mathbf{F}_{GB}$. But the question of whether the boy can move the wagon is decided entirely by the *balance of forces that act on the wagon alone.* If F_{GW}, which is actually the frictional force on the wagon that causes it to resist movement, is equal to F_{BW}, the force the boy exerts on the wagon, the wagon originally at rest will stay still. If F_{BW} is greater than F_{GW}, on the other hand, the wagon will start to move. Its acceleration, $\mathbf{a}_W$, can be obtained simply from the second law, remembering that $\mathbf{F}_{BW} - \mathbf{F}_{GW}$ is the net force acting on the wagon. If the wagon's mass is m_W, then

$$F_{BW} - F_{GW} = m_W a_W$$

or

$$a_W = \frac{F_{BW} - F_{GW}}{m_W}$$

Plainly, the statement that the boy could not possibly move the wagon is incorrect; it results from a superficial analysis of the problem that neglects to identify all the forces acting *on* the particular object of interest.

The question of whether the boy will move can be answered by summing up the forces acting *on* the boy. By means of his muscles, the boy exerts a horizontal force on the ground, $\mathbf{F}_{BG}$. The ground reacts to his push with an equal but oppositely directed push $\mathbf{F}_{GB}$, which is actually the force that propels the boy forward. The ground pushes the boy forward! Of course, if the boy is actually to move forward, this push of the ground *on him* must be larger than the force with which the wagon is pulling back on the boy.

3.7 APPLYING NEWTON'S LAWS

A major difficulty in solving problems that involve forces and Newton's laws is converting the words of the problem into mathematical information that can be used in precise equations. We must clearly identify all forces acting upon a body to apply Newton's second law correctly.

The best approach to problems that include forces is to follow this simple checklist:

1. Isolate the particular object or mass with which the problem is concerned.
2. Identify all the external forces that act on this object by drawing a sketch of the situation.
3. Calculate the net force on the object by working out the vector sum of all the individual forces. (This can be done by pictorial vector addition or by dividing each force into its x and y components. In three dimensions we would also have to add a z component perpendicular to both the x and y axes).
4. Apply Newton's second law $\mathbf{F}_{net} = m\mathbf{a}$. If necessary, this can be done individually for each of the three coordinates:

$$F_{x_{net}} = ma_x \qquad F_{y_{net}} = ma_y \qquad F_{z_{net}} = ma_z \qquad (3.4)$$

(In most problems we shall not require the z components.)

One important factor to bear in mind when considering the motion of an object is that if the net force acting upon the object is constant, then the resulting acceleration will be constant. Hence, in these restricted cases we can use the equations developed in Chap. 2, p. 35 that link velocity, displacement, and time to a constant acceleration.

The importance of identifying all the forces, with both magnitude and direction, which act *on* the particular object or system cannot be overemphasized. It is the net force acting *on* the object that determines its motion, not the forces it exerts on some other object. In many practical examples the forces are those resulting from the pull of a rope or wire. Such a force is often called a tension. Since the direction of all forces is important, it is useful, when dealing with a tension in a rope, to recall the old adage "you can't push a rope."

EXAMPLE 3.4

A home gardener uses an 800-lb garden tractor to pull a 200-lb cultivating machine to prepare her patch for planting. The tractor and cultivator accelerate at 8 ft/s². If the drag of the ground on the machine, that is, the frictional force, is 300 lb, as shown in Fig. 3.6, what is

Figure 3.6 Forces on a tractor and the cultivating machine that it is pulling.

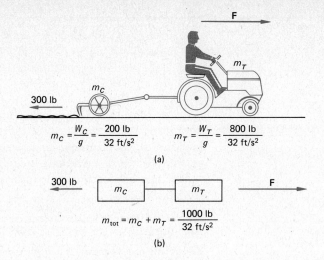

the total force the ground is actually exerting to accelerate both the tractor and the cultivator? This is the reaction force to the tractor's tires pushing against the ground.

Solution

First, we must isolate the system involved in the problem. This is the tractor and the cultivating machine. Then, we identify the external horizontal forces on this system in Fig. 3.6(b). The only horizontal forces external to the tractor-cultivator combination are the net forward force that the ground produces and the 300-lb drag force caused by friction acting on the cultivating machine. Applying Newton's second law to the system and being careful to find the masses from the weights given in the problem by the equation $m = W/g$, we have

$$\mathbf{F}_{net} = m_{tot}\mathbf{a}$$

$$(F - 300 \text{ lb}) = \left(\frac{1000 \text{ lb}}{32 \text{ ft/s}^2}\right) 8 \text{ ft/s}^2 = 250 \text{ lb}$$

$$F = 300 \text{ lb} + 250 \text{ lb} = 550 \text{ lb}$$

Thus, the tractor must supply 550 lb of driving force to accelerate the machinery with its large inertia and overcome the drag force. Realize that, although the driving force which moves the tractor comes ultimately from the tractor's engine, it is Newton's third-law reaction force to the tractor pushing against the ground that moves the tractor forward. That is, the tractor is pushed forward *by* the ground. Note that we have neglected the vertical forces because there is no motion in the vertical direction.

EXAMPLE 3.5

A 50-kg block on a table is attached to a 20-kg mass by a string that passes over a pulley in such a way that the 20-kg mass will descend and cause the 50-kg block to accelerate uniformly. The frictional force (f) impeding the motion of this 50-kg block is 10 N. Compute the acceleration of the blocks and the tension in the cord that connects them. Assume that the pulley has small mass and is frictionless.

Solution

Our first task is to draw a sketch of the situation (Fig. 3.7). Then, we must once again isolate the system actually involved in the problem. Because the two blocks are attached by a cord, they will move with exactly the same acceleration. And because the 20-kg block is hampered from undergoing free fall, we know that this acceleration is less than g.

Let us first approach the problem by assuming that the system consists of both blocks. In problems of this type, pulleys are assumed to have no effect on the motion apart from altering the direction of the string. We can therefore resketch this motion along a single line, as in Fig. 3.7(b).

We then apply Newton's second law, remembering that the weight of the 20-kg mass provides the force that actually moves the system. Thus,

$$\mathbf{F}_{net} = m_{tot}\mathbf{a}$$
$$m_{20}g - f = m_{tot}a = (m_{20} + m_{50})a$$

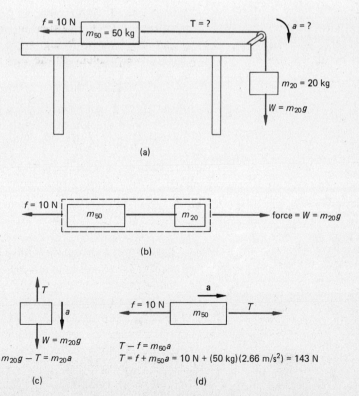

Figure 3.7 Block being pulled across a horizontal surface by a falling object.

In numbers, then,

$$(20 \text{ kg})(9.8 \text{ m/s}^2) - 10 \text{ N} = (50 \text{ kg} + 20 \text{ kg})a$$

$$a = \frac{(196 - 10) \text{ N}}{70 \text{ kg}} = \frac{186 \text{ kg} \cdot \text{m/s}^2}{70 \text{ kg}} = 2.66 \text{ m/s}^2$$

We can determine the tension in the rope during the acceleration of the two blocks by looking at either mass alone. For example, one might isolate the 20-kg mass and draw a force diagram for all forces acting on it, as shown in Fig. 3.7(c).

Applying Newton's second law to this single mass,

$$W - T = m_{20}a$$

Therefore

$$T = m_{20}g - m_{20}a$$

Using the value we have obtained for a above,

$$T = (20 \text{ kg})(9.8 \text{ m/s}^2) - (20 \text{ kg})(2.66 \text{ m/s}^2) = 143 \text{ N}$$

(Note that the value of the quantity $W - T$ must be positive because it represents a force that acts in a downward direction, which we have assumed to be the positive direction in this problem.)

We could also calculate the tension in the rope by doing a similar analysis on the 50-kg mass, as shown in Fig. 3.7(d). In fact, we could have solved the complete problem slightly more easily by isolating both masses at the start to determine their joint acceleration. This would have given us two simultaneous equations,

$$T - f = m_{50}a \quad \text{and} \quad m_{20}g - T = m_{20}a$$

from which T can be eliminated by simple addition of the two equations to give the identical result obtained earlier when we considered the system as the two blocks together:

$$m_{20}g - f = (m_{20} + m_{50})a$$

EXAMPLE 3.6

The apparatus known as the Atwood machine consists of a string over a pulley that links two unequal masses; it is often used as a simple experiment to determine the value of g. How does the machine produce a value for g?

Solution

The basic setup of the machine is shown in Fig. 3.8. We assume again that the pulley has small mass and is friction free. The equations can be solved simultaneously to eliminate T and thus give a value of g in terms of the known masses and their acceleration, which can be measured by

Figure 3.8 The Atwood machine.

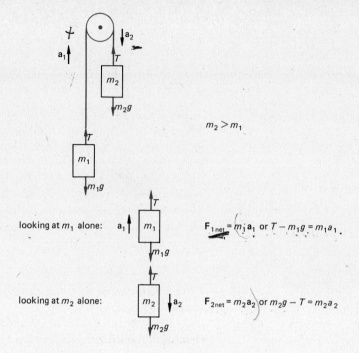

timing the fall of m_2 through a known distance. Note that m_2 is greater than m_1:

$$\mathbf{F}_{1\,net} = m_1 \mathbf{a}_1 \rightarrow T - m_1 g = m_1 a_1$$
$$\mathbf{F}_{2\,net} = m_2 \mathbf{a}_2 \rightarrow m_2 g - T = m_2 a_2$$

The magnitude of $\mathbf{a}_1$ equals the magnitude $\mathbf{a}_2$:

$$|\mathbf{a}_1| = |\mathbf{a}_2| = a = a_1 = a_2$$

Therefore, solving the two equations simultaneously by eliminating the T's (add the two equations),

$$m_2 g - m_1 g = m_2 a + m_1 a$$

or

$$a = \left(\frac{m_2 - m_1}{m_2 + m_1} \right) g$$

Obviously, the acceleration of the masses is always less than g because the denominator is always greater than the numerator; it equals g only when the value of the smaller mass is zero, which means that the remaining mass is falling freely. Equally, the smaller the difference between the two masses, the lower their acceleration. To measure g most conveniently when using the Atwood machine, the masses should be made quite similar; the acceleration will then be only a small

Translational Equilibrium

A particular case of the application of Newton's laws occurs when an object has no acceleration. If the acceleration of an object or a system of several objects is zero, we know from Newton's first law that the object or system will remain at rest or will continue to move with a constant velocity. A system or single object in this situation is said to be in **equilibrium** (or, more strictly, translational equilibrium), **which is the condition of the net force on the object or system being zero.** Simply stated, the condition for translational equilibrium is that the sum of the forces on a system is zero, or

$$\Sigma \mathbf{F} = 0$$

(The Greek capital sigma, Σ, is frequently used to indicate a summation.)

It may be instructive to examine the condition of translational equilibrium from the point of view of the addition of vectors. Forces are vector quantities. If we were to add the forces graphically by placing the vectors, representing all the forces on an object, with the head of one to the tail of the next at the appropriate angles, we would construct a closed polygon. The closure of the polygon indicates that the resultant of our vector addition is zero. A system of forces that is not in equilibrium may be placed in translational equilibrium by replacing the forces' resultant with an equal but oppositely directed force. For example, a woman standing at rest on the ground is in equilibrium. The force with which the ground pushes up on the woman will be equal in magnitude to the weight of the woman, which is of course a downward-acting force. Note that these two forces are not a Newton's third-law pair of forces but are equal because the acceleration of the woman is zero and by Newton's second law the $\Sigma \mathbf{F} = 0$. The woman is in translational equilibrium.

Examples 3.7 to 3.10 illustrate the application of Newton's laws to objects in translational equilibrium.

EXAMPLE 3.7

A 10-lb sign hangs from a pole by two wires arranged as shown in Fig. 3.9. Calculate the tensions in the two wires T_1 and T_2 if the angle θ_1 is 60° and the ange θ_2 is 30°.

Solution

Because the sign remains stationary, the net force acting on it is zero. Thus, the components of the forces acting on it in both the x and y axes also total zero separately. Redrawing the figure as a vector diagram and

Figure 3.9 Forces on a hanging sign.

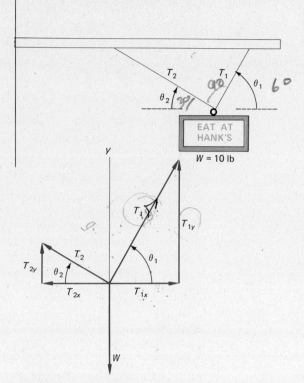

resolving the forces into their x and y components, we obtain the following equations:

$$T_{2y} = T_2 \sin \theta_2 \qquad T_{1y} = T_1 \sin \theta_1$$
$$T_{2x} = -T_2 \cos \theta_2 \qquad T_{1x} = T_1 \cos \theta_1$$

The x components of the forces add to zero; in other words,

$$T_1 \cos \theta_1 - T_2 \cos \theta_2 = 0$$

or

$$T_1 \cos \theta_1 = T_2 \cos \theta_2$$

Similarly, for the y components, which support the weight,

$$T_1 \sin \theta_1 + T_2 \sin \theta_2 - W = 0$$

where W is the weight of the sign.

Substituting the value we have for the angles θ_1 and θ_2 and the weight W, we find that

$$T_1 \cos 60° = T_2 \cos 30°$$

Thus,

$$T_1 = T_2 \frac{\cos 30°}{\cos 60°} = \frac{0.866}{0.5} T_2 = 1.73 T_2$$

Also,

$$T_1 \sin 60° + T_2 \sin 30° - 10 \text{ lb} = 0$$

$$T_1(0.866) + T_2(0.5) = 10 \text{ lb}$$

Substituting for T_1 in terms of T_2,

$$(1.73)(0.866)T_2 + 0.5T_2 = 10 \text{ lb}$$

$$2.0T_2 = 10 \text{ lb} \quad \text{or} \quad T_2 = 5.00 \text{ lb}$$

And since $T_1 = 1.73T_2$, $T_1 = 8.66$ lb.

One obvious point to note is that the total tension in the two wires is greater than the force of the weight they support. The sum of the tensions will only equal the weight when both wires are vertical, that is, $\theta_1 = \theta_2 = 90°$, and therefore $T_1 + T_2 - W = 0$.

As the wires approach a horizontal position, the tensions in them increase because a smaller proportion of the tension forces actually acts in the vertical direction necessary to support the weight of the sign. A warning here is in order. We should point out that the lengths of the wires in Fig. 3.9 do *not* determine the lengths of their force (tension) vectors.

EXAMPLE 3.8

The Russel system of traction for applying a longitudinal force to the femur is an example of an equilibrium situation, where $\Sigma \mathbf{F} = 0$. For the arrangement shown in Fig. 3.10, compute the resultant force on the femur.

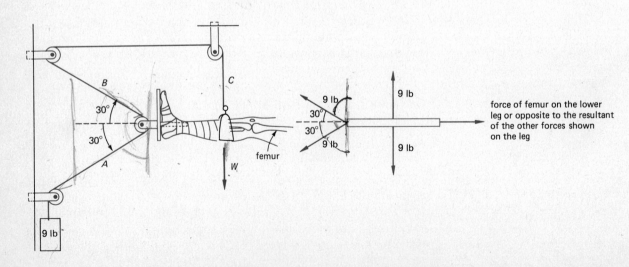

Figure 3.10 The Russell system of traction.

Solution

The leg and traction are obviously in equilibrium since the acceleration of the system is zero. Therefore, $\Sigma \mathbf{F} = 0$. Computing a resultant force on the femur requires the vector addition of all forces, which, of course, is an operation that can be carried out in nonequilibrium situations. However, it is important that the system is in equilibrium and no part of it is accelerating. If this is so, then the tension in the cord will be everywhere equal to the 9-lb weight force of the leg that hangs vertically from the cord. The pulleys serve only to redirect the forces. If we consider the forces in the vertical direction, we find, because of the symmetry of the traction with the leg midway between pulleys A and B, that the vertical component of the resultant force on the femur is zero:

$$B_y - A_y + C - W = R_y \qquad W = \text{weight of leg} = 9 \text{ lb}$$
$$B \sin 30° - A \sin 30° + C - W = R_y$$
$$9 \sin 30° - 9 \sin 30° + 9 - 9 = 0 = R_y$$
$$R_y = 0$$

Hence, the resultant force on the femur in this traction is horizontal. The sum of the forces in the horizontal direction will yield the horizontal component of the resultant—in this case, the magnitude of the resultant itself:

$$B_x + A_x = R_x$$
$$B \cos 30° + A \cos 30° = R_x$$
$$9 \cos 30° + 9 \cos 30° = R_x$$
$$R_x = 15.6 \text{ lb}$$

The resultant pull on the femur is horizontal and equal to 15.6 lb.

3.8 FRICTION

A sliding object experiences a frictional force that opposes the motion. The magnitude of the frictional force in any particular situation depends on a number of factors, among them the types of surfaces in contact, the physical condition of those surfaces, and the speed (or lack of it) of the moving object.

No physicist has been able to work out a reliable theory of friction from first principles.

No physicist has been able to work out a reliable theory of friction from first principles, but experiments over the years have produced a number of empirical rules regarding the action of frictional forces. A few simple experiments with blocks of wood and other material on a flat surface yield the following four conclusions:

1. The size of the frictional force is independent of the area of contact between the block and the surface.
2. The frictional force is greater when the object is not moving than when it is moving; in other words, a larger force must be applied to a block to start it in motion from rest than is necessary to keep it moving.

Sliding friction is, to a good approximation, independent of speed. However, air resistance and other fluid friction is closely proportional to speed at low speeds and depends on higher powers of the speed at greater speeds.

3. The frictional force is proportional to the force that presses the two surfaces together. This is known as the normal (perpendicular) force. It is equal to the component of the weight of the block that acts perpendicular to the surfaces in contact, *if* no other force acting on the block has a component perpendicular to the surfaces.

4. The constant factor that relates the frictional force and the normal force depends on the material of which both surfaces are made and their condition, such as wet, dry, or oiled.

Mathematically, the "simple law of friction" (3) is written

$$f = \mu N \tag{3.5}$$

where f is the frictional force; μ is the coefficient of friction, that is, the proportionality constant; and N is the normal force. Table 3.2 shows several typical coefficients of friction obtained experimentally. All depend on the conditions of the surfaces used to measure them.

Table 3.2 Coefficients of Static and Kinetic Friction

Material	μ_s	μ_k
Wood on wood	0.7	0.4
Steel on steel	0.5	0.4
Steel on steel (oiled)	0.1	0.03
Metal on leather	0.6	0.5
Wood on leather	0.5	0.4
Rubber on concrete (dry)	0.9	0.7
Rubber on concrete (wet)	0.7	0.6
Waxed wood on snow	...	0.05
Metal on ice	0.4	0.02

Note that there are many ways to reduce frictional forces; for example, rolling an object reduces considerably the frictional resistance which occurs when an object is slid along a surface. We have restricted our discussion of friction to static and sliding kinetic friction.

EXAMPLE 3.9

A 3200-lb automobile is speeding along an expressway at 50 mi/h in a straight line. The coefficient of static friction μ_s is 0.9, the coefficient of

Getting a Grip

Familiarity with friction is mandatory for designers of automobile tires. The designers must develop tires that, under widely different conditions, grip various surfaces well enough to give the driver control over the car but not so effectively that precious gasoline is wasted in getting the car moving against the frictional force.

Friction is not the only factor involved in designing a tire that will not be recalled; the experts must take into account such effects as the impact of cornering and other maneuvers on the tire, the tire's resistance to deflation, and the comfort of passengers. But the fundamental physics involved right from the beginning of tire making is frictional forces.

The first hurdle is to produce a tread profile that can cope with dry, wet, and tacky road conditions. On dry surfaces even bald tires—tires without any tread—can perform quite adequately. However, on wet surfaces the situation is quite different. A major problem is the buildup of water ahead of the moving tires; this water tends to reduce the tires' grip on the road. At high speeds, the effect known as aquaplaning often occurs. A thin film of water, perhaps a few millimeters thick, rests continuously between the bottom surface of the tires and the road. In this circumstance the coefficient of friction is so low that car wheels will not start to rotate after the brake is released, and so the unfortunate motorist is therefore unable to control the motion of the car. Skids and subsequent crashes are an all too-common result of aquaplaning.

The most obvious effect of wet weather—the buildup of water ahead of moving tires—is solved by cutting in the treads channels through which the water can drain and finer slits to help drainage and ensure that the tires grip the solid road. The identification of aquaplaning in the early 1960s persuaded tire manufacturers to widen the backward and forward drainage channels in the treads and to add new channels to disperse excess water sideways.

Another source of friction occurs when a tire deflates. The rim on which the tire is mounted settles against the flat tire, and the combined effects of friction caused by the rim and the road quickly tear the tire to ribbons. The Italian company Pirelli is now studying a new type of tire design that might prevent such loss and thus preclude the need for spare tires in the cars of the future. The DIP, as the experimental tire is known, has a triangular rather than an oval cross section, as shown in the diagram. When the tire is punctured, a large cushion of rubber is available between the rim and the road to prevent the tire from being destroyed. In early tests such tires ran comfortably for 100 mi without damage—distance enough for most motorists to reach a service station.

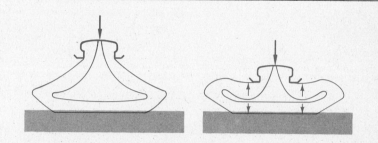

Time of tomorrow? The Pirelli DIP cross section inflated and deflated. Thin arrows indicate where friction forces act.

kinetic friction μ_k between the rubber and the concrete road is 0.7, and the air resistance is constant at 500 lb, as shown in Fig. 3.11. Calculate the (*a*) forward force that the automobile must generate to maintain the speed at 50 mi/h, (*b*) frictional braking force on the auto the instant the wheels are locked while the auto travels at 50 mi/h.

Figure 3.11 Forces on a moving automobile; F_R is the force of the road pushing the car forward. It is Newton's third-law reaction to the push of the turning wheels on the pavement via the force of friction.

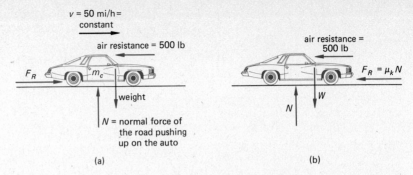

Solution

Friction is essential to driving, as any driver who has tried to maneuver on ice or a slick pavement can testify.

First let us determine the forces acting on the automobile. The weight force acts vertically downward, and the normal force of the road pushes directly upward on the car. Since there are no other vertically directed forces, the normal force on the auto is equivalent in magnitude to the weight of the auto. In the horizontal direction, the air resistance force of 500 lb acts on the auto opposing its motion. What other forces act *on* the auto? The engine of the auto turns the wheels. As a result friction between the tires and the road produces a Newton's third-law reaction force by which *the road,* through the frictional force, *pushes the car forward.* Friction in this case is the means by which we transmit the "driving force" to the car. Friction is in fact essential to driving, as any driver who has tried to maneuver on ice or a slick pavement can testify. (*a*) Now, to determine how large a force pushes the auto forward, we need only realize that the car is in equilibrium. It has a constant speed equal to 50 mi/h and hence is not accelerating. The sum of the horizontal forces must equal zero:

$$\Sigma F = 0$$
$$F_R - 500 \text{ lb} = 0$$
$$F_R = 500 \text{ lb}$$

Application of the 500-lb force is required to keep the automobile moving at constant speed because of the air resistance. Remember, however, that the car is in equilibrium, or, as Newton's first law expresses it, the automobile continues under the specified conditions in its state of uniform motion until another external force acts upon it. (*b*) When the brakes are slammed on, locking the wheels, the force of the road on the car now becomes a kinetic friction force that resists the motion of the auto and hence helps stop the auto. During the braking process, the auto slows down and the air friction force decreases. The amount of air resistance depends upon the speed of the car through the air, as you can easily show in a qualitative way by riding in an automobile at different speeds with your hand extended out of the window. If we consider only the stopping force at the instant the brakes are applied, the 500-lb air resistance force remains appropriate. The net resistive force acting on the car is the sum of the air resistance and the frictional force between the road and the tires:

Resultant $= F_R + 500$ lb
$R = \mu_k N + 500$ lb
$R = (0.7)(3200 \text{ lb}) + 500$ lb
$R = 2240 \text{ lb} + 500 \text{ lb} = 2740$ lb

As we can see from this example, frictional forces can be a help as well as a hindrance.

EXAMPLE 3.10

A fat child and a skinny child are on the opposite ends of a teeterboard. The fat boy decides to see how steeply the teeterboard must be inclined to cause his lighter companion to begin to slide down the teeterboard unless he hangs on with his hands, as shown in Fig. 3.12(a). What angle does the teeterboard make with the ground at this point?

We can translate this practical problem to a laboratory problem and then attempt to solve it: A block lies on an inclined plane. At what angle θ will the block just start to slide if the coefficient of static friction between the block and the plane is μ_s?

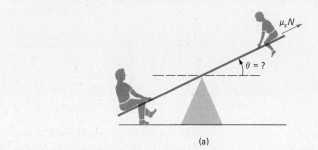

Figure 3.12 (a) A boy about to slide down a teeterboard, and (b) a block about to slide down an inclined plane.

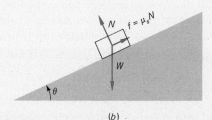

(b)

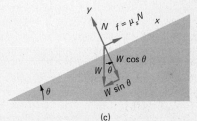

(c)

Solution

Figure 3.12(b) indicates the three forces that act on the block. These add to zero along both the x and y axes at all values of θ up to the value we require, where the block just starts to slide. The easiest way to approach the problem is to choose the x axis parallel to the inclined plane and the y axis perpendicular to it. The forces can then be resolved into x and y components, as shown in Fig. 3.12(b).

Summing the forces along the y axis,

$$N - W\cos\theta = 0 \quad \text{or} \quad N = W\cos\theta$$

Along the x axis,

$$\mu_s N - W\sin\theta = 0 \quad \text{or} \quad \mu_s N = W\sin\theta$$

If we link the two equations and solve them simultaneously, we obtain the relationship

$$W\cos\theta = \frac{W\sin\theta}{\mu_s}$$

In other words,

$$\mu_s = \frac{\sin\theta}{\cos\theta} = \tan\theta$$

Obviously, the problem can be reversed. By measuring the angle at which the block just starts to slip, that is, the angle at which the component of the block's weight along the plane is just sufficient to overcome the frictional force, we can determine the value of μ_s for any combination of materials, even jeans on wood.

SUMMARY

1. Forces can both move and distort objects on which they act.
2. According to Newton's first law of motion, objects continue to move at constant velocity, unless a net force is applied to them. The tendency to resist a change in motion is known as inertia.
3. Application of any net force accelerates an object; Newton's second law states that the amount of acceleration is proportional to the strength of the net force.
4. The proportionality factor between force and the acceleration it causes of any object is the object's inertial mass. This quantity indicates the object's inertia; it also gives a measure of the amount of material inside the object. The mass of an object never varies, whatever the location or time.
5. Weight is the force with which an object is attracted toward the earth, or whatever planet or star is closest to the object. An object's weight varies markedly with its location throughout the universe. Objects weigh about one-sixth as much on the moon as they do on earth.
6. Mass is the basic, defined physical quantity in metric systems of measurement. In the British engineering system, however, the pound of weight is the defined unit.
7. For every action, there is an equal and opposite reaction. This is Newton's third law.
8. Application of Newton's laws requires identification of all the forces that act on the system under examination.
9. The amount of frictional force between any two surfaces depends on the condition and type of those surfaces and whether they are moving relative to one another.

QUESTIONS

1. Express your weight in pounds, newtons, and dynes. What is your mass in kilograms?
2. In the following situations, how would you react to the motion: (*a*) standing in a train car that moves at a constant velocity, (*b*) standing in an enclosed truck as it moves around a right-hand turn with a uniform speed, (*c*) sitting in a roller coaster car just as it begins its journey at the top of the first hill?
3. Distinguish between mass, weight, and volume.
4. Is it necessary that the rocket pad from which satellite-bearing rockets take off be used in explaining how rockets take off in accord with Newton's third law of motion?
5. You are holding a bag of groceries while standing on ice in the middle of a pond. The ice is so slippery that no frictional force exists. How can you get off the ice?

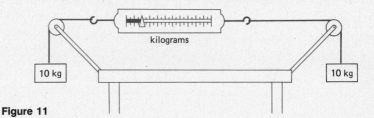

Figure 11

6. If a body is at rest, no force acts on it. What is wrong with this statement?
7. A body is whirled in a circle on the end of a string. The string snaps. What happens to the body?
8. In a tug of war, what is the net force acting on the rope when the two teams pull with opposite forces of 450 N? What is the tension force within the rope?
9. When you serve a tennis ball, consider the force of the racket on the ball as the "action" force and identify the "reaction" force. On which objects do these forces act?
10. A professor lies on the lecture table with a large block of concrete on her chest and asks a student to drive a nail into a board placed on the concrete. She does not expect to be hurt. Why?
11. Two 10-kg masses are attached to the spring illustrated in Fig. 11. What is the scale reading? Does it measure weight or mass?
12. If we identify one force acting on an object and yet the object is at rest, what can we conclude?
13. As you stand on the floor, the floor exerts an upward force against your feet. Why don't you move upward?

PROBLEMS

Newton's Second Law of Motion

1. What is the net force on an ape weighing 40 N sliding down a vine that exerts an upward force of 30 N?
2. A body that has a mass of 5 kg is acted upon by a net force of 45 N. Compute the resulting acceleration.
3. Calculate the net force required to give a body of mass 400 g an acceleration of 0.8 m/s².
4. What is the net force required to give a 0.2-kg baseball an acceleration of 20 m/s²?
5. What is the force required to give an automobile weighing 1600 N an acceleration of 9.8 m/s²?
6. Compute the mass of a body that is given an acceleration of 8 m/s² by a net force of 168 N.
7. The engines of a rocket whose initial mass is 10^5 kg provide a thrust of 10^6 N. What is the initial upward acceleration of the rocket?
8. An elevator is rising with an acceleration of 2.5 m/s². The elevator has a mass of 1000 kg. What is the tension in the cable?
9. A 100-kg sky diver "flies" with an acceleration of 1 m/s² before he opens his parachute. What is the resistive force of the air on the sky diver?
10. What force when applied to a 6-kg mass will reduce the mass' velocity from 20 to 15 m/s over a distance of 5 m?
11. A falling coconut has a mass of 0.25 kg, and the upward force of air resistance is 0.1 N. What is the acceleration of the coconut?
12. A child drags a small box across wet grass, which is so slippery that there is no friction between the box and the grass. The child pulls with a force of 3 N at 30°. (*a*) What is the upward force on the box by the cord? (*b*) What is the horizontal acceleration of the box?
13. A constant horizontal force of 10 lb acts on a body on a smooth, horizontal plane. The body starts from rest and is observed to move 250 ft in 5 s. What is the mass of the body? If the force ceases to act at the end of 5 s, how far will it move in the next 5 s?
14. A truck of mass 2000 kg is moving at 15 m/s and is acted on by two constant forces, the forward force of 800 N due to the engine and a 300-N retarding force due to friction. At what rate is the truck gaining speed? How far will it have gone in 6 s?

84 Problems

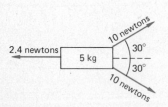

Figure 15

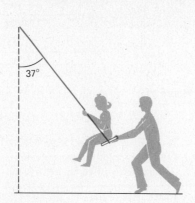

Figure 21

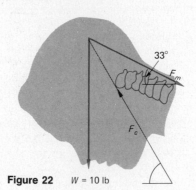

Figure 22 $W = 10$ lb

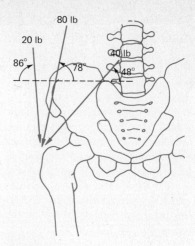

Figure 23

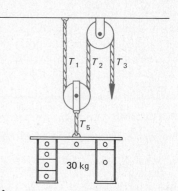

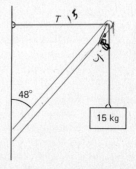

15. In Fig. 15 consider the top view of the 5-kg block on a table with a friction-free surface. Compute the resultant acceleration of the mass.
16. Two boys pull a 5-kg sled, each with his own rope. If they pull horizontally at an angle of 20° from one another with identical forces of 20 N, what will be the acceleration of the sled?
17. What is the acceleration of a block along the plane when it is sliding down a frictionless plane at angle θ?

Equilibrium

18. A bird of mass 170 g is gliding horizontally. What is the lift force on the bird?
19. A sky diver pulls the cord to open her parachute, and soon she is falling at terminal velocity, that is, at a constant speed. If she weighs 180 lb, what is the value of the frictional force of the air?
20. A 10-kg block of wood lies on the floor supporting a 70-kg person. Both are at rest. Draw diagrams to show the forces on the floor, the 70-kg person, and the 10-kg block.
21. A 60-lb girl sits in a swing and is pulled back by her father. Her father pulls horizontally to hold the swing at an angle of 37° with the vertical. What force must he apply? (See Fig. 21.)
22. When a student is bent over his desk, the forces exerted on his 10-lb head are shown in Fig. 22. The support comes from the neck muscle force F_m and a contact force F_c exerted by the atlantooccipetal joint. If the muscle force is 12 lb in the direction indicated, what is the magnitude and direction of F_c?
23. Figure 23 shows the hip abduction muscle, which connects the hip to the femur. The muscle actually consists of three independent muscles acting at the angles shown. Find the net force exerted by the three muscles.
24. Find the tensions T_1, T_2, T_3, and T_5 in the cords if the system is in equilibrium. Neglect the mass of the pulleys. (The top wheel is screwed to the ceiling.)
25. Find the tension in the horizontal cable and the force exerted on the boom by the pivot in the diagram shown below.

26. What is the tension in each wire of a 20-kg traffic light?

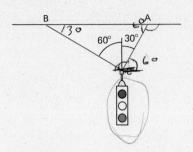

27. Dirty Dick is to be hanged at sunup. If Dick weighs 154 lb (685 N), what will be the tension in each rope?

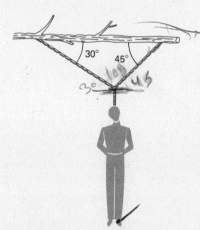

Friction and Newton's Laws

28. A force of 10 N is applied horizontally by a man to push a box. A resisting frictional force of 5 N opposes this motion. (a) What is the net force causing the box to move? (b) If the mass of the box is 2.5 kg, compute the resulting acceleration.

29. A ketchup bottle whose mass is 1.0 kg rests on a counter. If the frictional force between the counter and the bottle is a constant 3 N, what horizontal pull is required to accelerate the bottle from rest to a speed of 6 m/s in 2 s?

30. As shown in the following diagram, there is a coefficient of static friction equal to 0.866 between the block and the table top. Compare the force F required to just set the block in motion when you pull (a) with force F or push (b) with force F.

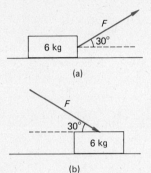

31. A physics professor pulls a 2-kg block across the lecture table by applying a force of 5 N to a scale. The scale has a mass of 1 kg. A frictional force of 1 N acts on the block. (See Fig. 31.) (a) What is the acceleration of the block and of the scale? (b) The scale reads the force pulling on the block. What is the reading on the scale?

32. If the coefficient of friction between tires and the road is 0.5, what is the shortest distance in which an automobile can be stopped when traveling 15 m/s? The automobile has a mass of 1000 kg.

33. A heavy block is dragged along a rough table by a spring scale attached to it. When the block is moving with a constant velocity, the scale reads 0.50 N, and when the block is dragged with a constant acceleration of 0.10 m/s², the scale reads 1.70 N. Find the (a) retarding force of friction and (b) mass of the block.

Challenges

In the following examples consider that the pulley has no effect on the motion except to change the direction of the pull of the string.

34. A force of 10 lb is exerted horizontally against an 80-lb block, which in turn pushes a 20-lb block, as in the figure below. If the blocks are on a frictionless surface, what force does one block exert on the other?

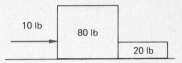

35. Three trailers, each weighing 980 N, form a *train* that is pulled along the floor of a factory by a tractor, which exerts a forward force of 750 N on the train. Find the (a) acceleration of the train, (b) tension in the coupling between the first and second trailers.

36. A 0.22 rifle bullet traveling at 36 000 cm/s strikes a block of soft wood and penetrates to a depth of 10 cm. The block does not move. The mass of the bullet is 1.8 g. Assume a constant retarding force. How long does it take the bullet to stop after hitting the block? And what is the force?

37. In Fig. 37, p. 86, $\mu = 0.25$ between all surfaces. What force will just set the 9-kg mass into motion?

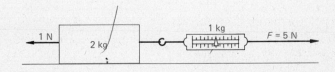

Figure 31

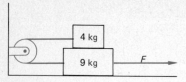

Figure 37

38. Two 1-kg blocks are hung over a small frictionless pulley by a light string. A 100-g mass is added to the block on the right. (a) What is the acceleration of the system? (b) How far will each block travel in 2 s? (c) What is the tension in the string?

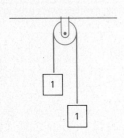

39. (a) For a frictionless surface, what is the acceleration of the following system? (b) For a coefficient of kinetic friction of 0.25, what is the acceleration of the system?

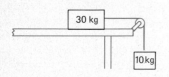

40. A 5-lb and a 2-lb block are sliding over a surface where the coefficient of friction $\mu = 0.20$. Compute the tensions T_1 and T_2 and acceleration of the 2-lb block.

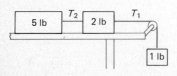

41. A 5-, 10-, and 20-kg set of blocks are attached as shown below. There is a frictional force on the two blocks, and this force is in proportion to the size of the two masses. The acceleration of the 10-kg block is 2 m/s^2. Find T_1 and T_2.

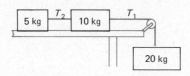

42. As shown in Fig. 42 below, a tugboat is pulling a line of three barges of unequal load. The forward acceleration is 0.1 m/s^2 and the pulling force $F = 900$ N. There is a frictional drag of the water on each barge which is independent of loading. (a) Calculate the frictional force acting on each barge. (b) Indicate in a diagram the forces acting on the first barge (i.e., the one of mass 1000 kg). (c) Calculate the tension T in the rope joining the first and second barges. (d) Calculate the weight of the first barge.

43. Consider the system shown in Figure 43, where the massless cords between the blocks pass over frictionless pulleys. Compute the acceleration of the system and the tension in the string between the 3- and the 4-M blocks. The coefficient of friction between the 4-M block and the table is 0.25.

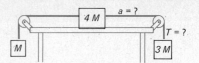

Figure 43

44. A driver of a 1200-lb sports car heading directly for a railroad crossing 1000 ft away applies the brakes in a panic stop. The car is moving at 90 mi/h and the brakes can supply a force of 300 lb. How fast will the car be moving when it reaches the crossing? Will the driver escape collision with a train that, at the instant the brakes are applied, is blocking the road and still needs 11 s to clear the crossing?

45. A skier, starting from rest, slides down a uniform slope that makes a $10°$ angle with the horizontal. If the frictional force between the skis and the slope is zero, how long will it take for the skier to travel a distance of 50 m? The skier has a mass of 70 kg.

46. Two blocks, A and B, are placed as shown in Fig. 46 and connected by ropes to block C. Both A and B weigh 20 lb, and the coefficient of sliding friction between each block and the surface is 0.5. Block C descends with a constant velocity. (a) Draw two separate force diagrams showing the forces acting on A and B. (b) Find the tension in the rope connecting blocks A and B. (c) What is the weight of block C?

47. Block A, of weight W, slides down an inclined plane S of slope

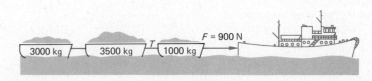

Figure 42

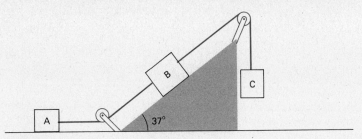

Figure 46

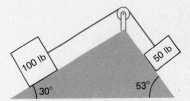

48. Two blocks are connected by a cord passing over a small frictionless pulley as shown below. Which way will the system move? What is the acceleration? The tension?

angle 37° at constant velocity, while the plank B, also of weight W, rests on top of A. The plank B is attached by a cord to the top of the plane. (See Fig. 47.) (*a*) Draw a diagram of all forces acting *on* block A. (*b*) If the coefficient of kinetic friction is the same between surfaces A and B and between S and A, determine its value. (*c*) Is block A in equilibrium? Why? Explain in words. (*d*) Is block B in equilibrium? Why? Explain in words.

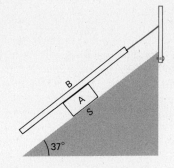

Figure 47

Special Topic
Return to a Small Planet

On the morning of December 27, 1968, the tension in the control room of Houston's Manned Spacecraft Center was at an all-time high. The events of the next few hours would decide the fate of the three men who were hurtling toward earth at thousands of miles per hour and in the process make or mar the $25 billion United States effort to land men on the moon by the end of the sixties. The tight-lipped controllers in front of the room's computerized instruments and television screens were preparing to bring the first crew ever to travel beyond the comparative safety of the earth's orbit down to a safe splashdown on the earth's ocean. The spaceship, Apollo 8, had already made history by orbiting the moon 10 times, and its crew of Frank Borman, James Lovell, and Edward Anders had struck a sympathetic chord by their reading of the first 14 verses of the Bible on Christmas Eve as their craft sped above the lunar landscape. Now they were returning to earth at a speed faster than anyone had ever traveled before.

The controllers' task was extraordinarily difficult to achieve. In cooperation with the three astronauts, those at mission control had to use the earth's atmosphere to reduce the spacecraft's speed from almost 25 000 mi/h (40 250 km/h) at an altitude of 400 000 ft (122 km) to just a few miles an hour at sea level.

This task demanded that the craft hit the top of the atmosphere at exactly the correct angle and location. If Apollo 8 arrived slightly above its target point atop the earth, it would skip off the top of the atmosphere like a stone skipping across a pond, taking the three pioneers on a lonely trip through space that would not return them to earth until long after their supplies of food and oxygen had run out. If the craft entered the atmosphere a little on the low side, it would plummet through the air at such high speed that the craft would be burned to a cinder despite its thick heat shield. The actual width of the "window" at which the spaceship was aiming at the top of the atmosphere was just 30 mi (48 km)—smaller than a needle in a haystack after a trip of 250 000 mi from the vicinity of the moon. That small size, and the fate of the ship once it arrived at the window, were entirely the consequences of the basic physics of forces.

The predominant force that controls the movement of any returning spacecraft is the earth's gravitational attraction. So precise were the demands on the spaceship, however, that the gravitational attraction of the moon, the sun, and other planets was also taken into account in the computerized equations of the craft's motion as it plunged toward the earth. Once the ship hit the atmosphere, another force took over the fate of its crew, a force that is normally neglected in simple physical problems: air resistance. The frictional force between the atmosphere and the outer skin of the Apollo craft was expected to reduce the ship's speed by about 25fold and warm the heat shield up to a fiery 5000°F ($\sim$2800°C). In fact, the limitations of the heat-shield material mandated a somewhat more complicated plunge through the atmosphere than was necessary according to a strict interpretation of the laws of forces. If the craft had plunged directly down toward the earth's surface, the heat buildup would have been so great as to destroy the heat shield. Thus, the path of the returning Apollo was designed to include two small jumps, which would disperse some of the heat. After these jumps, the craft reached its terminal velocity, at which the force of gravitational attraction was exactly balanced by the force of friction.

The frictional force between the atmosphere and the heat shield could not slow down the craft enough to bring it to a safe smooth landing. So, at an altitude of about 40 000 ft (12.2 km), a more direct form of air resistance came into play. Two small parachutes billowed out from the apex of the conical spaceship, followed, at an altitude of about 10 000 ft ($\sim$3000 m), by three larger chutes. The net effect of the parachutes was to reduce the spacecraft's speed to 19 mi/h (31 km/h) and deliver the craft safely into the sea.

Within minutes after the splashdown, the television screens in mission control showed Apollo 8 bobbing on the waters of the Pacific, and delighted space agency engineers lit up their traditional postmission cigars. The system had worked. Not only had Apollo made it to the neighborhood of the moon, its guidance systems had returned it with incredible precision to a pinpoint landing on earth. In many ways, the most amazing feature of the effort was that the combination of the laws of physics and the space agency's technology had slowed down the tiny craft by almost 25 000 mi/h during no more than 45 min.

Universal Gravitation

4.1 KEPLER'S LAWS OF PLANETARY MOTION

4.2 NEWTON'S LAW OF UNIVERSAL GRAVITATION

SHORT SUBJECT: On the Shoulders of Giants

Calculating G
G and g

4.3 MOVING ROUND CURVES
Going Around in Circles

4.4 GRAVITATION AND THE SOLAR SYSTEM

4.5 ARTIFICIAL SATELLITES

4.6 WHEN THE TIDE RUSHES IN

4.7 THAT FLOATING FEELING

SHORT SUBJECT: Getting Away from Gravity

4.8 FLEEING THE CENTER

4.9 EARTH'S GRAVITATIONAL FIELD

SHORT SUBJECT: Stars That Make Waves

Summary
Questions
Problems

SPECIAL TOPIC: An Open and Shut Universe

If you look up at the heavens on a clear night, you will have much the same view of the sky as did the ancient astronomers of China, Babylon, and Egypt. And if you were to spend a number of nights observing the relative movements of the shimmering stars and steadily shining planets, you would perhaps rediscover some of the sense of mystery that those observers experienced when they viewed the cosmos.

The ancient astronomers and astrologers contributed a great deal to what we now recognize as the sciences of astronomy and physics by identifying the planets in the solar system and carefully plotting the variations in their positions in the sky over periods of many years. These data eventually led to models and theories of the solar system. A mixture of improvements in data, modifications in theory, and large leaps in conceptual understanding finally led to the underlying principle that embodies the basic reason for all the movements of celestial objects. Known as the **law of universal gravitation,** the principle is another product of the immensely fertile mind of the English physicist Sir Isaac Newton.

Newton's law of gravitation is not the final word in our understanding of the universe. Albert Einstein modified Newton's concepts (although he did not really negate them) in his theory of relativity, and today a number of physicists are trying to modify even Einstein's theory. Furthermore, much of the most exciting and far-reaching research in current physics and astronomy concerns apparently incomprehensible objects undreamed of by Newton and Einstein, such as quasars, or "quasi-stellar objects," pulsars, and black holes. However, the core of our knowledge of how the universe works remains the gravitation law formulated by the seventeenth-century English academic.

The law of gravitation itself was actually the climax of a series of theories about the solar system. The first major theory, developed by the Greek philosophers Aristotle and Hipparchus, and elaborated by another Greek astronomer, Ptolemy, around 140 A.D., considered the earth as the center of the universe. According to the model, the sun and other stars supposedly revolved around the earth in simple circles, and the planets were believed to be in more complex orbits, revolving in small circles called epicycles that were imposed on their basic circular revolution around earth. This arrangement is shown in Fig. 4.1. Despite its inherent complexity, this model agreed so well with the naked-eye

Figure 4.1 Ptolemy's geocentric concept of the universe. Ptolemy in about 140 A.D. postulated that the moon moved in an orbit nearest the earth, Mercury in an orbit next farthest out, then Venus, then the sun, and then Mars, Jupiter, and Saturn. Ptolemy accounted for the long eastward motion of these heavenly bodies with respect to the stars by postulating counterclockwise motion along concentric circles called deferents. He ingeniously explained the loop (retrograde motion) formed by a planet when it moves westward for a short time by assuming that the planets move counterclockwise on smaller circles called epicycles as the centers of the epicycles move along the deferents.

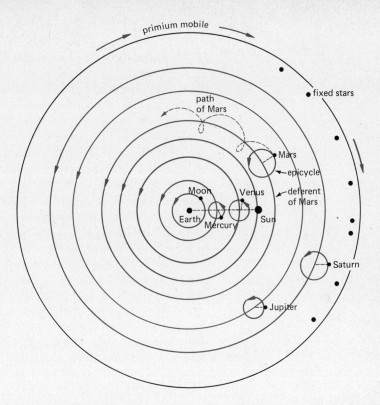

observations of the time that it was almost universally accepted for more than 14 centuries.

Eventually the Ptolemaic theory, as it was known, was superseded by the heliocentric theory, published in 1543 by Nicholas Copernicus, a Polish astronomer who had developed the ideas held many centuries earlier by Greek philosophers. The heliocentric theory precept was that each and every planet, including earth, revolves around the sun. In addition, Copernicus assumed that the earth rotated on its axis once each day. Two men—one an astronomical observer par excellence, the other a theorist with a splendid mind—brought rigorous scientific realism to the objective of understanding what makes the universe tick.

The observer was Danish astronomer Tycho Brahe, whose naked-eye observations of the planets during the last 25 yr of the sixteenth century were far more accurate than any performed previously. The theorist was the German Johannes Kepler, who, because of Brahe's data, found it necessary to modify the Copernican theory. Kepler was able to condense the motions of the planets into three relatively simple statements that enabled observers to produce more accurate mathematical predictions of planetary positions. Kepler's statements took the form of three empirical laws of motion that formed the basis of Newton's later statement of his law of universal gravitation.

4.1 KEPLER'S LAWS OF PLANETARY MOTION

The three laws are:

1. Each planet moves in an elliptical orbit, with the sun at one focus.

A simple way to draw a circle is to loop a piece of string (with its two ends tied together) around a thumbtack and a pencil, plant the thumbtack into a piece of paper or other suitable surface, and draw a line with the pencil, making sure that the loop of string remains at full extension. An ellipse can be drawn in much the same way, the only difference being that two thumbtacks are used instead of one (see Fig. 4.2). In the case of the circle, the thumbtack is the center; for the ellipse, the two tacks are at the two foci, or focal points. In fact, a circle is really a special case of an ellipse in which both foci occupy the same point.

In planetary motion, as shown in Fig. 4.3, the perihelion is the point P in the planet's orbit that is closest to the sun, and the aphelion, A, is the point in the orbit farthest from the sun. In the earth's orbit, the major axis—the line that joins the perihelion and aphelion—is about 186 million miles in length. At perihelion, the earth is 91.5 million miles from the sun; at aphelion, the distance is 94.5 million miles. Overall, the average distance between earth and sun taken over the course of any one year is 93 million miles. We might note that the orbit of the earth around the sun is relatively close to being circular. The separation of the two bodies never differs by more than 1.6 percent from the average.

Careful measurements of the apparent diameter of the sun on different days of the year indicate that the earth is closest to the sun (that is, at perihelion) about January 3 of each year and farthest from the sun around July 3. That our planet is at its greatest distance from our star at midsummer in the northern hemisphere is an apparent contradiction to those of us who live in the northern hemisphere. However, seasonal temperature variations are caused by a number of related physical factors, among them atmospheric rotation, angle of the sun's rays, ocean and air currents, and the relative positions of land and sea. Indeed, the very existence of clearly defined seasons on earth is an outcome of the axis around which the earth rotates not being perpendicular to the plane of earth's orbit (Fig. 4.4).

2. The imaginary line that joins a planet to the sun sweeps out equal areas in equal intervals of time.

This statement is known as the law of areas. It implies that the area swept out by the line which joins the earth and the sun between January 3 and January 31 is exactly the same as that swept out between July 3 and July 31, as shown in Fig. 4.5. We already know that the earth is closer to the sun in January than it is in July. Hence, the earth must move a greater distance around its orbit in January than in July, by traveling faster through space. In fact, the orbital speed of the earth varies from about 29.3 km/s (18.2 mi/s) on July 3 of every year to about 30.4 km/s (18.9 mi/s) on January 3. The average orbital speed of

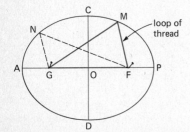

Figure 4.2 Construction of an ellipse. With pins placed at foci F and G, a loop of thread that is just big enough to cause a pencil point to fall at M is used. The loop will then guide the pencil successively through points C, N, A, D, P, and M, forming the ellipse.

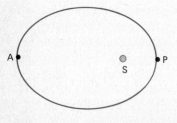

Figure 4.3 Elliptical orbit of a planet, with the sun (S) at one focus, to illustrate Kepler's first law of planetary motion. P is the perihelion, A the aphelion. For the earth moving about the sun the ellipse is much closer to a circle.

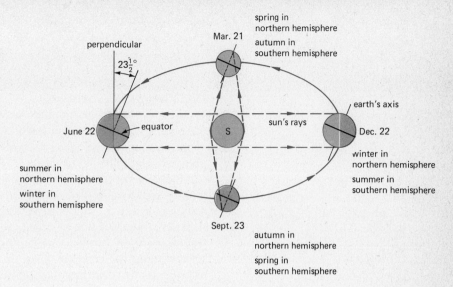

Figure 4.4 Main cause of seasons on the earth. The earth's positions with respect to the sun (S) at the beginning of the four seasons. The earth's axis points in the same direction with respect to the fixed stars in space—that is, to the north celestial pole. The axis is tilted $23\frac{1}{2}°$ away from the perpendicular to the plane of the earth's orbit.

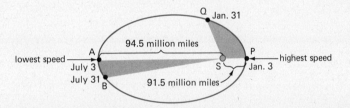

Figure 4.5 The earth in its orbit. The drawing, which is not to scale, represents the earth in orbit around the sun. The sun is at one focus of the elliptical path, with perihelion at P and aphelion at A.

our planet is close to 29.9 km/s (18.6 mi/s). This figure sounds much more impressive in miles per hour: it is an astonishing 67 000 mi/h. An important consequence of this observation is that the planets speed up as they fall toward the sun, just as objects accelerate as they fall toward earth.

3. The square of the time any planet takes to revolve around the sun is proportional to the cube of its average distance from the sun.

Mathematically, this is expressed as

$$T^2 \propto R^3$$

or

$$T^2 = (\text{constant}) \times R^3 \tag{4.1}$$

where **T is the time of one revolution, otherwise known as the period of the planet, and R is the average distance of that planet from the sun.** The value of the constant is determined by the units in which T and R are measured. In the case of the earth, for example, we can arbitrarily define our average distance from the sun as 1 astronomical unit (AU). Since T also has the value of 1 (in years), the constant is also 1. Using this value, and knowing that the planet Jupiter is 5.2 AU from the sun, we can calculate the time that the giant planet takes to revolve once around the sun in years:

$$T^2 = [1 \text{ yr}^2/(\text{AU})^3]R^3 = [1 \text{ yr}^2/(\text{AU})^3](5.2 \text{ AU})^3 = 140.6 \text{ yr}^2$$

Thus, Jupiter's period is 11.86 yr. In other words, a Jovian year would last 11.86 earth years.

Equation (4.1) also applies to other solar systems or to moons traveling about a particular planet. We shall show shortly that the value of the constant in the equation depends only on the mass of the body about which the planet or moon is circulating and the system of units.

Predictions of the positions of the planets according to Kepler's three laws agreed so well with observations that the Copernican hypothesis of a stationary sun and an orbiting earth was finally accepted despite a violent rearguard action by organized religion. The laws apply equally well to a variety of phenomena other than planetary motion. In the solar system, for example, they predict accurately the positions of asteroids, and of comets with elliptical orbits; in planetary systems, they allow scientists to calculate the positions of natural orbiting moons and artificial satellites.

EXAMPLE 4.1

If the period of revolution of an asteroid moving about the sun is 5 yr, what is its average distance from the sun in miles?

Solution

To solve this problem, we use Kepler's law, $T^2 = (\text{constant}) \times R^3$. Since the asteroid is moving about the sun, the constant can be taken to equal 1 if we use earth years as the units for T and astronomical units for the distance R:

$$T^2 = \left[\frac{1 \text{ yr}^2}{(\text{AU})^3}\right]R^3$$

$$R = \sqrt[3]{\frac{(T \text{ yr})^2(\text{AU})^3}{\text{yr}^2}} = \sqrt[3]{5^2} = 2.9 \text{ AU}$$

Since 1 AU is the average distance between the earth and sun, 93×10^6 mi, then

$$R = 2.9 \text{ AU} \times 93 \times 10^6 \text{ mi/AU} = 2.7 \times 10^8 \text{ mi}$$

4.2 NEWTON'S LAW OF UNIVERSAL GRAVITATION

Kepler's three laws represented a tremendous advance in astronomers' understanding of the universe. But important as they were in their own right, the laws' major role in the science of astronomy was to set the stage for a more powerful generalization about how the universe works. This was the recognition by Sir Isaac Newton, after he had formulated his relationships between force and motion, that the moon was held in its orbit around earth by precisely the same kind of force that causes objects near the earth to fall to the ground. Newton expressed this concept in a single sentence:

Every body (or mass particle) attracts every other body in the universe with a force that is directly proportional to the product of the masses of the two bodies, and inversely proportional to the square of the distance between them.

Mathematically, the law can be written as follows:

$$F = \frac{Gm_1 m_2}{R^2} \tag{4.2}$$

where F is the attractive force between two objects whose masses are m_1 and m_2, R is the distance between them, and G is a constant factor, known as the constant of universal gravitation.

What this means is that the gravitational force between any two objects is greater the more massive the objects are and the shorter the distance between them. Because the force is inversely related to the square of the distance, the law is known as an inverse square law. We shall encounter this type of law again when dealing with electrostatic forces and other situations in physics.

The numerical value of G, the constant of universal gravitation, depends on the units we use for the masses and distance involved in any particular equation. If, for example, we measure the masses in kilograms and the distance in meters, the force will be in newtons and G will have the value 6.67×10^{-11} N·m²/kg². In any one system of units, G has the same numerical value for every pair of objects in the cosmos.

The gravitational force between any two objects is greater the more massive the objects are and the shorter the distance between them.

EXAMPLE 4.2

Compute the gravitational force of attraction between the earth and the moon. The mass of the earth is 5.98×10^{24} kg, and the mass of the moon is about one-eightieth the mass of the earth. The distance between the earth and the moon is 3.8×10^5 km.

Solution

Using the given data, we need only substitute the known quantities into Newton's law of gravitation, Eq. (4.2):

$$F = \frac{Gm_1 m_2}{R^2}$$

$$= \frac{6.67 \times 10^{-11} \text{ N·m}^2/\text{kg}^2 (5.98 \times 10^{24} \text{ kg})(\tfrac{1}{80} \times 5.98 \times 10^{24} \text{ kg})}{(3.8 \times 10^8 \text{ m})^2}$$

$$F = \frac{(6.67)(5.98)^2 \times 10^{21} \text{ N}}{(80)(3.8)^2} = 2.1 \times 10^{20} \text{ N}$$

Needless to say, this is an extremely large force.

On the Shoulders of Giants

Sir Isaac Newton. (*National Portrait Gallery, London*)

Isaac Newton has been described as the greatest intellect of all time. His contributions to the science of physics are staggering. Yet in his early years at school, Newton was a loner who showed no hint of his future abilities.

Newton was born in 1642, the year of the death of Galileo, whose mantle as the foremost physical scientist of the age Newton eventually inherited. His ability was apparent only to an uncle who was a member of Trinity College, Cambridge, then as now one of the best colleges of one of the best universities. The uncle suggested that young Isaac study at his college, and Newton spent much of the rest of his life in the picturesque university city.

After graduating from Trinity College, Newton started a series of experiments on prisms and proved that white light actually consists of the spectrum of colors in the rainbow. So impressed were the Cambridge academics at this feat that they made Newton Professor of Mathematics—at the age of 27.

At the same time as he was starting his experiments on light, Newton was also looking into the subject for which he is most revered—gravitational attraction. Unlike most of the tales of scientific discovery, the story that Newton hit upon the universal nature of gravity when watching an apple fall to the ground is quite true—although the apple did not hit him on the head. The young academic speculated that the same force which pulled the apple down might also be responsible for keeping the moon in orbit. He theorized that the rate at which an object fell to earth was proportional to the gravitational force, which was in turn inversely proportional to the object's distance from earth, and he carried out a simple calculation for the case of the moon. Unfortunately, the calculation seemed to produce too low a figure for the effect of the earth's gravitation on the moon, and a dejected Newton left the subject alone for 15 yr.

When he did return to gravitation, Newton had the advantage of a powerful new mathematical tool—calculus—which he invented at about the same time as did the German mathematician Gottfried Wilhelm Leibnitz. The power of Newton's intellect can be seen from the fact that some aspects of his calculus, known as infinitesimals, are being rediscovered and extended by mathematicians today.

With calculus as his tool, Newton developed his three laws of motion and his law of universal gravitation, which survived intact for more than two centuries. He summarized these laws in his massive work *Philosophiae Naturalis Principia Mathematica,* which is still one of the greatest educational books.

In private life, Newton was the epitome of the absent-minded professor and surprisingly sensitive to criticism. Yet unlike Galileo, he was not without recognition in his lifetime. He was elected a member of Parliament in 1689, a role he fulfilled in near-total silence. He received a knighthood from Queen Anne in 1705 and was elected, fittingly, as President of the Royal Society, a post that he retained from 1703 to his death in 1727. One of Newton's most famous statements was his modest attribution of his discoveries to his scientific predecessors: "If I have seen further than other men, it is because I stood on the shoulders of giants." Subsequent generations of scientists have recognized that Newton's shoulders were the tallest and broadest of them all.

Calculating G

Newton obtained his law of universal gravitation by examining available data on the relative movements of the planets. And in a sophisticated form of double checking, he was able to derive Kepler's second law by combining the universal law with his own laws of motion, with which we dealt in Chap. 3. Even Newton could not work out the value of G, however, because he had no way of knowing the masses of the planets.

Not until 1797, more than a century after Newton formulated the gravitation law, was G determined experimentally. Henry Cavendish, the eccentric English scientist after whom the famous physics laboratory in Cambridge University is named, used a sensitive torsion balance (Fig. 4.6) to measure the force between two known masses in his laboratory.

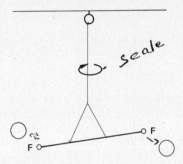

Figure 4.6 A simplified diagram of the sensitive Cavendish torsion balance used to measure the force between masses in the laboratory. The twist of the vertical fiber, produced by the two attractive forces between the small and large balls, can be used to determine the gravitation constant G.

EXAMPLE 4.3

Compute the gravitational force of attraction between two 5-kg balls 0.33 m apart (this is about the same as two 11-lb balls 1 ft apart).

Solution

$$F = \frac{Gm_1m_2}{R^2} = \frac{(6.67 \times 10^{-11}\ \text{N} \cdot \text{m}^2/\text{kg}^2)(5\ \text{kg})(5\ \text{kg})}{(0.33\ \text{m})^2}$$

$$F = 6.67(5)^2 \times (3)^2 \times 10^{-11}\ \text{N} = 1.5 \times 10^{-8}\ \text{N}$$

This is, of course, an insignificant force in relation to the weight of one 5-kg object, that is, $5 \times 9.8 = 49$ N, and hence we do not usually notice the gravitational attraction force between normal-sized objects.

The smallness of the force also explains the need to use a very sensitive torsion balance, such as that designed by Cavendish, to measure G. In fact, Cavendish's method is still used to determine this fundamental constant; its present-day value is more accurate than that measured by Cavendish himself because of technological improvements in instrumentation.

G and g One important outcome of the concept of the universal constant of gravitation is to give us a fresh understanding of *g*, *the acceleration due to gravity at the earth's surface,* which we discussed in Chap. 3.

Let us use Newton's law of universal gravitation to calculate the force F with which a cannonball whose mass is m is attracted to the earth. If the mass of the earth is M_e and the distance between the center of the earth and the center of the cannonball is R, then

$$F = \frac{GmM_e}{R^2}$$

We can simplify this equation slightly by assuming that the earth is a perfect sphere and that the cannonball is resting on its surface. Because the radius of the cannonball is negligible in comparison with the radius of the earth, we can imagine for the sake of argument that the distance between the centers of the earth and the cannonball is equal to the radius of the earth, R_e. The equation thus becomes

$$F = \frac{GmM_e}{R_e^2}$$

We already defined the force with which the cannonball is attracted to earth near the earth's surface as its weight $W = mg$. Equating these two expressions while neglecting the rotational effect yields

$$mg = \frac{GmM_e}{R_e^2}$$

Canceling out the mass of the cannonball on each side of the equation, we end up with the expression

$$g = \frac{GM_e}{R_e^2} \tag{4.3}$$

A few points are worth noting about this expression. First, the value of g is independent of the mass of any particular object experiencing the force of attraction by the earth. If the earth were a homogeneous perfect sphere, we see that g would be constant everywhere on the earth's surface, for the three factors G, M_e, and R_e are all constant in that idealized case. If we allow some reality to intrude on this picture, the equation shows that the value of g decreases if the cannonball is taken to the top of a mountain because at such a location the value of R_e is larger than it is elsewhere on the surface. Scientific measurements verify the dependence of g upon $1/R^2$ as one moves above the earth's surface. Other local variations in g are observed when one moves about the earth and are attributed to such factors as the rotation of the earth on its axis and the variation of the density just below the earth's surface in a particular locality.

The validity of Eq. (4.3) is not restricted to the earth and its environs. We can work out a similar series of calculations for the moon, making the same simplifying assumptions and substituting the mass and radius of the moon for those of earth. Such a calculation indicates what astronauts know from having been there—that the value of g on the moon is about one-sixth its value on earth.

EXAMPLE 4.4 Compute the gravitational acceleration at the surface of the planet Jupiter, which has a diameter of 88 000 mi compared with the 8000 mi for the earth's diameter, and a mass equal to 318 times that of the earth.

Solution This computation can be carried out directly using Eq. (4.3) and the values for the quantities for Jupiter. However, it is simpler to work out the problem in terms of the earth's characteristics. The radius of Jupiter is $(88\,000/2)/(8000/2) \times$ radius of the earth, that is, $11 \times R_e$, and the mass of the giant planet, M_J, is $318\, M_e$. Substituting into Eq. (4.3) yields

$$g_J = \frac{GM_J}{R_J^2} = \frac{G(318 M_e)}{(11 R_e)^2} = \frac{2.63\, GM_e}{R_e^2} = 2.63 g$$

The gravitational acceleration on Jupiter is 2.63 times that on earth; a 180-lb person on earth would weigh 2.63×180 lb, or 473 lb on Jupiter.

EXAMPLE 4.5 Using the approximate value of 4000 mi for the radius of the earth, compute the gravitational force on a 160-lb man positioned 4000 mi above the earth's surface on a small space platform.

Solution By definition, the man's weight is 160 lb on the earth's surface. When the man is 4000 mi above the earth's surface, his distance from the center of earth has increased to $2R_e$. Since the gravitational force depends on $g = GM_e/R^2$, the only change in this equation is in the value of R. The term $1/R^2$ has a value of $1/(2R_e)^2$, or $1/4R_e^2$, at an altitude of 4000 mi. Thus, the acceleration at this altitude becomes $GM_e/4R_e^2$, which equals $0.25g$. Thus, the man will experience a gravitational force one-fourth as much as on earth, or 40 lb, when he is 4000 mi above the earth's surface.

4.3 MOVING ROUND CURVES

One definition of scientific creativity is the ability to link apparently unrelated facts and observations in a single theory or hypothesis. This is what Newton achieved in the law of universal gravitation, combining the movements of the planets and those of objects on earth. The basic difference between the two types of motion is obvious: Planets revolve

Figure 4.7 Path of a barium sulfate crystal as it moves through the small intestine.

around their sun and satellites revolve around their planets in never-ending paths; yet objects on earth eventually fall to the surface. To understand the similarities that underlie these two types of motion, we must examine motion along curved paths.

Figure 4.7 shows the twisting and turning path of a barium sulfate crystal as it moves through the small intestine of an x-ray patient. The actual velocities of the crystal at various points along the intestine are represented by arrows pointing along the tangents to the curve at those points. We do not need to be told whether the crystal is traveling at a constant speed to know that its **velocity** is changing as it moves along the snaking path. The direction of the moving particle is changing from moment to moment, and a change in direction alters the velocity. Since acceleration is defined as change in velocity per unit time, we also know that the crystal is accelerating as it winds its tortuous path through the intestine. The average acceleration is given by the following simple equation:

$$\mathbf{a} = \frac{\mathbf{v}_2 - \mathbf{v}_1}{t_2 - t_1} = \frac{\Delta \mathbf{v}}{\Delta t}$$

where $\Delta \mathbf{v}$ is the vector difference between the two velocities $\mathbf{v}_1$ and $\mathbf{v}_2$ and is obtained by drawing a conventional vector diagram, remembering that one of the velocities in the diagram must have a negative value. This is shown in Fig. 4.8. The direction of $\Delta \mathbf{v}$ is the direction of the crystal's acceleration between the two points. Obviously, when we are dealing with a complicated curving path such as that through an intestine, the direction of $\Delta \mathbf{v}$ is also likely to change continually. Note that the quantity $\Delta \mathbf{v}/\Delta t$ can be applied in a realistic way only when one goes to small values of Δt. $\Delta \mathbf{v}/\Delta t$ then defines the meaningful quantity of interest, the instantaneous acceleration.

Figure 4.8 represents a special case of curved movement in another way; the magnitudes of the velocities $\mathbf{v}_1$ and $\mathbf{v}_2$ are roughly equal in the

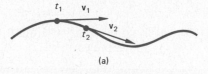

(a)

Figure 4.8 Motion along a curved path with constant speed.

(b) Vector determination of $\Delta \mathbf{v}$
$\Delta \mathbf{v} = \mathbf{v}_2 - \mathbf{v}_1 = \mathbf{v}_2 + (-\mathbf{v}_1)$

vector diagram. Plainly, the acceleration derives almost entirely from the change in the direction of the velocity. The simplest example of motion with a uniform speed but with a changing velocity caused by a directional change of the motion is illustrated by uniform circular motion.

Going Around in Circles

Uniform circular motion is simply the movement of an object at constant speed around a circle. A typical example of this occurs if one swings a stone on the end of a string in a horizontal circle above one's head. If R is the radius of the circle and T is the *period* of each revolution of the stone, then we know that the (constant) speed is equal to the circumference of the circle divided by the time of revolution, that is, expressed mathematically, $2\pi R/T$.

Because the velocity of the stone is continually changing direction, we also know that the stone is continually accelerating. Calculation of the value of acceleration is a relatively simple, although tedious, excursion into geometry, as shown in Fig. 4.9. The velocities of the stone, v_1 and v_2, at two different points P_1 and P_2 are tangential to the circle and hence perpendicular to the radius at the points P_1 and P_2. The vectors v_1 and v_2 are identical in length because the speed is constant. Basic geometry tells us that the isoceles triangle OP_1P_2 is similar to the vector diagram of the two velocities designed to yield the velocity difference Δv. Therefore,

$$\frac{\Delta v}{v} = \frac{\Delta s}{R} \quad \text{or} \quad \Delta v = \frac{v}{R}\Delta s$$

where v is the uniform speed of the stone at the end of the string. This equation is a scalar one, designed to calculate the *magnitude* of the stone's acceleration.

To take the next step to determine the acceleration (an instantaneous quantity), imagine that the points P_1 and P_2 are very much closer together, representing adjacent positions in the motion during an instant of time Δt. Under such conditions the chord of the circle, Δs, will almost equal the actual distance moved by the object on its curved (circular) path. Under this circumstance we can compute the magnitude of the acceleration

$$a = \frac{\Delta v}{\Delta t} = \frac{v}{R} \cdot \frac{\Delta s}{\Delta t} = \frac{v}{R} \cdot v = \frac{v^2}{R} \tag{4.4}$$

Here we use the fact that as Δt becomes very small, $\Delta s/\Delta t$ becomes the instantaneous speed v.

Having obtained the magnitude of the acceleration, we must now discover its direction. If we again imagine that the points P_1 and P_2 in Fig. 4.9 are moved very close together, a little thought will show that the direction of the vector difference between v_1 and v_2 will swing round in the same direction as the radius of the circle. And if the two points merge, the direction of Δv, and hence of the acceleration, becomes

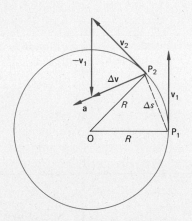

Figure 4.9 Uniform circular motion.

The nature of centripetal acceleration was first realized by Isaac Newton in the late seventeenth century.

identical with the radius at that point. Thus, the instantaneous acceleration of any object traveling in a circle of radius R at a constant speed v has a magnitude of v^2/R and is directed toward the center of the circle. Thus, it can be said that **an object moving in a circle at a constant speed is continually "falling" (accelerating) toward the center. This acceleration is called the centripetal, or center-seeking, acceleration.** Before returning to a discussion of gravitation we now present an example of uniform circular motion.

EXAMPLE 4.6

An object moves in a circle of radius 25 cm at a constant rate of 2 revolutions per second (r/s). Compute the linear speed and acceleration of the object.

Solution

The object is obviously undergoing uniform circular motion since its speed is constant and it moves in a circle of fixed radius. We just derived from geometrical considerations the expression for the magnitude of the acceleration of an object undergoing uniform circular motion. The acceleration given in Eq. (4.4) is equal to the speed squared divided by the radius of the circle $a = v^2/R$. In addition, the direction of a is along the radius toward the center of the circle. We do not know the speed v of the object in centimeters per second, but we do know that it completes 2 r/s. Since the object covers the distance of the circumference, $2\pi R$, in each revolution, then the speed must be given by

$$v = 2 \text{ r/s} \times 2\pi R \text{ cm/r} = 2 \times 2\pi(25) \text{ cm/s}$$
$$= 100\pi \text{ cm/s}$$

The centripetal acceleration will be

$$a = \frac{v^2}{R} = \frac{(100\pi \text{ cm/s})^2}{25 \text{ cm}}$$

$$= 400\pi^2 \text{ cm/s}^2 \quad \text{toward center of circle}$$

4.4 GRAVITATION AND THE SOLAR SYSTEM

We just saw that an object moving in any curved path must be accelerating. And if the object is moving in a circle with a constant speed, it will have a center-seeking or centripetal acceleration whose magnitude is given by $a_c = v^2/R$. If we apply Newton's second law, we see that this acceleration can only be produced by a net force on the circling object. This force is the centripetal force and is always required to keep an object moving in a curved path. This should be obvious from the first law. The centripetal force and the centripetal acceleration are both vector quantities and must be directed along the same line, namely, along a radius toward the center of the circle:

$$\mathbf{F}_c = m\mathbf{a}_c$$

The centripetal force = mass × centripetal acceleration,

$$F_c = m\frac{v^2}{R} \quad \text{in magnitude} \tag{4.5}$$

The stone in this example would still be subject to the vertical force of its own weight after release, and it would soon drop to earth. However, its horizontal movement would continue in a straight line.

To prove the existence of such a force in practice, one would merely have to set a stone on the end of a string swinging horizontally above one's head and then release the string. Without any force of tension acting on it through the string, the stone would move in a path consistent with Newton's first law of motion and fly off at a tangent in a straight line at a constant speed.

EXAMPLE 4.7

A 0.2-kg ball is tied to the end of a cord and whirled in a horizontal circle of radius 0.6 m. If the ball makes five complete revolutions in 2 s, determine the ball's linear speed, its centripetal acceleration, and the centripetal force.

Solution

The ball makes five revolutions in 2 s, traveling a distance of $2\pi R$ in each revolution. Hence it makes one revolution in 0.4 s. The linear speed of the ball is

$$v = \frac{2\pi R}{0.4 \text{ s}} = \frac{2\pi(0.6)\text{ m}}{0.4 \text{ s}} = 9.42 \text{ m/s}$$

The centripetal acceleration will be

$$a_c = \frac{v^2}{R} = \frac{(9.42 \text{ m/s})^2}{0.6 \text{ m}} = 148 \text{ m/s}^2$$

The centripetal force = $ma_c = (0.2 \text{ kg})(148 \text{ m/s}^2) = 29.6$ N.

EXAMPLE 4.8

Calculate the centripetal acceleration and centripetal force on a man whose weight is 176 lb (mass 80 kg) when resting on the ground at the equator.

Solution

Because of the rotation of the earth, the man at the equator moves in a circle whose radius is equal to the earth's radius, 6.4×10^6 m. The man makes one revolution in about 24 h; hence, his speed is given by

$$v = \frac{2\pi R}{T} = \frac{2\pi(6.4 \times 10^6 \text{ m})}{24(60)(60) \text{ s}} = 465 \text{ m/s}$$

This speed is approximately 1000 mi/h. The acceleration toward the center of the earth is

$$a_c = \frac{v^2}{R} = \frac{(465 \text{ m/s})^2}{6.4 \times 10^6 \text{ m}} = 3.37 \times 10^{-2} \text{ m/s}^2 \quad \text{or} \quad 0.0034g$$

The centripetal force equals

$$ma_c = 80 \text{ kg } (3.37 \times 10^{-2} \text{ m/s}^2) = 2.69 \text{ N}$$

Your weight on a scale is equivalent to the gravitational force only at the North and South poles.

With the results of Example 4.8 we can illustrate how a person's weight, as measured by a spring scale such as the normal bathroom scale, does *not* measure the true gravitational force except at two points on earth. These are the North and South poles, where there is no centripetal acceleration. For the man at the equator in Example 4.8, a spring scale would measure the gravitational force on him minus the centripetal force. Using the figures from Example 4.8, we can calculate the man's weight (i.e., what the scales indicate) at the equator (80 kg × 9.8 m/s²)N − 2.69 N = (784 − 2.69)N = 781 N. The 781 N or 175 lb is the force with which the man presses against the earth, as recorded by the scales at the equator. Only at the earth's poles will his measured weight (using a scale) be equivalent to the gravitational force.

Returning to the heavens, we can apply Newton's law of universal gravitation to the movements of the solar system. The sun and the earth are attracted to each other by the force

$$F = \frac{GM_e M_s}{R_{se}^2}$$

where M_e and M_s are the masses of the earth and sun and R_{se} is the distance between their centers. Now, if both bodies were stationary and the gravitational force were the only one acting on them, they would accelerate toward each other and collide disastrously. Since the solar system has been around for at least 4.5 billion years, we know that something is wrong with this picture. It is our neglect of the fact that the earth is revolving at high speed around the sun. This motion requires an inward, centripetal force. This centripetal force is provided by the mutual gravitational attraction between sun and earth.

The earth really moves in an elliptical orbit, but we can safely approximate the ellipse to a circle since the distance between the earth and sun varies from only 94.5 to 91.5 million miles, or 1.6 percent during the period of 1 yr. In this case, the force on the earth must be

$$F_{\text{net}} = M_e a_c = M_e \frac{v^2}{R_{se}} \tag{4.6}$$

Since this net force is the gravitational force,

$$\frac{GM_eM_s}{R_{se}^2} = \frac{M_ev^2}{R_{se}}$$

Cancellation of redundant terms leads to the following equation for the earth's speed in orbit:

$$v^2 = \frac{GM_s}{R_{se}} \qquad (4.7)$$

Both G and M_s are constants (at least over any time scale shorter than eons); since we simplified the earth's orbit to a circle with R_{se} being constant, the equation implies that the earth's orbital speed is also constant.

We can also relate the earth's orbital speed to the period of the earth's revolution, T, which is 1 yr:

$$v = \frac{2\pi R_{se}}{T} \qquad (4.8)$$

Substituting this value of v into Eq. (4.7), we find the following:

$$\frac{4\pi^2 R_{se}^2}{T^2} = \frac{GM_s}{R_{se}}$$

Rearranging the terms, we arrive at Kepler's third law:

$$T^2 = \frac{4\pi^2}{(GM_s)} R_{se}^3 = (\text{constant}) \times R_{se}^3 \qquad (4.9)$$

We now have a physical explanation for Kepler's empirical law.

The only factor actually related to the earth in this analysis is R_{se}, the distance between the centers of the sun and earth. Thus, if one knows the center-to-center distance between the sun and any other planet, one can apply the same set of equations to that planet. In addition, if the mass of the sun is replaced by the mass of a particular planet, one can use the process to work out the speed or period of any natural or artificial satellite in orbit around the planet.

One final point regarding mutual attraction: We have assumed to date that the earth moves around a stationary sun. A little consideration of Newton's third law tells us that this cannot be entirely accurate, for the same net force that acts on the earth is also acting on the sun. This effect is often referred to as the "mutual force of attraction."

To understand how this mutual force affects the relative movements of earth and sun, let us first consider an entirely hypothetical situation in which we have an earth and sun stationary relative to each other and separated by a mass of antigravity material (a fictitious material that eliminates the gravitational force between the earth and the sun). Once we remove the antigravity material, the gravitational attraction of the sun causes the earth to fall toward it. But equally, the gravitational attraction of the earth causes the sun to fall toward *it*. Now, Newton's second law tells us that when two objects are subject to the same force, the acceleration is inversely proportional to the masses of the two objects. Because the sun is about 300 000 times as massive as the earth, it will only move very slowly toward earth in our static example, whereas the light earth will plummet toward the sun.

Returning to reality, we find that the sun does revolve to a small extent around the earth or, more properly, that both sun and earth revolve around a point slightly separated from the center of the sun. Because of the sun's huge diameter, however, this point is still within our star.

The effect of mutual attraction is seen much more clearly in systems known as binary stars, which are visible in astronomical telescopes. In a number of such systems, the masses of the two stars revolving around each other are quite similar, and the stars revolve around a point roughly midway between their centers. This effect, along with the fact that the real paths of planetary objects are elliptical rather than circular, make our calculations approximate when we use the value R_{se} in Eqs. (4.6) to (4.9). In a more advanced treatment of this subject we could make an appropriate correction to R_{se}.

EXAMPLE 4.9

The moon travels in an approximately circular orbit about the earth at a distance of 3.8×10^5 km. Calculate the speed of the moon in its orbit and the period of the moon. The mass of the earth, M_E, is 5.98×10^{24} kg.

Solution

To determine the speed of the moon in its orbit about the earth, we set the gravitational attraction of the moon by the earth equal to the product of the mass of the moon and its centripetal acceleration:

$$F = M_m a_c$$

$$\frac{GM_e M_m}{R_{me}^2} = M_m \frac{v^2}{R_{me}} \quad \text{or} \quad v^2 = \frac{GM_e}{R_{me}}$$

$$v = \sqrt{\frac{GM_e}{R_{me}}} = \sqrt{\frac{6.67 \times 10^{-11} \text{ N} \cdot \text{m}^2/\text{kg}^2 \times 5.98 \times 10^{24} \text{ kg}}{3.8 \times 10^8 \text{ m}}}$$

$$v = 1.02 \times 10^3 \text{ m/s}$$

The period T of the moon can be calculated by using the data

$$v = \frac{2\pi R_{me}}{T}$$

$$T = \frac{2\pi R_{me}}{v} = \frac{2\pi \times 3.8 \times 10^8 \text{ m}}{1.02 \times 10^3 \text{ m/s}} = 2.34 \times 10^6 \text{ s}$$

$$T = \frac{2.34 \times 10^6 \text{ s}}{24 \times 60 \times 60 \text{ s/day}} = 27 \text{ days}$$

This is close to the observed value of 27.3 days. The difference results from such approximations as regarding the moon's orbit as circular instead of elliptical and from uncertainties in the average value for R_{me}.

4.5 ARTIFICIAL SATELLITES

For scientists in the space program, the problem of launching an earth satellite is a dual one of overcoming the earth's gravitational force and providing the satellite with proper velocity both in magnitude and direction to begin the desired orbit. Once placed into orbit, the satellite will be held there by the centripetal force provided by the gravitational attraction of the earth. The orbits of real satellites are always elliptical, but once again we can consider circular orbits, which we can analyze exactly, without losing any understanding.

Figure 4.10 shows the results of four separate efforts to launch satellites. Those taken up along paths 1 and 2 did not have enough speed in the right direction to make it into orbit, and those that went up along routes 3 and 4 did achieve stable orbits by getting above the earth's atmosphere. We can calculate the necessary speed by equating the centripetal force and the gravitational force of attraction:

$$\frac{GM_e m_s}{R^2} = m_s \frac{v^2}{R}$$

where M_e and m_s are the masses of the earth and the satellite, and R is the distance between their centers—the radius of the earth added to the altitude h required for the satellite. Canceling out redundant factors, we find that

$$v = \sqrt{\frac{M_e G}{R}}$$

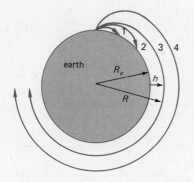

Figure 4.10 Paths of rockets and satellites. A rocket given sufficiently high speed in the correct direction goes into a stable circular orbit indicated by path 3 or 4. Rockets following path 1 or 2 return to earth. A satellite in orbit 3 must move more rapidly than a satellite in orbit 4. R_e is the radius of the earth. $R = R_e + h$ is the radius of the orbit of the satellite. To obtain orbits 3 and 4, correcting jet thrusts are required after launch.

This equation reveals that the speed required for any particular orbit is independent of the mass of the satellite but does depend on the height of the satellite's orbit above the earth. In fact, it is theoretically possible to put a satellite into orbit just 10 mi above the earth, rather than the normal altitude of more than 100 mi. However, the 10-mi orbit would require the satellite to be moving at a speed of about 7.9×10^3 m/s, or 17 700 mi/h. Although this speed can be attained, the drag of the air would quickly burn up the satellite. In Fig. 4.10, the satellite

108 Universal Gravitation

Newton showed the space community how to determine all the information necessary to launch satellites into orbit more than 200 yr before the experts were able to develop rockets powerful enough to achieve the task.

moving along path 3 requires a greater orbital speed than that traveling along route 4 because its orbit is closer to earth. Note that Newton showed the space community how to determine all the information necessary to launch satellites into orbit more than 200 yr before the experts were able to develop rockets powerful enough to achieve the task. This is characteristic of the development of science. Theories predict many things that cannot be tested for want of technology, and, conversely, observations are often made that existing theory cannot explain.

EXAMPLE 4.10

We wish to place in orbit a satellite that will always remain above the same point on earth. How far above the earth's surface will the satellite reside, and what orbital speed must we give the satellite?

Solution

Almost all communications satellites occupy "geosynchronous orbits" that keep them continuously above the same spot on earth.

The period of the satellite will be about 24 h, the same as the time it takes the earth to rotate on its axis. The orbital speed is given by Eq. (4.7), applied to the earth instead of the sun, and by $v = 2\pi R/T$. These two equations can be solved simultaneously for R, the distance between the satellite and the center of the earth. This resulting equation is of course Kepler's law and may be solved to determine R:

$$R^3 = \frac{M_e G T^2}{4\pi^2}$$

$$R = \sqrt[3]{\frac{M_e G T^2}{4\pi^2}}$$

$$= \sqrt[3]{\frac{(5.98 \times 10^{24} \text{ kg})(6.67 \times 10^{-11} \text{ N} \cdot \text{m}^2/\text{kg}^2)(24 \times 60 \times 60 \text{ s})^2}{4\pi^2}}$$

$$R = \sqrt[3]{75.4 \times 10^{21}} \text{ m} = 4.23 \times 10^7 \text{ m}$$

The height of the satellite above the earth's surface must be $h = R - R_e = 42.3 \times 10^6 - 6.37 \times 10^6$ m $= 35.9 \times 10^6$ m. This is about 5.6 earth radii above the earth's surface. The speed with which the satellite revolves can be computed most easily from

$$v = \frac{2\pi R}{T} = \frac{2\pi(4.23 \times 10^7 \text{ m})}{(24)(60)(60) \text{ s}} = 3.1 \times 10^3 \text{ m/s} = 6900 \text{ mi/h}$$

4.6 WHEN THE TIDE RUSHES IN

Natural philosophers had correlated the action of the oceans' tides with the movements of the moon long before Newton was born. However, the sage from Cambridge was the first to show the specific connection between the tides and the moon's gravitational pull. We have learned since Newton's time that the gravitational attraction of the sun also has a small effect on the tides and that the effects of the sun and moon are not restricted to the oceans; even solid rocks suffer a very small tidal effect.

The earth is not a perfectly rigid, spherical body. Consequently, the moon's gravitational attraction on different parts of the earth results in a differential or *tidal force* on the earth. This actually produces a slight distortion of the earth. The side of earth nearer the moon is attracted toward the moon more strongly than is the center of the earth, and the side of the earth farther from the moon is even less strongly attracted to the moon. The tidal force tends to stretch the earth slightly. Because the earth is solid, its actual deformation is only about 20 cm. Of course, since the earth is rotating upon its axis, different parts of the earth are continually coming "under" the moon. The direction and magnitude of the gravitational force of the moon on each point on the earth are continually changing. The earth readjusts its shape under this changing tidal force.

Note that this tidal effect upon the shape of the earth is superimposed upon the much larger distortion caused by the rotation of the earth upon its axis. The latter is responsible for the flattening of the earth at the poles.

The effect of the moon's gravitational pull upon the earth is most pronounced when we examine the tidal effect on the fluid part of the earth. We can start to understand the tides by first assuming that the earth is a sphere covered by a uniform depth of water. If we neglect the motions of the earth and the moon, it is clear that the pull of the moon on the water on the near side of the earth to the moon is greater than the pull on the far side. Recalling Newton's law of gravitation, we could actually calculate the difference in the force on the near and far side. The centers of the earth and moon are separated by about 3.8×10^5 km, and the radius of the earth is about 6.4×10^3 km. Since the force varies as 1 over the square of the distance, the difference between the force on the water on the near side and the force on the far side of the earth is about 6 percent.

How does this uniform sea on the earth's surface distort? In just the same way the solid earth is distorted. The tidal force stretches the earth and the sea of water surrounding it, producing a bulge on both the near and far sides of the earth relative to the moon. The water flows and piles up at the points under the moon and on the far side of the earth from the moon. This bulging effect is shown in Fig. 4.11.

These bulges are not static because the earth is rotating on its axis beneath the moon once every 24 h, and the moon and earth are moving around each other about once every 28 days. In just the same way that the earth and sun revolve about a point slightly displaced from the sun's center, the moon and earth revolve about a "center of gravity" that lies about 1600 km beneath the surface of the earth closest to the

Figure 4.11 The tides. *Top,* spring tides. When the earth, moon, and sun are in line (at the time of full moon and new moon), the pull of the sun (represented by black arrows) reinforces the pull of the moon (blue arrows), and the highest high tides and lowest low tides occur. *Bottom,* neap tides. When the moon is in a position at 90° to the sun's position relative to the earth (at half moon), the pulls of the sun and moon work against each other, so that high tides are of only medium amount and low tides are not as low. Relative sizes of the earth, moon, and sun, and their distances apart, are not drawn to scale, and tidal bulges are greatly exaggerated.

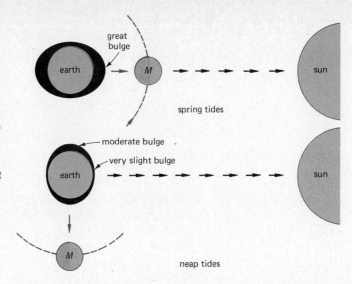

moon. (The center of the earth is at a depth of 6400 km.) The two bulges, or high tides, thus travel around the earth once each day. Since motion is relative, we could describe this motion as the earth's rotating once each day through the two bulges. Hence, if the earth were a uniform sphere, we would expect individual locations on the ocean-covered earth to pass through a sequence of high, low, high and low tides every 24 h. Thus a 6-h difference between high and low tides would be expected at our ocean shores.

In reality, the high tides do not occur exactly beneath the moon; owing to inertia and friction, they lag about 6 h behind the moon's motion. And the existence of the land masses on earth causes extremely complex tides. High tides occur once every 12 h 20 to 25 min on most coastlines. However, the variation in heights across the world is extraordinary. Such variations are caused by land and sea bottom topology. In many mid-Pacific islands, for example, the difference in height between high and low tides may be no more than 1 or 2 ft. In other areas, notably the Bay of Fundy between the Canadian provinces of New Brunswick and Nova Scotia, tidal variation can reach 70 ft. Tidal currents race into and out of narrow bays such as the Bay of Fundy at speeds up to 15 mi/h.

We have neglected the differential gravitational force of the sun on the near and far sides of the earth. Even though we are trapped in an orbit about the sun and it exerts a much larger gravitational attraction on the earth than the moon exerts on the earth, the tidal force of the sun upon the earth is weak compared to the moon's influence because the earth-sun separation is so much larger than the diameter of the earth, resulting in a very small difference between the tidal forces on the sides of the earth nearest to and farthest from the sun. The sun's influence on tides is less than half as strong as the moon's.

When the sun and moon are aligned (see Fig. 4.11), and thus pulling together, the earth's high tides are exceptionally high. These are the spring tides. On occasions when the sun is closest to earth and aligned with the moon, they can be catastrophically high. When the sun and moon are at right angles to each other relative to the position of the earth, their pulls do not reinforce. The high tides are then lower than normal and termed neap tides. High tides and neap tides each occur twice per lunar month.

4.7 THAT FLOATING FEELING

Any fan of space flight is used to hearing the term **weightlessness** describing the condition of an astronaut orbiting in a spaceship and to seeing vivid, real examples of this effect in telecasts from Skylab. The mathematical description of the state of weightlessness is somewhat more complicated than the visual evidence of television.

Weight is usually defined as the gravitational force on an object as observed on the surface of our rotating earth. Being a force, weight is proportional to the mass of an object. Mathematically we have set weight equal to mg, where m is the mass of an object and g is the acceleration due to gravity at the place where the object rests. From Newton's law of gravitation we know that g decreases as one moves away from the earth. However, because $g = GM_e/R^2$ (neglecting the rotational effect), g and our weight would not drop to zero unless R becomes infinitely large.

Weightlessness does not imply, therefore, that the value of the gravitational force becomes zero, i.e., that g has become zero. To illustrate what we commonly mean when we use the term weightlessness, here is an earthly example. Imagine the descent of an elevator covered all around with one-way mirrors, so that people outside the elevator can see inside but persons inside cannot see out. Say the elevator is descending with an acceleration of a, which is smaller in magnitude than g, and a man inside the elevator drops a ball. To the people outside the elevator the acceleration of the ball would be g, and to the man inside the elevator the acceleration of the ball would be $(g - a)$. The man in the elevator is in an accelerated system and measures an acceleration of the ball $g' = (g - a)$.

To examine weight in the accelerated system (the elevator), further imagine that a man in an elevator weighs himself on a spring scale. Let F_s be the force of the scale on the man acting upward. Then the force of the man on the scale—which is what the scale actually measures—is F_s acting downward. Because the man is descending with the elevator, we know that the scale will not measure the gravitational force on him. An observer from the outside who sees the accelerated elevator (downward) traveling uses Newton's second law, the scale reading (F_s), and her knowledge of the man's weight (mg) to set up the following equation (downward-directed quantities are negative):

$$-mg + F_s = -ma$$

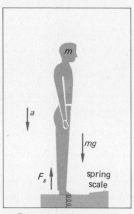

Figure 4.12 Forces on a person in a descending elevator as seen from the outside. Downward quantities are regarded as negative.

from which she can determine the scale reading

$$F_s = m(g - a) = W - ma$$

However, the man in the elevator who does not know about the acceleration of the elevator observes the scale reading as his weight:

$$F_s = W' = mg' = m(g - a) = W - ma$$

This equation is of course the same as the previous one. However, it has a different interpretation because of the two different points of view. The man in this downward-accelerating elevator sees his weight (W') as being less than that he would have when standing at rest on the surface of the earth. In fact, as the acceleration of the elevator approaches g, the man's weight from his frame of reference comes closer to zero. He approaches a state of weightlessness!

Despite the evidence of his eyes as he looks at the scale, the situation does not necessarily appear unusual to the man in the elevator. He might believe that he is in a region of the universe where the acceleration due to gravity is $(g - a)$. This is a simple example of Einstein's principle of equivalence, which asserts that there is no way of distinguishing between gravitational acceleration and acceleration through space produced by any net force. This principle is the cornerstone on which the general theory of relativity rests.

A practical way to obtain weightlessness, at least for a few seconds, is to ride in an airplane in a steep dive, that is, a plane "undergoing free fall." An alternative method is to ride in a satellite orbiting the earth and undergoing the centripetal acceleration GM_e/R^2, which is exactly the value of g at a distance R from the earth's center. Despite its circular motion, this satellite is undergoing free fall just as surely as the diving airplane is.

EXAMPLE 4.11

A woman with a mass of 45 kg (weight of about 100 lb) is standing on a scale in an elevator. The elevator accelerates upward with a constant acceleration of 1.2 m/s². What is the woman's weight as measured by her in the elevator?

Solution

Her weight as measured in the accelerating elevator is equal to the force F_s of the scale that is resting on the floor of the elevator pushing up on the woman's feet. We apply Newton's second law from our perspective outside the elevator, noting that the net force on the women is the push of the scale upward minus the pull of gravity (her weight), or

$$F_s - mg = ma$$
$$F_s = mg + ma = m(g + a)$$

Getting Away from Gravity

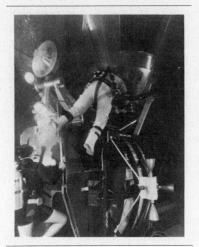
Underwater simulator. (NASA)

When it comes to space travel, being there is all the fun, at least as far as weightlessness is concerned. There is no easy way on earth that astronauts-to-be can fully experience the freedoms and restrictions of the weightless state.

In the early days of manned space flight, this state of affairs caused some concern in the National Aeronautics and Space Administration (NASA). The agency's simulators of spacecraft were perfect reproductions of the real vehicles, but if astronauts in training could not use them under the type of weightless conditions that they would experience in space, the value of the simulators would be small indeed. But NASA was undaunted and soon solved the problem.

The simplest means of experiencing the weightlessness associated with space flight is to travel in a plane in a free fall dive. The limitation of this method is that the state of weightlessness lasts no more than a few seconds at a time. More applicable for NASA's purposes in the early days of the Apollo moonflight program was a technique known as underwater simulation. The space agency constructed a large pool, 30 ft in diameter and 16 ft deep, at its astronaut center in Houston and mounted a mock-up Apollo spacecraft in the pool. Working inside the submerged mock-up, astronauts experienced much the same sort of buoyancy that they later came to know in space.

Even the water tank method is unsatisfactory for investigating the long-term effects of weightlessness, however, for an astronaut's time in the tank is limited by the amount of air in the breathing apparatus. The best method of determining how the months of weightlessness during manned flights to the planets will affect the salts and minerals of the body is to observe the effects of plain bed rest. Volunteers who spend months lying on their backs have shown many of the chemical changes experienced by astronauts on marathon Skylab flights.

$= 45 \text{ kg } (9.8 + 1.2) \text{ m/s}^2 = 495 \text{ N}$ or 111 lb

By riding in this upward-accelerated elevator the woman sees her weight as measured in the elevator increase. Not a happy experience for a weight watcher!

As an exercise, consider how your weight as measured in a moving elevator would change during a normal elevator trip to the top of a building and the subsequent return to the ground.

4.8 FLEEING THE CENTER

More critically minded readers may have noticed that this chapter has not had a mention of centrifugal (literally, fleeing from the center) force, although this term is much more common than centripetal force. The creation of the term centrifugal force is another example of the way in which identical phenomena appear different to observers in different frames of reference.

Imagine a traveler in a satellite that is orbiting the earth in uniform circular motion. When the traveler drops an object, it remains suspended. Now the observer knows that the traveler is pulled toward the earth by the force of earth's gravitational attraction. To account for the "obvious" lack of acceleration that the traveler experiences, the observer could invent a fictitious force that acts in the direction opposite to that of earth's gravity and exactly balances the gravitational pull. The observer calls this the centrifugal force.

Similarly, the concept of centrifugal force is used to account for the common experience of being thrown toward the outside of the curve when taking a sharp corner in a car. Actually, no force is "throwing" the car's occupants, but they are experiencing at first hand the consequences of Newton's first law—they continue to move in a straight line because of the *absence* of a force until the car, through the seat, exerts an inward, centripetal force on them, to set them moving in a circle. Most students are familiar with centrifuges, which have numerous applications in modern science, including separating red platelets and corpuscles from blood plasma. The separation process is a consequence of inertia, which is defined by Newton's first law; it is not the result of some mysterious force that pulls the material to the outside of the spinning centrifuge.

Centrifugal force is an invention used when applying Newton's laws in an accelerated frame of reference rather than in the inertial or nonaccelerated frame in which they actually hold. Since we avoid accelerated frames of reference in this text, the only proper use of the term centrifugal force would be an entirely different definition of centrifugal force. This new definition assigns the term centrifugal force as the Newton's third-law reaction to the centripetal force. Since the centripetal force is directed *toward* the center, its reaction would be directed *away from* the center—center fleeing! For example, when you whirl about a rock in a circle, the string pulling the rock inward toward the center of motion supplies the centripetal force. The reaction to this force, namely, the force the rock exerts on the string, can properly be termed the centrifugal force. Since we are nearly always concerned with the object in motion and the forces acting *on* the object that change its motion, it is strongly recommended that the term centrifugal be avoided.

4.9 Earth's Gravitational Field

The space around the earth in which a mass is affected by the force of the earth's attraction can be said to be modified by the earth's gravitational field. Theoretically, this field extends throughout the universe,

Stars That Make Waves

Stupendous as his achievements were in bringing the subject of gravitation down to earth, Sir Isaac Newton did not cover the whole subject of gravitational attraction. Although his law of universal gravitation accounts for all phenomena in the everyday world of physics, it is not quite accurate in the case of massive stars. Furthermore, Newton's law explained how gravitation worked, but not why. Countless scientists since Newton have asked the basic question: "What in the fundamental nature of gravitation?"

This question was answered partly by Albert Einstein, the man who inherited Newton's mantle as the synthesizer of physical theory. Einstein's general theory of relativity, published in 1916, extended Newton's concepts. Einstein modified Newton's law of gravitation in such a way as to account for the discrepancies in gravitational attraction involving stars and other cosmic objects, and he proposed that the essence of the gravitational force is a type of wave, not unlike light and sound. In simplest terms, Einstein's theory argues that waves pass, in effect, between two objects when they attract each other with the force of gravitation.

But while solving one problem, Einstein raised another. Large as it may appear to someone falling off a tree, the gravitational force is actually extremely small in comparison with other types of force in the universe. According to Einstein's theory, the waves associated with gravitational attraction are so minuscule as to make them all but indetectable. For 60 yr after the general theory was proposed, physicists believed that no equipment on earth could possibly spot a single gravitational wave from even the most massive stars. There was a shock for the gravitational experts in 1969, when Joseph Weber of the University of Maryland announced that he had detected several gravitational waves by using two ultrasensitive instruments more than 600 mi apart which were designed to oscillate only when penetrated by those specific waves. Most specialists, however, view the report with a great deal of skepticism.

Then in December 1978 came news from an unexpected quarter. A team of radio astronomers from the University of Massachusetts, headed by Joseph Taylor, announced that it had found indirect evidence of gravitational waves. The group had spent 4 yr focusing the giant antenna in Arecibo, Puerto Rico at a distant target consisting of two stars. One star was a pulsar, a peculiar type of dying star first discovered in 1968, that emits radio waves with the precision of a superaccurate clock. The pulsar's companion star was invisible even to radio telescopes.

If gravitational waves exist, their generation through the mutual attraction of two such stars must remove energy from the star system. Losing energy, the stars will drift closer together. The slow but inexorable approach of the stars in the system monitored by Taylor's team would make itself evident in a minute speeding up of the rate at which the pulsar emits radio waves. That is precisely what the University of Massachusetts astronomers found. The speeding up of 1 microsecond (μs) that they measured during their 4 yr of observation fitted exactly with that expected if gravitational waves were being emitted by the double-star system.

The measurement was not conclusive proof of the existence of gravitational waves; the physicists pointed out that other effects might have caused the pulsar to speed up. But the observation gave hope to the more than 10 teams of physicists worldwide who, at the end of 1978, were gearing up their own detectors designed to spot gravity waves directly.

---

but in practical terms it can really be noticed only in the solar system. **The strength of the field at any point is the force exerted by the earth on each unit of mass of any object placed at that point,** and the field is directed toward the center of the earth. This is a vector field defined by a stationary observer by the equation

$$\frac{F}{m} = \frac{mg}{m} = g \qquad (4.10)$$

For the earth the magnitude of g is the strength of the field and is given by $g = GM_e/R^2$.

The earth's gravitational field is represented graphically by imaginary lines in space, as shown in Fig. 4.13. Note in the figure that the lines are all directed toward the center of the earth. The closer the spacing of the lines at points above the surface, the stronger the field of these points.

Despite the foregoing body of knowledge on gravitation, note that the whole subject remains one of the most puzzling in all science. Scientists know how gravitation operates, but they have virtually no idea of the details of its origin except that it is a property of mass. Today, some exciting experiments in physics involve the search for "gravity waves," which might provide the key to just how objects are linked by gravitational attraction.

The distribution of electrostatic and magnetic fields, treated in Chaps. 13 and 15, is similar in many ways to that of gravitational fields.

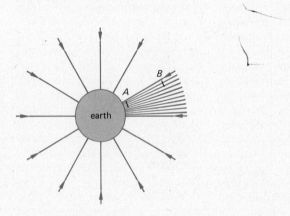

Figure 4.13 The strength of the earth's gravitational field depends on the number of lines per unit area. The unit area indicated by a blue line is everywhere perpendicular to the lines. The strength of the field is greater at A than at B.

SUMMARY

1. According to Kepler's laws of motion, each planet moves in an elliptical orbit around its sun. The line between the sun and the planet sweeps out equal areas in equal times during its movement. The square of any planet's time of revolution around its sun is proportional to the cube of its average distance from the sun.

2. Newton's law of universal gravitation states that every particle in the universe attracts every other particle with a force directly proportional to the masses of the two particles and inversely proportional to the square of the distance between them.

3. The constant G that relates the force of attraction to the masses of gravitationally attracting objects and the distance between them is a fundamental invariant that has the same value for all pairs of objects in the universe.

4. The acceleration due to gravity, neglecting the earth's rotation, for all objects at the earth's surface is proportional to the mass of the earth and inversely proportional to the square of the earth's radius at that point. The constant of proportionality is G.

5. Objects moving around curved paths are accelerating even when their speeds are constant.

6. The acceleration of any object moving in a circle at constant speed is directed toward the center of the circle. This is known as the centripetal acceleration.

7. The centripetal force maintains planets in orbit around their suns and satellites around their planets. The speed required for a stable orbit of any satellite can be obtained from the expression that equates the centripetal force with the gravitational force of attraction on the satellite.

8. As indicated by Newton's third law, planets and their suns always rotate around a common center, which is not the center of the sun. The same effect applies to satellites and their planets.

9. A combination of the gravitational pulls of both the moon and the sun produces the tides on the earth's waters and also tiny tides in solid rocks.

10. Weightlessness does not mean a value of zero for the earth's gravitational attraction. It implies that the object or person is described in a frame of reference whose acceleration matches the local value of the gravitational field g.

11. The space around earth in which objects are influenced by the earth's gravitational attraction is known as the earth's gravitational field. Its strength at any point is represented by the acceleration due to gravity at that point.

QUESTIONS

1. What is the Newton's third-law reaction force to your weight?

2. Once a value for G has been measured in the laboratory by Cavendish's method, it can be used to determine the mass of the earth. Explain how. How would you compute the "weight of the earth," and what does weight of the earth mean?

3. Would a 16-lb metal shot put used in the Olympic games weigh the same in Melbourne, Australia, as in Helsinki, Finland? Explain.

4. Why does the moon cause the water level to rise on both sides of the earth?

5. The force of gravity acts on all bodies in proportion to their masses. Why, then, in free fall doesn't a heavy body fall faster than a light body?

6. The earth is closer to the sun in January than in July. How does this affect the speed of the earth in its orbit about the sun?

7. High tide occurs not while the moon is overhead but several hours later. Can this be explained?

8. Why does a wobbly motion of a single star provide evidence that the star has a planet or a system of planets?

9. How does the force of a passenger on her seat in an airplane vary with the motion of the plane?

10. When satellites were first launched, it was not uncommon to hear on the radio or read in the newspaper that "the satellite will slow down and fall into a stable orbit closer to the earth." Is this consistent with Newton's laws? Explain.

11. Suppose you have two identical shoe boxes, one full of lead, the other full of feathers. You are in a spaceship where the two shoe boxes appear weightless. How could you tell which has the greater mass?

PROBLEMS

Useful constants and data:

$G = 6.67 \times 10^{-11}$ N·m^2/kg^2

$g = 9.8$ m/s^2

Radius of the earth = 3963 mi, or 6380 km

1 AU = 93 × 10^6 mi = 150 × 10^6 km

Mass of earth = 5.98×10^{24} kg

Mass of moon = 7.34×10^{22} kg

Radius of moon = 1.74×10^3 km

Mass of sun = 1.99×10^{30} kg

Radius of moon's orbit about earth = 3.8×10^5 km

Some useful planetary data is in the table on p. 118.

g and the Gravitational Field

1. What is the value of the gravitational acceleration when you are a distance of 1 earth radius above the earth's surface? 2 earth radii above the earth's surface?

2. If the radius of the earth were to expand by a factor of 5, by how much would g increase or decrease?

3. At what distance from the center of the earth does the gravitational acceleration have one-half the value that it has on the earth's surface?

4. At what distance from the center of the earth does the gravitational field of the earth have one-third the value it has on the earth's surface.

5. A 12-kg mass weighs 96 N on planet Goozoobie. What is the acceleration due to gravity on this planet?

6. Compute the surface gravity on Mars and your weight on Mars.

	Mercury	Venus	Earth	Mars	Jupiter	Saturn	Uranus	Neptune	Pluto
Distance from sun, mean (AU)	0.39	0.72	1.00	1.52	5.20	9.54	19.2	30.0	39.4
Equatorial diameter (km)	4849	12 122	12 760	6764	1.42×10^5	1.21×10^5	47 200	49 300	6300(?)
Rotation period	58.6 days	243 days	23 h, 56 min	1 day 43 min	9 h, 56 min	10 h, 14 min	10 h, 48 min	15 h, 48 min	6 days, 9 min
Revolution period, sidereal	88 days	225 days	$365\frac{1}{4}$ days	1.9 yr	11.9 yr	29.5 yr	84 yr	164 yr	247 yr
Mass, compared with earth	0.06	0.82	1	0.11	317.9	95.1	14.5	17.3	0.18(?)
Surface gravity compared with earth	0.39	0.91	1	0.38	2.64	0.88	0.99	1.1	0.44(?)
Number of natural satellites	0	0	1	2	12	10	5	2	0
Density, mean (g/cm^3)	5.5	5.27	5.52	3.95	1.33	0.69	1.7	1.6	4.86(?)

Newton's Law of Universal Gravitation

7. The gravitational force between Jupiter and one of its moons, when the two are separated by a distance R, has a value F. For the following changes in mass, distance of separation, or both, how will the gravitational force change? The factors not mentioned remain unchanged. (a) The moon's mass is doubled; (b) the moon's mass is halved; (c) Jupiter's mass is doubled; (d) Jupiter's mass is halved; (e) the distance between the two is doubled; (f) the distance between the two is halved; (g) the distance between the two is tripled; (h) both masses are doubled; (i) both masses are doubled and the distance is halved; (j) the mass of the moon is doubled and the distance is halved; (k) the mass of the moon is halved and the distance is doubled.

8. Compute the gravitational attraction between a college student and her male friend. He has a mass of 80 kg, she a mass of 50 kg, and they are 2 m apart. Is this force worth worrying about? Assume that Newton's law of gravitation applies to these nonparticle, nonspherical objects which are rather close together.

9. Compare the gravitational force of the sun on the earth to that of the moon on the earth. Note that the sun's mass is about 2.7×10^7 times the mass of the moon and the sun is about 400 times farther from the earth than is the moon.

10. The Apollo 8 space capsule orbited about the moon in an approximate circular orbit 112 km above the moon's surface. The period of the orbit was 120 min. Calculate the mass of the moon from these data and the radius of the moon.

11. Pluto was discovered in 1930, 14 yr after Percival Lowell had predicted its existence. How far has Pluto traveled in its path about the sun between 1930 and 1979?

12. The first orbital flight by a woman took place in 1963 when the Russians launched Vostok 6. The satellite was roughly 110 mi above the earth, and it took 88 min to complete one orbit. Compute the speed of the space capsule and the gravitational acceleration of the capsule.

13. The spectacular rings of Saturn are roughly at a distance twice the radius of Saturn from the center of the planet. What is the centripetal acceleration of the material in the rings? Compare this acceleration with g on earth.

Weight in an Accelerated Frame

14. A small boy of mass 30 kg is riding in an elevator whose acceleration is controlled by Dr. Wierdo. What is the boy's weight as measured on a scale in the elevator when the acceleration is (a) 2 m/s^2 upward, (b) 9.8 m/s^2 upward, (c) 2 m/s^2 downward, (d) 9.8 m/s^2 downward, (e) 19.6 m/s^2 upward?

15. You ride in an elevator that is accelerating and, as you stand on a scale, you find that your weight has been halved. What is the acceleration of the elevator, and in which direction is it accelerating?

16. An object hangs from a spring balance that is hung from the roof of an elevator. If the elevator has an upward acceleration of 1 m/s² and the balance reads 20 kg, what is the weight of the object while $a = 1$ m/s²? What is the weight of the mass when it is not accelerating? What is the mass of the object? Under what circumstance will the balance read 15 kg?

17. What is the acceleration of an elevator if you are standing on a scale in the elevator when it starts to rise and the reading jumps to 222 lb? Your weight on the scale is 192 lb when the elevator is at rest.

Uniform Circular Motion

18. (a) What is the centripetal acceleration of a person standing on the earth's equator due to the earth's rotation around its axis once each day? (b) What fraction of the acceleration due to gravity at the earth's surface is this centripetal acceleration?

19. A person in St. Paul, Minnesota, is at a latitude of 45° north of the equator. What is his centripetal acceleration due to the daily motion of the earth on its axis?

20. The earth is on the average 93 million miles (or 150 million kilometers) from the sun. What is the speed of the earth in its motion about the sun and its average centripetal acceleration during this yearly motion? Assume the earth moves in a circle at a uniform rate about the sun.

21. A Mickey Mouse watch has a second hand 1 cm long. (a) Compute the speed of the tip of this second hand. (b) What is the direction and magnitude of the acceleration of the tip of the second hand? Consider the motion uniform.

22. A small boy whirls a stone about in a slingshot in a circular motion at uniform speed. The rock moves at 60 r/min in a horizontal circle of radius 0.8 m. What is (a) the speed of the rock and (b) its centripetal acceleration?

23. How many revolutions per second are required to give a ball on the end of a 0.5-m string a centripetal acceleration of 4 m/s²? The string does not stretch.

24. A standard 12-in. diameter long-playing record moves at $33\frac{1}{3}$ r/min. (a) How fast does the outer edge move, and (b) what is its acceleration?

25. A high school runner does the mile in 4 min and 20 s. (a) What is his average speed? (b) What is his total displacement? (c) What is his centripetal acceleration? Assume a quarter-mile circularly shaped track.

Centripetal Force

26. A boy rides a roller coaster. With what force does he press against the seat when the car goes over a crest whose radius of curvature is 8.0 m at a speed of 4 m/s? Express the result as a fraction of his weight.

27. How will the weightlessness of space flight affect the growth of plants? Plants are to be grown in a spacecraft in a gravity field equal to that on earth, i.e., with $g = 9.8$ m/s². This is accomplished by rotating the plants on a drum in space. The plants in question are 0.41 m from the center of rotation. How fast must the platform rotate to provide the required acceleration?

28. Centrifuges are rated in "relative centrifugal force" (actually, relative centripetal force), which is equal to the ratio of the centripetal force to the acceleration of gravity. A centrifuge with a sample radius of 15 cm moves at 6000 r/min. What is the centripetal force on the sample and the relative centripetal force?

29. Calculate how fast a doughnut-shaped spaceship must revolve to introduce a radial acceleration of $0.5 g$. Assume that the radius of the spaceship is roughly the size of a football field, or 100 m.

30. A hammer thrower exerts a large centripetal force to keep the 7.2-kg (16-lb) ball in orbit. She swings the ball in a circular orbit with a radius of about 1.5 m. If it takes about 3 s to swing the hammer once around, what force must the thrower exert?

31. What centripetal acceleration is required to move a 16-lb weight in a horizontal circle of radius 6 ft with a speed of 42 ft/s? What is the tension in the wire holding the weight?

32. A test pilot dives at a speed of 600 m/s, a speed about twice the speed of sound at high altitudes (Mach 2). The pilot pulls out of his dive and goes into a curve of radius 900 m. (a) What is the acceleration he experiences at the lowest point in the dive? Express this in terms of the number of g's. (b) Compute the centripetal force the seat exerts on him.

33. A stone of mass 1 kg attached to one end of a string 1 m long of breaking strength 500 N is whirled in a horizontal circle on a frictionless tabletop. The other end of the string is kept fixed. Find the maximum velocity the stone can attain without breaking the string.

34. An electron in the atom according to the model developed by Niels Bohr moves at a uniform speed in a circular path of radius 5.3×10^{-9} cm about the positive nucleus. If the mass of the electron is 9.11×10^{-31} kg and its linear speed is 3.0×10^6 m/s, compute the centripetal acceleration and force on it.

35. A small boy (mass = 25 kg) sits on the edge of a merry-go-round but does not hold on. If the merry-go-round has a radius of 5 m and the coefficient of friction between his jeans and the metal floor is 0.20, what

is the maximum speed at which the merry-go-round can move if the boy is not to slide off?

36. A 20-kg packing crate rests on the back of a flatbed truck. The coefficient of friction between the crate and the surface is 0.6 for static friction and 0.3 for kinetic friction. What is the maximum speed the truck can move about a horizontal curve of radius 90 m without the box slipping? If the truck were to move at twice that speed, discuss the motion of the crate.

A Little More Difficult

37. We wish to place a satellite in an equitorial orbit about the earth so that it surveys the total circumference of the earth once each day. What is the period of such a satellite, and how far must it be above the earth's surface?

38. A girl whirls a pail of water in a vertical circle. What is the net force on the bucket at its highest and lowest points if the radius of the circle is 110 cm and it takes 2 s to complete one revolution? There is 1 kg of water in the bucket (mass = 1 kg also).

39. A rope with a mass m on its end is whirled about in a vertical circle. Write down an expression for the tension in the rope at the top and bottom of the swing. Is this motion uniform circular motion? Discuss.

40. Compute the distance to the point between the earth and the moon where the force on a satellite due to the earth would equal the tug of the moon on the satellite.

41. A conical pendulum consists of a bob of mass of 0.25 kg on the end of a 2-m-long string. (See Fig. 41.) (a) Compute the speed required to maintain the bob in a horizontal circle in uniform circular motion making an angle of $\theta = 9.6°$ with the vertical. (b) What is the tension in the string?

42. Bicycle racers in an indoor bowl 80 m in diameter are racing at a

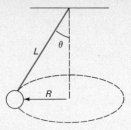

Figure 41

speed of 72 km/h on a portion of the curve where the banking is 60° with the horizontal. A racer and bike have a mass of 90 kg. Compute (a) the centripetal force and (b) acceleration.

43. In highway design, to make safe curves you generally have to bank the curve. Assume that the curve will provide only a small frictional force and that the centripetal force required to keep a vehicle on the curve is provided by the banking alone. Compute the angle of banking required on a curve of radius 90 m if the traffic will travel at up to 54 km/h around the curve.

Special Topic
An Open and Shut Universe

The part of physics known as cosmology is among the most philosophically interesting and intellectually challenging areas of all science. The philosophical interest derives from the subject matter of cosmology—the life and death, length and breadth of the universe. The challenge stems from the nature of the materials that cosmologists must use to define those terms: the stars, galaxies, and intergalactic matter which can never be sensed, measured, or observed directly. Relying on the extremely narrow view that they obtain through their instruments and the fundamental laws of gravitation, cosmologists attempt to determine the size of the universe and the nature and time of its birth and eventual death.

Cosmologists have long recognized that the universe is expanding, with distant stars and galaxies moving away from the earth and the rest of the universe at speeds approaching the speed of light. In the 1940s, the experts argued over two possible causes of the expansion. One school of thought believed that the universe started 5 billion years ago with an explosion of primordial matter—known as the big bang—that created the stars and galaxies and sent them speeding away from each other. Another group argued that the universe had existed from time immemorial and was expanding because new matter was continually being created. Most cosmologists now accept the big bang theory, and the great cosmological debate of the present concerns time rather than space.

The basic question is simple: "Will the universe continue to expand forever, or will its expansion slow down at some eon and change to a contraction, which will eventually bring all matter back to the state of the primordial fireball, ready for another big bang?" The first alternative—expansion forever—is known as the open universe; it is equivalent to sending up a rocket from earth with enough speed to escape earth's gravitational attraction and soar off into deep space. The second possibility—the closed, or oscillating, universe—is tantamount to sending up a rocket that does not have sufficient speed to continue to get farther away from earth's gravitational field; sooner or later this rocket returns to the surface of the earth.

The rocket analogy actually provides a glimpse of the methods cosmologists use in their effort to decide whether we inhabit an open or closed universe. The fundamental factors that determine whether the rocket escapes are its speed and the gravitational attraction between it and the earth. In just the same way, the eventual fate of the universe is decided by the speed at which its parts are receding from each other and by the total force of gravitational attraction between all the objects that it contains. Decades of observation and interpretation have given cosmologists a fair idea of the speed at which the universe is expanding, and thus the critical factor in determining the basic nature of our universe is the amount of gravitational force exerted by all the matter inside it. And since, according to Newton's law of gravitation, the strength of the gravitational force between any two objects is related to their masses, the problem of deciding between an open and a closed universe boils down to determining the total amount of mass in the universe. If the cosmos contains a certain critical amount of mass, the gravitational force inside it will be sufficient to slow down the motions of the outermost galaxies and reverse the direction in which they are moving. If the amount of mass is too small, the gravitational force will be insufficient to reverse the expansion.

Here we have a classic case of "easier said than done," for how can one possibly measure the mass of the universe? How, for that matter, can one measure the mass of an individual star? The cynical answer would be "very carefully"; the scientific answer is "very approximately."

The only star with which we have anything resembling first-hand experience is the sun. Its mass can be calculated from Newton's law of gravitation and the earth's period of rotation. The masses of other stars would be all but impossible to obtain were it not for the existence of slightly fewer than 100 combinations of stars known as binary systems. The binary systems consist of two stars that revolve about each other; the gravitational attraction of each individual star in a binary system influences the movement of the other star; and by observing the motions, astronomers can calculate the relative masses of the individual stars.

From there it is a short, although speculative, step to assign the same masses to single stars of similar brightnesses that are known to be at roughly the same distances from earth as particular binaries. Fortunately, stellar masses appear to vary relatively little throughout the observable universe. The most massive star observed has a mass no more than 10 times that of our sun, and the least massive comes in at about one-tenth the solar mass. One can thus take a "representative mass" for all stars, approximately equal to the mass of the sun. Extending the concept, astronomers can get a good idea of the masses of galaxies by studying the motion of stars inside them. Then, by counting all the stars and galaxies observed by optical and radio telescopes, and allowing a fudge factor for those too dim to be observed, astronomers can arrive at a very rough figure for the total mass of all the stars in the universe.

This is only part of the universe's total mass, however, for the spaces between the stars and galaxies are not entirely empty of mass; they contain clouds of interstellar gas and dust. And although the density of these clouds is unimaginably small—no more than a single atom for every few cubic meters—the clouds are so huge that they contain appreciable amounts of mass overall. This might amount to as much as 30 times the mass of all the stars.

A third source of mass in the cosmos is so mysterious that it literally defies description. This is the collection of black holes that theoreticians believe to exist throughout the cosmos. A black hole is essentially an invisible piece of the universe with a star-size mass. Theorists believe that certain ancient stars eventually collapse under the force of their own gravity, contracting into excessively dense pockets of mass. A star somewhat larger than our sun, for example, would collapse down to a black hole no more than a few miles in diameter. When matter is that dense, the normal laws of gravitation simply do not apply. The gravitational field of a black hole is so enormous that it attracts matter as if it were a cosmic vacuum cleaner. Nothing, not even light, can escape from the hole. Thus, black holes are by their very nature invisible reservoirs of huge amounts of mass. What evidence there is for the existence of black holes is highly circumstantial. It consists of observations of types of radiation that mass would be expected to emit as it is falling into a black hole.

Black holes in the argument over the open or closed universe are important because the entire mass of the stars, and the dust and gas between them, makes up only a small percentage of the total amount of mass required for a closed, oscillating universe. Because many cosmologists believe that an oscillating universe is philosophically more acceptable than one that expands eternally, they have put a great deal of effort into the search for black holes and any other unexpected type of cosmic object that might provide the "missing mass."

Yet at present, the open universe seems to fit the facts by a wide margin. In a report in December 1974, a group of cosmologists from the California Institute of Technology and the University of Texas concluded that, even with the addition of the masses of all the black holes imaginable, the total mass of the universe would still be below 10 percent of that required to halt the expansion. Only if black holes are scattered far more densely throughout the cosmos than theoreticians now believe to be possible could the total mass of the universe approach the figure necessary to close it. These figures are not enough to dismay the die-hard advocates of a closed universe, however. They continue to search for new sources of the extra mass that, if discovered, would mean another big bang billions of years in the future.

Energy

5.1 THE REAL MEANING OF WORK

5.2 KINETIC ENERGY: ENERGY OF MOVEMENT

   SHORT SUBJECT: The Dot Product

5.3 POTENTIAL ENERGY: ENERGY OF POSITION

5.4 CONSERVATION OF ENERGY

5.5 SWINGING THE CHANGES

   SHORT SUBJECT: Mean Machines

5.6 GRAVITATIONAL POTENTIAL ENERGY
   The Real Expression for Potential Energy
   Satellite Energy

5.7 POWER
   Summary
   Questions
   Problems

   SPECIAL TOPIC: Producing Energy for the People

Just about every American has personal memories of the energy crisis that started in the winter of 1973–1974. Waiting in lines for gasoline for 3 or 4 h, shivering in homes, schools, and office buildings with newly lowered thermostat settings of 68°F or less, and taking apparently interminable times to make journeys as a result of 50- or 55-mi/h speed limits reminded many Americans of sieges of wartime. The 55-mi/h speed limit has stayed with us, and as the summer of 1979 saw, we still have an energy crisis. But every experience has a good side, and the experience of living through energy crisises, along with the opportunity to learn of novel schemes to tap the sun, the oceans, and the wind to provide fresh sources of power, have given many people their first real understanding of what the word "energy" really means.

Energy is the most important concept in science and such a critical concept that it is studied in all sciences. It is not a simple concept. We cannot define what energy is, yet we know what it can do. We can identify a large number of different forms of energy, including kinetic energy, gravitational energy, heat energy, electrical energy, and chemical energy. The importance of what might be considered an abstract concept, energy, lies in one of the most basic laws of physics: the conservation of energy. The "law" of conservation of energy governs, as far as we know, all natural phenomena. The law simply states that the amount of the quantity we call energy does not change as changes occur in nature. The total energy remains constant even though in nature the amounts of various forms of energy may change as processes occur in nature. This is an abstract idea. We are saying that the total energy, which is something quantifiable by a number, remains constant even though the different parts of the total energy may be shifting among different forms.

Throughout the remainder of this text we shall attempt to develop an appreciation for the all-inclusive concept of energy by studying some of its forms and the transformations of energy from one form to another. To begin a study of the conservation of energy we restrict ourselves first to one small part of the energy picture: mechanical energy.

Since we "use" energy to perform tasks or "do things," it is common in physics to state that energy is the capacity to do work. This really does not help us define energy since we have not defined work. Hence we shall start our investigation of energy by defining the word "work."

5.1 THE REAL MEANING OF WORK

Whatever its meaning in daily life, **work** has a very precise definition in physics: **It is the product of a force that acts on an object and the distance which the object moves under the influence of that force, when both the force and the movement are in the same direction.** Stated slightly differently, **work is the product of the displacement of an object under the influence of a force and the component of the force in the direction of the displacement.**

Calculating the component of the force in the requisite direction is a simple exercise in trigonometry. One multiplies the force by the cosine of the angle between the direction in which the force acts and the direction of the motion that it causes. In general, work W of a constant force F producing a displacement s can be represented by the equation

$$W = Fs \cos \theta \qquad (5.1)$$

In Fig. 5.1, for example, which shows a railroad freight car being pulled along a straight, level track by a horse tugging a rope at an angle θ to the track, the work done by the horse on the freight car is $Fs \cos \theta$. Plainly, the amount of work will be greatest when θ is $0°$ because $\cos \theta$ then has its maximum value of 1; in this situation, the pull of the horse will be directed along the track. When θ is $90°$, $\cos \theta$ is 0, and hence no work is done in moving the car along its track; the horse would be pulling in a direction perpendicular to the track, thus providing no force at all in the direction of the track. Of course, if the horse performed an extraordinary feat of strength, it might manage to pull the freight car off the tracks and drag it away in the perpendicular direction; the work done in this case would depend on how far the freight car moved away from the tracks and the force exerted by the horse.

Another situation arises if θ is greater than $90°$; because the cosines of angles between $90°$ and $180°$ are negative, the equation tells us that the value of the work is negative. What this means is that the force acts in a direction opposite to that of the movement; it is, in other words, a retarding force. We might imagine that the horse in Fig. 5.1(b) tries to stop a moving freight car by pulling against the direction of the car's movement. A typical example of a retarding force is that of friction, which always acts at $180°$ to the direction in which an object is moving. Work produced by frictional force that opposes motion is negative.

Thus far we have implicitly assumed that the force that produces work is constant. As ever, this is a pie-in-the-sky assumption that rarely applies when we deal with real problems in the real world. If the force is not constant during the period in which an object moves under its influence, the calculation of work becomes obviously more difficult. One must compute the total amount of work done by summing the individual amounts of work performed over small portions of the displacement, each portion being so small that the force is effectively constant during that period. A summation of this type is best carried out analytically using integral calculus, but it can always be done graphically using the technique shown for computing the impulse of a

Figure 5.1 Horse pulling on a freight car in various directions.

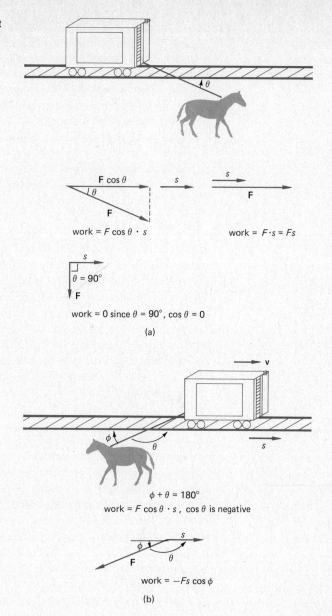

variable force on page 155. The basic concepts of work are the same for uniform and nonuniform forces, and it is therefore sufficient to use only those examples in which the force remains constant. The student need only be aware that forces do not remain constant in most practical situations.

The units of work are those of force multiplied by distance, such as dyne-centimeter (dyn · cm), newton-meter (N · m), and, in the English system, pound foot. To simplify matters somewhat, each unit has its own special name. The dyne-centimeter, for example, is known as the erg, which derives from ergon, the Greek word for work; the newton-meter is called the joule (J), after a nineteenth century British physicist who studied the conversion of various forms of energy to heat; and the pound foot is usually reversed and referred to as the foot pound (ft · lb).

EXAMPLE 5.1

A small girl drags a sled a distance of 50 ft along the horizontal level surface by applying a force of 15 lb at an angle of 30° with the horizontal. (See Fig. 5.2.) How much work does she perform in the process?

Figure 5.2 A girl pulling a sled along horizontal level ground.

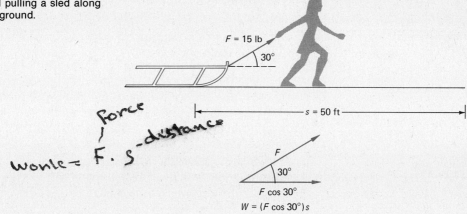

Solution

$$W = F(\cos \theta)s = 15 \text{ lb } (\cos 30°)(50 \text{ ft}) = 650 \text{ ft} \cdot \text{lb}$$

EXAMPLE 5.2

A man pushes a small boy on a sled up a hill. The man applies a horizontal force of 450 N, and a frictional force of 5 N opposes the motion of the sled, as shown in Fig. 5.3. The boy and sled together have a mass of 36 kg. If the angle of the hill is 30° and the man pushes for 20 m, (a) how much work does the man do on the sled? (b) How much work does the force of gravitation do on the sled? (c) How much work is done by friction on the sled?

Solution

(a) The man provides work on the sled since a component of his push is in the direction of the displacement. The component of the man's

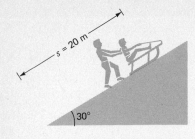

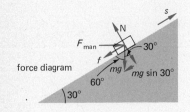

Figure 5.3 A man pushing a boy on a sled up a hill.

push up the incline actually does the work. This is given by

$$W = (F \cos \theta)s = +(450 \text{ N})(\cos 30)(20 \text{ m})$$
$$= +7.8 \text{ kilojoules (kJ)} = +7.8 \text{ kJ}$$

(b) Gravity opposes the displacement; hence, the gravity, or weight force, of the boy and sled does negative work on the sled. If we use our expression for work, noting from Fig. 5.3 that the weight force makes an angle of 60° with the plane, we obtain the gravitational work:

$$W = (F \cos \theta)s = -(mg \cos 60°) \times 20 \text{ m}$$
$$W = -(36 \text{ kg})(9.8 \text{ m/s}^2)(0.5)(20 \text{ m}) = -3.5 \text{ kJ}$$

We also could have computed the work by taking the product of the component of the weight along the plane ($-mg \sin 30°$) and the displacement (20 m), and produced the same result.

(c) The friction force also does negative work on the sled since it opposes the displacement. The friction force lies along the incline, so

$$W = Fs = (-5 \text{ N})(20 \text{ m}) = -0.1 \text{ kJ}$$

This is equivalent to computing $W = Fs \cos \theta$, where $\theta = 180°$, or $\cos \theta = -1$. We can compute the "total" work done on the sled by taking the sum of the work done on the sled by all the separate forces. The total work is actually the work done on the sled by the net force acting on it. In the present example, the total work will just equal the sum of the work done by the three forces acting along the inclined plane since the normal force equals, but is opposite to, the sum of the component of the weight of the sled and boy perpendicular to the incline and the component of the man's pushing force perpendicular to the incline. These forces "cancel" themselves, giving no net force perpendicular to the plane. The sum of the work done by the three forces is numerically (7.8, −3.5, −0.1) kJ, or 4.2 kJ. That we have associated plus and minus signs with work does not mean that work is a vector. It is not! Whether the work is negative or positive is related to whether the force responsible for the work opposes or supports the motion. As we shall see in the next section, positive work performed on an object increases the object's energy, and negative work on an object reduces its energy.

5.2 KINETIC ENERGY: ENERGY OF MOVEMENT

What are the consequences of the action of a force on an object? Let us imagine that a constant force F causes an object to move a distance x in the direction in which the force is acting. The work done by the force on the object is given by Eq. (5.1), that is, $W = Fx$. Because the force is constant, we know from Newton's second law that the object will undergo a constant acceleration a in the direction of the force. In other words, $F = ma$. Hence, $W = ma \cdot x$.

The Dot Product

Work is a scalar quantity that results from the product of two vectors, force **F** and displacement **s**. Such a product of two vectors in mathematics is called a scalar or dot product and is symbolized by

Work = **F** · **s**

The scalar product is obtained by multiplying the magnitudes of the force and the distance and the cosine of the angle θ between the vector force and vector displacement:

$$\mathbf{F} \cdot \mathbf{s} = F(s \cos \theta) = (F \cos \theta)s = Fs \cos \theta$$

Mathematically, the scalar product represents the product of the projection of one vector on the other vector times the magnitude of the other vector. The projection of one vector on the other is the product of the vector's magnitude and the cosine of the angle between the two vectors.

For example, $s \cos \theta$ is the projection of the displacement along the force vector. The projection of the force vector on the displacement vector is $F \cos \theta$.

Vectors can also be multiplied in another way to produce a vector rather than a scalar. This operation, called the vector product, is described in Chap. 7, where it is useful when dealing with angular motion.

Kinetic energy and work are related concepts, linked by the fact that the imposition of one—work done by a net force—on any object causes a change in the magnitude of the other—kinetic energy.

Now when the acceleration is constant, we also know from Eq. (2.9) that

$$v^2 = v_0^2 + 2ax$$

where v_0 is the starting speed and v is the speed after the object has traveled a distance x. Thus,

$$a = \frac{v^2 - v_0^2}{2x}$$

and, substituting,

$$W = \frac{mx(v^2 - v_0^2)}{2x} = \tfrac{1}{2}mv^2 - \tfrac{1}{2}mv_0^2 \tag{5.2}$$

The term $\tfrac{1}{2}mv^2$ is known as the kinetic energy (E_k) of the object. It is the energy that any object possesses by virtue of its motion alone. If the object is at rest, its kinetic energy is obviously zero because the speed v is zero. At any other speed v, the E_k is $\tfrac{1}{2}mv^2$.

Equation (5.2) tells us that the work done by a net force on an object causes the kinetic energy of that object to change. Kinetic energy and work are not identical quantities; they are related concepts, linked by the fact that the imposition of one—work done by a net force—on any object causes a change in the magnitude of the other—kinetic energy. This link demands that energy is measured in exactly the same units as work. In addition, the two concepts are both nondirectional, scalar

Although we have derived Eq. (5.2) for the case in which a constant force acts on the object, the result applies equally well to nonuniform forces.

quantities. That energy is a scalar quantity makes it much easier to deal with than, say, forces.

Equation (5.2) also shows that the kinetic energy of an object increases if positive work is performed on the object and decreases if the work is negative. Since a loss in kinetic energy means a loss in speed, negative work by a net force plainly means a slowing down.

EXAMPLE 5.3

Calculate the work done when a force of 0.4 N moves a model railroad coach of mass 0.50 kg a distance of 300 cm along a straight horizontal path in the direction of the force. What is the change in the kinetic energy of the coach? If the coach started from rest, what is its final speed?

Solution

Substituting for $\theta = 0$ in Eq. (5.1), $W = Fs \cos \theta$, we find that

$$W = Fs = 0.4 \text{ N} \times 3 \text{ m} = 1.2 \text{ N} \cdot \text{m}$$

Since 1 N·m equals 1 J, the work done on the coach is 1.2 J. Therefore, the change in kinetic energy of the model coach is also 1.2 J. If we set this equal to the change in kinetic energy and realize that the initial kinetic energy is zero, then we can calculate the resulting final speed after application of the 0.4-N force, using Eq. (5.2) and SI units:

$$\tfrac{1}{2} \times 0.50 \times v^2 = 1.2 \quad \text{and} \quad v^2 = 4.8$$

Hence, the coach's final speed is 2.2 m/s.

5.3 POTENTIAL ENERGY: ENERGY OF POSITION

Work can also alter the energy of objects without affecting their speed. Imagine that you pick up an object, say a book, and lift it very, very slowly without giving it any acceleration. To make the book move vertically upward, you must apply a force equal to the weight of the book. In other words, you must do work on the book. This work does not change the speed of the book, however, because after you have lifted it, the book has zero speed, just as it had before. Rather, the work increases the height of the book, or, more generally, it changes the book's separation from the earth.

We can, therefore, define a new type of energy, the energy of position. Technically, this is known as **potential energy**. With this definition in mind, we can expand our statement concerning the relationship between work and energy. **Work can change the kinetic energy of an object and/or the potential energy.** By defining as energy any quantity that results from the application of work, we have an extremely broad concept of what energy really is.

Returning to the problem of lifting a book, we see in Fig. 5.4 that the minimum force F required just to lift the book equals its weight mg.

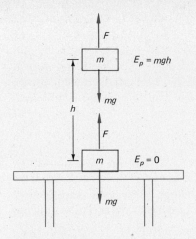

Figure 5.4 Change in potential energy of a book when it is lifted. The force **F** does work on the book equal to *Fh*, that is, *mgh*. Thus, the book's change in potential energy is *mgh*.

Application of any force greater than this will cause the book to accelerate upward and thus increase its speed. So, if we use just the minimum force and lift the book a distance h, we shall do an amount of work W on the book given by the equation

$$W = Fs \cos \theta = mgh \tag{5.3}$$

Since this work has increased the gravitational potential energy of the book, we can write the change in gravitational potential energy (E_p) as

$$\Delta E_p = mgh$$

The quantity mgh represents the gravitational potential energy near the earth's surface. It is the book's energy of position and the result of the presence of the force of gravity. In Eq. (5.3) the work done by our effort to lift the book produces a change in the potential energy since h represents a change in the book's vertical position. We can arbitrarily assume the value of h as zero at its starting point on the horizontal table and call the potential energy at height h above the table mgh.

We are now in a position to investigate the combined effect of work on the kinetic and potential energies. If, instead of just lifting the book by applying the minimum force, we give it an upward heave that accelerates it as well as lifting it to a new height, both the position and the speed of the book will change. Therefore,

$$\text{Work} = (\tfrac{1}{2}mv^2 - \tfrac{1}{2}mv_0^2) + mgh$$

or

$$\text{Work} = \Delta E_k + \Delta E_p \tag{5.4}$$

where the Greek letter Δ means, as before, the change in any particular quantity.

This equation is often called the work-energy principle, and it is an extremely useful relationship for solving a variety of problems involving mechanical energy.

If we consider a situation in which no external forces other than gravity act on an object, then no work can be done on the object. The effect of the force of gravity is contained in our expression for potential energy. The left-hand side of Eq. (5.4) is therefore zero. Hence, the sum of the change in kinetic energy and the change in potential energy will also be zero. In other words, the total mechanical energy of any system unchanged by external forces remains constant; any change in kinetic energy results from a change in potential energy, and vice versa. What this statement means is that mechanical energy is conserved. Its form may change, but its total amount does not. This is one statement of the principle of conservation of energy.

How can we measure potential energy? Consider a box of mass 2 kg resting on a tabletop that is 1 m above the second floor of a building. Imagine also that the second floor is exactly 4 m above the ground level of the first floor, as shown in Fig. 5.5. How do we know that the box on

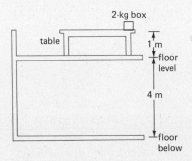

Figure 5.5 Varying values of potential energy due to position.

the table possesses potential energy? And how much does it possess?

Answering the second question first, we must point out that the amount of gravitational potential energy stored in any object depends on our plane of reference. If we consider the floor of the second story as our plane of reference, then the work that was originally done to lift the box from there to the tabletop, resulting in an increase in potential energy, would have been

$$W = Fh = mgh = (2 \text{ kg})(9.8 \text{ m/s}^2)(1 \text{ m}) = 19.6 \text{ J}$$

Hence, the box has 19.6 extra joules of gravitational potential energy while on the tabletop than it did on the floor. On the other hand, if we consider the ground—the first floor—as the reference plane, the work done to bring up the box from the first floor and put it on the tabletop would be

$$W = FH = mgH = (2 \text{ kg})(9.8 \text{ m/s}^2)(5 \text{ m}) = 98 \text{ J}$$

Here, the change in gravitational potential energy of the box is greater, and equal to 98 J. The box has different values of the gravitational potential energy when measured with respect to different reference planes. It is thus not very useful to speak of an absolute value of the potential energy unless one specifies the reference plane. The change in potential energy in lifting the box from the second floor to the tabletop, however, is the same no matter what reference plane has been chosen. **It is the change in potential energy that is the important factor in problem solving.** Note that potential energy could take on a negative value depending upon the reference taken for the zero of potential energy.

How does one convince a skeptic that the box on the table actually possesses potential energy whose value is changed by a change in position? One simply gives the box a slight horizontal push until it is teetering on the side of the table and then lets it fall to the floor. As it falls, the box picks up speed and, hence, kinetic energy. And since no force is acting on the box beyond its own weight (if we neglect air resistance), then the box can only gain kinetic energy at the expense of its potential energy.

Returning to Eq. (5.4) with the left-hand term equal to zero, we have the relationship

$$0 = \Delta E_p + \Delta E_k$$

or

$$\Delta E_k = -\Delta E_p$$

This means that any increase in kinetic energy is exactly equal to the decrease in potential energy. In the case of the 2-kg box falling to the second floor, its kinetic energy just before it hits the floor has increased over its short journey downward by exactly the amount of potential energy that it has lost, that is, 19.6 J.

5.4 CONSERVATION OF ENERGY

The previous example is a specialized case, involving just mechanical energy, of an all-embracing principle of conservation of energy that applies to all the physical universe. The principle can be stated two ways:

1. Energy can neither be created nor destroyed; it can be transformed from one type to another with exact equivalence.
2. The total amount of energy in the universe or in any isolated system remains constant.

The key word in statement 2 is *isolated.* The requirement that a system be isolated imposes a vital restriction when we deal with practical problems. Because in our practical applications we do not deal with the entire universe, but with "systems" that are some part of the universe, we must make sure that no energy gets into or out of our system. For example, a system is not isolated if it absorbs or emits heat during the time that we are dealing with it.

For now we are concerned with mechanical systems. To describe an isolated system in this context we must first distinguish between two types of force: conservative and nonconservative.

Conservative forces are those that can be represented by a potential energy. Gravity is a conservative force, for example; we just indicated how we can represent the influence of the gravitational force on a mass by specifying the gravitational potential energy. Elastic forces, such as those produced by a stretched or compressed spring, and electrostatic forces are other examples of conservative forces. The identifying characteristic of a conservative force is that the amount of work it performs is independent of the path taken by the object on which the force is operating and depends only on the initial and final position. This amount is equal to the difference between the final and initial values of the quantity that we have termed potential energy.

Forces that cannot be represented by a potential energy, such as dissipative or frictional forces, are called nonconservative because mechanical energy is not conserved. The work of a friction force depends on the actual path followed. In general, the longer the path, the more the amount of work performed.

No external force can act on the isolated system except those conservative forces, such as the force of gravity, that we describe by a potential energy term.

We shall now examine applications of the principle of conservation of energy to mechanical systems that are isolated because no external force acts on the system except gravity and for which we can neglect nonconservative forces such as friction and air resistance.

EXAMPLE 5.4

A ball with a mass of 0.50 kg (about 1 lb) is dropped from rest from a height of 10 m (above 33 ft) above the ground. How much mechanical energy does the ball have when it is 10, 7.5, and 2.5 m above the ground level? What form does the energy take at these heights?

Solution

At a height of 10 m above the ground, we know that the ball is at rest; its velocity is zero. Hence the kinetic energy is also zero. If we calculate the potential energy with respect to the ground, we have the equation

$$E_p = mgh = 0.50 \text{ kg} \times 9.8 \text{ m/s}^2 \times 10 \text{ m} = 49 \text{ J}$$

The total energy possessed by the ball is the sum of these two forms, that is, 49 J. We know from the principle of conservation of energy that this total remains constant during the ball's descent.

At the instant before the ball hits the ground, its potential energy is zero with reference to the ground. At this point, therefore, the ball's entire 49 J of energy is in the form of kinetic energy. For intermediate levels, we can calculate the relative amounts of potential and kinetic energy by first computing the value of mgh and then subtracting this quantity from 49 J. Table 5.1 shows the results of these calculations.

Table 5.1 Analysis of Example 5.4

Height Above Ground Level (m)	Form of the Energy	Amount of E_p (J)	Amount of E_k (J)	Total Energy $E_k + E_p$ (J)
10.0	All E_p	49	0	49
7.5	$\frac{3}{4}E_p, \frac{1}{4}E_k$	36.75	12.25	49
5.0	$\frac{1}{2}E_p, \frac{1}{2}E_k$	24.5	24.5	49
2.5	$\frac{1}{4}E_p, \frac{3}{4}E_k$	12.25	36.75	49
0	All E_k	0	49	49

Conservation of energy is also very useful when it comes to calculating speeds of objects. In Example 5.4 and similar cases we can use the potential energy at the point the ball is dropped to calculate the ball's speed v as it hits the ground:

$$mgh_{\text{top}} = \tfrac{1}{2}mv^2_{\text{bottom}}$$

or

$$v = \sqrt{2gh} \tag{5.5}$$

For Example 5.4, the speed of the ball just as it hits the ground is given by

$$v^2 = 2 \times 9.8 \text{ m/s}^2 \times 10 \text{ m} = 196 \text{ m}^2/\text{s}^2$$

Thus,

$$v = 14 \text{ m/s}$$

This method of working out the speed of a falling object is usually somewhat easier than that discussed in Chap. 3. It has the additional advantage of applying equally well to swinging objects.

5.5 SWINGING THE CHANGES

Changes of energy from one kind to another are continually taking place. One of the simplest examples of this is the change from potential to kinetic energy, and vice versa, that occurs in a rope swing, as shown in Fig. 5.6. At either end of the swing, the potential energy is at a maximum because the swing is highest, and the kinetic energy is zero because the swing's velocity is zero. At positions between these two extremes, the swing possesses both kinetic and potential energy. And when the swing is at its midpoint, the lowest position, the potential energy is at a minimum and the kinetic energy at a maximum value. Plainly, the energy changes back and forth between the two types as the swing swings. In the practical situation, we must consider friction at the support of the rope and friction caused by air resistance to the movement. Because of these types of friction, the mechanical energy is gradually changed into heat energy, which slowly disperses to the outside world. This loss eventually causes the swing to slow down. Thus the swing is not a completely isolated system, but in many examples we can neglect losses caused in this way. Note that if the person on the swing "pumps" the swing, the mechanical system is no longer isolated since the person puts the energy (essentially chemical energy) in his muscles into the swinging motion.

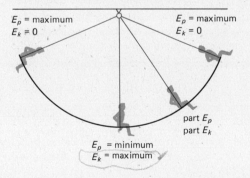

Figure 5.6 The continual transformation of energy between kinetic and potential forms in a rope swing.

EXAMPLE 5.5

Imagine that the swing in Fig. 5.6 rises to a height of 2 m (about 6 ft) above its lowest point. If the person on the swing has a mass of 70 kg (equivalent on the earth to a weight of 154 lb), compute the speed of the swing at its lowest point.

Mean Machines

Levers, wheels, screws, and even inclined planes represent a class of objects that is vitally important in keeping the world in motion. All are examples of simple machines, whose functions are to alter either the magnitude or the direction of forces applied to them. The devices appear most often as components of more complex machines, such as automobiles. But many common objects are fundamentally based on the action of one of the simple machines alone. Automobile jacks, for example, use either screws or levers to carry out their task of lifting heavy cars with a minimum expenditure of human effort. Human bones and joints are even more familiar instances of levers.

The four different types of simple machines work in the same way from the point of view of fundamental physics. One applies to each machine a force F_1, which does a certain amount of work W_1 on the machine. As a result of this input work, the machine applies its own output of work W_2, through a force F_2. The principle of conservation of energy requires that W_2 equals W_1 less any energy "lost" from the machine as heat, via friction. In equation form:

$W_2 = W_1 -$ energy lost as heat due to friction

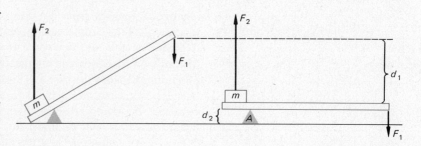

Figure A

The efficiency of the machine is simply the ratio of the output work to the input:

$$\text{Efficiency} = \frac{W_2}{W_1}$$

This figure cannot exceed 1, however much one lubricates or otherwise treats the machine to reduce friction. In practice, the efficiency never even reaches 1; every machine produces a net loss of work. So, one might well ask, what is a simple machine's value? We can get a glimmer of understanding by examining Fig. A, which shows the important details of a lever.

By applying a force F_1 to one end of the lever, we move that end downward a distance d_1. In the process, we perform an amount of work W_1 equal to $F_1 d_1$. As a result of our endeavors, the machine applies a force F_2 to raise the box of mass m a distance d_2, thereby performing an amount of work W_2 equal to $F_2 d_2$. If we make the somewhat unrealistic but nevertheless enormously helpful assumption that the machine is 100 percent efficient, then

$$W_1 = W_2 \quad \text{or} \quad F_1 d_1 = F_2 d_2$$

Solving for the output force,

$$F_2 = \frac{F_1 d_1}{d_2}$$

Plainly, if d_1 is larger than d_2, as is the case in this example, the lever does produce something useful: it helps us lift a box that requires more force to raise than the maximum we can apply directly. By moving the applied

Solution

For convenience, we take the gravitational potential energy of the swing to be zero at the bottom of its swinging. Thus, all the energy at this point is kinetic, in other words, $\frac{1}{2}mv^2$.

At the highest point, where speed and kinetic energy are both zero, force F_1 through a greater distance than the box moves, we actually apply a larger force to the box. This achievement is embodied in a term called the **mechanical advantage,** abbreviated M. The mechanical advantage is the ratio of the output and input forces F_2/F_1. In the case of the lever,

$$M = \frac{F_2}{F_1} = \frac{d_1}{d_2}$$

Obviously, the higher the value of M, the greater the force one can exert with the lever. By arranging the fulcrum of a lever in a suitable position, one that yields a large ratio of d_1 to d_2, one can set oneself up to perform Herculean tasks.

Similar reasoning applies to pulleys. A simple pulley is a suspended wheel that carries a rope. By pulling on one end of the rope, one can raise a mass attached to the other end. The force necessary to raise the mass equals the weight of the mass, assuming again no loss to friction, because a tug over a certain distance on the free end of the rope lifts the mass exactly the same distance. In other words, the mechanical advantage is 1.

This figure can be improved by producing more sophisticated pulleys. In the double arrangement in Fig. B, for example, an upward tug

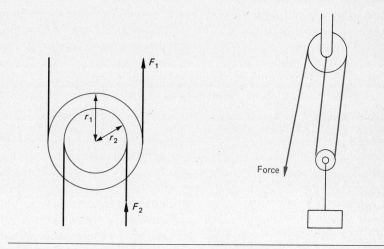

Figure B   Figure C

by force F_1 on the outer wheel of radius r_1 produces an upward force F_2 in the rope passing over the wheel of radius r_2. If the forces act for a full turn of the combined pulley wheel, the outside moves a distance $2\pi r_1$ and the inside $2\pi r_2$. Hence:

$$M = \frac{2\pi r_1}{2\pi r_2} = \frac{r_1}{r_2}$$

Thus the greater the contrast between the two radii, the larger the ratio of output to input force.

A final example of a simple machine is the block and tackle arrangement in Fig. C. Here, a force on the free end of the rope pulls up an object attached to the lower pulley. The geometry of the system is such that the force pulls the rope down twice the distance that the object rises. Hence the mechanical advantage of the system is 2, and the object can be lifted by a force equal to half its weight—once again ignoring any effect of friction.

the potential energy has increased to mgh, where h is the height of the topmost point above the bottommost point: Applying the principle of conservation of energy,

$$\tfrac{1}{2}mv^2_{\text{bottom}} = mgh_{\text{top}}$$

Thus,
$$v = \sqrt{2 \times 9.8 \text{ m/s}^2 \times 2 \text{ m}} = 6.26 \text{ m/s}$$

Note that this result is independent of the mass of the person on the swing. A 30-lb toddler and a 250-lb defensive tackle swing at the same speed under the same circumstances. This is not really too surprising when one realizes that the swing is really a falling object. Its speed at the bottom of its arc is the same as it would be if the swing simply dropped vertically through the 2 m between the top and bottom points of its movement.

A 30-lb toddler and a 250-lb defensive tackle swing at the same speed under the same circumstances.

EXAMPLE 5.6

A wooden block is placed at the top of an inclined plane 5 m above the ground. If the plane is friction free, compute the speed of the sliding object at the point where it reaches the ground.

Solution

Since the force of the inclined plane is at right angles to the motion, friction is neglected, and the force of gravity is taken into account by considering gravitational potential energy, the system may be considered isolated. Thus its mechanical energy remains constant throughout the slide. Substituting in $mgh_{\text{top}} = \frac{1}{2}mv^2_{\text{bottom}}$

$$v = \sqrt{2gh} = \sqrt{2 \times 9.8 \text{ m/s}^2 \times 5 \text{ m}} = 9.9 \text{ m/s}$$

Again, this slide is equivalent to a vertical drop of the block, over a distance of 5 m, because we have neglected the effect of friction. In fact, the block could have taken any sort of path from the top of the incline to the bottom and ended with the same speed. The incline could just as easily have been a curved hill as a flat inclined plane.

EXAMPLE 5.7

Reconsider Example 5.6 without ignoring the effect of friction. Imagine that the frictional force of the surface on the block is 3 N, the mass of the block is 5 kg, and the inclined plane is 15 m long.

Solution

Because of the frictional force, we can no longer regard the block as an isolated system. The external force of friction does work on the block, and we must apply the work-energy principle expressed in Eq. (5.4):

Work = $\Delta E_k + \Delta E_p$

Thus,

$$-Fs = \tfrac{1}{2}mv^2_{\text{bottom}} - \tfrac{1}{2}mv^2_{\text{top}} + mgh_{\text{bottom}} - mgh_{\text{top}}$$

We gave a negative sign to the work performed by the frictional force because the force acts in a direction exactly opposite to that of the block's displacement. The rest of the equation expresses the change in the block's kinetic and potential energies (which includes the effect of the gravity force) as a result of its trip down the inclined plane. And since $v_{top} = 0$ and $h_{bottom} = 0$ (the latter by our definition), the equation reduces to

$$-Fs = \tfrac{1}{2}mv_{bottom}^2 - mgh_{top}$$

Rearranging,

$$mgh_{top} - Fs = \tfrac{1}{2}mv_{bottom}^2$$

This equation tells us that the kinetic energy at the bottom of the incline is equal to the block's total energy at the top of the incline, less the work done on the block by friction as it slides down. Plainly, if the work term is larger than the total energy at the top, the object will not move, and v will have a value of zero; if we neglect friction, the equation will revert to the simpler formulation of Example 5.6. Substituting the numerical values into the equation,

$$5 \text{ kg} \times 9.8 \text{ m/s}^2 \times 5 \text{ m} - 3 \text{ N} \times 15 \text{ m} = (\tfrac{1}{2})5 \text{ kg } v_{bottom}^2$$

$$245 \text{ J} - 45 \text{ J} = (2.5 \text{ kg})v^2$$

$$v^2 = \frac{200 \text{ J}}{2.5 \text{ kg}} = 80 \text{ m}^2/\text{s}^2$$

$$v = \sqrt{80 \text{ m}^2/\text{s}^2} = 8.9 \text{ m/s}$$

5.6 GRAVITATIONAL POTENTIAL ENERGY

Thus far, we have considered only changes in the gravitational potential energy near the surface of the earth, where g is taken as a constant. Our expression mgh for the change in potential energy when an object of mass m was lifted a height h above the earth's surface raises a question when we consider the variation of g with distance. From Newton's law of gravitation, we know that

$$g = \frac{GM}{R^2}$$

where M is the mass of the earth, that is, the mass generating the gravitational field. If we reconsider the work done in lifting an object of mass m and compute $W = F \cdot s$ using the gravitational law $F = GmM/R^2$, we have a slight problem. The force changes as we move an object away from the earth. We need to compute the work done in moving an object with a variable force. This process is best done mathematically using calculus. It is sufficient here to state the result and examine its meaning.

The gravitational potential energy is a mutual property of a pair of objects by virtue of their mutual gravitational attractive force. Considered in the usual context of one body of mass m in the gravitational field of another body of mass M, we write the gravitational potential energy as

$$E_p = -\frac{GmM}{R} \tag{5.6}$$

Here G is the universal constant and R represents the separation between the centers of the two bodies.

Notice that the gravitational potential energy varies inversely as the separation R between the bodies, and the gravitational *force* between them varies as the inverse of the square of R. The negative before this scalar quantity arises because it is conventional to take the gravitational potential energy as zero when R approaches infinity, that is, when the two masses M and m are separated by an infinitely large distance. Having taken this reference point of $E_p = 0$ when $R = \infty$, then for all finite values of R the potential energy will be less and, hence, must take on negative values. Note that the negative potential energy is associated with an attractive force. Later, when we examine the electrical force, which can be repulsive or attractive, the negative potential energy will again be associated with the attractive force. Note that the kinetic energy $\frac{1}{2}mv^2$ is always a positive quantity, a fact guaranteed by the squaring of v.

We know that work is required to separate two bodies. It is this work that produces an increase in the potential energy of a pair of bodies as they are forcibly separated. As the two bodies move apart, their potential energy approaches zero—and yet we know that the potential energy is increasing. As we separate the two objects, we increase the potential energy from one negative value to another negative value that is numerically smaller. The smallest value of R comes about when the two bodies are touching each other; R equals the sum of the radii of the two bodies. This minimum value of R produces a maximum numerical value of the potential energy—the largest negative value, and the value of energy furthest below the maximum potential energy of zero, which occurs when R is literally infinite.

Our discussion of gravitational potential energy shows us that changes in potential energy are caused solely by alterations in the positions of objects. The positional form of potential energy is not restricted to simple alterations in the height of an object. Compression of a spring, for example, results from the performance of work on the spring. The result of the work is storage of potential energy in the spring, in the form of a change in the shape of the spring against the spring's elastic forces. Releasing the spring allows the spring to do work and hence reduces the potential energy in just the same way that knocking a book off a table reduces the book's gravitational potential energy.

5.6 Gravitational Potential Energy

The Real Expression for Potential Energy

We neglect the radius of the small object m since it is insignificant with respect to R_e.

The student might wonder how the potential energy expression mgh for *changes* in potential energy *near* the earth's surface can be consistent with our general expression for the potential energy $E_p = -GmM/R$. To answer this puzzle and to illustrate how Eq. (5.6) is employed, we can compute the change in potential energy near the earth's surface when we lift a small object a height h above the earth's surface (see Fig. 5.7). Initially, while the object is on the earth's surface, its potential energy is given by

$$E_p \text{ initially} = \frac{-GmM_e}{R_e}$$

When work is done on mass m to move it to a height h above the surface, which corresponds to a separation between the centers of the bodies equal to $R_e + h$, then the potential energy increases to

$$E_p = \frac{-GmM_e}{R_e + h}$$

To compute the change in the potential energy, we subtract the initial value from the final value, obtaining

$$\Delta E_p = \frac{-GmM_e}{R_e + h} - \left(\frac{-GmM_e}{R_e}\right) = GmM_e \left(\frac{1}{R_e} - \frac{1}{R_e + h}\right)$$

Taking a common denominator, we obtain

$$\Delta E_p = GmM_e \left[\frac{R_e + h - R_e}{R_e(R_e + h)}\right] = \frac{GM_e mh}{R_e(R_e + h)}$$

Now, if h is small, then $R_e + h$ is essentially R_e, and the product in the denominator $R_e(R_e + h) \rightarrow R_e^2$. Therefore,

$$\Delta E_p = \frac{GM_e mh}{R_e^2} = mgh$$

In the last step of our mathematics we substituted g for its value GM_e/R_e^2 and obtained an approximate expression for the change in potential energy *near* the earth's surface. The expression is good as long as h is very much smaller than the radius of the earth, which is the case for most everyday examples.

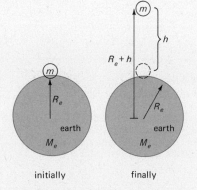

Figure 5.7 Diagram for computing the change in potential energy of an object whose mass is m when it is raised from the earth's surface.

Satellite Energy

The new equation for gravitational potential energy gives us the opportunity to take another look at the conservation of energy. Let us consider a system such as a satellite in orbit around the earth. Neglecting the forces of the sun, the moon, and other planets, because they are small in comparison to the earth's gravitation, we have an isolated system—the earth-satellite system. As a result, we know that energy will be conserved, and the total amount of energy in the system will remain constant. This total consists of three parts: the kinetic energies of the earth and the satellite and the mutual potential energy between the two. If the earth has a mass M_e and speed v_e, the satellite has a mass m_s and a speed v_s and their separation at any point is R, then

$$\text{Total energy} = \tfrac{1}{2}M_e v_e^2 + \tfrac{1}{2}m_s v_s^2 + \left(\frac{-GM_e m_s}{R}\right) = \text{constant}$$

Let us take a more specific look at the satellite itself, assuming that it is moving in the elliptical orbit about the earth shown in Fig. 5.8. If we assume that the massive earth is stationary, the energy of the satellite, like that of the earth-satellite system, also remains constant. In other words, $\tfrac{1}{2}m_s v_s^2 - (GM_e m_s/R) = $ constant. When the satellite is closest to the earth, meaning that its potential energy is at a minimum, its kinetic energy is at a maximum. Hence, its speed attains its highest value at the point closest to the earth. Similarly, the satellite travels at its slowest speed when it is farthest from the earth—in complete agreement with one prediction of Kepler's laws as we noted in Chap. 4 and with the commonsense notion that things speed up as they fall down.

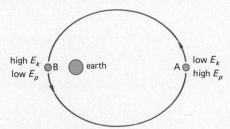

Figure 5.8 Transformation of energy during the motion of a satellite in orbit around the earth. The total energy (potential and kinetic) of the satellite is the same at all points in the orbit. When the satellite is farthest from earth, at A, the potential energy is a maximum, and hence the kinetic energy must be at a minimum. The satellite thus has its lowest speed at A. When the satellite is closest to the earth, at B, the potential energy is at a minimum and the kinetic energy is at a maximum. Hence, the satellite moves fastest at this point.

EXAMPLE 5.8

Examine how the speed of an earth satellite moving in an elliptical orbit varies with its distance from earth, given that this distance changes between a minimum value $R_{\min}$ and a maximum value $R_{\max}$.

Solution

Considering the satellite and the earth as an isolated system, we know that the total energy will remain constant during the satellite's orbit. To simplify the calculation, we can assume that the earth is stationary. Then, the total energy at the two extremities of the satellite's path is given by the equations

At $R_{\min}$:

$$E_{\text{tot}} = \tfrac{1}{2}m_s V_1^2 - \frac{Gm_s M_e}{R_{\min}}$$

At $R_{\max}$:

$$E_{\text{tot}} = \tfrac{1}{2}m_s V_2^2 - \frac{Gm_s M_e}{R_{\max}}$$

Since the energy is constant, these two quantities must be equal:

$$\tfrac{1}{2}m_s V_1^2 - \frac{Gm_s M_e}{R_{\min}} = \tfrac{1}{2}m_s V_2^2 - \frac{Gm_s M_e}{R_{\max}}$$

Even before we put numbers into this equation, it is clear that the speed at position 1, where R is at a minimum, is greater than that at position 2, where R has its maximum value. Note that the potential energy term is always negative; as the separation of the earth and the satellite decreases, it becomes more negative. In effect, the magnitudes of both kinetic and potential energy increase as the distance separating the earth and the satellite decreases. This becomes clear if we consider the expression for the total energy at any position:

$$E_{\text{tot}} = \tfrac{1}{2}m_s v^2 - \frac{Gm_s M_e}{R}$$

and then move the individual terms around:

$$\tfrac{1}{2}m_s v^2 = E_{\text{tot}} + \frac{Gm_s M_e}{R} = \text{constant} + \frac{Gm_s M_e}{R}$$

5.7 POWER

We have not yet mentioned time in our discussion of energy and work. Yet in a practical sense we are all aware that it is important whether a certain amount of work is done in an hour or a day. In physics a new quantity, **power, is defined as the time rate at which work is performed:**

$$P = \frac{\text{work}}{t} \quad \text{in ft} \cdot \text{lb/s or J/s} \tag{5.7}$$

The units of power are ft · lb/s in the British engineering system and J/s in the metric system. In the metric system a special name, watt (W), is defined as equivalent to 1 J/s. Most people are familiar with the watt or kilowatt in reference to electrical power. Note that when you pay your electric bill you pay for the number of kilowatthours (kWh), the product of power and time—that is, energy.

In the United States the horsepower (hp) is commonly used as a unit of mechanical power. One horsepower is equivalent to 550 ft · lb/s or 746 W.

Examples of interest often involve a constant force so that work is done in a uniform, continuous fashion. Under such conditions power can be expressed in the following way:

$$P = \frac{\text{work}}{t} = \frac{Fs}{t} = F\frac{s}{t} = Fv$$

where the velocity **v** is parallel to the direction of the force **F** doing the work.

EXAMPLE 5.9

Compute the (a) work done and (b) power in watts when an object of mass 100 kg is lifted 3 m in 10 s at a constant speed so that there is no increase in kinetic energy.

Solution

(a) Work $= F \cdot s = mg \times h = (100 \text{ kg})(9.8 \text{ m/s}^2)(3 \text{ m}) = 2.94$ kJ

(b) $P = \dfrac{\text{work}}{t} = \dfrac{2.94 \times 10^3 \text{ J}}{10 \text{ s}} = 294$ J/s $= 294$ W

This mechanical power corresponds to the electric power associated by a rather large light bulb.

EXAMPLE 5.10

A horse exerts a steady horizontal pull of 80 lb to move a wagon at a uniform speed along a horizontal road. The wagon, weighing 2500 lb including contents, is moved a distance of 0.25 mi (1320 ft) in 2.5 min. Compute the power associated with this work.

Solution

We are working with a uniform force of 80 lb to move the wagon horizontally. The weight of the wagon has no direct bearing upon our solution since the wagon is not being lifted, as would be the case if the wagon were moving up a hill.

$$\text{Power} = \frac{\text{work}}{t} = \frac{Fs}{t} = \frac{(80 \text{ lb})(1320 \text{ ft})}{2.5 \text{ min} \times 60 \text{ s/min}} = 704 \text{ ft} \cdot \text{lb/s}$$

or, in horsepower,

$$P = \frac{704 \text{ ft} \cdot \text{lb/s}}{550 \text{ ft} \cdot \text{lb/hp} \cdot \text{s}} = 1.28 \text{ hp}$$

SUMMARY

1. Work is the product of the displacement of an object and the component of the force acting on the object in the direction of that displacement.

2. Negative work occurs when a force acts in the direction opposite to that of the object's movement.

3. Kinetic energy is the energy that any object possesses by virtue of its motion. Its numerical value is half the product of the object's mass and the square of its speed.

4. Potential energy is the energy that objects possess as a result of their positions. The gravitational potential energy of an object is the product of its mass, its height above an appropriate reference level, and the local acceleration due to gravity.

5. The effect of work on any object is to alter the object's kinetic energy, or in the case of conservative forces, to alter the potential energy.

6. The total amount of energy in any isolated system remains constant whatever happens to the system. But note that energy can and does change from one form to another, notably from kinetic energy to potential energy, and vice versa.

7. Gravitational potential energy, as used in astronomical application, is a negative quantity that goes from a maximum value of zero when two objects are separated by an infinite distance to a minimum when the objects are touching.

8. Power, the time rate of doing work, is usually expressed in watts.

QUESTIONS

1. An orange hanging from a branch of a tree falls to the ground. While on the tree, it has gravitational potential energy. What happens to the gravitational potential energy just before the orange hits the ground? After it hits the ground?

2. Discuss a golf swing in terms of the conservation of energy. What is the form(s) of the energy at various points during the swing?

3. Water falls 100 ft down a waterfall. Where does the potential energy go when the water reaches the pool at the bottom of the fall? Has the state (or condition) of the water changed?

4. Consider a boy on a swing at the playground. During which part of his swing is the (a) potential energy a maximum, (b) potential energy a minimum, (c) potential energy zero, (d) kinetic energy a maximum, (e) kinetic energy a minimum, (f) kinetic energy one-half the total change in potential energy?

5. Consider an artificial satellite orbiting the earth in an elliptical orbit so that at one point the satellite is 200 mi above the earth and at the other end of the orbit it is 1000 mi from the earth. Apply the principle of conservation of energy to this situation and describe how the speed of the satellite changes in orbit.

6. Three skiers, Dick, Jane, and Sally, were together at the top of a hill. Dick went off a jump, Jane skied down the slope, and Sally rode the ski lift down. Compare the changes in their gravitational potential energy. Compare the speed at which Dick and Jane would arrive at the bottom of the hill, assuming they "lost" no energy due to frictional processes.

7. Is there a place between the earth and moon where an earth satellite will have zero gravitational energy? Is there a point where there is zero force on the satellite? Discuss your answers to these two questions. Are they in conflict?

8. A girl pushes a bicycle up a steep hill, increasing the gravitational potential energy of the bike and herself. Where did the energy come from? Follow back to the "source" of the energy as far as you are able. Is there an ultimate source of energy?

9. A track star runs once around a 0.25-mi track, returning to his starting point. Does he do any work?

PROBLEMS

Work-Energy Principle

1. A youth weighing 150 lb runs up a flight of stairs. The vertical distance between floors is 10 ft. Compute the work done against gravity.

2. An electric motor raises to a height of 20 m an elevator cage that, with contents, has a mass of 3000 kg. Compute the work done in (a) newton-meters, (b) joules.

3. An object having a mass of 5 kg is pulled with a uniform speed a distance of 4 m along a tabletop by a constant horizontal force of 3.0 N. Compute the work done in (a) newton-meters, (b) joules, and (c) ergs. (d) What is the value of the frictional force in this case?

4. In clearing a 6-ft, 6-in. bar, a high jumper who weighs 160 lb actually lifts his own weight a vertical distance of 3 ft. Compute the work done by the jumper.

5. You have a large box of mass 250 kg that you must move a distance of 20 m. How much work do you have to perform if the surface is level and frictionless?

6. You push a lawn mower at a uniform speed with a force of 8 N at an angle of 37° with the horizontal. In crossing your 30-m-wide lawn, how much work do you perform?

7. You have a small box of mass 250 kg that contains lead. You cannot conveniently move it the required 20 m over a surface where the kinetic friction is 0.3 unless you apply the force at an angle. Assume you use an angle of 45°. Which requires more work (or are they the same)—to (a) pull or (b) push the box? Calculate the work in both cases.

8. You have a large box of mass 250 kg that you move a distance of 20 m over a surface where the coeffi-

cient of kinetic friction is 0.3. How much work (minimum) must you do to move the box if you apply a horizontal force as a (*a*) push, (*b*) pull?

9. You carry a clothes basket containing 9 kg of dry laundry up the cellar stairs a height of 3 m, across the house a distance of 6 m, and then upstairs a height of 3 m. (*a*) How much work must you perform on the basket against the force of gravity? (*b*) What is the change in potential energy of the clothes basket and its contents?

10. A 4-kg book is slid across a table with an initial speed of 0.5 m/s. If it came to rest after 2.5 m, (*a*) what is the magnitude of the frictional force? (*b*) How much work is done by friction?

11. A force at an angle of 25° is applied to a 3-kg block. A frictional force of 3 N retards the motion of the block. (*a*) What is the work done by each force acting on the block? (*b*) What is the work done by the net force on the block? If the block starts from rest, (*c*) what will its speed be after it has been pulled 20 m?

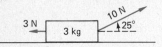

12. To remove logs from the woods in the winter, you use a tractor and a large chain. The chain is attached to a 1000-kg log and makes an angle of 30° with the horizontal. The tractor exerts 100 N on the log. How much work has been done in moving the log 500 m through the woods at a uniform speed?

13. A driver in a 1500-lb car moving at 45 mi/h applies his brakes in an erratic manner, causing the car to slow down to 15 mi/h. How much work is done by the frictional force acting on the car?

14. A baseball traveling at 40 m/s is caught by the catcher, and in the process the catcher's glove moves back 15 cm. Assuming a constant force, how large a force acts on the ball while it is being caught? The mass of the ball is 150 g.

15. A 2-kg box experiences a force of 8.3 N acting opposite to its motion on a horizontal surface. (*a*) What is the change in kinetic energy of the box if this force acts for 8 m? (*b*) How much work is done by this force?

16. A 250-g book is slid across a rough table with an initial velocity of 50 cm/s. (*a*) How much work is done by the retarding (frictional) force in bringing it to rest? If the magnitude of this retarding frictional force is 0.15 N, (*b*) how far does the book slide before coming to rest? (*c*) What power is associated with this work?

17. A mass *m* slides down a frictionless plane that is 10 m long. The top of the incline is 7 m above the horizontal surface on which the incline sits. How much work was required to (*a*) lift the mass to the top of the incline, (*b*) slide the mass up the incline to the top? How much potential energy is transformed into kinetic energy when the mass (*c*) slides down the incline, (*d*) falls off the top of the incline and drops vertically to the horizontal platform? (*e*) What is the speed of the mass as it reaches the horizontal in (*c*) and (*d*)?

18. You pull a wagon of mass 15 kg loaded with two children, each of mass 45 kg, up a 30° incline, pulling along the incline. If you move up the incline a distance of 40 m, how much work are you doing?

Kinetic Energy

19. An automobile moving at 15 m/s accelerates at 3 m/s² for 5 s. Compute its final kinetic energy if its mass is 1000 kg.

20. An automobile starts from rest and accelerates at 2 m/s². After it has traveled 10 m, what is its speed and kinetic energy if it "weighs" 1200 kg?

21. (*a*) How much kinetic energy does a 10-g mass moving at a speed of 2 cm/s have? Suppose the speed of this mass were to increase to 3 cm/s, 4 cm/s, and then 8 cm/s. (*b*) Compute the kinetic energy at each speed. (*c*) By what factor has the kinetic energy increased?

22. A bicycle rider whose combined mass with his bike is 80 kg and who is moving at 4 m/s accelerates to a speed of 8 m/s. What is the change in his kinetic energy?

23. A mass of 5 kg is given a speed of 2 m/s. What is its kinetic energy?

24. You are running around a track at 2 m/s and then increase your speed on the straightaway to 3 m/s. Compute your change in kinetic energy. Use your own mass. Does the fact that you are not in straight-line motion cause a problem?

Gravitational Potential Energy

25. A rocket is launched, attaining a velocity of 8×10^3 m/s just above the earth's surface. How high does it rise above the earth's surface?

26. Calculate the speed an object will have to be projected outward from the earth if it is just to escape the earth's gravitational field. This is called the escape velocity. Does this depend on the mass of the object?

27. At what distance from the center of the earth and moon will the gravitational potential energy of a satellite on a line between the earth and moon contain equal contributions due to the moon and the earth?

28. What is the gravitational potential energy of the earth-sun system? With what speed would the earth have to be projected away from the sun to escape the sun's gravitational influence?

Conservation of Energy

29. You are out bike riding and decide to relax and coast down a hill that has a vertical rise of 100 m. How fast will you be moving at the bottom in meters per second, kilometers per hour, and miles per hour, neglecting friction.

30. A hammer having a mass of 2 kg lies on a roof parapet 30 m above the ground. Compute its (*a*) potential energy, (*b*) kinetic energy. A workman accidentally knocks the hammer off the parapet. It passes a window ledge which is 10 m above the ground; compute its (*c*) potential energy, (*d*) kinetic energy at that level. Compute its (*e*) potential energy, (*f*) kinetic energy just as it reaches the ground level, then traveling at its maximum velocity. (*g*) What happened to the mechanical energy the hammer possessed after it struck the ground and came to rest?

31. You are on a ladder painting your house and you drop a bucket of paint weighing 10 lb a distance of 17 ft. (*a*) How much gravitational potential energy do you lose, along with your paint? (*b*) With what speed do the paint and bucket hit the rose bushes at ground level?

32. A bob ($m = 1$ kg) of a pendulum 0.8 m long is lifted 20 cm vertically above the equilibrium point and released, keeping the string taut. (*a*) What is the kinetic energy at the equilibrium position? (*b*) What is the form and magnitude of the energy when the pendulum is half-way vertically through its swing?

33. A 6-lb ax is used to split wood. If the ax head is raised to a height of 2.5 m vertically above the ground and then dropped, what will be its speed just before it hits the log on the ground?

34. If you ride on a roller coaster whose maximum height is 30 m, what will be the maximum speed you could attain on this roller coaster? The roller coaster passes the highest point with negligible speed.

35. A ball is thrown vertically to a height of 25 m. The ball hits the ground and bounces but loses one-half of its speed on the rebound. How high will it rise?

36. The system in the figure is released from rest with the 24-lb block 8 ft above the floor. Use the principle of conservation of energy to find the velocity with which the block strikes the floor. Neglect friction and the mass of the pulley.

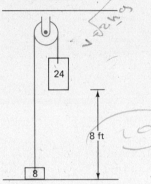

37. A 2-kg projectile is shot from a gun at an angle of 30° with the horizontal with an initial velocity of 300 m/s. (*a*) What is the form of its energy and how much energy does it have just as it leaves the gun and just before it hits the ground? (*b*) Using energy considerations, compute the maximum height to which the projectile rises. (*c*) What happens to the mechanical energy when the projectile hits the ground? Is energy conserved? Is mechanical energy conserved?

38. A skier takes off from a ski jump with a velocity of 50 m/s. The skier lands at a point whose *vertical* distance below the point of takeoff is 100 m. What is the speed just prior to landing? Ignore air friction.

39. The "hammer" of a drop pile driver weighs 4000 lb. When the hammer is stationary in a position 30 ft directly above the pile that is to be driven into the ground, compute its (*a*) potential energy with respect to the pile, (*b*) kinetic energy. The hammer is released. Considering it to drop freely, compute its (*c*) location 1 s after being released, (*d*) potential energy at that time, (*e*) kinetic energy at that time. Just as it reaches the pile, compute its (*f*) potential energy, (*g*) kinetic energy. (*h*) Assuming that 30 percent of the energy is transformed into heat energy on impact, compute the work that will be done by the hammer in forcing the pile into the ground.

40. Water over a falls 70 m high is to be used to generate electrical energy. How many 100-W bulbs could be lit if all the energy from 500 000 kg/s of water going over the falls could be converted to electrical energy?

41. A block of mass 1 kg (Fig. 41) is sent sliding down a frictionless, curved ramp with an initial speed of 6 m/s. The block reaches the horizontal and hits a rough surface, where a frictional force of 5 N acts for 5 m. The block continues on, going up an incline. (*a*) What is the total energy at point A just as the block is initially released? (*b*) What is the speed of the block at B? (*c*) What is the speed of the block at C? How much energy does the block have at C? (*d*) How high up the friction-free incline plane does the block go?

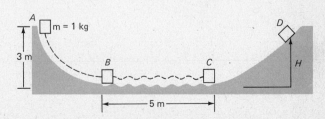

Figure 41

42. A skier travels down a curved, frictionless incline, then over another frictionless hill to a rough horizontal surface from D to E, where a constant frictional force of 50 N acts (see below). At E, the skier hits a snow drift. (*a*) What is the magnitude and form of the mechanical energy of the skier at A, B, C, D, and E just before he hits the snow and at E just after he hits the snow? (*b*) What is the speed of the skier at point B? (*c*) What is the speed of the skier just before he hits the snow at E? (*d*) What happens to the energy of the skier when he hits the snow?

43. A 3-kg block is at the top of a curved incline 20 m high. The block travels down the frictionless incline to point A, at which point it travels on a rough surface with a coefficient of friction $\mu_k = 0.20$. (*a*) What is the speed of the 3-kg block at A? (*b*) What is the speed of the 3-kg block at B? (*c*) How much farther beyond point B can the object travel before stopping?

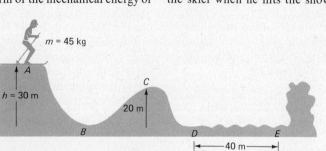

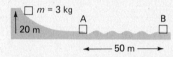

Special Topic
Producing Energy for the People

When people speak of energy, they normally refer to the energy that heats houses, lights offices, cooks meals, and keeps the television and other ingenious inventions in operation. This type of energy seems to be a far cry from the theories and equations of mechanical energy, which form the core of this chapter. However, closer examinations shows that mechanical energy is really a fundamental factor in keeping a person comfortable. Oil wells, natural gas fields, nuclear reactors, the sun, the wind, and the tides represent different methods of producing mechanical energy, which is then converted into practical forms of energy.

Energy is a particularly crucial matter in the United States because this country is extraordinarily energy intensive: It uses much more energy per capita than any other nation in the world. Only one-twentieth of the world's population is American, yet one-third of the energy used throughout the world is used by American consumers in some way or other. In another sense, the United States is using more than its portion because it possesses just one-fifth of the world's total energy resources in its oil, gas, and coal fields, its uranium ore, and other sources of energy. Still, as the rest of the world looks with envy on the American abundance of air conditioners, refrigerators that make their own ice, large, gas-guzzling cars, and other features of the nation's careless use of energy resources, its reaction is more a desire to emulate than a wish to avoid the American attitude to energy. Other nations want to catch up with this nation's freewheeling use of its energy—and that will only mean more pressure on the world's energy supplies. Experts foresee increasingly severe shortages unless conservation of energy is instituted (many observers think that the United States wastes half the energy it uses), and relatively unlimited sources of energy such as nuclear fusion and solar energy are developed to the stage of commercial production.

For the moment, conservation is politically unattractive, and both fusion and solar power appear unlikely to be economically feasible on the large scale—even if the technology works out—until well into the 2000s. Thus energy scientists are trying their best to produce a suitable mixture of the existing technology that will allow the United States to survive through the present century without drastically lowering its living standards. Much of the effort is aimed at converting various forms of energy to electricity that now accounts for slightly more than a quarter of the energy used in the nation. This figure is almost certain to increase over the years.

The key to the whole system is the turbogenerator—a piece of machinery that translates mechanical energy of motion into electricity. The electrical energy is generated when bundles of wires known as armatures are rotated in strong magnetic fields; this topic will be treated in more detail in Chap. 16. The armatures are forced to revolve by the spinning motion of a turbine under the stimulation of mechanical energy. A turbine contains a number of blades that are set into motion by the impact of a fluid flowing past them. Thus, the wide variety of different energy technologies are mostly concerned with one simple task: giving the fluid the necessary mechanical energy to set the blades in motion and keep them going.

The most common turbine fluid is steam, produced by the simple expedient of boiling water. In effect, then, most power plants are no more or no less than giant kitchen stoves, producing endless amounts of steam. Debates over priorities for development and use of different types of energy technology concern such issues as the cost, efficiency, cleanliness, extent of raw materials, and risk to the public of the alternative means of performing this large-scale household chore.

Coal, for example, one of the oldest fuels for producing the large amounts of heat necessary to produce steam on an industrial scale, is a relatively cheap resource that is likely to last for hundreds of years. However, obtaining it involves risk to miners when it is mined from deep in the earth and ruination of the landscape when it is strip mined.

Oil is in far shorter supply than coal. Estimates suggest that, at the present rates of usage, the world's supplies of the black gold are likely to start drying up in the twenty-first century, despite the exploitation of new fields on Alaska's north slope and beneath the North Sea. Furthermore, the world's present supply of oil is so concentrated in the Middle East as to make the resource a servant of diplomacy. And, when oil is used to stimulate the movement of cars rather than turbines, it has a catastrophic effect on the purity of our air.

The third "fossil fuel," natural gas, seems also to be in short supply, although it is an inherently cleaner method of generating power for both turbines and cars. In fact, the production of forms of natural gas from sewage is strongly promoted by environmental groups as one short-range contribution to alleviating the world energy crisis.

Nuclear fission was once hailed as the answer to all the world's energy problems, but over the years its reputation has slipped precipitously. Just like fossil fuel burners, nuclear reactors produce only one truly important product, heat, which converts water to steam. This heat is the result of a stringently controlled type of nuclear reaction produced by fission: the splitting of some nuclei of uranium atoms, which releases large amounts of energy. Environmentalists have long pictured nuclear energy as a hazard to the general public because of its production and use and the disposal of the resulting radioactive materials. Nuclear proponents have equally strenuously defended their technology on the grounds that the industry is more rigidly controlled by safety regulations and inspections than any other in the power business. However, the shocking disaster at the Three Mile Island reactor in Middletown, Pennsylvania, in the spring of 1979 brought the nuclear energy industry almost to a standstill. The near meltdown of the reactor's core and radiation leakage from the containment vessel awoke the public to the real dangers of nuclear power. This incident was further magnified by the almost simultaneous appearance in movie theaters nationwide of "The China Syndrome," a fictional story of a nuclear disaster at a plant. The resulting concerns about nuclear en-

Three Mile Island nuclear power plant. (*United Press International*)

ergy, such as frequent radiation spills in reactors, structural stability of reactors against earthquake shocks, safety design, "safe" radiation levels, and the relation of radiation to cancer, raise serious doubts as to whether the public will accept the risks of nuclear energy. Certainly, nuclear power has not spread as fast as its promoters once hoped, but it seems likely to continue to play a vital role in the world's energy future for many years to come.

Other methods of producing the energy necessary to drive turbogenerators are either more restricted or less well developed. Geothermal power is an example of a natural resource that has been tapped directly to drive turbines in a few parts of the world. Steam produced by underground heating is led through boreholes to power plants above ground in northern California, Italy, the north island of New Zealand, Iceland, and Mexico. However, the number of steam fields is so limited that this type of geothermal power could never satisfy more than a minute proportion of the world's energy needs. Engineers believe that geothermal fields containing hot water under pressure are both more numerous and more promising as large-scale sources of electric power. Experts are now developing the technology necessary to tap the hot pressurized water, which becomes steam once the pressure is released. They are also seeking to minimize a number of environmental problems associated with geothermal energy, among them extreme noise, the pres-

Geothermal plant at the geysers in northern California. (*U.S. Department of the Interior*)

Solar house built by engineers at the Massachusetts Institute of Technology. (From "Solar Energy—An Option For Future Energy Production." Peter E. Glaser, *The Physics Teacher*, **10**, no. 8, 443, November 1972)

ence of corrosive salts in the underground hot water and steam, and potential subsidence of the ground locally after the water and steam are removed.

Another natural source of power that has only been exploited minimally to date is the sun's heat. Its main use at present seems to be in domestic hot water systems and to a limited extent in heating houses, offices, and other buildings—a process which only involves mechanical energy insofar as water or air heated by the sun must be pumped around the installations. However, some engineers believe that suitably shaped arrangements of mirrors can focus the sun's heat sufficiently to convert water into steam for driving turbines. More futuristic proposals have been made to capture the sun's energy in huge orbiting satellites equipped with solar cells that convert the sun's light to electrical energy that can be beamed down to earth. At present such schemes are stymied by the very low efficiency (less than 20 percent) of solar cells, and even their proponents admit that such satellites are unlikely to get off the ground this century.

Some natural methods of generating electric power do not involve the production of steam. One of the most intriguing of these is known as thermal gradient energy conversion. In theory, it would exploit the temperature difference between the sun-warmed surface of the ocean and the cool depths of the water in tropical regions to boil and condense a suitable liquid, such as ammonia. The gaseous form of the fluid would be used to power a turbogenerator.

Other energy technologies are even more direct. Hydroelectric power stations use the energy of water that has spilled over waterfalls or has been piped into turbines to drive generators directly. A tidal power station in France exploits for the same purpose the huge difference between high and low tides that occur in that and a few other parts of the world. And windmills generate electrical energy when their vanes are sent spinning by the wind. All three methods suffer from a lack of suitable sites where power can be produced in really large amounts, although they have their uses on a small scale; in addition, hydroelectric stations tend to "self destruct" because they silt up over the years.

Perhaps the best hope of physicists for the energy of the future is nuclear fusion—a controlled version of the process that makes the hydrogen bomb boom. This is inherently safer than fission—the controlled version of the atom bomb—because it produces fewer radioactive products. However, the fusion reaction has not yet been perfected in the laboratory, and even the most bullish scientists do not foresee its wide-scale use until the next century. If technologic barriers arise, fusion may never be developed satisfactorily. Thus, experts are warning the politicians and bureaucrats who fund energy research not to put all their eggs into the fusion basket. For at least the next few decades it seems certain that our energy needs will be met by a combination of different technologies.

6 Impulse and Momentum

6.1 MOMENTUM

6.2 IMPULSE

6.3 CONSERVATION OF MOMENTUM

6.4 COMING TOGETHER

SHORT SUBJECT: Reaching for the Moon
Summary
Questions
Problems

SPECIAL TOPIC: The Impulsive Athlete

In much the same way that the name of George Washington appears again and again in early American history textbooks and the name of William Shakespeare is inseparable from any serious course in English literature, the work of Sir Isaac Newton is unavoidable in any discussion of force and motion. His scientific legacy returns once again in a new physical concept that forms the subject of this chapter.

If we recall Newton's original statement of the second law of motion, we find that his somewhat archaic wording contained the germ of this concept, which is of vital importance to a complete understanding of the physics of movement. Newton termed this concept the *motion* of an object. Today, we refer to it as *momentum*.

6.1 MOMENTUM

The momentum of any object is defined as the product of its mass and its velocity. Symbolically, the momentum, **p**, is given by

$$\mathbf{p} = m\mathbf{v} \tag{6.1}$$

Because the velocity is a vector quantity, momentum is also a vector. Its direction, not surprisingly, is exactly the same as that of the velocity.

Having gone to great lengths in Chap. 3 to convince you that an object's mass is constant throughout the universe, we might well ask the questions: Why bother with the product of mass and velocity? Would it not, after all, be adequate to deal alone with velocity, the variable quantity, in any case that involves momentum?

The answer to these questions is twofold. First, although the mass of any solid object remains constant under usual circumstances, there are a number of important cases in which masses do change during motion, as a result of the nature of the object. One simple example is a rocket. As it begins to lift off the launching pad, the rocket contains a great deal of fuel. By the time the rocket has risen a few hundred feet into the air, however, much of the fuel has been burned and released from the exhaust into the atmosphere, providing the rocket with the necessary force to get up those few hundred feet. The mass of the rocket's structural components has remained constant during the ascent, but the mass of the rocket and fuel together has obviously decreased in the period owing to the loss of burned fuel through the exhaust (Fig. 6.1). We should consider just the same effect when we are dealing with any moving vehicle that is powered by burning fuel; but in the case of, say, a typical automobile, the mass of the fuel is negligible in comparison with the mass of the vehicle's structure.

Figure 6.1 Saturn V rocket at lift-off.

Momentum is a quantity that is conserved.

The other reason for the importance of the concept of momentum is more profound. Momentum is a quantity that is conserved under all conditions, and as we shall see later in this chapter, it plays a vital role in solving a number of problems related to movement. By "conserved" we mean that if no *external* forces act on a system, the total momentum of the system remains constant.

Before considering this aspect of momentum, however, we must backtrack to study another physical concept related to momentum: **impulse**.

6.2 IMPULSE

Newton's original statement of his second law, which in mathematical shorthand is $\mathbf{F} = m\mathbf{a}$, was: *"The rate of change of motion (that is, the momentum) of a body is proportional to the net applied force, and the change takes place in the direction in which the force acts."* In symbols, this is expressed as follows:

$$\mathbf{F} = \frac{\Delta(m\mathbf{v})}{\Delta t} = \frac{m_2\mathbf{v}_2 - m_1\mathbf{v}_1}{\Delta t} = \frac{\mathbf{p}_2 - \mathbf{p}_1}{\Delta t} = \frac{\Delta \mathbf{p}}{\Delta t} \tag{6.2}$$

That is,

$$\mathbf{F}\,\Delta t = \Delta \mathbf{p} \qquad \text{where } \mathbf{p} = m\mathbf{v} \tag{6.3}$$

Here, Δt and $\Delta \mathbf{p}$ are the changes in time and momentum, respectively, over the period during which a constant force $\mathbf{F}$ is being applied.

The quantity $\mathbf{F}\,\Delta t$ is known as the impulse of the force. Equation (6.3), which is often referred to as the impulse-momentum theorem, is an alternative statement of Newton's second law that can be used in place of the more familiar formulation of Chap. 3. It states that **the impulse of a force equals the change in momentum which the force brings about.**

One area of interest is the type of force often called impulsive force. Such forces occur in a variety of situations, such as a golf club or a tennis racquet hitting a ball (Fig. 6.2), a hammer driving in a nail, and the explosion of gas in a pistol. These forces are not constant during their application to any particular object. Characteristically, they start

Figure 6.2 Impulsive forces: (a) a golf ball at the moment of impact with the head of a golf club; (b) a tennis ball being hit by a racquet. (*Wide World*)

(a)   (b)

Figure 6.3 An impulsive force as a function of time.

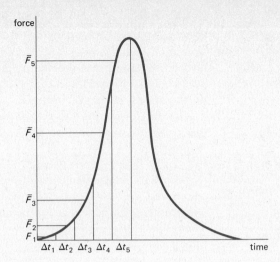

with a value of zero, rise to a high value, and then rapidly decrease back to zero, as shown in Fig. 6.3. Figure 6.3 shows us that the change in momentum caused by an impulsive force does not take place in a uniform fashion.

If we wish to calculate the change in momentum caused by a particular force using the equality between impulse and momentum, then we would first have to work out the value of $\mathbf{F}\,\Delta t$ over a series of very small time intervals (Δt_1, Δt_2, Δt_3, ..., in Fig. 6.3), each so short that the force is essentially constant during the period. Then, we would have to add all those small increments to obtain the total impulse of the force, and hence the change in momentum caused by the force. If the time intervals are not short enough, we can approximate our summation of impulses by using the average value of the force over each interval. These average values are indicated by $\bar{F}_1$, $\bar{F}_2$, $\bar{F}_3$, ..., in Fig. 6.3. The sum of the small increments of impulse actually turns out to be the area under the curve in Fig. 6.3 that relates the strength of the force to time. Newton himself derived this method of adding impulses and used it in his development of integral calculus.

Fortunately, we can calculate the change in momentum much more easily by remembering that momentum is the product of mass and velocity. All we need to know is the velocity and mass of the object in question before and after the action of the impulsive force. And because most masses remain constant during application of such forces, we generally require only the before and after velocities to obtain the change in momentum.

EXAMPLE 6.1

A batter hits a baseball of mass 0.25 kg that was thrown toward him with a speed of 10 m/s back toward the pitcher with a speed of 20 m/s.

If the process of hitting the ball takes 0.05 s, compute the impulse of the force and the average force exerted on the ball by the bat.

Solution

We can consider that the initial momentum of the ball is $+10$ m/s $\times$ 0.25 kg $= +2.5$ kg $\cdot$ m/s, and the momentum after the "collision" of bat and ball is -20 m/s $\times$ 0.25 kg $= -5$ kg $\cdot$ m/s. The positive and negative signs take account of the vector nature of momentum. To compute the impulse,

$$\mathbf{F}\,\Delta t = (m\mathbf{v})_{\text{final}} - (m\mathbf{v})_{\text{initial}}$$
$$F\,\Delta t = -5 \text{ kg}\cdot\text{m/s} - 2.5 \text{ kg}\cdot\text{m/s} = -7.5 \text{ kg}\cdot\text{m/s}$$

The negative sign just indicates that the impulse on the ball is opposite to the original direction of motion, which was chosen to be positive.

To compute the average force of the bat on the ball, we need only divide the impulse by the time of the collision, 0.05 s

$$\text{Average } F = \frac{-7.5 \text{ kg}\cdot\text{m/s}}{0.05 \text{ s}} = -150 \text{ N}$$

In reality, the force would not be constant but would resemble that shown in Fig. 6.3.

EXAMPLE 6.2

A small child is bouncing a tennis ball against the wall of a building. If the mass of the tennis ball is 60 g and the ball travels at 3 m/s, compute the impulse on the wall, assuming that the ball hits perpendicular to the surface of the wall and rebounds with the same speed of 3 m/s.

Solution

Since the motion of the ball changes, a force must have been exerted on the ball by the wall. Consequently from Newton's third law we know that an equal but oppositely directed force will be exerted on the wall by the ball. We can find the impulse using the impulse-momentum theorem, Eq. (6.3), if we evaluate the change in momentum. Initially, the ball has momentum $+mv = 0.06$ kg $\times$ 3 m/s $= +0.18$ kg $\cdot$ m/s toward the wall. Note that we have taken the positive direction to be toward the wall. After the collision, the ball is moving in the opposite direction with the same speed; hence, its momentum is -0.18 kg $\cdot$ m/s.

The change in momentum is the final momentum less the initial momentum. Hence, the impulse of the wall on the ball will be

$$\mathbf{F}\,\Delta t = m\mathbf{v}_{\text{final}} - m\mathbf{v}_{\text{initial}}$$
$$F\,\Delta t = -0.18 \text{ kg}\cdot\text{m/s} - (+0.18 \text{ kg}\cdot\text{m/s})$$
$$F\,\Delta t = 2(-0.18 \text{ kg}\cdot\text{m/s}) = -0.36 \text{ kg}\cdot\text{m/s}$$

The negative sign indicates that the impulse or force *on* the ball is away from the wall. The impulse of the ball on the wall will equal

+0.36 kg·m/s, an impulse toward the wall. Note in this type of collision the change in momentum of the ball is not zero but twice the value of the momentum because of the vector nature of momentum. The kinetic energy in this collision, on the other hand, does not change since it depends only on speed and not the vector nature of velocity.

More impulse is required in this process than if the ball stopped upon impact. In that case, the change in momentum would be just mv, or 0.18 kg·m/s instead of twice as much. When the ball rebounds, the wall in a sense must first stop the ball and then project the ball away from it, increasing the speed of the ball from 0 to 3 m/s.

6.3 CONSERVATION OF MOMENTUM

Whenever *two* objects are linked through the application of a force by one on the other—a punter kicking a football, a pistol propelling a bullet through the air, or a battering ram wielded by Islamic hordes against a Crusader's castle—we know from Newton's third law that each object is pushing on its complementary object with an equal but opposite force. Diagrammatically, we can express the situation as follows:

$$\xleftarrow{F_{BA}} \underset{M_A}{\textcircled{A}} \quad \cdots \cdots \cdots \cdots \quad \underset{M_B}{\textcircled{B}} \xrightarrow{F_{AB}}$$

Here, from Newton's third law,

$$\mathbf{F}_{BA} = -\mathbf{F}_{AB} \tag{6.4}$$

The diagram illustrates a collision in which the two objects meet and then separate. Clearly the two forces $\mathbf{F}_{BA}$ and $\mathbf{F}_{AB}$ last for the same time, because as soon as object A stops pushing on object B, B will cease pushing back. So, if Δt is the time during which the forces are acting, then

$$\mathbf{F}_{BA} \Delta t = -\mathbf{F}_{AB} \Delta t \tag{6.5}$$

This equation tells us that the impulses acting on the two objects are equal and opposite. Thus, the changes in momenta of the objects must also be equal and opposite. If we use the subscripts 1 and 2 to denote the velocities of the two objects before and after the collision, then

$$m_A \mathbf{v}_{A_2} - m_A \mathbf{v}_{A_1} = -(m_B \mathbf{v}_{B_2} - m_B \mathbf{v}_{B_1}) \tag{6.6}$$

(assuming, of course, that no net external force acts on the complete system).

In words, the change in momentum of object A is equal to but in the opposite direction to the change in momentum of object B.

By moving around the individual terms in Eq. (6.6), we can obtain two slightly different statements of this same truth. By transposing all the terms to the left-hand side of the equation, we have the expression

$$(m_A \mathbf{v}_{A_2} - m_A \mathbf{v}_{A_1}) + (m_B \mathbf{v}_{B_2} - m_B \mathbf{v}_{B_1}) = 0 \tag{6.7}$$

This means simply that the sum of the total change in momentum during the collision equals zero.

We can also put all the terms describing the situation before the collision on one side of the equation and all those relevant to the situation after it on the other side:

$$m_A \mathbf{v}_{A_1} + m_B \mathbf{v}_{B_1} = m_A \mathbf{v}_{A_2} + m_B \mathbf{v}_{B_2} \tag{6.8}$$

In other words, the total momentum of both objects taken together before the collision exactly equals the total momentum after it.

These three different statements are all expressions of a single physical law: the total momentum of any particular group of objects remains unchanged unless a net external force is applied to the system. Or: **The total momentum of any isolated system is a constant. This is the principle of conservation of momentum.** The word "isolated" in this context means that no net external force acts on the system.

The universal nature of this conservation law is best illustrated by Examples 6.3 to 6.5.

EXAMPLE 6.3

Air is forced through holes on the surface of the track to provide a cushion on which the gliders can slide with virtually no friction, in much the same fashion as in air hockey games.

Imagine two gliders on a level air track joined by compressed springs but kept from moving apart by a piece of string tying them together, as in Fig. 6.4. The two gliders push on each other, through the springs, with equal and opposite forces. If the string that prevents them from flying apart is severed, the two gliders will accelerate away from each other, gaining both speed and momentum. Show that the speeds of the gliders are inversely proportional to their masses.

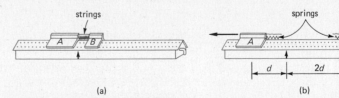

Figure 6.4 Conservation of momentum: two carts on an air track. In (a) two carts, B and A, of mass $m_B = \frac{1}{2} m_A$ are coupled together with a strong compressing the springs on each cart. The carts ride on a cushion of air that comes through small holes in the air track. After the string is cut (b), the two carts separate owing to the Newton's third-law pair of forces. Because the momentum of the two carts initially in (a) is zero, the sum of the momenta will also remain zero in (b). This means that the momentum of the cart B will be equal to and opposite that of cart A after the expansion of the springs (b). Note that because $m_B = \frac{1}{2} m_A$, cart A moves with one-half the speed of cart B in (b). Cart B will travel twice as far as cart A during the same time.

Solution

Because the force of gravity is canceled by the upward force of the air and the track is virtually frictionless, no external forces act on the system. We can apply the conservation of momentum principle to the two gliders:

$$m_A v_{A_2} - m_A v_{A_1} = -(m_B v_{B_2} - m_B v_{B_1})$$

Because both gliders started from rest, v_{A_1} and v_{B_1} are both zero; thus, the equation simplifies to

$$m_A v_{A_2} = -m_B v_{B_2}$$

or

$$\frac{v_{A_2}}{v_{B_2}} = \frac{m_B}{m_A}$$

The previous two equations show both the obvious—that the gliders will move away from each other in different directions because of the minus sign—and the slightly less obvious—that the speeds at which the two gliders separate are in inverse proportion to their masses. If cart A is twice as massive as cart B, it will move off with only half the speed of B. Coincidentally, this experiment provides a way to compare masses that is completely independent of gravity and weight.

EXAMPLE 6.4

A 5-g bullet is fired from a 500-g pistol with a muzzle velocity of 300 m/s. Calculate the pistol's recoil velocity.

Solution

Both pistol and bullet start off with zero velocity. Applying the principle of conservation of momentum, we know that the change of the pistol's momentum ($m_p v_p$) must be equal to but opposite in direction to the change in momentum of the bullet ($m_b v_b$). That is,

$$m_p v_p = -m_b v_b$$

Substituting all the known values, we have the relationship

$$v_p = -\frac{m_b v_b}{m_p} = -\frac{(5 \text{ g})(300 \times 10^2 \text{ cm/s})}{500 \text{ g}} = -300 \text{ cm/s}$$

Hence, the pistol moves at a speed of 3 m/s, which is one one-hundredth of the bullet's speed, and the pistol moves in the opposite direction to the bullet. The recoil velocity obviously accounts for the kick of any pistol, rifle, or cannon, and the negative sign in the expression indicates that the firearm moves in the opposite direction to the projectile it fires.

Recoil velocity accounts for the kick of any pistol, rifle, or cannon.

EXAMPLE 6.5

Consider the collision of two very sticky oranges, each with a mass of 0.50 kg. The oranges are moving toward each other in a straight line, one at 4 m/s, the other at 2 m/s. Assuming that the oranges stick together after the collision, compute the final velocity of the sticky messy mass.

Solution

We apply conservation of momentum to this collision. The sum of the momenta before the collision equals the momentum after the collision. Let us take the direction of the faster moving orange as positive. Hence,

$$m_A \mathbf{v}_A + m_B \mathbf{v}_B = (m_A + m_B)\mathbf{v}$$
$$(0.5 \text{ kg})(4 \text{ m/s}) - (0.5 \text{ kg})(2 \text{ m/s}) = (0.5 + 0.5) \text{ kg } (v)$$
$$(2 - 1) \text{ kg} \cdot \text{m/s} = (1 \text{ kg})(v)$$
$$v = 1 \text{ m/s}$$

in the direction in which the faster moving orange was initially moving.

One important point to remember when applying the principle of conservation of momentum is that both momenta and velocities are vector quantities. In most cases, including Examples 6.3 to 6.5, the momenta are in a straight line. However, one can expect to encounter more complex problems involving collisions that require one to sum the vectors.

6.4 COMING TOGETHER

The major topic of Chap. 5 was the conservation of energy; the major topic of this chapter is conservation of momentum. The two principles can be linked usefully in situations where objects collide via what are known as elastic forces. That means that they collide without losing mechanical energy to alternate forms such as heat and sound.

A typical example of elastic collision is the impact of a steel pendulum ball on an identical ball, as shown in Fig. 6.5. A similar situation occurs on an air track in which a glider collides with another glider of identical mass that starts off at rest before the collision. In both cases the moving object comes to a standstill after the impact, and the object at rest moves off with a speed identical to the original speed of the object that hit it. To understand why this happens, we must consider both energy and momentum.

Because the forces involved in the collision are horizontal, and the weights of the objects are supported in a way that effectively eliminates friction, we can neglect both the force of gravity and friction. Figure 6.6 illustrates the general situation in which ball B, whose mass is m, moves toward the stationary ball A, of identical mass, with a speed v_{B_i}. After

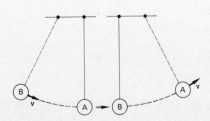

Figure 6.5 Collision of steel pendulum balls.

Reaching for the Moon

Like many scientific prophets before him, Professor Robert H. Goddard was without honor in his own country. The Clark University physicist's dream of sending rockets to the moon was scoffed at by his contemporaries in American laboratories during the 1920s, on what the critics thought to be strictly scientific grounds. They argued that Goddard's rockets would cease to move once they reached the vacuum of space because at that point the gases they emitted would have no atmosphere on which to push. Goddard answered these objections in the usual scientific way: by experimental demonstration.

The January 1925 issue of *Scientific American* reported that the rocketeer had fired a blank cartridge from a revolver inside a vacuum:

> Although the hot gases of the burning powder had no air to impinge against, the revolver recoiled on its prepared axis just as if it had not been fired in vacuum. Further experiments along this line were made with a small model of the rocket which it is planned to send to the moon. Professor Goddard never expected his rocket to get its kick from the impact of its escaping gases against air. On the contrary, its propulsion depends on Newton's Third Law, which states that "To every action there is an equal and opposite reaction."

Despite such demonstrations, Goddard's genius was never fully recognized in the United States. The Germans latched onto his ideas in their development of the fearsome V2 rockets of World War II and exported Goddard's expertise back to the United States in the minds of such technological leaders of the American space program as Wernher von Braun.

Robert Goddard: ahead of his time. (*Wide World*)

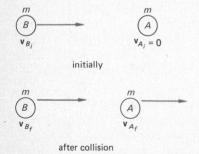

Figure 6.6 Elastic collision of two objects.

the impact, let us imagine that the two balls move off with speeds v_{B_f} and v_{A_f}.

Applying conservation of momentum, we have at any instant

$$m\mathbf{v}_{B_i} = m\mathbf{v}_{B_f} + m\mathbf{v}_{A_f}$$

Since the objects are of identical mass, the velocities are related:

$$\mathbf{v}_{B_i} = \mathbf{v}_{B_f} + \mathbf{v}_{A_f}$$

There are, of course, a large number of solutions for $\mathbf{v}_{B_f}$ and $\mathbf{v}_{A_f}$ in this equation. However, we reduce that uncertainty by specifying that the collision is "perfectly elastic." This means, as just noted, that *mechanical energy* is conserved in the interaction. This condition implies that the total kinetic energy of the system is the same just before and just after the collision; none is lost in the form of heat, sound, or deformation of one or both balls. Applying the equation for conservation of energy, therefore, we have

$$\tfrac{1}{2}mv_{B_i}^2 = \tfrac{1}{2}mv_{B_f}^2 + \tfrac{1}{2}mv_{A_f}^2$$

Since the masses are identical, the speeds are related:

$$v_{B_i}^2 = v_{B_f}^2 + v_{A_f}^2$$

If we solve this equation simultaneously with the previous equation obtained from conservation of momentum, then we can obtain the final speed of ball A in terms of the initial speed of ball B:

$$v_{B_f} = v_{B_i} - v_{A_f}$$
$$v_{B_i}^2 = v_{B_i}^2 - 2v_{B_i}v_{A_f} + v_{A_f}^2 + v_{A_f}^2$$
$$2v_{A_f}^2 - 2v_{B_i}v_{A_f} = 0$$

which yields two solutions:

$$v_{A_f} = v_{B_i} \quad \text{or} \quad v_{A_f} = 0$$

The only realistic solution of this is the one we noted originally: The first ball stops dead on impact, and the second shoots off with the same speed that the first ball possessed originally. The other solution of the equation, with $v_{A_f} = 0$ and $v_{B_i} = v_{B_f}$, simply means that no collision occurs at all.

This situation is artificial, insofar as kinetic energy is never conserved in practical situations of this kind. In real-life collisions, which are termed inelastic, where mechanical energy is transformed to sound or heat, both balls move after the collision. One important point to remember is that whereas kinetic energy is conserved only in perfectly elastic collisions, momentum is always conserved, whatever the elastic or inelastic nature of the impact.

Whereas kinetic energy is conserved only in perfectly elastic collisions, momentum is always conserved, whatever the elastic or inelastic nature of the impact.

EXAMPLE 6.6

An 8-kg ball moving at a speed of 16 m/s strikes head-on a 4-kg ball that is initially at rest. After the collision the larger ball is still moving in the same direction but at a speed of 8 m/s. Find the velocity of the 4-kg ball after impact, and compare the kinetic energies before and after the collision.

Solution

Momentum is always conserved in such a collision since no external forces act on the balls. After applying conservation of momentum we can compute the kinetic energies before and after the collision to determine whether or not kinetic energy is conserved. From conservation of momentum

$$m_8\mathbf{v}_8 + m_4\mathbf{v}_4 = m_8\mathbf{v}_{8f} + m_4\mathbf{v}_{4f}$$
$$8 \text{ kg} \times 16 \text{ m/s} + 0 = 8 \text{ kg} \times 8 \text{ m/s} + 4 \text{ kg } v_{4f}$$

$$4 \text{ kg } v_{4f} = (8 \times 16 - 8 \times 8) \text{ kg} \cdot \text{m/s} = 64 \text{ kg} \cdot \text{m/s}$$

$$v_{4f} = \frac{64 \text{ kg} \cdot \text{m/s}}{4 \text{ kg}} = 16 \text{ m/s}$$

After the collision the smaller ball moves off in the same direction as the larger ball but at twice the speed.

To compute the kinetic energy, one evaluates $\frac{1}{2}mv^2$ for both balls before and after the collision.

Before the collision:

E_k of m_8 ball $= \frac{1}{2}(8 \text{ kg})(16 \text{ m/s})^2 = 1024 \text{ J}$
E_k of m_4 ball $= 0$, since $v_4 = 0$

After the collision:

E_k of m_8 ball $= \frac{1}{2}(8 \text{ kg})(8 \text{ m/s})^2 = 256 \text{ J}$
E_k of m_4 ball $= \frac{1}{2}(4 \text{ kg})(16 \text{ m/s})^2 = 512 \text{ J}$
Total E_k after the collision $= 768 \text{ J}$
Loss of E_k during the collision $= 1024 \text{ J} - 768 \text{ J} = 256 \text{ J}$

Therefore $\frac{256}{1024} = 0.25$ or 25 percent of the kinetic energy is lost, mostly in the form of heat, during the collision. This is an inelastic collision.

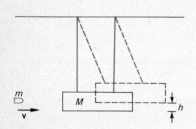

Figure 6.7 The ballistic pendulum.

One particular class of inelastic collision occurs when the two balls stick together on impact and move off in tandem. Such impacts are known as *completely* inelastic collisions. A practical example of completely inelastic collisions occurs in the ballistic pendulum. This is a device designed to measure the speed of a bullet. It may consist of a large wooden block, possibly filled with sand, that is suspended by two ropes 6 ft or more in length. A bullet is fired into the block and lodges there, creating a completely inelastic collision. As a result of the impact, the block rises by a height related to the speed of the bullet, as shown in Fig. 6.7.

EXAMPLE 6.7

If a bullet of mass 10 g strikes a ballistic pendulum of mass 2 kg and causes the pendulum to rise vertically 8 cm, what was the bullet's speed?

Solution

We know that momentum is conserved *during* the collision and mechanical energy is conserved *after* the collision as the pendulum swings.

During the collision, the magnitude of the initial momentum of the bullet mv is totally transformed to momentum of the bullet and block combined. If they move together with a speed V after the collision, then, from conservation of momentum, we can write a scalar equation

$$mv = (m + M)V$$

Therefore,

$$v = \left(\frac{m + M}{m}\right)V$$

Immediately after the collision, the kinetic energy of the block and bullet is given by

$$E_k = \tfrac{1}{2}(m + M)V^2$$

The pendulum now swings to the right, rising a height h. During this movement, the kinetic energy is transformed to gravitational potential energy. Hence,

$$\tfrac{1}{2}(m + M)V^2 = (m + M)gh$$

From the energy relation, we can obtain V, the speed of the block and bullet,

$$V = \sqrt{2gh}$$

Substituting this value of V into the equation for the speed of the bullet that we obtained from conservation of momentum,

$$v = \left(\frac{m + M}{m}\right)V = \left(\frac{m + M}{m}\right)\sqrt{2gh}$$

The speed of the bullet can now be calculated from the known masses $m = 10$ g, or 0.01 kg, $M = 2000$ g, or 2 kg, and the rise of the pendulum, $h = 8$ cm, or 0.08 m:

$$v = \left(\frac{0.01 \text{ kg} + 2.0 \text{ kg}}{0.01 \text{ kg}}\right)\sqrt{2(9.8 \text{ m/s}^2) \times 0.08 \text{ m}}$$

$$v = 201\sqrt{1.57 \text{ m}^2/\text{s}^2} = 2.52 \times 10^2 \text{ m/s}$$

It is instructive to compare the kinetic energy of the bullet prior to the collision, $\tfrac{1}{2}mv^2$, with the kinetic energy of the block and the bullet just after the collision. For the values given above,

$$\tfrac{1}{2}mv^2 = (\tfrac{1}{2})(0.01 \text{ kg})(2.52 \times 10^2 \text{ m/s})^2 = 3.17 \times 10^2 \text{ J}$$

$$\tfrac{1}{2}(m + M)V^2 = (m + M)gh = (0.01 \text{ kg} + 2 \text{ kg})(9.8 \text{ m/s}^2)(0.08 \text{ m})$$

$$\tfrac{1}{2}(m + M)V^2 = 1.57 \text{ J}$$

Note that over 99 percent of the energy is lost in the collision to heat and sound because the ratio of the mechanical energy after the bullet enters the block to the kinetic energy of the bullet before it strikes the block is $(1.57 \text{ J})/(3.17 \times 10^2 \text{ J}) = 4.9 \times 10^{-3}$, or 0.49 percent. The collision certainly is *inelastic*.

As another example of a collision, consider the impact of a cue ball C with an eight ball on a flat pool table. The motion is obviously restricted to the horizontal plane. If we neglect the ball's rotational motion (a fairly big if because such motions actually play vital roles in the strategy of real pool) and also assume that frictional forces are negligible, we can readily analyze the collision. Figure 6.8 shows a general impact, in which the cue ball strikes the stationary eight ball and both move off at different angles to the cue ball's original direction of approach. Applying the conservation of momentum, we obtain a vector equation:

$$m_C \mathbf{v}_C = m_8 \mathbf{u}_8 + m_C \mathbf{u}_C$$

Here, the term on the left of the equation refers to the momentum of the system before the impact and the terms on the right to the situation after the impact.

We can simplify the overall analysis by resolving the vectors into components in the x and y directions, parallel to and perpendicular to the cue ball's original motion, and obtain scalar equations:

For the x direction:

$$m_C v_C = m_C u_C \cos \theta + m_8 u_8 \cos \phi$$

For the y direction:

$$0 = m_C u_C \sin \theta - m_8 u_8 \sin \phi$$

Since the collision is perfectly elastic, the kinetic energy is also con-

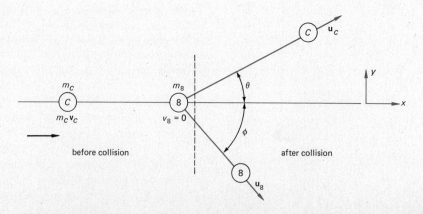

Figure 6.8 Collision of two pool balls.

served. We, therefore, have the additional condition that

$$\tfrac{1}{2}m_C v_C^2 = \tfrac{1}{2}m_C u_C^2 + \tfrac{1}{2}m_8 u_8^2$$

Taken together, these three equations provide the basic relationships between the final velocities of the two balls, the angles at which they part after the collision, and their masses. With the three equations, three unknowns can be determined; hence, all but three of the quantities must be known or determined from measurement. Of course, good pool players know such factors from experience.

The problem of the "scattering" of one ball off another is extremely important in other areas totally unrelated to pool—the determination of atomic and nuclear structures. By observing momentum changes resulting from collisions between atoms and subatomic particles, nuclear physicists can obtain information on the interior structure of regions too small to be seen in even the most powerful electron microscopes. In fact, the original nuclear model of the atom resulted from studies of this type in 1908 by British scientist Ernest Rutherford, who bombarded thin gold foils with subatomic entities known as alpha (α) particles, which are doubly ionized helium atoms—helium nuclei. Similar techniques are still used today in massive atom smashers by physicists attempting to probe the complex world of elementary particles and thus understand the true nature of matter.

SUMMARY

1. The momentum of any object is the product of the object's mass and velocity.
2. The impulse of a force is the product of the force and the time during which it acts.
3. The change of momentum produced by any force on an object equals the impulse of that force.
4. In elastic collisions, the total kinetic energy is conserved. In inelastic collisions, at least some of the kinetic energy is lost as other forms of energy, such as heat and sound. Momentum is conserved in all types of collisions, however.

QUESTIONS

1. Why is it important to extend the time of collision to minimize injury in a crash? Why, similarly, does a golfer or batter in baseball wish to extend the time of contact with the ball?
2. What is the difference between an elastic and an inelastic collision as far as conservation of momentum and energy is concerned?
3. When a firecracker explodes after having been lit and thrown into the air, how is momentum conserved? Is energy conserved? Explain.
4. How does the momentum of the earth in its orbit about the sun vary with time during the year?
5. As an exercise in analysis of hard-sphere collisions, play a little pool or billiards. Look at the collisions from the point of view of momentum. Before each collision, note the direction of the cue ball and use that as a reference to see if momentum is conserved.

PROBLEMS

Momentum

1. A bowling ball has a mass of 7.3 kg and is moving at a velocity of 3 m/s down the alley. Compute its momentum.
2. An airplane has a weight of 12 000 lb and is moving with a speed of 300 mi/h in a westerly direction. What is the momentum of the airplane?
3. What is the momentum of a 3-g bullet moving horizontally at 500 m/s in a northeasterly direction?

4. A 3-g bullet is moving at 500 m/s in a direction 60° away from the horizontal toward an unsuspecting sea duck. What are the components of the momentum of the bullet in a horizontal plane and in a vertical direction?

5. Compute the momentum of an oxygen molecule whose mass is 5.32×10^{-23} g if it is moving at a normal speed for a gas molecule of 500 m/s (over 1000 mi/h).

6. If an oxygen molecule of mass 5.32×10^{-23} g and $v = 500$ m/s hits a wall perpendicularly and bounces off at the same speed, what is its change in momentum? What is the change in momentum of the wall?

7. If our oxygen molecule of mass 5.32×10^{-23} g and $v = 500$ m/s hits a wall at an angle of 20° with the perpendicular to the wall, what is the momentum it transfers to the wall if it "reflects" off the wall at an angle also of 20° and with the same speed?

8. A raindrop of mass 3×10^{-4} g hits a window of a car at rest at an angle of 80° with the perpendicular to the window. The drop is moving at 5 m/s and stops when it hits the window. What is its change in momentum?

9. A pitcher delivers a 0.15-kg ball to the batter at a speed of 25 m/s in a horizontal direction. The batter fouls the ball off, and it leaves the bat at a speed of 10 m/s, moving directly upward. Compute the net change in momentum of the ball.

Impulse-Momentum

10. A 0.2-kg baseball arrives at the batter with a speed of 25 m/s, and, after being hit in the opposite direction, its speed is 40 m/s. (*a*) Compute the average force on the bat if the ball and bat were in contact for 0.002 s. (*b*) What is the average acceleration of the baseball? (*c*) Express the acceleration as a multiple of the acceleration of gravity g.

11. A child climbing a tree has a slight problem, and her watch gets caught by a branch, torn off, and falls to the ground 6 m below. If the watch has a mass of 25 g and it takes 0.005 s for the ground to stop the watch, compute the force on the watch. If the watch had hit some bushes, how might this "softening" of the shock have lowered the force on the watch?

12. A small child drops a 50-g tennis ball from a second-story apartment window 8 m above the ground. His friend below misses the ball and it hits the ground, rebounding and rising to a height one-half of that from which it was dropped. Assume free fall motion with no air resistance and compute the change in momentum of the ball.

13. Mr. Goodkid helps maintain an outdoor skating rink each cold winter night by flooding the rink using a fire hose. If he has a mass of 80 kg and the hose sends out 800 kg of water each minute at a speed of 15 m/s, (*a*) what is the force exerted on Mr. Goodkid and (*b*) what effect does it have on him as he tries to stand on the wet slippery ice?

Conservation of Momentum

14. Cart A of mass 300 g and cart B of mass 1500 g are tied together with a compressed spring between them on an air track, as shown in Fig. 6.4. One second after the string is cut, cart A is 120 cm from its starting point. At that instant, where is cart B?

15. A cannon has a weight of 12 000 lb. If an 80-lb projectile leaves the cannon horizontally with a muzzle velocity of 1500 ft/s, compute the velocity of recoil of the cannon.

16. A motorboat with contents has a mass of 900 kg, and a rowboat with contents has a mass of 300 kg. To transfer a passenger, a rope is thrown from one stationary boat to the other. If pulling on the rope results in a velocity of the rowboat of 1.2 m/s, compute the speed of the motor boat. Neglect the resistance of the water.

17. Two spacemen are both floating with zero velocity in a gravitation-free region of space. Spaceman A has a mass of 120 kg, and spaceman B has a mass of 90 kg. A pushes B away from him. If B's speed is 0.5 m/s, with what *velocity* does A recoil?

18. A boy skater and a girl skater are at rest in the center of a large ice rink. They push away from each other. How long does it take the boy to reach the edge of the ice if it takes 10 s for the girl to reach the opposite edge of the rink? The girl's mass is 45 kg, the boy's is 90 kg, and we assume no friction either by the air or ice.

19. A spaceman on a space walk is a golf nut and has a bucket of golf balls. If each ball has a mass of 40 g and the spaceman throws the balls in the same direction at a speed of 2.5 m/s and at a rate of one each 10 s, (*a*) what is his total change in momentum during 5 min of this action? The spaceman has a mass of 80 kg. Neglect the change in his total mass due to the loss of each ball as he throws it. (*b*) Compute his speed after 5 min.

20. Two girls in a rowboat decide to take a swim. One girl has a mass of 40 kg and the other a mass of 60 kg. The girls simultaneously dive off the boat, the larger girl diving from the stern, the other girl from the bow. If they leave the boat with a velocity of 5 m/s at an angle of 30° upward from the horizontal, in which direction and how fast does the boat move (neglect resistance)?

21. An astronaut of mass 100 kg is

separated from her spaceship and is at rest with respect to the spaceship 45 m away from her ship. Her oxygen tank contains 0.5 kg of gas, which can escape the valve at 50 m/s. If she releases one-half of the gas in a short burst at this speed in an appropriate direction, how long will it take for her to reach the spaceship?

22. Two sticky oranges, each of mass 0.25 kg, approach each other at right angles. One is moving at 1.5 m/s, the other at 2.0 m/s. If the oranges collide in a completely inelastic collision, what will be their final velocity after collision?

23. A bullet having a mass of 20 g is fired into a block of wood that has a mass of 1000 g. As a result, the block and the imbedded bullet start moving with a speed of 10.0 cm/s. Compute the speed of the bullet as it strikes the block. No frictional force acts on the block to oppose its motion.

24. A 200-lb hockey center moving at 7 m/s runs into the goalie, and the two players move together after the collision. What is their speed after collision if the goalie weighs 220 lb?

25. A 3600-kg railroad car moving at 2.5 m/s runs head-on into a 3200-kg railroad car moving in the opposite direction at 1.5 m/s. If the two cars couple together, what is the resulting *velocity* of the two cars? Is any mechanical energy lost in this collision?

26. A 20-kg wagon is moving with a velocity of 2 m/s. A boy of mass 40 kg jumps into the wagon. What is the resulting velocity of the boy and wagon together? Consider the circumstance (*a*) where the boy is initially at rest and also (*b*) where he moves initially at the same velocity as the wagon.

27. An explosion blows a rock into three parts. Two pieces go off at right angles to one another, a 1-kg piece at 12 m/s and a 2.0-kg piece at 8 m/s. The third piece of unknown mass goes off at 40 m/s. Determine the third mass and the direction it moves.

Conservation of Momentum and Energy

28. A 20-g bullet is fired horizontally from a 4-kg rifle with a 700-m/s muzzle velocity. What is the recoil velocity of the rifle? Compare the kinetic energy of the rifle and the bullet.

29. A child throws downward to his playmate 10 m below a 0.2-kg baseball from a second-story window at a speed of 6 m/s. The playmate, applying a constant force, catches the ball in his baseball mitt. In the process, the mitt moves 10 cm. (*a*) Compute the work done in catching the ball, (*b*) the change in momentum, the (*c*) impulse delivered to the mitt, and (*d*) the average force on the mitt.

30. A 3-ton truck just barely rolling at 3 mi/h along level ground runs into the back end of a larger truck weighing 5 tons. If the two trucks couple together, how fast do they move after the collision? Is there a change in kinetic energy in this collision?

31. An empty freight car with a mass of 1000 kg rolls at 3 m/s along a level track and collides with a loaded freight car with a mass of 2000 kg that is moving toward the first car at 1 m/s. These cars collide and couple together. (*a*) In what direction do they move after the collision? (*b*) What is the velocity after the collision? (*c*) What can be said concerning the energy before and after this collision?

32. A ball of mass $m_1 = 2$ kg is moving at a speed of 5 m/s toward a ball of mass $m_2 = 6$ kg, which is at rest. The mass m_1 collides with the mass m_2. After the collision, the smaller ball m_1 moves with a speed of 2 m/s along the same line. In which direction does this small ball (m_1) move after the collision? What is the speed of the larger ball after the collision? Is kinetic energy conserved during the collision (prove your answer)?

33. When a bullet of mass 10 g fired horizontally from a 4-kg rifle strikes a ballistic pendulum of mass 2 kg, the pendulum is observed to rise a vertical height of 10 cm. (See Fig. 6.7.) The bullet remains imbedded in the pendulum. Calculate the (*a*) velocity of the bullet, (*b*) recoil velocity of the rifle, (*c*) average force acting on the bullet to stop the bullet in the pendulum. The bullet travels a distance of 5 cm in the pendulum.

34. A rifle bullet of mass 0.01 kg is fired horizontally with a velocity of 800 m/s into a ballistic pendulum. The pendulum has a mass of 5 kg and is suspended from a cord 1 m long. The bullet is fired into the pendulum and is imbedded in the pendulum, causing the pendulum to swing. (*a*) Compute the recoil velocity if the rifle had a mass of 12 kg. (*b*) Compute the vertical height to which the pendulum rises. (*c*) Compute the change in mechanical energy of the system during the process of the bullet hitting the pendulum.

35. A block travels down a curved, frictionless incline, then over another frictionless hill where it collides with a stationary block twice as massive (see Fig. 35, p. 170). The first block continues with a speed one-third its incoming speed. The second, more massive block moves to the right over a rough surface with a coefficient of friction equal to 0.5. (*a*) What is the speed of the first block at points *B*, *C*, and *D* (just before it hits the 4-kg block)? (*b*) What is the kinetic energy of the 4-kg block just after the collision? (*c*) How far will the large block move toward point *E* at the base of a wall? (*d*) How much kinetic energy

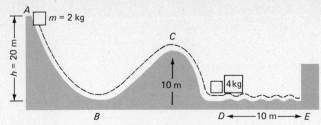

Figure 35

was lost in the collision at D? (e) How much momentum was lost in the collision at D?

36. Tarzan (mass of 80 kg) swinging on a vine through the jungle leaves a branch 10 m above the ground, and at the low point of his swing hits a bad guy (mass 120 kg) who is standing at rest on the ground. If the two move together, (a) what is their final speed? If Tarzan had hit the bad guy and stopped short, (b) what would have been the bad guy's final speed?

37. A neutron runs into a proton initially at rest, making a direct hit. The neutron and proton both have essentially the same mass: 1.67×10^{-24} g. If the collision is elastic and the neutron's initial speed is v_0, with (a) what speed will the proton leave? (b) What happens to the neutron? (c) If you replaced the protron by a nucleus 10 times more massive, what would happen to the neutron in a collision?

38. A 100-g ball makes an elastic head-on collision with a ball of unknown mass that is initially at rest. If the 100-g ball rebounds at one-fourth its original speed, what is the unknown mass?

39. A proton with a mass of 1 on the atomic mass unit (u) scale collides head-on with a helium atom of mass 4 u. If the initial speed of the proton is v_i and the collision is elastic, (a) what is the forward speed of the helium atom and (b) what fraction of the E_k of the proton is transferred to the helium atom?

Special Topic
The Impulsive Athlete

In common with the rest of physics, momentum is more than a concept to be studied diligently by professors in stuffy laboratories. An understanding of the impulse-momentum theorem and the conservation of momentum, for example, has almost as much value for the student on an athletic scholarship as it does for a science major. A few minutes of meditation on mv and $F\,\Delta t$ before a big game might give the sports competitor that small but important edge over less intellectually minded opponents.

The most obvious patch of green on which the laws of momentum play a vital role is the pool table. Any would-be Minnesota Fats should know by experience, even if he or she cannot work out the equations, that the combined momentum of the cue ball and the ball it hits will be identical to the momentum of the cue ball alone before the impact. Furthermore, as we shall see in Chap. 7, conservation of momentum applies to all axes and to the angular rotation or spin of the balls. If the target ball moves off after impact on one side of the cue ball's original line of travel, the cue ball will move away from the collision on the other side of that line. The laws of momentum simply do not allow pool ingenues to achieve some of the more ambitious placements they try.

Those same laws do allow the occasional small football player to star in the major leagues, however. A typical example is running back Mack Herron. In the fall of 1974, when he joined the hapless New England Patriots, Herron had little going for him. Not only had he fallen out with his college coach and the Canadian Football League Club that he joined after college, he also stood a mere 5 ft 5 in. and weighed just 170 lb—hardly the perfect physique

Minnesota Fats: a natural understanding of momentum. (*Wide World*)

for blasting gaps between 260-lb linemen and running through the crunching tackles of 220-lb linebackers. Yet Herron made the grade and helped the Patriots to a respectable season. The reason: his intuitive understanding of the principles of momentum.

Since momentum is the product of mass and velocity, Herron made up for his relative lack of mass with sheer speed. By building up enough speed fast enough before he reached the line of scrimmage, Herron and other small players could and can withstand the body checks and tackles of players far heavier than they.

At a less physical level, tennis players use the impulse-momentum theorem whenever they play a shot. The conventional ground shot in tennis is designed to give the ball as much momentum as possible. The player uses a flowing motion with a long follow-through that allows the racket to stay in contact with the ball as long as possible. The longer the contact, the greater impulse given to the ball, and hence the greater the ball's change in momentum. When a player wants to play a stop shot that just drops the ball over the net, she or he stabs at the ball with the racket. This action reduces the time of contact between racket and ball to a minimum. The ball receives a relatively small impulse and thus takes off from the racket with little momentum. If the shot is played correctly, the ball has just enough momentum to make it over the net before it "dies," leaving the player's opponent stranded near the service line.

Combination of the impulse-momentum theorem and conservation of momentum is also used to ensure the well-being of athletes. When high jumpers and pole-vaulters come down to earth after

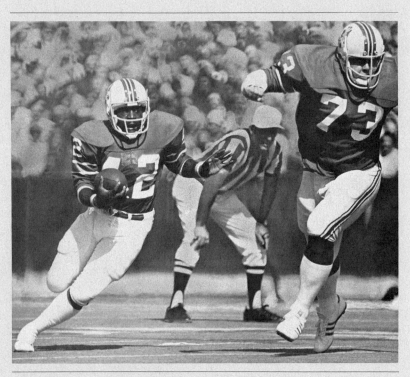

Mack Herron: good despite the lack of mass. (*Wide World*)

soaring to great heights, their momentum is suddenly reduced to zero. (Conservation of momentum is obeyed, because the earth itself moves imperceptibly as a result of the collision.) If the reduction of the leapers' momentum occurs in too short a time, that is, the term Δt in the expression for impulse $F \Delta t$ is very small, the retarding force on the body may be great enough to damage ligaments, tendons, and even bones. Thus, the jumpers and vaulters land on materials with a certain amount of give, such as foam rubber. By stretching out the time of impact, these materials reduce the amount of retarding force the athlete must undergo. Fitness buffs performing squats and similar jumping exercises at home automatically take similar precautions to reduce bodily damage. By flexing their knees on landing, they increase the time of impact with the floor and thus reduce the amount of force on their bodies.

7 Systems of Many Particles

7.1 THE TWO CENTERS
Locating the Center of Gravity

7.2 TAU IS FOR TORQUE

7.3 GETTING AT THE CENTER OF GRAVITY

SHORT SUBJECT: Vector Product

7.4 MOVING AROUND

SHORT SUBJECT: Kinematical Equations for Rotational Motion

7.5 ROTATIONAL DYNAMICS

7.6 THE GYROSCOPE

SHORT SUBJECT: Come Fly with Me
Summary
Questions
Problems

SPECIAL TOPIC: A Revolutionary Means of Staying on the Straight and Narrow

A predominant theme of this book so far has been the need to approximate to real-life situations in order to treat them mathematically with the techniques of physics at our disposal. This type of simplification is essential for students because we do not wish to muddy the clear waters of underlying physical principles with the dirt of extraneous factors such as air resistance and friction. In reality, this type of approach is not restricted to a limited category labeled "problems for students," for most theoretical physics is conducted by approximation and estimate. Only by improving their estimates and approximations do working physicists slowly approach a more exact description and a more complete understanding of everyday experience.

Newton's laws, for example, accurate as they are in our everyday experience on earth, turn out to be no more than fairly good approximations to real life when they are applied to cosmic phenomena on the scale of huge galaxies or to the ultrasmall world of subatomic particles. In these cases Einstein's theory of relativity and quantum theory prove to be more accurate approximations to the real situations. A number of scientists believe that even relativity theory has its limitations in some circumstances; they are, therefore, trying to refine the theory to take account of such possibilities. What this means is that relativity is a better approximation than the collection of Newton's laws to the observed facts about the universe as we understand them now.

Advances in physics, then, often come about as a result of improved approximations. We are now in a position to make such an advance by moving from our initial set of idealized approximations to a set that is rather less gross. In most of the previous discussions we dealt with objects as if they have no physical size. This is a highly dubious assumption tenable for only the simplest of mechanical problems. As it happens, it is a simple matter to extend our understanding of the mechanical world to take into account the finite size of objects.

One example of the deficiencies of the "no size" approach is to consider a book resting on a table near the edge, as shown in Fig. 7.1.

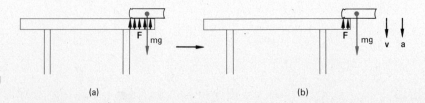

Figure 7.1 A book balanced on the edge of a table. If it is moved beyond the edge, it rotates and falls.

Owing to its weight, the book presses down on the table and the table pushes up on the book. Since the book is in equilibrium, the net force on the book is zero. Hence, the table pushes up on the book with a force equal to the weight of the book. Now imagine that one pushes the book horizontally across the edge of the table, as illustrated. The book still weighs the same, and the table continues to push up on it with the same force. Yet, sooner or later the book falls.

Why does it fall? Clearly, the points at which the two forces, the weight and the upward push of the table, are applied play an important role in the action, as shown in Figs. 7.1(a), (b). In these figures the weight of the book is indicated as acting at a single point, exactly in the book's center—a perfectly valid assumption, as it turns out. While that central point remains above the surface of the table, the upward force of the table on the book acts through that point, balancing the book's weight. As soon as the center of the book passes over the table's edge, however, this particular balancing act is impossible. The book's weight and the upward force of the table now act at different points on the book. Instead of canceling each other, they act in concert to rotate the book until the table no longer exerts an upward force and the force of gravity sends the book down to the ground, still rotating. Investigation of rotation and its cause will allow us to complete our look at motion, at least as far as it applies to objects that do not deform when external forces act on them.

7.1 THE TWO CENTERS

We noted, when studying the forces acting on the book at the table's edge, that the weight of the tome could be regarded as acting just as if it were all concentrated at the book's geometric center, if we assume the mass of the book is uniformly distributed. For a rough proof of this assertion, try to balance a meter rule, held horizontally, on one finger or on a knife edge. If you support the rule off center, it will rotate and fall. Only if your finger or the knife edge runs exactly along the 50-cm line will the stick stay in balance, albeit a somewhat unstable balance. If you were to extend the experiment and try to support the rule on, say, the point of a thick pin, you could only balance the stick if the pin were both on the 50-cm line and exactly halfway across the ruler's width. This is the only point at which the weight of the ruler is exactly balanced by the upward force of the pin; the downward force of the weight and the opposite upward force of the pin coincide at this point alone, thus negating each other.

If you completed the experiment by finding the position on the edge of the ruler for which the same condition of balance applied, you would discover that the critical point of balance in three dimensions is at the geometric center of the rule. This point is known as the ruler's center of gravity, and in a constant gravitational field this is also the center of mass.

The center of mass is identical with the geometric center in this case, but it can be calculated by simple mathematics only in symmetrical

The center of mass of a table with sturdy legs is located in the space beneath the tabletop.

objects, such as uniform books and rulers. For irregularly shaped objects such as buildings, rocks, and plants, the center of mass is well-nigh impossible to calculate geometrically. In some cases it may not even be located inside the object at all. The center of mass of a table with sturdy legs, for example, is located in the space beneath the tabletop.

The importance of the concept of the center of mass is that if a net external force is acting on the object, then the **center of mass** of the extended object moves in the same way that a particle of identical mass would move if the same net force were acting on it. Hence, when applying Newton's laws to an extended object, the center of mass is that point at which all the mass of the body could be considered concentrated. Newton's first law should then be stated as follows: "The center of mass of a body remains at rest or moves with uniform velocity in a straight line, unless a net external force acts on the body." And Newton's second law states that "the net external force acting on the object equals the product of the total mass of the object and the acceleration of the center of mass." The motion of the center of mass can be captured using flash photography, as Fig. 7.2 illustrates. The camera has caught the motion of a wrench sliding across a smooth, horizontal table in such a way that it continually rotates during its trip. All the individual points on the wrench revolve, except for the point marked with the dark spot in the series of photographs. This point represents the wrench's center of mass. It moves across the table in uniform motion in a straight line. If the wrench were thrown through the air, its center of mass would follow a simple parabolic path and the wrench would rotate about the center of mass.

The center of mass is often confused with a similar term: **center of gravity.** The two terms are so similar in many respects that one can use them interchangeably without introducing any wild inaccuracies into problems. However, there is an important difference between the two centers.

The center of gravity is the point associated with an object where all the weight could be concentrated when considering the motion of the object. If the object is in a completely uniform gravitational field, that

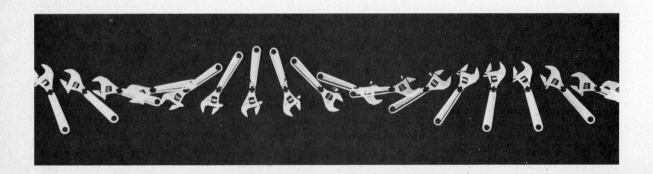

Figure 7.2 A multiple-flash photograph of a wrench sliding across a smooth, horizontal surface. The black cross marks the center of mass, which moves in just the same way a particle with all the mass of the wrench would. (From *PSSC Physics*, D. C. Heath & Company, Lexington MA, 1965)

is, *g* is constant, the center of gravity is the same point as the center of mass. In almost all real situations, however, the center of gravity is slightly displaced from the center of mass. We can understand just why this is so by considering a book standing upright on a bookshelf. Assuming that the book is uniformly distributed, its center of mass is at its geometric center. Its weight, however, is not uniformly distributed because the value of *g*, the gravitational pull of the earth, varies with height above the earth. Because the lower half of the book is closer to the earth than the upper half, it is pulled toward the earth with a slightly stronger force of attraction; hence, the lower half weighs more than the upper half. The book's center of gravity is thus very slightly below its center of mass.

In most cases, such as that just described, this discrepancy is so small that it can be neglected. Physicists can determine an object's center of mass by experimentally discovering its center of gravity. But there are some cases, such as large astronomic objects, in which the difference between the two centers is significantly large. The moon, for example, experiences the force of earth's gravitational attraction appreciably more on the side closer to the earth than on the opposite side because the near side is about 3000 mi closer to its parent planet. As a result, the moon's center of gravitation is closer to earth than is its center of mass.

Locating the Center of Gravity

Returning to the efforts to balance a meter rule, it is obvious in this case that we were really concerned with center of gravity rather than center of mass because we sought to balance the *weight* of the rule against the upward force of its support. Because the weight acts downward from the center of gravity, balance is only possible when the line along which the upward force acts also passes through the object's center of gravity. To determine just where an object's center of gravity is, then, we merely have to balance the object in three different orientations and find the meeting point of the three lines through which the upward forces act.

In the special case of thin, flat objects, which are essentially two-dimensional, we need only balance the objects in two orientations.

This is simple enough for objects whose centers of gravity are internal. But what of articles so strangely shaped that the center of gravity is outside the object? Here, we must try a different approach. The technique is to suspend the object in question from two or three different points (depending on whether the object is two- or three-dimensional). Figure 7.3(b) illustrates the method for a two-dimensional boomerang. In this example the line of action of the force holding up the suspended object, in both positions, passes through the object's center of gravity. All we have to do is find the intersection of these lines for the two different orientations. It is also possible to compute the centers of mass of objects given knowledge of the forces acting on them. To achieve this, however, we need to take into account the effects known as torque and rotation.

7.2 TAU IS FOR TORQUE

Getting back to the example that opened this chapter, that of the book toppling off the end of the table, we can now analyze why the fall

Figure 7.3 Two methods of locating the center of gravity.

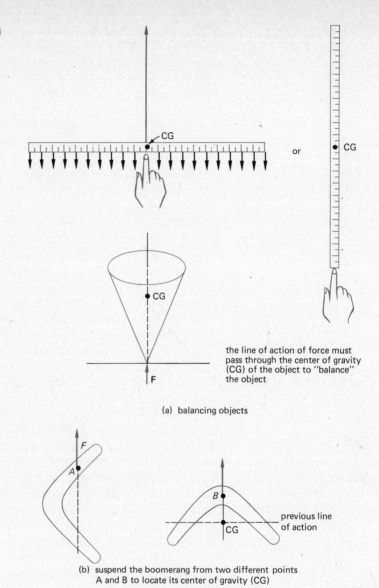

(a) balancing objects

(b) suspend the boomerang from two different points A and B to locate its center of gravity (CG)

occurs. The book drops once the downward force of its weight, acting through the book's center of gravity, is no longer balanced by the upward force of the table, because the center of gravity has moved past the table's edge. With the center of gravity hanging in space beyond the edge of the table, the two forces acting on the book can no longer balance each other. Instead, one force the weight, pulls the book downward at its center of gravity, and the other force, the upward force of the table, pushes the book up at the point immediately above the table's edge. The net results is that the two forces act together to rotate

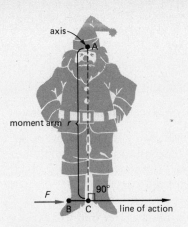

Figure 7.4 Christmas tree ornament undergoing torque.

the book as it falls to the ground. Technically, one says that the two forces create a net torque on the book. **Whereas a force is a "push" or "pull" on an object, a torque is a "twist" on an object.**

Strictly speaking, **a torque is the product of the magnitude of a force that tends to rotate an object and the perpendicular distance of the line through which the force acts from the axis about which the rotation occurs.** Figure 7.4, for example, shows an irregularly shaped, planar Christmas decoration pinned to a tree at point A. Pushing on the decoration at point B with a force F causes the decoration to rotate about an axis at A. To obtain the torque of this force, we must extend the line along which the force acts to point C, at which a perpendicular dropped from the line passes directly through A. This perpendicular distance from A to C is known as the moment arm of the force. If points A and C are separated by a distance r, then the torque of the force F is simply Fr. The Greek letter tau, τ, usually denotes torque. Thus, the equation is

$$\tau = Fr \tag{7.1}$$

The role of torque in rotational motion is analogous to that of a force in translational motion. Just as the action of a net force on an object causes the object's center of mass to accelerate, so a resultant torque causes the object to change its rotational state—to undergo *angular acceleration* around an appropriate axis. It is, therefore, convenient to assign a negative sign to counterclockwise torques and a positive sign to clockwise torques. This small elaboration allows physicists to treat torques according to the methods of simple algebra, just as they treat forces in translational motion along a straight line.

The analogy between forces and torques also extends to cases of equilibrium. In the same way that an object is undergoing uniform motion if the net force acting on it is zero, **any object is in rotational equilibrium when the sum of the torques acting upon it is zero.** Because both force and torque are vector quantities, these two conditions actually yield six separate equations for an object that is in complete equilibrium. The equations are

$$\Sigma F_x = 0 \quad \Sigma F_y = 0 \quad \text{and} \quad \Sigma F_z = 0$$

for the forces along each of the three mutually perpendicular axes, and

$$\Sigma \tau_x = 0 \quad \Sigma \tau_y = 0 \quad \text{and} \quad \Sigma \tau_z = 0 \tag{7.2}$$

or the torques around each of three mutually perpendicular axes. The sign Σ represents the sum of all the force components acting in a particular direction, or the sum of all the torque components taken about a particular axis. Note that one can compute the torque about any axis. The axis can be quite arbitrarily chosen either through the object under consideration or through some point outside it. When an object is not in equilibrium, the sums of the torques taken about different axes will not be equal. However, the sum of the torques taken

EXAMPLE 7.1

We wish to lift a 200-lb object with a lever, by applying a force in the way shown in Fig. 7.5(a). The lever is 12 ft long and rests on a fulcrum (or knife edge) support 4 ft from the end that holds the 200-lb load. What is the minimum force that will lift the load?

Solution

The force that will just lift the 200 lb is really the force that will exactly balance the lever. When the lever is exactly balanced, the sum of the forces in the vertical direction, ΣF_y, equals zero. Unfortunately, this equation is of no immediate use because the vertical forces include two unknowns: the downward force tending to lift the weight and the upward force of the fulcrum on the lever. We can avoid the latter force, however, if we consider the sum of the torques around the fulcrum when the lever is balanced and, for example, in a horizontal position. The torque equation is

$$\Sigma \tau = (8 \text{ ft})(F) - (4 \text{ ft})(200 \text{ lb}) = 0$$
$$8 \text{ ft } F = 4 \text{ ft}(200 \text{ lb})$$
$$F = 100 \text{ lb}$$

The choice of the fulcrum as the axis for the torque equation was entirely arbitrary and made only because it eliminated the second unknown force. We could have set up a torque equation around any other axis—either end of the lever, for example—but all such equations would have included both unknown forces, in the same way as the rejected summation of vertical forces.

In this problem the lever was horizontal, and hence the moment arms were simply the distances between the axis and the points along the lever at which the forces acted. If the lever were not horizontal, as in Fig. 7.5(b), the moment arms would be different and would require elementary trigonometry to be calculated.

The moment arm of any force is the perpendicular distance of the force's line of action from the axis around which the force tends to cause rotation. The choice of axis is of course arbitrary. In this case, the moment arm of the force F is $8 \cos \theta$, and that of the 200-lb weight will be $4 \cos \theta$. Using our condition for equilibrium that $\Sigma \tau$ is zero, we have

$$\Sigma \tau = (8 \cos \theta)(F) - (4 \cos \theta)(200 \text{ lb}) = 0$$

or

$$F = \frac{(4 \cos \theta)(200 \text{ lb})}{8 \cos \theta} = 100 \text{ lb}$$

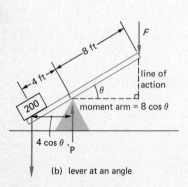

Figure 7.5 Using a lever.

(a) horizontal lever

(b) lever at an angle

This is exactly the same as the result for the horizontal lever. Plainly, the angle is unimportant in calculating the force, for although the moment arms change at different angles, they change by the same relative amount. Nevertheless, it is important to compute the moment arms correctly in cases that involve different shapes.

EXAMPLE 7.2

A hulking boy who weighs 120 lb wishes to balance on a seesaw with his small sister, who weighs only 60 lb. The sister sits on the very end of a 12-ft seesaw. To confuse matters, their 10-lb pet cat settles down on the same side as the sister, 2 ft from the center of the seesaw, as shown in Fig. 7.6. Where should the brother sit to ensure balance?

Solution

The sum of the forces in the vertical direction tells us only the force that the fulcrum exerts on the seesaw to balance the combined weight of the two children, the cat, and the plank. To work out the distances, we must once again take the sum of the torques, which is zero for the balanced situation. And by taking this sum around the fulcrum we avoid the need to introduce the weight of the plank into the equation because the fulcrum is at the seesaw's center of gravity. The torque equation is, therefore,

$$\Sigma \tau = (+120 \text{ lb})(x) - [(10 \text{ lb})(2 \text{ ft})] - [(60 \text{ lb})(6 \text{ ft})] = 0$$

or

$$(120 \text{ lb})(x) = (20 + 360) \text{ ft} \cdot \text{lb}$$

$$x = \tfrac{380}{120} \text{ ft} = 3.17 \text{ ft}$$

This is of course a rigorously calculated quantitative answer to a problem that most 5-yr-olds have worked out by experience on the school playground.

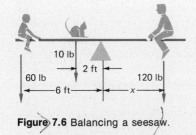

Figure 7.6 Balancing a seesaw.

7.3 GETTING AT THE CENTER OF GRAVITY

In our examples so far we have either neglected the weight of such objects as levers and seesaws or managed to avoid introducing it by mathematical manipulation. Sooner or later, however, we have to include such weights in our analysis of situations. As it happens, this is relatively simple once we know the location of the center of gravity of the object whose weight is involved. Because the center of gravity of any object is the point through which all the weight acts, the object's total weight taken at its center of gravity has the same torque around any axis as the sum of the torques of each individual element of mass in the object. We can use this fact to calculate the center of gravity of unusually shaped objects, as Examples 7.3 to 7.6 show.

Vector Product

In problems that involve many vectors and their mathematical relationships, it is often helpful to make use of the vector product, or "cross" product. This is particularly useful when dealing with vectors representing angular quantities such as torque, angular velocity, and angular momentum. The vector product is most useful in helping us find the direction of the new vector that results from a product of two other vectors. Mathematically, the vector product is the product of one vector and the projection of the other perpendicular to it.

The vector product of two vectors **A** and **B** is defined as a vector of magnitude $AB \sin \theta$, where θ is the angle between **A** and **B**. If **C** represents this product, then the vector or cross product is written $\mathbf{C} = \mathbf{A} \times \mathbf{B}$, resulting in a relationship, $C = AB \sin \theta$, between the magnitudes of the vectors.

In this chapter, we mention two such vector products, namely,

Torque $\boldsymbol{\tau} = \mathbf{r} \times \mathbf{F}$
Angular momentum $\mathbf{L} = \mathbf{r} \times \mathbf{p}$

or

$\mathbf{r} \times m\mathbf{v}$

In many instances the relationship between the vectors is simplified, so we do not need to use the power of the vector product. Whenever the

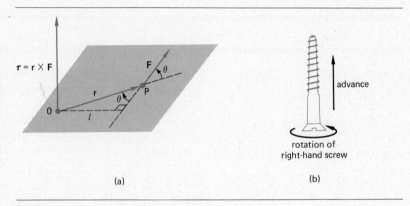

Figure A. The right-hand screw rule.

two vectors forming the cross product are at right angles, for example, the magnitude of the resulting vector is simply the product of the magnitudes of the two vectors, since $\sin 90° = 1$. Note that the cross product of two vectors which lie along the same line is zero, for in that case $\theta = 0°$ or $180°$, and $\sin \theta = 0$.

We can further examine the magnitude of the vector product by examining the definition of torque. In our approach to torques on page 179 we define a lever arm that is the perpendicular distance from the axis of rotation to the line of action of the force. This distance is, in fact, $r \sin \theta$, where r is the distance from the axis to the point of application of the force. Figure A shows the general case of a force acting on a plane at an angle θ with the vector that joins the axis of rotation (0) and the point of application P of the force vector. Since the lever arm $l = r \sin \theta$, the magnitude of the torque is $\tau = Fr \sin \theta$. The vector product mathematically forms the product of one vector and the projection of the other vector perpendicular to the first.

What about the direction of the vector formed by the product of two others? Perhaps the best way to determine the direction is to use the "right-hand screw rule." First, the vector product, such as the torque $\boldsymbol{\tau}$

EXAMPLE 7.3

Find the center of gravity of a system consisting of a long pole, weighing 19.6 N (2-kg mass), which has attached to it a 39.2-N weight (4-kg mass) at one end and a 58.8-N weight (6-kg mass) at the other. The centers of the two objects are 3 m apart.

Special Subject: Vector Product

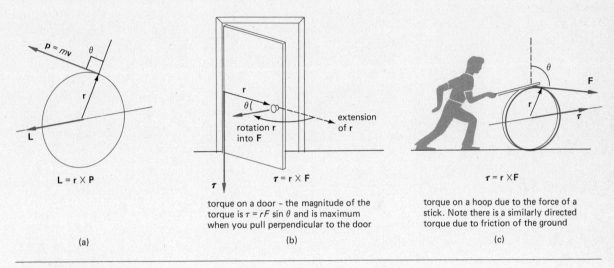

Figure B. Vector products.

or **C** in our generalized definition, is always perpendicular to the plane formed by the other vectors, **r** and **F** or **A** and **B**, respectively. There are, however, two possible directions that are perpendicular to a plane. To select the appropriate one, rotate the first vector (**r** in our product to find **τ**) through the smaller angle (θ) between the vectors so that it approaches being parallel to the other vector. The resulting vector (**τ**, for example) points in the direction that a right-hand screw would advance when rotated through the same angle. For example, rotating **r** into **F** in Fig. A produces a torque **τ**, which points upward perpendicular to the plane containing **r** and **F** since a right-hand screw [Fig. A(b)] would advance upward if it were rotated in the same manner. Of course, to use this rule you must realize that you turn a right-hand (threaded) screw clockwise to advance it forward into the wood.

Figure B(a) is another example of vector cross products. This shows the direction of the vector angular momentum $\mathbf{L} = \mathbf{r} \times m\mathbf{v}$. To form the rotation of **r**, the first vector, into the second vector **p**, the radius vector was extended and then rotated into **p** through angle $\theta = 90°$. The resulting angular momentum is along the axis of the circle and perpendicular to the plane of the circle in direction and has a magnitude of mvr, since $\theta = 90°$ and $\sin 90° = 1$.

Examine the other diagrams in Fig. B to see if you can apply the right-hand screw rule.

Solution

Assuming, as we generally do, that the pole's weight is distributed uniformly throughout its length, we know that its center of gravity is at its geometric center, as shown in Fig. 7.7. To calculate the center of gravity of the whole system, then, we need merely to compute the sum of the torques of the three components of the system about any particular axis and set that sum equal to the torque of the total weight of the system acting at its center of gravity. The choice of an axis is arbitrary; it could, in fact, be outside the system. The simplest choice is the geometric center of the pole:

$$\Sigma \tau \text{ about center} = -(39.2 \text{ N})(1.5 \text{ m}) + (58.8 \text{ N})(1.5 \text{ m})$$
$$= 29.4 \text{ N} \cdot \text{m}$$

This equals the torque of the total weight about the geometric center of the pole:

$$(39.2 + 19.6 + 58.8) \text{ N} \cdot (x) = 29.4 \text{ N} \cdot \text{m}$$

$$x = \frac{29.4 \text{ N} \cdot \text{m}}{117.6 \text{ N}} = 0.25 \text{ m} \quad \text{or} \quad 25 \text{ cm}$$

The center of gravity of the system is 25 cm from the center of the pole toward the end of the pole where the heavier weight has been attached.

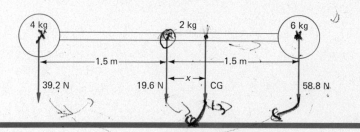

Figure 7.7 Locating the center of gravity of an unsymmetrical object.

EXAMPLE 7.4

Find the center of gravity of a flatbed, tractor-trailer truck loaded with two autos, as shown in Fig. 7.8. The weights of the tractor, trailer, and cars are as indicated and act at the center of gravity of each vehicle. All distances are measured with respect to the rear axle of the tractor.

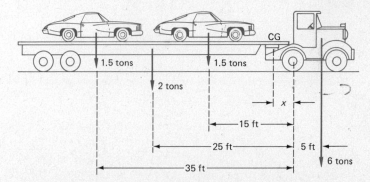

Figure 7.8 How to determine the center of gravity of a loaded truck.

Solution

Here, we must take the sum of the torques of the component parts, for example, about the rear axle of the tractor and set that equal to the product of the total weight of the system and the distance between the center of gravity of the system to the rear axle:

Total weight = (1.5 + 1.5 + 2 + 6) tons = 11 tons

If x is the distance from the center of gravity (CG) to the rear axle, then

$$-(11 \text{ tons})(x) = -(1.5 \text{ tons})(35 \text{ ft}) - (2 \text{ tons})(25 \text{ ft}) - \\ (1.5 \text{ tons})(15 \text{ ft}) + (6 \text{ tons})(5 \text{ ft})$$

$$-(11 \text{ tons})(x) = -95 \text{ tons} \cdot \text{ft}$$
$$x = 8.64 \text{ ft} \quad \text{behind rear axle}$$

EXAMPLE 7.5

A 20-ft ladder weighing 80 lb leans at an angle of 30° against a vertical, frictionless wall, as shown in Fig. 7.9. Calculate the magnitudes and directions of the forces that the ground and wall exert on the ladder.

Solution

Since the wall is smooth and frictionless, it can only push horizontally on the ladder. Hence, the force F_2 is horizontal. The force of the ground on the ladder, F_1, has two components; F_{1_y} acts vertically and F_{1_x} acts horizontally. The resultant ground force does not necessarily act straight up the ladder. In fact, we can calculate its direction only by first working out the magnitudes of the two perpendicular components and then combining them in vector addition.

Applying the first condition of equilibrium $\Sigma \mathbf{F} = 0$,

$$\Sigma F_x = -F_2 + F_{1_x} = 0$$

or

$$F_2 = F_{1_x}$$
$$\Sigma F_y = -80 \text{ lb} + F_{1_y} = 0$$

or

$$F_{1_y} = 80 \text{ lb}$$

Applying the second condition of equilibrium $\Sigma \tau = 0$, we must first select an axis. A simple choice is an axis through the lower end of the ladder. Then, to compute the torques of the 80-lb force and F_2, we must compute the moment arms as indicated in the force diagram, Figs. 7.9(c), (d). For the 80-lb force,

$$r = (10 \text{ ft})(\cos 60°) = 5 \text{ ft}$$
$$\Sigma \tau = +(80 \text{ lb})(5 \text{ ft}) - (F_2)(17.32 \text{ ft}) = 0$$

A smooth, frictionless wall is a figment of a physicist's imagination. By definition, no force can exist parallel to such a wall.

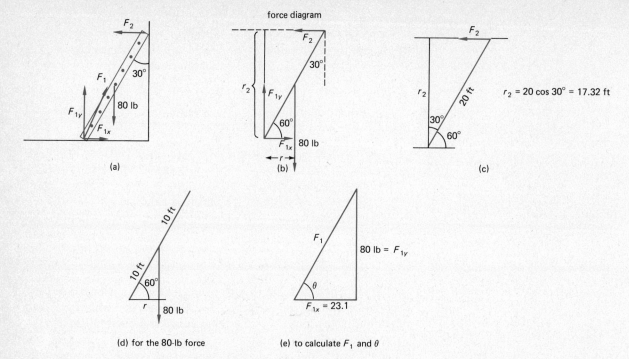

Figure 7.9

or

$$F_2 = \frac{(80 \text{ lb})(5 \text{ ft})}{17.32 \text{ ft}} = 23.1 \text{ lb}$$

Since $F_{1_x} = F_2$, the magnitude of $F_{1_x} = 23.1$ lb.

To compute F_1 and θ, one makes use of the triangle with sides F_{1_y}, F_{1_x}, and F_1, as shown in Fig. 7.9(e), and the Pythagorean theorem:

$$F_1^2 = F_{1_x}^2 + F_{1_y}^2 = (23.1 \text{ lb})^2 + (80 \text{ lb})^2 = 6934 \text{ lb}^2$$

$$F_1 = \sqrt{6934 \text{ lb}^2} = 83.3 \text{ lb}$$

$$\tan \theta = \frac{F_{1_y}}{F_{1_x}} = \frac{80 \text{ lb}}{23.1 \text{ lb}} = 3.46$$

$$\theta = \tan^{-1}(3.46) = 73.9°$$

EXAMPLE 7.6

The center of gravity of a shot-putter's forearm and hand is 7 in. from his elbow joint, and his hand and forearm together weigh 4 lb. Compute the force he must exert on his forearm to keep forearm and hand horizontal while holding a 10-lb ball, if the force is exerted by a muscle 1.5 in. from the elbow joint, as illustrated in Fig. 7.10.

Solution

Taking the sum of the torques about point O in the figure yields

$$\Sigma \tau = -(F)(1.5 \text{ in.}) + (4 \text{ lb})(7 \text{ in.}) + (10 \text{ lb})(14 \text{ in.}) = 0$$

or

$$F = \frac{(28 + 140) \text{ lb} \cdot \text{in.}}{1.5 \text{ in.}} = 112 \text{ lb}$$

7.4 MOVING AROUND

What happens to the objects in the previous examples if the sum of the torques does not equal zero? The objects will undergo angular acceleration, just as objects undergo translational acceleration when the sums of all forces on them are not zero. Before we examine the dynamics of such problems it is useful to define several angular characteristics. For convenience, physicists use angular quantities to describe rotational motion, rather than the linear quantities s and v that we encountered in Chap. 2. Each linear quantity has its rotational analog; the latter are generally symbolized by Greek letters.

The archetypal example of circular motion is that of a ball rotating in a circle on the end of a string, illustrated in Fig. 7.11. Say that at the initial time t_1 the ball is directly to the right of the center of the circle as we visualize it and that at a later time t_2 it has traveled along its circular arc a linear distance of s. During this time the ball has also moved through an angular displacement θ. Angles are generally measured in degrees, each degree being defined as one-360th of a complete circle. The size of the degree agreed upon internationally is completely arbitrary, however, because we could just as easily have divided the circle into any number of parts.

There is a more natural measure of angular displacement with which to attack rotational problems; it is known as the radian (rad). **A radian is the angle that subtends an arc of a circle whose length exactly equals the radius of the circle.** Because the circumference of a circle is 2π times its radius, there are 2π rad in a full circle. Hence, 2π rad equal 360°, and 1 rad is $360/2\pi$, or 57.3°.

In radian measure, the relation between the linear distance along an arc subtended by an angle θ in a circle of radius r is

$$\theta = \frac{s}{r}$$

where r is the radius of the circle. Hence, with θ in radians,

$$s = r\theta \tag{7.3}$$

Because θ is the ratio of two lengths, it is actually a unitless quantity. These two equations are valid *only* when the angle is measured in radians.

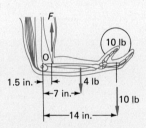

Figure 7.10 Force on a shot-putter's forearm.

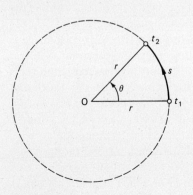

Figure 7.11

With radians now defined, we are in a position to define a characteristic of rotating objects: the magnitude of the angular velocity which is indicated by the Greek letter omega (ω). **Angular velocity is the rate at which the angular displacement θ of a rotating object changes in time.** That is,

$$\omega = \frac{\Delta\theta}{\Delta t} \text{ in rad/s} \tag{7.4}$$

If Δt is a finite quantity in this equation, the expression gives us the average angular velocity. If the time becomes extremely small, the relationship defines the instantaneous angular velocity. The analogy with the expression $\Delta s/\Delta t$ for linear velocity v is plain, and the relationship between angular velocity and linear velocity can be obtained in straightforward fashion by using the equations for ω in terms of θ and for θ in terms of s and r; these yield the expression

$$v = r\omega \tag{7.5}$$

In this case, ω is in rad/s, r in units of length, and v in units of length/s, the units of length obviously matching for v and r. The instantaneous velocity is perpendicular to the radius r.

The angular acceleration, denoted by the Greek letter alpha (α), can be defined in a similar fashion

$$\alpha = \frac{\Delta\omega}{\Delta t} \tag{7.6}$$

The angular acceleration's units are rad/s^2; the term represents the rate at which angular velocity changes. Angular acceleration can be related to linear acceleration by using Eqs. (7.5) and (2.3). The relationship is

$$a_t = r\alpha \tag{7.7}$$

Angular acceleration α is different from the centripetal acceleration with which we dealt in Chap. 4. The angular acceleration α produces an acceleration *perpendicular* to the radius of the circle around which the object is moving. For this reason it is sometimes called the tangential acceleration a_t. Centripetal acceleration, by contrast, acts along the radius toward the center of the circle. Angular acceleration is only present when the *angular* velocity of the object moving in a circle is changing. Centripetal acceleration is *always* present whenever an object moves in a circle or any curved path. The **centripetal acceleration gives the rate of change of direction of the velocity, and the tangential acceleration a_t gives the rate of change of the speed.**

Centripetal acceleration is related to angular velocity by a simple equation. From Chap. 4 the centripetal acceleration a_c is equal to v^2/r, where v is the linear speed of the moving object, r is the radius of the circle around which it moves. Since $v = r\omega$, clearly $a_c = r\omega^2$.

In the general case of angular motion in which angular velocity is not constant, the moving object undergoes both tangential and centripetal acceleration. Since these two components are at right angles, the total acceleration is the vector sum of the two.

One base of linear motion is the set of equations relating displacement, time, and velocity at constant acceleration considered in Chap. 2. These equations, slightly rewritten, can also be applied to angular motions (see p. 191).

It is a common practice when dealing with practical problems involving angular motion to present the angular velocity in units of revolutions per second (r/s) or hertz (Hz) (the name now used to denote cycles per second), rather than rad/s.

The **frequency** ν (Greek letter nu) *is measured in hertz and* **is equal to the number of times an object completes a trip around the circle in a second, that is, the number of revolutions per second.** The reciprocal of the frequency, **the time to make one complete revolution, is called the period** and is usually denoted by T. Note that $\nu = 1/T$. The relationship between frequency ν and the angular velocity ω is formally written $\omega = 2\pi\nu$. This relation arises from the fact that one revolution corresponds to an angular displacement of 2π rad, so a frequency of 1 r/s equals 2π rad/s. Care must be taken not to confuse frequency ν with angular speed ω when employing kinematic equations.

The relationship between frequency and period will reappear in Chap. 12, in connection with waves.

EXAMPLE 7.7

A boy swings a ball on the end of a 1-m string. Ten seconds after starting from rest, the ball has an angular speed of 5 r/s. Compute the (*a*) angular speed after 10 s, (*b*) linear speed after 10 s, (*c*) angular acceleration, assuming it has been constant, (*d*) total angular displacement during the 10 s, (*e*) total distance during the 10 s, (*f*) centripetal acceleration after 10 s.

Solution

(*a*) After 10 s the ball is moving at a rate of 5 r/s. Since there are 2π rad in each revolution, we know that $\omega = 2\pi\nu = 2\pi(5 \text{ r/s}) = 10\pi$ rad/s.
(*b*) From Eq. (7.5), $v = r\omega = (1 \text{ m})(10\pi \text{ rad/s}) = 10\pi$ m/s.
(*c*) From the definition of angular acceleration given in Eq. (7.6),

$$\alpha = \frac{\omega - \omega_0}{\Delta t} = \frac{10\pi \text{ rad/s} - 0}{10 \text{ s}} = \pi \text{ rad/s}^2$$

(*d*) The total angular displacement during the 10 s is computed by multiplying the average angular speed by the time interval. Then,

$$\theta = \left(\frac{\omega + \omega_0}{2}\right)(\Delta t) = \left(\frac{10\pi \text{ rad/s} + 0}{2}\right)(10 \text{ s}) = 50\pi \text{ rad}$$

One can also use Eq. (7.10) [see the second Short Subject, p. 191, for Eqs. (7.8) to (7.10)], the rotational analog to $s = v_0 t + \frac{1}{2}at^2$, to solve for θ:

$$\theta = \omega_0 t + \tfrac{1}{2}\alpha t^2$$
$$\theta = (\tfrac{1}{2}\pi \text{ rad/s}^2)(10 \text{ s})^2 = 50\pi \text{ rad}$$

One can easily convert this value of 50π rad to degrees, remembering that 2π rad correspond to one revolution, which, in turn, is equal to 360°:

$$\frac{50\pi \text{ rad}}{2\pi \text{ rad/r}} = 25 \text{ r} \quad \text{or} \quad (25)(360°) = 9000°$$

(e) The distance is found by multiplying the angular displacement in radians by the radius of 1 m:

$$s = r\theta = (1 \text{ m})(50\pi \text{ rad}) = 50\pi \text{ m}$$

Note that the radian is not actually a unit of measure but just refers to a fraction or part of a circle.

(f) The centripetal acceleration is given by

$$a_c = \frac{v^2}{r} \quad \text{or} \quad r\omega^2 = (1 \text{ m})(10\pi \text{ rad/s})^2 = 100\pi^2 \text{ m/s}^2$$

Note that a_c is not constant but changes as ω increases from 0 to its maximum value; a_c also changes direction continuously.

EXAMPLE 7.8

A grinding wheel rotating at 4 r/s receives a constant angular acceleration of 3 rad/s² for 3 s. (a) What will be its final angular speed? (b) What angular displacement will it describe in 3 s? (c) How many revolutions will it make?

Solution

(a) The wheel starts rotating with a frequency of $v = 4$ r/s. Since the angular speed $\omega = 2\pi v$,

$$\omega_0 = (2\pi)(4 \text{ r/s}) = 8\pi \text{ rad/s} = 25.1 \text{ rad/s}$$
$$\omega = \omega_0 + \alpha t = (25.1 \text{ rad/s}) + (3 \text{ rad/s}^2)(3 \text{ s}) = 34.1 \text{ rad/s}$$

(b) We can use Eq. (7.10) to compute θ; alternatively, we could use the approach of Example 7.7, working out the average value of ω:

$$\theta = \omega_0 t + \tfrac{1}{2}\alpha t^2 = (8\pi \text{ rad/s})(3 \text{ s}) + \tfrac{1}{2}(3 \text{ rad/s}^2)(3 \text{ s})^2$$
$$\theta = (24\pi + \tfrac{27}{2}) \text{ rad} = 88.9 \text{ rad}$$

(c) The total angular displacement is 88.9 rad. There are 2π rad/r. Thus,

$$\text{Number of revolutions} = \frac{88.9 \text{ rad}}{2\pi \text{ rad/r}} = 14.1 \text{ r}$$

Kinematical Equations for Rotational Motion

The equations of kinematics developed for constant acceleration in Chap. 2 for linear variables can be rewritten in terms of the angular variables. These relations, summarized here, are excellent aids when solving problems involving angular motion. Equations (7.8) through (7.10) do not necessarily require units of rad/s; r/s can be used if θ is expressed in revolutions and α in r/s^2.

		Linear motion		*Angular motion*	
$s = r\theta$	(7.3)	$v = v_0 + at$	(7.5)	$\omega = \omega_0 + \alpha t$	(7.8)
$v = r\omega$	(7.5)	$v^2 = v_0^2 + 2as$		$\omega^2 = \omega_0^2 + 2\alpha\theta$	(7.9)
$a = r\alpha$	(7.7)	$s = v_0 t + \frac{1}{2}at^2$		$\theta = \omega_0 t + \frac{1}{2}\alpha t^2$	(7.10)

7.5 ROTATIONAL DYNAMICS

Replete with this brief background in angular motion, we are now in a position to rewrite some of the basic laws of motion in such a way as to take account of the special circumstances of angular movements. To start, we can convert the translational term for kinetic energy, $\frac{1}{2}mv^2$, to a rotational equation.

If a particle is moving in a circle of radius R with an angular velocity ω, we know that its actual speed around the circumference is given by

$$v = R\omega$$

Thus, its kinetic energy is $\frac{1}{2}m(R\omega)^2$, or $\frac{1}{2}mR^2\omega^2$.

If an object of finite size is rotating, each particle of matter that makes up the object is rotating with it. Because of their differing distances from the axis of rotation, the different particles move around circles of different radii, but all have exactly the same angular speed ω. The total kinetic energy of the object is the sum of the individual kinetic energies for the individual particles. In mathematical terms,

$$E_k = \Sigma(\tfrac{1}{2}mr^2\omega^2)$$

Now, because ω is the same for each and every piece of the object, we can write this equation in a slightly different form:

$$E_k = \tfrac{1}{2}(\Sigma mr^2)\omega^2 = \tfrac{1}{2}I\omega^2 \tag{7.11}$$

I in this equation is defined as the sum of all the mr^2 terms and is called the moment of inertia of the object. It is an important concept because it plays the same role in rotational motion that inertial mass plays in translational movement. It is just the conventional term for what might

Moment of inertia substitutes for mass and angular velocity substitutes for straight-line velocity when we calculate kinetic energy for angular motion.

be called "rotational inertia." Thus, the resemblance between the kinetic energy terms for translational and rotational motion, $\frac{1}{2}mv^2$ and $\frac{1}{2}I\omega^2$, is more than a coincidence. Moment of inertia substitutes for mass and angular velocity substitutes for straight-line velocity when we calculate kinetic energy for angular motion.

The moment of inertia of a single particle of mass m moving in a circle of radius R is simply mR^2. For objects of extended size, particularly those possessed of unusual shapes, things are not that simple. The moment of inertia can only be obtained experimentally, or by summing the products mr^2 for each individual particle in the object using calculus. Fortunately, physicists have already done the hard work on a number of objects with common shapes. A selection of their results is shown in Fig. 7.12.

One interesting point to note is that the moment of inertia of a thin hoop rotating about an axis through its center is exactly the same as that of a single rotating particle of the same mass. It is worth noting that a solid disk has a smaller moment of inertia than a hoop of the same radius and mass because of the different distribution of mass in the two objects. The mass near the center of the disk contributes very little to that object's moment of inertia, whereas each unit of the mass of the hoop contributes exactly the same amount.

One final point must be remembered when one uses the moment of inertia to calculate the kinetic energy of a rotating object. For all objects the value of I depends on the axis of rotation. The moment of inertia of a rod whose mass is m and length L, for example, is $\frac{1}{12}mL^2$ if

Figure 7.12 Moments of inertia for some commonly shaped objects.

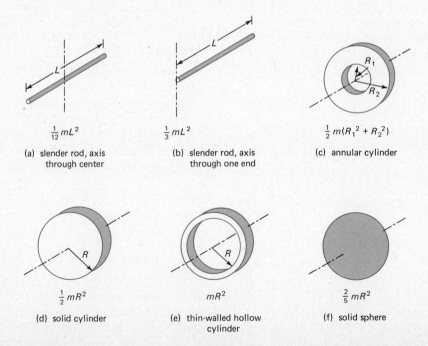

$\frac{1}{12}mL^2$

(a) slender rod, axis through center

$\frac{1}{3}mL^2$

(b) slender rod, axis through one end

$\frac{1}{2}m(R_1^2 + R_2^2)$

(c) annular cylinder

$\frac{1}{2}mR^2$

(d) solid cylinder

mR^2

(e) thin-walled hollow cylinder

$\frac{2}{5}mR^2$

(f) solid sphere

the rod is revolving about its geometric center but $\frac{1}{3}mL^2$ if it is rotating about one end.

Having defined moments of inertia, we can now relate torque and rotational motion. The two come together through Newton's second law, which is $F = ma$ for the case of translational motion.

The torque τ on a particle of mass m is the product of the force F that acts on the particle and the perpendicular distance r between the force's line of action and the axis about which it rotates the particle. In algebra,

$$\tau = Fr$$

Substituting the second-law equation,

$$\tau = mar$$

The tangential acceleration ($a = a_t$) is given by $a_t = r\alpha$, where α is the angular acceleration. So, the torque becomes

$$\tau = mr^2\alpha$$

This is the expression for the accelerating effect of a torque on a single particle. Generalizing this for an extended object,

$$\tau = (\Sigma mr^2)\alpha = I\alpha \tag{7.12}$$

As in the case of kinetic energy, the moment of inertia in an angular equation is equivalent to mass in a straight-line example. Note that when an object is not in equilibrium, the sums of the torques taken about different axes will not be equal.

EXAMPLE 7.9

A belt applies a constant force of 20 lb tangentially to a solid cylinder with a radius of 2 ft and a weight of 320 lb. Find the (a) angular acceleration and (b) kinetic energy of the cylinder after 4 s.

Solution

(a) To find the angular acceleration, one need only apply Newton's second law, $\tau = I\alpha$. The torque $\tau = Fr$, where F is the tangentially applied force, 20 lb, and $r = 2$ ft. The moment of inertia for a solid cylinder is given in Fig. 7.12:

$$I = \tfrac{1}{2}mr^2$$
$$\tau = I\alpha$$
$$Fr = \tfrac{1}{2}mr^2\alpha$$

$$\alpha = \frac{2F}{mr} = \frac{(2)(20\text{ lb})}{\left(\dfrac{320\text{ lb}}{32\text{ ft/s}^2}\right)2\text{ ft}} = 2.0 \text{ rad/s}^2$$

(Note that the mass was obtained by dividing the weight by $g = 32$ ft/s^2.)

(b) To find the kinetic energy after 4 s, one needs to compute the angular velocity ω after 4 s:

$$\alpha = \frac{\omega - \omega_0}{t} \quad \text{or} \quad 2.0 \text{ rad/s}^2 = \frac{\omega - 0}{4 \text{ s}}$$

$$\omega = 8.0 \text{ rad/s}$$

The kinetic energy is given by

$$E_k = \tfrac{1}{2}I\omega^2 = \tfrac{1}{2}(\tfrac{1}{2}mr^2)\omega^2$$

Hence $E_k = \dfrac{1}{2}\left[\dfrac{1}{2}\left(\dfrac{320 \text{ lb}}{32 \text{ ft/s}^2}\right)2^2 \text{ ft}^2\right]8^2 \text{ rad/s}^2 = 640 \text{ ft}\cdot\text{lb}$

EXAMPLE 7.10

Imagine that a hoop and a solid cylinder of identical mass and radius are released at the same time from the top of a hill of height h. Which will reach the bottom of the hill first?

Solution

We need to apply conservation of energy in this example, equating the potential energy at the top with the kinetic energy at the bottom for the two objects:

$$mgh = \tfrac{1}{2}mv^2 + \tfrac{1}{2}I\omega^2$$

For the cylinder, $I = \tfrac{1}{2}mr^2$, and thus,

$$mgh = \tfrac{1}{2}mv^2 + \tfrac{1}{2}(\tfrac{1}{2}mr^2)\frac{v^2}{r^2} = \tfrac{3}{4}mv^2$$

or

$$v = \sqrt{\tfrac{4}{3}gh} \quad \text{at the bottom}$$

For the hoop, $I = mr^2$; therefore,

$$mgh = \tfrac{1}{2}mv^2 + \tfrac{1}{2}mr^2\frac{v^2}{r^2} = mv^2$$

or

$$v = \sqrt{gh} \quad \text{at the bottom}$$

Hence, the speed of the center of mass of the cylinder will be larger than the speed of the hoop regardless of the radius and mass at the bottom of the hill. The cylinder will get to the bottom first. If the objects had skidded frictionlessly down the incline rather than rotating, both would have moved at the higher speed $v = \sqrt{2gh}$. Some of this energy is used in the task of keeping the objects rotating when they do not skid.

Finally, we can rewrite the equations for momentum in angular terms. Consider, for example, a ball of mass m that is moving in a circle of radius r. Its linear momentum at any instant is the product mv. **We define the angular momentum of the ball, L, as the product of its linear momentum and the perpendicular distance between the axis of rotation and the velocity vector of the rotating ball.** In other words,

$$L = mvr \tag{7.13}$$

This is the moment of the linear momentum. Its relation to momentum is exactly the same as that of the torque to force. Using the relation $v = r\omega$, Eq. (7.13) can be written

$$L = mr^2\omega$$

Generalizing this expression for angular momentum in the case of an extended object in which all particles have the same angular velocity,

$$L = I\omega \tag{7.14}$$

The angular momentum of a system remains constant as long as no net external torque is applied to it.

Just as in the case of linear momentum, the importance of this expression is that it is a conserved quantity. **The total angular momentum of an isolated system remains constant,** or, stated slightly differently, the angular momentum of a system remains constant as long as no net external torque is applied to it.

Once again, the simple system that involves a ball revolving on the end of a string in a horizontal plane provides a good example of the application of this conservation law. In the setup illustrated in Fig. 7.13, the string passes into a vertical tube at the center of the circle around which the ball rotates; this arrangement allows adjustment of the length of the string, and hence the circle's radius, with a tug.

Suppose the ball, whose mass is m, whirls around with a linear velocity v_1 and an angular velocity ω_1 when the radius is r_1, after application of a torque to start it up. Then, suppose that one pulls the string down the tube, reducing the radius to r_2, and observes what happens to the linear and angular velocities.

Because the force on the string acts toward the center, the action does not apply torque to the ball. Hence, we can apply conservation of angular momentum. The angular momentum before the string was jerked must equal the angular momentum after the string was jerked:

$$L_1 = L_2$$
$$mv_1r_1 = mv_2r_2$$
$$v_2 = v_1 \frac{r_1}{r_2}$$

Since $r_1 > r_2$, v_2 must be larger than v_1. The object speeds up.

Anyone who has been to an ice show or a ballet performance or who has had the opportunity to look in on a satellite control room has seen this effect in action. Ballet dancers and ice skaters start spinning slowly

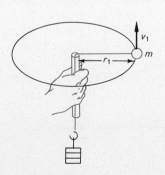

Figure 7.13

with their arms outstretched and then speed up their spins by pulling their arms close to their bodies. Satellites moving in orbit about the earth and planets drifting around the sun speed up when they get closer to the body around which they rotate—another confirmation of the cosmic phenomenon we discussed in Chap. 4.

If angular momentum is analogous to momentum, it must be a vector quantity whose direction is obtained by employing the vector cross product (see page 182) of the momentum **p** and **r**, the "lever" or moment arm. In vector notation, Eq. (7.13) is written

$$\mathbf{L} = \mathbf{r} \times \mathbf{p} = \mathbf{r} \times m\mathbf{v} \tag{7.15}$$

Furthermore, we find that a more generalized expression of the second law, applied to angular motion, is not $\tau = I\alpha$ but torque equals the time rate of change of the angular momentum,

$$\tau = \frac{\Delta \mathbf{L}}{\Delta t} \tag{7.16}$$

This latter expression is the rotational analog to the impulse-momentum theorem.

EXAMPLE 7.11

A physicist with a constant moment of inertia equal to 20 kg · m² stands on the axis of a massless, rotating platform with frictionless bearings. He holds two 4-kg bricks, one in each hand. The man and bricks are set in rotation with an angular speed of 1 r/s, with the bricks held out at a distance of 1 m from the axis of rotation. Compute the angular speed or frequency in revolutions per second when the man brings both of the bricks toward his body, a distance of 0.5 m from the axis of rotation.

Solution

Assuming no external torques act upon the system, angular momentum will be conserved; that is,

$$(I\omega)_{\text{final}} = (I\omega)_{\text{initial}}$$
$$I = I_{\text{man}} + I_{\text{masses}} = 20 \text{ kg} \cdot \text{m}^2 + 2mr^2$$

Initially,

$$I_i = 20 \text{ kg} \cdot \text{m}^2 + 2 \times 4 \text{ kg} \times (1 \text{ m})^2 = 28 \text{ kg} \cdot \text{m}^2$$
$$\omega_i = 2\pi \times \nu = 2\pi \times 1 = 2\pi \text{ rad/s}$$

Finally,

$$I_f = 20 \text{ kg} \cdot \text{m}^2 + 2 \times 4 \text{ kg} \times (\tfrac{1}{2} \text{ m})^2 = 22 \text{ kg} \cdot \text{m}^2$$
$$I_i\omega_i = I_f\omega_f$$
$$\omega_f = \frac{I_i}{I_f}\omega_i = \frac{28 \text{ kg} \cdot \text{m}^2}{22 \text{ kg} \cdot \text{m}^2} \times 2\pi \text{ rad/s} = 8.0 \text{ rad/s}$$

or

$$\frac{8.0 \text{ rad/s}}{2\pi \text{ rad/r}} = 1.27 \text{ r/s}$$

Note that the force required to produce this change is that of the man's arms—an internal force from the standpoint of the system. Furthermore, the man pulls in along a radius, meaning that no torque is applied.

7.6 THE GYROSCOPE

In the examples of angular momentum, we restricted ourselves to cases in which the angular momentum vectors are along the same axis. To examine the gyroscope and many other practical angular motions such as the stability of a bicycle, we must include the vector nature of angular momentum, torque, and angular velocity.

Perhaps the simplest explanation for the rather strange motion of the gyroscope, or top, comes from the application of Newton's first law to the behavior of a small element of mass on the rim of a rotating wheel, gyro, or rotor rim. That law says that a body in motion tends to remain in motion at uniform velocity unless acted upon by some external force. Since velocity is a vector quantity, both the speed and direction of the body tend to remain constant.

Examine the behavior of a particle A on the outer edge of the wheel or rotor shown in Fig. 7.14. At a particular instant, the particle is moving in the direction indicated by the velocity vector shown in the figure. This velocity vector is along a tangent to the circumference of the rotor at the point where the particle A exists. Now, suppose we freeze the rotor at this instant of time and simultaneously rotate the rotor about a vertical axis, as indicated in Fig. 7.14(b). The external rotation force, or torque, tends to displace the original direction of the velocity vector of point A. According to Newton's first law, the law of

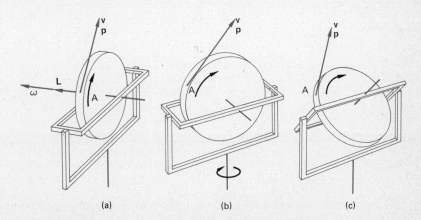

Figure 7.14 The principle of the gyroscope—rotational inertia.

Come Fly with Me

Whoever invented the potter's wheel was one of the first people to put the law of conservation of energy into practice. By giving the wheel a quick kick, the discoverer provided it, and the piece of art mounted on it, with a store of energy that only dissipated gradually as a result of the frictional force of the person's hands. Today, engineers are looking at flywheels with a weightier purpose in mind: storing energy in power plants, cars, trains, and other machines.

Any large flywheel has a large moment of inertia. It requires a large torque to set it in motion quickly, but once in motion it continues to revolve for long periods because such forces as friction and air resistance are minimal. While rotating, the flywheel acts as a reservoir for $\frac{1}{2}I\omega^2$ of energy. Engineers are, therefore, investigating the possibility of coupling large flywheels—with diameters up to 15 ft—to electrical power generators. Energy diverted to the flywheels during off-peak hours would be reclaimed during peak periods of use simply by reversing the process used to speed up the flywheels. Instead of feeding energy into the flywheels to speed them up, the energy of the flywheel's motion is extracted, thus causing the flywheels to slow down. Other engineers are developing smaller types of flywheel for installation in trolleys, trains, and even cars as means of avoiding the loss of energy that occurs in conventional vehicles in the process of braking.

The fundamental limitation on the development of bigger and/or better flywheels in the past has been a material one. The maximum speed at which a flywheel of any particular diameter can revolve is restricted by the strength of the material of which it is made. If a flywheel is driven too fast, it flies apart. Two developments in recent years have increased the capacity of flywheels to store energy and thus made them serious candidates for inclusion in the power production process. The replacement of the traditional metals with fiberglass and other synthetic products has strengthened flywheels by as much as 1000 percent. Improvements in flywheel geometry by researchers in applied physics at Johns Hopkins University and other institutions have made the simple devices even more shatterproof.

According to research by these experts, a flywheel whose center is slightly thicker than its rim is inherently stronger than the traditional flat flywheel. This strength is gained at the expense of some stored energy, because the moment of inertia of a flywheel with a thickened center is smaller than that of a flat wheel. A "multirim" design, consisting of cylindrical sandwiches of fiberglass and rubber, appears to be the most effective design for storing energy in a restricted space, such as the area under the hood of an automobile. Where size is no object, a "superflywheel" that looks like a spoked wheel without a rim can store more energy than any other variety.

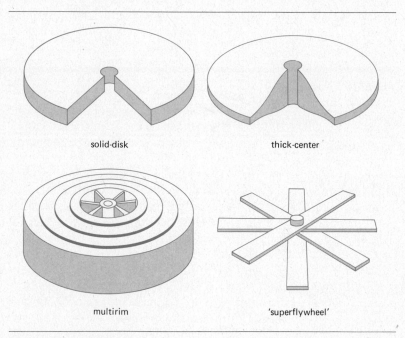

The geometry of flywheels.

inertia, particle A will tend to continue traveling in the direction of the original tangent. In this case we are dealing with a rotational force, a torque, and with rotational inertia. If the rotor axis is fixed to the frame to which we apply the torque about the vertical axis, then the whole device would just twist about the vertical axis as shown in Fig. 7.14(b). We would find that twisting this whole system requires more torque when the rotor is spinning on its axis than when it is not spinning because the rotating wheel or rotor has angular momentum. The result of this angular momentum can be seen very vividly if we mount the rotor in such a way that it is free to rotate about a horizontal axis perpendicular to the rotor spin axis. Then, as we apply the external torque about the vertical axis, particle A tries to continue in its original direction—that is, the velocity vector tries to maintain its original orientation in space. To do so, the whole rotating wheel must rotate about a horizontal axis. The rotor will slowly tip, or precess, as shown in Fig. 7.14(c). Of course, the rigid construction of the rotor and its mounting is such that particle A cannot actually travel in a straight line at any instant, but the tendency to do so exists.

With this qualitative description, we can use the vector nature of torque, angular velocity ω, and the analog of Newton's law of motion for rotational motion to analyze this and other gyroscopic devices. Newton's law states that the net external torque produces a time rate of change in angular velocity.

It is the directional changes with which we need be concerned when considering gyroscopic devices. As long as no torques are produced by forces applied tangentially to the rotating wheel, the wheel will not change its angular speed. We must carefully define the directions of torques and angular velocities.

We can analyze the spectacular top shown in Fig. 7.15 by application of Newton's laws. A top is simply a symmetric object that spins with

Figure 7.15 Precession of a top.

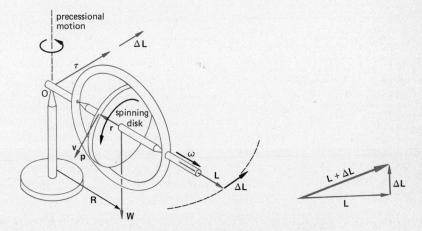

one point fixed. It differs from a gyroscope only in that the fixed point of a gyroscope is its center of gravity. The fixed point of the top in Fig. 7.15 is the point O, which is clearly not the center of gravity of the spinning object. The disk of the top is initially spinning with an angular velocity ω whose direction acts along the axis of rotation in the direction indicated. The initial angular momentum **L** will be along the same line as ω since $\mathbf{L} = I\omega$. The direction **L** is determined from $\mathbf{L} = \mathbf{r} \times \mathbf{p}$, where **r** is the radius of the disk. Check the diagram with the right-hand screw rule to see that the diagram is correct.

Now, when the top is released with this initial angular momentum **L**, it begins to rotate about a vertical axis through point O. This motion is called precession and can be explained if we consider the force of gravity acting upon the top. The weight force **W** acts vertically downward and produces a torque about O given by

$$\tau = \mathbf{R} \times \mathbf{W}$$

Using the right-hand screw rule, we see that this torque is directed at right angles to **W** and to **L**, which lies along the axis of spin. Now, from the analog of Newton's second law for rotational motion, the torque can be related to a change in the angular momentum:

$$\tau = \frac{\Delta \mathbf{L}}{\Delta t}$$

The gravity torque produces a change in the angular momentum $\Delta \mathbf{L}$, which must lie along the same direction as the torque. To find the new angular momentum, we must add this change in angular momentum vectorially to the original angular momentum **L**. We do this by transposing the vector $\Delta \mathbf{L}$ to the tip of **L** and constructing the resultant. The resultant angular momentum $\mathbf{L} + \Delta \mathbf{L}$ is no longer along the initial axis of spin. Hence, the top must have rotated about the vertical axis through point O. The top precesses. Note that $\Delta \mathbf{L} = \tau \Delta t$ and hence depends on time. Also note that the top is supported vertically by the pivot at O so that Newton's second law $\Sigma F_y = 0$ is not violated by the apparently amazing fact that the top does not fall. The result of the gravity torque causes precession if the top is spinning initially. If not, then the top just falls. A similar analysis of a more conventional child's top in Fig. 7.16 will explain its precession.

This discussion is certainly not complete; it is meant primarily to illustrate the complexities associated with the angular quantities τ, **L**, and ω when one is dealing with real-life, three-dimensional examples. In these three-dimensional examples analysis is complicated because more than one axis of rotation is required. The positions where the forces, including the weight, are applied are very important in determining the resulting motion. It is critical in all these examples whether the object is rotating, thus providing angular momentum. For example, isn't it much more difficult to balance a bicycle while sitting on it when it is stationary than when the bike is rolling along?

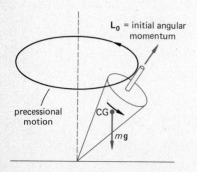

Figure 7.16 A toy top in precession.

SUMMARY

1. The center of mass is the point associated with an object that will move in the same way whether forces are applied to various points on an extended object or whether the resultant of those forces is applied to this point, which is assumed to have all the mass of the object.
2. The center of gravity of an object is the point associated with an object where all the weight could be concentrated when considering the motion of the object.
3. A torque is a twist that tends to rotate an object. Torque is the rotational equivalent of force.
4. The magnitude of a torque is the product of the force that tends to rotate an object and the perpendicular distance between that force's line of action and the axis of rotation.
5. Rotational motion produced by torques is studied in terms of angular quantities such as angular displacements in radians, angular velocities, and angular accelerations.
6. Angular acceleration is quite distinct from centripetal acceleration. Centripetal acceleration is directed inward along the radius of a circle and arises from the directional change of an object moving along a curved path. The angular acceleration is proportional to the tangential acceleration of an object in circular motion and expresses the fact that the speed of the object in circular motion is changing.
7. The moment of inertia of any object about a particular axis is the sum of the product of the masses and the square of the distances from the axis of all particles that make up the object.
8. When the basic laws of motion are rewritten in angular terms, the moment of inertia substitutes for the mass of translational motion and angular velocity for straight-line velocity.
9. The gyroscope illustrates the application of the laws of rotational motion to three-dimensional problems.

QUESTIONS

1. Write down the angular quantities discussed in this chapter and note their linear counterparts.
2. Give several examples of objects or groups of connected objects whose center of mass lies outside of the body.
3. Could the angle used in θ, ω, and α be expressed in degrees or revolutions in the equations of kinematics for angular variables?
4. Why doesn't the centripetal force acting on a rotating object produce a torque?
5. Before power-assisted steering was invented, why was it almost a necessity to have a very large steering wheel on a large truck?
6. Analyze the motion of a spool or yo-yo rolling on a table. If you pull the cord at a low angle, the spool rotates one way. If you pull at a higher angle, the spool rotates the opposite way. At a particular angle, no rotation occurs. Try this and explain.

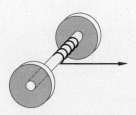

7. Figure out your center of gravity when you are in a sitting position.
8. Why is it easier to balance a bicycle while it is moving?
9. Consider a rolling sphere, hoop, and solid disk, all of the same diameter and identical masses. How do their kinetic energies compare when their centers of mass are all moving at the same speed?
10. If the ice at the poles of the earth were to melt and add additional mass to the middle latitudes of the earth, how would this affect the length of the day?
11. The earth does not travel at a uniform speed throughout the year in its orbit about the sun. How are the "orbital" acceleration and the centripetal acceleration related?
12. You are riding along on a bicycle with one of your arms extended and you pick a book bag off a wall and hence have an added weight in your extended arm. What would you do to keep your balance?
13. How would you find the center of mass of an irregularly shaped work of art?
14. Discuss the motion of a diver doing a dive involving a rotation. How does the diver's center of mass move, and what is the status of his energy?

PROBLEMS

Torque

1. A pole 2 m in length has one end attached to a hinge mounted on a wall. (*a*) Compute the net torque acting on the pole around the attached end if the weight of the uniform pole is 40 N and you apply a force of 100 N, as shown in the diagrams. (*b*) In which direction is the torque vector—into or out of the paper?

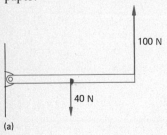
(a)

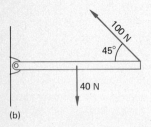
(b)

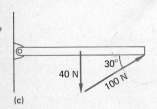
(c)

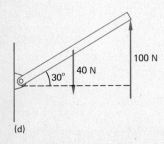
(d)

2. You wish to lift a 500-lb rock using a lever 10 ft long. You rest the board on a smaller rock 2 ft from the large rock, placing one end of the board under the rock. If you weigh 200 lb, how far up the board must you sit to lift the rock?

3. Compute and compare the different torques required by the muscle at the elbow for the two positions shown in the figure.

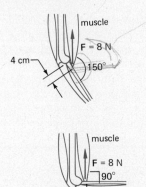

4. A common exercise for leg and stomach muscles is to lie on your back and lift your legs off the ground by only a small angle, for example, 20°. If the mass of each leg is 11 kg and the center of mass is 38 cm from the hip joint, find the force exerted by the muscles in the stomach if these muscles have a lever arm of 10 cm.

Center of Gravity and Center of Mass

5. Two masses are hung on a meter stick, which itself has a mass of 100 g: 250 g at the 20-cm mark and 500 g at the 75-cm mark. Find the center of gravity of this arrangement.

6. Find the center of gravity of a human leg and foot if the center of gravity of the thigh (mass 6 kg) is 20 cm above the knee, the center of gravity of the lower leg (mass 3 kg) lies 20 cm below the knee, and the center of gravity of the foot (mass 1 kg) lies 50 cm below the knee.

7. A man in a "football stance" has a particularly stable position. Relate this observation to the location of center of gravity by finding his center of gravity. In the following diagram, something a little better than a stick diagram will enable you to make a rough estimate of the center of gravity. The forces on the legs and arms have been doubled to include the fact that there are two arms and two legs.

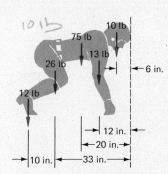

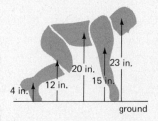

8. The mass of the earth is about 80 times that of the moon, and the distance from the moon's center to the earth's center is about 60 times the radius of the earth. How far is the center of mass of the earth-moon system from the center of the earth? Assume the earth is a sphere of radius, 6400 km.

9. Calculate the center of mass of a water molecule that is a planar molecule of dimensions shown in Fig. 9. The mass of an oxygen atom is 16 times the mass of a hydrogen atom.

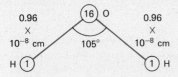

Figure 9

10. Locate the center of mass of three particles at the corners of an equilateral triangle with side 1 m (a) when the two masses at the base are 3 kg and the third is 6 kg, (b) when the three masses are 1 kg, 2 kg, and 3 kg.

Equilibrium

11. A 65-lb boy and a 45-lb girl are on opposite ends of a seesaw supported at its center. The seesaw is 13 ft long. If the girl sits exactly on one end, where should the boy sit to cause a balance?

12. Two workmen are carrying a 100-kg load of bricks on a uniform plank of mass 12 kg and length 5 m. One man is at each end, and the bricks are centered 2 m from Mr. Big. How much of the load is carried by Mr. Big and the other man, Mr. Small?

13. Three painters are on a scaffold that is 8 m long and has a mass of 15 kg. They are situated as shown in Fig. 13. Compute the tensions in the two ropes.

14. A uniform plank 4 m long of mass 30 kg is supported on two sawhorses 1 m from each end. How close to the end of the plank can a 90-kg painter walk?

15. The forces on the lower arm as shown in Fig. 7.10 could be represented by the following diagram. If the weight is a mass of 10 kg and the mass of the forearm is 2 kg, find the force F the biceps must provide.

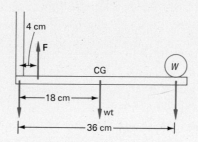

16. What force **F** must be applied by the Achilles tendon and at the ankle joint for a person of mass 90 kg to stand on the balls of his feet? To be in equilibrium, the center of gravity will lie over the ball of the feet. Consult Fig. 16. The weight can be assumed to be shared equally by the 2 feet.

17. A marcher in a parade (Fig. 17) carries a flag on 3-m pole. The flag and pole together have a mass of 8 kg. The center of gravity of the flag and pole is 1 m from the top end.

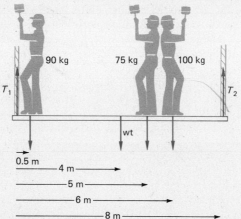

Figure 13

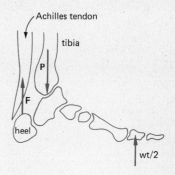

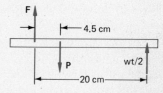

Figure 16

The flagpole makes an angle of 60° with the horizontal and is held by the arm pulling horizontally 1 m from the lower end, with the lower end held by a belt about the waist. Compute the forces exerted by the arm and by the stomach. Neglect the force of the wind on the flag.

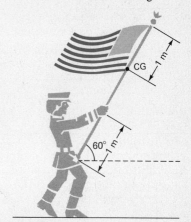

Figure 17

18. A light fishing pole 9 ft long (Fig. 18, p. 204) is supported by a horizontal string. A 10-lb fish hangs from the end of the pole. (a) What is the tension of the supporting string, and (b) what are the components of the force of the pivot on the pole?

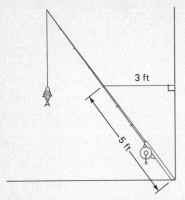

Figure 18

19. In the following figure, the distance along the boom from the hinge to the cable is one-half the length of the boom. The boom weighs 500 lb and is uniform. (*a*) Draw a free-body diagram indicating where and in what direction the forces *on* the boom act. (*b*) Calculate the tension in the cable. (*c*) Calculate the horizontal and vertical components of the force on the wall by the boom.

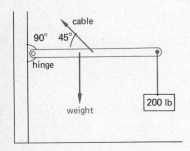

20. Compute the tension in the cable and the force magnitude and direction on the pole at the base of the pole shown in Fig. 20.
21. A uniform ladder leans at an angle of 60° at rest against a rough vertical wall. The ladder weighs 50 lb. The coefficient of static friction between the ladder and the wall and the ladder and the floor is 0.268. The ladder is 20 ft long. (*a*) Draw a dia-

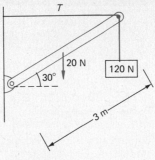

Figure 20

gram indicating all the forces acting on the ladder. (*b*) Apply the conditions of equilibrium to this problem. (*c*) Determine all the forces on the ladder when it is on the verge of slipping.

Angular Kinematics

Assume constant acceleration where appropriate in Probs. 22–31.
22. A fan that normally operates at 130 r/min is turned on and takes 8 s to reach its final speed. (*a*) What is the angular acceleration of the fan? (*b*) Compute the angular displacement of one blade during the 8-s time. (*c*) Compute the angular displacement after 20 s. (*d*) If the diameter of the fan is 0.5 m, compute the distance the end of the one blade has traveled in the first 8 s.
23. A pulley on a motor turns at 400 r/min. (*a*) What is its angular speed? (*b*) What is its angular displacement after 3 s? (*c*) How far has a point in the rim of the pulley moved in 3 s if the radius of the pulley is 5 cm?
24. The top of a tennis racket during a serve moves through 0.33 m with a radius of 1.6 m. Compute the angular displacement in radians, revolutions, and degrees.
25. A wheel has a constant acceleration of 4 rad/s². If it starts from rest, what is its angular displacement and angular velocity after 4 s?
26. A record on a turntable moves at 78 r/min. After being turned off, it stops after 20 rotations. What is the angular acceleration of the turntable?
27. The fan on an automobile rotates at 2 r/s; at 5 s later the speed has increased to 3.5 r/s. (*a*) What is the average angular acceleration over this period? (*b*) How many revolutions occur during this period?
28. A 33-r/min record on a record player comes to rest 1 min after the machine has been shut off. (*a*) What is the tangential acceleration of the outer edge of a 10-in. record? (*b*) What is the centripetal acceleration of the outer edge of the 33-r/min record before it is turned off? Is there any reason that these two should be related?
29. A grinding stone moves at a rate of 240 r/min at full speed. It takes 2 min to reach this speed, and then it is used for 5 min. It is then turned off and takes 3 min to stop. (*a*) Compute the angular acceleration of the stone during the first 2 min. (*b*) Compute the speed 1 min after start-up. (*c*) Compute the total angular displacement over the 10 min of operation. (*d*) If the diameter of the stone is 12 cm, what is the linear distance covered by the outside edge of the wheel in 10 min? (*e*) What is the linear speed of the edge of the stone halfway through the operation?
30. A bike rider is moving along at a linear speed of 4 m/s. If her bike has wheels 0.7 m in diameter, what is the angular speed of her wheel?
31. A 0.7-m-diameter bike wheel is turning at 125 r/min. (*a*) What is the linear speed of a point on the outer edge of the wheel? (*b*) How is this speed related to the speed of the center of the wheel along the road? (*c*) If the bike wheel is accelerated to a speed of 200 r/min over a distance of 300 m, what is the linear acceleration of the wheel?

Rotational Dynamics

32. What is the moment of inertia of the wheel (mass of 1 kg) and axle (mass of 3 kg) shown in the following figure?

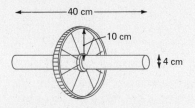

33. Compute the moment of inertia of a baton of dimensions shown in the figure about a vertical axis through CG.

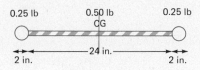

34. A ball on the end of a 0.6-m string is whirled about in a horizontal plane at 90 r/min. Compute the (a) moment of inertia, (b) kinetic energy, (c) angular momentum, (d) torque required to initiate this motion from rest in 20 rotations of the ball. m = 0.2 kg

35. A 12-in.-diameter 33-r/min record on a superbly good turntable is allowed to come to rest when the turntable is turned off and the needle is left in place 8 cm from the center. It takes only 20 s for the record and turntable of mass 140 g to stop. (a) What is the average angular acceleration of the record? (b) What are the average torque and force applied by the needle in the stopping process? (c) What is the initial kinetic energy of the record? (d) Compute the angular displacement during the 20-s stopping time using knowledge of torque and kinetic energy.

36. A 6-g mass is attached to a 30-cm-long string to form a pendulum. If the mass is raised 5 cm above its equilibrium position, compute the angular momentum of the mass when it passes by its original equilibrium position (lowest point of swing).

37. A racing car of mass 800 kg is traveling around a circular track of radius 200 m with a steady speed of 30 m/s. Find its angular momentum.

38. A block of mass 50 g is attached to a cord passing through a hole in a horizontal frictionless surface as in the following figure. The block is originally revolving at a distance of 20 cm from the hole with an angular velocity of 3 rad/s. The cord is then pulled from below, shortening the radius of the circle in which the block revolves to 10 cm. (a) What is the new angular velocity? (b) Find the change in kinetic energy of the block. (c) What is the acceleration of the block initially?

39. A 10-g mass swings about horizontally at a rate of 1 r for every 2 s at a distance $r_1 = 1$ m from the glass tube shown. The string is pulled in to make $r_2 = 0.5$ m. What is the final angular speed ω_2 of the 10-g mass?

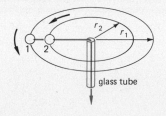

40. A merry-go-round with a moment of inertia of 10^4 kg·m² is rotating at 0.5 rad/s. A small boy of mass 30 kg standing on the outer edge 3 m from the center jumps radially out and up off the merry-go-round. What is the resulting angular velocity of the merry-go-round?

41. What is the angular momentum of the moon about the earth assuming the moon takes 27.3 days to orbit about the earth in a circular orbit? If the moon were pulled suddenly toward the earth so that the radius of its orbit were reduced to one-fourth its original value, what would happen to the angular momentum?

42. The sun has a mass of 1.99×10^{30} kg. Apparently, the sun makes one rotation on its axis roughly every 25 days. If the sun should suddenly expand to twice its present diameter, 13.9×10^8 m, compute its new rate of revolution. Assume the sun rotates in one unit like a solid object and that it is a sphere of uniform density.

43. A 200-lb man decides to ride on a merry-go-round in a city park. The merry-go-round is 6 ft across and is a disk of weight 800 lb. Initially, the angular speed established by pushing is 15 r/min when the man sits on the outer edge of the merry-go-round. (a) What is the angular speed of the merry-go-round if the man crawls into the center of the merry-go-round? (b) What is the speed if he decides to hang out over the edge of the merry-go-round so that his center of mass is 5 ft from the center of the merry-go-round? Consider the man as a point mass and that friction does not affect the motion.

44. A skinny physicist with zero moment of inertia stands on the axis of a massless rotating platform with frictionless bearings. He has an 8-kg lead brick in each hand, held out a distance of 1 m from the axis of rotation of the system. The initial speed of rotation of the system is 6 r/min. (a) What speed of rotation will result if the lead bricks are brought in to a distance of 0.3 m from the axis of rotation? (*b*) What is the increase in rotational kinetic energy of the system? (*c*) What force is responsible for the work done to increase the kinetic energy? Does this force produce a torque?

45. A hoop of radius 1 m weighing 1400 N rolls along a horizontal surface with a speed of 0.2 m/s. How much work must be done to stop the hoop in a distance of 20 m by a tangentially applied force?

46. A solid cylinder rolls down an inclined plane that is 10 m long and makes an angle of 30° with the horizontal. Compute the speed of the cylinder at the bottom.

47. A tall uniform pole 4 m long hinged at its base and of mass 5 kg is released from rest while standing upright. Compute the speeds with which the end and the center hit the ground.

48. A uniform ladder of mass 50 kg and length 5 m leans against a vertical wall with its base 3 m from the wall. The wall is smooth and can only exert a force perpendicular to its surface on the ladder. Find (*a*) this force and (*b*) the horizontal and (*c*) vertical components of the force of the floor on the ladder. (*d*) If a 40-kg boy attempts to climb the ladder, how far up the ladder can he go before the ladder slips if the coefficient of static friction at the floor is 0.4?

49. A hoop, a disk, and a spherical ball all with the same radius and mass roll down an inclined plane. Which reaches the bottom first; that is, which has the highest speed at the bottom?

50. A constant force of 15 N is applied tangentially to a pulley of radius 8 cm. The pulley has a mass of 2 kg and can be considered a solid cylinder. Compute the (*a*) moment of inertia of the pulley, (*b*) angular acceleration of the pulley, (*c*) angular speed after 3 s, (*d*) angular momentum of the pulley after 3 s, (*e*) kinetic energy of the pulley after 3 s, (*f*) work done during the 3 s. Compare (*e*) and (*f*).

51. In the Atwood machine discussed in Chap. 3, we neglected the inertia of the pulley. Determine the rotational inertia of the pulley if one mass is 500 g, the other is 460 g, the radius of the pulley is 5 cm, and the heavier block falls 75 cm in 5 s when released from rest. Neglect friction.

52. Two masses are on opposite ends of a string hung over a solid cylindrical pulley of mass 3 kg. (*a*) Compute the speed of the 4-kg mass just before it strikes the floor. Assume the cord does not stretch and there are no frictional losses. (*b*) Compute the torque on the pulley and its angular acceleration. (*c*) With knowledge of the final speed of the 4-kg block, compute the final angular momentum of the pulley and the time for the 4-kg mass to drop the 15 m.

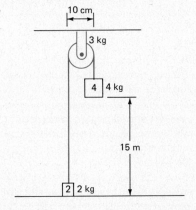

Special Topic
A Revolutionary Means of Staying on the Straight and Narrow

If you have ever been aggravated by a pesky fly that seems uncannily able to avoid all your attempts to swat it, if you have ever been intrigued by the type of spinning top that seems to defy gravity, if you have ever traveled on a plane or a ship, then you have experienced the principle of conservation of angular momentum. The common link between all these scenarios is the gyroscope, a rotating device that consistently tries to point in the same direction whatever is done to change its pointing.

The name of the instrument, which derives from the greek words *scope* (to see) and *gyro* (rotation), was first coined in the 1860s by the French physician-physicist Jean Foucault. The term actually applies to any object that is rotating with a fairly large angular speed and, hence, an appreciable angular momentum. We are most familiar with the gyroscopic top, the gravity-defying child's toy, but we also have direct contact with gyroscopic action by existing on our native planet. The earth, as an object that spins in space, is also an example of a gyroscope.

The basic property of gyroscopic action is that the gyroscope stays spinning in exactly the same direction in space over both short and long periods of time. The spinning portion of the gyroscopic top, for example, remains pointed in the same direction however much one moves the frame that encircles and holds it. If left spinning long enough, a toy gyroscope does, indeed, subtly change its pointing. Far from disproving the basic principle of gyroscopic action, this change actually confirms it, for the tiny toy is actually compensating for the rotation of the earth beneath it. The apparent alteration of its pointing occurs because the observers are moving with the earth, whereas the gyroscope itself stays pointed at exactly the same spot in space.

Astronomical measurements show that the gyroscopic earth indulges in exactly the same behavior. Despite the gravitational attractions of the sun, the moon, and the planets, the axis of the earth's rotation remains pointed in almost the same direction in space throughout the passage of the seasons and the years. (This direction does change very slightly at a very slow rate.)

Actually, gyroscopes are influenced to some extent by outside forces; they remain perfectly pointed, according to the conservation of angular momentum, only when external forces are absent. However, any force applied to a gyroscope produces an unusual effect on the device. It sets it moving in an awkward-looking type of rotation about its original direction of pointing that is known as precession. (This type of motion is discussed formally on page 199, in reference to Fig. 7.15.) The extent of this precession gradually decreases with time as the instrument tends to return to its original direction. The cause of the precession is

The gravity-defying gyroscope. (*Fundamental Photographs from Granger*)

the combination of two torques, one produced by the force that is intended to shift the gyroscope, the other by the gyroscope's weight, the torques acting around the pedestal or similar device on which the gyro is mounted.

A gyroscope's accuracy and consistency in pointing in a specific direction make it far superior to magnets as a basis for navigational compasses. Magnetic north, after all, is not in the same location as true north and is somewhat variable over periods of just a few years. And spacecraft traveling beyond the earth obviously cannot rely for their navigation on a purely local phenomenon in planetary terms. Thus, the gyroscopic compass has become part of modern aerospace technology.

Gyroscopes are incorporated into luxury liners and other vessels that travel on the high seas as aids in minimizing the effect of the swells of the ocean on the stability of the ships. Any incipient roll produces a force that acts on a small gyro inside a ship, causing the gyro to start precessing. By itself, this gyro would be too small to be able to dampen down the ship's roll. Hence, its precession is used to start up an artificial precession in a much larger gyroscope located in the bowels of the ship. It is the precession of the monster gyro that smooths out the roll of the vessel and keeps the passengers from instant seasickness.

In recent years experts have devised much more sophisticated types of gyroscopes for space navigation. The gyros are based on lasers and similarly advanced technology. Even these developments are primitive, however, in comparison with the type of instrument that steers flies. Two tiny devices that resemble gyros mounted behind a fly's wings consist of miniature rods that vibrate continually during flight, keeping the insect moving in exactly the direction it chooses.

8 Matter Under Stress

8.1 SOLIDS, LIQUIDS, AND GASES

8.2 DENSITY AND RELATIVE DENSITY

SHORT SUBJECT: Liquid Crystals: The In-Between State of Matter

8.3 UNDER PRESSURE

8.4 STRESSES AND STRAINS

SHORT SUBJECT: The Real Difference Between Mice and Elephants

8.5 SURFACE TENSION
Summary
Questions
Problems

SPECIAL TOPIC: Bone As a Building Material

Ever since prehistory people have wondered about the nature of the matter of which they themselves and everything in the world around them consists. What makes a rock different from a tree? How does running water differ from the soil it replenishes? What gives stone its toughness and animal skin its flexibility? These were among the questions pondered by the philosophers of ancient times. About 25 centuries ago in ancient Greece the debate over the nature of matter narrowed down to a simple fundamental issue: is matter indefinitely divisible (continuous) or ultimately discrete? This question embodies a philosophic conflict that still occupies scientists today. What, it asks, is the end product when one starts to cut a piece of matter—be it wood, stone, or iron—into two pieces and then divides the two pieces each into two more pieces, and so on? Can one continue cutting ad infinitum and still produce entities that contain all the characteristics of the original material? Or does one eventually reach a limit of size beyond which the material can no longer be subdivided? Although that particular question has long been settled, a philosophical variation of it remains of interest to physicists at the frontier of their subject today. Are the fundamental particles known as quarks real physical objects, the theoretical physicists ask, or merely mathematical abstractions without any true physical existence?

The first set of questions was complex enough for the ancient Greeks. About 2300 yr ago the predominant view among the Greek academic establishment was that of Aristotle, who argued that matter was by its very nature continuous. Some philosophers, however, among them Democritos and Leucippus, argued the reverse. They believed that matter consisted ultimately of tiny, invisible, indivisible particles to which they gave the name atoms. One substance differed from another, the early atomists surmised, because their atoms were of different shapes and sizes.

Although hardly accepted with approbation when it was first formulated, the atomic theory remained a part of the intellectual heritage into and beyond the Middle Ages, partly as a result of the eager proatomistic writings of the Roman philosopher and poet Lucretius.

However, neither the Greeks nor the peoples of early medieval Europe believed in experimentation. These philosophers generally believed in the efficacy of *pure* reason and so did not feel it necessary to dirty their hands with testing these ideas by scientific experiments. Not until the advent of Galileo, sometimes called the father of the scientific method, and Sir Francis Bacon in the early 1600s did philosphers feel

The word atom derives from the Greek term *atomos,* which means uncuttable.

experimentation to be necessary for the establishment of scientific theories. For example, in 1620 Bacon wrote, "Nothing duly investigated, nothing verified, nothing counted, weighed, or measured, is to be found in natural history; and what in observation is loose and vague, is in information deceptive and treacherous."

Not until the nineteenth century did the theory come under any sort of critical confirmation by experimentation. Then the work of such fathers of modern chemistry as Dalton, Gay-Lussac, Avogadro, and Cannizzaro led to the firm conclusion that molecules of matter validly could be regarded as the atoms that the early Greeks had in mind. Molecules are not, it must be emphasized, the ultimate constituents of matter. They can be split into atoms, which in turn can be subdivided into a whole array of even more fundamental particles. *But molecules are the smallest units of individual substances*. Splitting a molecule of, say, carbon dioxide or water leads to components that can in no way be regarded as small units of carbon dioxide or water.

The early scientists made another vital discovery: *all molecules are moving to some extent or other*. The nature of matter can thus be summarized as follows: *All matter, whether in the solid, liquid, or gaseous state, is made up of very small particles, called molecules, that are continuously in motion*. Because it involves motion, this theory is referred to as the kinetic theory of matter.

8.1 SOLIDS, LIQUIDS, AND GASES

The exact form of motion depends on the state of matter in which the substance exists, whether it is a gas, a liquid, or a solid. If you remove the stopper from a bottle of perfume, for example, the gaseous scent will quickly be sniffed by you and other people in the same room. If you release a drop of colored liquid from a medicine dropper into a glass of water, by contrast, the color will spread outward relatively slowly, perhaps taking several minutes to spread throughout the water. There is no way of directly observing the movement of molecules in solids; one is forced to use indirect methods such as subjecting them to forces that cause the molecules to move faster.

Molecules of gas, such as those of the perfume vapor, move very fast but collide frequently with each other and with air molecules. Measurements show that the individual molecules of gases move in straight lines at speeds similar to those of speeding bullets. However, the frequency of collisions between molecules in air causes the movement of the molecules of the gas, when they are released into air from a perfume bottle or any other container, to be much slower than this. This type of motion, which is common to both gases and liquids, is known as **diffusion.**

The random motions of molecules was first observed in a liquid. In 1827, the Scottish botanist Robert Brown noticed by viewing with a miscroscope that fine grains of pollen suspended in water underwent continual zigzag movements. Later studies established that the movements, referred to as Brownian motion, resulted from bombardment by the multitude of moving molecules of water on the tiny, light grains of pollen. Because molecules of water are extremely small and light themselves, and thus have relatively little momentum despite their high speeds, their collective effect can only be noticed on very lightweight objects such as the pollen. One modern method of viewing Brownian motion is to peer through a microscope into a vessel containing puffs of cigarette smoke or smoke from a match, as illustrated in Fig. 8.1. A strong light will reveal that the smoke particles move hither and thither under impact of the molecules of air.

As it happens, random molecular movements are only partly responsible for the mixing of one liquid with another. Just as important a factor in the mixing is convection: the tendency of hot liquids or gases to rise upward through colder regions.

The existence of motion of molecules of solids is much harder to establish than it is for liquids and gases because of the strength of the atomic forces that bind solids together. Even the most frantic movements of the molecules cannot usually tear them out of their basic positions in solids; thus, the normal motion that occurs takes the form of small oscillations or vibrations of the molecules about their positions. The eighteenth century physicist Count Rumford first used the kinetic theory to account for the large amount of heat that appeared in a cannon boring tool and borings whenever a cannon was bored. From the boring process, Rumford correctly surmised that heat was related to motion. Later experiments have shown that the kinetic energy of the individual molecules in the metals is increased by the boring process.

Of course, the differences between the three states of matter are more obvious than those we have outlined on the basis of kinetic theory alone. The two most most obvious means of distinguishing one from the other two involve the boundary surfaces of the three states and their compressibility. Solids have complete boundary surfaces of their own that do not change unless a force is applied to them. Even then they often require a considerable force to make much impression on them. A

According to one's perspective, Benjamin Thompson, Count Rumford, was either a vile traitor or a patriot supreme. Born in Woburn, Massachusetts, in 1753, he sailed to England as a loyalist in 1776. He received the title of Count from the Elector of Bavaria and Palatine in 1800 for scientific and technical services rendered. Whatever Americans think of his political views, they admire him as a superbly skilled scientist and engineer.

Figure 8.1 The way to view Brownian motion.

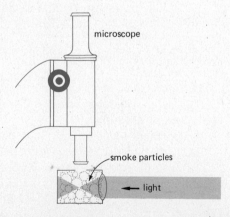

chunk of iron or a rock, for example, must be subjected to very strong forces before it will change shape, and it requires immense forces to change its volume, that is, compress it.

The boundary surfaces of a liquid at rest on the earth's surface are identical to those of the vessel containing it, apart from the uppermost surface, which is always horizontal. Liquids offer almost no resistance to changes in shape, as one can easily demonstrate by pouring water, milk, or any other liquid from a pitcher into a glass. Like solids, though, liquids are very hard to compress into a smaller volume. Forces on the order of thousands of pounds applied to each square inch of the surface are necessary to alter the volume of a typical liquid by any appreciable amount.

Owing to the speed of their molecules and their intermolecular spacing, gases behave in an entirely different manner. Expansion is their natural characteristic, but they can be compressed under the impact of relatively small forces. Left to their own devices, gases tend to expand indefinitely and thus fill any container into which they are channeled. Their surfaces are always identical to those of their container.

8.2 DENSITY AND RELATIVE DENSITY

One further area of difference between solids, liquids, and gases concerns the amount of matter—that is, the number of molecules of each state of matter—that occupies a given volume. The differences are expressed numerically by the quantity known as **density, which is the mass of a substance per unit of volume.**

Giving density the symbol rho (ρ), we have

$$\rho = \frac{m}{V} \tag{8.1}$$

Knowing that the mass of 10 cm^3 of a sample of steel is 78 g, one can readily calculate that the density of that particular steel is 78 g/10 cm^3, or 7.8 g/cm^3.

For simplicity, the density of water was originally assigned the value of 1 g/cm^3 as 1 g was defined as the mass of 1 cm^3, and 1 kg as the mass of 1000 cm^3 (1 L) of pure water at 4°C, the temperature at which water is most dense. Unfortunately, subsequent measurements have shown that the original standards were not quite accurate. However, the density of water can be taken as 1 g/cm^3 with reasonable accuracy for most purposes, although the exact density depends on the temperature of the water and, to a lesser extent, the pressure.

A cubic foot of water weighs 62.4 lb. To find the density of the water in English units, we must first compute the mass of water in a cubic foot. Recalling that weight is the force of gravity on the mass mg near the earth's surface, then

$$m = \frac{62.4 \text{ lb}}{32.2 \text{ ft/s}^2} = 1.94 \text{ slugs}$$

Liquid Crystals: The In-Between State of Matter

One characteristic of science is that it is rarely a black-and-white type of subject. Exceptions to rules always seem to arise after the rules are formulated, and, according to one's viewpoint, these exceptions can present either intractable problems or manifest opportunities. An example of the opportunistic application of an exception to the normal rules is provided by the burgeoning applications of the types of material known as liquid crystals.

Liquid crystals are essentially fluids that possess certain properties of crystals. Thus, they fall in the cracks between the classically defined solid and liquid states. They are composed of rod-shaped organic molecules that line up in three different types of parallel array, as shown in the accompanying figure. These have been given the names nematic, smectic, and cholesteric liquid crystals.

The molecules of nematic crystals always remain parallel to each other, but they are otherwise free to move

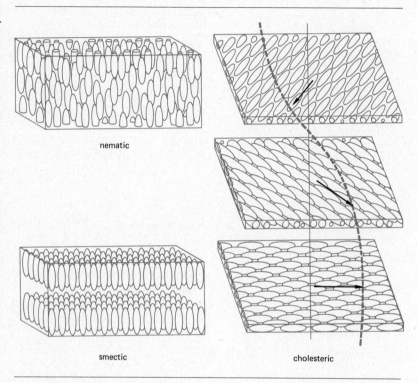

The three types of liquid crystal.

The density of water in the English system is

$$\rho_{\text{water}} = \frac{m}{V} = \frac{1.94 \text{ slugs}}{1 \text{ ft}^3} = 1.94 \text{ slugs/ft}^3$$

Because engineers tend to work with weight and forces, engineers in the United States often use a term known as the **weight density, D, which is simply the weight of a substance per unit volume.** It is the product of the mass density and the acceleration due to gravity. For water, which from side to side and up and down at will. Molecules of the other two types have slightly less freedom. In smectic materials the molecules line up vertically in horizontal layers (see illustration); these layers can move sideways but not up and down relative to each other. In cholesteric crystals the molecules line up horizontally in horizontal layers, with the general orientations of adjacent layers forming an overall helical pattern. In general, liquid crystals reflect light extremely well; furthermore, many of the cholesteric variety are brightly colored.

Physicists have recognized the existence of liquid crystals since the 1890s, but not until the late 1960s did applications for these unusual substances start to emerge. Now, however, liquid crystals have become a growth industry. The applications arise basically because of the reactions of certain liquid crystals to changes in temperature, pressure, and electric field. By slightly altering the ordering of the molecules within the sample these changes produce substantial differences in their ability to reflect light.

Thus, very simple, cheap, liquid crystals form the heart of the mood rings that were briefly popular a while back. The rings were publicized as indicators of one's psychic state; all they actually did was change color in response to alterations in the temperature of the part of one's skin in contact with them. In a more serious application of the same sort, liquid crystals are being used as convenient throwaway thermometers in hospital nurseries. Their colors indicate the temperatures of babies accurately enough to tell the nursing staff whether or not all is well with the infants.

The ability of liquid crystals to reflect light that can be clearly seen even in bright sunlight is being combined with their reaction to electric fields to produce spectacular outdoor display devices, which are turned on with just a small electric field. Digital watches use the same combination, and researchers are now using liquid crystals to develop Dick Tracy-type wristband television sets that can produce clear pictures despite their small size.

---

weighs 62.4 lb/ft³,

$$D_{\text{water}} = \frac{62.4 \text{ lb}}{1 \text{ ft}^3} = 62.4 \text{ lb/ft}^3$$

A liter (1000 cm³) of air under normal atmospheric pressure and at a temperature of 0°C has a mass of 1.293 g. Hence,

$$\rho_{\text{air}} = \frac{1.293 \text{ g}}{1000 \text{ cm}^3} = 0.001\ 293 \text{ g/cm}^3 = 1.293 \times 10^{-3} \text{ g/cm}^3$$

This value indicates the relatively low density of gases as compared with solids and liquids.

In some cases it is more convenient to compare the density of a substance with the density of a standard, such as water, rather than to know the numerical value of the substance's density. Hence, we have the quantity known as specific gravity. **The specific gravity of any particular substance is the ratio of its density to the density of water.** Plainly, the specific gravity compares the mass of a certain volume of the substance with the mass of the same volume of water, and it tells us how many times more dense the substance is than water. In fact, "relative density" would be a more appropriate term than specific gravity.

As an example, the specific gravity of steel is the density of steel (7.8 g/cm^3) divided by the density of water (1 g/cm^3), that is, 7.8. Because it is simply a ratio, the specific gravity has no units. In the cgs metric system, in which the density of water is 1 g/cm^3, the specific gravity is always numerically equal to density. In the mks units, by contrast, the numbers differ; the density of water is 1000 kg/m^3, and the specific gravity of water remains, of course, 1. Table 8.1 shows the specific gravities of a number of common materials.

One extra point worthy of note: the mass and weight of any specific object, when taken at the same location, are proportional to one another. Plainly, the specific gravity compares the weight of a certain volume of a substance with the weight of the same volume of water when the substance and the water are weighed and measured at the same place.

Specific gravity compares the weight of a certain volume of a substance with the weight of the same volume of water.

8.3 UNDER PRESSURE

If one wishes to analyze exactly how a force deforms any particular material—what effect a gusty wind will have on a rattling windowpane, or whether a 200-lb man will manage to walk across a pond covered with thin ice without plunging into the freezing water—the actual strength of the force is relatively unimportant. The critical factor in determining how much damage or deformation occurs is the force over each unit of area. We therefore define a quantity called **pressure.** This **is the force applied perpendicularly to a surface divided by the area of the surface over which it is applied.** Algebraically,

$$P = \frac{F}{A} \tag{8.2}$$

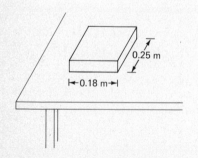

Figure 8.2 Pressure exerted on a tabletop by *Webster's New Collegiate Dictionary,* which weighs 3 lb, or 13.36 N.

As an illustration of pressure, imagine that a Webster's New Collegiate Dictionary is closed and resting on a tabletop, as shown in Fig. 8.2. The book weighs 13.36 N; this is the force it exerts downward on the tabletop. The book is 0.25 m long and 0.18 m wide, giving it an area of 0.045 m^2. This is the area over which the force of the book's weight acts. Hence the pressure of the book on the tabletop is 13.36 N/4.5 × 10^{-2} m^2, or 297 N/m^2. This is equivalent to 2.97 × 10^{-2} N/cm^2, or 4.3 × 10^{-2} lb/in.^{2}

Table 8.1 Specific Gravity

Substance	Specific Gravity
Aluminum	2.70 *gm/cm³*
Bone	1.7–2.0
Brass	8.44
Concrete	2.7
Copper	8.93
Gold	19.3
Ice	0.92
Iron	7.86
Steel	7.8
Lead	11.3
Magnesium	1.75
Wood	0.2–0.9
Liquids	
Ethyl alcohol	0.81
Carbon tetrachloride	1.60
Gasoline	0.67
Glycerin	1.26
Seawater	1.025
Blood plasma	1.03
Water	1.00
Mercury	13.6
Gases	
Air	1.293×10^{-3}
Carbon dioxide	1.977×10^{-3}
Helium	0.1785×10^{-3}
Hydrogen	0.0899×10^{-3}

Taken at 0°C and 1 atm of pressure. Density of water $\rho = 1 \text{ g/cm}^3 = 1000 \text{ kg/m}^3$; $D = 62.4 \text{ lb/ft}^3$ (wt density).

Application of pressure is a vital method in deciding the strength of materials. Imagine, for example, a man who weighs 200 lb standing on a floor. Let us further imagine that his shoes are rectangular in shape. If the shoes are each 1 ft long and 4 in. ($\frac{1}{3}$ ft) wide, each shoe has an area of $\frac{1}{3}$ ft². Hence, the man is in contact with the floor over an area of $\frac{2}{3}$ ft² and the pressure he exerts on the floor is

$$\frac{200 \text{ lb}}{\frac{2}{3} \text{ ft}} \quad \text{or } 300 \text{ lb/ft}^2$$

To support the man when he is standing flat-footed in his shoes, the floor must withstand a pressure of at least 300 lb/ft². But what if he now changes to a pair of cowboy boots that have a 1-in. square heel and for some mysterious reason decides to stand with all his weight on one of the heels? In this case, the pressure on the floor will be

$$P = \frac{F}{A} = \frac{200 \text{ lb}}{\frac{1}{12} \text{ ft} \times \frac{1}{12} \text{ ft}} = \frac{200 \text{ lb}}{\frac{1}{144} \text{ ft}^2} = 28\,800 \text{ lb/ft}^2$$

Obviously, the flooring must withstand a huge amount of pressure in this case, and often tile floors or carpets are simply not up to a task of that magnitude. In the 1950s and early 1960s, floor surfaces were frequently dented and damaged by the ultrathin stiletto heels that were then the vogue for women, and aircraft manufacturers went through plenty of headaches in their efforts to ensure that their cabin floors would not give way under the impact of a stewardess or passenger who momentarily put all her weight on one heel. Today, tubular kitchen furniture that has lost its protective feet and small wheels on couches can produce the same damaging effect on carpets.

Two wintry examples give us further insight into the technical meaning of pressure. If we wish to walk on a carpet of fresh snow without getting our feet wet, it is best to don snowshoes. By spreading our weight over a large area, these shoes exert relatively little pressure on the delicate snow surface and thus allow us to walk without sinking in too deeply. Ice skates, by contrast, have a very small area. Skaters exert a great deal of pressure on the surface of the ice through their skates, causing the ice to melt immediately beneath each skate. Therefore, skaters move smoothly and easily on a thin layer of water (Fig. 8.3).

Figure 8.3 Two wintry applications of the principles of pressure. (*Wide World*)

Figure 8.4 Stretching or compressing a rod with a normal stress.

8.4 STRESSES AND STRAINS

When a sleepy driver lets the steering wheel slip from her grasp and her car ends up smashed against a bridge abutment, the metal parts of the vehicle deform in rather drastic fashion. A car crash is an extreme, real-life example of an area of physics that crops up constantly: deformation of solids. By looking first at simple examples, we can gain an understanding of the topic, along with an insight into the elastic properties of matter.

As a starting point for studying deformation, we must define two more physical quantities: **stress and strain. Stress is force per unit area.** Pressure, which we have just defined, is one type of stress known as a normal stress. In Fig. 8.4, for example, the stress produced by a force F pulling or pushing on a metal rod whose area is A is simply F/A. The units of stress are force/length^2. If the stress tends to stretch the rod, it is known as a tensile stress; if it tends to squash it, it is a compressive stress.

In physics, the term "normal" usually means perpendicular. In this case a normal stress indicates one that is applied perpendicular to the cross-sectional area of an object.

Strain is the relative deformation that is caused by application of a stress to the object involved. In Fig. 8.4, for example, the strain is $\Delta L/L$. Because it is a ratio of lengths, strain has no units. The amount of strain produced by a stress in any particular material is basically determined by the forces between the atoms in the material. The stronger these forces are, the greater stress will be required to produce a given amount of strain.

The numerical relationship between stress and strain was first discovered in the late seventeenth centry by Robert Hooke, an accomplished British physicist and astronomer. He found that the extension of an iron wire caused by a force was proportional to the force up to a particular value known as the elastic limit. Below that limit, Hooke discovered, the wire returned to its original length when the force was removed; above it, the wire would stay stretched even after the stretching force was removed. Furthermore, the wire stretched much farther for each unit of force above the elastic limit than it did below it, as shown in Fig. 8.5.

The part of the graph below the elastic limit is now known as the Hooke's law region. In this portion, we see the real meaning of Hooke's observation that stress is proportional to strain. This is written as

$$\frac{F}{A} = Y \frac{\Delta L}{L} \qquad (8.3)$$

where Y is the proportionality constant. It is known as Young's modulus, named after Thomas Young, a British physicist who was also a

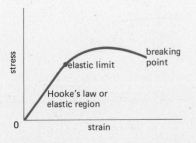

Figure 8.5 The stress-strain diagram.

language expert. Rearranging the terms in the equation, we see that

$$Y = \frac{F/A}{\Delta L/L}$$

Young's modulus is plainly the ratio of the tensile or linear compressive stress to the tensile or linear compressive strain. Because the strain is a dimensionless ratio, the units of Y are identical with those of stress, namely, force/length^2. The value of Y obviously differs for different materials. In individual materials it also varies to some extent with the temperature at which it is measured.

More algebraic manipulation gives us a slightly different way in which to write the basic Hooke equation:

$$F = \frac{YA \, \Delta L}{L} \quad \text{or} \quad F = k \, \Delta L \quad (8.4)$$

In this case the quantity k is constant for any particular object such as a particular piece of wire or a spring. The quantity k is known as the force constant and is useful when one deals with oscillatory motion produced by the elastic restoring forces in materials such as a bob on a spring. Note that k varies for each different wire or spring since it depends on YA and $1/L$.

EXAMPLE 8.1

A telephone wire 120 m long and 2 mm in diameter is stretched by a force of 628 N. (*a*) Compute the stress. (*b*) If the length after stretching is 120.24 m, what is the strain? (*c*) Determine Young's modulus for the wire. Can you identify the material of which the wire is made?

Solution

(*a*) The stress is defined as F/A, where F is the tension in the wire and A is the cross-sectional area of the wire. The cross-sectional area of the wire, in square meters, is

$$A = \pi r^2 = \pi \left(\frac{2 \text{ mm}}{2} \times \frac{1}{10^3 \text{ mm/m}}\right)^2 = 3.14 \times 10^{-6} \text{ m}^2$$

$$\text{Stress} = \frac{F}{A} = \frac{628 \text{ N}}{3.14 \times 10^{-6} \text{ m}^2} = 2.0 \times 10^8 \text{ N/m}^2$$

(*b*) The strain is given by

$$\frac{\Delta L}{L} = \frac{0.24 \text{ m}}{120 \text{ m}} = 2.0 \times 10^{-3}$$

(*c*) Young's modulus is the proportionality factor between stress and strain

$$Y = \frac{\text{stress}}{\text{strain}} = \frac{2 \times 10^8 \text{ N/m}^2}{2 \times 10^{-3}} = 1.0 \times 10^{11} \text{ N/m}^2$$

The value of Young's modulus is similar to that for copper, as Table 8.2 indicates.

Table 8.2 Elastic Moduli

Material	Young's Modulus Y (N/m^2)	Shear Modulus S (N/m^2)	Bulk Modulus B (N/m^2)[a]
Tungsten	36×10^{10}	15×10^{10}	20×10^{10}
Steel	20×10^{10}	8.4×10^{10}	16×10^{10}
Copper	11×10^{10}	4.2×10^{10}	14×10^{10}
Brass	9.1×10^{10}	3.6×10^{10}	6.1×10^{10}
Aluminum	7.0×10^{10}	2.4×10^{10}	7.0×10^{10}
Glass	5.5×10^{10}	2.3×10^{10}	3.7×10^{10}
Lead	1.6×10^{10}	0.56×10^{10}	0.77×10^{10}
Bone	1.6×10^{10} (tensile)		
Bone	0.9×10^{10} (compressive)		
Ethyl alcohol			0.10×10^{10}
Water			0.20×10^{10}
Glycerin			0.45×10^{10}
Mercury			2.5×10^{10}
Air			1.01×10^5
Carbon dioxide			1.01×10^5
Hydrogen			1.01×10^5
Helium			1.01×10^5

[a] B at 1 atm of pressure.

Thus far we have dealt with normal stresses in which the force acts in a direction perpendicular to the area. A different form of stress is known as shear stress. This occurs when a force acts along a surface. If you place your hand flat down along the surface of a household sponge, a book, or a deck of cards lying horizontally on a table, for example, and push horizontally, the object will deform in two directions, as shown in Fig. 8.6. The *shearing stress* in this case is F_t/A, where F_t is the horizontal (or tangential) component of the hand's force along the sponge's surface coupled with an equal but opposite force acting on the

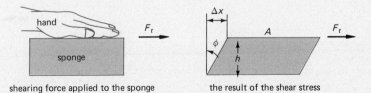

Figure 8.6 Shear stress on a sponge.

bottom. The shearing *strain* for the deformation of the sponge is the ratio of the displacement caused by the force in a horizontal direction x to the height of the surface *perpendicular* to the horizontal h. This is also the tangent of ϕ, the angle of deformation.

One important point to note in this example is that because the hand is planted on the sponge, say from above, it exerts a normal, compressive stress in addition to the shearing stress. The shearing stress is developed through the frictional force between the surface of the sponge and your hand. Of course, a normal force is required to maintain the frictional force.

Shearing stress and strain are proportional the same way that normal stress and strain are. Not surprisingly, the constants of proportionality are different in the two cases. For shear, the constant S is known as the shear modulus, or rigidity modulus. Its value is given by the equation

$$S = \frac{F_t/A}{\Delta x/h}$$

The higher the rigidity modulus of any material, the greater is that material's resistance to a shearing stress. Materials with a large value of S are recommended for such devices as trailer hitches, which experience huge shearing stresses in action.

EXAMPLE 8.2

A short steel rod 3 cm in diameter and 6 cm long projects out from a wall. The rod supports a pulley that is used to lift heavy objects. The pulley exerts a shearing force of 30 000 N on the rod. Compute the downward deflection of the rod. The length h of the rod is exaggerated to show the angle ϕ. The pulley is as long as the rod, and hence it is assumed that the rod does not bend but undergoes only shearing.

Figure 8.7 Shear stress on a rod. The length of the rod has been exaggerated to show the small angle; that is, h should be about twice the diameter of the rod.

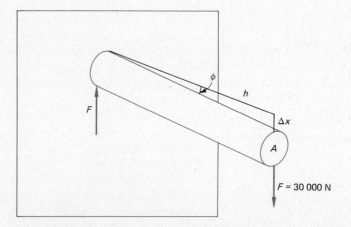

Solution

We must determine both ϕ, the angle of deformation, and Δx, using the expression for the shear modulus

$$S = \frac{F_t/A}{\Delta x/h}$$

Solving for Δx, using $A = \pi r^2$, $r = \frac{3}{2}$ cm, and $S = 8.4 \times 10^{10}$ N/m^2 from Table 8.2 yields

$$\Delta x = \frac{F_t h}{SA} = \frac{(3 \times 10^4 \text{ N})(0.06 \text{ m})}{(8.4 \times 10^{10} \text{ N/m}^2)(\pi)(0.015 \text{ m})^2} = 3.03 \times 10^{-5} \text{ m}$$

$$= 3.03 \times 10^{-2} \text{ mm}$$

The types of deformation with which we have dealt so far apply only to solids. Because their molecules or atoms are not bound tightly together and are relatively free to move about, fluids (that is, liquids and gases) cannot support shearing stresses. The only type of stress fluids encounter is that resulting from the application of pressure—a normal, compressive stress that leads to a change in volume of the fluid. The strain, or relative deformation, involved in this type of stress, which can also be applied to solids, is the ratio of the change in volume to the original volume, $\Delta V/V$. The stress and strain are once again proportional to each other, the stress being simply the pressure applied to the solid or fluid; the constant of proportionality is known as the bulk modulus and has the symbol B. The equation is

$$B = -\frac{P}{\Delta V/V}$$

The negative sign indicates that an increase in pressure always reduces the volume of the solid or liquid involved. Materials scientists occasionally refer to the compressibility K, which is the reciprocal of B. This is not to be confused with another k, the force constant!

Table 8.2 gives some examples of Young's modulus, shear modulus, and bulk modulus of a number of common solids, and the bulk moduli of certain liquids and gases. The table contains a number of points of interest. For a start, solids with large values of Young's modulus also have relatively high bulk moduli. In addition, liquids as a whole have bulk moduli that are not far below those of solids, indicating, as we noted previously, that liquids are largely incompressible. The bulk modulus for mercury, the densest of all liquids at room temperature, is actually higher than that of solid lead. The most striking feature of all about the table, however, is that all the gases have exactly the same bulk modulus. This has a very strong bearing on the behavior of gases at different pressures and temperatures, as we shall see in Chap. 11.

The Real Difference Between Mice and Elephants

A zoologist deciding to make a model of the skeleton of an elephant on the same scale as that of a mouse would immediately be struck by a disparity between the two sets of bones. The bones of the scaled-down elephant would be appreciably thicker than those of the real mouse. The reason stems directly from the effects of forces on objects of different sizes.

The ability of any particular object to withstand a force depends on the material of which it is made. A metal wire, for example, resists efforts to break it using pulling forces far better than does a rope of the same diameter. Resistance to such forces among objects of the same material depends essentially on cross-sectional area. If a 0.5-in.-diameter piece of wire just breaks when suspending a certain weight, we know that a 0.4-in.-diameter piece of the same wire will snap more easily when subjected to the same weight, and a 0.6-in.-diameter wire will probably manage to support the weight. The situation is much the same in the case of bones. The ability of any particular bone to withstand the types of forces that its owner puts on it by moving around, lying down, or sitting still depends mainly on the bone's cross-sectional area. Because a bone's area is proportional to the square of its diameter, we say that the ability of any bone to withstand a force without breaking is proportional to the square of its radius.

What of the forces that bones are expected to withstand? The major force acting on the skeleton of any animal is that of the creature's weight—the force produced by earth's force of gravity. If we allow the simplifying assumption that all animals have the same density, we can say that the weight of any animal is proportional to its volume, that is, the unit of length cubed.

A typical elephant is about 10 ft tall, a typical mouse about one-tenth of a foot; hence, the elephant's length is 100 times as great as that of the mouse. The force on the elephant's bones produced by the beast's weight is thus 100^3, or 1 million, times the force on the mouse's bones caused by its weight. But the ability of the elephant's bones to withstand force would be only 100^2, or 10 000, times as much as the mouse bones' resistance to force if the two sets of bones were directly proportional in cross-sectional area. Plainly, the elephant's weight would crush its bones under these circumstances. To survive, the elephant must have bones that are proportionately thicker than those of the mouse.

Perhaps the best illustration of

Beached whale: crushed by its own weight. (*Wide World*)

this physical principle is provided by an apparent exception to the rule: the blue whale. The largest creature that has ever lived on earth, the blue whale weighs about 40 times as much as an elephant. Yet its bones are no thicker in proportion to the rest of its body than elephant bones because the whale's great weight is supported over its entire lower surface by its watery environment, and the bones have to withstand far less force than they would if the whale were a land creature. Only when whales are beached do their bones suffer the burden of the creatures' enormous

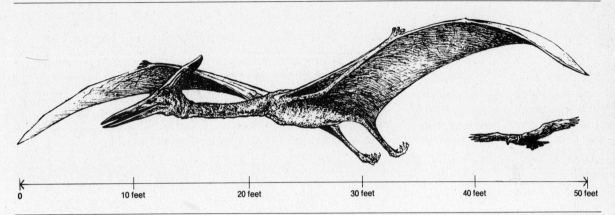

Giant pterodactyl: sky diver or flapper? (*Ib Ohlsson, Newsweek*)

weight. The result is predictable: beached whales are crushed by their own weight.

In practical terms, zoologists have advanced beyond comparisons of the sizes of mice, elephants, and whales. However, this type of force analysis is relevant to a long-running controversy among paleontologists. The argument concerns the flying abilities of pterodactyls, the large flying reptiles that lived in the age of dinosaurs. Their wings consisted of membranes of skin that were supported by two arm bones and the reptiles' hands. Some experts believe that the membranes were too thin to support the flapping motion that modern birds use in flight. In the view of these paleontologists, the pterodactyls had to walk up to elevated perches, like sky divers going to the top of a hill, and take off from the perches in glider fashion. Other scientists believe that the pterodactyls could manage to take off from level ground by flapping their wings in much the same ungainly way employed today by albatrosses and gooney birds. The key to this problem is an accurate knowledge of the weight of pterodactyls, whose wingspans sometimes reached 50 ft. But evidence of pterodactyl weights is buried in the dim and distant past.

Closer to the present, the study of bone reaction to tensile forces indicates that anatomy is really a function of gravity. If we and the animals around us lived on another planet, with a different value of g, our skeletal structures would be markedly different from what they are. For science fiction buffs, it may come as a shock to learn that the anatomy of such a creature as "The Ant That Lifted the Washington Monument" is unsupportable on this earth.

EXAMPLE 8.3   What increase in pressure is required to decrease the volume of a block of aluminum by 1 percent?

Solution   Since the problem involves pressure and a change in volume, we can employ Hooke's law, which relates pressure to volume strain by the bulk modulus:

$$B = -\frac{P}{\Delta V/V}$$

We are asked for the increase in pressure P, which will produce a 1 percent change in volume. That is,

$$\frac{\Delta V}{V} = -1.\% \quad \text{or} \quad -0.01$$

$B = 7 \times 10^{10}$ N/m² for aluminum. Therefore,

$$P = \frac{-\Delta V}{V} B = (0.01)(7 \times 10^{10} \text{ N/m}^2) = 7 \times 10^8 \text{ N/m}^2$$

$$= 6.8 \times 10^3 \text{ atm}$$

8.5 SURFACE TENSION

One of the most important properties of a liquid is that its surface is continually trying to decrease its area. A result of this tendency for the surface to contract is the formation of liquids into droplets as spherical as possible, considering the constraint of the ever-present gravity force. Surface tension arises because the attractive forces between molecules deep inside a liquid are symmetrical; molecules situated near the surface are attracted from the inside but not the outside. The surface molecules experience a net inward force. The energy ΔE required to move additional molecules to the surface and increase the area by ΔA is proportional to the surface area; therefore,

$$\Delta E = \sigma \, \Delta A$$

where σ, **the porportionality factor, is called the surface tension,**

$$\sigma = \frac{\Delta E}{\Delta A} \tag{8.5}$$

and is measured in J/m². **The surface tension is also defined as the force F exerted per unit length of interface (l),**

$$\sigma = \frac{F}{l} \tag{8.6}$$

where σ is expressed in N/m. Since a N/m equals a J/m², it is not surprising to find that these two expressions are one and the same.

Figure 8.8 Surface tension. The liquid surface pulls inward with a tension σ, which is a force per unit length of the interface.

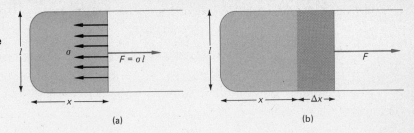

To obtain some feeling for surface tension, we can examine the work required to increase the surface area of a liquid. Consider a film stretched across a U-shaped wire, with one end a sliding bar as shown in Fig. 8.8. The force needed to hold the bar in position F is σ, the force per unit length along it, times the length,

$$F = \sigma l$$

If the bar is moved a short distance Δx, the work required to stretch the surface is

$$W = F \Delta x = \sigma l \, \Delta x = \sigma \, \Delta A$$

or

$$\sigma = \frac{W}{\Delta A} = \frac{2 \, \Delta E}{2 \, \Delta A} = \frac{\Delta E}{\Delta A}$$

Thus, the surface tension can be expressed in two ways that are actually the same. The surface tension is the force per unit length that the surface of the liquid exerts on any line in the surface. The force lies parallel to the surface and is perpendicular to the line. In a sense, it is a linear pressure.

A simple way to determine the surface tension is to examine the forces on a wire ring laid upon the surface of a liquid. As the ring is pulled out of the liquid, it carries with it a circular film. At the point when the wire is just about to break free of the surface tension forces, the liquid is vertical and the surface tension acts essentially vertically downward. There are actually two surface tension forces acting on the ring, one due to each side of the surface of the liquid, as indicated in Fig. 8.9. Therefore, the force F required to hold the ring is given by the equation

$$F = \text{weight of ring} + 2L\sigma$$

where σ is the surface tension and L is the circumference of the ring, $2\pi R$. The factor of 2 is required because both surfaces pull downward on the ring. If we consider only the additional force due to surface tension, then the added force required to support the ring is

$$\Delta F = 2(2\pi R)\sigma$$

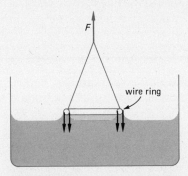

Figure 8.9 Measuring surface tension. A wire ring is laid on the surface of the liquid and then pulled with force F. The pull of the surface on the wire, downward, is the result of surface tension.

228 Matter Under Stress

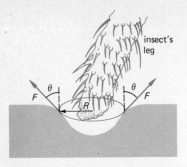

Figure 8.10 Surface tension force on an insect's leg. The component $F \cos \theta$ acts upward. F equals σL, where L is the circumference of the top of the depression in the water. To determine R, the radius of the top of the depression, the force vectors **F** are drawn in an approximate fashion along the walls of the depression, assuming there is a sharp break at the surface from horizontal to a depression wall at angle θ.

Table 8.3 Surface Tension

Liquid in Contact with Air	Temperature (°C)	Surface Tension (N/m)
Water	20	7.28×10^{-2}
Water	100	5.9×10^{-2}
Acetone	20	2.37×10^{-2}
Ether	20	1.7×10^{-2}
Mercury	20	4.7×10^{-1}
Lead	350	4.5

A delicate balance can measure the surface tension in this manner.

The surface tensions of several liquids are listed in Table 8.3. The consequence of surface tension is that small objects such as a needle, a razor blade, and a bug do not sink on an undisturbed surface. When a small insect walks on water, the surface tension supports the insect, as shown in Fig. 8.10. Here we can approximate the result of the surface tension force acting upward on the leg as being due to a ring of surface of circumference $2\pi R$. Since the surface tension force acts parallel to the surface, only the component in the vertical direction acts to support the insect. This component will be $F \cos \theta$. The force is the product of the surface tension σ and the length of the circumference L or $2\pi R$. So, the total upward force is

$$\sigma L \cos \theta = 2\pi R \sigma \cos \theta$$

Surface tension can be reduced by addition of a detergent, thus allowing water to creep more effectively into small gaps between fibers of material. Detergents are examples of "surfactants," which supposedly make water wetter. Ducks and other water fowl have an oil covering on their feathers that prevents wetting—the reverse effect of that due to a detergent.

The same effect accounts for the children's party trick in which one floats a needle on the surface of water in a bowl after rubbing the needle between one's forefinger and thumb. The rubbing coats the needle with body oil, which reduces the water's wetting effect on the needle. Surface tension does the rest of the job of keeping the needle afloat.

EXAMPLE 8.4

Another example of the results of surface tension is capillary action, shown in Fig. 8.11. The force of adhesion that attracts the liquid molecules to the glass is greater than the cohesive forces between the water molecules themselves. The result is that the liquid rises in the small capillary tube. Ultimately, the surface tension forces acting upward must support the column of liquid in the capillary tube. Determine an expression for the surface tension in terms of the height h to which the liquid rises in the capillary tube.

Solution

If the radius of the uniform capillary is small, then the top surface is a hemisphere of radius R, and the upward surface tension force will be essentially vertical. The total upward force of surface tension will be $\sigma L = \sigma 2\pi R$. This force must ultimately support the weight of the column of fluid when the system is in equilibrium. This downward force is

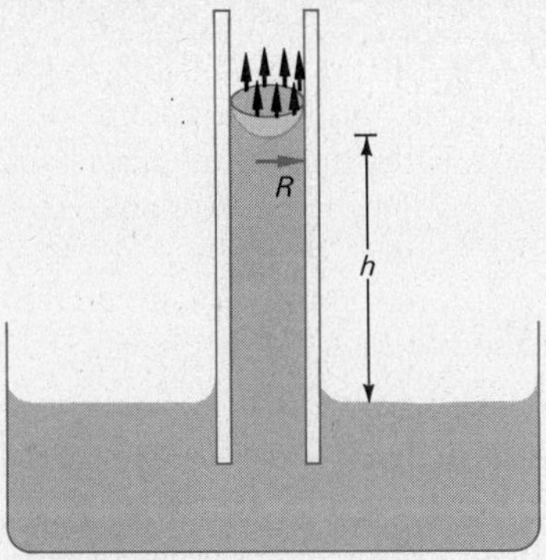

Weight of liquid $= mg = \rho V g = \rho(\pi R^2) h g$

Equating the surface tension to this weight, we have

$$\sigma(2\pi R) = \rho(\pi R^2) h g$$

$$\sigma = \frac{\rho h g R}{2}$$

Figure 8.11 Capillary action.

The quantities on the right side of the equation are all readily measured and hence provide a way to measure the surface tension.

The previous analysis could have been carried out more generally in terms of pressure; this will be more apparent after the next chapter.

There are many examples in nature where the competition between the forces of adhesion of the liquid molecules to another material and the cohesive forces between the liquid molecules (responsible for surface tension) lead to a rather different appearance of the fluid, for example, water "beads" on the surface of a newly waxed automobile hood, because cohesive forces between the water molecules are much stronger than the force of adhesion between the water molecules and the waxed surface. The surface tension is intimately connected to water-droplet formation due to condensation and water-droplet formation in clouds.

SUMMARY

1. All matter consists of very small particles, known as molecules, that are always in motion.
2. Solids, liquids, and gases are differentiated by the forms of motion in which their molecules indulge. Molecules of gases move hither and thither with speeds similar to those of bullets. Molecules of liquids move just as freely but more slowly if the liquid is at a lower temperature than the gas. Molecules of solids just oscillate around fixed positions.
3. The density of any substance is the ratio of its mass to its volume. The specific gravity of a particular substance is the ratio of its density to the density of water.
4. Pressure is the ratio of the magnitude of a force to the area over which that force is applied perpendicularly.
5. Pressure is one example of the general type of action known as stress, which is defined as force per unit area.
6. Application of a stress to an object produces a relative deformation called strain. The ratio of the change in length to the original length is known as the longitudinal strain.
7. The strain produced in wires and similarly shaped objects is proportional to the normal stress that causes it for a large range of values of stress from zero upward. The constant of proportionality is known as Young's modulus.
8. The limit of the simple stress-strain relationship is known as the elastic limit. Below that limit, a stretched wire returns to its original length once the stress is removed. Above it, the wire does not spring back to the original length.
9. Shearing stress is produced when a force acts along a surface. Shearing strain is the ratio of the tangential displacement due to a pair of equal

but opposite forces divided by the distance between the forces. Shearing stress and strain are directly proportional over a wide range of stresses; the constant of proportionality is known as the shear, or rigidity, modulus.

10. Tensile and shear stress can be applied only to solids.

11. A stress from all directions that acts to change the volume of an object can be applied to liquids and gases as well as solids. The strain produced by this type of force is defined as the ratio of the change in volume to the original volume. This strain is proportional to the stress that produces it over a wide range of stresses; the constant of proportionality is known as the bulk modulus.

12. Surface tension is the force produced on the surface molecules of liquids by the attraction of molecules in the interior of the liquid. As a result of this force, liquid surfaces tend to contract.

QUESTIONS

1. How can you account from a molecular point of view that animals such as dogs can track other animals even after many hours have elapsed since the first animal passed by?

2. Opening a perfume bottle releases vapor molecules that move at about 800 m/s, yet the perfume odor travels much more slowly. Placing a teaspoon of dye into calm water eventually results in coloring the water completely. What is this process, and how is it explained from a molecular point of view? How do the gaseous and liquid states differ here?

3. How do the volume, mass, and density of a loaf of bread change if the loaf is squeezed?

4. Examine Table 8.2. Do you notice any striking information here? How do the bulk moduli of solids and liquids compare? What is strange about the bulk modulus of gases?

5. What parameter determines whether or not a material is in the gaseous, liquid, or solid state?

6. How can specific gravity be used to tell how much antifreeze is in a car radiator? A similar measurement is used to determine how much acid is mixed with water in a car battery.

7. Using a graduated cylinder and a balance, how could you determine the specific gravity of a small rock?

8. To supplement your income, you decide to make and sell candy apples about the campus. Should you use large or small apples to minimize the quantity of candy coating, and hence your cost?

9. If you double the size of a bone, what will happen to the breaking strength? Will it be twice as large?

10. What is the difference between stress and strain? Translate the physical meaning to the everyday meaning. Are the words used to mean the same thing?

11. Which is more compressible, water or steel?

12. If you turn a screw too hard it may break. Which elastic modulus is important in this case?

13. Can you determine the diameter of a capillary tube from knowledge of surface tension and observation of the height water rises in the tube? How?

PROBLEMS

Density, Specific Gravity, and Pressure

1. A small bottle has a mass of 25 g when empty. When filled with water it has a mass of 75 g. When filled with an unknown liquid its mass is 77.4 g. What is the specific gravity of the liquid?

2. If an object "weighs" 75 kg and its specific gravity is 1.05, what is its volume?

3. A liter (1000 cm^3) bottle of alcohol (just over a quart) contains what mass of alcohol? If the bottle were filled with mercury, what would be the mass of the mercury in the bottle?

4. If a graduated cylinder of capacity of 50 cm^3 is filled with glycerin, what will be the increase in the mass of the cylinder?

5. If you fill a spherical balloon 6 cm in diameter with alcohol, what will it weigh? Neglect the mass of the balloon.

6. A girl of mass 50 kg is standing on one blade of her ice skates. The blade is 2.5 mm wide and 25 cm long. What pressure does she exert on the ice?

7. A fat man with a mass of 160 kg sits in a kitchen chair. The feet on the end of each leg are 3 cm in diameter. What is the pressure the chair exerts on the floor? Remember there are four legs!

8. A child sits in a kitchen chair whose legs are made of tubular aluminum. The walls of the tube are 2 mm thick, and the outer diameter is 2 cm. Compute the pressure on the floor if the child has a mass of 35 kg.

Stretch and Compression

9. Continuous pounding on a metal spike changes its length from 15 to 14.82 cm. What is the magnitude of the strain, and what is the strain called?

10. A tendon originally 8 cm long is stretched to 8.2 cm. Compute the strain. What type of strain is this?

11. The elastic limit of a steel elevator cable is 35 000 lb/in.^{2} The cable is 0.50 in.² in cross-sectional area and supports an elevator weighing 1500 lb. If we allow a safety factor 4, what is the maximum upward acceleration that can be safely allowed for the elevator?

12. A 45-N weight hangs on a vertical copper wire that is 0.8 m long and 0.006 cm² in cross section. What is the tensile stress, strain, and elongation?

13. How much will a piece of hair 20 cm long and 0.1 mm in diameter stretch if a 0.2-kg mass hangs from it? Young's modulus for hair is 0.2×10^{10} N/m².

14. A piano wire 60 cm long is stretched 0.9 mm. (a) What is the strain? If the wire is made of steel with a Young's modulus of 20×10^{10} N/m², (b) compute the stress responsible for this strain.

15. The elastic limit for a steel is 36 000 lb/in.² If a wire is 0.25 in. in diameter, what is the maximum load the wire can lift to avoid exceeding the elastic limit?

16. A steel post used in construction is 10 cm in diameter and 5 m long. It supports a load of 80 000 N. What is the (a) stress in the post, (b) strain, and (c) compression?

17. The cross-sectional area of a bone such as the tibia is a thick ring of inner radius of about 0.6 cm and outer radius of about 1.3 cm. The compressional force on a tibia due to the weight of the person and muscle contractions can be of the order of 2500 N. (a) Compute the stress in the bone, and, using data from Table 8.2, (b) determine the strain in this tibia. If the bone is 37 cm long, (c) what is the change in length for this strain?

18. A piece of bone 12 cm long with a cross-sectional area of 5 mm² is loaded in a test with 22 kg. As a result, the bone is compressed 0.05 mm. Calculate Young's modulus for the bone.

19. A boy with a slingshot whirls a 2-kg stone in a vertical circle of radius 0.6 m at 60 r/min. (a) At which point in the circle do you expect the cord is most likely to break? If the string the boy uses has a cross-sectional area of 0.02 cm² and its maximum elongation in the circular swing is 2 cm, (b) what is the Young's modulus of the string?

20. A 45-N keg hangs from the midpoint of a 4-m-long steel wire of cross-sectional area equal to 0.2 cm². How much does the wire stretch due to the weight of the keg?

21. A coil spring of the sort used in a spring balance stretches 1 cm for each 50-g mass hung upon it. What is the spring constant?

Shear and Bulk Modulus

22. A force of 600 lb is applied to a piston of radius 2 in. What pressure results? If the cylinder contains alcohol, by how much will the volume of the alcohol decrease?

23. The skate blades on a pair of ice skates are held on by four rivets each 4 mm in diameter. If the skates are worn by a 100-kg man, what is the shear stress on each rivet if he stands on one skate and each rivet supports an equal share of the load?

24. A woman pushes horizontally with the palm of her hand on the top of a rectangular solid sponge that is 1.5 in. thick. She applies 10 lb, and the top moves 0.50 in. horizontally over the bottom. What is the shear strain? Using the estimated surface area of your palm, determine the shear stress and then the shear modulus.

25. A copper block is placed under 100 atm of pressure (1×10^7 N/m²). Determine the fractional change in volume of the copper block.

26. If you place an aluminum cube into a vacuum—that is, reduce the pressure from 10^5 N/m² in the atmosphere to 0—by what percentage will the cube expand?

27. What pressure would be required to increase the density of water by 0.1 percent?

Surface Tension

28. To what height will acetone rise in a tube of radius 0.05 mm at 20°C due to surface tension effects? Specific gravity = 0.88.

29. To what height will water rise in a tube of radius 0.05 mm at 100°C due to surface tension effects?

30. An unknown liquid rises 10 cm in a tube of diameter 0.1 mm at 20°C. If the surface tension of the liquid is measured by independent means to be 5.9×10^{-2} N/m, what is the density of the liquid?

31. If blood has a surface tension of 5.8×10^{-2} N/m, how high can blood rise in a capillary vessel of radius 2×10^{-3} mm?

32. A water spider of mass 2 g is sitting on the surface of the water supported by eight legs. What is the radius of the top of the depression made by each leg if the surface tension makes an angle of 45° with the vertical?

Special Topic
Bone As a Building Material

The relationship between stress and strain is the starting point for any architectural effort to design a new building or structure. The amounts and types of stresses expected to be borne by individual members, such as rods, bars, and girders, the weights of those members themselves, and the stresses that they will exert on the rest of the structure are fundamental factors in determining whether a specific design is inherently rigid and, if it is, which materials should be used for the different parts.

Nowhere does the architect have a better example of the perfect balance of such factors than in the human skeleton. The interlinking and internal construction of different bones provide a perfect example of versatile engineering, which combines strength with lightness and rigidity with flexibility.

The relative lightness of bones in human beings is perhaps their most impressive quality. A man who weighs 160 lb, for example, is completely supported in all his activities, from sleeping on a soft mattress to scaling a vast mountain, by a collection of just over 200 bones that weigh 29 lb altogether. The significance of this figure can best be seen when one compares bones with other strong materials such as steel; the tensile strength of steel is about four times that of bone, but the density of steel is three times as great as the density of bone. Hence, in terms of strength per unit weight—an important criterion in any type of construction—bones are almost as strong as steel. And when it comes to compressive stress, bones appear even more favorable in comparison with steel.

The reason for the high strength-to-weight ratio of bones, and the difference between their compressive and tensile stresses, is a combination of their basic structures and their chemical composition. About half of any bone's weight consists of such inorganic substances as calcium, phosphorus, and other minerals; another eighth is made up of an organic fiber known as collagen; and the rest consists of water. The great strength of bone derives largely from the way in which collagen and the minerals are combined. They are cemented together in much the same way as concrete that is reinforced with the addition of steel rods, a method of construction that gives them great rigidity.

Bones are relatively light because not all their material contributes to their strength. The long bones in the limbs, such as the femur and tibia in the legs and the radius and humerus of the arms, consist of tubes of stress-bearing material filled with very light bone marrow; the marrow contributes nothing to the actual strength of the bone, as can be seen by calculating the effective cross section of a human thigh bone.

Studies on bone have shown that the maximum compressive stress of the human femur (the thigh bone) is about 16.5×10^7 N/m². Tests on samples of femurs also show that the bone breaks under a compressive force of 5×10^4 N for men and 4×10^4 N for women. Since stress equals force per unit area, the effective cross-sectional areas for the male and female bones are given by the relations $(5 \times 10^4 \text{ N})/(16.5 \times 10^7 \text{ N/m}^2)$ and $(4 \times 10^4 \text{ N})/(16.5 \times 10^7 \text{ N/m}^2)$, that is, 3×10^{-4} m² or 300 mm² for men and 2.4×10^{-4} m² or 240 mm² for women. In reality, the cross sections are closer to 1.25 times these values. The difference represents the bone marrow, thus illustrating that it has no effect on bone strength.

Over the years, surgeons and physicians have developed a number of inert materials that can replace shattered or malformed bones. However, no alternative structural material can equal the outstanding quality of living bone—its ability to adjust continually to new stresses. When human beings undertake unaccustomed, strenuous exercise, such as starting a program of push-ups after a life of avoiding any physical activity, their bones strengthen up in such a way that they can easily support the extra load. Conversely, when normally active individuals spend long periods of time in relative inactivity, owing to extended bed rest or space flight, for example, their bones tend to become weaker.

Now, researchers are trying to encourage this natural ability of bones to speed up healing rates of fractured bones. Medical experts have generally believed that a certain amount of compressive stress can improve the healing of broken bones, and in recent years they have set about gathering the necessary data to confirm that impression. In one typical investigation, researchers at Battelle's Columbus Laboratories in Columbus, Ohio, have studied the effect of compressive stress on the healing and remodeling of rabbits' bones. The procedure is to inject dyes whose colors depend on the amount of calcium present into the broken bones of the rodents and then stress the same bones mechanically. After an appropriate period, 3 to 5 weeks in normal cases, the bones are examined for traces of fresh calcium, which indicate areas in which new bone is being formed. By tracing the healing process in this way, researchers gain some insight into the value of compressive stress in the process and the optimum amount of stress required for particular types of bones.

9 Fluids Under Pressure

9.1 THE PRESSURE OF THE ATMOSPHERE

9.2 ARCHIMEDES AND HIS PRINCIPLE

SHORT SUBJECT: Introducing the Atmosphere

9.3 FLUIDS IN EQUILIBRIUM

9.4 THE BERNOULLI EFFECT

9.5 VISCOSITY: A REAL DRAG
Summary
Questions
Problems

SPECIAL TOPIC: The Circulatory System

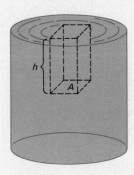

Figure 9.1 The pressure exerted by a fluid on an immersed surface.

The characteristic shared by liquids and gases is the freedom of movement that their molecules enjoy—a freedom that contrasts with the relative imprisonment of molecules in solids, which can only oscillate about fixed positions. For this reason **liquids** and **gases, which are known collectively as fluids,** lack the ability to withstand tensile and shearing stresses. As a result, the physics of liquids is quite different from that of solids.

That pressure is far more convenient than force in dealing with the deformation of fluids can be seen by considering the force and pressure exerted at some point below the surface of the liquid in the jar shown in Fig. 9.1. The column of area A that stretches downward from the liquid surface to a distance h beneath the surface exerts the force of its own weight on the liquid beneath it. This weight can be simply calculated from the density of the liquid and the quantities A and h: the mass of the column is the product of its volume Ah and its density ρ, and the weight is the product of the mass and g, the acceleration due to gravity, that is, ρAhg. We know that pressure is force divided by area, and hence in this case the downward pressure is $\rho Ahg/A$; that is,

$$P = \rho hg \tag{9.1}$$

The pressure is, therefore, independent of the area; it varies with the height of the column of liquid under consideration and the density of the liquid. We can validly speak of the pressure at any point at a particular depth beneath the surface of a fluid because the pressure is exactly the same whatever area is considered.

Pressure is a scalar quantity, devoid of any specific direction. The pressure at any particular level beneath the surface of a liquid is exactly the same in all directions: up, down, sideways, and diagonally. This fact is obvious when one considers that liquids can exist without internal movement. If the pressure at any depth were not equal in every direction, the inequality would imply the existence of a net force in some direction, which would cause the fluid to move that way. The stability of fluids denotes the lack of any net force and thus the equality of pressures in all directions at each point.

Thus far in the discussion we have considered liquids to be incompressible. Hence, their density does not alter with changing depth. We shall continue to make this assumption, which is almost true to reality, for the rest of the chapter. The assumption does not, however, apply to gases.

EXAMPLE 9.1

The fish tank in Fig. 9.2 is filled with alcohol of specific gravity 0.8. Compute the force of the alcohol on (a) the bottom surface (F_b), (b) the front face (F_f), and (c) one end (F_e). Also compute (d) the pressure on the bottom (P_b) and (e) the pressure at a point on the front face 5 cm below the surface (P_e). Assume the alcohol to be incompressible.

Solution

(a) Force on bottom = pressure × area of bottom

$$F_b = \rho A h g$$
$$= (0.8 \times 1000 \text{ kg/m}^3)(0.40 \text{ m} \times 0.15 \text{ m}) \times 0.20 \text{ m} \times 9.8 \text{ m/s}^2$$
$$= 94.1 \text{ N}$$

(b) The force on the front face equals the product of the average pressure on the front face and the area of the front face; it tends to push the face outward. The pressure due to the alcohol is zero at the top and increases in proportion to the depth h. It is clear that the average pressure would be one-half the maximum pressure; that is, $P_{av.} = \rho H g/2$, where H is the depth of the tank. The force on the front face is, therefore,

$$F_f = \frac{\rho H g A}{2}$$
$$= \tfrac{1}{2}(0.8 \times 1000 \text{ kg/m}^3)(0.20 \text{ m})(9.8 \text{ m/s}^2)(0.40 \text{ m} \times 0.20 \text{ m})$$
$$= 62.7 \text{ N}$$

(c) The force on the end is computed in a similar manner, using again the average pressure and the area of the end:

$$F_e = \frac{\rho H g A}{2}$$
$$= (0.8 \times 1000 \text{ kg/m}^3)\left(\frac{0.20}{2} \text{ m}\right)(9.8 \text{ m/s}^2)(0.20 \text{ m} \times 0.15 \text{ m})$$
$$= 23.5 \text{ N}$$

(d) The pressure at the bottom depends only on H, the depth of the tank:

$$P_b = H\rho g = (0.20 \text{ m})(0.8 \times 1000 \text{ kg/m}^3)(9.8 \text{ m/s}^2)$$
$$= 1.57 \times 10^3 \text{ N/m}^2$$

(e) The pressure 5 cm below the surface anywhere across the tank is

$$P_e = h\rho g = (0.05 \text{ m})(0.8 \times 1000 \text{ kg/m}^3)(9.8 \text{ m/s}^2)$$
$$= 3.92 \times 10^2 \text{ N/m}^2$$

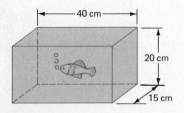

Figure 9.2 A happy fish in a tank full of alcohol.

EXAMPLE 9.2

Compute the (a) force exerted by water against the vertical face of a dam 60 ft long and 40 ft high if the water comes just to the top of the dam, and (b) pressure against the dam 15 ft below the surface.

Solution

(a) $F = A \times \dfrac{H}{2} \times \rho \times g$ where $H =$ total depth of water

$F = (60 \text{ ft} \times 40 \text{ ft}) \times \tfrac{40}{2} \text{ ft} \times 62.4 \text{ lb/ft}^3 = 3.0 \times 10^6 \text{ lb}$

using the weight density of water, 62.4 lb/ft³, for the product of mass density and g.

(b) $P = h \times \rho \times g$

$P = 15 \text{ ft} \times 62.4 \text{ lb/ft}^3 = 936 \text{ lb/ft}^2$

9.1 THE PRESSURE OF THE ATMOSPHERE

So far we have assumed that the densities of the fluids involved in our problems are independent of depth. This is patently untrue for our atmosphere because the weight of the atmosphere above tends to compress the gas bringing the molecules closer together near the surface of the earth. In liquids, this effect is small enough to be neglected, but in gases such as air it plays a vital role in determining basic behavior.

The earth is surrounded by a layer of gas that extends several hundred miles above the planet's solid surface. The density of this atmosphere is far greater close to the surface of the earth than it is at the highest altitudes. About 90 percent of the air and fully 95 percent of the water vapor in the atmosphere lies within 10 mi of the earth's surface, in the region known as the troposphere. By contrast, there is hardly any evidence of the atmosphere above an altitude of 250 mi.

The atmosphere exerts a measurable pressure in all directions at any point above the surface of the earth as a result of the weight of the column of air that stretches above it. The existence of this atmospheric pressure was first shown in 1654, in a classic experiment by Otto von Guericke, mayor of the German town of Magdeburg and the man credited with the invention of the vacuum pump. He had a sense of showmanship that would do credit to a Hollywood executive.

Von Guericke constructed a pair of hollow bronze hemispheres 2 ft in diameter, fitted them together, and pumped as much air as he could out of the combined vessel. Each hemisphere had a handle on its outside, and to each handle von Guericke attached a team of four powerful horses. He and his assistants then did all they could to urge the teams to pull apart the hemispheres, but to no avail. The pressure of the atmosphere pushing on the outside of the hemispheres containing air at a reduced pressure was simply too much for the animals.

It is the rare physics laboratory that is equipped with even a pair of horses, let alone two teams of four, but many laboratories do possess scaled-down versions of von Guericke's hemispheres, which are universally known as Magdeburg hemispheres. If the pair of 6-in.-diameter hemispheres illustrated in Fig. 9.3 is joined and evacuated of air, the seal is likely to be strong enough to defy the efforts of two people to pry apart the hemispheres by pulling on the handles. We can

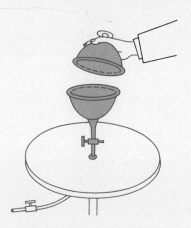

Figure 9.3 Magdeburg hemispheres adapted for the laboratory.

Von Guericke's sense of scientific efficiency lagged somewhat behind his feeling for the spectacular. According to Newton's third law of motion, we now know that he could have obtained twice the pulling power from his horses by attaching all eight to one hemisphere and tying the other to a tree or similarly stable foundation. But the civic-minded scientist can hardly be faulted; Newton did not formulate his laws of motion until more than a quarter of a century after the Magdeburg experiment.

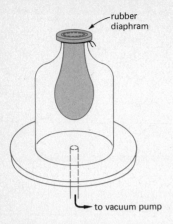

Figure 9.4 Pressure of the air against a rubber diaphragm.

Fluids tend to move from regions of higher pressure toward regions of lower pressure.

calculate the force of the atmosphere on the evacuated spheres very simply from the relationship

$$F = PA$$

where P is the pressure of the atmosphere, to which we assign a typical value of 14.7 lb/in.², and A is the cross-sectional area of the hemispheres. Hence, assuming $P = 0$ inside:

$$F = PA = P(\pi R^2) = 14.7 \text{ lb/in.}^2 \times \tfrac{22}{7} \times (\tfrac{6}{2}\text{ in.})^2 = 416 \text{ lb}$$

We use the cross-sectional area in this case because the definition of pressure demands that the force be always perpendicular to the area. In fact, the efforts to pry apart the two hemispheres also involve forces perpendicular to the cross section; the pullers are pulling against the components of the atmospheric pressure that act along one specific axis.

Other methods of demonstrating the pressure of the atmosphere are somewhat simpler and require less muscle. One might, for example, evacuate an empty can using a mechanical vacuum pump or a bicycle pump modified to run in reverse. As the air is removed, the sides of the can will start to buckle inward as a result of the difference between the pressure exerted by the atmosphere and the reduced pressure inside. Similarly, one can connect a vacuum pump to a jar with a rubber diaphragm stretched over its opening, as shown in Fig. 9.4. As the air is pumped out of the jar, the excess pressure of the outside air causes the diaphragm to bulge downward spectacularly. A similar demonstration employs a sealed balloon partially filled with air and placed in a bell jar, which is subsequently evacuated. The balloon expands as the air pressure on its outside decreases.

All these experiments lead to a broad generalization: Fluids tend to move from regions of higher pressure toward regions of lower pressure. Practical applications of this fundamental physical fact are in vacuum cleaners, medicine droppers, milking machines, siphons, drinking straws, and barometers.

A simple barometer is put together by filling a glass tube, at least 80 cm in length and closed at one end, with mercury. After heating the liquid metal to drive out any air bubbles clinging to the glass wall of the tube, one turns over the tube and places it in a dish of mercury, as illustrated in Figs. 9.5(a), (b). The level of mercury inside the tube drops somewhat, leaving an almost perfect vacuum above it (it contains nothing more than a few molecules of mercury vapor). The level of the mercury in the tube does not, however, fall to the same level as that in the dish. This allows us to use the device to measure the pressure of the atmosphere.

Figure 9.5(c) shows why the mercury remains suspended in the tube above the dish. The surface of the mercury in the dish is subjected to the downward force of the atmosphere, indicated by the arrows; the top of the column of mercury in the tube suffers no such pressure. The only pressure on the mercury in the dish right beneath the tube is due to the

Figure 9.5 The mercury barometer. A glass tube (a) 80 cm long is filled with mercury and inserted in a dish of the liquid element (b). A cross-sectional view of the barometer appears in (c).

weight of the column of liquid mercury. Because the system is in equilibrium, and mercury is essentially incompressible, this pressure must be exactly equal to the pressure of the atmosphere on the rest of the mercury in the dish. In other words, the pressure of the column of mercury is equal to the pressure of the atmosphere. When the atmospheric pressure increases, mercury rises up the column; when it decreases, the level of the mercury in the column falls accordingly.

The actual atmospheric pressure varies with a large number of factors, including altitude, latitude, and weather conditions. However, meteorological studies over the course of many years, using barometers that record their readings automatically, have shown that the average height of mercury in a barometer at sea level and latitude 45°N is 76 cm, or 29.92 in. From this figure, one can readily calculate the "standard atmospheric pressure." Using the relationship $P = h\rho g$, and given that the density of mercury (Hg) is 13.6 g/cm^3, the pressure is found to be

$$P = 76 \text{ cm} \times 13.6 \text{ g/cm}^3 \times 980 \text{ cm/s}^2$$
$$= 1\ 013\ 000 \text{ dyn/cm}^2 \quad \text{or} \quad 101\ 300 \text{ N/m}^2$$

The corresponding value in English units is 14.7 lb/in.^{2}, but because the concept of standard pressure is a nebulous one at best, meteorologists often use the approximate value of 15 lb/in.^{2} The effect of altitude on atmospheric pressure is fairly obvious. The higher the altitude above sea level, the less air exists in the atmosphere above that location. Hence, there is a reduction in pressure with increased altitude. Near the

The athletic advantage of living at high altitude was demonstrated in the 1968 Olympic Games in Mexico City (altitude 8000 ft). Athletes reared at or above the altitude of the Olympic stadium made almost a clean sweep of the middle- and long-distance running events.

surface of the earth, there is generally a reduction of about 2.5 mmHg for each extra 30 m of elevation. From the point of view of human physiology, the importance of such a reduction is that the amount of oxygen decreases. Thus, when normal human beings used to living close to sea level venture into mountainous terrain above an altitude of about 8000 ft, they experience shortness of breath, headaches, and other physical discomforts brought about by a relative lack of oxygen. At the peak of Mount Everest, the highest point on earth, the normal atmospheric pressure in only about 27 cmHg (Fig. 9.6). There, as elsewhere on earth, the actual pressure never varies by more than 3 cmHg from the average value.

Weather reporters tend to avoid such terms as dynes and newtons in conversations among themselves and with the public. They prefer to use the term bar as their basic unit of pressure. A bar is defined as 1 million dyn/cm^2, or 10^5 N/m^2. For most practical purposes, the experts refer to millibars, abbreviated mbar, which are thousandths of a bar. The standard atmospheric pressure (76 cmHg) from Eq. (9.1) is 1013 mbar. Often, the weather maps shown on television contain pressure readings in both inches of mercury and millibars. Note, however, that such units as inches, centimeters, millibars, and atmospheres have no place in equations that involve atmospheric pressure. The pressure in these equations must be expressed in N/m^2, dyn/cm^2, or $lb/in.^2$, that is, as force per unit area. In the SI system of units, pressure is specified in pascals (Pa) or kilopascals (kPa); 1 Pa equals 1 N/m^2.

Figure 9.6 The low pressure—just 27 cmHg—on the summit of Mount Everest forces mountaineers aiming for the world's highest peak to wear oxygen masks. Edmund Hillary and Tenzing Norgay, the first conquerors of the mountain, show off their oxygen equipment.

9.1 The Pressure of the Atmosphere

The starting point for this discussion of the earth's atmosphere was the introduction of the barometer. At this point we must backtrack to ask why the barometer requires mercury as its working liquid. After all, any environmentalist knows that mercury is an extremely poisonous substance when it is present in any appreciable amount. It is also cumbersome and difficult to handle in an instrument such as the barometer. But these problems are insignificant if we consider the difficulty of using any liquid other than mercury to construct a barometer. Let us calculate the height of the column of liquid if we substitute water for mercury in a simple barometer. Using the standard atmospheric pressure 14.7 lb/in.² in the standard equation for pressure, we have

$$14.7 \text{ lb/in.}^2 = H\rho g$$

or

$$H = \frac{14.7 \text{ lb/in.}^2}{\rho g} = \frac{14.7 \text{ lb/in.}^2}{62.4 \text{ lb/ft}^3}$$

$$= \frac{14.7 \text{ lb/in.}^2 \times 144 \text{ in.}^2/\text{ft}^2}{62.4 \text{ lb/ft}^3} = 34 \text{ ft}$$

A water barometer would require a tube longer than 34 ft.

Thus, a water barometer would require a tube longer than 34 ft. Certainly such an instrument would be extraordinarily accurate in measuring small changes of pressure, but it would also be quite impractical. Hence, meteorologists and physicists are forced to use the ultradense liquid mercury in their barometers.

The effective exclusion of water as a working liquid does not preclude all types of barometers apart from the mercury-column variety. One practical instrument that is more portable than the mercury barometer is the aneroid barometer. It consists of a thin metallic box from which some air—but not all—has been removed. As the air pressure increases or decreases, the top of the box moves up and down in response; a series of levers and gears magnifies this motion and transfers it to a needle on a direct-reading scale. Pressure gauges in gas stations are generally based on this principle. The user should be cautioned, however, that the scales are generally calibrated to read zero at normal atmospheric pressure. They therefore record not the actual pressure in a car or bicycle tire but the amount by which the pressure in the tire exceeds atmospheric pressure. Such scales or gauges indicate what is termed the gauge pressure. In effect, these devices measure differences in pressure. An automobile tire inflated to a gauge pressure of 28 lb/in.², for example, actually contains air at 28 lb/in.² plus 1 atm of pressure, 15 lb/in.², or 43 lb/in.²

The relation between atmospheric pressure and the height of a column of water that the pressure supports tells us that we cannot pump water beyond a height of 34 ft with a simple pump that creates a vacuum in the top end of a tube whose bottom is immersed in water. This is a basic physical limitation on the depth of wells using vacuum pumps at the earth's surface. We can use powerful pumps at water level to push water up to greater heights.

A more sensitive device for measuring pressure differences is the open-tube manometer, which is a U-shaped tube partly filled with an appropriate liquid, such as mercury, water, or colored alcohol. The instrument has medical applications in measuring pressures greater

Figure 9.7 U tube manometers. (a) A water-filled manometer used to measure the pressure of the lungs during exhalation. (b) Demonstration of the use of a mercury-filled tube to measure blood pressure.

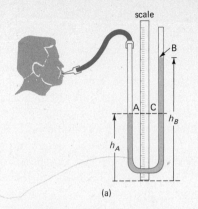
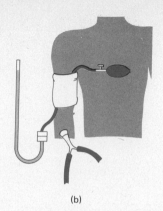

than that of the atmosphere, such as the pressure in the lungs of a wheezing patient and the blood pressure of a subject suspected of having hypertension.

Figure 9.7(a) shows the use of the device to measure the gauge pressure in a patient's lungs. In this case the safest and most appropriate liquid is obviously water. At the start of the measurement, the liquid is at the same height in the two arms of the U tube, under the influence of the pressure of the atmosphere, which we shall call P_0. Then the patient exhales as much breath as he can, driving the water around the tube to the position shown in Fig. 9.7(a). The pressure at point C in the diagram is given by the equation

$$P_C = P_0 + \rho g(h_B - h_A) = P_0 + \rho g h$$

where $h = h_B - h_A$, or the height of water in the right arm in excess of the height in the left-hand arm. Now, because C is at the same vertical level as A, the top of the lower column, we know that the pressures at both points are equal. The pressure at A is the quantity we seek to measure—the absolute pressure of the lungs, P_a. Hence,

$$P_a = P_0 + \rho g h$$

When water is the liquid in the tube, a good breath of air normally produces a difference in height between the two arms of about 60 cm. If mercury were used as the working liquid—putting aside for a moment its extraordinary toxicity—it would only rise to 1/13.6 of that height because mercury is 13.6 times as dense as water. Obviously, the water manometer is the more sensitive instrument of the two.

EXAMPLE 9.3

Using the data in the previous paragraph, compute the absolute pressure developed by the lungs.

Solution

The absolute pressure of the lungs is equal to the atmospheric pressure plus the pressure due to the difference in the heights of the water on the two sides of the manometer. This difference was given as 60 cm. Hence, the absolute lung pressure P_a is

$$P_a = P_0 + \rho gh$$
$$P_a = 101\ 300\ \text{N/m}^2 + 1000\ \text{kg/m}^3\ (9.8\ \text{m/s}^2)(0.60\ \text{m})$$
$$P_a = 101\ 300\ \text{N/m}^2 + 5880\ \text{N/m}^2$$
$$= 107\ 180\ \text{N/m}^2 \quad \text{or} \quad 1.06\ \text{atm}$$

The use of a manometer to measure blood pressure is shown in Fig. 9.7(b). The nurse or physician pumps air into a bag strapped around the patient's arm until the pressure is great enough to stop the flow of blood through the arm's main arteries. The operator then releases air from the bag, thus reducing the pressure, and listens with a stethoscope for the pulse to return. Its return is known in medical terminology as the first sound; the pressure indicated in the manometer at this point equals the systolic blood pressure—the peak pumping pressure of the blood. After hearing the first sound, the operator continues to release air from the bag, waiting for the pulse to vanish and then reappear. The pressure at the return of the pulse is the diastolic pressure—the blood pressure when the heart is relaxed.

9.2 ARCHIMEDES AND HIS PRINCIPLE

In the last section we studied the downward pressures of fluids on objects beneath them. Fluids also exert upward, buoyant pressures on things immersed in them, and now is the time to reverse our way of looking at pressures in fluids. The name most commonly associated with buoyancy is that of Archimedes, the Greek mathematician of the third century B.C. who invented the planetarium and reputedly burned a Roman fleet that was besieging his home town of Syracuse by using an array of mirrors.

The best-known story about Archimedes concerns the task he was given by the local king: to determine whether a crown the monarch had received consisted of pure gold, as advertised, or was tainted with silver. Direct measurement of the crown's density was out of the question because of the crown's complex geometry, and for a while the problem seemed intractable. Then, the story goes, one day when Archimedes was in his bath he realized a fundamental principle: **the upward force on any object floating or immersed in a fluid is equal to the weight of fluid that the object displaces.** By weighing the crown in air and then in water, Archimedes reasoned, he could determine its specific gravity, and hence its purity. Flushed with his insight, the tale continues, Archimedes rushed naked out of the bath and out of his house shouting "Eureka," "I've found it."

However apocryphal the legend may be, the principle is a cornerstone of the physics of fluids. We can derive it today by simple reasoning rather than the flash of inspiration that struck the worthy Archi-

Introducing the Atmosphere

The layer of mixed gases with locally included particles of dust and moisture is called the atmosphere, or air, and extends upward several hundred miles. Close to the surface of the earth, the atmosphere is about 0.0013 times as dense as water, but this density gradually decreases to nearly 0 at 3000 mi up. This falling off in density is not directly proportional to elevation. About half the total weight of the atmosphere is located within 3.5 mi of the surface of the earth, half of the remaining half is located in the next layer of 3.5 mi, and so on. Accordingly, almost 90 percent of the atmosphere is within about 10 mi of the surface of the earth. A few years ago, scientists defined the boundary between the earth's atmosphere and interplanetary space as an altitude of about 600 mi, where they believed that atmospheric gases were free to escape from the earth's gravity into space. Recently, however, using artificial satellites, scientists have discovered an envelope of charged solar particles trapped in the earth's magnetic field and have therefore extended the earth-space boundary to between 37 000 and 62 000 mi.

Until the advent of the earth satellites and space and lunar probes, it was customary to divide the earth's atmosphere into three main zones: the troposphere, stratosphere, and ionosphere (Fig. A). Because of varying thermal and electrically charged atomic conditions at different heights, as determined by satellites and probes, experts named additional zones: the mesosphere, thermosphere, exosphere, and magnetosphere.

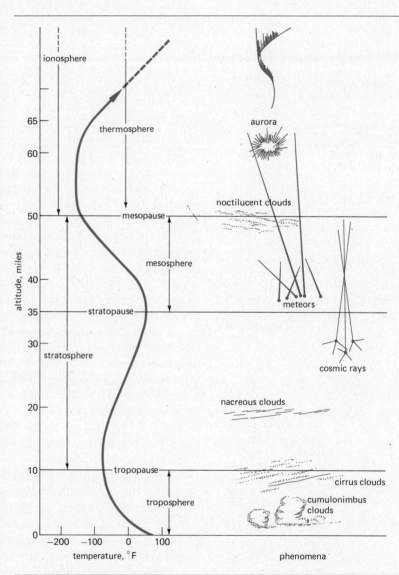

Figure A. The earth's atmosphere. Temperature changes, indicated by the dark line, define the various regions of the lower atmosphere. Altitudes are shown at left, natural phenomena typical of the regions at right.

The troposphere extends to about 10 mi above the earth's equator and diminishes to about 5 mi above the poles. This zone contains over 75 percent of our air and nearly all the water vapor. Hence, our weather—the movement of air currents, condensation of water vapor to form clouds, rain, snow, and so on—is almost entirely a phenomenon that occurs in the troposphere.

A reasonable idea of the variation of pressure with altitude of the earth's atmosphere can be obtained by assuming that the density ρ is proportional to the pressure. This would be exactly true if the temperature of the air were to remain constant at all altitudes. It is easy to see from the derivation of Eq. (9.1) that the variation of pressure with respect to a change in vertical direction (altitude, Δh) is given by

$$\frac{\Delta P}{\Delta h} = -\rho g$$

The minus sign indicates that the change in pressure, ΔP, decreases with height, Δh. The assumption that pressure is proportional to density can be written

$$\frac{P}{P_0} = \frac{\rho}{\rho_0}$$

where P_0 and ρ_0 are, respectively, the atmospheric pressure and density of the atmosphere at sea level.

Combining these two equations and making use of elementary integral calculus, scientists can show that the pressure at any height h above sea level is given by the following exponential function:

$$P = P_0 e^{-g(\rho_0/P_0)h}$$

The value of $g(\rho_0/P_0)$ can be evaluated using the known values at sea level. Then, the pressure can be written as

$$P = P_0 e^{-(0.116 \text{ km}^{-1})h}$$

or

$$P = \frac{P_0}{e^{(0.116 \text{ km}^{-1})h}}$$

The quantity e is the base of natural logarithms and is equal to 2.72. The pressure P equals the pressure P_0 when $h = 0$, since any number (e in (*continued*)

Composition of Troposphere and Stratosphere

Material	Percentage by Volume of Dry Air
Nitrogen molecules	78
Oxygen molecules	21
Argon atoms	1
Carbon dioxide molecules	0.03
Ozone (largely in stratosphere)	Traces
15 or more other gases (including pollutants[a])	0.04[b]
Particulate matter (dust)	Traces[b]
Water vapor	Trace to 4

[a] Pollutants include carbon monoxide (nearly half of all created pollutants), sulfur oxides, hydrocarbons, nitrogen oxides, particulate matter.
[b] Locally at much higher concentrations in the troposphere during high wind storms, near volcanic eruptions, or over industrial areas.

this case) raised to the zero power is 1. As h increases, the denominator increases, predicting a decrease in the pressure. For example, when $h = 16$ km (about 10 mi), then

$$P = \frac{P_0}{e^{(0.116 \text{ km}^{-1})16 \text{ km}}} = \frac{P_0}{e^{1.86}}$$

$$= \frac{P_0}{(2.72)^{1.86}} = \frac{P_0}{6.40} = 0.156 P_0$$

At $h = 16$ km, $P = 0.156 P_0$, or the pressure has decreased to about 15 percent of the pressure at sea level.

This variation in pressure is an exponential function and is plotted in Fig. B. It is far from a linear relationship and indicates a slower and slower decrease in pressure per change in altitude at higher altitudes. This type of function, the exponential, appears in many natural phenomena; we shall examine other exponentially varying functions later in this book.

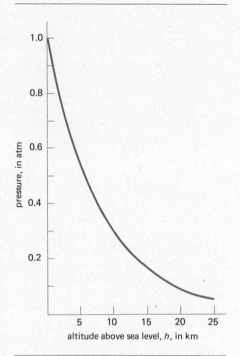

Figure B. Variation of atmospheric pressure with altitude.

medes. Consider a cube of some solid material immersed beneath the surface of a fluid of density ρ, as illustrated in Fig. 9.8. Imagine the area of each face of the cube to be A, and the length of each side of the cube to be $h_2 - h_1$, where h_2 and h_1 are the distances between the surface of the fluid and the bottom and top of the cube, respectively. Because the fluid is at rest, there is no net force on its sides. However, because of the variations in depth, the downward force on the top of the cube, F_1, is different from the upward force on its bottom, F_2. The external forces are given by the equations

$$F_1 = P_1 A = h_1 \rho g A \quad \text{downward}$$
$$F_2 = P_2 A = h_2 \rho g A \quad \text{upward}$$

Figure 9.8 The buoyant force on a cube.

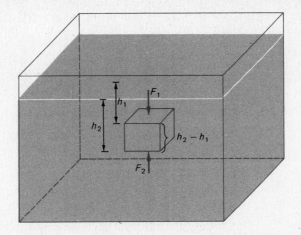

The net applied force in the vertical direction will be the vector sum

$(F_2 - F_1)$ upward $= (h_2 - h_1)\rho g A = V\rho g$

Thus, the net upward force, otherwise known as the **buoyant force, BF, is the product of the volume of the object (or equivalently the volume of the displaced fluid), the density of the fluid, and the acceleration due to gravity.** Remembering that the fluid's density is actually its mass per unit volume, we see that

$$\text{BF} = V\rho g = V\frac{m}{V}g = mg \tag{9.2}$$

In other words, **the buoyant force equals the weight of the displaced fluid.** If the object weighs less than the buoyant force, it will rise like a helium balloon through air once it is released. An object weighing less than its buoyant force will end up, after release in a liquid, floating on the surface in such a way that the weight of the liquid it finally displaces exactly equals its own weight. Of course, if the object weighs more than the buoyant force, it will sink like a stone in a pond or a pearl in Prell shampoo.

One means of demonstrating Archimedes' principle is shown in Fig. 9.9. A solid cylinder and an empty bucket into which the cylinder exactly fits are both suspended from one side of a balance and then brought to equilibrium. The cylinder is then immersed in water, Fig. 9.9(b). The balance pan on that side rises, indicating the effect of a buoyant force on the cylinder. Then the bucket is filled to the brim with water, Fig. 9.9(c), and the balancing act repeated with the cylinder in the water. This time the balance is perfect, indicating that the buoyant force exactly equals the weight of water in the bucket. Because this water occupies just the same volume as the cylinder, the experiment shows that the buoyant force equals the weight of the water displaced by the cylinder.

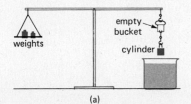
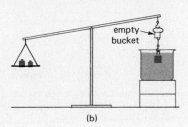
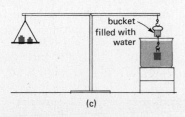

Figure 9.9 The bucket and cylinder experiment that demonstrates Archimedes' principle for immersed objects.

Figure 9.10 Weight loss! In air, the object weighs 4.5 kg × 9.8 m/s². In water, the object weighs 2 kg × 9.8 m/s².

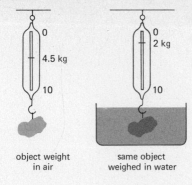

object weight in air | same object weighed in water

Figure 9.10 shows the use of Archimedes' principle to determine specific gravity. One first weighs an object, using a spring balance, and then weighs it in water. The difference between the two weights is the buoyant force; according to Archimedes' principle, this equals the weight of water displaced by the object. This leads to a new expression for specific gravity (sp. gr.):

$$\text{sp. gr.} = \frac{\rho_{\text{object}}}{\rho_{\text{water}}} = \frac{\text{weight of object}}{\text{weight of equal volume of water}}$$

$$= \frac{\text{weight of object}}{\text{weight of water displaced}}$$

$$= \frac{\text{weight of object}}{\text{loss of weight in water}}$$

Note that the spring balance is calibrated in mass units—kilograms. To convert these to weight units, one must multiply them by the acceleration due to gravity, g, whose value is 9.8 m/s².

Applying this relation to the object in Fig. 9.10,

$$\text{sp. gr.} = \frac{4.5 \text{ kg} \times 9.8 \text{ m/s}^2}{(4.5 - 2.0) \text{ kg} \times 9.8 \text{ m/s}^2} = \frac{4.5}{2.5} = 1.8$$

EXAMPLE 9.4

A stone weighs 500 N and 375 N in water. Compute its specific gravity.

Solution

$$\text{sp. gr.} = \frac{\text{weight}}{\text{loss of weight in water}}$$

$$= \frac{500}{500 - 375} = \frac{500}{125} = 4$$

One important modern application of Archimedes' principle is the submarine, whose basic modus operandi is pictured in Fig. 9.11. When

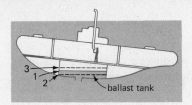

Figure 9.11 Archimedes' principle in action: how a submarine dives and surfaces.

the submarine is riding steadily beneath the surface of the ocean, the sea water in the open compartment, or ballast tank, is at level 1. In this position, the downward force—the total weight of the submarine and its contents—exactly equals the buoyant force, which is the net weight of the ocean water displaced by the submarine. To make the submarine rise, the crew pumps compressed air into the compartment, forcing the water down to level 2. This displaces more ocean water and thus produces an increased buoyant force (the weight of the displaced seawater), which lifts the vessel. When the skipper gives the order to submerge, the crew releases air from the compartment, allowing water to enter up to level 3. The action reduces the buoyant force to the point at which it is smaller than the downward force of the vessel's total weight, and the submarine sinks. (The submarine's crew also has a finer resolution means of altering the submarine's depth when it is moving forward in the water, in the form of the horizontal rudder, which controls the vessel's up-and-down movements to a small extent.)

9.3 FLUIDS IN EQUILIBRIUM

Submarines, of course, move through liquids—the waters of the ocean. But what of the movements of liquids themselves? An understanding of the laws that guide the movements of liquids and gases is essential to the study of such practical events as the pumping of blood through veins and arteries, the ebb and flow of currents in rivers and streams, the inexorable progress of the Gulf Stream from the eastern shores of North America to Northern Europe, the slow and steady buildup of storms, and the stark fury of hurricanes and tornadoes. Before moving on to such specific items, however, we must pause to consider fluid statics.

Moving at a rapid clip through almost 2000 yr of the history of science, we pass from Archimedes (third century B.C.) to the seventeenth century French mathematician and philosopher Blaise Pascal. Pascal might be best remembered by this generation of calculator-conscious students as the inventor of the first rudimentary adding machine, but physicists have attached his name to an important principle that underlies fluids. **Pascal's principle states that a change in pressure applied at any point in a liquid at rest in a closed vessel is transmitted undiminished to all parts of the liquid and to the walls of the vessel that contains it.**

The meaning of the principle can be demonstrated by the modified syringe shown in Fig. 9.12. Instead of a plain orifice at its end, the syringe contains a spherical glass bulb with a row of holes around the bulb in one plane. If one fills the syringe with water, holds it in such a way that all the holes are in a horizontal plane, and then squeezes in the piston, an unexpected effect will be apparent: the water will squirt out of the openings in streams of equal length, without any detectable variation. This confirms Pascal's observation, showing that the pressure applied to the piston is carried equally to all parts of the liquid in the bulb.

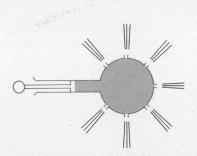

Figure 9.12 The Pascal principle. Water emerges from all the horizontal openings of the syringe, which is viewed from above, at the same speed.

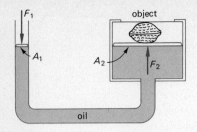

Figure 9.13 World's biggest nutcracker: a hydraulic press.

This analysis ignores atmospheric pressure, but for a very good reason: the air pressure is present on both sides of the device and thus cancels itself.

The hydraulic press is a simple, practical application of Pascal's principle. It uses a liquid to convert a relatively small force into a much larger one. The press illustrated in Fig. 9.13 is a rather intricate and unorthodox nutcracker. It consists of two pistons, one small in diameter, the other large, with a quantity of oil in the space between them. If one applies a downward force F_1 on the small piston, whose area is A_1, then the oil suffers a pressure P equal to F_1/A_1. Pascal's principle states that this pressure of the liquid against the large piston, which is to crack the walnut, is also P. If A_2 is the area of the second piston, the force F_2 exerted on the walnut is given by the equation

$$\frac{F_1}{A_1} = \frac{F_2}{A_2}$$

This equation tells us that the force applied to the first piston is enlarged by the time it gets to the second piston by the ratio of the areas of the two pistons. If the larger piston has a cross section appreciably greater than the smaller one, the press can magnify a small force quite spectacularly.

Note that even though force can be multiplied by the hydraulic press, the press is not an exception to the conservation of energy. Actually, the work input on piston 1 equals the work output on piston 2. Of course, this means that piston 1 will travel farther than piston 2. As piston 1 is pressed down in Fig. 9.13, piston 2 must rise to accommodate the liquid transfer.

The hydraulic nutcracker is a device that shows more inventiveness than commercial appeal. But a variety of devices used in real life rely on exactly the same principle: automobile lifts, barbers' and dentists' chairs, hydraulic automobile brakes, and elevators, for example.

In the automobile brake shown in Fig. 9.14 the pressure exerted on the foot pedal by the driver is translated unchanged to the brake pistons. These exert a large decelerating force on the car by pushing the brake shoes against the drums of the wheels.

Figure 9.14 How hydraulic brakes work.

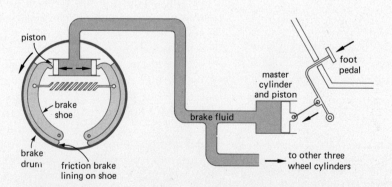

EXAMPLE 9.5

A dental assistant can apply, by means of a lever, a force of 30 lb to the piston of the dentist's chair, which has a patient in it. The force moves a small piston whose cross section is 2 in.² If the cross section of the large piston, which supports the chair and the patient, is 20 in.², compute the (a) pressure of the small piston against the fluid, (b) pressure of the fluid against the large piston, and (c) maximum total weight—of the larger piston, chair, and patient—that can be raised by the assistant.

Solution

(a) $P_1 = \dfrac{F_1}{A_1} = 30 \text{ lb}/2 \text{ in.}^2 = 15 \text{ lb/in.}^2$

(b) Hence, by Pascal's principle,

$P_2 = P_1 = 15 \text{ lb/in.}^2$

(c) $F_2 = P_2 A_2 = 15 \text{ lb/in.}^2 \times 20 \text{ in.}^2 = 300 \text{ lb}$

Because this weight includes the hefty chair, the dentist's assistant must either restrict herself to lightweight patients or be prepared to use more force on the chair's hand lever.

EXAMPLE 9.6

The gauge of a hydraulic hoist (Fig. 9.15) in a gas station indicates a compressed air pressure of 100 lb/in.² Compute the force exerted on a car on the hoist if the area of the piston that supports it is 1 ft in diameter.

Solution

The pressure against the bottom of the supporting piston under the car is 100 lb/in.² Thus, the force on the car, F, is given by

$F = PA = P\pi r^2 = (100 \text{ lb/in.}^2)\pi(6 \text{ in.})^2 = 11\,300 \text{ lb}$

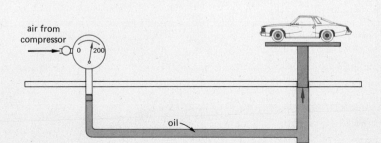

Figure 9.15 The hydraulic hoist.

9.4 THE BERNOULLI EFFECT

Having studied the statics of fluids, we can now move on to free flow, or streamline flow as it is often called. The term applies to the simple movement of liquids or gases from one point to another without any rotation or turbulence—again, somewhat of an idealized situation. The underlying principles of streamline flow are Newton's laws. However, the basic mathematics are more involved because the motion of fluids is inherently more complex than that of solids. One major reason for the mathematical complexity is the existence of viscous forces that retard the flow of liquids in much the same way that frictional forces slow down solid objects; another reason is the dependence of the pressure of a fluid on the velocity with which it is traveling. We shall examine this latter effect first.

Each year when the hurricane season arrives in the southern United States or the twisters start to whistle up tornado alley in the midwest, newspapers contain gruesome reports of high winds taking off roofs of houses and blowing out windows of stores, homes, and factories. The general impression from these reports is that the huge pressures of the whistling winds cause the intense damage. But that impression is entirely wrong. Certainly, the high winds lead to damage, but in an indirect fashion. The pressure in the winds is actually lower than that of the still air in the houses before they are damaged (Fig. 9.16). It is actually the pressure *inside* the houses that pushes off (explodes) the roofs and blows out the windows. This unexpected relationship between velocity and pressure in moving fluids was first outlined by the eighteenth century Swiss scientist Daniel Bernoulli.

A simple picture of how movement alters the pressure of liquids can be obtained from the two sets of glass tubes shown in Fig. 9.17. The two parts of the figure show water flowing from left to right along a

It is the pressure inside the houses that pushes off the roofs and blows out the windows.

Figure 9.16 Hurricane damage. The buildings actually explode. (*UPI*)

horizontal pipe. The heights to which the water rises in the vertical pipes attached to it provide a measure of the pressure of the water at those horizontal points.

In the top picture the water flows through a pipe whose cross section remains constant: the water levels indicate the reduction in pressure as the water moves farther along the pipe. This reduction is caused, of course, by the water's viscosity. It is experienced frequently in practical situations; the relatively low pressure of water from a city supply in houses far from the pumping station is an example.

The situation in the lower picture of Fig. 9.17 is somewhat different. Here the pipe has a narrow, constricted section in it, attached to its own vertical tube. To make its way through the constriction, the water must speed up when it reaches it. To speed up the fluid from the wide region into the narrow part of the pipe, a force has to be applied; hence the pressure must be larger in the wide part. For the fluid to slow down when entering the wide part, an opposing pressure difference must develop. As the figure shows, the water in the center tube, above the constriction, occupies the lowest level of the three tubes; the pressure at this point, where the horizontal flow is fastest, is plainly the least of the three.

Bernoulli used the principles of conservation of energy and mass to explain this phenomenon. Assuming an ideal fluid, which is incompressible and moving in streamline flow, Bernoulli stated that the sum of the fluid's energies per unit volume—its kinetic energy, its potential energy as represented by the fluid pressure, and its gravitational potential energy—is constant, as long as no energy is added to or removed from the fluid. Putting this statement into algebraic terms,

$$\tfrac{1}{2}\rho v_1^2 + h_1 \rho g + P_1 = \tfrac{1}{2}\rho v_2^2 + h_2 \rho g + P_2 \tag{9.3}$$

The subscripts refer to two different but arbitrarily chosen points such as A, B, or C illustrated in Fig. 9.18. The first term in the equation is the kinetic energy per unit volume, where $\rho = m/V$. The quantity $h\rho g$ is the gravitational potential energy per unit volume; this is constant in the case under study because the pipe is horizontal and $h_1 = h_2$. The third term in Eq. (9.3) is the pressure in the fluid, neglecting any viscosity or turbulence. If viscous forces are present, they do work on the fluid, reducing both the internal pressure in the fluid and its kinetic energy. It is clear from the equation that as long as Bernoulli's principle is applicable, flow in a horizontal pipe is such that the speed increases when the pressure decreases because the sum of $\tfrac{1}{2}\rho v^2$ and P must stay constant. The pressure in Fig. 9.18 is indicated by the level of the water in the small, open-ended vertical tubes. The velocity in the same figure is indicated by streamlines that indicate smooth paths of steady flow. These streamlines are crowded in narrower regions, indicating a greater velocity. If the velocity becomes too great, forms of turbulence such as vortices and eddies will develop; Bernoulli's principle will then no longer apply.

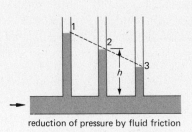

reduction of pressure by fluid friction

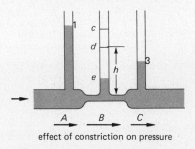

effect of constriction on pressure

Figure 9.17 Bernoulli's principle.

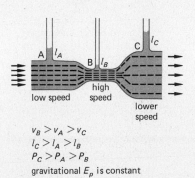

$v_B > v_A > v_C$
$l_C > l_A > l_B$
$P_C > P_A > P_B$
gravitational E_p is constant

Figure 9.18 The total energy in a flowing fluid.

EXAMPLE 9.7

A storage tank (Fig. 9.19) contains water to a depth of 4.9 m. A small hose is connected to a tap in the side of the tank at the very bottom. (*a*) At what speed does the water come out of the tank initially? (*b*) If the hose were pointed upward, to what height would the water rise, assuming the water height in the tank remained constant at 4.9 m?

Solution

According to Bernoulli's equation, the total energy remains constant. If we write Bernoulli's equation for the water at the top of the tank and at the bottom of the tank as the water leaves the opening, we have

$$\tfrac{1}{2}\rho v_{\text{top}}^2 + E_{p_{\text{top}}} + P_{\text{top}} = \tfrac{1}{2}\rho v_{\text{bottom}}^2 + E_{p_{\text{bottom}}} + P_{\text{bottom}}$$

At the top of a wide container the speed is negligible; hence the first term is zero. The potential energy per unit volume required here is given by $E_{p_{\text{top}}} = h\rho g$. The pressure at both the top and the opening in the bottom is equal to the atmospheric pressure P_{atm}.

Hence,

$$P_{\text{atm.}} + h\rho g = \tfrac{1}{2}\rho v_{\text{bottom}}^2 + P_{\text{atm.}}$$

Canceling the densities, and P_{atm}.

$$v_{\text{bottom}} = \sqrt{2gh} = \sqrt{2(9.8 \text{ m/s}^2)(4.9 \text{ m})} = 9.8 \text{ m/s}$$

This is just the same result as one obtains for an object falling a distance *h*. If the water comes out of the tank in an upward direction and the process occurs at atmospheric pressure, application of the conservation of energy indicates that the $E_{p_{\text{gr}}}$ at the top of the stream equals the original E_k as the water just leaves the tank; that is,

$$\tfrac{1}{2}\rho v^2 = h\rho g \quad \text{or} \quad h = \tfrac{1}{2}v^2/g = 4.9 \text{ m}$$

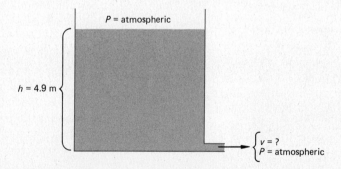

Figure 9.19 Speed of the water passing out of a tank containing water 4.9 m high.

A more spectacular demonstration of Bernoulli's principle is an apparent gravity-defying performance that can be put together using a

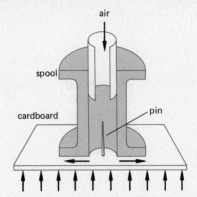

Figure 9.20 The cardboard and spool trick. The pin restricts the lateral motion.

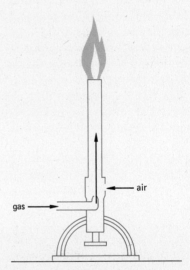

Figure 9.21 The action of a Bunsen burner.

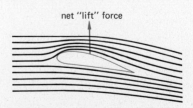

Figure 9.22 Lift of an airplane wing. The wing is shaped in such a way that it produces streamline flow, with the air moving faster above it than below it.

simple cotton spool and a large piece of cardboard. If you blow through the cylindrical opening in the spool (see Fig. 9.20), you can hold the piece of cardboard in place beneath the spool; as soon as you stop blowing, the cardboard falls to the floor.

The reason for the fall is obvious. When no air is blown through the spool, the pressure of the atmosphere downward on the cardboard is just the same as atmospheric pressure upward on it. The weight of the piece of cardboard causes it to drop. When air is blown through the spool, it rushes out of the thin opening between the bottom of the spool and the cardboard. According to Bernoulli's principle, this high-velocity stream of air is at a reduced pressure. The result is a net upward force that is sufficient to overcome the downward force of the cardboard's weight. In action, the cardboard seems to be touching the spool; however, it undergoes small movements, illustrating the presence of a thin layer of air above it.

Another example of the Bernoulli principle occurs in the chemistry laboratory. If the side opening of the Bunsen burner in Fig. 9.21 is closed, a smoky flame results; if it is opened, a hot, blue flame results. The hot flame is caused by the admission of air into the burner tube, and we might ask how the air manages to get in. If the orifice is open before the gas is turned on, the air pressure is the same inside and outside the opening. When the gas is turned on, a stream of it rushes up at high speed through the small hole at the bottom of the shaft. Its effect is to reduce the pressure inside the burner tube. The excess pressure of the outside atmosphere forces air into the burner hole, where it mixes with the gas and produces a hot blue flame.

The Bernoulli principle is also vital to the success of aviation. An airplane wing is designed as shown in Fig. 9.22, and as a result the air moves with a greater velocity above the wing than below it. The normal air pressure above the wing is reduced somewhat, whereas the pressure below is unchanged, producing a net upward force. This force, together with the force of the air moving against the bottom of the wing, provides the lifting force on the wing.

The Bernoulli effect also explains a number of everyday occurrences. The feeling that car drivers have of being sucked toward large, passing vehicles is more than feeling, for example. It is real because the action of passing reduces the pressure between the two vehicles. A shower curtain is tugged in toward the shower because the fast-moving shower water lowers the pressure inside the shower stall. The vacuum cleaner sales display uses the exhaust air blowing upward to "support" a beach ball, keeping the ball confined in the air column. And a canny curve-ball pitcher instinctively uses the principle.

9.5 VISCOSITY: A REAL DRAG

If we used dye markers to follow the motion of any liquid through a tube, they would reveal that the movement occurs in a far from uniform manner. Looking at the fluid's cross section, we would see that motion takes place along a U-shaped front. The liquid at the center of the tube moves fastest, and that close to the inside walls of the tube moves slowest. Closer examination of the flow would show the reason for this strange profile. The thin layer of liquid at the very edge that is in direct contact with the inner wall of the tube remains stationary, stuck to the wall by molecular forces. This thin layer exerts a drag on the liquid layer immediately inside it, which, in turn, has some retarding effect on the layer inside it. **The overall dragging effect is known as the viscosity of the liquid.**

Mathematical analysis shows that, in the case of liquid flowing without turbulence or rotation through a cylindrical pipe of length L, the strength of the viscous retarding force F_v is related to the peak velocity v_m, the velocity in the center, by the equation

$$F_v = 4\pi \eta L v_m \tag{9.4}$$

The Greek letter η (eta) stands for the coefficient of viscosity. It is a constant for any specific fluid at a particular temperature. Its magnitude indicates the ease, or lack of it, with which the fluid flows. Water, for example, which has a low viscosity, flows very easily. By contrast, molasses, with a very high viscosity, is a very sluggish-moving liquid. Motor oils are rated according to viscosity.

The viscous force F_v opposes the motion of the liquid. To keep the liquid moving with a constant speed, therefore, we must apply some kind of driving force equal in size to F_v. This driving force is measured by a difference in pressure of the liquid. We could say that the liquid enters the section of pipe at a high pressure P_1 with a force $F_1 = P_1 A$ and leaves the section after traveling a distance L with a lower pressure P_2, which exerts a force $P_2 A$ in the same direction. This force provides the drive necessary to move the liquid farther along the pipe. F_2 and P_2 are smaller than F_1 and P_1 because of the conversion of mechanical energy to heat as a result of viscosity.

When the peak velocity of the liquid is constant, we know from Newton's first law that no net force acts on the liquid. Therefore, the driving force equals the viscous force. Put in the form of an equation,

$$P_1 A - P_2 A = 4\pi \eta L v_m$$

If r is the radius of the pipe, the area $A = \pi r^2$. Hence, by substituting for A and canceling the π's, we have

$$\Delta P = P_1 - P_2 = \frac{4\eta L v_m}{r^2} \tag{9.5}$$

EXAMPLE 9.8   Blood is traveling through a small capillary whose radius is 1.5×10^{-6} m. If the maximum speed of the flow is 0.5 mm/s and the viscosity of the blood 0.004 N · s/m², calculate the pressure drop of the blood in traveling along 2 mm of the capillary.

Solution   In Eq. (9.4), $L = 2 \times 10^{-3}$ m, $r = 1.5 \times 10^{-6}$ m, and $v_m = 0.5 \times 10^{-3}$ m/s; therefore,

$$\Delta P = P_1 - P_2 = \frac{4\eta L v_m}{r^2}$$

$$= \frac{4(0.004 \text{ N} \cdot \text{s/m}^2)(2 \times 10^{-3} \text{ m})(0.5 \times 10^{-3} \text{ m/s})}{(1.5 \times 10^{-6} \text{ m})^2}$$

$$= P_1 - P_2 = 7.11 \times 10^3 \text{ N/m}^2$$

This can be converted to mmHg, recalling that 1 atm = 760 mmHg = 1.013×10^5 N/m²:

$$P = 7.11 \times 10^3 \text{ N/m}^2 \times \frac{760 \text{ mm}}{1.013 \times 10^5 \text{ N/m}^2} = 53.3 \text{ mmHg}$$

One critical factor in all problems involving the flow of fluids is the actual amount of liquid that passes through a pipe in a specific time. To determine the volume of liquid flowing through a pipe per second, per minute, or per hour, one must know the average velocity of the liquid. Because of the uneven profile of a moving liquid front, this is a somewhat difficult quantity to calculate. In the simple cases of liquid flow through a pipe with a circular cross section, in which the velocity varies from zero at the inside wall of the pipe to a maximum v_m in the very center, we can consider the average velocity $\bar{v}$ to be simply $v_m/2$. Extending Eq. (9.5), we have

$$\bar{v} = \frac{v_m}{2} = \frac{(P_1 - P_2)r^2}{8\eta L}$$

The quantity of liquid flowing through the pipe in a given time has the symbol Q. It represents the volume of liquid traveling through the pipe in a particular length of time divided by that time. One can think of Q as the rate at which the fluid pours out the end of the pipe. If we consider liquid traveling along a pipe whose cross section is A, the area

A can be regarded as generating a volume $\Delta V = A\,\Delta s$ when a particular front of liquid moves along the length Δs. If this movement occurs in time Δt, then the average velocity $\overline{v}$ is simply $\Delta s/\Delta t$. Thus, $\Delta t = \Delta s/\overline{v}$, and our equation for the time rate of flow becomes

$$Q = \frac{\Delta V}{\Delta t} = \frac{A\,\Delta s}{\Delta s/\overline{v}} = A\overline{v} = \pi r^2 \overline{v}$$

Hence, the quantity of liquid flowing per unit time is readily determined from the previous two equations:

$$Q = \frac{\pi r^4 (P_1 - P_2)}{8L\eta} \tag{9.6}$$

The quantity of liquid that flows per unit time is thus proportional to the fourth power of the radius of the pipe and directly proportional to the pressure drop in the pipe. This is known as Poiseulle's law, after the French scientist Jean Léonard M. Poiseuille.

> Because ΔP depends on $1/r^4$, constrictions in blood vessels clearly put an extra large load on the body's pump, the heart; increased pressure is required to push blood through the constrictions.

Movement through pipes is just one simple example that involves the concept of viscosity. Another set of problems related to viscosity is the movement of spherical drops of a solid or liquid at low speeds through a gas or a liquid. The descent of raindrops through the air is one example of this type of flow through a gas; the famous advertisement in which a pearl drops through Prell shampoo is another.

The viscous force in these cases acts to slow down the plummeting objects, preventing them from attaining free fall. Instead, they reach a constant velocity—the terminal velocity—when the viscous force exactly equals the weight reduced by the buoyant force. The size of the viscous force, F_v, is given in these cases by Stokes' law:

$$F_v = 6\pi \eta r v \tag{9.7}$$

> The law is named for British mathematician Sir George Gabriel Stokes (1819–1903), who represented Cambridge University as its member of Parliament from 1887 to 1891.

where r is the radius of the object, η the coefficient of viscosity, and v is the terminal speed. Stokes' law is only applicable to situations of low-speed, streamline flow; it does not apply to sky divers because of the turbulence they encounter and because they are not spherical.

EXAMPLE 9.9

A raindrop with a radius of 0.1 mm is dropping through air that has a viscosity of 1.8×10^{-5} N·s/m². Calculate the raindrop's terminal speed.

Solution

At terminal velocity the weight of the drop equals the Stokes' law viscous force

$$mg = 6\pi \eta r v$$

Hence,

$$v = \frac{mg}{6\pi \eta r}$$

One of the most ingenious applications of this type of situation is found in a totally unrelated field of physics—the determination of the fundamental electric charge by the Millikan oil-drop experiment. We shall encounter this classic experiment in Chap. 13.

We know the value of all quantities except the velocity and mass of the drop. Since mass is the product of density and volume,

$$v = \frac{\rho(\frac{4}{3}\pi r^3)g}{6\pi \eta r} = \frac{2\rho g r^2}{9\eta}$$

$$= \frac{(2)(1000 \text{ kg/m}^2)(9.8 \text{ m/s}^2)(10^{-4} \text{ m})^2}{9 \times 1.8 \times 10^{-5} \text{ N} \cdot \text{s/m}^2} = 1.21 \text{ m/s}$$

SUMMARY

1. The pressure of a liquid is independent of the area over which it acts.
2. A liquid's density varies to a very small extent with depth, but the density of a gas varies markedly with altitude. This effect controls the basic behavior of the earth's atmosphere.
3. Fluids tend to move from regions of higher pressure to regions of lower pressure.
4. As stated in Archimedes' principle, the upward, or buoyant, force on any object floating on or immersed in a fluid equals the weight of fluid that the object displaces.
5. Pressure applied to a liquid is transmitted unaltered to all parts of the liquid. This principle is applied in the hydraulic press.
6. The pressure of a fluid moving in streamline flow falls with increasing speed. This is the Bernoulli principle, which states that the sum of the energies per unit volume—kinetic energy, the potential energy caused by the pressure of the fluid, the gravitational potential energy—of a fluid is constant during streamline flow.
7. Forces of viscosity retard moving fluids in much the same way that frictional forces slow down moving solids. Viscosity gives a characteristic U-shaped profile to the front of moving fluids in a pipe.
8. The quantity of liquid flowing through a cylindrical pipe per unit time is directly proportional to the pressure drop across the liquid in the pipe and proportional to the fourth power of the pipe's radius.
9. A drop of a solid or liquid that is falling through a fluid eventually reaches a constant "terminal velocity" when its weight equals the viscous force acting on it.

QUESTIONS

1. Explain how to make a simple mercury barometer. What is a barometer supposed to measure? Does it do this directly or indirectly? Discuss briefly.
2. Explain how a drinking straw works.
3. Assume that the water in a lake is at the top of the dam but not flowing through the spillway. How would the pressure at any point on the immersed surface of the dam be affected (increased, decreased, or remain the same) if the water is backed up from the dam a short distance (a few miles) or a longer distance (many miles)? That is, does the horizontal distance of water behind a dam affect the pressure on the dam, assuming that the water remains up to the top of the dam? Give reason(s) for your answer.
4. Diagram and explain at least one application, not given in this textbook or by your instructor, of Pascal's principle, Archimedes' principle, Bernoulli's principle.
5. With the aid of a properly labeled diagram, explain in terms of Bernoulli's principle the action of a Bunsen gas burner, an atomizer or sprayer, nondraft ventilation of an automobile.
6. Do Pascal's principle, Archimedes' principle, and Bernoulli's principle apply to both liquids and gases? Explain your answer and give illustrative examples for each principle.
7. Differentiate between force and pressure.
8. Why does a boat float higher in salt water than in pure lake water?
9. Which displaces a greater quantity of water when submerged, a cubic centimeter of aluminum or a cubic centimeter of lead?
10. A toy boat containing a rock is placed in a bowl of water. If the rock is removed from the boat and put into the bowl, does the water level in the bowl rise, stay the same, or go down?
11. An ice cube is placed in a glass of water that is completely full. When the ice melts, will the glass overflow? Does the water go down or remain the same? Explain.
12. Fill a glass with water, place a piece of cardboard over the glass, and invert the glass holding the cardboard in place. What keeps the water

in the glass? For how tall a glass would this work?

13. To remove the contents of a can of frozen orange juice or cranberry jelly, you remove one end, invert, and, if you punch a hole in the closed end on top, the contents "plop" out quickly. Why? Explain.

14. A filled air balloon that rests on the floor is dropped into a mine shaft. What will happed to the balloon if the shaft is very deep?

15. About how much weight of air in the atmosphere does your body support?

16. Why is your car attracted to a large truck or bus when you pass the truck or bus at high speed?

17. You wish to monitor a chemical reaction, e.g., the fermentation process. Could you easily determine the specific gravity of your brew using a small spring balance, a bucket of water, and a piece of copper? Once determined, could the specific gravity be used to determine the alcohol content in the brew? Explain.

PROBLEMS

Hydrostatics

Unless otherwise specified, use 76.0 cmHg = 1.01×10^5 N/m² = 14.7 lb/in.² for atmospheric pressure.

1. A fish tank 0.6 m long, 0.3 m wide, and 0.4 m high is full of pure water. Compute the (a) force and (b) pressure on the bottom of the tank, (c) pressure 0.2 m from the top surface of the water, and (d) force on the larger vertical face.

2. If the air pressure in a respirator is 25 cmH$_2$O, express this pressure in cmHg, lb/in.², N/m², and Pa.

3. What depth of water is equivalent to 1 atm of pressure?

4. At what altitude above the earth will the barometric pressure decrease to one-half its value at sea level?

5. If the gauge pressure of blood leaving the heart is 100 mmHg, what is the absolute pressure?

6. If a diving sphere is lowered 1 km under the sea, what would be the pressure on it? Express in terms of atmospheres.

7. What is the pressure on a diver who is 10 m below the surface of a lake?

8. Estimate the total force on your chest if you dive 7 m under water.

9. A submarine dives to a depth of 30 m. What is the net force on a hatch on the sub if the hatch has a diameter of 0.8 m? The inside of the sub is kept at atmospheric pressure.

10. If you attempt to plug a hole in a dam 5 m below the water surface, (a) what pressure must you exert? If the hole you wish to plug is the size of your thumb, (b) estimate the force on your thumb if it acts as the plug.

11. If someone opens the drain of a swimming pool and you place your foot over the drain, what force will be exerted on your foot? Consider the drain 8 cm in diameter and the water 2.5 m deep. The pressure in the empty drain below your foot is 1 atm.

12. A class of physical science students did not believe that the maximum height to which water could be "sucked" at normal atmospheric pressure was 34 ft. Therefore, a tuba player in the class decided to experiment with this question. Standing on the lecture table, he was easily able to "suck" water up a straw 9 ft long. The straw was made of three sections of glass tube with a small rubber tube attached. What was the pressure differential created by his lungs between his lungs and the atmosphere? Eventually, in a stairwell he "sucked" water over 20 ft. What was the differential then? (*Note:* Caution must be used to do this experiment safely.)

13. If the difference in height of the two columns of mercury in a mercury manometer is 15 cmHg, what is the absolute pressure in the vessel?

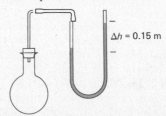

14. Estimate the hydrostatic blood pressure in your foot artery, assuming the density of blood is 1 g/cm³.

15. Two hollow hemispheres 10 in. in diameter are placed together and the air removed to create a pressure of 0.10 atm. inside. What force is required to separate the two halves?

16. A barber's chair is connected with a large cylinder that has a cross-sectional area of 15 in.² The chair is raised by applying a force of 45 lb to a small piston that has a cross-sectional area of 1.5 in.² The two pistons are part of a hydraulic press arrangement, with a heavy oil in the chamber between them. Compute the (a) pressure of the small piston against the fluid; (b) pressure of the fluid against the large piston; (c) maximum weight, including the weight of the chair, large piston, and occupant, that can be lifted.

17. A passenger car is driven on the rack of a hydraulic hoist. The car weighs 4000 lb, and the rack, large cylinder, and piston of the hoist weigh 1000 lb. The compressed air is turned on, and a gauge shows that the pressure is 110 lb/in.² The bottom of the hoist piston has a cross-sectional area of 90 in.² (a) Compute the upward force against the hoist piston. (b) If the car is replaced by a truck that weighs 5 tons with contents, what happens when the compressed air is turned on?

18. The cross-sectional area of a piston in a water pump is 10 in.² What force is required to raise the water

90 ft with the piston?

19. If the heart pumps blood into an artery with a gauge pressure of 110 mmHg and the artery's diameter is 0.8 cm, what is the force exerted by the heart on the blood in this artery?

20. A hypodermic syringe acts as a hydraulic device. If a syringe with a 1.5-cm-diameter piston is pushed with a force of 5 N, what is the pressure of the fluid in the syringe above atmospheric pressure (the gauge pressure)? If the needle has a 0.8-mm diameter and is inserted into a vein to inject fluid into the blood, with what force would you have to push the syringe if the blood pressure (gauge pressure) is 16 mmHg?

Archimedes' Principle

21. A block of copper has a mass of 1.5 kg. What will it "weigh" under water? The specific gravity of copper is 8.5.

22. An object "weighs" 100 g in air, 88 g in water, and 90 g in alcohol. Compute the (a) density of the object, (b) density of the alcohol.

23. What fraction of an iceberg is below the surface of the water? The density of ice is 917 kg/m^3.

24. A stone "weighs" 40 kg in air, and its "apparent weight" under water is 36 kg. What is the volume of the stone and its specific gravity?

25. A piece of wood with a specific gravity of 0.5 and a volume of 0.05 m^3 floats on water. What volume is below the water surface, and what weight can be placed on the wood to just completely submerge the wood?

26. What are the volume and average specific gravity of a 45-kg human who floats with 10 cm^3 of her volume above the surface of pure water?

27. A tanker barge is 30 m long and 20 m wide. If the tanker is filled with 1200 m^3 oil of specific gravity 0.8, how far will it sink into the water?

28. A rectangularly shaped barge 60 by 30 ft with vertical sides sinks into the water an additional 6 in. when loaded with trucks and cars. Compute the weight of the trucks and cars that the barge ferried.

29. A piece of metal floats in mercury with 84 percent of its volume submerged. What is the specific gravity of this metal? The specific gravity of mercury is 13.6.

The Bernoulli Effect

30. A horizontal tube 2 cm in diameter is constricted to 0.5 cm. If the flow rate in the large tube is 8 cm/s, what is the rate in the narrow tube?

31. Water flows through a horizontal tube at a speed of 0.5 m/s when the pressure is 2×10^6 N/m^2. The tube diameter of 4 cm is reduced to 0.4 cm. (a) Compute the speed of the water in the constricted part of the tube. (b) What is the pressure in the narrow part of the tube?

32. Water flowing at 3 m/s in a 15-cm-diameter pipe is connected to a 7.5-cm-diameter pipe. Compute the speed of the water in the smaller pipe.

33. What is the speed of alcohol running out of a small hole in the side of a can full of alcohol if the hole is 20 cm below the top of the can, which is 5 cm in diameter?

34. A crack occurs in a stand pipe 6 m below the top surface of the water. If the area of the crack is 1.3 cm^2, what is the force required to plug the leak, and with what speed would the water initially spurt out of the crack?

35. If a 96 km/h (60 mi/h) wind blows across the roof of your house, compute the reduction in pressure outside the house. The density of air is 1.25 kg/m^3, and the pressure in still air is 1 atm (10^5 N/m^2).

36. An airplane wing is 7 m long, 2 m wide. If the airplane's lift is 8000 N, half on each wing, what is the pressure difference between the wing's upper and lower surfaces?

37. Water is pumped at a rate of 9 m/s from a river through a 10-cm-diameter pipe up over the bank a height of 3 m, then through a constricted pipe of one-half the original diameter. The water finally is discharged into the air, falling into an irrigation ditch. (a) What is the rate of flow in the upper pipe, and what are the pressures in the (b) upper and (c) lower pipes (neglect viscosity)?

Viscosity

38. Compute the terminal speed of a pearl 2 mm in radius if it drops through glycerin whose viscosity is 1.49 N · s/m^2. The specific gravity of the pearl is 1.2.

39. What is the average speed of blood in capillaries of length 0.05 cm and radius 0.004 mm if the pressure change across the capillary is 10 mmHg? $\eta = 0.004$ N · s/m^2 for blood.

40. What is the pressure difference of water in a glass capillary 25 cm long of radius 0.5 mm if the average speed is 0.7 m/s? The viscosity of water is 0.0008 N · s/m^2.

41. What pressure is required to send water through a hypodermic needle of length 2 cm and diameter 0.3 mm at an average rate of 0.9 m/s? The viscosity of water is 0.001 N · s/m^2.

42. If the average speed of blood flow through a small artery of length 0.2 cm and radius 2×10^{-2} cm is 0.1 cm/s, what is the fluid flow through the capillary and the pressure drop across the artery? The viscosity of blood is 0.004 N · s/m^2.

43. Determine the terminal speed of a water drop whose radius is 0.02 cm if the viscosity of air is 1.9×10^{-5} N · s/m^2 at 40°C.

Special Topic
The Circulatory System

Anyone who has tried to hook up a washing machine or dishwasher or even repair a leaky faucet can testify that the real motion of fluids is far less simple that one might expect on the basis of the fundamental laws that form the bulk of this chapter. In fact, plumbing systems actually obey the laws with a fair degree of accuracy, except when the water flows through the pipes so fast that its flow is no longer streamline. The difficulties for do-it-yourself plumbers arise mainly from the complex, curving paths that water pipes generally take. But when put against another type of plumbing—that of the blood's circulatory system—the difficulties of the domestic variety are quite insignificant. The flow of blood through the vessels of the circulatory system differs in so many ways from the simple forms of fluid motion on which the fundamental laws of fluid dynamics are based that physicists and physiologists must virtually rethink the rules when they try to understand it.

The blood circulation of human beings—and of all other mammals, for that matter—is a closed system in which blood is pumped throughout the body by the heart to deliver oxygen and nutrients to the tissues via the arteries and to remove such waste products as carbon dioxide through the veins. The complexities in understanding this flow system arise right from the start in the very nature of the fluid. Blood can in no way be considered as a simple fluid; it contains many solids, most notably the red blood cells that are actually responsible for the transport of oxygen around the body.

The flow of blood is also different from that involved in theoretical studies of fluids because the pumping of the heart sends the red liquid coursing around the circulatory system in spurts or, as doctors say, pulses. The maximum pressure of the pulse is known medically as the systolic pressure, and the lowest pressure between pulses is known as the diastolic pressure. Even if one incorporates these two unusual effects into the equations of fluid flow, one still has difficulties fitting the expressions to the facts for the blood only just manages to fulfill the conditions for streamline flow—the only conditions under which Bernoulli's and Poiseulle's principles apply. Indeed, in some types of circulatory ailment, such as aortic stenosis, the flow of blood becomes definitely turbulent.

To add further to the complications of studying the circulation, blood vessels are anything but rigid tubes of fixed diameter. The arteries and veins are actually flexible pipes that dilate in response to any surge of blood through them, not only when their owner partakes of vigorous exercise, but also during the normal pumping action of the heart. Since the flow rate of fluids, according to Poiseulle's law, is proportional to the fourth power of the radius of the tube through which they flow, even a small dilation of this sort has a marked effect in the blood flow. The flexibility of blood vessels serves an important purpose. As the vessels return to their normal radii once the peak of the pulse has passed, they exert a small pressure of their own on the blood that prevents the rate of flow through the system from dropping right down to zero between successive pumping beats of the heart. This small pressure assures all parts of the body of a continuous supply of oxygen.

Once pumped out of the heart, at a typical pressure of about 110 mmHg, the blood moves through a series of increasingly narrow vessels that take it to the tissues which are its target. First comes the aorta, the body's main artery, whose radius of about 0.9 cm is so large that the pressure of the blood is only reduced by a few millimeters of mercury in transit through it. Branching off the aorta are the major arteries that reduce the blood pressure by slightly more than that amount. Next, the blood flows into the large arteries, which reduce the pressure by slightly less than 20 mmHg, and smaller vessels known as arterioles that take the pressure down by a further 55 mm. Finally, the blood pressure is reduced by another 20 mmHg in the capillaries, hair-thin vessels that deliver the oxygen-loaded blood to the tissues. The pressure drop through the capillaries is less than that in the wider arterioles because there are more of the former than of the latter, and hence less actual flow of blood through any specific capillary.

By the time the blood has yielded up its load of oxygen and started on its way back to the heart through the venules, the venal system's equivalent of the capillaries of the arterial system, its pressure is down to no more than a few millimeters of mercury. This drops even further as the blood moves on through increasingly wide small veins, large veins, and major veins. However, one-way valves in the veins ensure that the blood continues to flow in the correct direction; and in the final stages of the circulation, the flow is helped along by muscles that contract and squeeze the blood toward the heart.

Given this fluid dynamic look at the circulatory system, we are in a position to obtain a simple idea of the reasons for certain circulatory ailments, such as heart disease. If, for example, an artery wall is coated by internal deposits of materials such as cholesterol, then the speed at which blood flows through the constriction is increased. As a result, according to Bernoulli's principle, the potential energy of the blood decreases, thus reducing the blood pressure in the area of the constriction. In some parts of the arterial system, this drop may be so great that external pressure on the artery is sufficient to close up the vessel completely, thus blocking the supply of blood. If this effect occurs in the coronary artery, which supplies blood to the heart muscle, the heart simply stops; death quickly follows unless heroic medical measures are instituted rapidly.

10 Thermal Energy and Thermal Properties of Matter

10.1 TAKING THE TEMPERATURE

SHORT SUBJECT: Measuring the Mercury: The Origins of Temperature Scales

10.2 EXPANDING ON EXPANSION

10.3 MOVING IN ALL DIRECTIONS

10.4 A QUESTION OF CALORIES

SHORT SUBJECT: The Hot, the Cold, and the Thermometer

10.5 HIDDEN HEAT

10.6 THE MECHANICAL EQUIVALENT

10.7 HEAT, TEMPERATURE, THERMAL ENERGY, AND KINETIC THEORY

10.8 THE TRANSFER OF HEAT
Conduction
Convection

Summary
Questions
Problems

SPECIAL TOPIC: Holding Down Those Heating Bills

The scene is as American as apple pie. Hundreds of thousands of spectators at the Indianapolis 500 react with various amounts of shock, horror, and curiosity as two cars collide and cause a sickening pileup of mangled metal on the short straight in front of the pits. Drivers rounding the turn before the straight slam on their brakes in a frantic effort to avoid the deadly melee. As they do, their vehicles add clouds of smoke and the acrid smell of burning rubber to the chaotic, fiery scene. The smoke and the heating of the rubber obviously arise from the tires as the drivers fight with the brakes of their cars. But we might ask more specifically from where does the heat come that produces the smoke and burns the rubber and where do the huge amounts of kinetic energy of the speeding autos go during their spectacular slowdowns?

More mundane events raise similar questions. What, for example, happens to the kinetic energy of a falling object, such as a book, pen, or parachuted package, when it hits the ground and stops still? Where does the kinetic energy of a hammer go when it hits a nail on the head? The simple answer to these questions is from Chap. 5: the mechanical energy of all the objects in question is converted entirely to other types of energy, most notably heat and sound. In this chapter we shall explore this answer in some detail by zeroing in on the nature of heat and temperature, studying how they are related to energy in general and conservation of energy in particular, and examining some of the thermal properties of matter.

One of the earliest demonstrations of the link between heat and mechanical energy was in 1798, as noted in Chap. 8. American-born Count Rumford, the Bavarian minister of war, noticed that the process of boring a cannon heated up the cannon itself, the boring tool, and the chips that the tool extracted. One year later, the young English chemist Sir Humphry Davy showed the connection in a different fashion. He melted two pieces of ice simply by rubbing them together. The mechanical energy of the rubbing motion provided enough thermal energy—or heat—to melt the pieces.

The two examples illustrate the conversion of mechanical energy into thermal energy. The reverse process—conversion of thermal energy into mechanical energy—can also be demonstrated quite simply by using the well-established fact that most objects expand to a greater or lesser extent when they are heated. As it expands, matter can be made to perform mechanical work. And as might be expected, more heat is required to increase the temperature of a sample of any particular gas if the gas is free to expand against external pressure, thus

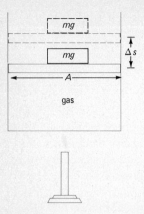

Figure 10.1 Expansion of gas by heating under constant pressure.

performing mechanical work, than if the sample is confined in a specific volume and therefore performing no work as it is heated.

Figure 10.1 shows a laboratory demonstration of the conversion of heat into work. A gas in a vertical cylindrical vessel can expand, moving a piston on which a weight is perched. For simplicity, the piston is regarded as being friction free. Obviously, the weight on the top of the piston remains constant whatever happens to the gas. If we assume that the weight of the piston itself is so small as to be negligible, then the pressure on the gas—also constant—is that of the weight above the piston, mg, distributed over the area of the piston, A, plus the atmospheric pressure. Mathematically,

$$P = \frac{mg}{A} + P_{atm}.$$

We now imagine that the gas is heated by a flame beneath the vessel. As it heats, the gas expands, moving the piston and weight above it up by, say, a distance Δs. The work performed by this action is the product of the force and the distance:

$$W = F\Delta s = PA\Delta s = P\Delta V$$

where ΔV is the change in the volume of the gas that results from heating it.

Work is not the only product of the process. The flame also raises the temperature of the gas. This is a universal observation. As we shall see later in this chapter, heat can never be converted entirely into work or mechanical energy. However, mechanical energy or work can be converted 100 percent into thermal energy. Before considering the reasoning that leads to this situation, we must stop to define several basic concepts we use to measure heat and its impact.

10.1 TAKING THE TEMPERATURE

The thermal concept that is entirely different from any quantity in mechanics is called **temperature.** It is, of course, **the numerical indication of hotness and coldness.** Scales by which temperature is measured are just as arbitrary as scales of other basic quantities, such as length and weight. Over the years a number of different temperature scales emerged, were tested, and either survived or declined into near uselessness. Three different scales are in common use today: the Fahrenheit, Celsius (or centigrade), and Kelvin scales.

The Fahrenheit scale, which is easily the most familiar of the three to residents of the United States, divides the interval between the freezing point of water at atmospheric pressure and the boiling point of water at the same pressure into 180 divisions, or degrees. Because of the history of the scale, these freezing and boiling points are themselves marked by somewhat incongruous numbers: 32°F and 212°F (F, of course, standing for Fahrenheit).

Much more rational is the Celsius scale, which has 100° between the

Figure 10.2 The three most widely used scales of temperature.

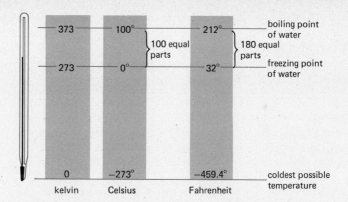

freezing point of water at atmospheric pressure (0°C) and the boiling point of water at the same pressure (100°C). The disadvantage of both the Fahrenheit and Celsius scales is that colder temperatures have negative numbers. The Kelvin scale avoids this particular pitfall. Zero on the Kelvin scale is defined as the coldest temperature than can be obtained theoretically. The theoretical background of this "absolute zero" must wait until Chap. 11. For the moment, it is sufficient to note that the temperature of this point is about $-273°$ on the Celsius scale. The degree divisions of the Kelvin scale are identical to those of the Celsius scale. Thus, the freezing point of water (again at atmospheric pressure because the exact freezing temperature varies with pressure) is 273 K, and the boiling point of water at atmospheric pressure is 373 K. Obviously, the Kelvin temperature is the Celsius temperature plus 273°. Note that by convention the degree sign (°) is not used with kelvin (K).

Figure 10.2, which relates the three major temperature scales, enables us to calculate the relationship between Fahrenheit and Celsius temperatures. The liquid range of water covers 100°C and 180°F. The ratio of these two numbers is $\frac{9}{5}$, indicating that 9°F are equivalent to 5°C; the degree Celsius is, therefore, a larger unit than the Fahrenheit degree. To move from this ratio to a general formula that links the two systems, we must remember that zero on the Celsius scale is equivalent to 32°F. Hence, the Celsius temperature t_C is equal to $\frac{5}{9}$ of the Fahrenheit temperature t_F, less 32°.

$$t_C = \tfrac{5}{9}(t_F - 32) \tag{10.1}$$

Using algebra, this can be rewritten for t_F as follows:

$$t_F = \tfrac{9}{5}t_C + 32$$

EXAMPLE 10.1

Assume that room temperature is 70°F and compute the same temperature on the Celsius and Kelvin scales.

Measuring the Mercury: The Origins of Temperature Scales

On a chilly winter's morning, a scientist walking to her laboratory might notice that the temperature reading atop a local high building is 20°. When she reaches the warmth and comfort of her laboratory, she might glance at the thermometers inside and find that they are also reading 20°. The paradox is simply explained; the two thermometers are operating on different scales of temperature. The Fahrenheit scale, according to which the temperature is 20° on an icy morn, is commonly used in America. The Celsius (or centigrade) scale, which registers 20° in a warm room, is used by the worldwide scientific community.

These two temperature scales, and others now forgotten, were established in the early eighteenth century. One man who turned his attention to the subject was Gabriel Daniel Fahrenheit, a maker of meteorological instruments who emigrated to Amsterdam, Holland, from his birthplace in Danzig (now Gdansk, Poland) early in his life. In 1724, the émigré set about devising a scale that would preclude the need for negative values of temperature. To this end he assigned the value of 0° to the lowest temperature that he could obtain, using a freezing mixture of water, ice, and salt. As his other reference point, Fahrenheit chose normal human body temperature. He did not, however, as is widely misreported, assign the value of 100° to that benchmark; instead, he found it convenient to divide the distance between the two reference points of his scale into 12 major pairs, each in turn split into 8 divisions. Thus, body temperature was set at 96°.

Fahrenheit later adjusted his scale to make the temperature of boiling pure water exactly 212°. This set the freezing point of pure water at 32°; we now know that body temperature is 98.6°F. The new scale was adopted in the Netherlands and Great Britain as soon as Fahrenheit announced its invention in 1724; today, however, it is in general use in only a few countries of the English-speaking world, such as the United States, Canada, and Britain (although Canada and Britain have officially converted to the Celsius scale).

The rest of the world's general population, and all its scientists, read their thermometers in degrees Celsius. This scale was devised in the early 1740s by Anders Celsius, a Swedish astronomer who divided the difference between the freezing and boiling points of pure water into 100 steps. Celsius placed the boiling point of water at 0° and the freezing point at 100°, but he later reversed them. Until recently, the scale was known as centigrade, from the Latin for 100 steps, but now scientists favor the use of its inventor's name.

Fahrenheit and Celsius were not the only men to produce temperature scales in the first half of the eighteenth century. A French physicist, Antoine Ferchault de Réaumur, invented a system that placed the freezing point of pure water at 0° and the boiling point of water at 80°. This scale was widely accepted at first but gradually faded into scientific oblivion.

Neither Fahrenheit nor Celsius

Solution

Strictly speaking, the answer to this problem is 21°C because we should avoid gaining significant figures in converting from one scale to another; 21.1°C is the valid Celsius equivalent of 70.0°F.

The Celsius temperature corresponding to 70°F is, from the formula,

$$t_C = \tfrac{5}{9}(70 - 32) = \tfrac{5}{9}(38) = 21.1°C \simeq 21°C$$

This temperature on the Kelvin scale is obtained by adding 273° to the Celsius temperature:

$$T = 21° + 273° = 294 \text{ K}$$

Anders Celsius. (*Granger*)

Antoine Ferchault de Réaumur. (*Culver*)

Lord Kelvin. (*Culver*)

overcame the problem that Fahrenheit tried to avoid: negative temperatures. Not until more than a century later, after physicists had come to grips with the concept of absolute zero, did an alternative scale emerge. The Scottish physicist Lord Kelvin invented a scale that started at absolute zero and went upward in degrees Celsius. According to this Kelvin scale, an absolute scale, the freezing point of pure water is 273.15 K and the boiling point 373.15 K.

The Fahrenheit equivalent of Kelvin's scale was devised by another Scotsman, William John Macquorn Rankine, and is known as the Rankine scale, which is used by engineers.

EXAMPLE 10.2

On a winter's day in Bangor, Maine, the temperature varied from -20 to $+40\,°F$. Express this temperature range in degrees Celsius. Compute the extreme temperatures of $-20.0\,°F$ and $+40.0\,°F$ in degrees Celsius.

Solution

The temperature interval in °F is

$$\Delta t = 40\,°F - (-20\,°F) = 60\,°F$$

5°C are equivalent to 9°F, so this temperature interval in °C is

$$\Delta t = 60°F \times \frac{5°C}{9°F} = 33.3°C$$

To compute the temperature $-20°F$ on the Celsius scale, we can use the relationship

$$t_C = \tfrac{5}{9}(t_F - 32) = \tfrac{5}{9}(-20 - 32) = \tfrac{5}{9}(-52) = -28.9°C$$

The temperature of 40°F is 33.3°C above this temperature, or 40°F is equivalent to $-28.9°C + 33.3°C = 4.4°C$. We can check our result by converting 40°F directly to °C:

$$t_C = \tfrac{5}{9}(t_F - 32) = \tfrac{5}{9}(40 - 32) = 4.4°C$$

Measuring temperature involves detecting and measuring a particular property of matter that varies consistently with temperature. Given the large range of temperatures which must be measured and the variety of circumstances under which the measurements must be taken, it is not surprising that a large number of different types of thermometers have been perfected, from the simple mercury thermometer that parents turn to when their children get the sniffles to sophisticated detectors of tiny changes in electrical resistance. The most common property used in thermometry, however, is the most obvious effect of the application of heat: thermal expansion.

10.2 EXPANDING ON EXPANSION

Almost all substances—solids, liquids, and gases—expand when the temperature increases and contract when the temperature decreases. The effects of expansion, and precautions to avoid these effects, are visible everywhere. The bubbles on asphalt roads and parking lots in very hot weather, for example, arise from the expansion of the asphalt. The gaps between contiguous sections of railroad track are designed to allow for expansion of the tracks without catastrophic buckling. Stretches of roadway over bridges are designed in sections for just the same reason, thus accounting for the bumps that drivers experience when they travel over them.

All materials that expand do so in every direction when the temperature is increased. For simplicity, however, we shall first deal with linear expansion—the expansion of solids in a single direction. The amount of such expansion is proportional to both the original length of the object in that particular direction and the change in temperature. In the case of a brass rod of length L whose temperature is increased by Δt degrees, the amount of expansion ΔL is given by the equation

$$\Delta L = L\alpha\,\Delta t \tag{10.2}$$

The constant factor α, the Greek alpha, is known as the **coefficient of linear expansion** of the brass. It **represents the fractional change in**

length per degree of increase or decrease in temperature and has the units of 1/temperature. Because the value of α stays fairly constant over large ranges of temperature, linear expansion offers an obvious means of measuring temperatures.

EXAMPLE 10.3

An iron pipe, 300 ft long at room temperature (20°C), is used as a steam pipe. How much allowance must be given for expansion? What will the length of the pipe be when steam is in the pipe?

Solution

The temperature of steam is 100°C and the coefficient of thermal expansion α for iron is $1.2 \times 10^{-5}/°C$, $L = 300$ ft:

$$\Delta L = L\alpha \Delta t = (300 \text{ ft})(1.2 \times 10^{-5}/°C)(100°C - 20°C) = 0.288 \text{ ft}$$

Therefore, the final length of the pipe, $L_{100°}$, will be

$$L_{100°} = L + \Delta L = (300 + 0.288) \text{ ft} = 300.29 \text{ ft}$$

Table 10.1 shows the value of the coefficient of linear expansion for a number of different substances. One noticeable point about the list is that values of α for metals are much larger than those for most nonmetallic substances, such as glass. Housewives make use of this observation when they put the metal top of a tightly sealed jar of jelly or marmalade under hot water to loosen the seal. The hot water expands the metal cap more than it expands the jar, thus making it easier to unscrew the cap.

Modern thermostats often use a bimetallic strip to activate a mercury switch; the mercury actually makes the electrical contact that turns on the furnace.

A more technological application of differences between the coefficients of expansion of different substances occurs in the bimetallic strip that is used to trigger thermostats. The strip consists of thin sheets of different metals welded together, as shown in Fig. 10.3. If the coefficients of linear expansion of the two metals are different, the metals will expand and contract at different rates when their temperature is increased or decreased. Because the metals are welded together, this differential expansion or contraction will cause the bimetallic strip to bend when it is heated or cooled. In Fig. 10.3, which shows a simple thermostat switch, metal B has a higher coefficient of linear expansion than metal A. On cooling, B will therefore contract more than A, forcing the strip toward the switch. At a preset temperature, the strip will touch the switch, complete the electrical circuit, and start up the furnace.

Another effect of expansion, this time detrimental, occurs if a solid is heated in confinement without having room available to accommodate the expansion. The result is a very large stress being applied to the solid; the magnitude of this stress can be calculated using Young's modulus for the solid. The amount of *thermal* strain, $\Delta L/L$, is calcu-

272 Thermal Energy and Thermal Properties of Matter

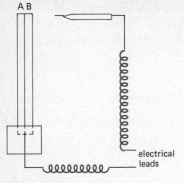

(a) at temperature t, switch open

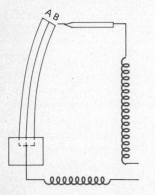

(b) at a lower temperature, switch closes

Figure 10.3 Bimetallic strip used as a thermostat. The coefficient of expansion of metal B is greater than that of metal A.

Table 10.1 Coefficients of Thermal Expansion at 0°C

Substance	Coefficient of Linear	Coefficient of Volume
Solids	Expansion α (10^{-5}/°C)	Expansion β (10^{-5}/°C)
Aluminum	2.4	7.7
Brass	1.9	5.6
Concrete	0.7–1.2	2.1–3.6
Copper	1.7	5.1
Glass, ordinary	0.85	2.6
Glass, Pyrex	0.33	1.0
Iron	1.2	3.6
Lead	3.0	9.0
Quartz	0.04	0.1
Silver	2.0	6.0
Steel	1.2	3.6
Zinc	2.6	7.8
Liquids		
Ethyl alcohol		110
Glycerin		51
Mercury		18
Petroleum		90
Water		21
Gases		
Air		367
Hydrogen		366
Helium		367

lated from the change in temperature and the solid's coefficient of expansion. Thermal stress can also result when one subjects solids to sudden, large temperature changes, such as putting a tea cup into a hot oven. Often these stresses are so severe that the object cracks, just as it would if struck with a hammer. This effect can be largely overcome if glassware is made with a very low coefficient of linear expansion, thus preventing large changes in temperature from producing excessively large thermal stresses; Pyrex brand glassware is a prime example of this defensive strategy.

EXAMPLE 10.4

A piece of steel used on a roadway bridge is foolishly set tightly between the bridge abutments 20 m apart on a cold day when the temperature is -2°C. The next day the temperature rises to 30°C. Compute the force of compression in the steel, assuming that it has a cross-sectional area of 100 cm².

Solution

The steel will tend to expand as the temperature rises, but because it is between the two bridge abutments, it will not be able to do so. It will, therefore, be under compression. We can first compute the increase in length the steel would experience if it were free to expand and then using that number determine from Hooke's law the compression force required to compress the steel back this same amount ΔL:

$$\Delta L = L\alpha\,\Delta t = (20\text{ m})(1.2 \times 10^{-5}/^{\circ}\text{C})[30\,^{\circ}\text{C} - (-2\,^{\circ}\text{C})]$$
$$= (20\text{ m})(1.2 \times 10^{-5}/^{\circ}\text{C})(32\,^{\circ}\text{C}) = 7.68 \times 10^{-3}\text{ m}$$

Stress is proportional to strain, with Young's modulus Y being the proportionality constant. For steel, $Y = 20 \times 10^{10}\text{ N/m}^2$:

$$Y = \frac{\Delta F/A}{\Delta L/L}$$

$$\Delta F = \frac{Y\,\Delta L\,A}{L}$$

$$= \frac{(20 \times 10^{10}\text{ N/m}^2) \times (7.68 \times 10^{-3}\text{ m}) \times (100\text{ cm}^2 \times 10^{-4}\text{ m}^2/\text{cm}^2)}{20\text{ m}}$$

$$= 7.68 \times 10^5\text{ N}$$

This is a force in excess of 170 000 lb.

10.3 MOVING IN ALL DIRECTIONS

Mica and graphite are two examples of substances whose coefficients of expansion vary according to the direction in which they are measured. Both have layered molecular structures, with distances between molecules in adjacent layers differing from distances between molecules in individual layers.

All substances, whether solid, liquid, or gaseous, experience a change in volume as their temperature changes.

Linear expansion, as we noted, is a somewhat artificial example of thermal expansion in general because nearly all materials expand outward in every direction when they are heated and contract inward in every direction when they are cooled, at least unless they are constrained. A long rod, for example, increases in cross section as well as length when it is heated. Of course, the actual amount of expansion of the cross section is much smaller than the expansion of length because the length is greater than the diameter to start with. In this example the linear coefficient of expansion of the rod is exactly the same in every direction. There are certain compounds, however, whose coefficients of linear expansion vary according to the directions in which they are measured. This effect is normally caused by unusual molecular structures; the molecules might be arranged in layers, for example, with strong atomic forces between molecules inside each layer and much weaker forces between molecules in adjacent layers. In this type of compound the coefficient of linear expansion parallel to the layers is smaller than that perpendicular to them.

Whatever the molecular peculiarities of individual compounds, however, it is an established fact that all substances, whether solid, liquid, or gaseous, experience a change in volume as their temperature changes. For most substances the change is an expansion that is directly proportional to the change in temperature over a wide range of temperature. Mathematically, this is expressed as follows:

$$\Delta V = \beta V \Delta t \tag{10.3}$$

where ΔV is the change in volume of an original volume V on heating through a temperature change Δt. The Greek letter beta (β) stands for the coefficient of volume expansion; like the coefficient of linear expansion, its unit is the reciprocal of temperature, 1/temperature.

Interestingly enough, $\beta = 3\alpha$ for most solids, as can be seen from the data in Table 10.1. We can also show this using Eqs. (10.2) and (10.3) and some simple algebra. Assume for simplicity that we have a cube of side L. Then,

$$V = L^3$$

and after expansion, $V + \Delta V = (L + \Delta L)^3$. If we expand the right side of the equation, we have

$$V + \Delta V = L^3 + 3L^2 \Delta L + 3L \Delta L^2 + \Delta L^3$$

Neglecting powers of ΔL greater than 1, since ΔL is so small that ΔL^2 and ΔL^3 are negligible in size, we have

$$V + \Delta V = L^3 + 3L^2 \Delta L$$

Since $V = L^3$,

$$\Delta V = 3L^2 \Delta L$$

Substituting for ΔV and ΔL from Eqs. (10.3) and (10.2), we arrive at the result

$$\beta V \Delta t = 3L^2 \alpha L \Delta t \qquad \beta L^3 \Delta t = 3L^3 \alpha \Delta t$$

or

$$\beta = 3\alpha$$

Table 10.1 also shows that the coefficients of volume expansion are generally higher for gases than liquids and higher for liquids than solids. This is not unexpected if one considers the atomic forces that hold together the three different states of matter. The strong binding forces in solids oppose the tendency to expand on heating far more effectively than the weaker binding forces of liquids, in just the same way that they greatly restrict the movement of molecules in solids. In gases there are essentially no binding forces between atoms or molecules.

Another noticeable point in Table 10.1 is that all the gases have a nearly identical coefficient of volume expansion: 0.00366 per degree Celsius. This figure applies to all gases that are well above the temperature at which they condense to liquids—a condition that includes most common gases at room temperature. Strange as this may seem, the value 0.00366 is the source of an even more peculiar effect. The reciprocal of this figure, 1/0.00366 per degree Celsius, is 273°C, the exact difference between the Celsius and Kelvin scales of temperature.

Is this merely a strange coincidence, or does it tell us something about the basic physics of gases? To investigate this question, rewrite Eq. (10.3) as follows:

$$\frac{1}{\beta} = \frac{V}{\Delta V}\Delta t$$

Since the original volume, V_0, was measured when $t = 0°C$, then $\Delta V = V - V_0$ and $\Delta t = t - t_0 = t$. Hence this expression becomes

$$\frac{1}{\beta} = \frac{V_0}{V - V_0}(t - t_0) = \frac{V_0}{V - V_0}t$$

However,

$$\frac{1}{\beta} = 273°C = \frac{V_0}{V - V_0}t$$

If we assign the value of zero to V, this equation tells us that the equivalent temperature is $-273°C$. This relationship seems to tell us that all gases contract to a volume of zero at $-273°C$ and that, therefore, there is something special about this temperature. We know from the laws of thermodynamics that, indeed, $-273°C$ is a special temperature: it is the lowest temperature theoretically possible. In practice, it can be approached only in the laboratory but never attained. Every gas, including hydrogen and helium, the lightest of all gases, liquifies above this theoretical limit. We shall examine this concept of "absolute zero" in more detail in Chap. 11, which is devoted entirely to gases.

The volume expansion of liquids is the property used in the most familiar type of thermometer, the liquid-in-glass thermometer. Most typically, a small amount of mercury is placed in a bulb attached to a capillary tube. The tube itself, above the mercury, is evacuated and then sealed. The mercury expands when it is heated and flows up the tube; the narrowness of the tube magnifies this expansion by channeling it effectively in one direction. Because the glass expands slightly on heating, the length of the column of mercury in the tube is proportional to both the temperature and the difference between the coefficients of volume expansion of the mercury and the glass.

EXAMPLE 10.5

A glass bulb is filled with 50 cm³Hg at 20°C. What volume will overflow if the system is heated to 60°C? (Note that the volume enclosed by the glass expands as the glass expands outward.)

Solution

The initial volume of the glass bulb, 50 cm³, will increase on heating, with a volume expansion $\beta_G = 3\alpha_G = 2.6 \times 10^{-5}/°C$. At the same time, the volume of mercury will increase with a volume expansion $\beta_M = 18 \times 10^{-5}/°C$. The net result will be an expansion rate per degree Celsius equal to the difference between these two expansion coefficients. That is,

$$\begin{aligned}
\text{Vol. overflow} &= \text{vol. increase in mercury} - \text{vol. increase in glass} \\
&= \Delta V_M - \Delta V_G \\
&= \beta_M V \Delta t - \beta_G V \Delta t = (\beta_M - \beta_G) V \Delta t \\
&= (18 \times 10^{-5}/°C - 2.6 \times 10^{-5}/°C)(50 \text{ cm}^3) \\
&\quad (60°C - 20°C) \\
&= 0.31 \text{ cm}^3
\end{aligned}$$

Most liquids, like mercury, expand in an essentially uniform manner; in other words, their coefficients of volume expansion stay constant over wide spans of temperature. One remarkable exception to this rule is water, the most common liquid. For the top 90°C of its liquid existence, the coefficient of volume expansion of water behaves much like that of any other liquid. But below 10°C water becomes an entirely different liquid, at least as far as its reaction to heating and cooling is concerned.

Between 10 and 4°C, the rate at which water contracts in volume with falling temperature slowly decreases, until at 4°C exactly, the rate is zero (Fig. 10.4). As the temperature drops further, the water actually starts to expand, until the freezing point, 0°C, is reached. What this means is that the coefficient of volume expansion of water is negative

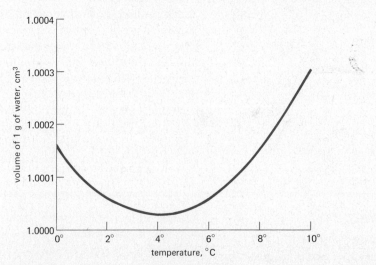

Figure 10.4 Changes in the volume of 1 g of water between 0 and 10°C.

between 4° and 0°. The maximum density of liquid water occurs at 4°C rather than at the freezing point, as is the case with most liquids. Adding to the peculiarity of water, the liquid is denser than its solid, ice—another rare situation in the world of science.

These strange events have great importance in a number of aspects of daily life. Water pipes would not burst after freezing and thawing if ice were denser than water at its freezing point. Cocktails and mixed drinks would look very peculiar if the ice in them were denser than water and sank to the bottom. Even stranger, and for winter sports enthusiasts highly frustrating, would be the appearance of ice at the bottom of lakes and ponds—a process of "icing under" rather than the "icing over" with which we are so familiar.

Cold air in the atmosphere cools the water at the top of a pond. Above 4°C cooler water is denser than warmer water, and hence the cold water at the surface sinks slowly to the bottom of the pond, mixing the water and exposing warmer water to the cold air. This process continues until all the pond is at a uniform temperature of 4°C. As the water at the surface cools below this temperature, however, it stays on the top of the pond for it is less dense than the warmer water beneath it. Eventually the top layer of water freezes to ice that, because it is less dense than the water beneath, forms a crust on top of the pond. If water behaved like other liquids, it would freeze much more uniformly, and the ice would fall to the bottom of the pond, making skating possible only on the rare occasions when the entire pond was converted to ice. Were it not for the strange properties of water, there is some question as to whether life could exist on earth, at least away from the equator.

If water behaved like other liquids, it would freeze much more uniformly, and the ice would fall to the bottom of the pond, making skating possible only on the rare occasions when the entire pond was converted to ice.

10.4 A QUESTION OF CALORIES

Temperature is not the only fundamental quantity involved in the physics of thermal phenomena. Just as important is the amount of heat necessary to raise the temperature of any particular object a certain number of degrees. The difference between quantity of heat and temperature is obvious immediately when one realizes that a gallon of water requires more exposure to a fire to reach boiling temperature than does a tea cup of water.

The fundamental unit of heat is the calorie, a familiar term to diet faddists and students of nutrition. Actually, the heat calorie (with a lowercase c) is just one-thousandth of the Calorie (with a capital C) or kilocalorie (kcal) that is the basic currency of the dietary trade. But the dimensions of both types of calories are those of energy, illustrating again the interconvertibility of different forms of energy.

The heat calorie is defined as the amount of heat required to raise the temperature of 1 g of water 1°C between 14.5 and 15.5°C. The reason for specifying the exact temperature range in this definition is that the amount of heat required to raise a gram of water a degree Celsius varies slightly with temperature.

Not surprisingly, equal masses of different materials require differ-

The Hot, the Cold, and the Thermometer

For the average person, measuring temperature is simply a matter of reading the scale on a mercury-in-glass thermometer. By contrast, in the scientific laboratory temperature measurement is a sophisticated branch of physics that requires ingenuity, skill, and extraordinary technical care. Although members of the public are concerned at best with the temperatures of the air and of their own immediate families, scientists must monitor temperatures ranging from within a few thousandths of a degree of absolute zero to thousands and even millions of degrees.

A typical example of a thermometric adaptation concerns the conventional mercury thermometer. The clinical variety of this device contains a kink in its internal tube that prevents the mercury from contracting back into the bulb once the instrument is removed from the patient's mouth. The purpose is simple: by holding the reading steady, this kink allows nurses, doctors, and harried mothers to read off the temperatures of feverish patients without exposing themselves to infections in the patient's breath.

Mercury has one major restriction as a temperature sensor: it freezes at $-39°C$. To measure temperatures lower than this, scientists are forced to substitute colored alcohol for the quicksilver metal in their capillary tubes or to turn to one of the more advanced means of temperature measurement, which has the extra advantage of yielding more accurate readings than any liquid-in-glass device can produce.

One of the most accurate laboratory thermometers is the gas thermometer shown in the accompanying figure. It is a highly accurate thermometer that is often used in scientific laboratories as a standard against which other thermometers can be calibrated. The gas thermometer employs a volume of gas whose pressure is monitored as a function of temperature. When the gas container is heated in a uniform manner, the gas expands, pushing the mercury in the U-tube manometer. The flexible U tube is adjusted to return the volume of the gas to its original volume, and h is measured. The pressure in the right-hand sealed tube is zero, so the height h determines the absolute pressure of the constant volume of gas. Such an apparatus is bulky but very accurate if care is taken in the measurements and corrections are made for the expansion of the glass.

Expansion is not the only characteristic of materials that alters with temperature, and over the years scientists have incorporated a number of other basic properties of matter into their thermometers. For example, two dissimilar metals produce an electrical voltage when they are joined together in two places and the joints kept at different temperatures. This property is utilized in the device

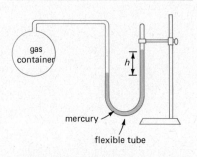

The gas thermometer.

known as a thermocouple. When one joint is kept at a standard temperature, the amount of voltage across the two joints depends on the temperature of the second joint. The electrical resistance of metals also varies in a consistent way with temperature, and resistance thermometers using such metals as platinum are extremely accurate temperature sensors for a wide range of temperatures.

The most challenging regions for temperature measurement are the very hot and the very cold. The only practical method for measuring temperatures of a few thousand degrees or more involves comparison of the colors of objects at the unknown temperatures with the colors of objects at standard temperatures. Researchers investigating the lowest of low temperatures, within a degree of absolute zero, must rely on atomic effects, such as the magnetization of atoms.

Table 10.2 Specific Heat Capacities

Substance	Specific Heat Capacity cal/g·°C or kcal/kg·°C Btu/lb·°F
Water	1.00
Air	0.25
Aluminum	0.22
Ethyl alcohol	0.60
Brass	0.094
Copper	0.093
Glass	0.20
Gold	0.03
Helium	1.24
Ice	0.50
Iron	0.11
Lead	0.031
Mercury	0.033
Silver	0.056
Steam	0.48
Steel	0.11
Zinc	0.092

ent amounts of heat to increase their temperatures by an identical number of degrees. This is quantified by defining yet another property of materials: the specific heat, or specific heat capacity. **The specific heat of a material,** normally given the symbol c, **is the amount of heat required to raise the temperature of 1 g of that material by 1°C** and its units are calories per gram per degree Celsius:

$$c = \frac{Q}{m \, \Delta t} \tag{10.4}$$

where Q is the quantity of heat required to heat the mass m of the substance by Δt degrees.

Given the definition of the calorie, the specific heat of water is clearly 1 cal/g·°C, at least in the temperature range between 14.5 and 15.5°C. At other temperatures the value is very close to 1. The specific heats of a number of common substances are listed in Table 10.2.

It is clear from Table 10.2 that water is again an unusual material by this criterion. It has the highest heat capacity of any substance except helium. This ability of water to hold a large quantity of heat is extremely important to our climate and weather. It explains, for example, how the Gulf Stream is able to carry heat all along the East Coast of the United States and across the North Atlantic to Europe. The energy released by the waters of the Gulf Stream near Europe heats the air that moves across Europe, substantially moderating its temperature.

The Gulf Stream is a relatively narrow ocean current that flows from the warm coast of Florida across the Atlantic Ocean to the British Isles and other parts of western Europe. It moderates the British weather markedly, reducing the amount of snow that the country receives well below the amounts that drop elsewhere at the same latitude.

A new use for water, primarily because of water's large heat capacity, is the storage of heat in a solar heating system. Water heated by the sun in solar "collector panels" is stored in a large insulated tank. Heat can be removed from the water with a heat pump at a later time, at night or on a cloudy day.

In the British system of units, the British thermal unit (Btu) is defined as the quantity of heat required to raise the temperature of 1 lb of water by 1°F from 63 to 64°F. The specific heat of any material is the quantity of heat necessary to raise the temperature of 1 lb of that substance by 1°F.

Measurements of the quantities of heat involved in these definitions are made using containers called calorimeters. These are essentially insulated cans that neither gain nor lose heat. Actually, there are no perfect calorimeters; the best ones slow the rate of heat flow but do not stop it entirely.

EXAMPLE 10.6

Compute the heat required to raise the temperature of a 2-kg aluminum cooking pan from room temperature, 20°C, to 100°C. The specific heat of aluminum is 0.22 kcal/kg · °C.

Solution

$$Q = mc\,\Delta t = (2\text{ kg})(0.22\text{ kcal/kg} \cdot \text{°C})(100\text{°C} - 20\text{°C}) = 35.2\text{ kcal}$$

EXAMPLE 10.7

A handful of lead shot whose specific heat is 0.031 cal/g · °C is heated to 90°C and then dropped into 80 g of water at 10°C. If the final temperature of the mixture is 14°C, what is the mass of the lead shot?

Solution

If we assume that no heat is given to the container, then the heat lost by the hot lead shot will equal the heat absorbed by the water. If we let the subscript l denote the lead and the subscript w denote the water, then

$$m_l c_l \,\Delta t_l = m_w c_w \,\Delta t_w$$

The specific heat of lead, c_l, is 0.031 cal/g · °C and the specific heat of water is 1. Hence,

$$m_l(0.031\text{ cal/g} \cdot \text{°C})(90\text{°C} - 14\text{°C})$$
$$= 80\text{ g}\,(1\text{ cal/g} \cdot \text{°C})(14\text{°C} - 10\text{°C})$$

Note that the temperature differences were computed as positive on both sides of the equation since the heat from the lead on the left side equals the heat absorbed by the water on the right side:

$$m_l = \frac{(80)(1)(4)\text{ g}}{(0.031)(76)} = 136\text{ g}$$

In a practical calorimetry experiment, the can holding the water would absorb heat. This fact would require the inclusion of another term in the equation.

10.5 HIDDEN HEAT

A thermometer placed in a teakettle on a stove indicates surprising behavior in the temperature of the water as it comes up to boiling temperature and then lets off steam. The temperature increases consistently from its starting point to 100°C, at which temperature the water starts to boil. But then, instead of increasing further, the temperature stays constant. It rises again only after all the water is converted into steam. The heat from the stove has no effect on the temperature of the water while it boils. Instead, the heat from the stove is devoted entirely to changing the state of the water from its liquid form into gaseous steam. Only when the teakettle boils dry will the temperature of the metal begin to rise.

The heat required to change the state of any substance without altering its temperature is known as the latent heat, latent meaning "hidden." All changes of state require the addition of a certain amount of latent heat (or the removal of such heat if the change goes in the reverse direction) between a state of matter that exists at a higher temperature and one that exists at a lower one.

The latent heat of vaporization, L_v, for any substance is defined as the amount of heat necessary to convert a gram of the substance in liquid form to the gaseous state. The latent heat of fusion, L_f, is the amount of heat required to convert a gram of the solid form to liquid. The latent heat of sublimation, L_s, is the heat needed to change a gram of the solid directly to the gaseous state. The latent heat of vaporization of water, for example, is 540 cal/g. This means that 540 cal of heat must be added to 1 g of water at 100°C to convert it all to steam and that 540 cal are given off by 1 g of steam as it condenses entirely to water at 100°C. The latent heat of fusion of water is 80 cal/g; 1 g of water at 0°C turns entirely to ice if 80 cal of heat are removed from it, and the addition of 80 cal to 1 g of ice at 0°C converts it all to liquid water.

To illustrate how a substance absorbs heat, we can combine the equations involving specific heat and latent heat. Figure 10.5 shows the variation in temperature of 10 g of water, starting in the form of ice at -25°C and heated at a constant rate until it is all in the form of steam at 100°C. The conversion occurs in four separate steps. The heat first warms up the ice, raising the temperature to 0°C. Assuming that the heat is distributed uniformly throughout the ice, and noting that the specific heat of ice, 0.5 cal/g · °C, is different from that of water, the amount of heat required for this stage of the warming is

$$Q = mc\,\Delta t = 10\text{ g }(0.5\text{ cal/g} \cdot °C)[0°C - (-25°C)] = 125\text{ cal}$$

The ice then melts at 0°C. Since water's latent heat of fusion is 80 cal/g,

Figure 10.5 Graph of temperature versus heat for 10 g of ice originally at −25°C, to which heat is added at a constant rate. The ice first rises in temperature, then turns to water, which itself rises in temperature until it boils and turns to steam.

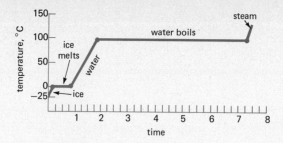

the amount of heat absorbed in this part of the process is

$$Q = L_f m = (80 \text{ cal/g}) \times 10 \text{ g} = 800 \text{ cal}$$

Stage three is the heating of 10 g of water at 0°C to 100°C. For this,

$$Q = mc\,\Delta t = 10 \text{ g} (1 \text{ cal/g} \cdot °C)(100°C - 0°C) = 1000 \text{ cal}$$

Finally, the water at 100°C is converted to vapor (steam) at the same temperature, using latent heat at a rate of 540 cal/g. The total amount of heat used in this part of the process is

$$Q = L_v m = (540 \text{ cal/g}) \times 10 \text{ g} = 5400 \text{ cal}$$

Thus, the total amount of heat absorbed is 7325 cal. We could continue to heat the vapor (steam) to a higher temperature and add the actual amount of extra heat required by specifying the actual temperature reached by the steam and knowing the specific heat of steam, which, like that of ice, is different from the specific heat of water.

One point to note in the above calculations is that almost three quarters of the heat involved is needed to convert water to steam at 100°C. It is the extremely high latent heat of vaporization of water that makes steam heat a very attractive means of heating houses, office buildings, and other installations; each gram of steam in the pipes yields 540 cal of heat as it condenses entirely into water.

The reverse side of this coin is that steam causes very severe burns because of its high latent heat of vaporization. Each gram of vapor (steam) that condenses on one's person—and because steam is gaseous it can make good contact over large areas—plunges 540 cal into the victim's skin.

Of course, whenever a gram of water evaporates or a gram of water vapor condenses, about 540 cal of heat are absorbed or released. This process is exremely important in determining the weather, and it is responsible for cooling the human body.

EXAMPLE 10.8

Suppose 10 g of steam at 100°C are condensed into a mixture of 200 g of water and 100 g of ice in a 50-g aluminum calorimeter. Find the final temperature and composition of the mixture.

Solution

One could set up a conservation of energy equation with the sum of the heat input from the condensing steam and the change in temperature of the resulting 10 g of water at 100°C equal to the heat absorbed by the ice, cold water, and cold calorimeter. However, such an equation does not clearly indicate if the heat released from the steam and cooling of the resulting water will be sufficient to melt all the ice. If it is not, some ice will remain, and the final temperature will be 0°C. To check this, we first compute the heat given up by 10 g of steam, which condenses at 100°C and cools to 0°C:

$$Q_{\text{input}} = m_s L_v + m_w c_w \Delta t$$
$$= 10 \text{ g } (540 \text{ cal/g}) + (10 \text{ g})(1 \text{ cal/g} \cdot {}^\circ\text{C})(100{}^\circ\text{C} - 0{}^\circ\text{C})$$
$$= 5400 \text{ cal} + 1000 \text{ cal} = 6400 \text{ cal}$$

To melt the ice would require

$$Q_{\text{melt}} = m_i L_f = (100 \text{ g})(80 \text{ cal/g}) = 8000 \text{ cal}$$

Hence, the 6400 cal from the steam is insufficient to melt all the ice. The final temperature will be 0°C. The final composition can be determined by computing the amount of ice that 6400 cal will melt:

$$m_i L_f = 6400 \text{ cal}$$

$$m_i = \frac{6400 \text{ cal}}{80 \text{ cal/g}} = 80 \text{ g}$$

The ice remaining will be equal to $100 - 80 = 20$ g. The amount of water will be equal to the sum of the original water (200 g) plus the condensed steam (10 g) and the melted ice (80 g):

$$m_w = 200 \text{ g} + 80 \text{ g} + 10 \text{ g} = 290 \text{ g}$$

In this example we assumed no heat loss to or gain from the perfect aluminum calorimeter.

10.6 THE MECHANICAL EQUIVALENT

The first man to measure the ratio of mechanical energy to heat energy was Count Rumford. Shortly after his observation of the heat produced by cannon boring, he set up an experiment in which a horse rotated a cylinder that was located in a large wooden bucket full of water. After 2.5 h, the turning cylinder had raised the water to boiling temperature. From the mass of the water and the cylinder and the change in temperature, Rumford calculated that 1.2×10^3 kcal of heat had been generated in the effort. Knowing that his horse was an average horse, Rumford credited it with delivering 1 horsepower (hp), which we now define at 7.5×10^2 J/s. The total work done by the horse to heat the water was readily obtained by multiplying this figure by the total time of working, 2.5 h, expressed in seconds. The result is 6.7×10^6 J. Assuming that all the horse's energy was translated into heat for the

Power is defined as the rate at which work is done. Thus, the product of power and the time during which it is generated gives the actual amount of work performed in any period.

water, Rumford postulated the ratio of the mechanical energy to the heat energy would be the mechanical equivalent of heat; the figure worked out to 5.6 J/kcal.

This experiment was extremely crude. In the mid-1800s the English physicist James Joule took this type of work a stage further, with a large number of experiments designed to measure the mechanical equivalent of heat with extreme accuracy. In one experiment Joule raised the temperature of a cup of water by rotating a paddle wheel in it; in another experiment he heated water electrically. Common to all his experiments was his great care in insulating the apparatus and measuring all the heat produced in the experiments. Joule's reward was the consistent finding in all the experiments that the ratio of energy input to quantity of heat produced was constant. This ratio, known as Joule's constant, or J, is an extremely important universal constant. Its accepted value is 4186 J/kcal, or 4.186 J/cal, and its constancy illustrates that mechanical energy is converted to heat energy at a constant ratio. This equivalence retains the concept of conservation of energy.

EXAMPLE 10.9

Water in a waterfall drops 120 m into a pool. If all the gravitational potential energy of the falling water is transformed to thermal energy that only raises the temperature of the falling water, by how much would the temperature of the water at the bottom of the falls rise?

Solution

If no energy is lost and all the change in gravitational potential energy is transformed to heating the water, then we can equate these two energies since the gravitational work can go 100 percent into thermal energy. We must, however, make use of Joule's equivalent:

Loss in $E_p = mgh$ J

Gain in heat energy $= mc\,\Delta t$ kcal

Conversion $= 4186$ J/kcal

$$\frac{mgh\text{ J}}{4186\text{ J/kcal}} = mc\,\Delta t\text{ kcal}$$

$$\Delta t = \frac{gh}{c\,4186\text{ J/kcal}} = \frac{(9.8\text{ m/s}^2)(120\text{ m})}{1\text{ kcal/kg}\cdot{}^\circ\text{C} \times 4186\text{ J/kcal}} = 2.81\,^\circ\text{C}$$

EXAMPLE 10.10

How much work is done by a 75-kg man who climbs a 1500-m-high mountain? If the man consumes 5 kcal of food energy for each 1 kcal of work performed, how much food energy does he consume to climb the mountain? Consider only the work done to lift the man vertically.

Solution We assume that the work done by the man goes only into lifting himself to the top of the mountain and hence changing his potential energy:

$$W = mgh = 75 \text{ kg } (9.8 \text{ m/s}^2)(1500 \text{ m}) = 1.10 \times 10^6 \text{ J}$$

This energy can be converted into kilocalories using Joule's equivalent, 4186 J/kcal:

$$W = \frac{1.10 \times 10^6 \text{ J}}{4186 \text{ J/kcal}} = 263 \text{ kcal}$$

If the man consumes 5 kcal of food energy to produce 1 kcal of work, then the man would have to consume 5×263, or 1320 kcal, to climb the mountains.

10.7 HEAT, TEMPERATURE, THERMAL ENERGY, AND KINETIC THEORY

According to the kinetic and molecular theory of matter, heat added to any object, not used for changing a state or for doing work, but to raise the temperature, represents an increase in the total kinetic energy of the random motions of the individual molecules that make up that object. When an object is heated, its molecules move faster and therefore have more kinetic energy. It is important to remember, however, that this extra kinetic energy, and hence momentum, is distributed among the molecules as random motion; if it were not, the molecules, and therefore the object they make up, would move in a particular direction, just as if a force had been applied to the object.

We can describe the conversion of mechanical energy to thermal energy, according to the kinetic theory, by noting that the purposeful ordered motion of molecules characteristic of mechanical energy is changed to the disorderly molecular movements typical of thermal energy. When a hammer hits a nail on the head, heating both the hammer and the nail, for example, the orderly motion of the hammer falling is converted to the more chaotic motion of the molecules in the hammer and the nail.

This model readily accounts for the expansion of objects when they are heated. In solids, for example, heating results in an increase in the frequency and amplitude with which the molecules vibrate around their basic positions. The increased vibration results in the molecules' moving out farther from their basic position. The effect is to produce expansion of the distances between molecules in local regions, and thus expansion of the solid as a whole. At the melting point of the solid, this increased random kinetic energy becomes great enough to overcome the forces that hold the molecules together. The latent heat of fusion finally allows the molecules to become disordered, and the solid melts.

From the point of view of kinetic theory, **temperature is a measure of the average kinetic energy of the individual molecules in their random motion.** For although some molecules move faster than others, the total

The amplitude of a vibration is the maximum distance from the center that the vibrating object reaches. This concept will be dealt with more thoroughly in Chap. 12.

energy of all the molecules in a solid increases when the temperature of the solid rises and decreases when the temperature falls. We shall supply concrete figures for this argument in the next chapter.

In the discussion in this chapter we frequently used two terms—heat and thermal energy—without formally defining them. **Thermal energy is that form of energy associated with the random motion of the molecules and atoms** which make up an object. The greater this motion, the more thermal energy is contained in the object. The higher the temperature, the greater the thermal energy of the object. Thermal energy is a form of energy that an object "possesses," just as kinetic energy is energy an object "possesses" because of its motion, and just as potential energy is energy an object "possesses" by the position it occupies above the earth. **Heat is** now the term used to describe **energy in transit** from one place to another. If a flame is held under an object, heat flows from the hot flame to the colder object, increasing the thermal energy of the object. It might seem that heat and thermal energy are the same; however, in the next chapter the difference will become very important. The importance of the distinction between thermal energy and heat is apparent when we note that the thermal energy—the energy of random motion of the atoms or molecules—can be changed without any flow of heat into or out of an object. Mechanically rub two objects together or strike one object on the other, and the mechanical work will increase the thermal energy, yet no heat has moved from one object to the other. When we examine thermodynamics in Chap. 11, it will be very important to distinguish between the two terms heat and thermal energy.

10.8 THE TRANSFER OF HEAT

We developed a picture of heat as a form of energy and temperature as a measure of the average kinetic energy of the individual molecules in their random motion within a material. Whenever there is a difference in temperature between two points in a material, thermal energy moves from the higher to lower temperature. Physicists recognize three types of heat flow, or heat transfer: conduction, convection, and radiation (see Fig. 10.6).

In this chapter we shall restrict ourselves mainly to conduction and convection, leaving a detailed discussion of radiation for later in the book. **Radiation refers to the transfer of energy by electromagnetic waves.** Radiative transfer of energy is extremely important to us on earth because it is the mechanism by which energy is transmitted from the sun to the earth through the vacuum of space. Of the three means of heat transfer, radiation is unique in that it is the only one that can transmit thermal energy through a vacuum. Although it does not require a material medium, radiation can and does transmit energy through materials also. All objects radiate and absorb heat energy in the form of electromagnetic waves. Considering radiation alone, an object that emits more radiation that it absorbs "cools down," and a

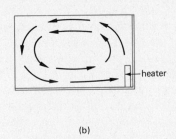
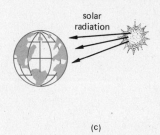

(a)   (b)   (c)

Figure 10.6 Mechanisms of heat transfer: (a) conduction, (b) convection, (c) radiation.

We use the terms heat up and cool down to indicate both the usual result of a temperature change and the other processes resulting from the emission or absorption of heat, namely, changes in state. For example, ice at 0°C could absorb more heat than it emits by radiation and thus undergo a change of state. The ice would melt but the temperature would not rise.

body that absorbs more radiation than it emits "heats up." The details of radiation are best described in terms of electromagnetic waves and will be discussed in Chaps. 17 and 21.

The other two methods of heat transfer, convection and conduction, can be described from a molecular point of view using kinetic theory. Molecular motion is important in both processes, and hence both must involve a material medium. However, only convection involves the actual transport of large numbers of molecules from one locality to another within the medium. The two processes are defined as follows:

Conduction is the heat-transfer process in which thermal energy is transmitted through the material by molecular collisions or vibrations. The molecules undergoing random motion remain in one local region, transmitting some of their kinetic energy of random motion to adjacent molecules through collisions or through vibrations of the molecules.

Convection is the process whereby thermal energy is transmitted by the movement of mass involving fluid flow in a gas or liquid.

All three heat-transfer processes are important, and they often take place simultaneously, which makes the study of heat transfer involved and complicated. In fact, real-life heat-transfer processes can be described quantitatively only in an empirical manner. We shall describe conduction and convection under a set of ideal conditions to indicate the important factors that influence these processes. More detailed and practical discussions are left to engineering analyses, which usually involve actual measurements on the particular materials in the physical situation in which the materials are employed.

Conduction

From the point of view of kinetic theory, temperature is a measure of the average kinetic energy of the random moving molecules. The higher the temperature, the greater the internal energy. This internal, molecular energy is transferred by conduction as one molecule collides with its neighbors. If a temperature difference exists in a material, then heat is conducted from hot to cold. The ability of a material to conduct heat depends a great deal on its structure. In general, gases are poor conductors of heat because of the large spaces between the molecules. Solids, on the other hand, are much better conductors. However, some solids, such as the metals, are extraordinarily good conductors, whereas others, including cork, asbestos, brick, concrete, and glass, are relatively poor conductors of heat. (The latter solids are often known as insulators.) Metals are the best conductors of both heat and electricity because they possess relatively "free" electrons that can readily move about inside them. The ability of a material to conduct heat is expressed by an experimentally determined quantity called the **thermal conductivity,** several values of which are in Table 10.3. The units of the thermal conductivity are best understood by examining the dependence of the thermal conduction process on several variables for a particular simple geometry.

Consider the slab of material of thickness d and cross-sectional area A shown in Fig. 10.7. If one face of the slab is kept at a high temperature t_2 and the other at a lower temperature t_1, the quantity of heat Q that flows through the slab in a time τ—that is, the rate of heat flow Q/τ—is given by

$$\frac{Q}{\tau} = \frac{kA(t_2 - t_1)}{d} = \frac{kA\,\Delta t}{d} \tag{10.5}$$

The rate of heat flow, expressed as the number of kilocalories per unit time crossing an area A of material, depends directly upon the thermal

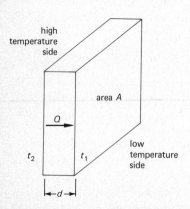

Figure 10.7 Measurement of thermal conductivity by determining the heat flow through a slab of material.

Table 10.3 Thermal Conductivities

Material	k, kcal/m·s·°C	k, Btu·in./ft^2·h·°F
Aluminum	5.1×10^{-2}	1.5×10^3
Brass	0.25×10^{-2}	0.75×10^3
Copper	9.2×10^{-2}	2.7×10^3
Iron and steel	1.1×10^{-2}	0.32×10^3
Silver	10×10^{-2}	2.9×10^3
Asbestos	1.4×10^{-4}	4.0
Brick	1.7×10^{-4}	5.0
Concrete	4.1×10^{-4}	12.0
Corkboard	0.1×10^{-4}	0.3
Fiberglass	0.1×10^{-4}	0.3
Glass	1.7–2.4×10^{-4}	5–7
Sawdust	0.14×10^{-4}	0.4
Wood	0.25–0.4×10^{-4}	0.75–1
Water	1.4×10^{-4}	4.2
Air	0.055×10^{-4}	0.16
Hydrogen	0.4×10^{-4}	1.2

conductivity of the material and the temperature difference across the slab of material, and inversely upon the thickness of the slab. The constant k is known as the thermal conductivity of the material. Its value is obtained by appropriate manipulation of Eq. (10.5)

$$k = \frac{Qd}{A\tau \Delta t} \tag{10.6}$$

All these quantities can be measured experimentally, thus enabling physicists or engineers to determine values of thermal conductivity. The units of the thermal conductivity k are derived from the units of the quantities used in Eq. (10.6). For example, in the metric system Q would be measured in kilocalories, A in square meters, d in meters, time τ in seconds, and the change in temperature in degrees Celsius. In the British engineering system the units are usually mixed in the following way: Q in British thermal units, A in square feet, d in inches, time τ in hours, and the change in temperature in degrees Fahrenheit. In calculating heat flow one must be careful to use units appropriate to the units in which the thermal conductivity has been measured.

EXAMPLE 10.11

A house window is constructed of a pane of glass 2.5 ft wide, 3 ft high, and $\frac{3}{16}$ in. thick. If the inner surface of the window is at 72°F and the outside temperature is 20°F, how much heat is conducted through the window during 1 h? During 1 day?

Solution

Since the quantities are given in the British system, it is appropriate to use the thermal conductivity for glass from Table 10.3. In British units $k = 6.0$ Btu · in./ft² · h · °F—an average value for glass. For 1 h the heat loss will be

$$Q = \frac{kA\tau \Delta t}{d} = 6.0 \, \frac{\text{Btu} \cdot \text{in.}}{\text{ft}^2 \cdot \text{h} \cdot \text{°F}} \times \frac{(2.5 \text{ ft} \times 3 \text{ ft})(1 \text{ h})(72\text{°F} - 20\text{°F})}{\frac{3}{16} \text{ in.}}$$

$$= 1.25 \times 10^4 \text{ Btu/h}$$

For a 24-h day the heat loss will be

$$Q = 24 \times 1.25 \times 10^4 \text{ Btu} = 3.00 \times 10^5 \text{ Btu}$$

Walls are often made of layers of different materials whose conductivities are different. Such composite walls can also be treated by Eq. (10.5) if one realizes that the rate at which heat is conducted through each material must be constant, since heat cannot "build up" at any point in the composite. A steady heat flow will be maintained across the slab or wall of composite materials if the temperature change across the wall remains constant.

Figure 10.8 Forced convection currents circulate the hot water from the furnace to the radiator in the room; air heated by the radiator expands and rises, producing natural convection currents in the air.

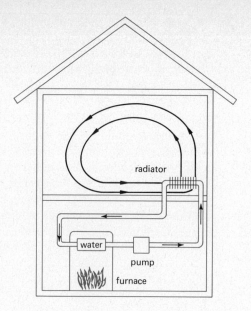

Convection

Perhaps the best way to describe heat transfer by convection is to discuss an everyday example—at least an example familiar to those who live in the northern latitudes. Home heating systems, whether wood stoves in kitchens or central furnaces fired by oil or gas, require convection to transport heat. The hot water heating system pictured in Fig. 10.8 involves *forced convection* in the transport of heat from the boiler of the furnace to the "radiator" or baseboard heating pipes placed in the rooms throughout the home. The water, heated in the firebox of the furnace, is pumped by a circulating pump through the pipes. In this way hot water is physically transported to the upstairs rooms.

In these rooms the hot water flows through a radiator or, particularly in modern heating systems, pipes to which fins are soldered at right angles. Air in the vicinity of these pipes and fins is heated and expands. The expanded air is less dense than the air in the remainder of the room, and hence it rises. The hot air rising results in the circulation of air (Fig. 10.8). The cold air circulates to the lower regions of the room, is heated, and rises. This is *natural convection*.

Natural convection plays an important role in our weather. For example, when the sun rises and its radiation is absorbed by the earth, the air near the earth is heated. The air expands and rises, producing circulation by convection. But because these processes are very complicated and depend on many variables, it is difficult to write even empirical formulas to describe convection under differing conditions.

One empirical observation is, however, worthy of note. A hot object loses heat to the surrounding air at a rate approximately proportional

The name radiator for the piping arrangement through which the hot water circulates in a room is a misnomer since the major function of the device is to heat the air by contact or conduction, thereby setting up the convection currents in the air. Some heat is indeed radiated from the hot pipes, but the major room heating is by convection. Note that most baseboard heating units are placed too close to the floor of the room for efficient heating. Actually, engineers have shown that the efficiency of the pipes with fins in heating a room is greatly increased by raising the pipes to about 2.5 ft above the floor. This provides for a much more efficient convection pattern—but it gives interior designers Excedrin headaches.

to the temperature difference between the object and the air. This empirical observation is called *Newton's law of cooling*. Unfortunately, even this simple "law" does not apply if the temperature differences are large. Engineers have established a number of empirical rules of thumb or guidelines to describe convection under more or less controlled circumstances. However, because of the multitude of restrictions and variable conditions, we shall not examine any of these in detail.

SUMMARY

1. Temperature is the numerical indication of hotness and coldness. It is measured on various different scales.
2. Almost all substances expand when their temperature increases and contract when their temperature falls. The coefficient of expansion of any substance indicates the relative amount of expansion produced when the temperature of the substance is changed by a specific amount.
3. Coefficients of expansion are generally higher for gases than for liquids and higher for liquids than for solids.
4. The expansion of water goes "the wrong way" between its freezing point of 0 and 4°C. This effect makes possible many familiar features in our daily lives.
5. The amount of heat, as opposed to temperature, is measured in units of calories. The heat capacity of a substance indicates how many calories are required to raise its temperature one degree.
6. Changes of state, from gas to solid, for example, or from solid to liquid, are accompanied by a release or intake of "latent heat."
7. An amount of mechanical energy equal to 4.186 J is always equivalent to 1 cal of heat.
8. According to the kinetic theory, the temperature of an object is a measure of the average kinetic energy of the molecules that make up the objects.
9. The transport of thermal energy occurs by radiation, conduction, and convection. Radiation is the process by which we obtain energy from the sun through the vacuum of space. Conduction is the heat-transfer process in which thermal energy is transmitted through material objects by molecular collision or vibration. Convection is the process whereby thermal energy is transmitted by the movement of mass involving fluid flow in a gas or liquid.
10. The rate at which heat flows through a slab of material depends upon the area of the slab, the thermal conductivity of the material, the temperature difference across the slab of material, and inversely upon the thickness of the slab.
11. Heat is energy in transit. Thermal energy is the energy associated with the random motion of the molecules that make up matter.

QUESTIONS

1. Distinguish between heat and temperature. How are they related? How are they different? Distinguish between heat and thermal energy.
2. Discuss in terms of internal molecular motion the expansion of solids, liquids, and gases upon heating.
3. By examining Table 10.1, explain why placing a hot, tightly fitting iron ring over a cold brass cylinder makes a bond that cannot be separated.
4. If we wish to make a thermometer to operate from just over 0°C to 80°C, why would water be a poor choice for the working substance?
5. Is water a "typical" substance? Name and explain at least two characteristics that make water the "model" or the "exception."
6. How is it possible for hot water from the Gulf Stream to move all the way from the eastern United States to England and the coast of Europe and still have a tremendous warming effect on the climate?
7. When you go on a camping trip and use an icebox or large cooler containing ice to keep your perishables, is it wise to wrap your block of ice with a number of newspapers? What does this do for you?
8. Why does blowing on a spoonful of hot soup cool the soup?
9. A soldier stationed near the Arctic Circle washes his trousers and hangs them out to dry. They freeze quickly, but after a few hours the ice disappears and the trousers are dry. Explain.
10. Ice on the sidewalk disappears even though no one walks on it and the temperature does not rise above the freezing point. Explain.

11. A steam burn is usually more serious that a hot water burn. Explain.

12. Frost forms on the sheet metal of cars outside in the early morning even when the temperature is above freezing. The metal is a good conductor. How can this be explained?

13. On a desert it is very, very hot in the daytime but very cold in the evening. Explain this change.

14. It is possible to heat water in a paper cup with a flame and not have the paper burn. Explain.

15. When you touch a rock, a piece of wood, and a metal plate that are all outside on a cold day at the same temperature, the rock and the metal feel colder than the wood. Explain.

16. Which process is faster, convection or conduction of heat? Give examples to justify your answer.

17. Two rectangular blocks, one made of copper and the other of aluminum, have identical masses and are placed on a large cake of ice after having been heated to 200°C. If the surface area of the base of each block on the ice is the same, which will stop sinking first and which will sink deeper? Consult Tables 10.2 and 10.3.

PROBLEMS

Temperature Scales

1. The temperature of the human skin is approximately 34°C while resting. What is this temperature in degrees Fahrenheit?

2. Mercury freezes at −38.9°C and vaporizes at 357°C. What are these temperatures on the Fahrenheit scale?

3. At what temperature do the Celsius and Fahrenheit scales coincide?

4. The boiling points of water on top of Mt. Everest and at sea level are 158.4°F and 212.0°F, respectively. What are these temperatures on the Celsius scale?

5. The sun has a temperature of about 20 000 000 K. What is this on the Celsius scale?

6. On the moon, when the sun is overhead the temperature is about 117°C; at sunset it is a mild 14°C; and at night a cool −163°C. Express the maximum change in temperature in degrees Fahrenheit.

Expansion

7. A steel rod is 15 cm long at −30°C. The rod is heated to 200°C. What is the new length?

8. A steel rod expands 0.98 mm for every 1°F. How much does it expand for 1°C?

9. How long is a steel bridge on a hot summer's day of 37.8°C (100°F) if its length in mid-December at −4°C is 92 m?

10. A hole 8.9 cm in diameter is bored in a brass plate at a temperature of 24°C. What is the diameter of the hole at 100°C?

11. A rectangular sheet of aluminum is 6 × 8 cm at 25°C. What is its area at 50°C?

12. To ensure a tight fit, the aluminum rivets used in construction are made slightly larger than the rivet holes and cooled by using dry ice before being driven. If the diameter of a hole is 0.6400 cm, what should be the diameter of a rivet at 20°C if its diameter is to equal that of the hole when the rivet is cooled to −78°C, the temperature of dry ice?

13. The diameter of a zinc sphere is 5 cm at −24°C. The sphere is heated to a temperature of 78°C. What is the change in volume?

14. The volume of a copper ball bearing is 2 cm³. The temperature is increased by 96°C. How much has its volume increased?

15. A motorist buys 12 gal of gasoline on a day when it is −25°C. What is the new volume on a day when it is +10°C outside? $\beta = 96 \times 10^{-5}/°C$.

16. A glass gallon jug is filled to the top with alcohol at −12°C. How much alcohol is lost if the jug is placed where it is 37°C?

17. An aluminum sphere 4 cm in diameter is 0.002 cm too large to fit through a brass loop at 25°C. At what temperature will the sphere and the loop just fit together? Is this a heating or cooling process?

18. A steel rule correct at 20°C is used to measure the size of a gold bar. If the length of the gold bar is 20.00 cm as measured by the rule when they are both at 0°C, what is the true length of the bar?

19. A brass block of volume 1 cm³ is heated from 0 to 400°C when it is confined in one dimension so that it cannot expand. (a) What is the resulting strain in the brass? (b) What stress must be applied to maintain the brass block at its original size?

20. An aluminum wire 0.8 m long is part of a metal sculpture. The wire is at 20°C. When the wire is placed outside on a cold day at −20°C, what strain is produced in the wire? How much stress must the sculpture be able to withstand to resist the contraction of this wire?

Heat and Change in State

21. It requires 490 cal to melt 10 g of a certain substance at its melting temperature. What is the heat of fusion of this substance?

22. What is the initial temperature of 140 g of water if the final temperature is 50°C and 980 cal of heat are given off?

23. There are 100 g of water at 25°C. If a lead weight of mass 200 g and

temperature 95°C is dropped into the water, what will be the final temperature?

24. A 100-g copper calorimeter holds 25 g of water and 8 g of ice at $t = 0$°C. If a 200-g brass weight at 100°C is added to the system, what is the final temperature?

25. It is observed that 600 cal of heat are required to raise the temperature of 50 g of a substance from 10 to 40°C. What is the specific heat of this substance?

26. A brass calorimeter of mass 30 g holds 100 g of water at 200°F. A 104-g lead weight at -40°C is added. What is the final temperature of the system?

27. A kilogram of water with 100 g of ice is in a calorimeter at $t = 0$°C. A kilogram block of aluminum at 100°C is added. Calculate the final temperature of the system.

28. A bathtub contains 100 000 g of water at 25°C. How much water at 60°C must be added to provide a hot bath of 40°C?

29. Compute the heat lost by an aluminum kettle of mass equal to 3000 g in cooling from 99 to 19°C, given that the specific heat of aluminum is 0.22 cal/g · °C.

30. Compute the amount of heat required to raise the temperature of a 3-lb electric flatiron from 75 to 275°F, given that the specific heat of the iron is 0.12 Btu/lb · °F.

Heat Conduction

31. A steel rod is 50 cm long and has a cross-sectional area of 4 cm². One end of the rod is placed in an ice bath, and the other end is placed in a steam bath. (*a*) How much time in minutes is required to transfer 100 kcal from one end to the other? (*b*) Compare this time to the time it would take if the steel rod were replaced by a glass rod of the same dimensions.

32. The bottom of a steel pan is 8 in. (20.3 cm) in diameter and 0.25 cm thick. How many kilocalories per minute pass through the pan if the surfaces of the pan are 104°C and 100°C?

33. What thickness of copper is equivalent to 2 cm of cork as far as insulating value is concerned?

34. What thickness of wood has the same insulating value as a 4-mm-thick plate of aluminum?

35. Compare the heat loss for an equal time interval through equal areas of a 5-cm-thick wooden wall and a 0.5-cm-thick windowpane.

36. How much heat in British thermal units is conducted through an 8-in.-thick concrete basement wall 10 ft high and 20 ft long if the temperature inside is 70°F and the temperature outside is 40°F?

37. A copper-bottom teakettle is used to boil water on a stove burner. (*a*) If the stove maintains a temperature of 300°F and the pan is 0.1 in. thick and 6 in. in diameter, compute the rate of heat flow into the water. (*b*) If 50 percent of the heat goes into boiling the water, how much water boils in 10 min? Latent heat of vaporization = 970 Btu/lb.

38. A copper rod is rigidly attached to a brass rod. Each rod is 40 cm long and 0.25 cm² in cross-sectional area. The free end of the copper rod is maintained at 100°C, and the free end of the brass rod is maintained at 20°C. (*a*) What is the temperature at the point where the two rods are connected? (*b*) What is the rate of heat flow through these rods?

39. A wall is constructed from a layer of wood 0.75 in. thick and of fiberglass 3 in. thick. If the wall is 10 ft high and 25 ft long and the outside temperature is -5°F, what is the amount of heat flowing through the wall in 12 h? The inside temperature is 70°F.

40. A 15-ft³ freezer is kept at 0°F inside with an outside temperature of 70°F. Consider the freezer a cube constructed of 2-in.-thick walls made of a composite of metal with fiberglass insulation with a thermal conductivity of 0.2 Btu · in./(ft² · h · °F). Determine the rate of heat flow into the freezer. If the freezer is turned off, how long would it take before 1 lb of ice in the freezer completely melted? Consider the freezer to contain only this ice.

41. A picnic cooler has an overall area of 1.5 m² and is made of a 1-cm-thick material with a thermal conductivity of 0.08×10^{-4} kcal/(m · s · °C). How many grams of ice will melt in 20 min if the inside temperature is 4°C and the outside is 30°C?

Special Topic
Holding Down Those Heating Bills

Determine how much money is being wasted through inadequate weather protection of your house or any other residence you choose, and devise a realistic plan to recoup at least some of that cash by simple, economic improvements.

Impossible? Not according to the Department of Energy. Engineers from the University of Maine at Orono who prepared a document titled Project Retrotech say that it is a relatively simple matter to identify heat leaks, translate the data into wastage of dollars and cents, and plot the improvements that will pay for themselves rather quickly.

The Maine engineers suggest four steps to new, improved home weatherization and the cash savings that it can produce:

1. Inspect the building to identify sources of heat loss.

2. Calculate how much heat actually escapes, bearing in mind such factors as the positions of obvious leaks and the thickness of insulation used for the house.

3. Evaluate the data, to decide what improvements in weather protection really make sense economically.

4. Install the appropriate materials to bring about the improvements.

The formula for economic satisfaction in these days of rising energy prices is based on a simple equation. If the heat given off by the furnace—or whatever other source of heat the home possesses—equals the total amount of heat lost through the walls, up the chimney, under the doors, and so on, then the temperature of the house will stay comfortably constant. Therefore, to keep the heating bills down, it is obviously prudent to minimize the amount of heat lost from the house.

Detailing the sources of heat loss is the purpose of step 1. After writing a basic description of the size and construction of the house—noting such features as double doors, windows, and the general condition of the outside of the house—examine potential sources of heat loss. Doors that fit well, and thus cause little loss of heat, resist the effort to be opened quickly because the seal between door and jamb creates a kind of vacuum. Windows are well sealed if one cannot push a 25-cent piece between the window and its casing. If two quarters can be forced into the gap, the window is a veritable heat sink. Check walls for signs of draughts coming through electrical outlets.

The second stage of the process, calculating how much heat is lost from the house, requires simple mathematics along with rough knowledge of the types of insulation in walls, ceilings, and the like and its effective insulating value. This latter, known as the R value, is really available from manufacturers. One other essential quantity is the "district heating function," an easy-to-obtain figure that indicates the amount of heating energy that should be required in a particular geographic location in a period of 1 yr.

The third step is to outline the obvious improvements at locations identified in stage 2 as the worst heat wasters. Such measures might include fitting storm windows, adding extra thicknesses of insulation to walls, installing weatherproofing around doors, and caulking window frames. Then calculate the amount of heat that such proposed changes should conserve inside the house, work out how much the materials for the improvements would cost, and figure whether the additions would pay for themselves in terms of the fuel savings that they impel. If the payoff time is just a few years, or "heating seasons," the investment is obviously worthwhile, and you can proceed to step 4—actually undertaking the improvements—in the knowledge that you are both saving money and doing your duty for the economy.

Ideal Gases, and Thermodynamics

11.1   CHEMICAL INTERLUDE

      SHORT SUBJECT: The Skyptical Chymyst

11.2   THE GAS LAW

11.3   THE IDEAL GAS

11.4   MAXWELL'S SPEED DISTRIBUTION

      SHORT SUBJECT: A Real Application for a Real Gas

11.5   VAPOR PRESSURE—THE KEY TO GOOD BREATH

11.6   RELATIVE HUMIDITY—THE COMFORT BARRIER

11.7   LAWS OF THERMODYNAMICS
The Efficiency of Engines

      SHORT SUBJECT: The Heat Death of the Universe

Summary
Questions
Problems

SPECIAL TOPIC: The Cruel Air, the Cruel Sea

We took then a long glass-tube, which, by a dexterous hand and the help of a lamp, was in such a manner crooked at the bottom, that the part turned up was almost parallel to the rest of the tube, and the orifice of this shorter leg of the siphon (if I may so call the whole instrument) being hermetically sealed, the length of it was divided into inches (each of which was subdivided into eight parts) by a straight list of paper, which containing those divisions, was carefully pasted all along it. Then putting in as much quicksilver as served to fill the arch or bended part of the siphon, that the mercury standing in a level might reach in the one leg to the bottom of the divided paper, and just to the same height of horizontal line in the other, we took care, by frequently inclining the tube, so that the air might freely pass from one leg into the other by the sides of the mercury (we took, I say, care) that the air at last included in the shorter cylinder should be of the same laxity with the rest of the air about it. This done, we began to pour quicksilver into the longer leg of the siphon, which by its weight pressing up that in the shorter leg, did by degrees straighten the included air; and continuing this pouring in of quicksilver till the air in the shorter leg was by condensation reduced to take up by half the space it possessed (I say, possessed, not filled) before; we cast our eyes upon the longer leg of the glass, on which was likewise pasted a list of paper carefully divided into inches and parts, and we observed, not without delight and satisfaction, that the quicksilver in that longer part of the tube was 29 inches higher than the other.

(*Treasury of World Science,* ed. Dagobert D. Runes. Philosophical Library, New York, 1962, p. 89.)

Such was the description—at once more folksy and more explanatory than typical scientific papers today—of a pioneering experiment by the brilliant seventeenth century chemist Robert Boyle. Like most scientists of his time, Boyle was a generalist, and in 1662 he set out on the series of experiments outlined above; the experiments were designed to investigate "the spring of the air," what we now recognize as air pressure. The experiments, carried out with the help of Robert Hooke, whom we encountered in Chap. 8 in connection with the "spring" of solids, led to a fundamental principle of the physics of gases: the volume of a particular mass of gas is inversely proportional to gas pressure. Expressed symbolically,

$$P \propto \frac{1}{V} \quad \text{or} \quad PV = \text{constant} \tag{11.1}$$

This relationship, shown graphically in Fig. 11.1, applies only to a constant mass of gas at a constant temperature, although the latter condition escaped the attention of Boyle. The first man to realize that the relationship does require constant temperature was Edmé Mariotte, a French contemporary of Boyle. Nevertheless, the relationship is universally known in America as Boyle's law.

The product of pressure and volume for a certain sample of gas at a given temperature remains constant; therefore, if V_1 is the volume of a particular gas at pressure P_1 and V_2 is the volume of the same mass of gas at pressure P_2 all at the same temperature, then

$$P_1 V_1 = P_2 V_2$$

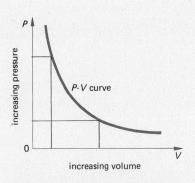

Figure 11.1 Relationship between the volume V of a gas and its pressure P, as summarized by Boyle's law. As the pressure increases, the volume decreases, and vice versa.

Used in this form, the equation does not require any specific set of units; obviously, however, the units used on the two sides of the equation must be consistent. One caution: the pressure must be the total or absolute pressure and not the gauge pressure. Recall that gauge pressure refers to measurements made with instruments calibrated to indicate zero when exposed to atmospheric pressure.

EXAMPLE 11.1

A 600-cm³ volume of nitrogen gas is at a temperature of 25°C and a pressure of 75 cmHg. Compute the volume of this gas if the pressure is decreased to 73 cmHg while its temperature remains at 25°C.

Solution

Because the temperature remains constant, and if we assume that no gas leaks from the container, we can use Boyle's law:

$$P_2 V_2 = P_1 V_1$$

$$(73 \text{ cmHg})(V_2) = (75 \text{ cmHg}) \times (600 \text{ cm}^3)$$

$$V_2 = \frac{75 \text{ cmHg} \times 600 \text{ cm}^3}{73 \text{ cmHg}} = 616 \text{ cm}^3$$

Boyle's law defines the specific connection between the pressure and volume of any constant mass of gas in the particular case of constant temperature. But what if we keep the pressure constant and alter another of the variables, say, the temperature? We can achieve this situation experimentally by placing a particular mass of gas in the cylinder shown in Fig. 11.2. The piston above the cylinder is free to move up and down, and the constant weight on the piston produces a constant pressure on the gas. By measuring the volume of the gas at a variety of different temperatures, we find that volume and temperature vary linearly with each other, as illustrated in Fig. 11.3. This is exactly the relationship we encountered for expansion of volume in Chap. 10.

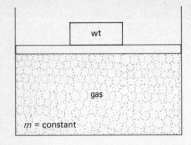

Figure 11.2 Gas under a constant pressure. The gas is held in a cylinder; the pressure on it is maintained by a movable piston that supports a weight.

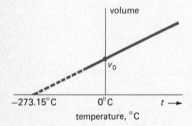

Figure 11.3 Thermal expansion of an ideal gas when the pressure is constant. The dotted line indicates the area in which the gas departs from ideal nature.

From Eq. (10.3),

$$\Delta V = \beta V_0 \Delta t \tag{10.3}$$

Taking $V = V_0$ at a temperature of 0°C,

$$V - V_0 = \beta V_0 (t - 0)$$

which can be rewritten as:

$$V = V_0(1 + \beta t)$$

This is known as Charles' law, after the French hot air balloonist and physicist Jacques Charles, who discovered it in the late eighteenth century.

The value of β, the volume coefficient of expansion for any gas, is readily obtained from the slope of the graph of volume plotted against temperature. The slope actually equals $V_0 \beta$. Substituting the value of 0.00366 (that is, 1/273.15) for β in the equation, we find that

$$V = V_0 \left(1 + \frac{t}{273.15}\right) = V_0 \left(\frac{273.15 + t}{273.15}\right)$$

That is,

$$\frac{V}{V_0} = \frac{273.15 + t}{273.15}$$

If we now recall that temperatures on the Kelvin scale equal degrees Celsius +273.15 (and note in passing that we have substituted the more accurate value of 273.15 in place of the 273 we used previously), then the above equation becomes a simple ratio:

$$\frac{V}{V_0} = \frac{T}{T_0} \tag{11.2}$$

at constant pressure and mass, where V is the volume of the gas at temperature T on the Kelvin scale and V_0 is the volume at temperature T_0. In this specific case, T_0 is 0°C because we arbitrarily chose V_0 as the volume of the gas at 0°C. But the general meaning of the equation is that **the volume of a constant mass of any gas is directly proportional to the absolute temperature of the gas when the pressure is constant.**

It is necessary to repeat the caution of Chap. 10: despite its appearance, Eq. (10.3) does not state that the volume of a gas drops to nothing at absolute zero. The reason is that the relation between temperature and volume deviates from a straight line above absolute zero because whatever gas is used liquefies at some finite temperature above absolute zero. Equations (11.1), (11.2), and (10.3) and many of the other equations in this chapter are strictly obeyed only by gases of relatively low density that are significantly above their liquefaction temperatures. Under these conditions, real gases approach the state defined as an ideal gas.

EXAMPLE 11.2

A 500-cm³ volume of hydrogen gas is at a temperature of 20°C and a pressure of 74 cmHg. If this quantity of gas is heated to 40°C while the pressure remains unchanged, compute the resulting volume.

Solution

We can use Charles' law since both the mass of the gas and the pressure are kept constant. The temperatures must be converted to the Kelvin scale ($T = t°C + 273$, using the approximate value):

$$\frac{V_1}{V_2} = \frac{T_1}{T_2}$$

or

$$\frac{500 \text{ cm}^3}{V_2} = \frac{20° + 273}{40° + 273} = \frac{293 \text{ K}}{313 \text{ K}}$$

$$V_2 = \tfrac{313}{293} \times 500 \text{ cm}^3 = 534 \text{ cm}^3$$

Note that the Celsius temperature is doubled but the Kelvin temperature is not, resulting in a final volume which is not double the initial volume.

Yet another alternative is to keep the volume of a certain mass of gas constant and check how the pressure of the gas varies with its temperature, using a constant-volume gas thermometer. This yields an equation for pressure identical to that in which volume varies with temperature; that is,

$$P_t = P_0(1 + \beta t) = P_0(1 + 0.00366t)$$

where P_0 is the pressure at $t = 0°C$, and P_t is the pressure at temperature t in degrees Celsius. Repeating the same procedure used to obtain Charles' law, we can obtain

$$\frac{P}{P_0} = \frac{T}{T_0} \quad \text{or} \quad \frac{P_1}{P_2} = \frac{T_1}{T_2} \quad \begin{array}{l} V = \text{constant} \\ m = \text{constant} \end{array} \quad (11.3)$$

Because the statement's form is so similar to that of Charles' law, one is tempted to make the statement that the pressure of a gas falls to zero at absolute zero of temperature. But again, this argument is misleading because no substances exist as gases at absolute zero.

EXAMPLE 11.3

In a laboratory, 200 cm³ of oxygen are collected at a pressure of 73 cmHg and a temperature of 27°C. Compute the pressure of the gas if this volume of gas is heated to 100°C while the volume and the mass of the gas remain constant.

Solution

We use the proportionality between pressure and absolute temperature here; since we are working with proportions, the pressure may be expressed in cmHg even though it is not a "proper" unit of pressure:

$$\frac{P_1}{P_2} = \frac{T_1}{T_2}$$

or

$$\frac{73 \text{ cmHg}}{P_2} = \frac{27° + 273}{100 + 273} = \frac{300 \text{ K}}{373 \text{ K}}$$

$$P_2 = \frac{373 \text{ K}}{300 \text{ K}} \times 73 \text{ cmHg} = 90.8 \text{ cmHg}$$

11.1 CHEMICAL INTERLUDE

Before taking the next obvious step—combining the three gas laws described so far in this chapter into one all-embracing statement of principle that is applicable to gases—we must delve into physical chemistry. Our purpose is to define a quantity related to mass that has particular application in linking the three laws of gases. The quantity is the gram-molecular weight, or mole, as it is known for short.

A mole (abbreviated mol) of any particular element or compound is defined as the numerical value of the substance's atomic or molecular weight expressed in grams. Atomic and molecular weights (which, to add to the confusion of physics students, are actually masses) are normally related to the atomic mass unit (u). The atomic mass unit was originally defined as the mass of the hydrogen atom. It is now defined, in the interests of greater accuracy, as one-twelfth the mass of the atom of the most common isotope of carbon, carbon 12. Its value is 1.66×10^{-24} g, an almost unimaginably small mass.

Moles represent much larger and more practical quantities that enable scientists to deal with large-scale processes. A mole of oxygen, whose atomic weight is 16, is 32 g, because oxygen is a diatomic gas. Similarly, 1 mol of diatomic nitrogen, whose atomic weight is 14, is 28 g. Helium, whose atomic weight is 4, is a monatomic gas, by contrast; thus, its mole has a mass of just 4 g. And 1 mol of steam, which consists of molecules containing one atom of oxygen and two atoms of hydrogen, has a mass of 18 g. The number of moles in any particular mass of gas is found by dividing the atomic or molecular weight of the gas into the mass; 10 g of hydrogen, for example, are $\frac{10}{2}$, or 5 mol.

What is significant about this apparently cumbersome means of measuring the masses of gases is that a mole of any gas contains the same number of particles, whether these be molecules or individual atoms; furthermore, 1 mol of gas always occupies the same volume at

That all gases at a given temperature and pressure contain the same number of particles per unit volume was advanced by the Italian scientist Amadeo Avogadro in 1811. Observations in the years since have shown that each mole of gas contains 6.02×10^{23} molecules; this number is known as Avogadro's number.

The Skyptical Chymyst

By any criterion, Robert Boyle is one of the giants of scientific thought. A pioneer in both chemistry and physics and a charter member of the Royal Society, which is still one of the most prestigious scientific clubs in the world, Boyle was at the center of the seventeenth century trend to advance science through experimentation rather than robotlike reliance on the word of previous authorities.

Born in 1627, the fourteenth son of the Earl of Cork, Boyle was a child prodigy who, at the age of 14, traveled to Italy to study the writings of the just deceased Galileo. His first major contribution to science was his book *The Skyptical Chymyst*, published in 1661, which established chemistry as an independent science and effectively led to the death of alchemy. Boyle argued that chemical elements were substances that could not be broken down into smaller fragments and that compounds consisted of combinations of elements. He did not shut the door entirely on alchemy, however, for he believed in the possibility of transmuting base metals into gold—and eventually, with the discovery of nuclear reactions, he was proved right!

The religiously devout bachelor demonstrated his scientific versatility a few years before the book was published when he designed an air pump and embarked, with the help of Robert Hooke, on the series of gas experiments outlined at the start of this chapter. These led to the law named after Boyle and to a revolution in scientific understanding of the nature of gases.

Robert Boyle: a giant of scientific thought. (*Culver*)

the same temperature and pressure, whatever the chemical composition of the gas. **A mole of any gas at the condition known as standard temperature and pressure (STP)—that is, 0°C and 76 cmHg—occupies 22.4 L,** for example.

11.2 THE GAS LAW

Having made this diversion, we can now return to our development of general law that connects the volume, pressure, temperature, and mass of gases. We can achieve this goal by the simple expedient of expressing the mass of gas in terms of the number of moles, which is given the symbol n.

Then, by combining the three laws with which we opened the chapter, we come up with a more general relationship

$$\frac{PV}{T} = \text{constant}$$

for a given quantity of gas. However, since the volume depends on the amount of gas, conveniently expressed as the number of moles (n) of gas present, we can write the law as

$$\frac{PV}{nT} = R \quad \text{where } R \text{ is a constant}$$

This, of course, says that

$$\frac{P_1 V_1}{n_1 T_1} = \frac{P_2 V_2}{n_2 T_2} = \frac{P_3 V_3}{n_3 T_3} \cdots \tag{11.4}$$

These two equations, known as the ideal gas law, hold for all gases of low density at temperatures well above their liquefaction temperature. This general gas law incorporates the previous three laws, which are just special cases of the general gas law. This procedure provides a good illustration of the direction in which science proceeds: to find more general laws or theories with which to replace many.

The constant R in the first equation is a truly universal constant whose value is exactly the same for all gases and is completely independent of pressures, temperatures, volumes, and other variables. Ensuring that R is a fundamental constant is actually the reason for introducing the concept of moles into the chapter.

Of course, the actual numerical value of R does depend on the units used for pressure, volume, and temperature. If P is in atmospheres, V in liters, T in K, and n in gram-moles (or simply moles), then

$R = 0.0821$ L·atm/mol·K

With P in N/m^2, V in m^3, and T in K,

$R = 8.314$ J/mol·K

The latter value may be converted from joules to calories using the Joule equivalent, yielding

$R = 1.987$ cal/mol·K

In practice, the second form of the general gas law is much simpler to use because it involves far less worry with units. One needs only to ensure that the units of volume and absolute pressure are the same on each side of the equation, although one must also remember that the **temperatures can only be expressed in absolute degrees** rather than in degrees Celsius or Fahrenheit. The constant R is the product of PV/nT in any one set of experimental conditions. One convenient means of calculating R mentally is to remember that at STP conditions—1 atm of pressure at 273 K—a mole of any gas occupies 22.4 L.

EXAMPLE 11.4

There are 8 g of helium gas at room temperature 20°C confined to a volume of 10 L. Under what pressure must the gas be held?

Solution

Since we know only one state of the gas, we must use the first form of the general gas law in Eq. (11.4). Because the volume is in liters, it is appropriate to use the value $R = 0.0821$ L·atm/mol·K. The pressure will then come out in atmospheres. We have to convert the mass of helium to moles, dividing the mass of the gas, 8 g, by the atomic weight of helium, which is obtained from App. F:

$$n = \frac{8 \text{ g}}{4 \text{ g/mol}} = 2 \text{ mol}$$

$$\frac{PV}{nT} = R$$

$$P = \frac{nTR}{V} = \frac{(2 \text{ mol})(273 + 20)\text{K}(0.0821 \text{ L} \cdot \text{atm/mol} \cdot \text{K})}{10 \text{ L}}$$

$$= 4.81 \text{ atm}$$

EXAMPLE 11.5

On a day when the temperature is 19°C and the barometer registers 74 cmHg, a balloon is inflated with helium gas to a volume of 100 000 ft³. On release from the earth's surface, it rises to an altitude of 3.5 mi, where the pressure is just 37 cmHg and the temperature is −10°C. Compute the volume of the helium at the higher level.

Solution

Because we are comparing the gas under two sets of conditions and we have such an odd collection of units, it is best to use the proportional form of the general gas law. We can assume that the balloon does not leak and the number of moles—not given—remains constant:

$$\frac{P_1 V_1}{n_1 T_1} = \frac{P_2 V_2}{n_2 T_2}$$

Canceling the n's and assuming that the effect of the balloon is to simply confine the gas, we obtain, upon substituting the given data,

$$\frac{(74 \text{ cmHg})(10^5 \text{ ft}^3)}{(273 + 19) \text{ K}} = \frac{(37 \text{ cmHg}) V_2}{(273 - 10) \text{ K}}$$

This example illustrates why weather balloons burst at very high altitudes. The expansion of the gas in the balloons eventually proves too much for the balloon material.

Solving for V_2,

$$V_2 = \frac{(74 \text{ cmHg})(263 \text{ K}) \, 10^5 \text{ ft}^3}{(37 \text{ cmHg})(292 \text{ K})} = 1.8 \times 10^5 \text{ ft}^3$$

The number of moles of gas in the balloon could be calculated from the given data using the gas constant. Convert the values of the variables measured in one state, say, when the balloon was on the ground, to a consistent set of units and use the first form of the gas law to calculate n.

EXAMPLE 11.6

A 200-cm³ volume of oxygen gas is collected under outside atmospheric pressure. In the laboratory, a thermometer indicates 25°C, and the barometer indicates 73 cmHg. Compute the amount of gas collected and the volume of the gas under standard conditions (0°C and 76 cmHg).

Solution

Using the data given, together with an appropriate value for R, one can directly calculate the number of moles of gas. However, the number of moles could also be determined by computing the volume of the gas at standard temperature and pressure and then comparing this volume to 22.4 L, the volume that 1 mol of gas occupies at STP.

To calculate n directly, we must choose a set of units and match a value of R. Let us choose $R = 0.0821$ L · atm/mol · K. The pressure of 73 cmHg is (73 cmHg) ÷ (76 cmHg/atm) = 0.96 atm. The original volume was 200 cm³. A liter is 1000 cm³, so our original value is (200 cm³)/(1000 cm³/L) = 0.20 L. Using the gas law, remembering to change the temperature to degrees Kelvin, we have

$$\frac{PV}{nT} = R$$

$$n = \frac{PV}{RT} = \frac{(0.96 \text{ atm})(0.20 \text{ L})}{(0.0821 \text{ L} \cdot \text{atm/mol} \cdot \text{K})(273 + 25) \text{ K}} = 0.00785 \text{ mol}$$

To compute the volume under standard conditions, we can use the proportional form of the gas law:

$$\frac{P_1 V_1}{n_1 T_1} = \frac{P_2 V_2}{n_2 T_2}$$

The n's remain unchanged, and thus cancel. Substituting the data given yields

$$\frac{(73 \text{ cmHg})(200 \text{ cm}^3)}{(273 + 25) \text{ K}} = \frac{(76 \text{ cmHg})(V_2)}{273 \text{ K}}$$

$$V_2 = \frac{73 \text{ cmHg} \times 273 \text{ K} \times 200 \text{ cm}^3}{76 \text{ cmHg} \times 298 \text{ K}} = 176 \text{ cm}^3$$

As noted, we can compute the number of moles of gas by realizing that under STP the volume of the gas is 22.4 L. Converting 176 cm³ to liters yields

$$\frac{176 \text{ cm}^3}{1000 \text{ cm}^3/\text{L}} = 0.176 \text{ L}$$

$$n = \frac{\text{no. of liters}}{22.4 \text{ L/mol}} = \frac{0.176}{22.4} = 0.0079 \text{ mol or } 7.9 \times 10^{-3} \text{ mol}$$

11.3 THE IDEAL GAS

By developing a molecular model of what is termed an "ideal gas" we can get a clearer understanding of the large-scale concepts of temperature and pressure.

Having introduced the concept of moles, which is related to large numbers of individual molecules, we might ask if there is any way of understanding the general gas law in strictly molecular terms. It turns out that there certainly is. By developing a molecular model of what is termed an "ideal gas" we can get a clearer understanding of the large-scale concepts of temperature and pressure.

The model of the ideal gas involves a number of simplifying assumptions, including:

1. Gases are made up of small molecules (or atoms) that are assumed to be point particles occupying no volume. Each individual gas consists of vast numbers of identical particles that are relatively far apart.
2. The molecules are moving in random motion—the basic assumption of kinetic theory.
3. No molecular forces exist between the particles. The molecules interact with each other only when they collide. Furthermore, collisions between molecules, and collisions between molecules and the walls of any vessel containing the gas, are assumed to be perfectly elastic (no kinetic energy is lost).

These conditions seem quite restrictive. However, real gases at sufficiently low densities and high enough temperatures fulfill them surprisingly well.

Having made the assumptions, we can apply Newton's laws to a simple assemblage of ideal gas molecules, represented, say, by the gas in the cubical box whose sides are each of length l shown in Fig. 11.4.

The pressure on the walls of the box results from the bombardment of the individual molecules. Imagine a single molecule traveling along the x axis with a velocity v_x and colliding with the wall. Because the collision is perfectly elastic, according to one of the specifications above, we know that the molecule rebounds from the wall with exactly the same speed, v_x, in the opposite direction. The change in momentum as a result of the collision is $mv_x - (-mv_x)$, that is, $2mv_x$, where m is the mass of the molecule (see Example 6.2, page 157). Now we know that the change in momentum of the molecule equals its impulse on the wall and the impulse is the product of the force and the time during which it acts. If we can define the time of collision, therefore, we can calculate the force the molecule exerts on the wall during the collision.

Stretching our imagination a little, we can define the time of each collision as the time between two successive collisions between the wall and the same molecule. If we further imagine that the individual molecule simply travels back and forth along the x axis between the two walls of the box, with a constant speed v_x, then the time t between successive collisions is obviously given by the equation

$$v_x = \frac{2l}{t} \quad \text{or} \quad t = \frac{2l}{v_x}$$

Using the impulse-momentum theorem, we can compute the force

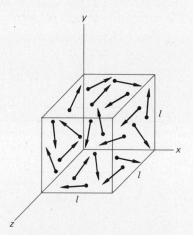

Figure 11.4 Kinetic model of a gas. The molecules are confined in a cubic box, through which they move in a random manner. Their collisions with the walls of the box produce the pressure that the gas exerts on the walls.

imparted to the wall during a single collision:

$$Ft = \Delta mv_x = 2mv_x$$

Hence,

$$F\left(\frac{2l}{v_x}\right) = 2mv_x \quad \text{or} \quad F = \frac{mv_x^2}{l}$$

The pressure is given by

$$P = \frac{F}{A} = \frac{mv_x^2}{lA} = \frac{mv_x^2}{V}$$

(Here we have recognized that lA is the volume of the cube.)

We have considered only one molecule when, in fact, the box contains some large number N of molecules. The pressure on the one wall due to the N number of molecules is

$$P = \frac{Nmv_x^2}{V}$$

> The average *velocity* of the molecules must be zero in all directions because all molecules move back and forth across the box. If this average of the velocity were not zero, one of the walls would experience a greater pressure than the others—and might be moving. Note that the average of interest here is the average of the square of the speeds of all the molecules, which is obtained by squaring all the speeds first and then averaging these values. The square root of this average is called the root mean square (rms) value of the speed. Furthermore, this average, appropriate to our problem, is different from the square of the average value of the speed, at least for a group of molecules with a random distribution of speeds.

What about the other walls of the box? Obviously, the molecules do not move in any preferential direction because we have already stated that their motion is random. To answer this question it is convenient to turn our attention to the kinetic energy of the individual molecules.

If N is large, we can assume that the average kinetic energy associated with the molecules moving in the x direction is equal to the average kinetic energy associated with the y direction, and likewise for the z direction. This means that

$$\tfrac{1}{2}\overline{mv_x^2} = \tfrac{1}{2}\overline{mv_y^2} = \tfrac{1}{2}\overline{mv_z^2}$$

In effect, this condition says that average value of the square of the speed is the same in each direction, or

$$\overline{v_x^2} = \overline{v_y^2} = \overline{v_z^2}$$

where v_x, v_y, and v_z are the speeds of molecules in the x, y, and z directions, respectively, and the average is denoted by the bar over the quantity.

Since the average value of the velocity squared is given by the sum of the squares of its components,

$$\overline{v^2} = \overline{v_x^2} + \overline{v_y^2} + \overline{v_z^2}$$

in other words,

$$\overline{v^2} = 3\overline{v_x^2} \quad \text{therefore} \quad \overline{v_x^2} = \tfrac{1}{3}\overline{v^2}$$

Substituting this value into our expression for pressure, we obtain

$$P = \frac{Nm\overline{v^2}}{3V}$$

or

$$PV = \tfrac{1}{3}Nm\overline{v^2} \tag{11.5}$$

This can be rewritten as

$$PV = (\tfrac{2}{3}N)(\tfrac{1}{2}\overline{mv^2}) = \tfrac{2}{3}N \times \text{(average molecular kinetic energy)}$$

In determining this expression, we considered N molecules moving in the box in a random way. We would have determined the same expression by considering one-third of the molecules moving in each of the three mutually perpendicular directions.

This expression for PV is interesting in that it defines the product of two macroscopic variables in terms of the average kinetic energy of the individual particles or molecules. It connects macroscopic and microscopic quantities. We can gain further insight into the connection between our microscopic model and the macroscopic parameters by recalling the gas law:

$$PV = nRT$$

This gas law expresses what is observed for real gases under conditions we have called ideal, meaning that the gas is far from conditions of pressure and temperature which would cause liquefaction. Our microscopic gas law resulted from a simple kinetic model of the gas and the application of Newtonian mechanics. We can now combine these two results for PV and make some predictions. If our predictions withstand experimental tests, we shall know that our model is reasonable.

Setting the right side of the two equations for PV equal to one another, we obtain the result

$$nRT = \tfrac{2}{3}N(\tfrac{1}{2}\overline{mv^2})$$

The number of molecules N in a sample is usually obtained from a measurement of the mass of the gas, which can then be expressed as the number of moles of gas present. Since each mole of a substance contains Avogadro's number of molecules, N_A, the number of molecules present can be written as the number of moles present times Avogadro's number:

$$N = nN_A$$

Substituting this value for the number of molecules and solving for the average kinetic energy, we obtain

$$nRT = \tfrac{2}{3}(nN_A)(\tfrac{1}{2}\overline{mv^2})$$

so

$$\tfrac{1}{2}\overline{mv^2} = \frac{3}{2}\left(\frac{R}{N_A}\right)T = \frac{3}{2}kT \tag{11.6}$$

The constant $k = R/N_A$ is the gas constant per molecule (R is the gas constant per mole). This new constant k is known as Boltzmann's constant, after the Austrian physicist Ludwig Boltzmann, who derived it, and its value is $k = 1.38 \times 10^{-23}$ J/K.

How do we interpret Eq. (11.6)? It shows in quantitative terms what we previously assumed in a qualitative fashion in dealing with the kinetic theory of gases in previous chapters—that the **Kelvin temperature of any gas is proportional to the average kinetic energy of the molecules in the gas,** and hence provides **a measure of average molecular kinetic energy.** The equation represents a microscopic definition of a macroscopic quantity: temperature. As the temperature of a gas rises, the kinetic energy of its molecules increases; thus the momentum and its pressure increase. The **concept of temperature is** plainly **a statistical one that depends on large numbers of molecules.** The idea of the temperature of a single molecule is meaningless; indeed, the above argument breaks down if we try to apply it to just a few molecules.

In real situations, even with vast numbers of molecules, the reactions between the molecules and the walls will not take place in exactly the way we assumed. Individual molecules will collide with others and thus be unable to complete the round-trip journeys between opposite walls, for example. However, the presence of huge numbers of molecules, and the fact that all collisions are elastic, will overcome this difficulty. The individual molecules with which we are concerned may collide with one another and bounce backward, but the other molecules with which they collide will also be driven back in their paths, conserving momentum. Furthermore, the large number of molecules ensures that each molecule knocked out of the x direction is replaced by another knocked into that direction. Overall, the average velocity along any particular line remains constant at a value of zero.

EXAMPLE 11.7

What is the average kinetic energy of the oxygen molecules and nitrogen molecules in normal air on a warm summer day (27°C)? From knowledge of the kinetic energy compute the rms speed of the nitrogen and oxygen molecules in this air.

Solution

The average kinetic energy of an individual gas molecule of mass m is determined by temperature (only) and is given by Eq. (11.6):

$$\tfrac{1}{2}\overline{mv^2} = \tfrac{3}{2}kT$$

Note that the average kinetic energy of any or all gases will be the same at the same temperature. At $t = 27°C$, $T = 27 + 273 = 300$ K, and the average kinetic energy will equal

$$\tfrac{1}{2}\overline{mv^2} = \tfrac{3}{2}(1.38 \times 10^{-23} \text{ J/K})(300 \text{ K})$$
$$= 6.21 \times 10^{-21} \text{ J}$$

This is an extremely small amount of energy, but, after all, it is the average kinetic energy of a *single* molecule that has an extremely small mass.

Now to compute the rms speed, $\sqrt{\overline{v^2}} = v_{\text{rms}}$, we must have a value for the mass of the molecules in the gas. The value of v_{rms} depends inversely upon the size of the mass of the molecule. Hence, at a particular temperature, the massive molecules will move more slowly than the lighter, less massive molecules. The mass of a molecule can be determined by dividing the molecular weight (mass) M of the gas by Avogadro's number N_A since a mole of gas "weighs" one molecular weight and contains N_A molecules:

$$m = \frac{M}{N_A}$$

$$\tfrac{1}{2}\overline{mv^2} = 6.21 \times 10^{-21} \text{ J}$$

$$\tfrac{1}{2}\frac{M}{N_A}\overline{v^2} = 6.21 \times 10^{-21} \text{ J}$$

$$v_{\text{rms}} = \sqrt{\overline{v^2}} = \sqrt{\frac{2N_A}{M} \times 6.21 \times 10^{-21} \text{ J}}$$

For oxygen the kilogram molecular weight (kmol) is 32 kg, and for nitrogen the molecular weight is 28 kg. The rms speed at 27°C for molecules of the two gases can now be evaluated:

For oxygen:

$$v_{\text{rms}} = \sqrt{\frac{2 \times 6.023 \times 10^{26}(\text{kmol})^{-1} \times 6.21 \times 10^{-21} \text{ J}}{32 \text{ kg/kmol}}}$$

$$= \sqrt{2.34 \times 10^5} = 4.84 \times 10^2 \text{ m/s}$$

For nitrogen:

$$v_{\text{rms}} = \sqrt{\frac{2 \times 6.023 \times 10^{26}(\text{kmol})^{-1} \times 6.21 \times 10^{-21} \text{ J}}{28 \text{ kg/kmol}}}$$

$$= \sqrt{2.67 \times 10^5} = 5.17 \times 10^2 \text{ m/s}$$

These speeds are over 1000 mi/h! The results indicate that the lighter nitrogen molecules move, on the average, more rapidly than do the more massive oxygen molecules.

11.4 MAXWELL'S SPEED DISTRIBUTION

Extending the statistical argument, it is useful to get some idea of the range of speeds of the individual molecules in an ideal gas. A typical distribution is shown in Fig. 11.5. This is known as the Maxwell distribution, after Scottish physicist James Clerk Maxwell, who first calculated this curve on theoretical grounds in 1860. So far was Max-

A Real Application for a Real Gas

The kinetic theory, Boyle's law, and other fundamental principles of the gaseous state all refer to ideal gases—the theoretical entities in which the forces of attraction between molecules are at a minimum. In practice, of course, no gas is completely ideal; the very fact that all gases liquefy precludes this. And although most gases approach fairly close to the ideal state when their densities are low and their temperatures well above their boiling points, the nonideal characteristics of gases are generally those most useful in practical applications.

The departures from nonideal behavior can best be visualized graphically by plotting a series of "isotherms"—curves that relate pressure and volume at constant temperature, as seen in the accompanying figure. Isotherms for gases well above their boiling points are almost perfectly hyperbolic in shape, as shown in Fig. 11.1, thus illustrating the validity of Boyle's law. At lower temperatures, however, the isotherms first develop definite kinks, then show an extremely disturbed region, consisting

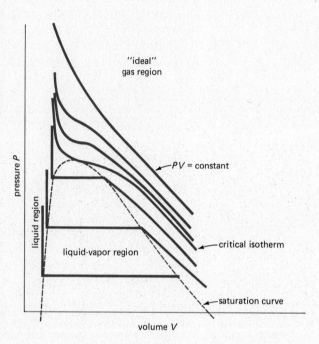

Isotherms for a pure substance. The isotherms indicate the various constant temperatures. The dashed curve that separates the gas and liquid-vapor regions is the vapor saturation curve, and is continued to the left to separate the liquid region from the liquid-vapor region.

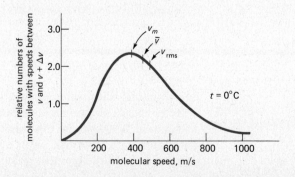

Figure 11.5 Distribution of molecular speeds in a gas or liquid.

of a nearly horizontal line in which pressure remains almost constant with changing volume, and then show an almost vertical line that shows virtually no change in volume in response to large applications of pressure.

These two areas represent the liquefaction of the gas. The constant-pressure region corresponds to condensation of the gas, which reduces its volume considerably at constant temperature and pressure. The constant-volume region represents the all-but-incompressible liquid state of the condensed gas.

The change of state that occurs along the horizontal section of the isotherms requires heat supply or removal accompanying the volume change. The temperature at which the process takes place depends on the pressure (as explained in Sec. 11.5). The principles embodied in the curves are utilized in the design of that most common and necessary household device, the refrigerator.

The purpose of the refrigerator is to lower the temperature of a certain volume of space. According to the second law of thermodynamics, this can be achieved only by introducing work into the system. The laws of thermodynamics also tell us that any liquid at its boiling point requires an input of heat to convert it to gas. In the case of the refrigerator, heat is removed from the inside compartment in just this way; it is used to boil the refrigerator fluid, which is generally a chemical such as ammonia or Freon, whose boiling point is relatively low. The gas is then taken to another part of the device, where it is condensed back to a liquid, thus giving off heat that is dispersed to the atmosphere. In refrigerators, this boiling-condensation cycle is accomplished by variations in pressure rather than temperature.

Two types of systems are in common use for refrigerators: compression and absorption devices. In the compression model, the liquid is first evaporated, thus drawing the requisite heat from the machine's interior storage space. A small compression pump draws the vapor off, compresses it, and leads it to a condenser. By this time the vapor is at sufficiently high pressure to liquefy readily. The newly liquefied refrigerant is then expanded to a lower pressure, at which it boils more readily once it is returned to the evaporator to continue the cooling cycle.

The absorption refrigerator is somewhat more complicated. In this device, water containing a large amount of dissolved ammonia is heated; the heat drives off the ammonia in the form of vapor, leaving almost pure water behind. The pressure of the outgoing ammonia vapor gradually increases until it is high enough to cause it to condense in a condenser. This liquid is expanded using a special valve, and then evaporates again, withdrawing heat from the inside of the refrigerator in the process. The gaseous ammonia from this process is mixed with the pure water from the boiler, which has been cooled in a heat exchanger; in fact, the device is set up in such a way that the cooling is carried out by mixing the pure water with the new water-ammonia mixture, after which it is ready to return to the boiler and start the cooling process again.

---

well ahead of the experimenters that it took applied physicists more than 65 yr to confirm his thinking. They finally did so by using a complex experimental procedure that involved channeling gaseous mercury vapor through rotating disks with small slits in them.

The Maxwell distribution curve, which shows the number of gas molecules possessing speeds from v to $v + \Delta v$, has a number of points of interest. First, most of the molecules cluster around the speed v_m, where the curve has its peak. More of the molecules are traveling with speed v_m than with any other speed; v_m is, therefore, termed the most probable speed. As the curve shows, it is slightly smaller than the average speed of all the molecules, $\bar{v}$. Examination of the graph shows why: although very few of the molecules possess a speed of zero, a significant but still relatively small number are traveling at twice the most probable speed, and some are traveling with much greater speeds than that. Hence, the average speed of all the molecules must be higher than the most probable speed. One other marker on the graph indicates v_{rms}, the root-mean-square speed, which we have already encountered. It is also larger than the average speed.

Use of the Maxwell distribution is not restricted to problems involving gases. We can invoke it, for example, to understand why liquids cool as they evaporate because molecules in liquids have a similar spread of speeds. A liquid tends to remain in the liquid state because of the attraction between its molecules. The molecules that move fastest are least likely to remain bound by this type of attraction and hence most likely to break out of the liquid during evaporation. The loss of the fastest moving molecules means that the average speed of the molecules remaining in the liquid is lower than before. This, in turn, means that the average kinetic energy, and hence the temperature, of the liquid is lower as a result of evaporation.

11.5 VAPOR PRESSURE—THE KEY TO GOOD BREATH

The molecules that evaporate from a liquid exert their own pressure above the liquid. To understand the real meaning of this "vapor pressure," let us imagine the evaporation of a dish of liquid such as water in three different situations.

First, imagine that the dish of water is placed inside a closed container which has been evacuated of all its air. The fastest molecules in the liquid water immediately start to evaporate into the vacuum above the water. At the same time, some of the liberated molecules find their way back into the liquid, by collision, for instance. However, as long as more molecules leave the liquid than return, the amount of vapor increases.

Eventually, the rate at which vapor molecules return to the liquid water exactly matches the rate at which liquid molecules spring out into the vapor. At this time we say that an equilibrium exists between the water and the water vapor and that the space above the liquid water is saturated with water vapor because the quantity of vapor no longer increases. The pressure of the water vapor above the dish at this equilibrium point is known as the saturated vapor pressure. The saturated vapor of any particular liquid varies only with temperature, increasing quite rapidly as the temperature increases up to the liquid's boiling temperature, which is the temperature at which the pressure of the saturated vapor of the liquid is equal to the external pressure. Table 11.1 shows some values of the saturated vapor pressure of water at different temperatures.

For our second situation, imagine that the dish of water is placed in a large room so carefully sealed from the outside that it is completely

Table 11.1 Saturated Vapor Pressure of Water at Temperatures from 0 to 150°C

Temperature (°C)	Pressure (atm)
150	4.69
100	1.00
80	0.467
60	0.196
40	0.0728
35	0.0555
30	0.0418
20	0.0231
15	0.0168
10	0.0121
5	0.00861
0	0.00626

11.5 Vapor Pressure—The Key to Good Breath

gas tight. Once again, molecules of liquid water escape and molecules of vapor return to the liquid. But because of the size of the room and the presence of air molecules, a great deal of time passes before equilibrium is established. Once it is established, however, the molecules of water vapor contribute their part—equal to the saturated vapor pressure of water at that temperature—to the overall pressure of the room. As a result, the overall pressure in the room is increased. If the air pressure before the dish of water is placed into the room is 1 atm, and the temperature remains at 30°C throughout the experiment, the pressure in the gas-tight room once equilibrium is reached would be 1.0418 atm—the original 1 atm plus 0.0418 atm, which is the saturated vapor pressure of water at 30°C. This latter pressure is called the partial pressure caused by the water vapor.

The third scenario is closest to reality. We place the dish of water in a room that is not air tight, a condition which describes most of the rooms around the world. In this circumstance the total pressure remains at 1 atm even after equilibrium is set up between the liquid water and water vapor. However, 0.0418 atm of that 1 atm is now contributed by the pressure of the water which has evaporated into the air. This contribution is known as the partial pressure of the water vapor at that particular temperature. In fact, whenever gases are mixed, each exerts a partial pressure equal to the pressure that it would produce if the same amount of gas as there is in the mixture were to occupy the container alone.

Whenever gases are mixed, each exerts a partial pressure equal to the pressure that it would produce if the same amount of gas as there is in the mixture were to occupy the container alone.

Although the majority of the population is blithely unaware of the fact, the concept of partial pressure is vitally important to our continued healthy existence on earth. Neglecting small amounts of carbon dioxide, water vapor, and the rare gases helium, neon, and argon, the air consists roughly of 80 percent nitrogen and 20 percent oxygen. Our lungs and circulatory systems, as well as those of all other air-breathing creatures on earth, are adapted to the partial pressure of oxygen.

At high altitudes, as we recall from Chap. 9, atmospheric pressure is appreciably lower than it is at sea level. More importantly for mountaineers, parachutists, and other high-altitude adventurers, the partial pressure of oxygen is also lower at elevated altitudes than it is at sea level. The fall in the partial pressure of oxygen is roughly proportional to the overall drop in atmospheric pressure. For lungs and circulation systems used to oxygen at its "normal" sea-level partial pressure, exposure to the partial pressure of the gas at the altitude of Mexico City or the Himalayas can be quite a shock. At Mexico City's height of about 8000 ft it takes most people who normally live at sea level a number of days of acclimation before they can breathe comfortably and carry out any strenuous work.

The reverse situation occurs in the case of deep sea diving. The pressure in the ocean depths is much higher than that of the atmosphere at sea level; to keep the pressure in the diver's lungs in equilibrium with the outside pressure, the breathing apparatus must deliver air at a pressure higher than that at sea level. But if the apparatus delivers "normal" air, consisting of 20 percent oxygen, the partial pressure will be too high for the unfortunate ocean explorer. The solution is an appropriate mixture of gases in which oxygen has its normal sea-level partial pressure, as detailed in the Special Topic at the end of this chapter.

11.6 RELATIVE HUMIDITY—THE COMFORT BARRIER

One consistent ingredient of weather forecasts, along with temperature, wind velocity, and barometric pressure, is the **relative humidity** of the local region. The term **is defined as the ratio of the actual vapor density to the density of saturated vapor at the particular temperature involved** and is normally expressed as a percentage. When it is foggy, the air generally contains as much water vapor as it can take, and the relative humidity hovers between 95 and 100 percent. On exceptionally dry days in desert areas, such as those of the southwest United States, the relative humidity can fall to 20 percent or even as low as 5 percent.

The maximum amount of water vapor the air can contain obviously represents the saturated vapor density of the atmosphere. It is not surprising, therefore, that this quantity, which is normally measured in grams per cubic meter, also rises rapidly as the temperature increases. Figure 11.6 shows a graph of the saturation density.

Another new concept related to that of relative humidity is the dew point, which is the temperature at which the air becomes just saturated with water vapor when it is being cooled. Let us imagine that the atmosphere on a rather muggy summer's day contains 20.5 g/m³ of water. A glance at Fig. 11.6 shows that as long as the temperature remains in the high seventies, the atmosphere is not completely saturated with water vapor. But once the temperature drops to 73.4°F (23°C in the figure), the 20.5 g of water in each cubic meter is enough to saturate the vapor. At this point, which is the dew point for the particular conditions we specified, drops of water condense on metal surfaces and other objects.

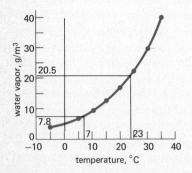

Figure 11.6 Changing humidity. The graph indicates the density of saturated vapor at different temperatures.

EXAMPLE 11.8

What are the dew point and relative humidity for the air of a home whose room temperature is 23°C on a day during which moisture starts to condense on the outside surface of a bright can of water at 7°C?

Solution

The temperature at which water vapor condenses from the air on the metal can is 7°C. We can assume that this is the dew point. To compute the relative humidity of the air at 23°C we use the relation

$$\text{Relative humidity} = \frac{\text{density of vapor present}}{\text{density of vapor for saturation}}$$

This is equivalent to

$$\text{Relative humidity} = \frac{\text{density to saturate at dew point}}{\text{density to saturate at room temperature } 23°C}$$

To determine the amounts for saturation we use Fig. 11.6:

$$\text{Relative humidity} = \frac{7.8 \text{ g/m}^3}{20.5 \text{ g/m}^3} = 0.38 = 38\%$$

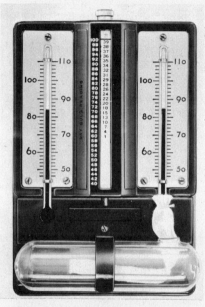

Figure 11.7 A wet-and-dry bulb hygrometer. Both thermometers indicate the same temperature because there is no water in the wet bulb reservoir. (*Taylor Instrument Consumer Products Division, Sybron Corporation*)

We feel most comfortable when the relative humidity is between about 35 and 60 percent.

A simple way to determine relative humidity is based on the fact that evaporation causes cooling. The instrument used is called the wet-and-dry bulb hygrometer. It consists, as illustrated in Figure 11.7, of two similar thermometers: one open to the air and the other surrounded by a wick kept wet by a container of distilled water. The presence of the wick means that the water evaporates unless the air is saturated. If the surrounding air is saturated with water vapor, both the wet and dry thermometers will register the same temperature, indicating no net evaporation. If the air is not saturated, however, the cooling produced as water evaporates from the wet thermometer will cause this instrument to register a lower temperature than the dry thermometer. The wet reading is not the dew point, however.

Over the years, the experts in standards laboratories have carried out experiments to interpret wet-and-dry bulb readings in terms of relative humidities. These experiments have resulted in a series of standard tables that anyone owning a wet-and-dry bulb hygrometer can use to obtain relative humidity at a glance. The tables show, for example, that when the dry thermometer reads 20°C and the wet one reads 15°C, the relative humidity is 59 percent. Hygrometer owners should take one precaution to ensure accurate results: they must keep the air around the thermometers in motion with a fan, to prevent the air in the vicinity of the bulbs from becoming more saturated than the rest of the air in the room.

For most people, the important thing about relative humidity is not its measurement but its effect on bodily health and comfort. We feel most comfortable when the relative humidity is between about 35 and 60 percent. In hot weather our bodies cool themselves down through evaporation of sweat, but when the relative humidity is greater than 60 percent, the amount of moisture in the air greatly reduces the rate at which sweat can evaporate. We therefore feel generally uncomfortable, complain of the muggy weather, often suffer from sinus trouble, and comfort ourselves with such clichés as "It isn't the heat, it's the humidity."

On the other hand, if the air inside a home is so dry in winter that the relative humidity is less than 35 percent, sweat evaporates from our bodies at too fast a rate for comfort, cooling us down so effectively that we feel chilly even when the thermometer reads 73°F or more. In fact, we feel just as warm at 73°F and a relative humidity of 50 percent as we do at a temperature of 76°F and a relative humidity of 25 percent. Furthermore, physicians agree that we are more susceptible to colds when the relative humidity of our homes is too low, owing to the excess evaporation of moisture from the membranes of the mouth, nose, and throat. The penalty of too high a relative humidity indoors in winter is more directly financial; in addition to causing condensation of moisture on cold windows, which fog up, it may also leave moisture on windowsills and curtains, ruining the paint jobs and destroying the curtain material.

Obviously, then, householders should keep the relative humidity of their homes at a moderate level in winter. This is more difficult than it might appear because warm air can hold much more water vapor than cold air. Thus, although the relative humidity of outside air at a temperature around the freezing mark may be a respectable 80 percent, simply heating that air in a furnace to 20°C reduces greatly the relative humidity. The same amount of water is a far smaller proportion of the amount necessary to saturate the air at higher temperatures. A rough reading of the humidity graph in Fig. 11.6 shows that heating saturated air at 0 to 20°C reduces the relative humidity of the air to about 22 percent—just about the same as in the dry air over the Sahara Desert.

The solution to this dilemma is to add water vapor to the air as it emerges from the furnace. One simple device consists of a pan of water attached to the side of a hot-air furnace. Another solution involves the use of a device that drips water on to a metal trough which extends across the hot-air chamber at the top of the furnace. Experience shows that either type of humidifier uses several quarts of water daily during winter to keep the relative humidity inside a typical house at a comfortable 40 to 50 percent.

In summer, unfortunate house owners encounter the reverse difficulty: the relative humidity of the inside air is too high, particularly in basements. Moisture then tends to condense on cold water pipes and, more catastrophically, on leather and other materials, which mildew. Dehumidifiers are designed with cold coils on which moisture condenses preferentially, and a major function of air conditioners is to remove moisture from the air.

11.7 LAWS OF THERMODYNAMICS

Thermodynamics is the area of physics that is concerned with the large-scale properties such as pressure, volume, and temperature of systems which exchange heat with their surroundings. The laws of thermodynamics are, in their way, the heat equivalents of the basic conservation laws of motion. Before looking at these laws, we must

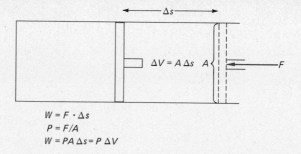

Figure 11.8 The work ($F \cdot \Delta s$) done against pressure on a gas equals $P \Delta V$.

define a new quantity: **the internal energy of a substance.** This factor, generally given the symbol U, **is the total amount of energy inside the substance, stored in the forms of chemical binding energy and translational, rotational, and vibrational kinetic energy of the molecules or their parts.** The internal energy of any substance can be altered by the application of either work or heat. According to the kinetic picture of matter, we expect that internal energy increases with temperature.

The first law of thermodynamics is simply an expression for the conservation of energy in processes involving heat. It states that the amount of heat added to any system equals the sum of the increase of the system's internal energy, defined as thermal energy in Chap. 10, and the work done by the system against its surroundings. Put into the form of an equation, the first law is

$$JQ = U_2 - U_1 + W \tag{11.7}$$

where Q is the amount of heat in kilocalories put into the system; J is the mechanical equivalent of heat; U_1 and U_2 are the internal energies, in joules, of the system before and after the application of heat; and W is the amount of work, in joules, done by the system against its surroundings when it is heated. This equation is general enough to encompass both the addition and removal of heat. If heat is taken from the system, Q becomes a negative quantity. Similarly, W is *positive* if the system actually *does work against* its surroundings and *negative* if work is *done on the system*. The law applies to any type of work, but in most practical applications involving a gas, the work is given by the equation $W = P \Delta V$. This can be obtained from our definition of work, $W = F \Delta s$, the definition of pressure, and a look at Fig. 11.8.

One particular class of processes that illustrates the application of the first law of thermodynamics is the group known as adiabatic processes. Scientifically, **an adiabatic process is one in which no heat is added to or removed from the system.** This condition is attained by either isolating the system thermally from its surroundings or carrying out the process very rapidly. This does not mean that the temperature of the system remains constant, for even in the absence of external interactions with the surroundings the system is free to exchange its fund of energy between thermal (internal energy) and mechanical forms. Giving Q the value zero in Eq. (11.7),

$$0 = U_2 - U_1 + W$$

or

$$U_2 - U_1 = -W \tag{11.8}$$

This equation means that when the system performs work, its internal energy (and therefore its temperature) decreases; when work is done *on* the system, its internal energy (and therefore its temperature) increases.

One common example of an adiabatic process is the rapid compression of a gas in a hand pump. If the compression is carried out fast enough, no heat can be added to, or subtracted from, the system. Work is plainly done on the air in the pump, and hence its internal energy increases. This increase manifests itself in an increase in temperature on the compressed air.

The reverse process, which is also adiabatic, is the expansion of a compressed gas through a nozzle into a region of lower pressure than the initial pressure of the compressed gas. Here, the gas expands and, therefore, does work on its surroundings. This is positive work according to our definition, and substitution in Eq. (11.8) shows that its effect is to decrease the internal energy of the gas, thus lowering its temperature. A typical example of this type of process is the release of carbon dioxide from a fire extinguisher. Often enough, expanding carbon dioxide starting at room temperature becomes so cold that it forms crystals of dry ice. The container also becomes cold since the average kinetic energy of the molecules left behind is lower. (Demonstration of this particular process for or by students is guaranteed to give gray hairs to college administrators!)

Fortunately, there is a much simpler means of demonstrating adiabatic heating and cooling. If a rubber band is stretched quickly, work is done on it, causing its internal energy to rise. This rise shows itself as a small increase in temperature, which one can detect by touching the rubber band to one's lips. If a stretched rubber band is allowed to relax quickly, it performs work on itself as it restores itself to its equilibrium site and suffers a reduction in internal energy. The resulting lower temperature can be monitored in the same way.

EXAMPLE 11.9

A gas absorbs 500 cal of heat, doing 200 J of work in the process. What is the change in internal energy of the gas?

Solution

We can use the first law of thermodynamics:

$$JQ = U_2 - U_1 + W$$

where $J = 4.186$ J/cal; $Q = +500$ cal since the heat is absorbed by the system; and $W = +200$ J since work is done by the system.

Therefore,

$$U_2 - U_1 = JQ - W = (500 \text{ cal})(4.186 \text{ J/cal}) - 200 \text{ J} = 1893 \text{ J}$$

The internal energy will increase.

The first law of thermodynamics is a general expression regarding the conservation of energy when it is interchanged between the mechanical and thermal forms. However, observations over the centuries have shown that although mechanical energy can be converted entirely into thermal energy, the reverse is not the case. Thermal energy cannot be converted entirely into mechanical work; when the conversion takes place in this direction, a certain amount of thermal energy remains unconverted as a kind of exhaust. In Fig. 11.2, for instance, it is impossible to heat the gas and simply raise the piston above it without also increasing the temperature of the gas. This restriction is neither stated nor implied in the first law of thermodynamics. As a result, we must, therefore, formulate a second law to take account of the restriction.

The chances of changing disorderly motion entirely into the orderly motion of work are slim indeed.

The second law of thermodynamics can be expressed in a number of different ways, each statement corresponding to some observation of the way that events occur in the natural world. The oldest statement of the law is that heat cannot flow spontaneously from material at a lower temperature to material at a higher temperature. We recall that work is orderly energy that can be changed completely into random, or disorderly, motion in a natural way. Thermal energy, by contrast, is disorderly motion, and the chances of changing disorderly motion entirely into the orderly motion of work are slim indeed. In fact, such processes as the flow of heat from hot areas to colder ones and the spontaneous mixing of liquids with the absence of spontaneous unmixing are all examples of irreversible processes. They are all examples of the one-sidedness of nature with which the second law of thermodynamics is concerned. These observations are summarized in a more general statement of the law, as follows:

In any energy transformation, some of the energy is altered to heat or internal energy that is no longer available for further energy transformation. In simple language, this says that all processes, no matter how efficient, produce exhaust heat and fall short of 100 percent efficiency. Put another way, **the law states that any system containing a large number of molecules will, when left to itself, assume a state as disordered as possible.** The extent of disorder is measured by the quantity known as **entropy**.

Entropy is a measure of the probability that a particular type of change will occur. When we are working with a large number of molecules, it is appropriate to speak in terms of statistical probabilities. The disordered situation is the more probable in nature, and entropy is a measure of the disorder. The slightest perturbation will cause an ordered situation to change rapidly to a more disordered situation. For example, an ordered deck of cards when dropped becomes disordered. Furthermore, if the cards are picked up one at a time, without their faces being looked at, there is only a very, very small chance that the cards will be picked up in order. Humpty Dumpty falling off the wall will be smashed into a disordered state. What are the chances (in the absence of outside influence) that Humpty Dumpty will return to his ordered state?

To return to a more physical example, consider water spilling over a waterfall into a pool. Here the more orderly potential energy is transformed partly into an increased disorderly motion of the water molecules in the pool below the falls. The thermal or disordered energy of the water has increased. What is the chance that all this random thermal energy will "get together" and provide the energy required to lift the water back to the top of the falls, that is, reverse the falls? There is certainly nothing in the first law of thermodynamics (conservation of energy) that would *not* allow this process. The first law would require only that the water cool down as its thermal energy was "used" to raise the water to the top of the falls. The probability that the water will flow uphill must be very small because we do not observe it happening. Such a process would be from a disordered to a more ordered state; this is not nature's way. These are examples of the very general law, the second law of thermodynamics. In terms of entropy, the second law means that when left to itself (isolated), a system with a given amount of thermal energy moves to a condition where entropy or disorder is maximized.

Practically, the second law of thermodynamics places restrictions on the operation of machines in general, predicting, for example, that no machine can run with ideal efficiency. The law also plays an important part in our environmental future, predicting an event with a science fiction cachet: the heat death of the universe. (See page 322.)

The Efficiency of Engines

An engine is a device that converts some form of energy such as heat, chemical, or electrical energy into work. An important application of the second and first laws of thermodynamics is the computation of the efficiency of an engine. Thermodynamics is useful here because it provides a generalized approach to problems. Any results obtained can be applied to specific situations. For example, by treating engines in a general way, we gain insight into such varied devices as gasoline motors, steam engines, electric motors, and muscles. In all cases, the process by which energy is changed into work is quite complex and difficult to analyze in detail.

We shall consider for simplicity an engine as working in a cycle and converting heat to work. We shall then examine in more detail the most efficient or ideal heat engine, called the Carnot engine after the French engineer Sadi Carnot. Carnot was the first to examine the heat engine

Carnot's contribution to thermodynamics promised to be immense, but the French physicist died in 1832 during a cholera epidemic at the early age of 36.

from the fundamental thermodynamic point of view. Carnot considered only the significant features of the heat engine, disregarding the details of the operation. The engine would take energy in the form of heat and convert it to work. Then, the engine would reject some heat to a lower temperature. The processes would necessarily have to obey the first and second laws of thermodynamics.

In an engine, some working substance, say, a gas, at one temperature expands against a piston doing work and then is brought back to its original condition in a cyclic manner. For example, in a steam engine, steam at some high temperature expands against a piston doing work at the expense of its internal energy. The steam cools and leaves the expansion chamber, returning perhaps as hot water to a boiler where it is heated again back to steam, and the process can then repeat itself. We can represent such a situation by a simple diagram shown in Fig. 11.9. Here, heat Q_2 is supplied to the engine that does work W and rejects exhaust heat Q_1 at a lower temperature. The work done in thermal units by the engine is, from the first law of thermodynamics (conservation of energy), just the difference between the heat input and the heat output:

$$W = Q_2 - Q_1$$

The efficiency of the machine is the ratio of the work output to the work input. In thermal energy units,

$$\text{Efficiency} = \frac{\text{work output}}{\text{heat energy input}}$$

$$= \frac{W}{Q_2} = \frac{Q_2 - Q_1}{Q_2} = 1 - \frac{Q_1}{Q_2}$$

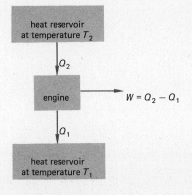

T_2 is greater than T_1

Figure 11.9 Energy flow in a heat engine.

Obviously, the most efficient engine is one that exhausts as little heat as possible. We already noted in our discussion of the second law that there will always be some exhaust heat Q_1, so that no engine can be 100 percent efficient. What can be the maximum attainable efficiency? Can we proceed further without the details of a specific engine?

Carnot answered the questions by devising an idealized engine which involved the cyclic process outlined in the pressure-volume graph shown in Fig. 11.10. The cycle is a four-step process. First, the engine absorbs heat Q_2 from a constant-temperature heat reservoir held at temperature T_2. During the process, the volume of the gas (assumed to be the working substance) expands and the pressure is reduced while the temperature is held constant at temperature T_2. In the second step, the gas expands adiabatically with no heat loss to or gain from the outside. Then, in step 3 the gas is compressed along an isotherm; in other words, the temperature is held constant at T_1. Heat of amount Q_1 leaves the working gas and is transferred to the lower temperature reservoir. In step 4, the gas is compressed adiabatically to its original state. In the whole process, the engine does an amount of

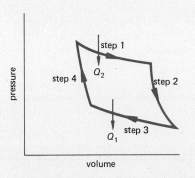

Figure 11.10 The Carnot cycle. Steps 1 and 3 are isothermal; steps 2 and 4 are adiabatic.

The Heat Death of the Universe

Imagine a dark, silent universe, populated by billions upon billions of cold stars and planets, all at the same temperature as the vast clouds of gas between them, and without a glimmer of life in its vase volume. This was the picture of the eventual fate of our cosmos predicted by physicists at the end of the last century, as a direct result of their interpretations of the second law of thermodynamics.

The startling, if remote, possibility of a dead universe came about as a consequence of a concept termed **entropy,** from the Greek word for transformation. Entropy represents a means of quantifying the amount of disorder in any particular isolated system. The entropy of any particular state in which a system finds itself is related to the probability of that state. Rudolph Clausius, the German scientist who invented the term entropy, carried the concept of entropy to its logical conclusion in his alternative statement of the second law of thermodynamics (written in 1865):

> We can express the fundamental laws of the universe, which correspond to the two fundamental laws of the mechanical theory of heat, in the following simple form:
>
> **1.** The energy of the universe is constant.
> **2.** The entropy of the universe tends toward a maximum.

We can see this statement of the second law of thermodynamics in action in daily life. A piece of precious china dropped on the floor maximizes its entropy by shattering. If the individual pieces are dropped together, they do not immediately re-form into the original article. A cut of meat cooks when heated because its molecules break down into assemblages of greater disorder. The meat does not return to the more ordered raw state when it is stored in a freezer. And the water in swimming pools does not spontaneously convert into a mixture of ice and hot water.

So realistic did this form of the second law of thermodynamics appear to physicists a century ago that Lord Kelvin made the daring but apparently justifiable extension to the largest isolated system imaginable: the universe itself. According to the Clausius principle, Kelvin reckoned, all objects in the cosmos will eventually reach the same temperature by exchanging heat with each other (an action that increases a system's entropy). When this situation comes about, Kelvin reasoned further, it will be impossible to produce useful work from heat because the creation of work by heat engines mandates the flow of heat from a hot object to a cooler one. Thus, the uniform temperature universe will also be a dead universe, the victim of "heat death."

In its day, this theory was as controversial among scientists and nonscientists as Darwin's theory of evolution. Weighing in to argue for the theory or against it were not only the scientific worthies of Victorian times, such as Kelvin, Boltzmann, and Clausius, but also mathematicians of the caliber of the Frenchman Henri Poincaré, philosopher-physicists of the quality of the Austrian Ernst Mach, honest-to-goodness philosophers of the towering repute of Nietzsche, and even such purveyors of popular prose as H. G. Wells. A French astronomer, Camille Flammarion, wrote a tract about the ways in which the heat death might end the world, and an American historian named Henry Adams attempted to apply the Clausius version of the second law of thermodynamics to human history. Among other things, the whole controversy served to cast a grave shadow of doubt over the kinetic theory of gases, which was relatively new at the time.

Nevertheless, the kinetic theory survived, and the prospect of the heat death slowly became less horrific to physicists. But even now, it appears that entropy may signal the end of the living universe as we know it—although this event will certainly not occur for many billions of years.

---

work W that is actually represented graphically by the area inside the closed cycle in Fig. 11.10. The Carnot cycle is completely reversible. In the reverse direction the engine acts as a refrigerator that takes in heat at temperature T_1 and rejects heat to the hot reservoir. Of course, in this case, work is done on the device.

Starting with the second law of thermodynamics, it can be shown that a Carnot engine is the most efficient engine which can operate between the two temperatures T_1 and T_2, and, in addition, the efficiency of the Carnot engine is independent of the working substance. Since the efficiency of an engine in given by

$$\text{Efficiency} = 1 - \frac{Q_1}{Q_2}$$

the efficiencies of all Carnot engines working between the same two temperatures must depend only on the two temperatures T_2 and T_1 and nothing else. Lord Kelvin proposed that this fact be used to define a temperature scale such that the ratio of the temperatures of the two reservoirs is defined by

$$\frac{Q_2}{Q_1} = \frac{T_2}{T_1}$$

This temperature scale turns out to be identical to the absolute temperature scale discussed previously.

The efficiency of the Carnot cycle is therefore given by

$$\text{Efficiency (Carnot engine)} = 1 - \frac{T_1}{T_2}$$

The efficiency can be maximized by raising the intake temperature T_2 as high as possible and lowering the exhaust temperature T_1 as low as possible. Usually, T_1 is the temperature of the air or of a lake or river, and T_2 is limited by the fact that most working substances at high temperature are gases whose pressure increases with temperature. High pressures of course require strongly built boilers. In practice, T_1 and T_2 have limits. Theoretically, we must reduce T_1 to 0 K, or $-273.15\,°\text{C}$ to obtain 100 percent efficiency. To make $Q_1 = 0$, we must use a low-temperature reservoir at absolute zero. This line of argument leads to the third law of thermodynamics: **although absolute zero may be approached as close as we like, it is actually not possible to reach the absolute zero of temperature.**

EXAMPLE 11.10

In a steam engine steam is heated to 200°C; because it is above the boiling point, this is called superheated steam. The steam expands against a piston doing work and then condenses into water in thermal equilibrium with an icy river at 4°C. Compute the maximum efficiency of this engine.

Solution

The maximum efficiency would occur if the steam cycle operated in a Carnot cycle. If we assume this, then the efficiency can be calculated from the temperatures of the two reservoirs. The temperatures must be in degrees Kelvin:

$$T_2 = 200°C + 273° = 473 \text{ K}$$
$$T_1 = 4°C + 273° = 277 \text{ K}$$
$$\text{Efficiency} = 1 - \frac{T_1}{T_2} = 1 - \frac{277 \text{ K}}{473 \text{ K}} = 1 - 0.59 = 0.41$$

The efficiency is 0.41, or 41 percent. Therefore, for every 100 J of heat energy taken in from the high-temperature reservoir, 41 J of useful work will be done by the machine. Of course, any real machine will be less efficient because of heat loss during operation and frictional processes.

SUMMARY

1. According to Boyle's law, the product of the pressure and volume of a gas is constant at constant temperature.
2. According to Charles' law, the volume of a gas is directly proportional to the absolute temperature at constant pressure.
3. A "mole" of gas always contains the same number of particles.
4. The general gas law states that the product of pressure and volume of any gas divided by the product of the number of moles and the absolute temperature is constant for all temperatures well above the liquefaction temperature.
5. The concept of the ideal gas explains macroscopic characteristics of gases in microscopic terms.
6. Speeds of molecules or atoms in a gas are distributed in characteristic fashion, with the most probable speed being slightly lower than the average speed of all the molecules.
7. Whenever gases are mixed, each gas exerts a pressure in the mixture equal to that it would produce if the amount in the mixture were present alone in the container.
8. The relative humidity at any temperature is the ratio of the amount of water vapor in the air to the maximum amount of water vapor that the air can contain at that temperature. This quantity decreases markedly with increasing temperature.
9. The first law of thermodynamics states that the amount of heat added to a system equals the sum of the increase in the system's internal energy and the work done by the system against its surroundings. This law extends the conservation of energy to processes that involve heat.
10. Although mechanical energy can always be converted entirely into heat, the reverse is not true. Attempts to convert heat completely into energy always produce some waste heat. This observation is the basis of the second law of thermodynamics.
11. The second law of thermodynamics limits the efficiency of engines to below 100 percent. The most efficient engine is one based on the Carnot cycle.

QUESTIONS

1. How does the kinetic theory explain the fact that gases exert pressure on all surfaces with which they come in contact?
2. A sample of hydrogen gas is compressed to half its original volume while its temperature is held constant. What happens to the average speed of the hydrogen molecules?
3. One explains Brownian motion of large smoke particles as being due to the bombardment of the smoke by the randomly moving, very much smaller air molecules. The air and the smoke must be at the same temperature. What must be true of the relative speeds and the relative kinetic energies of the smoke particles and the air molecules? Discuss.
4. How does perspiration provide the body with a means of cooling itself?
5. Evaporation is said to be a cooling process; that is, the liquid remaining is cooler. Explain in terms of molecular motion.

6. Estimate the number of molecules in 1 cm³ of air and the distance between them.
7. A container of gas is placed into a "gravity-free" place in outer space. Since the molecules of the gas exert a pressure and hence a force on the walls, why doesn't the container accelerate steadily?
8. Why must absolute, or Kelvin, temperatures be used in the gas law equation?
9. What effect do elastic collision between gas molecules have upon our derivation of the gas law?
10. When air in a bicycle pump is compressed, it heats. Why?
11. Why does the boiling point of a liquid depend upon atmospheric pressure?
12. Canteens usually come with a canvas cover. Why does wetting the cover on a hot day help keep the contents of the canteen cool?
13. If a piston is pushed rapidly into a container of gas, what happens to the kinetic energy of the molecules of the gas? What happens to the temperature of the gas?
14. On a very hot day in your home, you decide to cool off your kitchen by opening the refrigerator door and closing all the kitchen doors and windows. Would this process cool off the kitchen? Explain why or why not.
15. If you put a pinhole in a helium-filled balloon in a laboratory, the helium diffuses throughout the whole lab. How does this illustrate the second law of thermodynamics?
16. The ocean contains a tremendous amount of heat energy. (a) Under what conditions could a ship use this heat to propel it? (b) How would this be done? (c) What would be the change in the temperature of the water? (d) Does this violate the laws of thermodynamics?
17. Give several examples of everyday processes in which work is converted to heat and heat is converted to work.
18. Give examples of everyday processes in which entropy is increasing.
19. Do the following processes obey the first law of thermodynamics: warming a poker in a fire; the boiling away of water in a saucepan on a stove; lighting a campfire by rubbing two sticks together or striking a metal with flint? Explain.
20. Does the growth of an animal throughout its life violate the second law of thermodynamics?
21. Discuss the application of the laws of thermodynamics to social interactions and interactions of people and nations.

PROBLEMS

Macroscopic Gas Law

1. How many moles of gas are contained in 0.3 kg of carbon dioxide gas (CO_2)?
2. Show that 1 u in grams is equal to the reciprocal of Avogadro's number.
3. How many moles of gas occupy 1 L at STP?
4. What volume does 1 mol of gas occupy at STP?
5. The pressure of a fixed volume of oxygen at 25°C is 735 torr (mmHg). What is the temperature if the final pressure is 700 torr?
6. (a) Compute the volume of air, starting at 1 atm of pressure (14.7 lb/in.²), which would be required to fill a 10-ft³ tank with air to a pressure of 8 atm. (b) What will a pressure gauge, which measures excess of inside pressure over outside pressure, register for this compressed air?
7. The volume of CO_2 gas at 10°C is 2 L. The temperature is increased to 100°C at constant pressure. What is the new volume?
8. Ten liters of air at 20°C are confined in a vertical cylinder by a piston resting on top of the gas. As the air in the cylinder is heated, the piston rises, thus keeping the pressure constant. Compute the temperature in degrees Celsius to which the enclosed air must be heated to increase the volume to 20 L.
9. The pressure of 6 L of gas at 25°C is 4 atm. What is the new pressure if the temperature is decreased to 10°C and volume increased to 7 L?
10. Calculate the ideal gas constant if 1 mol of gas at 0°C and 6 atm occupies 3.7 L.
11. What is the pressure of 4 mol of gas at 35°C and occupying 2 L?
12. What volume does 4 mol of oxygen gas occupy at 25°C and 760 torr (mmHg)?
13. A cylinder contains 100 cm³ of air at a total pressure P. What will the total pressure become if the (a) volume is reduced by half to 50 cm³ while the temperature remains constant, (b) volume is doubled to 200 cm³ while the temperature remains constant, (c) volume is doubled and the absolute temperature is also doubled, (d) volume is halved and the absolute temperature doubled?
14. An ideal monatomic gas at 400 K exerts a pressure of 10^4 N/m². If it is compressed to half its original volume and, as a result, its temperature changes to 480 K, what will be its new pressure?
15. Two grams of nitrogen at 27°C occupy a volume of 2 L. If the pressure is doubled and the temperature raised to 127°C, calculate the final volume.
16. If the pressure in a lecture room 30 × 20 × 5 m is to remain constant as the temperature rises from 17 to 27°C, what volume of gas must escape?

17. A balloon is to be filled to a volume of 10 000 ft³ with helium at atmospheric pressure. If each cylinder of gas contains 2 ft³ at a gauge pressure of 13 atm, how many cylinders are required?

18. A 14-L helium tank at 25°C and a pressure of 27 atm is used to fill a balloon. The gas in the balloon is at 1 atm and initially at −20°C due to the cooling caused by the expanding gas. What is the volume of the balloon? When the balloon returns to the outdoor temperature of 25°C, what will its volume be?

19. A very good vacuum corresponds to a pressure of about 10^{-10} N/m². How many molecules remain in 1 cm³ at 27°C? Compare this number with the number in our normal atmosphere.

20. A gauge on a tank of oxygen reads 212 lb/in.² Apparently the tank has a leak because a day later the unused tank reads 122 lb/in.² What fraction of the original mass of gas contained in the tank has been lost?

21. A bubble of air rises from the bottom of a lake, where the pressure is 3.03 atm, to the surface, where the pressure is 1 atm. The temperature at the bottom of the lake is 7°C, and the temperature at the surface is 27°C. What is the ratio of the size (volume) of the bubble as it reaches the surface to the size of the bubble at the bottom? Neglect the vapor pressure of the water.

22. An air bubble 2 cm in diameter at the bottom of a lake 27 m deep where the temperature is 4°C rises to the surface where the temperature is 20°C. Compute the diameter of the bubble at the surface. Neglect the vapor pressure of the water.

23. It is important for proper tire wear to keep the pressure of the air in the tires at the proper value. Consider the volume of the tire as fixed. If you inflate the tires to gauge pressure of 24 lb/in.² on a day when the temperature is 50°F, (a) what will be the pressure when the temperature rises to 90°F on a summer's day? (b) When the temperature drops to 15° below 0°F in December?

Kinetic Theory

24. What is the rms speed of CO_2 molecules in a tank where the pressure is 4 atm and the temperature is 200°C?

25. What is the rms speed of helium atoms in the atmosphere of a star where the temperature is 100 000 K?

26. Helium remains a gas down to 4.2 K under atmospheric pressure. Compute the average speed and kinetic energy of helium atoms at this temperature.

27. A gas is enclosed in a cubical box 0.5 m on a side at a pressure of 4×10^3 N/m². What is the force exerted by the gas on one wall of the box?

28. An inert gas, helium, is monatomic. If a sample of helium is confined inside a box of volume 500 cm³ at a pressure of 5×10^3 N/m² and a temperature of 200 K, what is the average kinetic energy per atom?

29. The rms speed of oxygen molecules is 400 m/s when the pressure is 700 mmHg. What is the density of the gas?

30. Some helium gas is contained in a box of negligible weight at room temperature (300 K). If all the internal energy could be abstracted from the gas and used to lift the box, through what height could it be raised? Express your answer in miles.

31. One mole of an ideal monatomic gas A at 200 K is mixed with 2 mol of a different ideal monatomic gas B at 400 K. They come into thermal equilibrium with one another without losing any internal energy. What is their final common temperature?

32. At what temperature will the rms speed of oxygen molecules be twice their rms speed at 27°C?

33. At what temperature will the rms speed of nitrogen molecules equal the rms speed of oxygen molecules at 27°C? Mass of oxygen atom = 16 u; mass of nitrogen atom = 14 u.

34. Show that the ratio of the rms speeds of molecules of two different gases is inversely proportional to the square root of their masses.

35. (a) At what temperature would the rms speed of oxygen molecules equal the escape velocity from the earth (11.3 km/s)? (b) Calculate the corresponding temperature for the moon. Discuss your results in terms of atmospheric conditions on the moon and the earth.

36. One mole of an ideal gas is warmed from 300 to 350 K without changing the volume of its container. How much heat is added? What is the final pressure?

37. Two moles of helium at 300°C are mixed with 2 mol of argon at 100°C. What will be the final temperature of the mixture assuming that no heat is lost and the mixing takes place at constant volume?

Vapor Pressure

38. A container holds 12 g of methane (CH_4) and 34 g of carbon monoxide (CO). What are the partial pressures of the two gases if the total pressure is 2 atm?

39. A container with a volume of 1 m³ is filled with a cubic meter of hydrogen (H_2) at 1 atm and nitrogen (N_2), which occupies 3 m³ at 0.7 atm of pressure. Calculate the partial pressures and the total pressure.

40. Air in the lungs usually contains a partial pressure of carbon dioxide (CO_2) of about 40 mmHg. What is the percentage of CO_2 in the air in the lung?

41. A gas is composed of 2.6 g of

oxygen and 1.3 g of helium in a 2.5-L container at a temperature of 22°C. What is the partial pressure of each gas and the density of the mixture?

42. If the room temperature is 24°C and the relative humidity is 67 percent, what is the dew point?

43. If the temperature of the air is 20°C and the dew point is 12°C, what is the relative humidity?

44. Air at 0°C on a winter day when the relative humidity is 45 percent comes into the house through an open door and is warmed to 23°C. What is the relative humidity in the house?

45. On a hot day in the summer, each cubic meter of air contains 28 g of water vapor. If the temperature is 32°C, what is the relative humidity?

Thermodynamics

46. A gas expands from a volume of 10^3 cm^3 to a volume of 4×10^3 cm^3 at a constant pressure of 5×10^5 N/m^2. How much work is done?

47. How many joules of energy are used when a force of 500 N moves an object through 10 m?

48. If water at a waterfall drops 150 m into a pool, how much is the temperature of the falling water raised if 100 percent of the potential energy is converted into heat?

49. A lead bullet moving at a speed of 400 m/s strikes a target and stops. If 40 percent of the heat generated in the stopping process were absorbed by the bullet, what would be the final temperature of the bullet? The original temperature was 25°C.

50. A spacecraft reentering the earth's atmosphere must slow down from a speed of 29 000 km/h to essentially zero. If all this kinetic energy is absorbed as heat energy by the spacecraft, by how much would the temperature rise? Assume the specific heat capacity of the spacecraft is 0.5 kcal/kg · °C. What is done to avoid this temperature rise?

51. What is the change in internal energy of a system in which 200 cal are added and 20 J of work are done by it?

52. What is the change in internal energy of a system in which 200 cal are given off by the system and 40 J of work are done on it?

53. A Carnot engine is operated between two heat reservoirs of temperature 400 and 300 K. If the engine receives 1200 cal from the 400-K reservoir, how many calories does it reject to the lower?

54. If an engine does 400 J of work and 800 J of heat are absorbed while doing this work, what is the efficiency of this engine?

55. If an engine is 35 percent efficient and does 560 J of work, how much heat must have been supplied?

56. If an engine is 55 percent efficient and is supplied with 1600 J of heat energy, how much work will be accomplished by the engine?

57. What is the efficiency of an engine that receives heat from a reservoir at 981°F and rejects it to a reservoir at 300 K?

Special Topic
The Cruel Air, the Cruel Sea

It is not uncommon for visitors to the Soviet scientific base of Vostok in the heart of Antarctica to take one step out of their pressurized plane and drop to the ice in a faint. The cause is not the extreme cold of the lonely base, whose temperature rarely goes above −40°F even in the southern summer, but the thinness of the air. The base itself, on the icy Antarctic plateau, is at an altitude of 11 000 ft. In addition, as a result of the earth's rotation, the atmosphere dips somewhat near the earth's poles, and the air there is naturally thinner than air everywhere else on the globe. Thus, the atmospheric pressure at Vostok is as low as that at altitudes of 14 000 ft in equatorial regions. More important for hapless visitors, the partial pressure of oxygen at Vostok is also much lower than that at sea level throughout the more temperate regions of the globe.

The section in the *Guinness Book of Records* that deals with deep diving by adventurers breathing air and gas mixtures is replete with asterisks which refer to the gloomy notation "Died on the ascent." The reason is also related to the partial pressure of oxygen; many of the deaths were caused by a much higher partial pressure than normal.

These two examples illustrate the vital importance of the concept of partial pressure in human endeavor and exploration. Only through a full understanding of the concept can people fulfill their desire for studying the remotest regions of their environment in relative safety.

The effects of reduced partial pressure of oxygen on human beings start to become evident at altitudes well below 10 000 ft. The effects include labored breathing, a heart rate above 100 beats per minute, headaches, nausea, vomiting, insomnia, and a general feeling of bodily malaise.

A number of balloonists in the nineteenth century died by ascending to altitudes above 8000 m, at which the partial pressure of oxygen is no more than 63 mmHg, appreciably less than the 160 mmHg that is normal at sea level. The nature of this type of disaster was best described by French balloonist M. Lortet, who ascended to an altitude reckoned at 8600 m in the mid-1800s, together with two companions who died on the trip:

> Towards 7500 meters the numbness one experiences is extraordinary, the body and the mind weaken little by little, gradually, unconsciously, without one's knowledge. One does not suffer at all; on the contrary, one experiences inner joy, as if it were an effect of the inundating flood of light. One becomes indifferent, one no longer thinks of the perilous situation or of the danger; one rises and is happy to rise. Vertigo of lofty regions is not a vain word. But as far as I can judge by my personal impressions, this vertigo appears at the last moment; it immediately precedes annihilation, sudden, unexpected, irresistible.

The first man to show that such effects as these were caused by alteration of the partial pressure of oxy-

Low-pressure life: The Double Eagle II, crewed by Ben Abruzzo, Maxie Anderson, and Larry Newman, over the Normandy coast during the Eagle's pioneer Atlantic crossing. (*United Press International*)

gen rather than simple reduction in atmospheric pressure was Paul Bert, Professor of Physiology of the Faculté des Sciences in Paris between 1869 and 1886. In Bert's experiments, volunteers in low-pressure chambers suffered no ill effects when the pressure fell as long as the partial pressure of the oxygen in the chamber was maintained at about 160 mmHg. In fact, it was extension of such studies at lower than normal partial pressures of oxygen that led to an understanding of the way in which hemoglobin carries oxygen through the blood and releases it to the tissues of the body.

Any student of pulmonary physiology quickly learns that too much oxygen in the lungs, that is, oxygen at too high a partial pressure, can be just as dangerous as too little oxygen. The use of high-pressure oxygen to treat premature babies was halted in the mid-1950s when evidence emerged that it could lead to blindness; indeed, in a celebrated malpractice case in 1975, a 21-year-old New York woman who had been treated with high-pressure oxygen at birth was awarded $900,000 for her subsequent sufferings—although she settled for $165,000, just before learning of the jury's decision.

Outside the delivery room, the most important sites for demonstrating the dangers of oxygen in excess are the watery depths of the oceans. In the main body of Chap. 9 we dealt with the relationship between pressure and depth beneath the surface of a liquid. In the case of sea water, it turns out that the surrounding pressure increases roughly by 1 atm for each extra 10 m of depth. If the pressure at the surface is 1 atm, that at a depth of 10 m is 2 atm, that at 20 m is 3 atm, and that at 100 m is 11 atm.

If divers are to survive at these depths without their lungs collapsing and their blood vessels rupturing, the pressure of gas inside their lungs must equal the pressure outside their bodies. Thus, divers in 20 m of water must breathe gas at a pressure of 3 atm to retain equilibrium inside and outside their bodies. If they are breathing pure oxygen, its partial pressure in these circumstances is 3×760, that is, 2280 mmHg—more than 10 times the normal partial pressure of 160 mmHg (0.209 atm).

This partial pressure is also well above the danger level; researchers have established that if divers breathe oxygen at a partial pressure of 2 atm for any appreciable length of time, they go into convulsions, probably because the excess oxygen oxidizes certain body enzymes. After 2 or 3 min of writhing, the oxygen-filled divers become unconscious; unless the oxygen pressure is rapidly lowered, they soon die.

An obvious alternative to this grisly progression is for deep divers to use normal air, consisting of about 20 percent oxygen and 80 percent nitrogen. In this case, the partial pressure of oxygen is only one-fifth that of pure oxygen. The use of air creates another partial pressure problem, however: at high pressures, the nitrogen in the air dissolves in divers' blood and tissues.

This effect is unremarkable during descents; after a period at depth, divers' bodies become "saturated" with nitrogen. The problem arises when divers try to ascend to the surface. If they move up too fast, the decrease of pressure outside and inside their bodies occurs too rapidly for the nitrogen to move out of solution smoothly. The gas forms bubbles in the blood and tissues, producing the intensely painful and dangerous ailment known as the bends. The only treatment for this effect is to increase

High-pressure existence: divers working and living in an undersea habitat. (*General Electric*)

the pressure on a diver as quickly as possible and then decompress the diver over several hours in a special chamber whose pressure is reduced very slowly to atmospheric level.

One approach that reduces the difficulty to some extent is the substitution of helium for nitrogen in divers' breathing mixture. Helium is less soluble in the blood than nitrogen; thus, less gas is dissolved during diving, and a shorter decompression time is required. The major disadvantage of helium is its effect on speech; it gives divers' voices their characteristic Donald Duck sound—but that seems a small price to pay for relative safety.

12 Making Waves

12.1 WHY WAVES WAVE

12.2 SIMPLE HARMONIC MOTION

12.3 WAVE SPEED
SHORT SUBJECT: Boom!

12.4 PERIODIC WAVES

12.5 REFLECTION
The Superposition Principle

12.6 STANDING WAVES AND RESONANCE

12.7 AND THE BEATS GO ON

12.8 SOUNDS—PHYSICAL VERSUS SUBJECTIVE
Intensity and Loudness
Frequency and Pitch
Wave Shape and Quality

12.9 THE DOPPLER EFFECT

SHORT SUBJECT: Ultrasound: Sound Medicine

Summary
Questions
Problems

SPECIAL TOPIC: Hear, Hear: The Inside Story of the Ear

Question: What is the connection between a patient receiving a chest x-ray in a mobile x-ray unit, the water of the blue Caribbean lapping up on a white beach in the West Indies, the clumsy descent of a Slinky spring down a staircase, the movie cliché in which an Indian puts his ear to the ground to listen for evidence of the approach of buffalo or the white man's army, the up-and-down movement of laughing children on a merry-go-round in an amusement park, and the collapse of the Tacoma Narrows Bridge?

Answer: All the incidents involve some type of wave motion. X-rays, like light, radio, and other forms of electromagnetic radiation, move invisibly through the air and space in the form of waves. The ocean provides a more visible example of waves. A Slinky moves down a staircase because its coils undergo the type of wave motion that alternately extends and contracts them. The canny Hollywood Indian listens for the sound waves produced by the buffalo's or horse's hooves; this evidence of oncoming animals is carried more clearly through the solid earth than it is through the fluid atmosphere. The up-and-down movement of the children on a merry-go-round is the epitome of wave motion. And the Tacoma Narrows bridge in the state of Washington was destroyed during a mild gale that set the central span waving in resonance (Fig. 12.1). As most of these examples show, wave motion is actually a major method of transferring energy and information from one place to another.

Figure 12.1 Collapse of the Tacoma Narrows Bridge as a result of resonance. (*Wide World*)

Figure 12.2 Motion of a wave pulse in transverse fashion. The particles of the medium, a rope, vibrate at right angles to the direction in which the wave propagates.

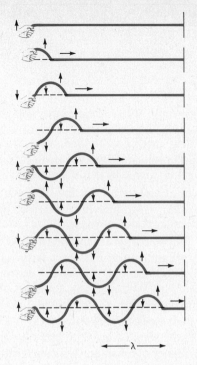

Wave motion is extremely important in our everyday lives and in scientific studies, primarily because we can transmit information and energy by means of waves. The movement of a wave through a medium is actually the movement of a disturbance or deformation that represents the transmission of energy and momentum from one point to another through the medium.

You can produce one of the simplest types of wave motion by fastening one end of a clothesline to a post, holding the other end taut, and then giving it a quick jerk. A wave pulse travels from the jerked end to the fixed end of the line, as shown in Fig. 12.2. By continuing to move the free end of the rope up and down, you can produce a series of wave pulses.

This simple illustration tells us a great deal about the nature of waves. First, we see that these waves in particular, and as it happens waves in general, are produced by some kind of vibration. Second, the waving clothesline provides us with **a definition of wave motion: the movement of a disturbance or deformation through a medium.**

The deformation may take the form of a change in pressure, density, extension, twist, or surface profile, and the medium can be solid, liquid, or gaseous. In the special case of electromagnetic waves, the strength of the local electric and magnetic fields is the disturbance that actually travels through empty space. Most of the discussion in this chapter will be devoted to mechanical waves. We shall leave a more complete discussion of electromagnetic waves, which include such important waves as infrared (heat), microwaves, radiowaves, light, ultraviolet, x-rays, and γ rays, until after the chapters on electricity and magnetism. At times in this chapter we shall point out how electromagnetic waves do and do not differ from mechanical waves.

12.1 WHY WAVES WAVE

In microscopic terms, the movement of a mechanical wave through any particular medium is the transmission of the vibration of one individual particle in the medium to an adjacent particle, after which the particles return to their original positions. In other words, there is no permanent displacement of the particles of the medium through which a wave travels. In the example of the waving clothesline, the material through which the wave moves is the rope, and the disturbance is a change in the rope's position in space. As an individual wave or wave pulse passes along the rope, each individual molecule in the rope moves up, returns to its original position, moves down, and returns once again to its starting point just once or many times in a repetitive manner.

One particular example of wave motion that has wide applicability is the movement of sound waves through air. Figure 12.3 illustrates the production of a sound wave using a tuning fork. For simplicity, let us deal with just the right-hand prong of the tuning fork, which starts to move as soon as the fork is struck. As the prong swings from its equilibrium position A to one extremity, position B, it pushes molecules of air ahead of it. As a result, this region of air temporarily contains many more than the normal number of air molecules. In fact, the forward motion of the prong drives many millions of extra molecules into the small region in front of the prong because each cubic centimeter of air contains more than a billion billion molecules. **The squeezing effect is called a compression.** The compression moves rapidly outward from the tuning fork as molecules arriving in the compressed regions push molecules already there farther forward, by collisions.

The prong then swings back from B to C. Immediately to the right of the prong, in the region between B and C, the air contains fewer molecules than normal. **This partial vacuum is called a rarefaction.** Once the rarefied region is created, air molecules from the region beyond it move into it, in an effort to equalize the pressure. As those molecules move out, they reduce the density of molecules in the area from which they move, thus causing the rarefaction to move outward from the tuning fork, right behind the compression. As the fork continues to vibrate, a series of condensations and rarefactions moves out from it with a speed of about 1100 ft/s and produces the sensation of sound in a hearer's ears as discussed in the Special Topic at the end of this chapter.

Although the variety of wave motions is almost limitless, all waves are one or both of two simple types: transverse and longitudinal. **In transverse wave motion, the individual particles of the medium vibrate at right angles to the direction in which the wave is moving,** as shown in Fig. 12.2. The movement of a wave pulse along a clothesline is an example of transverse wave motion, as are all types of electromagnetic waves. **In longitudinal wave motion,** illustrated in Fig. 12.4, **the particles of the medium vibrate back and forth along the direction in which the**

Most of the discussion in this chapter excludes shock-type waves, such as sonic booms and ocean waves breaking on a beach. The particle motion in these waves is large and produces permanent deformation of the medium that carries them.

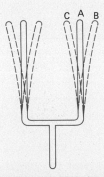

Figure 12.3 Vibrating prongs of a tuning fork, producing compressions and rarefactions in the air.

Figure 12.4 A horizontal wave on a spring, produced by waving the spring back and forth with one's hand. This is a longitudinal wave in which the particles of the medium (the spring) vibrate back and forth along the same direction as that in which the wave is propagating. The wavelength λ is the distance between two adjacent points of the same phase.

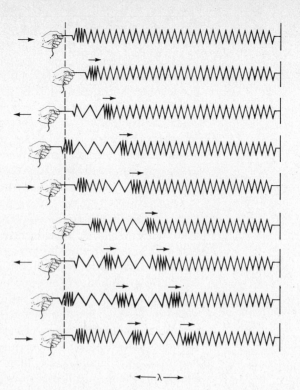

Note that a Slinky can also support and transmit a transverse wave if it is extended and shaken at right angles to its length.

wave is moving. A typical example of this type of motion is the Slinky spring. The deformation here is a compression of the coils of the spring; as this compression moves along the spring, the particles that make up the Slinky's coils oscillate to and fro. When the medium supporting a wave motion is a gas, the motion is always longitudinal. The most common example of longitudinal wave motion in a gas is the transmission of sound through air.

Many forms of wave motion involve both transverse and longitudinal waves. Sound waves and other signals travel through rocks in both forms. This is an extremely useful phenomenon for geologists and geophysicists, who try to interpret the differences between earthquakes and underground nuclear explosions as well as locate such events by measuring exactly when the different seismic signals from the incidents reach their measuring equipment. Ocean waves are even more complicated. As Fig. 12.5 shows, the water rotates in circles as the wave disturbances pass through it. This circular motion can be regarded as a combination of transverse and longitudinal motion because the individual water molecules vibrate in both the vertical and horizontal directions. These are surface waves.

Waves of every type are defined by measuring certain physical characteristics. These can be best understood by referring to one of the simplest types of oscillatory motion, the swing of a pendulum. Imagine

Figure 12.5 A water wave. The particles move in circular motion. The envelope of their position at any specific time is the wave illustrated.

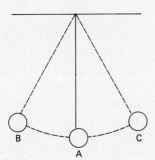

Figure 12.6 Oscillating pendulum. The potential energy is a maximum at B, and the kinetic energy is a maximum at A.

that the pendulum bob in Fig. 12.6 is pulled up from its equilibrium position A to the point B and then released. If no frictional force slows down the bob, it will pass through point A and swing up to another point C, whose height is equal to that of point B, then retrace its path back to B and start the whole process over. The net effect will be an infinite vibration of the pendulum because in the absence of friction the system will not lose any mechanical energy. In real life, of course, friction is never absent. As a result of the loss of kinetic energy to heat via friction, the pendulum bob will lose mechanical energy. Each successive swing will leave the bob at a slightly lower height, and the motion will eventually die down and stop.

This oscillatory motion, with or without friction, is known as vibration. Each complete vibration, or cycle, consists of a round trip of the pendulum from any starting point to each extreme of the motion (points B and C in Fig. 12.6) and back to the starting point. The journey BACAB in the figure is one complete vibration, as is the trip ABACA. **The time the bob takes to make that round trip is known as the period of the vibration or of one cycle.** It is normally given the symbol T. **The inverse of the period (that is, $1/T$) is called the frequency of the vibration** and is symbolized by the Greek letter nu (ν). The units of frequency are vibrations per second (or vibrations per hour, per day, or any other unit of time selected). If one vibration of the pendulum takes 3 s, for example, the frequency is $\frac{1}{3}$ vibration per second. If each vibration takes 0.20 s, the frequency is 5 vibrations per second. And for a vibration whose frequency is 10 vibrations per second, the period is plainly 0.10 s. The terms hertz and vibrations per second are used interchangeably.

In the International System of units, which we are using in this book, frequency is specified in units of hertz (Hz) after Heinrich Hertz, the discoverer of radio waves. One hertz equals one cycle per second. A few years ago, frequencies on a radio dial were given in cycles per second, or kilocycles per second. Now they are specified by hertz (Hz), kilohertz (kHz), and megahertz (MHz).

Two other definitions important when discussing wave motion can be best understood by reference to the wave on a clothesline shown in Fig. 12.7. **The wavelength of a wave is the total length of a complete wave along its direction of movement**—the distance from B to C or from D to E in Fig. 12.7. The symbol for wavelength is the Greek lambda (λ).

Figure 12.7 A traveling periodic wave on a clothesline.

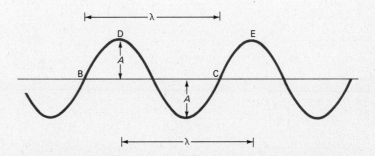

The amplitude of the wave is referred to by the letter A. Amplitude measures the maximum displacement of the individual particles of the medium through which a wave is traveling from their equilibrium positions. For the transverse wave shown in Fig. 12.7, the amplitude is the maximum height that the wave reaches *above* or *below* the horizontal line.

12.2 SIMPLE HARMONIC MOTION

One of the least complicated types of mechanical vibration is that known as simple harmonic motion. Study of simple harmonic motion, as exhibited by a vibrating spring, gives us a fundamental insight into the types of forces that control many forms of wave motion.

Imagine we stretch a spring from its equilibrium position a certain distance s by applying to it a specific force of magnitude F. The force will tend to restore the spring to the equilibrium position from which it started before we pulled it. As long as we do not pull so forcefully that we permanently deform the spring, we know that s and the restoring force F are related by Hooke's law equation from Chap. 8:

$$F = -ks \qquad (12.1)$$

where k is a constant for any particular spring and is determined from the normal length of the spring and Young's modulus by Eq. (8.4). This equation is the fundamental force law of simple harmonic motion and tells us that the restoring force for such motion, that is, the force that tends to bring the medium back to its original position, is proportional to the displacement of the medium from its equilibrium, or original, position. More precisely, simple harmonic motion is the type of vibration that is produced by a linear restoring force. The negative sign in the equation indicates that the restoring force is in the opposite direction to the displacement.

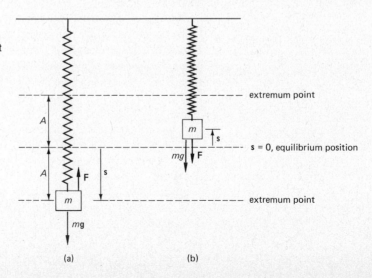

Figure 12.8 Oscillating spring. In (a) the mass is at the end of its downward trip; thus, the displacement $s = A$, where A is the amplitude of the motion. In (b), the mass is displaced upward. The displacement is always in the opposite direction to the restoring force.

Figure 12.9 A vibrating spring. A is the amplitude.

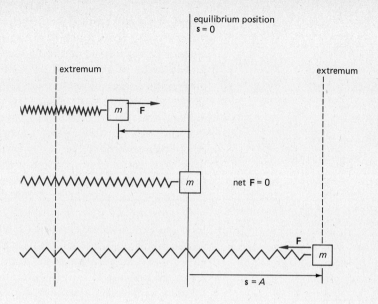

In the case of a mass vibrating vertically on the end of a spring, as shown in Fig. 12.8, the restoring force is a maximum at the maximum extension or compression of the spring, each a distance A from the equilibrium position, and the restoring force is zero at the equilibrium position. In this case A is the amplitude of the vibration. The value of the constant k in Eq. (12.1) is determined by adding a known force to stretch the spring and measuring the extensions that the force produces.

We can obtain further insight into simple harmonic motion by turning Fig. 12.8 on its side, as illustrated in Fig. 12.9. Here the spring is oscillating on a frictionless surface, and the weight of the mass at the end of the spring acts downward, in a direction perpendicular to that of the spring's restoring force. We know from Eq. (12.1) that the restoring force F is equal to $-ks$; because Newton's second law also tells us that $F = ma$, we have

$$ma = -ks$$

or

$$a = -\frac{ks}{m} \tag{12.2}$$

We realize from Newton's second law, $\mathbf{F} = m\mathbf{a}$, that since the restoring force is directed toward the central point of the motion, or equilibrium position of the spring, the acceleration will be likewise directed toward the equilibrium position. Furthermore, the acceleration varies with displacement from a value of zero at the equilibrium position ($s = 0$) to a maximum at the extremities of the motion where the magnitude of the displacement equals the amplitude of the vibra-

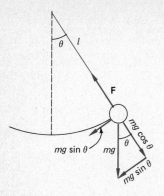

Figure 12.10 Analysis of the forces acting on a pendulum.

tion. The ratio of the acceleration to the distance from the center point is a constant since it equals k/m and both m and k are constants. This is another characteristic of simple harmonic motion.

The swing of a pendulum is another example of simple harmonic motion, as examination of Fig. 12.10 readily illustrates. This figure shows the two forces that act on the pendulum bob: the bob's weight, mg, and the tension of the string that supports the bob, **F**. Let us say that the bob has swung out an angular distance of θ rad from its equilibrium position. The restoring force that tends to bring the bob back to its equilibrium position is the component of its weight in the direction of that spot—$mg \sin \theta$. If the length of the pendulum string is l, then we know from our study of circles that the displacement of the bob along the arc from equilibrium is $l\theta$. Hence, the ratio of restoring force to displacement is $(mg \sin \theta)/l\theta$. A quick examination of trigonometric tables yields the information that the sine of an angle is roughly equal to the radian value of the same angle ($\sin \theta = \theta$) when the angle is less than about 15°. In this circumstance, the ratio of the restoring force to displacement becomes mg/l. This ratio is a constant, which is the basic condition for simple harmonic motion.

A somewhat unexpected example of simple harmonic motion is the movement of the projection on a diameter of a particle that is whirling around in uniform circular motion. Study of this particular case gives us an equation from which we can calculate the period of any simple harmonic motion.

The situation is illustrated in Fig. 12.11. Here, the shadow of a knob that is moving on a uniformly rotating disk is displayed on a flat screen. Figure 12.12 is a schematic diagram of the same effect. The radius of the disk, R, is the amplitude of the simple harmonic motion on the screen. As the knob moves from point 1 to point 2 along its circular route, the shadow moves from its furthest extremity to the equilibrium position 3. If the knob moves through an angle θ in getting from 1 to 2, then the shadow moves a distance $A \cos \theta$. We know from Eq. (7.10) that the angle through which the knob moves is related to its angular

Figure 12.11 The projection, or shadow, of a ball attached to a rotating disk moves with simple harmonic motion.

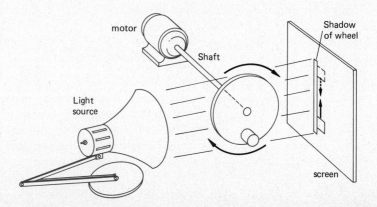

velocity ω and the time t by the equation $\theta = \omega t$. Hence, the equation for the displacement of the shadow can be written as follows:

$$s = A \cos \omega t$$

Now ω is simply the ratio of the total angular displacement of the knob in undertaking one whole revolution (that is, 2π) to the period of the revolution, T, which is exactly the same as the period of the oscillatory motion on the screen. Hence,

$$s = A \cos \frac{2\pi t}{T} \tag{12.3}$$

This equation indicates that the oscillation of the shadow (representing the projection of the uniform circular motion on the vertical diameter in Fig. 12.12) is represented by a cosine function as time passes. This displacement, s, varies with time in the manner shown in Fig. 12.13.

Extending our examination to the velocity of the shadow, we find a slightly different type of variation. The velocity of the shadow at point 1 is zero, but it increases as the knob progresses around the circle up to a maximum at point 3. From then it declines until it reaches zero at point 4. Plainly, the displacement is at a maximum when the velocity of the projection is at a minimum, and vice versa. A little trigonometry will quickly show that the velocity of the shadow varies according to a sine function given by

$$v_{\text{shadow}} = -v \sin \theta = -A\omega \sin \omega t \tag{12.4}$$

where v_{shadow} is the velocity of the shadow and $v = A\omega$ is the speed of the object m undergoing uniform circular motion. The minus sign is appropriate because the projection of v is downward. Note that had our analysis of this motion begun ($t = 0$) as the motion was directed upward at point 5, the displacement would vary with time as the sin ωt; the corresponding velocity would vary as cos ωt. Of course, the shapes of the sine and cosine functions are exactly the same; the only difference between them occurs in the displacement at the starting point $t = 0$. For a sine wave this point is zero; for a cosine wave it is a maximum value.

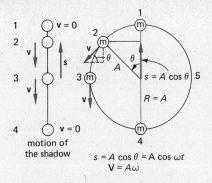

Figure 12.12 Reference circle to analyze an oscillating object in simple harmonic motion. As the mass m rotates in a circle, its projection along a diameter moves in SHM. The projection's distance from the center of the motion at any point is $A \cos \theta$, where A, the amplitude, equals the radius of the circle.

Figure 12.13 Variation with time of the displacement of an object undergoing simple harmonic motion. Refer to Fig. 12.12.

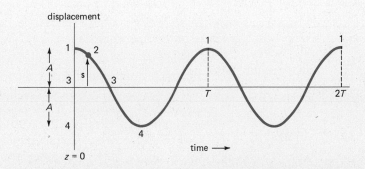

Continuing our analysis of the motion of the shadow, let us consider how the acceleration varies with time. The acceleration of the shadow is the projection of the centripetal acceleration of the object m as it moves in uniform circular motion. The centripetal acceleration of magnitude $a_c = v^2/R$ acts inward toward the center of the circle along the radius vector. This center-seeking acceleration is directed in the opposite sense to the radius vector. Hence the projection of the acceleration along the diameter, the acceleration of the shadow, varies with time in a similar manner to the displacement s but in the opposite direction. Applying trigonometry to a figure like Fig. 12.12, but in this case for the acceleration, we find that the acceleration of the shadow is given by

$$a_{\text{shadow}} = -a_c \cos \theta = -a_c \cos \omega t$$

Since $a_c = v^2/R$ and $A = R$

$$a_{\text{shadow}} = -\frac{v^2}{A} \cos \omega t \tag{12.5}$$

We can easily relate the acceleration of the shadow to the displacement s since $s = A \cos \omega t$. Substituting for $\cos \omega t$ from the displacement equation into the acceleration yields:

$$a_{\text{shadow}} = -\frac{v^2}{A} \frac{s}{A} = -\frac{v^2}{A^2} s$$

Since v and A are constant, the acceleration of the shadow is equal to the negative of a constant times the displacement of the shadow. This is exactly the condition for simple harmonic motion Eq. (12.2):

$$a = -\frac{k}{m} s \quad \text{where} \quad \frac{k}{m} = \frac{v^2}{A^2}$$

Since $v = R\omega$ or $v = A\omega$ in our notation

$$\frac{k}{m} = \omega^2 \quad \text{or} \quad \omega = \sqrt{\frac{k}{m}} \tag{12.6}$$

But ω is also equal to $2\pi/T$, where T is the period of the simple harmonic motion. Thus,

$$\frac{2\pi}{T} = \sqrt{\frac{k}{m}} \quad \text{or} \quad T = 2\pi \sqrt{\frac{m}{k}} \tag{12.7}$$

Since we have shown that the motion of the shadow of the object m undergoing uniform circular motion is just an example of simple harmonic motion (SHM), we can use Eqs. (12.3) to (12.7) when examining SHM in general. If we interpret A as the amplitude, or the maximum displacement, of the SHM and v as the maximum speed of the SHM, then we can summarize the relationships describing SHM by the following:

$$s_{SHM} = A \cos \omega t = A \cos \frac{2\pi t}{T} \tag{12.3}$$

$$v_{SHM} = -v \sin \theta = -A\omega \sin \omega t \tag{12.4}$$

$$a_{SHM} = -\frac{v^2}{A} \cos \omega t = -A\omega^2 \cos \omega t = -\omega^2 s_{SHM} \tag{12.5}$$

$$\omega = \sqrt{\frac{k}{m}} \tag{12.6}$$

$$T = 2\pi \sqrt{\frac{m}{k}} \tag{12.7}$$

This expression gives us the period of an object moving in simple harmonic motion in terms of the mass of the object and the force constant of the medium supporting the motion. The equation tells us, for example, the period of a mass m that is oscillating on the end of a spring whose force constant is k. Alternatively, we can measure the period of oscillation of such a system and the mass of the vibrating object. This way we can therefore obtain the force constant of the spring.

Extending this analysis to a simple pendulum swinging in simple harmonic motion, we recall that the force constant of the pendulum is equal to mg/l, where m is the mass of the pendulum bob and l is the length of the string. Substituting into Eq. (12.7),

$$T = 2\pi \sqrt{\frac{m}{k}} = 2\pi \sqrt{\frac{m}{mg/l}} = 2\pi \sqrt{\frac{l}{g}} \tag{12.8}$$

Note from the equation that the period of the pendulum is independent of mass.

The type of understanding of wave motion that we derive from studying simple harmonic motion, the simplest example of vibrational motion, gives physicists an insight into many more complex oscillating systems. In fact, most of the natural driving forces for mechanical waves, among them the twang of a guitar string, the vibration of vocal cords, the vibration of a tuning fork, and the vibration of an elastic solid, produce simple harmonic motion because matter generally tends to restore itself to equilibrium according to Hooke's law when it is stretched—as long as the deformation does not exceed the elastic limit of the material.

12.3 WAVE SPEED

Having calculated the period of simple harmonic motion, we can now take the next step and look at the speeds of wave motions. In the type of wave motion that we encounter in daily life, **wave disturbances travel at definite speeds which depend only on the nature of the medium**

Figure 12.14 Analysis of the restoring forces on a rope that is supporting the movement of a wave.

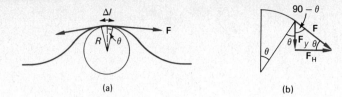

Specifying the mass per unit length turns out to be more convenient than dealing with the total mass of the string.

through which the disturbance is traveling, at least as long as the amplitude of the wave is small. This condition once again restricts us to those waves covered by the Hooke's law restoring forces.

Let us first imagine a pulse of transverse waves traveling from left to right at a speed v along a string that has a mass per unit length of μ (Greek mu). Figure 12.14 shows the pulse at some point in its progress along the string. Let us focus on the motion of the small piece of the string whose length is Δl. As a result of the tension in the string, two equal forces, each of size F, pull on the small increment of the string that we are considering. Each force acts in a tangential direction to the string, and Fig. 12.14(a) shows a circle constructed in such a way that the length Δl is actually an arc of a circle whose radius is R. This arc subtends an angle of 2θ. We can now analyze these forces by using vectors.

The horizontal components of the two forces are clearly equivalent, and because they act in opposite directions, they cancel out. (If these horizontal components did not cancel each other, the string would obviously be moving horizontally—an impossibility in this case because we have specified transverse wave motion.)

If F_y is the component of the force **F** acting in a vertical direction, Fig. 12.14(b) shows that

$$F_y = F \sin \theta$$

The total vertical force acting on the small length Δl is twice this amount, since two separate F forces act on the length. Hence the total restoring force on the segment of string is $2F \sin \theta$. And since $\theta \simeq \sin \theta$ at small angles, we have the following approximate equation:

$$\text{Total restoring force} = 2F\theta = 2F\left(\frac{\Delta l/2}{R}\right) = \frac{F \Delta l}{R}$$

To continue the analysis, it is convenient to alter our perspective. Suppose that instead of standing at rest watching the pulse in the shape of an arc of a circle moving along the rope, we move along the rope at the speed with which the pulse moves (the wave speed) just under the pulse. It would appear as if the particles of the rope were moving in a direction toward us in an arc of a circle, as if they were undergoing uniform circular motion with a centripetal acceleration $a_c = v^2/R$. This same situation could also be visualized by an observer at rest watching a pulse travel on a rope that extends from one end of a flat car to the other. If the car is moving as fast in one direction as the pulse travels in the opposite direction, then the rope goes through a circular path, as seen by the observer at rest. The restoring force $F\,\Delta l/R$ acts as the centripetal force producing this centripetal acceleration, so that application of Newton's second law yields

$$\frac{F\,\Delta l}{R} = \frac{mv^2}{R}$$

where m is the mass of the length of string. Now, from our definition of μ, the string's mass per unit length, it is plain that the mass of the small element of the string is $\mu\,\Delta l$. Thus,

$$\frac{F\,\Delta l}{R} = \frac{\mu\,\Delta l v^2}{R}$$

from which

$$v^2 = \frac{F}{\mu} \quad \text{and} \quad v = \sqrt{\frac{F}{\mu}} \tag{12.9}$$

That is, the speed of the wave pulse equals the square root of the tension in the string divided by the string's mass per unit length. This particular derivation is actually somewhat approximate, but it does yield a result that is observed experimentally. The method also shows how the tension of a string provides the restoring force that produces the transverse oscillation of the string when a wave passes along it.

In general, the speed of a wave passing through a solid medium equals the square root of the appropriate elastic modulus of the material divided by its density. The speed of a longitudinal wave traveling through a solid rod that has a density ρ and Young's modulus Y is given by the equation

$$v = \sqrt{\frac{Y}{\rho}} \tag{12.10}$$

For a transverse wave $v = \sqrt{(\text{shear modulus}/\rho)}$.

In the case of liquids and gases, the velocity of longitudinal waves depends on the value of the bulk modulus B:

$$v = \sqrt{\frac{B}{\rho}} \tag{12.11}$$

A brief examination of the relative bulk moduli and densities of solids, liquids, and gases in Tables 8.1 and 8.2 show that the speed of sound or any other longitudinal wave is much greater in a solid than it is in a fluid. Example 12.1 illustrates this difference.

EXAMPLE 12.1

Compute and compare the speed of sound in air and in a steel rod.

Solution

The speed of a longitudinal wave in a steel rod is given by Eq. (12.10), where Young's modulus is 20×10^{10} N/m² (see Table 8.2) and the density of steel is 7.8×10^3 kg/m³ (see Table 8.1):

$$v = \sqrt{\frac{Y}{\rho}} = \sqrt{\frac{20 \times 10^{10} \text{ N/m}^2}{7.8 \times 10^3 \text{ kg/m}^3}} = 5.06 \times 10^3 \text{ m/s}$$

The speed of sound in air is determined by using Eq. (12.11). The bulk modulus for air is 1.01×10^5 N/m² (see Table 8.2), and the density of air is 1.29 kg/m³ (see Table 8.1). The velocity of sound in air will be

$$v = \sqrt{\frac{B}{\rho}} = \sqrt{\frac{1.01 \times 10^5 \text{ N/m}^2}{1.29 \text{ kg/m}^3}} = 2.80 \times 10^2 \text{ m/s}$$

Because the elastic modulus of solids is about 10^5 times larger than that of air or other gases, the speed of sound in the solid is much greater than the speed of sound in air.

Note that this result for the speed of sound in air is only approximate because we used an inappropriate bulk modulus. For a gas, the most applicable bulk modulus is the adiabatic bulk modulus given by γP, where P is the pressure and γ has a value of 1.67 for monatomic gases, 1.40 for diatomic gases, and 1.28 for many triatomic gases. For air, which is a diatomic gas, the appropriate γ is 1.40. Using the ideal gas law we can reevaluate Eq. (12.11) for a gas. The density of the gas is the mass per unit volume, which can be written in terms of the number of moles of the gas n and the molecular weight M.

$$\rho = \frac{nM}{V}$$

Using this value for ρ, Eq. (12.11) becomes

$$v = \sqrt{\frac{B}{\rho}} = \sqrt{\frac{\gamma P}{nM/V}} = \sqrt{\frac{\gamma PV}{nM}}$$

From the ideal gas law we know that

$$PV = nRT$$

so

$$v = \sqrt{\frac{\gamma n RT}{nM}} = \sqrt{\frac{\gamma RT}{M}}$$

(Do not confuse this T, which denotes absolute temperature, with the T used to denote the period of oscillation.) To evaluate the sound speed in air we require a value for the molecular weight of air. Since air is essentially 80 percent nitrogen of molecular weight 28 g and 20 percent oxygen of molecular weight 32 g, the molecular weight of air is $M = (28 \text{ g})(0.8) + (32 \text{ g})(0.2) = 28.8$ g. We can now evaluate the speed of sound in air at $0°C$ ($T = 273$ K) by using the values $\gamma = 1.40$, $R = 8.31$ J/mol·K, and $M = 28.8$ g or 28.8×10^{-3} kg.

$$v = \sqrt{\frac{\gamma RT}{M}} = \sqrt{\frac{(1.40)(8.314 \text{ J/mol·K})(273 \text{ K})}{28.8 \times 10^{-3} \text{ kg}}} = 332 \text{ m/s}$$

This value agrees with the measured value of the speed of sound in air at $0°C$ of 331.6 m/s or 1088 ft/s.

An evaluation of the temperature dependence of the speed using this equation shows that the speed of sound increases by about 2 ft/s or 0.6 m/s for each $1°C$ rise in temperature. Hence the speed of sound in air at $25°C$ will be 331.6 m/s + 25×0.6 m/s = 346.6 m/s, which is equivalent to 1138 ft/s.

EXAMPLE 12.2

Using the values of the speed of sound in air given above, compute the time delay between sighting a lightning flash and hearing the sound of thunder on a warm summer's day when $t = 30°C$. The lightning bolt strikes 0.50 mi away, and we assume that light travels so fast it arrives at the instant the lightning occurs.

Solution

We can resolve the problem into one of computing the time of travel of the sound wave a distance of 0.50 mi in air when the temperature of the air is $30°C$. Since the speed of sound increases 2 ft/s for each degree rise of temperature, the speed at $30°C$ will be the speed at $0°C$, 1088 ft/s, plus 2 ft/s × 30, or 60 ft/s. Hence,

$$v_{t=30°C} = (1088 + 60) \text{ ft/s} = 1148 \text{ ft/s}$$

To travel 0.50 mi, or 0.50×5280 ft = 2640 ft, will require a time

$$t = \frac{\text{distance}}{\text{speed}} = \frac{2640 \text{ ft}}{1148 \text{ ft/s}} = 2.30 \text{ s}$$

Therefore, for every 5-s delay of thunder behind the lightning, the lightning is 1 mi away.

Boom!

When the British Aircraft Corporation started supersonic flight tests of the Anglo-French Concorde airplane in the summer of 1970, protests and insurance claims flowed into the company from residents of the remote areas of Wales, Scotland, and England over which the pioneering plane flew. Farmers lamented that their animals had aborted as a result of experiencing the shock of the Concorde's passage. Householders complained of broken windows and falling masonry in the wake of the vehicle's flight. Authorities at St. David's Cathedral expressed fears that the 900-yr-old house of God was being slowly shaken to pieces by the twentieth century plane. And residents throughout the area protested about the aural assault to their senses by the Concorde's sonic boom, which is an excellent example of a shock-type wave that does not obey Hooke's law.

Concorde taking off. (*Wide World*)

Investigations of the incidents revealed a somewhat different pattern of events. Many of the claims involved purported damage that occurred on days when the Concorde was not flying, or when it was cruising hundreds of miles away. Many of the verbal protesters had plainly

---

12.4 PERIODIC WAVES

If one end of a rope or Slinky is moved back and forth in a regular manner, a periodic wave train results. A vibrating tuning fork produces a periodic sound wave in air. Periodic waves are common, and a study of such waves enables us to understand more complex mechanical waves in general.

Perhaps the simplest type of periodic wave is one that is generated by a driving force which is itself vibrating in simple harmonic motion—a string that is made to wave by having one of its ends moved up and down in simple harmonic motion, for example. In this case, the end of the string moves up and down according to the equation

$$y = A \sin \frac{2\pi t}{T} = A \sin 2\pi \nu t$$

heard nothing resembling Concorde's sonic boom. Yet in many of the doubtful cases the Corporation had cheerfully paid up, convinced that their tests had shown the supersonic passenger plane to be acceptable to, if not loved by, most of the population.

Of course, the Concorde engineers realized that many of the claims of property damage and mental aggravation were perfectly valid. One can understand the reason by analyzing the cause of sonic booms.

We can best approach the problem by considering a related type of case: a speedboat cruising on a lake. When the boat travels slowly, the waves that it produces on the surface of the water emanate from it in all directions, including forward. But once the pilot opens up the engine, the boat starts to travel at the same speed as the waves. As a result, the fronts of the waves pile up at the bow of the boat, forming a kind of barrier that can be overcome only by an extra burst of power from the engine. Once the pilot had made it past this "wave barrier," however, she has smooth movement the rest of the way. Knifing through the smooth water, she leaves her waves behind her boat. As they spread out and combine, the waves produce a V-shaped disturbance in the water behind the boat.

Supersonic aircraft produce exactly the same type of "shock wave" once they have passed the sound barrier—the speed of sound through air, which is about 1100 ft/s. Instead of producing a V-shaped wave, though, the supersonic plane produces a conical disturbance in the air, which trails the speeding plane. Just as the wake from a speedboat rocks any vessel it passes, the wake from a supersonic plane affects both the people and property it encounters. The effects, of which the sharp crack of the sonic boom is the most obvious, are caused because a large number of sound waves from the plane reach the listener at exactly the same time.

The extent of the area affected by a sonic boom depends on the altitude of the plane producing it. The cone of disturbance from an aircraft flying supersonically at 1000 ft will plainly cover less ground than one from a similar plane at 30 000 ft. Moreover, the amount of damage to both nerves and structures decreases as the plane's altitude increases because the intensity of the waves is reduced. In practice, the much-criticized Concorde disturbs a 50-mi-wide swath of ground when flying at its cruising altitude of 65 000 ft and, according to most of the evidence, produces relatively little physical damage—although the boom puts environmentalists through mental torture.

This equation relates the position y of the end of the string to the frequency of its vibration ν and the passage of time t by a sine wave function. As the disturbance moves along the string, every point along it oscillates up and down in SHM.

We can introduce and define the phase of the wave by examining Fig. 12.15. The continuous line in the figure shows the position of the whole string at an instant of time at which its driven end is exactly at its equilibrium point. The wavelength of the wave motion is easily obtained by measuring the distance between successive peaks or troughs of the sine wave. More generally, **the wavelength is the distance between any two points along the string that are in the same relative position and moving in the same direction at any point in time.** Points A and G, or B and H, or F and L, for example, are separated by a single

Figure 12.15 Propagation of a traveling wave.

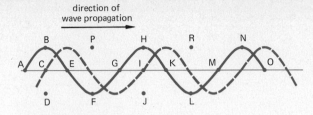

transverse wave motion
(dashed line shows the position
of the wave one-quarter of a period later)

wavelength. Physicists describe such points as being *in phase* or, alternately, possessing the *same phase*. The term applies to all points along a particular wave motion that are separated by a finite number of wavelengths. Points B, H, and N in Fig. 12.15 are all in phase, for example. Not surprisingly, points not in phase are described qualitatively as being out of phase with each other. However, the term *completely out of phase* has a more specific meaning, implying that the points in question are separated by $\frac{1}{2}$ wavelength, $1\frac{1}{2}$ wavelengths, $2\frac{1}{2}$ wavelengths, or any other integral number plus $\frac{1}{2}$ wavelength. This represents 180° of the angle determining the sine function. In Fig. 12.15, for example, point F is completely out of phase with points B, H, and N, as is point L. Note also that the points A and E, although apparently in the same relative position on the wave, are not in phase. Because the rope at A is moving down while at E it is on its way up, the two points are completely out of phase with each other.

The dashed line in Fig. 12.15 shows the position of the rope one-quarter of a period after the situation illustrated by the continuous line. The peaks, troughs, and all other parts of the wave have progressed $\frac{1}{4}$ wavelength along the rope. Comparison of the two sets of positions clearly illustrates that the displacement of any point along the rope perpendicular to the direction of the wave motion depends on both the actual position of the point on the rope and the time that has elapsed since the wave motion started to flow along it.

The speed of the wave in a string is a constant whose magnitude depends on the linear density and the tension of the string. Because the speed is constant, we know that 1 wavelength passes any specific point on the string during one period. In mathematical terms,

$$v = \frac{\lambda}{T}$$

And since

$$\nu = \frac{1}{T}$$

we have

$$v = \lambda\nu \tag{12.12}$$

In other words, **the speed of a wave is the product of its wavelength and frequency.** This speed, known as the **phase velocity,** is a fundamental formulation of wave motion that applies to any periodic wave, including electromagnetic waves.

EXAMPLE 12.3

Compute the wavelength of the sound emitted by a tuning fork of frequency 256 vibrations per second when the temperature is 20°C.

Solution

$$\lambda = \frac{v}{\nu} = \frac{1088 \text{ ft/s} + (20 \times 2 \text{ ft/s})}{256 \text{ vib/s}} = \frac{1128 \text{ ft/s}}{256 \text{ vib/s}}$$

$$= 4.41 \text{ ft/vib} \quad \text{or} \quad 4.41 \text{ ft}$$

Because the wavelength is by definition the distance a wave travels during one complete vibration, the units of wavelength are simply units of length.

EXAMPLE 12.4

Compute the wavelength of the wave sent out by a radio station that broadcasts on an assigned frequency of 1500 kHz, or 1.5 MHz.

Solution

All electromagnetic waves, including light, radio waves, television waves, x radiations, and so on, move through a vacuum with a speed of 186 000 mi/s, or 3×10^{10} cm/s. From $v = \lambda\nu$,

$$\lambda = \frac{v}{\nu} = \frac{30 \times 10^9 \text{ cm/s}}{15 \times 10^5 \text{ Hz}} = 2 \times 10^4 \text{ cm} = 2 \times 10^2 \text{ m}$$

12.5 REFLECTION

Thus far in this chapter we have concerned ourselves with simple, unhindered pulses of waves. Unfortunately, such uncomplicated wave pulses rarely occur in nature because every wave sooner or later interacts with surfaces or boundaries that reflect it. The sound from a vibrating tuning fork returns to the listener after reflection from a wall, the floor, or the ceiling. It is these reflections that lead to complications in analyzing wave motion, although they also provide information of interest to seismologists and radar or sonar operators.

Mechanical waves are reflected whenever they come to a boundary. A simple illustration of such reflection is shown in Fig. 12.16, which sketches the reflection of a wave pulse in a rope attached to a wall as the pulse comes into contact with the wall. Observation shows that the reflected pulse is exactly out of phase with the original pulse. This effect

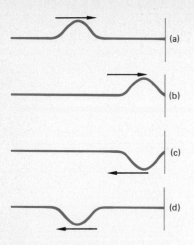

Figure 12.16 Reflection of a transverse pulse at a fixed boundary. (a) and (b) Incoming positive wave pulse. (c) and (d) Reflected wave at the point of reflection is always 180° out of phase with the incident wave.

results from Newton's third law. As the pulse reaches the wall, the rope exerts an upward force on the wall. In reaction, the wall exerts a downward force on the rope, causing the rope to move downward. If the wall had not been in the way of the pulse, the rope would be starting to move upward at the point at which the pulse hits the wall. As it is, with the wall in place, that same point on the rope is just starting to move in the downward direction. Hence, it is 180° out of phase with the original wave at the end of the rope.

A slightly different type of reflection occurs if a rope is held vertically, with its lower end dangling free. A wave sent down the rope by giving the rope a sharp jerk at the top will be reflected from the free end to produce a return wave that is in phase with the original. Here, the only force on the rope's free end is the tension of the rope pulling the free end. The end of the rope sways back and forth in a vibration produced by the original wave; that vibration, in turn, generates a return wave in phase with the original one at the end of the rope.

Reflection of sound waves normally takes the form of the first example because such waves come up against a definite solid boundary, such as a wall, a house, a rock, or the sea bed.

However, sound waves are reflected when the sound strikes a region where the density of the air changes. Such density variations result from temperature differences and are common in our atmosphere. The reflection of sound waves, if strong and clearly heard, is often termed an echo.

Interesting phenomena occur when reflected waves mix with the original wave train. Such phenomena give specific examples of wave interference. We shall examine a number of different ways to produce interference.

The Superposition Principle

The most obvious difficulty in dealing with mixtures of wave motions is that any specific wave pulse loses its own identity when it comes into contact with another. When waves on a string attached to a wall meet their own reflection from the wall, the result is a new series of waves that are different in shape and amplitude from the original and the pure reflected wave. Exactly the same effect occurs in all types of wave motion, including sound, water waves, and electromagnetic waves. In fact, whenever two or more pulses, or wave trains, meet at any particular point in the medium through which they are traveling, the displacement of the medium at that point is related to the individual displacements of the two waves. In cases in which the Hooke's law situation holds—a condition that actually covers many common forms of interference between waves—the displacement at any point produced by the combination of two waves is simple to calculate: it is merely the algebraic sum of the displacements that each individual wave would produce by itself at that point. This statement is known as the principle of superposition. Its effect is to provide a relatively simple

Figure 12.17 The superposition principle. The figures represent snapshots of a string ($t = $ constant) and show how the displacements of two waves add to produce a resultant wave.

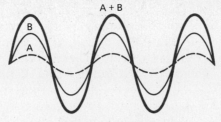

(a) constructive interference

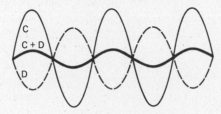

(b) destructive interference

means of starting out on analysis of inherently complicated types of interference between waves.

The application of the principle can be best understood by reference to the two parts of Fig. 12.17, each of which shows the mixing of two waves of the same frequency on the same string. In Fig. 12.17(a), lines A and B represent the positions that two waves would occupy if each were traveling along the string alone. The line labeled A + B represents the sum of the displacements of the two waves, taking as the starting point the straight line that represents the equilibrium position of the string at rest. The A + B line shows the shape of the string caused by the interference of the two waves A and B at a particular instant of time. Because the two waves are in phase at this instant of time, their interference produces a wave that has larger displacement than either individual wave at every point except the equilibrium points. This effect is termed **constructive interference.** By contrast, in Fig. 12.17(b) the summation of the displacements of the waves C and D at the instant of time shown illustrates **destructive interference.** At this instant waves C and D are out of phase with one another, and the summation of the two waves results in a wave that has smaller displacements all along the rope than would be the case if either wave C or wave D were on the rope all alone. One extra point should be noted: if the two waves in Fig. 12.17(b) had been of the same amplitude as well as the same frequency, they would have canceled out each other entirely.

Note that each sketch in Fig. 12.17 signifies a snapshot of the rope representing an instant of time. In the usual condition of observation, we see the results of the interference of waves traveling on the string over a period of time. For example, a positive pulse traveling along a rope, on meeting a negative pulse of identical amplitude traveling in the opposite direction, would produce destructive interference during the time period while they pass one another. After passing each other, the individual pulses would continue on as before their meeting.

12.6 STANDING WAVES AND RESONANCE

One particularly interesting type of interference pattern produces what are known as standing waves, which occur whenever two sets of periodic waves of equal wavelength and equal amplitude are traveling in opposite directions. Often the two wave trains are an ongoing wave train and a reflected wave train.

A simple example of a standing wave pattern appears in Fig. 12.18. This shows the interference of two wave trains of equal amplitude and wavelength moving in opposite directions along a string. Figures 12.18(a) to (d) picture the positions of the two individual waves and the resultant interference patterns at intervals of one-quarter of a period, starting with an instant at which the two waves are completely out of phase with each other. The continuous line in the diagrams denotes the wave that is moving from left to right, and the dashed line indicates the wave traveling in the opposite direction. Figure 12.18(e) shows the overall result of the interferences, with a few additional interference patterns from times between the quarter-period intervals added. This diagram clearly shows that the string oscillates as a unit. Certain points (N) along the string, separated by $\frac{1}{2}$ wavelength, never move from their equilibrium positions at any time during the interference of the two waves. Other points, midway between successive motionless spots, swing up and down between two extremes of height, these extremes being twice the amplitude of the individual interfering waves. This type of pattern is known as a standing wave. The **motionless points** along it are called **nodes,** indicated by N in Fig. 12.18; the **points** of **maximum amplitude** between the nodes are called, not surprisingly, **antinodes,** indicated by A in Fig. 12.18. The distance between adjacent nodes and antinodes is $\frac{1}{4}$ wavelength.

Actual standing wave patterns on a string result from the interference of an ongoing wave train and a reflected wave train while the string is vibrating at one of its natural frequencies, a condition called **resonance.** Any object composed of an elastic material will vibrate at its own special frequency when disturbed. The natural frequency or frequencies depend upon the size, shape, and elasticity of the object.

When a string fixed at each end is vibrating at a natural frequency, waves traveling down the string are reflected, undergoing a 180° phase shift. After traveling back down the string, they are reflected again with another 180° phase shift, with the result that they travel along with the incoming waves, in phase. Hence, the twice-reflected wave adds to the incoming wave, which increases the amplitude of the standing wave pattern. Standing wave patterns that occur when the medium is oscillating under the resonance condition at a natural frequency have an

Figure 12.18 Interference of moving waves on a string produces a standing wave. The solid line represents a wave traveling to the right; the dashed line represents a wave moving to the left.

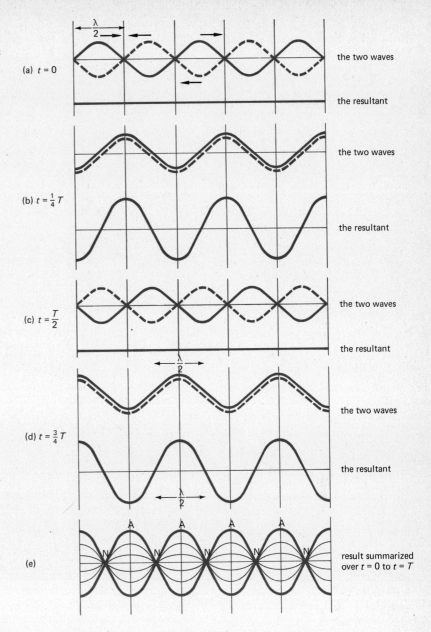

Standing waves are actually responsible for the tones emitted by musical instruments.

increased amplitude. Standing waves and resonance are actually responsible for the tones emitted by musical instruments.

A typical real-life example of resonance is the effect of pumping or pushing a child on a swing. If each successive push or pumping action is made at exactly the right moment—in phase with the vibration of the swing—the amplitude of the swing's swing increases. A spectacular illustration of resonance at the beginning of this chapter was the

Figure 12.19 Standing wave patterns on a string as a function of tension.

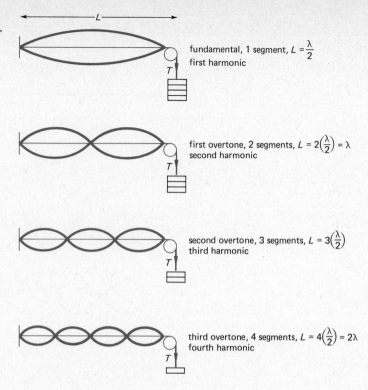

resonance of torsional waves in the Tacoma Narrows Bridge. Resonance is also involved in the production of music, as we shall see later in this chapter.

We can produce resonance in a string by the method illustrated in Fig. 12.19. This shows a string that is held firmly at both ends. One end is attached to a vibrator, and the other is weighed down by a weight that holds it over a pulley. By varying this weight, one can alter the tension in the string. Let us first set the frequency of the vibrator at a specific value, say, 120 vibrations per second. The vibrator makes waves that travel along the string to the pulley at a rate of 120 vibrations per second before being reflected back 180° out of phase with the original waves.

To produce a standing wave in this system, we must adjust the magnitude of the weight on the end of the string until we find a particular value that gives us the simplest resonance possible. This is a standing wave with two nodes, one at either end of the string, and a single antinode in the string's center. This is the least complex type of standing wave that can be produced in the string. It is known as the string's fundamental frequency, or first harmonic.

We already mentioned that the distance between successive nodes in a standing wave is half the wavelength of the waves that produce the pattern. In this example, the distance between the nodes is L, the length of the string; the wavelength of the fundamental standing wave is, therefore, $2L$.

If we now continue the experiment by removing some of the weights while the vibrator is still operating, we lose the fundamental wave. However, we quickly come upon another standing wave. This so-called first overtone has a third node in the center of the string in addition to the fixed nodes at either end. The distance between successive nodes in this case is $L/2$, and the wavelength of this first overtone is half that of the fundamental; the overtone is also known as the second harmonic of the string.

Removal of more weights, as illustrated in Fig. 12.19, produces the second, third, and subsequent overtones, with three, four, and more standing segments of string. Because the wavelengths of these standing waves are $\frac{1}{3}, \frac{1}{4}$, and so on times that of the fundamental, we know these patterns as the third, fourth, and so forth harmonics. We can generalize this series of wave patterns to produce the basic condition for production of a standing wave in a string of length L: L must equal $n\lambda/2$, where λ is the wavelength and n is any integral number.

How is the tension in the string related to the order of the harmonic it produces? Reversing the wave equation, we have the formulation

$$\lambda = \frac{v}{\nu}$$

where v is the speed of the wave. We produce the differing wavelengths of the various harmonics, therefore, by altering the speed of the traveling waves that produce the standing wave. Now we have already related the speed v, tension F, and mass μ, per unit length of the string in Eq. (12.9):

$$v = \sqrt{\frac{F}{\mu}}$$

Combining the two equations,

$$\lambda = \frac{1}{\nu}\sqrt{\frac{F}{\mu}}$$

Substituting our condition for standing waves,

$$L = \frac{n\lambda}{2}$$

we see that a standing wave will be produced in a string of length L whose mass per unit length is μ which is under tension F and carrying interfering waves of frequency ν whenever

$$L = (n/2\nu)\sqrt{F/\mu}$$

The longest wavelength, fundamental, standing wave occurs when the tension is greatest, that is, when $n = 1$ in the equation. As the tension is reduced, the equation is satisfied for higher integral values of

n, which produce the higher harmonics of the fundamental standing wave.

EXAMPLE 12.5

A horizontal string is 60 cm long and has a mass of 5 g. One end is attached to a vibrator whose frequency is fixed at 120 vibrations per second (that is, 120 Hz). The other end of the string passes over a pulley and is held down by a 50-g hanger. What mass must be placed on the hanger to cause the string to vibrate in four segments?

Solution

When the string vibrates in four segments we know that $4(\lambda/2) =$ length of the string; thus $\lambda = L/2 = 0.6 \text{ m}/2 = 0.3 \text{ m}$. To find the required tension in the string that equals the weight of the hanger and the added masses, we must compute the velocity of the wave by using Eqs. (12.9) and (12.12):

$$\lambda = \frac{1}{\nu}\sqrt{\frac{F}{\mu}}$$

Rewriting this relationship to determine F,

$$F = \mu\lambda^2\nu^2 = \frac{(5 \times 10^{-3} \text{ kg})(0.3 \text{ m})^2(120 \text{ s}^{-1})^2}{0.60 \text{ m}} = 10.8 \text{ N}$$

The mass in excess of the 50-g hanger can be calculated by the following:

$$mg + (50 \times 10^{-3} \text{ kg})(g) = 10.8 \text{ N}$$

$$m = \frac{10.8 \text{ N} - (50 \times 10^{-3} \text{ kg})(g)}{g} = \frac{10.8 - (50 \times 10^{-3})(9.8)}{9.8}$$

$$= 1.05 \text{ kg}$$

$$= 1050 \text{ g}$$

In normal circumstances strings are not driven with fixed frequencies. Rather, the tension is fixed and the strings are free to vibrate with any frequency. This is the case with musical instruments, as Example 12.6 illustrates.

EXAMPLE 12.6

A 50-cm length of steel piano wire has a mass of 5 g and is under a tension of 500 N. What are the wavelengths and frequencies of its fundamental and first two overtones?

Solution

The wavelengths can be determined by examining Fig. 12.19. The nodes must be spaced $\frac{1}{2}\lambda$ apart, and the length of the piano wire, L, equals $n\lambda/2$. The wavelengths will be given by $\lambda = 2L/n$; hence,

λ of fundamental $= 2L = 2(50 \text{ cm}) = 1$ m
λ of 1st overtone $= 2L/2 = L = 0.5$ m
λ of 2d overtone $= 2L/3 = 2/3(50 \text{ cm}) = 0.33$ m

To compute the corresponding frequencies, we use the fact that $\nu = v/\lambda$ and $v = \sqrt{F/(m/L)}$:

$$\nu = \frac{1}{\lambda}\sqrt{\frac{F}{m/L}} = \frac{1}{\lambda}\sqrt{\frac{500 \text{ N}}{5 \times 10^{-3} \text{ kg}/0.5 \text{ m}}} = \frac{1}{\lambda}\sqrt{5 \times 10^4} = \frac{224}{\lambda} \text{ Hz}$$

fundamental frequency, or 1st harmonic $\nu_1 = \dfrac{224}{1} = 224$ Hz

1st overtone, or 2d harmonic $\nu_2 = \dfrac{224}{0.5} = 448$ Hz

2d overtone, or 3d harmonic $\nu_3 = \dfrac{224}{0.333} = 672$ Hz

One point to note is that the fundamental has the lowest frequency and the longest wavelength of all the harmonics. Furthermore, the frequencies of the higher harmonics in a wire under constant tension are simple multiples of the fundamental frequency. In this example we have an instrument with strings whose lengths and tensions are fixed. Several different frequencies can set up standing waves in resonance with the wire. In striking the piano wire, one actually produces many frequencies; however, only the natural resonant frequencies set up standing waves, which are the only waves of sufficient amplitude to be heard clearly.

The existence of nodes and antinodes in strings and wires is a simple matter to demonstrate. Standing waves also occur in pipes and tubes containing air, as can be shown by the Rube Goldbergian apparatus in Fig. 12.20. In this experiment a longitudinal sound wave produced by a speaker sets a rubber membrane in vibration. The membrane in turn

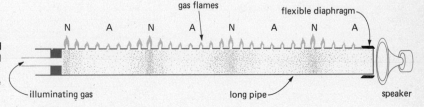

Figure 12.20 Experimental illustration of standing waves in a tube filled with illuminating gas. Letters A and N refer to displacement antinodes and nodes. The density pattern shows the distribution at a specific time.

generates a new sound wave that travels through a flammable gas in a long tube. The tube contains a row of pinholes through which small amounts of gas can escape. If one lights the gas escaping from each hole before turning on the speaker and then observes the heights of the different flames once the speaker is excited with a resonance frequency, one will see the result of a standing wave in action.

At points of high pressure variation, corresponding to the positions of nodes, the flames will rise appreciably higher than those around them. Above points of no pressure variation in the tube, corresponding to antinodes, the flames will hardly manage to stay above the pinholes. For the analytically inclined, the distance between adjacent high- and low-pressure points will be one-fourth the wavelength of the standing wave.

This demonstration is just one example of a phenomenon that is responsible for a wide range of physical and musical instruments. A standing wave occurs in a column of air whenever air rushes past an open end of the column and starts the air inside vibrating at a natural resonant frequency. As in the case of the piano wire, the air inside a tube actually vibrates with many different frequencies; however, only the few resonant vibrations that are present set up audible standing waves. The musical notes produced by woodwind and brass instruments, including flutes, trumpets, and organs, are all examples of standing waves.

The air inside a tube actually vibrates with many different frequencies; however, only the few resonant vibrations that are present set up audible standing waves.

Although musicians categorize an orchestra's wind and brass sections into a large number of different instruments, physicists, for simplicity, divide air columns into just two types: open and closed. The classic example of standing waves in an open column occurs in the pipe organ with both ends open, as illustrated in Fig. 12.21. The basic patterns of these longitudinal standing waves are the same as in the case of a vibrating string, and the wavelengths of the various harmonics in an open resonating air column of length L are given by the equation $L = n\lambda/2$, just as in the case of the string. However, the arrangement of nodes and antinodes in an open air column is exactly opposite that in

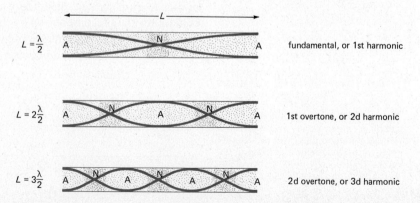

Figure 12.21 Standing waves in a resonant air column: an open pipe. Displacement nodes and antinodes are indicated. The envelope of a time-varying situation is indicated by the solid blue line; the density pattern shows the distribution at a specific time.

a string. Because the air is free to expand at each end of the pipe, antinodes occur at the two ends, rather than the nodes that are located at each fixed end of a string. The simplest standing wave possible in the pipe organ, therefore, consists of a node in the center of the pipe and antinodes at each open end. This is the fundamental, or first harmonic, of the pipe; because $\lambda/2 = L$, its wavelength is twice the length of the pipe, just as the wavelength of the first fundamental of a string is twice the string's length.

The first overtone occurs when the column contains two nodes, as shown in Fig. 12.21. Here, the length of the column is two $\frac{1}{2}$ wavelengths; the wavelength of the overtone is obviously one-half the wavelength of the fundamental, making this overtone the second harmonic. The second overtone involves three nodes inside the column, giving the standing wave a wavelength of two-thirds the length of the pipe. Because this is one-third the fundamental's wavelength, this second overtone is the pipe's third harmonic.

The pattern of standing waves in a closed column—which is not a closed column at all, but a column with one end stoppered and the other end open to the air—is slightly more complex. The source of the complexity is that the conditions are different at the two ends of the column. Because air is trapped at the closed end of the pipe, unable to move, this end must be the site of a node. The open end is an antinode, however, because the air is free to move there just as it is at both ends of an open column. Hence, the fundamental standing wave in a closed pipe consists of a node at the closed end and an antinode at the other end, as pictured in Fig. 12.22. We know that the distance between adjacent nodes and antinodes is $\frac{1}{4}$ wavelength. Therefore, the wavelength of the fundamental standing wave in a closed column is four times the length of the column. This is twice the wavelength of the fundamental in an open pipe because the wavelength is inversely proportional to frequency at constant speed. It is clear that the fundamental frequency of a closed pipe is half the fundamental frequency of an open one of the same length.

Figure 12.22 Standing waves in a resonant air column: a closed pipe. Displacement nodes and antinodes are indicated. The envelope of a time-varying situation is indicated by the solid blue line; the density pattern shows the distribution at a specific time.

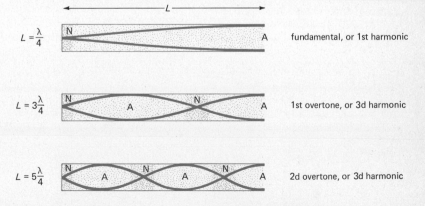

Closed pipes can produce only odd harmonics.

The next simplest standing wave in a closed pipe is that containing a node at the closed end and another node inside the pipe, as illustrated in Fig. 12.22. In this case, $\frac{3}{4}$ wavelengths spread out along the length of the pipe. The wavelength of this first overtone is $4L/3$. This figure is one-third the fundamental wavelength, and hence the overtone is the *third*, not the second, harmonic. Similarly, the second overtone has a wavelength of $4L/5$ and is therefore the fifth harmonic. In fact, closed pipes can produce only odd harmonics because of the difference between their two ends.

EXAMPLE 12.7

What are the frequencies of the fundamental and the first overtone in a closed organ pipe that is 2 m long at a temperature of 20°C?

Solution

The fundamental wavelength of a closed pipe is $4L$. In this example the wavelength is obviously 8 m. The first overtone is the third harmonic, whose wavelength is one-third the fundamental wavelength. This is $\frac{8}{3}$ or 2.67 m. To calculate the frequencies of these harmonics we use the wave equation

$$v = \lambda \nu \quad \text{or} \quad \nu = \frac{v}{\lambda}$$

The speed of sound in air at 0°C is 331.6 m/s, and the speed increases by 0.6 m/s for each degree Celsius rise in temperature. Hence, the speed at 20°C is

$$v = 331.6 \text{ m/s} + (20)(0.6 \text{ m/s}) = 343.6 \text{ m/s}$$

The frequency of the fundamental is

$$\nu = \frac{v}{\lambda} = \frac{343.6 \text{ m/s}}{8 \text{ m}} = 42.95 \text{ Hz}$$

The frequency of the first overtone (third harmonic) is three times the fundamental:

$$\nu = 3(42.95) \text{ Hz} = 128.85 \text{ Hz}$$

or

$$\nu = \frac{v}{\lambda} = \frac{343.6 \text{ m/s}}{\frac{8}{3} \text{ m}} = 128.85 \text{ Hz}$$

EXAMPLE 12.8

What length of organ pipe open at both ends produces a note of frequency 1.2 kHz as its first overtone? The speed of sound in air can be taken as 340 m/s.

Solution

The first overtone in an open pipe is the second harmonic, and the length of the pipe is given by $L = 2(\lambda/2)$ for the open pipe; hence $\lambda = L$. Using $v = \lambda \nu$, then

$$\nu = \frac{v}{\lambda} = \frac{v}{L}$$

or

$$L = \frac{v}{\nu} = \frac{340 \text{ m/s}}{1.2 \text{ kHz}} = 0.28 \text{ m} \quad \text{or} \quad 28 \text{ cm}$$

EXAMPLE 12.9

A stretched wire is placed near the open end of a tube 1.0 m long that is closed at one end. The wire is 0.30 m long and has a mass of 0.01 kg. The wire is fixed at both ends and vibrates in its third harmonic, setting the air column in the tube into vibration at its fundamental frequency. The velocity of sound in air can be taken as 330 m/s. Compute the frequency of oscillation of the air column and the tension in the wire.

Solution

The vibrating wire via resonance has set the air column into resonance. Both the wire and the air column must oscillate with the same frequency. However, the wavelengths of the transverse wave in the wire and the longitudinal sound wave in air are different because the speeds of these two waves are different in air and the wire. First, we compute the frequency of the sound wave in the air column using $v = \lambda \nu$ and the fact that the wavelength of the fundamental of a closed pipe can be determined by $L = \lambda/4$. Hence,

$$\lambda = 4L = 4(1 \text{ m}) = 4 \text{ m}$$

$$\nu = \frac{v}{\lambda} = \frac{330 \text{ m/s}}{4 \text{ m}} = 82.5 \text{ Hz}$$

To compute the tension in the string, we make use of $v = \lambda \nu$ and $v = \sqrt{F/(m/L)}$. Hence

$$\lambda \nu = \sqrt{\frac{F}{m/L}} \quad \text{or} \quad F = \frac{\lambda^2 \nu^2 m}{L}$$

For a wire 0.3 m in length oscillating in its third harmonic (three segments), the length of the string is

$$L = 3\frac{\lambda}{2} \quad \text{or} \quad \lambda = \frac{2}{3}L$$

Using $L = 0.3$ m, $\nu = 82.5$ Hz (the same as the air column), and $m = 0.01$ kg:

$$F = \lambda^2 v^2 \frac{m}{L} = \frac{(2L)^2}{3^2} v^2 \frac{m}{L} = \frac{4L}{9} v^2 m = \frac{4}{9}(0.3 \text{ m})(82.5 \text{ Hz})^2(0.01 \text{ kg})$$

$$= 9.1 \text{ N}$$

12.7 AND THE BEATS GO ON

Varied as it may have appeared so far, our analysis of interference has been strictly limited to one specific condition: the interfering waves are all of the same frequency. Also, standing waves are the result of interference that varies with position but is constant in time. Another interference effect that is very significant in music involves interference between two waves whose frequencies differ just a little. This phenomenon is known as **beats,** and it is an interference effect that varies with time at a fixed position in space.

Imagine two tuning forks close to each other that are vibrating with slightly different frequencies, as diagrammed in Fig. 12.23. The interference pattern of these two waves, obtained by the principle of superposition, gives a sound whose frequency is about the same as those of the two forks but whose amplitude goes up and down regularly with time as shown in the same figure. Any listener within range of the two simultaneously vibrating tuning forks hears a characteristic "beating" sound whose volume goes up and down between near zero and a maximum as a function of time. This so-called sound is termed the beat note.

The frequency of the beats, that is, the number of beats heard per second, equals the difference between the frequencies of the two individual sources of sound. This type of variation in amplitude is

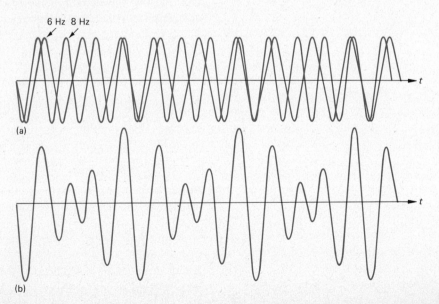

Figure 12.23 The origin of beats. (a) Shows the waves, one of 6 Hz and one of 8 Hz. (b) Shows the sum of the displacements of the two waves in (a). The result produces a beat of 2 per second.

known as amplitude modulation (AM), for the amplitude of the resulting wave varies with time. It is the type of wave that carries the signal of AM radio stations. Piano tuners and musicians use beats to tune their instruments. They listen simultaneously to the sound from a tuning fork, or the standard A string bowed by the concertmaster of the orchestra, and to the same note played on their instruments. If any instrument is out of tune, the combination of the two notes will produce beats. The musicians adjust their strings or reeds until the beats disappear because at that point the frequencies of the instruments and the standard are exactly equal.

12.8 SOUNDS—PHYSICAL VERSUS SUBJECTIVE

The discussion of standing waves on strings and in air in open pipes and closed pipes leads the way to a more general field of musical sounds. Most sounds that we hear are in reality noises. The sounds of music have more sustained notes than the sounds of noise. Obviously, it is difficult for us to be specific in our distinction between music and noise. The reason of course is that we have moved into a qualitative or subjective area similar to art; that is, what appears as great painting or music to one person may be scribbling or noise to another person.

Sounds can be characterized by the following three different subjective attributes: loudness, pitch, and quality. These three subjective attributes correspond to physical terms: intensity, frequency, and shape of the waves producing the sound caused by the number of harmonics.

Intensity and Loudness

The intensity of sound depends upon pressure variations in a sound wave. The intensity is proportional to the square of the amplitude of a wave. It is a physical quantity that can be measured for sound waves in air or for any type of wave using appropriate instruments. **The intensity of a wave is a measure of the power or energy per unit time carried by the waves across a unit area.** We can write a general expression of the intensity as follows:

$$I = \frac{E}{At} = \frac{P}{A} \tag{12.13}$$

The energy E might be the energy incident upon a detector, such as a microphone for sound in air, and A the area of the detector. The units of intensity are watts per square meter.

It can be quite readily shown that the intensity at a given frequency is proportional to the square of the amplitude of the wave. For sound waves in air, the proportionality factor depends on the density of the air and the speed of the sound in the air at that particular temperature. The amplitude of a sound wave in air is the difference between the maximum pressure the sound wave produces and the pressure of the undisturbed air.

The intensity of sound varies with distance from the source. In general, this variation is complicated by reflection and interference effects. However, in the case of a relatively small source, such as a small bell in a very large room, a fire whistle on top of a building heard by a person far from the building where no other buildings obstruct the sound, or sound from an airplane engine from a plane high in the sky moving at slow speed, the intensity varies inversely as the square of the distance from the source to the observer.

This variation in intensity is caused by sound from a point source moving out in all directions. Hence, the total energy of the wave is spread out over the surface of a sphere centered on the point source. As the sound moves further away from the source, the same energy (assuming no loss of sound energy to other forms) spreads out over a larger sphere. Since the surface area of a sphere is $4\pi R^2$, the energy crossing a unit area of the sphere $E/4\pi R^2$ will decrease as $1/R^2$.

The intensity thus obeys an inverse square law similar to the dependence on distance shown by gravitational and electrical forces. The intensity of light and other electromagnetic waves from a point source also obeys this inverse square law dependence on distance.

EXAMPLE 12.10

You can barely hear a small single-engine airplane flying rather slowly when it is 4 km away. Some time later, it is 0.50 km away from you. By how much has the intensity of the sound increased as the plane moves from 4 to 0.50 km?

Solution

Let the intensity at 4 km be I_0 and the intensity when the plane is 0.50 km away I_2. Since the intensities vary inversely as the distance squared, we can write

$$R_0^2 I_0 = R_2^2 I_2 \quad \text{or} \quad (4 \text{ km})^2(I_0) = (0.50 \text{ km})^2(I_2)$$

The intensity at $R_2 = 0.50$ km, I_2, is given by

$$I_2 = \frac{(4 \text{ km})^2}{(0.50 \text{ km})^2}(I_0) = (16 \times 4)(I_0) = 64 I_0$$

The intensity of the sound, the physical quantity, will increase 64 times as the plane moves from 4 to 0.50 km. However, the loudness, the subjective measure of "intensity," will not change by 64 times, as we shall soon see.

Bell became interested in communicating sounds mechanically as a result of his work with deaf children. Eventually, the inventor's studies led to the creation of the telephone, which he patented in 1876.

Loudness ($\mathcal{L}$) is a physiologic sensation that depends upon the intensity in a rather complicated way. A measure of loudness, or more exactly the relative loudness, is called the intensity level and is measured by bels, or more usually decibels (abbreviated dB), a unit named after Alexander Graham Bell. The intensity level of sound is related

12.8 Sounds—Physical Versus Subjective

to the intensity by the relationship, as shown in the following equation:

$$\mathcal{L} = 10 \log \frac{I}{I_0} \tag{12.14}$$

The number of decibels is 10 times the log to the base 10 of the ratio of two intensities. The decibel is not an absolute unit; it expresses the ratio of two intensities. However, the zero decibel level is defined by convention by taking as $I_0 = 10^{-12}$ W/m². The number was selected because it is the minimum sound intensity of a 1-kHz tone detectable by the normal human ear.

A loudness scale for sounds from the lower limit of the ear's detection to the initial levels of pain is shown in Fig. 12.24. This scale is of course approximate since the human ear differs from person to person. However, the threshold of pain occurs at an intensity of about 1 W/m². Using Eq. (12.14), we can convert this intensity to decibels if we take $I_0 = 10^{-12}$ W/m²:

$$\mathcal{L} = 10 \log \frac{1 \text{ W/m}^2}{10^{-12} \text{ W/m}^2} = 10 \log 10^{12} = 120 \text{ dB}$$

On our scale in Fig. 12.24 we can see that amplified rock music is above the pain level for most people. In fact, damage to hearing actually begins at exposure to about 85 dB. The degree of damage depends upon the time of exposure to the loud sound. Note that the intensity level from the limit of sound detection to the threshold of pain spans 120 dB, whereas the intensity of the corresponding sound levels spans 12 orders of magnitude, a very large range.

Extensive studies of the hearing of millions of people have been carried out by Public Health Services. Experts have used measurements to produce statistical data on the human ear; these data show how the ear's response depends upon the frequency of the sound.

Figure 12.24 Approximate sound levels in decibels produced by different sources, compared with the intensities of the sounds in W/m².

Frequency and Pitch

Pitch is the sensation that a sound produces in the ear of the listener and is closely related to the frequency of the sound. The frequency is defined as the number of vibrations that occur per second. Frequency and pitch are both measured in hertz (Hz). The average human ear can detect sounds over a vast range of frequencies, from about 20 to 30 vibrations per second (Hz) in the lowest ranges to an upper limit of around 20 to 50 kHz. The piano, based on an international pitch of 440 Hz for the standard note A, extends from 32.7 to 4186 Hz. A bass singer typically harmonizes between 130 and 350 Hz, and a soprano trills between 250 Hz and 1.2 kHz.

The pitch of sounds produced by various musical instruments usually depends upon the natural resonant frequencies of the instrument. This means that smaller instruments usually produce higher pitch sounds. Large instruments, dense wide strings, and large air columns all produce lower pitch sounds.

Figure 12.25 Fundamental overtones and their addition.

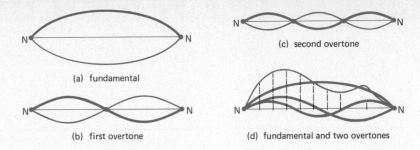

Wave Shape and Quality

The third, and most important, subjective attribute of sound in music is the quality, or timbre. Even the most tone-deaf person can tell the difference between a middle C struck on a piano and the same note bowed with the same intensity on a violin. The reason for this obvious difference is the existence of differing numbers and relative intensities of overtones accompanying the fundamental tone in the two cases. The combination of fundamental and overtones is the quality, or timbre, of the sound.

We can demonstrate this effect using a violin bow, a cork, and a length of wire under sufficient tension to emit a musical note. When it is bowed, the wire tends to vibrate in a fundamental standing wave, as shown in Fig. 12.25(a). If one touches the cork to the exact center point of the wire while it is sounding off, one hears a fainter note of higher pitch. This is the second harmonic shown in Fig. 12.25(b), otherwise known as the octave; its frequency is twice that of the fundamental. An octave results because the cork stills the center of the wire, thus creating a permanent node there that suppresses the fundamental. Similarly, if one touches the cork against the string one-third of the way along its length, it suppresses all but the third harmonic, whose frequency is three times the fundamental, as shown in Fig. 12.25(c). *Musical instruments very rarely produce pure tones of a single frequency.* It is the combination of fundamentals and overtones that gives individual instruments their musical character or quality.

We can physically analyze the sounds from instruments to try to determine just which harmonics are contained in a given sound. This is done by detecting the sound with a microphone, where the longitudinal sound wave is changed to an oscillating electrical signal that can be displayed in an oscilloscope. Figure 12.26 shows three waveforms, from a tuning fork, noise, and music. The musical sound and the noise are qualitatively different in that the music has a more sustained tone. Both noise and musical sound can be broken down mathematically into a series of sine waves of different frequencies by a process known as Fourier analysis. A note played on a musical instrument can be resolved into a few sine waves of varying intensity. The fundamental would usually be the most intense, and the first few higher harmonics or overtones in reduced intensity would add to the fundamental, producing the more complex pattern observed on the oscilloscope.

Figure 12.26 Waveforms produced on an oscilloscope by three types of sounds.

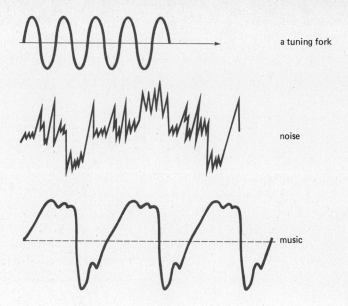

a tuning fork

noise

music

12.9 THE DOPPLER EFFECT

When the base of a tuning fork is placed on a table, or a tuning fork is placed near an air column and the loudness of the sound increases, we know that the table or air column has begun to vibrate in resonance with the tuning fork at the same frequency. The wavelength of the sound waves and the speeds of the waves in the air and wooden table will be different. However, the frequency remains constant. It is the frequency of vibration of the source that determines the frequency of the waves, no matter through what medium the wave is transmitted. There is, however, a very important situation in which the pitch of the sound perceived by an observer changes even though the actual frequency of the vibration of the *source* remains constant. This phenomenon occurs when the source of sound and the observer move relative to one another. It is known as the Doppler effect, after Christian Doppler, an Austrian physicist who explained the commonly observed phenomenon and worked out the mathematical description in 1842.

We can best understand the effect of relative motion upon the perceived frequency in a qualitative manner by examining Fig. 12.27. Part (a) shows a source and the observer of the wave at rest with respect to one another. Since the Doppler effect occurs for any kind of wave, the source might be a small tuning fork sending out a note of single frequency, a musical instrument, a light source, or a small beetle beating one leg up and down in the water in a regular manner. The wave travels out in all directions, producing expanding wave fronts, which are drawn as circular lines in Fig. 12.27. These circular wave fronts represent points of the same phase. For example, the beetle wiggling its foot sends out circular ripples, or "wave fronts." The

Like von Guericke before him, Doppler had a touch of the scientific showman. He tested his mathematical description in a series of experiments involving instrumentalists on trains moving at certain speeds, playing one note while musicians gifted with perfect pitch stood beside the track and recorded their perceptions as the train approached and receded.

Figure 12.27 The Doppler effect.

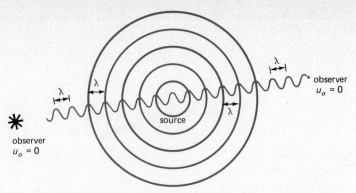

(a) the source and observers are all at rest in the medium where the wave of wavelength λ travels at a speed v

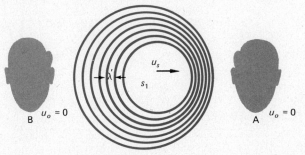

(b) the source moves toward observer A and away from observer B at speed u_s

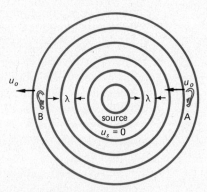

(c) observer A moves toward the stationary source at speed u_o, while observer B moves away from the source at the same speed

separation of the wave fronts equals a wavelength λ, since by definition λ is the separation of adjacent points in the direction of wave propagation that are in phase.

What happens if either the source or the observer is moving? Let us assume that the speed of propagation—the wave speed $v = \nu\lambda$—is constant, as we have seen is the usual case. We shall also restrict our analysis to relative motion of source and observer along a straight line. (This assumption merely simplifies our analysis, for with a little geometry we could generalize our results.)

Consider first the case of a moving source approaching observer A and leaving observer B, as shown in Fig. 12.27(b). Both observers are at rest in the medium, in which the sound is moving at speed v. As is apparent from the figure, the wavelength of the sound appears shorter in front of the moving source and longer behind the moving source. This effect can be readily observed in a water wave tank or ripple tank with a moving rippler. The water waves, or ripples, actually bunch up in front of the moving source, as shown in Fig. 12.27(b).

We can derive a relationship for the frequency ν_s detected by observers A and B. When the source and observer are at rest, the wavelength of the wave is λ, the frequency is ν, the period is T ($T = 1/\nu$), and the speed of the wave is v through the medium. We know that $v = \lambda\nu$. Now, if the source moves through the medium at a speed u_s toward observer A, the new wavelength detected by observer A will be given by

$$\lambda_A = vT - u_sT = (v - u_s)T = \left(\frac{v - u_s}{\nu}\right)$$

The faster the source moves, the shorter will be the wavelength λ_A detected by observer A. Observer B, behind the moving source, will hear a longer wavelength:

$$\lambda_B = vT + u_sT = (v + u_s)T = \frac{v + u_s}{\nu}$$

We can convert the wavelengths λ_A and λ_B observed from the moving source to frequencies ν_A and ν_B, the apparent frequencies of the moving source detected by observers A and B. The speed of the wave is constant and equal to v, so

$$\lambda_A \nu_A = v$$

or

$$\nu_A = \frac{v}{\lambda_A}$$

For observer A, as the source approaches,

$$\nu_A = \frac{v}{\lambda_A} = \frac{v}{\frac{v - u_s}{\nu}} = \nu\left(\frac{v}{v - u_s}\right)$$

For observer B, as the source moves away,

$$\nu_B = \frac{v}{\lambda_B} = \frac{v}{\frac{v + u_s}{\nu}} = \nu\left(\frac{v}{v + u_s}\right)$$

Ultrasound: Sound Medicine

x-rays have been a major diagnostic tool of medicine for so long that one is tempted to think of them as the only form of radiation that can be applied to probing the human body. But recent years have seen the emergence of an alternative that is inherently safer than x-rays: ultrasound. Unlike x-rays, ultrasound does not damage cells. It does, however, possess remarkable versatility. Measurements of internal reflections of ultrasound can be used in such varied forms of diagnosis as spotting breast cancer and taking the heartbeats of fetuses and newborns.

One simple ultrasonic instrument, called a Hemosonde by its manufacturer, Parke-Davis, uses the Doppler effect of ultrasonic waves reflected from moving internal masses in the patient. The device is a very sensitive one for detecting blood flow and works extremely well for faint heartbeats and in a very noisy environment where the stethoscope (a listening device that acts like a microphone) occasionally may not function reliably.

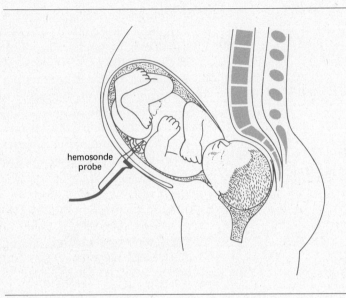

The use of the device to monitor a fetus in the womb.

This ultrasonic device measures movement rather than the sound of the heartbeat. Its principle of operation is similar to that of devices employing radar (electromagnetic waves with frequencies in the gigahertz range) to detect motion of automobiles or devices employing sonar (ultrasonic waves) in the detection of objects underwater. The instrument is a hand-held device consisting of a bull's-eye-shaped ultrasonic trans-

We can combine this expression for the frequency v' as heard by either observer when the source is moving:

$$v' = v \left(\frac{v}{v \mp u_s} \right) \qquad (12.15)$$

The upper sign ($-$) before u_s indicates that the source is approaching the observer [A in Fig. 12.27(b)] and the lower sign ($+$) before u_s shows that the source is moving away from the observer [B in Fig. 12.27(b)].

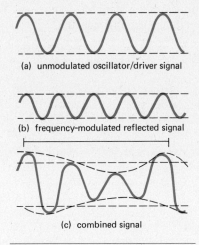

The hemosonde in action. (a) to (c) The reflected signal is combined with the original signal to produce the combined effect shown.

ducer driven by a crystal oscillator in the region of 5 MHz. The transducer is a crystal that converts an oscillating electrical signal into a mechanical sound wave in the ultrasonic region above the limits of the human ear. The transducer is placed with a gel against the skin of the patient, to provide a close coupling between the transducer and the patient. A portion of the ultrasonic signal transmitted into the patient is reflected whenever it strikes an internal mass. If the internal mass is moving, then the frequency of the reflected wave will be shifted by the Doppler effect. The transducer detects the rather complicated ultrasonic signal that consists of various amplitudes of the original frequency reflected from stationary matter and Doppler-shifted frequencies reflected from moving internal matter. It then converts the ultrasonic signal to an electrical signal that is amplified and eventually monitored using a headset or an oscilloscope. The accompanying figure shows (a) the unmodulated 5-MHz sine wave produced by the oscillator that originally was radiated into the body tissue and some of which is picked up by the receiver along with the reflected signals, and (b) the return (reflected) signal at a different frequency due to the Doppler effect of the moving target. The signals (a) and (b) will be added linearly by the transducer, resulting in an amplitude-modulated signal with a low-frequency envelope as shown in (c). The frequency of the envelope is limited to the difference in frequency between (a) and (b) and appears in the audio region. This signal, which is an electrical output from the transducer, can be amplified and used to drive a headset so it can be "heard," or to drive an oscilloscope.

One great advantage of this device over the stethoscope is its increased sensitivity and nonreliance upon the direct audio signal from the heartbeat. Room noises do not hinder the detection sensitivity. Of course, the device is not inserted into the body as are other types of blood flow monitors, which is an added advantage. This ultrasonic device is perhaps the simplest of the ultrasonic devices that have been employed to scan the brain or abdomen to produce a map much like an x-ray photograph.

There are two other conditions we can readily treat: when the source is at rest and the observer is moving either toward the source or away from it. These situations are illustrated in Fig. 12.27(c). For the observer moving at speed u_o toward the source that emits waves of frequency ν and wavelength λ, the frequency of the wave appears to rise because the waves reach the observer more rapidly than if the observer were at rest. That is, the observer in motion toward the stationary source receives more waves per unit time. These additional waves are contained in the distance the observer travels at speed u_o. Since λ is the wavelength, the additional number of waves received by the observer per unit time is u_o/λ. Since $v = \lambda \nu$, we can express u_o/λ as

$$\frac{u_o}{v/\nu} = \nu \left(\frac{u_o}{v}\right)$$

This represents the increase in frequency of the wave as observed by observer A. Hence, as observer A approaches the stationary source, he detects a frequency ν_A that is greater than ν by this amount:

$$\nu_A = \nu + \nu \left(\frac{u_o}{v}\right) = \nu \left(1 + \frac{u_o}{v}\right) = \nu \left(\frac{v + u_o}{v}\right)$$

For observer B moving away from the source, the frequency ν_B will *decrease* to:

$$\nu_B = \nu \left(\frac{v - u_o}{v}\right)$$

We can again combine the two cases in the following equation:

$$\nu' = \nu \left(\frac{v \pm u_o}{v}\right) \tag{12.16}$$

Here, the upper, or $+$ sign, before u_o is for the observer approaching the source, and the lower, or $-$ sign, is for the case of the observer moving away from the source.

Now, Eqs. (12.15) and (12.16) can be combined into a general relationship that expresses the observed frequency for any of these four cases in which the source and/or the observer move relative to one another. The general expression for the Doppler effect for all but electromagnetic waves is as follows:

$$\nu' = \nu \left(\frac{v \pm u_o}{v \mp u_s}\right) \tag{12.17}$$

The upper signs indicate that the source and observer are approaching one another, and the lower signs show that the two are separating from one another.

How do we experience the Doppler effect in practice? The most common example is when an express train with its whistle blowing passes through a station. As the train approaches, the sound waves from its whistle are compressed into the diminishing space between it and passengers waiting at the station. After the train has passed through, the same sound waves are stretched out over an increasing distance. The result is that the stationary commuters in the station hear a sudden drop in the pitch of the whistle. Of course, the loudness of the sound also reaches a peak as the train is immediately opposite the

The difference between the ways in which the Doppler effect affects light and sound arises from the unusual nature of light, and of all other forms of electromagnetic radiation for that matter. Such radiation requires no medium to transmit it, and its measured speed is constant whatever the speed of the observer making the measurement. Such peculiar properties will be dealt with in our discussion of Einstein's theory of relativity in Chap. 20.

waiting crowds, but the change in pitch is obviously a different type of effect.

The Doppler effect can be observed for all wave phenomena. It is readily demonstrated in a laboratory ripple tank. However, the most important illustration of the Doppler effect is in astronomy, where the Doppler effect is used to interpret the "red shift."

The Doppler effect with light is similar to that involving sound but does involve significant differences. Most important is that whereas the individual speeds of the observer, u_o, and the source, u_s, with respect to the medium determine the frequency change for sound waves, the frequency change of light waves depends solely on the relative speed of source and observer.

It is this relative speed that explains the red shift in the collection of wavelengths of light, or "optical spectra," from stars. Spectral lines from most stars and galaxies appear to be shifted toward the red, or longer, wavelength. This indicates a decrease in frequency, which is explained according to the Doppler effect if the star is receding from the observer. Measurement of the red shift of spectral lines from stars allows astronomers to calculate the speeds with which the stars are moving away from us. Since almost all stars and galaxies exhibit a red shift, it is believed that the universe is continually expanding, at least over the time we have observed it. This statement should be considered in view of the fact that the light you see tonight from the more distant stars and galaxies has taken millions and millions of years to reach the earth. Therefore, this light does not show what is happening *now* on these stars and galaxies, but what happened in the past. Astronomical observations allow us to look back in time—something historians are always wishing very hard to accomplish. The Doppler shift is the basis for the interpretation of many astronomical observations and is extremely important to the development of cosmological theories.

Before terminating the discussion of the Doppler effect, we should examine what happens to a sound or water wave if the source moves through the air or water at the speed of the wave. Or one might ask "What happens when a bug or a beetle moves through water faster than the speed of the water waves?" Or, "What happens when an airplane that generates sound waves moves through the air faster than the speed of sound in the air?" The result is perhaps best seen by extending Figure 12.27(b) to higher and higher source speeds, as shown in Fig. 12.28. As the speed of the source reaches the speed of sound in the medium, the waves pile up in front of the source, producing a V-shaped bow wave in water and a cone-shaped wave in air. The superposition of these many waves produces an extremely large-amplitude "shock wave," resulting in a large bow wave in water and a sonic boom in air. When a supersonic aircraft passes by, this cone of waves compressed upon one another produces a high-pressure shock wave that passes over the ground as the plane flies along. Once an

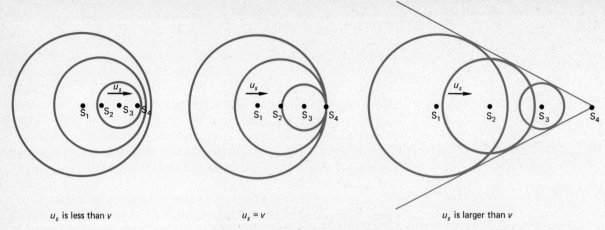

u_s is less than v $u_s = v$ u_s is larger than v

Figure 12.28 Production of waves by a source moving through a fluid when the speed of the source is less than, equal to, and greater than that of the waves. In the third case, a bow wave, or shock wave, develops as the waves build up in front of the source. In each example, S_1, S_2, and S_3 indicate successive positions of the moving source.

The sonic boom is not a one-shot affair that occurs just at the instant the speed of sound is exceeded.

object moves faster than the speed of sound it will continue to drag a shock wave (sonic boom) along with it. The sonic boom is not a one-shot affair that occurs just at the instant the speed of sound is exceeded; it continues as long as the aircraft is moving faster than the speed of sound.

EXAMPLE 12.11

A note of frequency 500 Hz is emitted from a sound-broadcasting truck. (*a*) If the truck moves toward you at a speed of 60 km/h, what is the frequency you will detect? (*b*) After the truck passes you and moves away from you at 60 km/h, what frequency will you detect? The speed of sound is 340 m/s.

Solution

First, the speeds should all be put into the same units. The speed of the truck is 60 km/h, which can be converted to m/s:

$$60 \text{ km/h} \times 10^3 \text{ m/km} \times \frac{1}{3600 \text{ s/h}} = 16.7 \text{ m/s}$$

(*a*) To compute the frequency when the source approaches we use Eq. (12.17), where $u_s = +16.7$ m/s, $v = 340$ m/s, $\nu = 500$ Hz, and $u_0 = 0$. The observed frequency is

$$\nu_A = \nu \left(\frac{v + u_0}{v - u_s} \right) = 500 \text{ Hz} \left(\frac{340 \text{ m/s}}{340 \text{ m/s} - 16.7 \text{ m/s}} \right)$$

$$\nu_A = 500 \text{ Hz} \left(\frac{340}{323.3}\right) = 525.8 \text{ Hz}$$

(b) When the source moves away from you

$$\nu_B = \nu \left(\frac{v - u_0}{v + u_s}\right) = 500 \text{ Hz} \left(\frac{340 \text{ m/s}}{340 \text{ m/s} + 16.7 \text{ m/s}}\right)$$

$$\nu_B = 500 \text{ Hz} \left(\frac{340}{356.7}\right) = 476.6 \text{ Hz}$$

SUMMARY

1. Wave motion is the movement of a disturbance or a deformation through a medium. In transverse motion, the individual particles of the medium vibrate at right angles to the direction in which the wave moves; in longitudinal motion, the particles move back and forth along the wave's direction of motion.
2. The time that a particle takes to make a single round trip of oscillatory motion is known as the period of vibration. The reciprocal of the period is the vibration frequency.
3. The wavelength of a wave is the length of a complete wave along the direction in which it is moving. The amplitude is the maximum displacement of the material supporting the wave from its equilibrium position.
4. Simple harmonic motion is the type of mechanical vibration in which the restoring force that tends to bring the medium back to its original position is proportional to the displacement of the medium from that position.
5. The speed of a wave along a string is the square root of the string's tension divided by its mass per unit length.
6. Points along the path of a periodic wave that are in phase are separated by an integral number of wavelengths.
7. The speed of a periodic wave is the product of its wavelength and frequency.
8. When two wave trains meet, or interfere, the resulting displacement at any point is the algebraic sum of the displacements of the individual waves at that point. This is known as the superposition principle.
9. Standing waves arise when two sets of periodic waves of equal amplitude and wavelength traveling in opposite direction interfere. Such waves are characterizd by fixed nodes, where the displacement of the medium is always zero, and antinodes, where the displacement of the medium reaches a maximum. Standing waves are responsible for the tones produced by musical instruments.
10. Interference of two waves of slightly different frequencies produces the throbbing effect known as beats. The frequency of the beats equals the difference between the frequencies of the two individual sources of waves.
11. Physical characteristics of sound, such as intensity, frequency, and the mixture of harmonics, are interpreted subjectively by listeners as loudness, pitch, and quality of sound.
12. Relative movement of a source of waves and an observer causes the observer to detect a frequency different from that actually produced by the source. This is the Doppler effect.

QUESTIONS

1. How is the frequency of a wave related to the frequency of the vibrating source?
2. What is the difference between a traveling wave and a standing wave?
3. What type of mechanical wave can propagate in a solid? In a fluid?
4. What kind of motion does the medium exhibit at nodes of a standing wave? At antinodes?
5. The speed of sound in air or any gas is dependent on temperature of the medium. Explain in terms of kinetic theory.
6. How are different notes produced on a single string?
7. A pendulum clock is found to run too fast, gaining time; what adjustment should be made to the pendulum to correct this?
8. Can you think of an experiment to detect that waves transport energy?

9. How is the speed of a wave on a string altered if (a) the tension is tripled, (b) the tension is cut by one-fourth, (c) the string is replaced by a string of twice the mass per unit length?
10. Graphically, show what results if two waves of equal frequency are at a point (a) 180° out of phase but equal in amplitude, (b) in phase but one has twice the amplitude of the other.
11. If you were walking on the moon, would you hear a moon monster stomping up behind you? Explain.
12. When purchasing a stereo set, should you spend extra money for a system with frequency response higher than 20 kHz?
13. Which notes on a stereo or radio are transmitted best from one part of a building or from one apartment to another? Does the answer here indicate anything concerning the natural resonance of building materials?
14. If you walk by a ringing bell, its pitch does not change. Yet, if you move by in a rapidly moving car, the pitch does change. Why?
15. The intensity of a sound of 40 dB is how many times greater than the threshold of hearing?
16. Inhaling helium gas increases the pitch of the sounds from the vocal cords. Explain.
17. How could the Doppler effect and reflection be used to detect speeding automobiles?

PROBLEMS

Simple Harmonic Motion

1. A force of 20 N is required to stretch a spring 4 cm. What is the spring constant?
2. What force is required to stretch a spring 4 in. with a force constant of 100 lb/in.?
3. How far can a force of 14 N stretch a spring with a force constant of 20 N/m?
4. The scale of a spring balance reads from 0 to 450 g and is 10 cm long. What is the force constant of the spring? How many grams correspond to 1 mm of stretch of the spring?
5. A force of 6 N stretches a spring 18 cm. What is the force constant of this spring?
6. A force of 7 lb stretches a spring 8 in. What weight must be hung from it to make it oscillate with a period of $\pi/2$ s?
7. A 3-kg mass on a spring moves in simple harmonic motion at frequency ν. What mass will cause the same system to vibrate at twice the frequency, 2ν?
8. When a 1-oz letter is placed on a pan spring balance, it goes down $\frac{1}{16}$ in. What is the force constant of the spring?
9. What is the period of a simple pendulum whose length is 9.8 m and holds a bob of mass 6 g?
10. What is the frequency of a 4.9-m-long pendulum?
11. A pendulum is held 10 cm from its equilibrium position and let go. What is its amplitude?
12. The length of a simple pendulum is 6.2 cm. What is the period? The frequency?
13. What must be the length of a simple pendulum to have a period of 1 s?
14. What is the period of a simple pendulum of length 6.5 m on the moon, where the acceleration due to gravity is one-sixth that on the earth?
15. The scale of a spring balance reading from 0 to 40 lb is 8 in. long. A body suspended from the balance oscillates vertically at 1.5 Hz. What is the weight of the body?
16. A mass vibrates in simple harmonic motion with an angular frequency of 4 rad/s. The amplitude is 10 cm. (a) What is the maximum acceleration? (b) What is the maximum speed? $\omega = 2\pi\nu$
17. If an object in simple harmonic motion has a maximum acceleration of 20 m/s² and a maximum speed of 3 m/s, what are its (a) period and (b) amplitude?
18. Two girls on a rubber raft bob up and down in simple harmonic motion a total distance of 40 cm with a period of 2 s. (a) Compute their maximum acceleration. (b) Where does this occur? (c) What is the maximum and minimum weight of one girl if her mass is 45 kg and she sits on a scale on the raft?
19. Examine a pendulum in its motion. Draw a graph of the displacement of the pendulum as a function of time.
20. Show, using the reference circle, that the velocity of the projection of a rotating object on the diameter of the orbit is given by

$$v = \pm 2\pi\nu \sqrt{R^2 - x^2}$$

where ν is the frequency, $2\pi\nu$ being the angular frequency of the circular motion; R is the radius of the circle and equals A, the amplitude of the simple harmonic motion of the projection; and x is the displacement of the projection from the center.
21. An object moves in simple harmonic motion at the end of a spring. The amplitude is 12 cm, and the frequency is 4 Hz. (a) What is the maximum acceleration and where does it occur? (b) What is the maximum velocity and where does it occur? (c) What is the displacement after 4 s if the motion began when $x = 0$. (d) What is the displacement after $\frac{1}{8}$ s?

22. A test tube 2 cm in diameter has a mass of 6 g, contains 10 g mercury, and is floated in water. Calculate the period of oscillation of the test tube. Neglect the inertia of the water.

23. The equation of motion for a mass ($m = 200$ g) attached to a spring is $y = 10 \cos(4\pi t)$; y is in cm and t in s. Find the (a) amplitude, (b) frequency, (c) period, and (d) force constant of the spring. (e) Calculate the velocity at $t = 3$ s.

24. A 10-g mass vibrating with simple harmonic motion is pulled out 10 cm and let go. The force constant of the spring is 1000 g/s^2. Where is the mass after (a) 1 s? After (b) 2 s?

25. A body vibrates with simple harmonic motion of amplitude = 14 cm and frequency = 4 Hz. Compute the (a) maximum acceleration and (b) acceleration when $y = 4$ cm. (c) What is the speed when $y = 4$ cm?

Frequency and Wavelength

26. The speed of waves on the coast is 14 ft/s and the wavelength is 3 ft. With what frequency do the waves hit the beach?

27. A boy on an inflatable raft observes two waves pass in 1 s. The distance between crests is 1 m. How fast are the waves moving?

28. If the wavelength of a sound in air is 6 m and the speed of sound is 340 m/s, what is the frequency of this sound when it moves into sea water?

29. Compute the wavelength of the sound emitted by a G fork (384 Hz) when the temperature is 32°C.

30. Compute, in centimeters and meters, the wavelength of the broadcast wave from an AM radio station broadcasting on a frequency of 800 kHz. How long is this wavelength in terms of some familiar object?

31. Assume that the lowest audible note has a frequency of 16 vibrations per second and the highest audible note has a frequency of 20 000 vibrations per second. Taking the speed of sound in air as 1100 ft/s, find the corresponding wavelength in air. In each case think of some familiar object that has a size comparable with the wavelength in question.

32. Write the equation of a sine wave with amplitude 10 cm and wavelength 20 cm.

Speed of Waves

33. An explosion occurs 20 km away. How long will it take for the sound to reach you through the air?

34. Sound waves in a sonar system are emitted from a submarine and reflect from a group of whales back to the sub in 0.5 s. How far away are the whales? Use data from Tables 8.1 and 8.2 to assist you.

35. Compute the speed of sound in steel or iron railroad rails using data from Tables 8.1 and 8.2. Compare with the speed of sound in air.

36. The specific gravity of copper is 8.8 and Young's modulus for copper is 11×10^{10} N/m^2. Compute the speed of sound in a copper rod. Compare this speed with that of sound in air.

37. If the speed of sound in a piece of wood is 4.3×10^3 m/s and the density of the wood is 890 kg/m^3, what is the value of Young's modulus? This is a good way to determine the modulus and thus gain some information about the strength of the wood.

38. Compute the speed of sound in bone, where Young's modulus is 1.6×10^{10} N/m^2 and the average specific gravity is 1.8. How does this compare with the speed of sound in tissue (1500 m/s)?

39. What is the speed of propagation of transverse waves in a string of tension 400 N if its length is 1 m and its mass 1 g?

40. If the speed of a transverse wave on a bass string 1.5 m in length is 3 m/s when the string is stretched to a tension of 6 N, what is the mass of the string?

41. What is the speed of a wave on a guitar string 1 m long with a mass of 0.01 kg, stretched with a tension of 16 N?

42. Compute the speed of sound in helium at 0°C and compare with the speed of sound in air.

43. A string 7 m long is hooked to a vibrator at one end, and the other end, hanging over a pulley, has a 10-kg mass hanging on it. The string vibrates, forming a standing wave of three antinodes. Compute the frequency of the vibrator. The mass of the string is 7 g.

44. A guitar string is 80 cm long and has a fundamental frequency of v. To produce other fundamentals, the length of the string must be shortened by pressing it. What length is required to produce a fundamental of $1.5\,v$?

45. A string 4 m long of mass 8 g is under a tension of 50 N. Compute the frequency of the fundamental and the first two overtones.

46. What are the fundamental and the first four overtones of a closed-end organ pipe 20 cm long when the speed of sound in the pipe is 330 m/s?

47. If you blow into a soda bottle that is 20 cm tall, what is the fundamental frequency? Speed of sound = 344 m/s. Assume the bottle is a cylinder.

48. An apparatus often used in elementary laboratories to demonstrate resonance in a column of air is shown in the accompanying figure. As the water column is lowered in the resonance tube, a resonance column open at one end is created. The first (shortest length) resonance occurs at the fundamental. Resonance occurs

for the single frequency at other discrete but longer air columns. If resonance occurs with a tuning fork at 15 and 45 cm, what is the frequency of the fork?

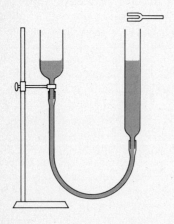

49. An open-end organ pipe 1.5 m long has a fundamental frequency of 116 Hz. Calculate the wavelength and the speed of sound in the air in the pipe.

50. Find the fundamental frequency and first four overtones in a 40-cm-long pipe containing air. Consider both cases: (*a*) the pipe is open at both ends, and (*b*) the pipe is closed at one end.

51. An open pipe (open at both ends) 75 cm long is sounding in its fundamental. What length pipe closed at one end would resonate in its fundamental with this open pipe?

52. A closed pipe 40 cm long is filled first with air and then with helium. What is the frequency of the fundamental heard in each case?

Doppler Effect and Beats

53. If two frequencies emitted from two sources are 49 and 56 vibrations per second, how many beats per second are heard?

54. Two whistles emit sounds of wavelengths 2.7 m and 3 m. If the speed of sound is 340 m/s, how many beats are heard each second when these two sounds are heard simultaneously?

55. Compute the wavelength of the sound emitted by a tuning fork of frequency 460 Hz when the temperature is 31°C. If you were moving toward this tuning fork at a speed of 60 km/h, what frequency would you hear?

56. A whistle has a frequency of 380 Hz. If a car is moving toward the whistle at a speed of 25 m/s, what frequency is heard?

57. A train whistle is blowing at 280 Hz. The train is moving at 80 km/h as you approach in a car at a speed of 48 km/h on a road alongside the railroad track. If the speed of sound is 340 m/s, what is the frequency you will hear?

58. A 512-Hz tuning fork is moved away from a group of students in a lecture hall toward a wall at a speed of 3 m/s. What is the apparent frequency of the waves coming directly to the students? What is the frequency of the waves reflected from the wall? How many beats will the students hear?

59. A 512-Hz note is emitted from a stationary source. If you are traveling toward this source at 32 km/h as the wind blows toward you at 16 km/h, what is the frequency of the note heard by you? The speed of sound is 340 m/s in air.

Intensity and Loudness

60. If the intensity of sound is tripled, by how much does the loudness increase?

61. What is the loudness level of a sound of intensity 4.5×10^{-3} W/m²?

62. What is the sound intensity 10 m from a small speaker at a football stadium if the speaker sends out 8 W of power over a hemisphere? If you move twice as far from the speaker, by how much is the sound intensity reduced?

63. If 4 W of sound spread out in all directions, what is the intensity 5 m away? What loudness level does this correspond to?

64. The loudness level of a sound detected by your ear is 60 dB. If the area of your eardrum is 0.6 cm², what is the power-level incident on the eardrum? If the sound lasts for 2 min, what energy was incident on the eardrum?

Special Topic
Hear, Hear: The Inside Story of the Ear

The cry of a baby, the throb of a rock group, the roar of a crowd in a sports stadium, the whine of a jet plane, the insistent dripping of water from a leaky faucet, the boom of an exploding bomb: For all their variety, these sounds travel through exactly the same channel from their sources to the brains of their hearers. Having passed through the atmosphere, the sounds reach the listener's ear and then embark on an incredibly complicated journey through the compressed, winding channels that comprise the organ responsible for hearing.

The human ear is a marvelously efficient instrument whose compactness impresses any electronic engineer. The ear is practically shatterproof, quick to shrug off infections, and capable of responding to sounds as diverse as the rustle of leaves in the early fall and the crack of a thunderstorm in summer. The loudest sound that the ear can detect without pain is a huge 10^{12} times as intense as the faintest whisper it can hear. And to add icing to the cake, our ears are also responsible for giving us our sense of balance.

Each ear is divided into three distinct parts. The vibrating waves that make up any sound first find their way into the outer ear. This area contains the ear canal, which is, in effect, a closed pipe about 1 in. long. One end is open to the atmosphere, the other stoppered by the membrane known as the eardrum. Having passed through the ear canal, the sound sets the eardrum in vibration. An area of condensation in the sound

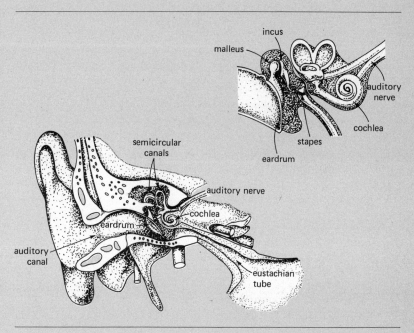

The human ear.

pushes the drum in slightly, and the subsequent region of rarefaction draws it out.

From the eardrum, the sound passes to a delicate lever system of three bones in the middle ear, the second region of the unique listening device. Because of their distinctive shapes, these three bones are named the hammer, stirrup, and anvil. They carry the vibrations to another interface, this one known as the oval window. The three bones transmit a roughly constant amount of force, but because the area of the eardrum is appreciably larger than that of the oval window, the pressure of the sound vibrations is somewhat magnified by a factor of about 20. Owing to the shapes of the three bones, the intensity is magnified by a factor of more than 30 000 at best.

The oval window is the threshold of the inner ear, where the experience of hearing really goes into action. The functional part of the inner ear, as far as hearing is concerned, is the snail-shaped cylinder known as the cochlea. If unrolled, this part of the body is no more than $\frac{1}{4}$ in. long. Yet it contains an ultrasophisticated system for converting the pressure waves that come from the middle ear into tiny electrical signals which pass on to the brain and allow that organ to detect and discriminate between sounds.

The cochlea is filled with a liquid known as perilymph and is split into two areas along its whole length by a membrane known as the basilar membrane. This part is attached to a conglomeration of nerve fibers and minutely fine hairs called the organ of Corti. The basilar membrane has a variable thickness and tension. It is thin and extremely taut close to the oval window but thicker and looser at greater distances from the window. Each intensity sound, in effect, finds its own level in the cochlea. Although every sound causes the basilar membrane to sway up and down, the maximum amplitude of the swaying is located in a particular portion of the membrane, according to the frequency of the sound. The taut sections of the membrane near the oval window react preferentially to high-frequency sounds, such as the shriek of an ambulance siren. The slacker areas at the other end of the membrane vibrate with the greatest amplitude in response to low-frequency noises, such as the croak of a frog.

These membrane movements produce shearing stresses in the contiguous organ of Corti, thus deforming some of the tiny hairs that the organ contains, in much the same way that one might pluck a piano wire. The deformations stimulate the ends of the auditory nerves; then, in some way not yet fully understood, this action provides the nerves with coded electrical signals that contain all the necessary information about the pitch, intensity, and quality of the incoming sounds. The signals reach the brain, where they are translated into joy or aggravation on behalf of the listener.

13 Electrostatics

13.1 GETTING A CHARGE OUT OF ELECTRICITY

13.2 CONDUCTORS AND INSULATORS

SHORT SUBJECT: Copying by Electrostatics

13.3 INTRODUCING THE ELECTROSCOPE

13.4 COULOMB'S LAW

13.5 ELECTRIC FIELDS

13.6 POTENTIAL DIFFERENCE AND POTENTIAL ENERGY

SHORT SUBJECT: The Electrostatic Accelerator

13.7 POTENTIAL DIFFERENCE AND POTENTIAL GRADIENT

13.8 EQUIPOTENTIALS AND ELECTRIC FIELD LINES
Summary
Questions
Problems

SPECIAL TOPIC: How Millikan Measured the Ultimate Electric Charge

It all started with Thales, a sage who held court in the town of Miletus around 600 B.C. Thales observed that the yellowish resin known as amber attracted light objects, such as small pieces of dry grass, thread, and the pith from the centers of weed stalks, after it was rubbed with silk. Without the preliminary rubbing, the resin had no effect at all on the objects. That observation set the stage for more than 25 centuries of probing into a phenomenon that affects virtually every facet of our lives today: electricity.

Looked at dispassionately, electricity is truly an amazing effect. Like gravity, it is an example of a force that seems to act at a distance; there is no obvious connection between, say, the amber rod and the pieces of grass that it pulls toward itself. On the other hand, the differences between electrical and gravitational forces are striking. For a start, electricity involves both attraction and repulsion, rather than the simple attraction of the gravitational force, because there are two different types of electrical objects, which we term positive and negative according to the convention introduced by Benjamin Franklin. Even more impressive is the size of the electrical force between objects—roughly one thousand trillion trillion trillion times stronger than the gravitational force! Finally, we do not usually experience the electrical force directly in our daily lives because the balance between positive and negative electric charge in our bodies is absolutely perfect. If it were not, any individual would suffer an overwhelming force of attraction to or repulsion from anyone else standing nearby—a force strong enough to lift the entire earth if there were only a 1 percent imbalance in electric charge on the two people. We do, of course, see the result of even a very slight imbalance of electric charge when, after walking across a carpet in a dry room, we go to grab a metal door knob and a spark jumps to the knob.

13.1 GETTING A CHARGE OUT OF ELECTRICITY

The word electricity—from the Greek word for amber—was coined about 1600 by Dr. William Gilbert, physician to Queen Elizabeth I. Gilbert was the first person to discover that the amber effect could be reproduced by rubbing together a number of other materials. One can repeat and extend the experiments of Thales and Gilbert by rubbing together a variety of pairs of substances and testing each one for the ability to attract lightweight objects. Rubber and fur, glass and wool, and wood and flannel are good examples of the frictional separation of electric charge to create an imbalance. One can also produce charged objects by scraping one's shoes across a carpet (assuming that the shoes

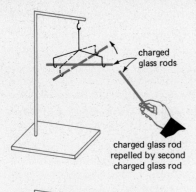

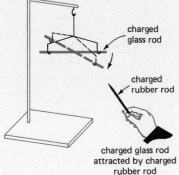

Figure 13.1 Like electric charges repel each other, the unlike electric charges attract.

are made of leather and the carpet of wool in this age of plastics) and even by simply running a comb through one's hair.

Having demonstrated the existence of electric charge in this way, one turns to the next obvious questions. Are there different kinds of electric charge? And if there are, how many exist? A few more experiments quickly yield the answers. As a starter, rub a glass rod vigorously with a piece of silk cloth, thus "charging" the rod, and then place the rod in the holder suspended from a bar by a piece of thread, as illustrated in Fig. 13.1. Now rub a second glass rod with silk and bring the rod up toward the first rod; the latter rod swings away from the second charged rod, demonstrating a force of repulsion between the two charged objects of the same type.

Next, charge a rubber rod by applying fur and elbow grease to it, and bring this rod close to the suspended, charged glass rod. The reaction here is the reverse of the first case; the glass rod swings toward the rubber rod, as shown in Fig. 13.1, demonstrating both that the charge on the rubber rod is different from that on the glass and that the two different types of charges attract each other. The two charges are arbitrarily distinguished by calling the charge produced on the glass when glass is rubbed with silk positive and that produced on a rubber rod by the frictional action of fur on the rod negative. Vast numbers of similar experiments involving pairs of materials have proved that any charged object either attracts or repels the suspended, charged glass rod. There are two, and only two, different kinds of electric charge.

A note of caution is perhaps appropriate here. A neutral object, as we shall discuss later in this chapter, is attracted to either a positively or negatively charged object. This means that when you carry out electrostatic experiments, you must be very careful in how you interpret them.

Basic experiments have yielded six ground rules for **the study of simple attraction and repulsion between charges that, because it does not involve continuous flow of electric charge, is known as electrostatics.** These six fundamental rules are:

1. Opposites attract; that is, objects of unlike charge exert pulling forces on each other.
2. Like repels like; in other words, objects with the same type of charge push each other apart.
3. Charged objects can attract uncharged objects.
4. Whenever two different substances are rubbed together, one becomes positively charged and the other negatively charged. The quantity of charge on the two objects, which is extremely difficult to measure, is exactly the same. This leads to the law of conservation of charge.
5. Any electric charge in certain materials known as conductors automatically distributes itself throughout the exterior surface of the material. Charges remain localized in other materials, which are called insulators.

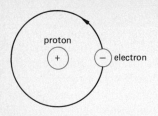

Figure 13.2 The Bohr concept of the hydrogen atom. The mass of the proton is 1836 times greater than the mass of the electron.

Only negative charges can move easily through a solid material because the negatively charged electrons are both less massive and closer to the outside of the atom than are the positively charged protons.

6. Just as the atom is the smallest unit of matter, so a specific charge (the charge on the electron and on the proton) is known to be the smallest unit of electric charge.

These basic rules and regulations of electrostatics can be understood in terms of simple atomic theory. To explain them, we shall take a sneak preview of the atomic theory of the Danish physicist Niels Bohr, which is analyzed in broader terms in Chap. 21.

Bohr proposed this theory in 1913, when physicists' excitement about such new discoveries as x-rays, radium, and radioactivity was at a heady peak. Bohr conceived his model of the simplest of all atoms, the hydrogen atom, in analogy to the solar system. Instead of gravitation however, Bohr argued that the electrical force of attraction held the atomic system together. He envisioned the hydrogen atom as consisting of a central nucleus containing a single positive charge, known as a proton, around which a body with a single negative charge, known as an electron, circled like a planet around a sun. The theoretical system, which has since been superseded in a number of important ways, is illustrated in Fig. 13.2. The proton and electron are very different in many ways, but as far as we know their charges are of *exactly* equal magnitude but opposite sign. We now know that the mass of the proton is 1836 times that of the revolving electron. The centripetal force that keeps the electron in orbit is the result of the electrostatic attraction between the positive and negative charges, which is immensely greater than the gravitational force between them.

The difference between atoms of various elements, according to Bohr's theory and all theories that have succeeded it, arises from differences in the number of protons and electrons. Note that under ordinary circumstances the number of protons in an atom's nucleus always equals the number of electrons in orbit. Matter is electrically neutral unless we do something to "charge it electrically."

This approach leads to the significant idea that only negative charges can move easily through a solid material because the negatively charged electrons are both less massive and closer to the outside of the atom than are the positively charged protons. Thus, when glass is rubbed with silk, some of the negative electrons are transferred from the glass to the silk, leaving the glass with a net positive charge and the silk with a net negative charge. Similarly, when rubber is stroked with fur, some of the negative electrons jump from the fur to the rubber, giving the rubber a net negative charge and the fur a net positive charge. Plainly, the charging of a substance by rubbing it with another is just a matter of redistributing charges between the two substances. Thus, it follows that an uncharged, electrically neutral object is not one that contains no electric charges; rather, the object contains an equal number of positive and negative charges, which produce the same external effect as if no charges are present. It is as simple as the equation

$$(+x) + (-x) = 0$$

One point worth noting is that the definition of the charge on glass as positive turned out to be unfortunate at best, and totally confusing at worst. It means, as we shall see in the next chapter, that electric charge is carried around circuits by negative particles—electrons. Geared as we are to the definition of current as the flow of positive charge, we must perform mental contortions to understand the movement of electric charge in the "wrong" direction. Of course, Benjamin Franklin defined rubbed glass as electrically positive in all good faith. He picked the wrong horse because he made the definition many years before the existence of such a particle as the electron was even suspected.

13.2 CONDUCTORS AND INSULATORS

Chemical materials are divided into two broad groups according to the ability of electrons to move through them. In one class of materials, which includes all metals, the negative electrons move readily from one atom to another. Individual atomic nuclei have little grasp on several of their electrons in these conductors, although the atoms en masse do retain their hold on the electrons. In the other class of substances, which contains such materials as glass, hard rubber, plastics, and dry wood, the individual atomic nuclei or molecules have a much firmer hold on their electrons. Even the outermost electrons of insulators have a tough time getting away from the atoms to which they belong.

This difference is the reason that wires of copper, aluminum, and other metallic materials are used to transport electric charge, and the supports that attach the wires to the poles consist of glass, ceramic, or some other insulator. It also explains why a charge remains at a specific point on an insulator without spreading, whereas an additional charge introduced to a neutral metal object that is insulated from its surroundings spreads throughout the surface of the metal object and remains there. Any excess charge placed on a conductor will move to the outer surface of the conductor and spread out over the surface because of the repulsive force between like charges.

This approach to conductors and insulators also explains observation 3 on page 383—the attraction between electrically charged and electrically neutral objects. When a strongly charged rod is brought close to a neutral insulator, the charges in the neutral object rearrange themselves. They become displaced, thus producing a "polarization," as demonstrated in Fig. 13.3. To envision this process on a relatively simple level, let us imagine that the atoms in the neutral object all consist of just one positive and one negative charge, bound to each other by the electrical force. This is shown in Fig. 13.3(a). When a positively charged object comes near a neutral insulator, such as the

The real situation is not quite as simple as we have pictured it. In fact, nature has provided us with materials whose ability to conduct charges spans the whole range from these two extremes: the conductor and the insulator. A particular group of materials, which lies somewhere between the good conductors and insulators, is known as semiconductors. Semiconductors are vital components of the modern electronics industry.

Figure 13.3 A charged object attracts a neutral object. The charges of a neutral body (a) are rearranged in the presence of a positively charged rod (b) and a negatively charged rod (c). Such a rearrangement is called polarization and results in a net attractive force between the neutral object and the charged object.

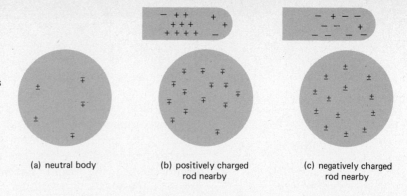

(a) neutral body   (b) positively charged rod nearby   (c) negatively charged rod nearby

pith ball in Fig. 13.3(b), electrostatic attraction will cause the negative parts of its atoms to move slightly toward the rod, repelling the positive parts and thus causing the polarization. Now, the electrostatic force is similar to the gravitation force insofar as the force of attraction between opposite charges, and the force of repulsion between identical charges, increases as the distance between the charges decreases. Thus, the attraction between the positive charges of the rod and the nearby negative regions of each atom in the uncharged object is greater than the repulsive force between the positive charges in the rod and the more remote positive nuclei of each atom. The net result is that the polarized neutral object is attracted to the positively charged rod.

A similar effect occurs when a negatively charged rod is placed near the uncharged insulator, as illustrated in Fig. 13.3(c). In this case the rod repels the negative parts of each atom, after which it attracts the positive parts with more force than it repels those negative parts that it has already pushed away. Again, the rod attracts the pith ball. In both cases we say that the charged rod has induced an electric charge on the neutral ball.

This polarization effect is much more pronounced when a charged insulator is brought near a metallic conductor. In this case the negative charges on the metallic conductor are free to move to the far end of the metal rod if the insulator has been charged negatively or to the near end if the insulator has been charged positively.

Returning to insulators, what happens when the attracted object comes into contact with the charged rod? It may fly right off again, as can be shown by using a (negatively) charged rubber rod to pick up the very small pieces of paper. The papers become polarized and are pulled toward the rod by the process outlined above. But once the pieces touch the rod, some of the negative charge drains off onto the paper, charging the paper negatively. According to the principle that like repels like, the papers immediately scatter away from the rod. This observation can also be used to show that there has been some charge tranfer to the paper.

Copying by Electrostatics

Science thrives on the distribution of knowledge. A theoretical advance or experimental finding by an individual scientist is of little value until her or his scientific peers have had the chance to review the piece of work and try to duplicate it. On a formal level, new knowledge is passed from one scientist to the rest of the community via science conferences, scientific journals, preprints, and reprints. But in recent years these formal methods have been joined by a much faster means of transporting knowledge from brain to brain—through copies of scientific papers and reports. So widespread has the use of duplication become that publishers of some small scientific journals have called for laws to restrict such copying.

The most common duplicating technique, known universally as the Xerox process after the company that markets it, relies on electrostatics. The method, invented in the 1930s, involves a type of material known as a photoconductive semiconductor. This is a substance that acts as a conductor of electric charge when it is exposed to light but acts as an insulator in the dark. The photoconductive semiconductor used today generally consists of the elements selenium, arsenic, and tellurium, in a coating on the Xerographic plate.

The plate is first charged in the dark, after which the pattern to be reproduced is projected onto it. Dark regions of the pattern project as dark regions on the plate, and because the coating is an insulator in the dark, these regions retain their electrostatic charges. Light regions on the original become light regions on the coating; the light makes these parts of the coating conductive, and the electric charges in the light regions are conducted away.

The next stage is to develop the dark areas, the only ones now charged, to make them visibly dark. This is achieved using a special dark powder that is charged electrostatically with charges opposite to those on the Xerox plate. These charges, and thus the flakes of powder, are attracted electrically to the charged regions of the plate, where they stick and outline the original pattern. Finally, the plate produces individual paper copies of itself, and hence of the original.

13.3 INTRODUCING THE ELECTROSCOPE

It follows from this analysis that the only sure test to determine whether an object is electrically charged is electrostatic repulsion. If an object whose electrical state is under investigation is attracted by a (positively) charged glass rod, the object may be either negatively charged or *neutral*. One can differentiate between the two possibilities only by bringing up a negatively charged object. If the object is neutral, it will again be attracted, but if it is negatively charged, it will be repelled. This principle is put to quantitative, as well as qualitative, use in an instrument called the electroscope.

The electroscope, shown diagrammatically in Fig. 13.4, consists of a metal knob atop a metal rod that is joined at its bottom to two parallel leaves of very thin gold foil. The rod is supported by insulating material in a vessel that also covers the gold leaves. If a negatively charged rod touches the instrument's metal knob, some of the negative charge moves into the electroscope. Because all the instrument's functional parts are metallic, this excess charge distributes itself throughout the

Figure 13.4 Enlarged view of the events in an electroscope charged by contact. In (a), the knob, stem, and leaves have an equal number of positive and negative charges and are neutral as a unit. In (b), charges move off the negative rubber rod that contacts the knob; some electrons are also repelled from the knob to the leaves that diverge. In (c), after the removal of the negative rod, the electrons regroup themselves toward the knob, leaving a net negative charge over the entire electroscope; the leaves will remain separated. Note that the sign of the net charge on the electroscope is the same as that on the charging rod, which is negative in this example.

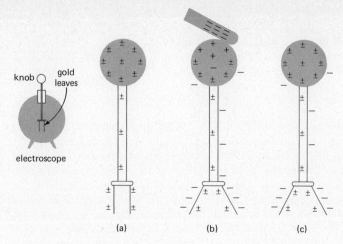

electroscope, including both leaves. The charges on the two leaves repel, causing the leaves to separate. They remain separated even when the charging rod is removed from the knob of the electroscope because the negative charge remains in the instrument. The result of touching a positively charged glass rod against the electroscope knob and then removing it appears exactly the same. Here, however, electrons move *from* the knob *to* the rod, leaving the instrument with a *net positive charge*.

Electroscopes can be charged by induction as well as by contact, as Fig. 13.5 demonstrates. When a negatively charged rod is brought close to the instrument's knob—but not so close that a spark leaps from the rod to the knob—it forces electrons away from the top of the electroscope. These electrons distribute themselves over the parts of the instrument well away from the negative rod, including the gold leaves. Hence, the leaves have an "excess" charge and separate. The electroscope remains neutral as a whole, but its charges are redistributed.

One can bring the leaves back to normal temporarily by touching the knob with one's fingers while the negatively charged rod remains close by. This action "grounds" the electroscope by providing an easy channel for the electrons at the knob to travel away from the leaves and knob to the earth, which is effectively an infinite sink for electrons (as well as an infinite source, when necessary). When one lets go of the knob and then removes the rod, as in Fig. 13.5, the electroscope stays *positively* charged, and its leaves separate once more. Note that this charging by induction involved no contact between the charging rod and the instrument and that the electroscope received a charge opposite to that on the rod. Another way to illustrate charging by induction is to place two metal spheres supported by insulating posts in contact with one another and then bring up a negatively charged rod. Separate the two spheres and you will find that the sphere that had been close to the

Figure 13.5 Enlarged view of the events in an electroscope charged by induction. In (a), the knob, stem, and leaves of the electroscope have an equal number of positive and negative charges and are neutral as a unit. In (b), the negative rod brought near the knob repels electrons toward the stem and leaves, giving a net negative charge below, as indicated by the divergent leaves. If the knob is grounded by touching the knob as in (c) while the negative rod remains nearby, the surplus electrons on the leaves will flow to earth, causing the leaves to collapse; electrons in the knob are still repelled by the nearby rod. After the finger is removed, followed by the withdrawal of the rod, the electrons redistribute themselves throughout the electroscope, leaving a net positive charge on the electroscope and the leaves diverged. Note that the sign of the net charge on the electroscope is the opposite to that on the charging rod.

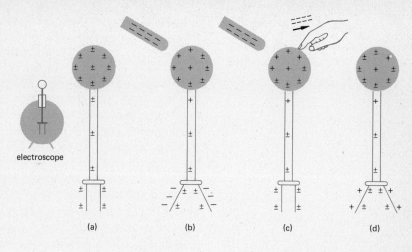

negatively charged rod is now positively charged, and the other sphere has a negative charge of equal magnitude. Here, the first sphere is equivalent to the electroscope knob, the other, to the earth.

13.4 COULOMB'S LAW

Experiments designed to measure the extent of forces of electrical attraction and repulsion, using electroscopes and other instruments, have established the basic law of electrostatics that was first advanced more than 190 yr ago by the French scientist Charles Augustin Coulomb. Inspired by Newton's law of universal gravitation, Coulomb looked for an inverse square relation; he found that **two unlike charges attract each other, and two like charges repel each other, with a force that is directly proportional to the product of the magnitude of the charges and is inversely proportional to the square of the distance between them.** Symbolically,

$$F = \frac{kQ_1Q_2}{r^2} \tag{13.1}$$

where F is the force between charges Q_1 and Q_2 that are separated by a distance r, and k is a proportionality constant whose value depends on both the units used for the other quantities and the nature of the medium between the charges.

This law is, of course, another example of an inverse square law. In fact, it is identical in form to Newton's law of universal gravitation, which we met in Chap. 4. Just as in the gravitational case, the value of r is the distance between the centers of the two objects. Equation (13.1) applies to real charged objects of significant size as long as their separation is large in comparison with their individual sizes. Coulomb found it necessary to introduce another quantity, or concept—that of

Describing the charge on the individual proton or electron as the smallest known is possibly simplistic. Theorists believe that quarks, which are possibly among the true building blocks of matter, possess charges one-third and two-thirds as strong as those of the proton and electron. No one has experimentally proved that belief.

electric charge—to describe electricity. One could no longer get by with mass, length, and time to study this force.

Coulomb's law clearly gives us the chance to define the unit of charge. The obvious choice would be the charge on the individual electron or proton because this is the smallest unit of charge known. Unfortunately, the electron was discovered too late to allow this rational decision to be made. Thus, we are stuck with a definition of the fundamental charge measured in coulombs, at least in the mks system, which is based on the unit for electric current, the ampere; we shall reach this concept in Chap. 14. For the moment, we need just note that this somewhat forced definition requires that k in air have a value of approximately 9.0×10^9 N·m²/C² (C stands for coulomb) in mks units. Life is simpler in the cgs units, for which k has the value of 1. The constant is also defined, on occasion, in terms of a property called permittivity, which measures the electrical properties of a medium in which electrical forces are acting. The relationship is

$$k = \frac{1}{(4\pi\epsilon_0)}$$

where ϵ_0 is the permittivity of free space.

The exact amount of the charge on the electron was the subject of much research in the first 30 yr of this century. The classic means of measuring this fundamental "electronic" charge is the Millikan oil drop experiment, which is outlined in the Special Topic at the end of this chapter. The most accurate application of this technique has produced a value for the electronic charge of 1.602×10^{-19} C. Inverting this figure, we see that a charge of 1 C contains $1/(1.602 \times 10^{-19})$, that is, 6.232×10^{18} electronic charges.

As mentioned, a major difference between electrical and gravitational forces is the existence of repulsion as well as attraction in the former. Note that we take account of this by attaching the appropriate positive or negative signs to the charges in the Coulomb force law. Thus, when like charges interact, their product is positive; when unlike charges interact, the product is negative. This means that **a negative sign indicates an attractive force** and **a positive sign a repulsive force.** However, force is a vector quantity whose *direction* is also symbolized by positive and negative signs. The whole situation is ripe for confusion. The only sure means of avoiding it is to bear in mind the two different meanings of the negative and positive signs and to try to avoid mixing them, as some of the following examples illustrate. If the underlying physics is kept in mind, the sign problem should be eliminated.

EXAMPLE 13.1

Two balls 2 cm apart each possess a positive charge of 6 μC (or 10^{-6} C). What is the force of repulsion between them?

Solution

$$F = \frac{kQ_1Q_2}{r^2} = \frac{(9 \times 10^9 \text{ N} \cdot \text{m}^2/\text{C}^2)(6 \times 10^{-6} \text{ C})(6 \times 10^{-6} \text{ C})}{(2 \times 10^{-2} \text{ m})^2}$$

$$= 810 \text{ N}$$

EXAMPLE 13.2

A charge $Q_2 = +4\,\mu\text{C}$ is placed midway between two charges $Q_1 = -8\,\mu\text{C}$ and $Q_3 = +10\,\mu\text{C}$, which are 8 cm apart, as shown in Fig. 13.6. Compute the net force on Q_2.

Solution

To solve this problem, we consider that the charges act in pairs, and compute the force F_1 of Q_1 on Q_2 and the force F_3 of Q_3 on Q_2:

$$F_1 = \frac{kQ_1Q_2}{r^2} = \frac{9 \times 10^9 \text{ N} \cdot \text{m}^2/\text{C}^2 (-8 \times 10^{-6} \text{ C})(4 \times 10^{-6} \text{ C})}{(0.04 \text{ m})^2}$$

$$= -180 \text{ N}$$

The negative sign tells us that this is an attractive force. In this example, it acts to the left toward Q_1 in the figure. The force on Q_2 owing to Q_3 is

$$F_3 = \frac{kQ_3Q_2}{r^2} = \frac{9 \times 10^9 \text{ N} \cdot \text{m}^2/\text{C}^2\,(10 \times 10^{-6} \text{ C})(4 \times 10^{-6} \text{ C})}{(0.04 \text{ m})^2}$$

$$= 225 \text{ N}$$

This is a positive force and hence a repulsive force to the left.

The total force $\mathbf{F} = \mathbf{F}_1 + \mathbf{F}_3$. Since $\mathbf{F}_1$ and $\mathbf{F}_3$ are in the same direction, the two can be added arithmetically:

$$F = 180 \text{ N} + 225 \text{ N} = 405 \text{ N} \quad \text{to the left}$$

Note that even though we did put the $+$ or $-$ signs into the individual expressions of Coulomb's law, we had to consider the direction in which these forces acted on Q_2 when it came time to add the force vectors. We assigned directions to the two forces consistent with attractive and repulsive forces as shown in Fig. 13.6

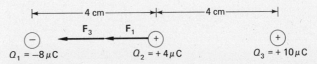

Figure 13.6 Force on a central charge Q_2 due to two unequal charges of the opposite sign, Q_1 and Q_3. The three charges are on a line.

EXAMPLE 13.3

Three charges, each of magnitude $+4 \times 10^{-6}$ C, are placed at three adjacent corners of a square, as shown in Fig. 13.7. If the side of the square is 8 cm, compute the net force on Q_2, the midcharge.

Solution

Let $\mathbf{F}_1$ be the force of Q_1 on Q_2 and $\mathbf{F}_3$ be the force of Q_3 on Q_2. Both are repulsive forces as indicated in the figure, and their magnitudes are given by Coulomb's law. Since the pairs of charges are both separated by the same distance and the charges all have the same value, F_1 must equal F_3 in magnitude:

$$F_1 = F_3 = \frac{kQ_1Q_2}{r^2}$$

$$= \frac{(9 \times 10^9 \text{ N} \cdot \text{m}^2/\text{C}^2)(4 \times 10^{-6} \text{ C})(4 \times 10^{-6} \text{ C})}{(0.08 \text{ m})^2}$$

$$= 22.5 \text{ N}$$

Clearly, $\mathbf{F}_1$ is the x component of the resultant force $\mathbf{F}$, and $\mathbf{F}_3$ is the y component of $\mathbf{F}$. Using the Pythagorean theorem, we can write the resultant force F as

$$F^2 = F_1^2 + F_3^2 = 2(22.5 \text{ N})^2 = 1010 \text{ N}^2$$

$$F = \sqrt{1010 \text{ N}^2} = 31.8 \text{ N}$$

$\mathbf{F}$ will be directed at 45° from the horizontal.

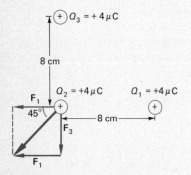

Figure 13.7 Three identical charges placed on adjacent corners of a square. The resultant force on the "central" charge acts along a diagonal of the square toward the outside of the square.

EXAMPLE 13.4

Compare the electrical force with the gravitational force between a proton and electron in the hydrogen atom. The mass of the proton is 1.67×10^{-27} kg, and it carries one fundamental charge, $Q_p = +1.602 \times 10^{-19}$ C. The mass of the electron is 9.11×10^{-31} kg, and it carries a charge of $Q_e = -1.602 \times 10^{-19}$ C. The two charges are separated by 0.53×10^{-10} m.

Solution

To compare the forces, we need only compute the ratio of the electrostatic force to the gravitational force and not worry about the sign:

$$\frac{F_e}{F_G} = \frac{\dfrac{kQ_pQ_e}{r^2}}{\dfrac{Gm_em_p}{r^2}} = \frac{kQ_pQ_e}{Gm_em_p}$$

$$= \frac{(9 \times 10^9 \text{ N} \cdot \text{m}^2/\text{C}^2)(1.602 \times 10^{-19} \text{ C})^2}{(6.67 \times 10^{-11} \text{ N} \cdot \text{m}^2/\text{C}^2)(1.67 \times 10^{-27} \text{ kg})(9.11 \times 10^{-31} \text{ kg})}$$

$$= \frac{2.31 \times 10^{-28}}{1.01 \times 10^{-67}} = 2.28 \times 10^{+39}$$

Plainly, the electrostatic force is such an immense amount larger than the gravitational force that gravitational attraction may be neglected in calculations of forces between elementary charges. Only if we were to compute the forces between very massive charged objects, such as astronomical bodies, would the gravitational force become important.

13.5 ELECTRIC FIELDS

One aid to our understanding of gravitational force in Chap. 4 was the introduction of the concept of the gravitational field, which has a strength of GM/r^2 at a distance r from a mass M. Because the form of the Coulomb's law equation for electrostatic force is mathematically identical to that for gravitational force in Newton's law of universal gravitation, it is not surprising that the concept of an electric field is both valid and useful in electrostatics.

According to this concept, any charged object modifies all space around it, extending to infinite distances. This modification leads to an electrical force on any other charge placed anywhere in space. We say that the first charge sets up an electric field that affects the second charge. Of course, the reverse also applies: the second charge sets up its own electric field that influences the first charge.

Pictorially, we can view the electric field, as British scientist Michael Faraday did, in terms of lines of force emanating from an individual charge (in the case of positive charges) or impinging upon the charge (in the case of negative charges), as shown in Fig. 13.8. This is an obvious visual aid, as are lines of gravitational force. In both cases the strength of the field is represented by the number of lines of force that cross a unit area that is at right angles to the lines. The concept of the electric field is also extremely important in helping us understand what is really happening, in an electrical sense, in situations that involve complicated distributions of charge, such as the charged parallel plates shown in Fig. 13.18.

The strength of the electric field E in the vicinity of a charge Q is defined in much the same way as the gravitational field, that is, as the electric force (F) per unit charge. The strength of the gravitational field was defined by the value of **g**, where $\mathbf{g} = \mathbf{F}/m$.

For the electric field:

$$\mathbf{E} = \frac{\mathbf{F}}{q} \tag{13.2}$$

The electric field is measured in newtons per coulomb. As we already implied in referring to the electric field lines, **the direction of an electric field is defined by convention to be the direction in which the electric force acts on a positive test charge.**

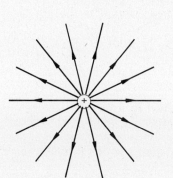

Figure 13.8 Electric field, indicated by lines of force in the vicinity of (a), an isolated positive charge, and (b), an isolated negative charge.

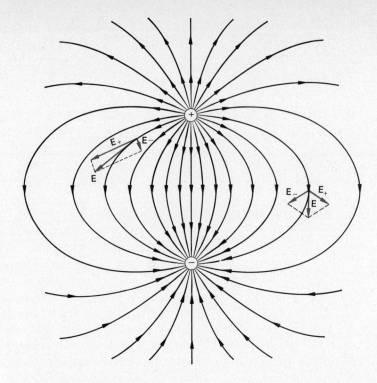

Figure 13.9 Electric field in the vicinity of two equal charges of opposite sign.

We can extend Eq. (13.2) by combining it with the Coulomb's law equation for electrical force. The coulomb force F produced on a charge q by the charge Q is given by $F = kQq/r^2$. Hence, the strength of the electric field at the point occupied by the charge q—the force per unit charge—is

$$E = \frac{F}{q} = \frac{kQq}{qr^2} = \frac{kQ}{r^2} \tag{13.3}$$

The strength of the electric field at any point, therefore, depends only on the magnitude of the charge Q that is responsible for the field and the distance which is between that charge and the point under investigation.

The vector nature of electric field becomes obvious when we encounter situations that involve more than one electric charge. The net electric field in such cases can only be obtained by adding the electric fields caused by the separate charges in a vector manner. The effect of mixing two opposite charges is shown in Fig. 13.9, and that of mixing two like charges appears in Fig. 13.10. In mixing two opposite charges the electric field lines all follow different routes between the positive and the negative charge, by leaving the positive charge and entering the negative charge. In mixing two like charges in which the two charges repel, not a single electric field line joins one charge to the other. Note

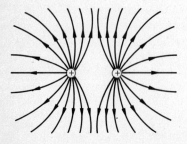

Figure 13.10 Electric field in the vicinity of two equal positive charges.

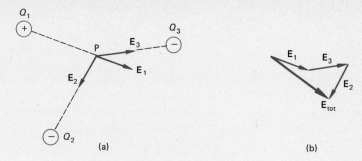

Figure 13.11 (a) Electric field at a point P in the vicinity of three charges. The electric field vectors due to each of the three charges must be added in a vector manner. (b) Here the three vectors $\mathbf{E}_1$, $\mathbf{E}_2$, and $\mathbf{E}_3$ are added graphically.

that the electric field at the midpoint between the two identical charges is zero because of superposition of equal and opposite fields.

Of course, one is not restricted to just two charges in calculating the value of the electric field at a point in space. Figure 13.11 provides an example of how to calculate the net intensity of the electric field at any point in space that is influenced by three separate charges. Note that we could reexamine Example 13.3 and find the net force on Q_2 by first finding the electric field at the point where Q_2 resides owing to the other two charges. The net force on Q_2 would then be given by

$$\mathbf{F} = \mathbf{E} Q_2$$

EXAMPLE 13.5

What is the intensity of the electric field at a distance of 3 m from a charge of $-18\ \mu C$?

Solution

Notice that we did not refer to the sign of the charge before doing the calculation. Instead, we applied it before doing the mathematics to determine the direction of $\mathbf{E}$. Putting the sign of the charge in the equation usually confuses the vector nature of the problem, as Examples 13.6 and 13.7 show.

Because the charge is negative, the electric field intensity is directed toward the charge as shown in Fig. 13.12. The magnitude is given by

$$E = \frac{kQ}{r^2} = \frac{(9 \times 10^9\ \mathrm{N \cdot m^2/C^2})(18 \times 10^{-6}\ \mathrm{C})}{(3\ \mathrm{m})^2}$$

$$= 1.8 \times 10^4\ \mathrm{N/C} \quad \text{toward the negative charge}$$

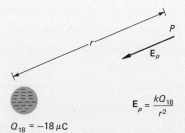

Figure 13.12 Computation of the electric field intensity at a distance r from the center of a $-18\text{-}\mu C$ charge.

EXAMPLE 13.6

Compute the electric field at the point midway between two charges of $+20\ \mu C$ and $+40\ \mu C$ that are 4 m apart. If a charge of $-5\ \mu C$ were placed at this point, what would be the force on this charge?

Solution

As shown in the sketch in Fig. 13.13, we must add the electric field intensities produced at the midpoint by these two charges. First, we calculate the magnitude of the electric field at the midpoint due to each separate charge:

$$E_{40} = \frac{kQ_{40}}{r^2} = \frac{(9 \times 10^9\ \text{N} \cdot \text{m}^2/\text{C}^2)(40 \times 10^{-6}\ \text{C})}{(2\ \text{m})^2} = 9 \times 10^4\ \text{N/C}$$

$$E_{20} = \frac{kQ_{20}}{r^2} = \frac{(9 \times 10^9\ \text{N} \cdot \text{m}^2/\text{C}^2)(20 \times 10^{-6}\ \text{C})}{(2\ \text{m})^2}$$

$$= 4.5 \times 10^4\ \text{N/C}$$

The electric fields in both cases are directed away from the charges producing them since that is the direction of the force on a positive charge placed at the midpoint. To "add" these electric field intensities in a vector manner, therefore, **we must actually subtract their magnitudes.** Taking the positive direction to the right,

$$E_{\text{net}} = E_{40} - E_{20} = [(9 \times 10^4) - (4.5 \times 10^4)]\ \text{N/C}$$
$$= 4.5 \times 10^4\ \text{N/C}$$

E_{net} has direction to the right at the midpoint.

To compute the force on the charge of $-5\ \mu C$ placed at the midpoint, we need only use the definition of the electric field intensity:

$$\mathbf{F} = Q\mathbf{E}$$
$$\mathbf{F}_{\text{net}} = Q\mathbf{E}_{\text{net}} = (-5 \times 10^{-6}\ \text{C})(4.5 \times 10^4\ \text{N/C}) = -0.225\ \text{N}$$

The direction of the force on this negative charge is to the left, opposite the direction of the electric field.

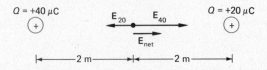

Figure 13.13 Electric field intensity at the midpoint between two positive charges of unequal magnitude.

EXAMPLE 13.7

Compute the electric field intensity at a point 3 m above a charge of $-6\ \mu C$ if a charge of $+46\ \mu C$ is also placed 6 m to the right of the $-6\ \mu C$ charge.

Figure 13.14 Computation of the net electric field due to two charges, $+46\ \mu C$ and $-6\ \mu C$, placed 6 m apart. The field is computed at point P, 3 m directly above the $-6\text{-}\mu C$ charge. The two electric fields must be summed in a vector manner. (a) Diagram. (b) Vector summation.

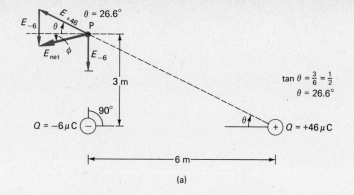

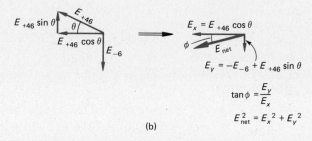

Solution

First, construct a diagram of the problem, as demonstrated in Fig. 13.14. Next, compute the magnitude of the electric field at point P due to each separate charge:

$$E_{+46} = \frac{kQ_{46}}{r_6^2} = \frac{(9 \times 10^9\ \text{N} \cdot \text{m}^2/\text{C}^2)(46 \times 10^{-6}\ \text{C})}{(36 + 9)\ \text{m}^2}$$

$$= 9.2 \times 10^3\ \text{N/C}$$

$$E_{-6} = \frac{kQ_{-6}}{r_6^2} = \frac{(9 \times 10^9\ \text{N} \cdot \text{m}^2/\text{C}^2)(6 \times 10^{-6}\ \text{C})}{(3\ \text{m})^2} = 6 \times 10^3\ \text{N/C}$$

To compute the net electric field intensity, we must add the two vectors. Their directions are shown in the diagram, which indicates the direction of the force on a positive charge placed at point P. The easiest approach to the addition is to take the x and y components of $\mathbf{E}_{+46}$, add these to the y component of $\mathbf{E}_{-6}$, and then determine the resultant electric field intensity (negative signs mean to the left or downward):

x component:

$$E_{+46} \cos \theta = (-9.2 \times 10^3)(\cos 26.6°) = (-9.2 \times 10^3)(0.894)$$
$$= -8.23 \times 10^3\ \text{N/C}$$

y component:

$$E_{+46} \sin \theta = (9.2 \times 10^3)(\sin 26.6°) = (9.2 \times 10^3)(0.447)$$
$$= 4.12 \times 10^3\ \text{N/C}$$

The net x component is -8.23×10^3 N/C to the left, and the net y component is $(+4.12 \times 10^3)$ N/C $-$ (6×10^3) N/C $= -1.88 \times 10^3$ N/C downward. The value of E_{net} is given by the Pythagorean theorem:

$$E_{net}^2 = E_x^2 + E_y^2 = (-8.23 \times 10^3)^2 + (-1.88 \times 10^3)^2$$
$$E_{net} = 10^3 \sqrt{67.7 + 3.53} = 10^3 \sqrt{71.26} = 8.44 \times 10^3 \text{ N/C}$$

E_{net} makes an angle ϕ with the horizontal:

$$\tan \phi = \frac{E_y}{E_x} = \frac{1.88 \times 10^3}{8.23 \times 10^3} = 0.228 \quad \text{or} \quad \phi = 12.9°$$

13.6 POTENTIAL DIFFERENCE AND POTENTIAL ENERGY

One important simplifying factor we introduced as an aid to our understanding of mechanical problems in Chap. 6 was the concept of energy and the law regarding its conservation. The use of energy, which is a scalar quantity, allowed us to deal simply with problems that could be described by the transformations between kinetic and potential energy and that would otherwise have required the complex addition of vector forces. We can again capitalize on the similarities between gravitational and electrical forces by introducing the concept of energy into our study of electrostatics. Continuing the analogy with gravitation, we move from electrical force to electrical energy via the concept of work. We must remember, of course, that the electrical case is slightly more complicated because of the existence of repulsive forces as well as attractive ones. Nevertheless, work is still defined as force times distance, $W = \mathbf{F} \cdot \mathbf{s}$.

Figure 13.15 illustrates the method of calculating electrical work. To move a positive charge of $+Q$ from point B to point A in the uniform electric field shown in this figure, along an electric field line in a direction opposite to that of the electric field, requires a quantity of work $\Delta W = \mathbf{F} \cdot \mathbf{s}$. This amount of work alters the energy of the charge, and if the charge moves so that the applied force which balances the electrical field produces no effective change in its speed, we know from our mechanical examples that only its potential energy—the energy of position—is altered. We can say that the work has produced a change in potential energy, which we can symbolize as $\Delta W = W_{BA}$. This action is analogous to lifting a mass from point B to a higher point A against the gravitational field of the earth.

In just the same way that physicists tackling mechanical problems deal with *differences* in potential energy rather than absolute values, they find it convenient to introduce a term known as the *potential difference* when they tackle electrical problems. **Potential difference is defined as the difference in potential energy per unit charge** and is given the symbol V. In the example illustrated by Fig. 13.15, the potential difference V_{AB} is the work per unit charge to move charge Q from B to

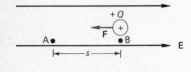

Figure 13.15 Work to move a charge $+Q$ a distance s from B to A against the electric field **E** (indicated by the parallel vectors). The force **F** is required to do the work $W_{BA} = \mathbf{F} \cdot \mathbf{s}$.

A, that is, W_{BA}/Q. Thus, as can be seen in the following equation:

$$W_{BA} = QV_{AB} \tag{13.4}$$

Since W_{BA} represents work, or energy, it is measured in units of joules. The potential difference is energy per unit charge and obviously requires a new unit of measurement. We use the term volt (V), a familiar household word named after the Italian electrical pioneer Alessandro Volta. According to its definition, **the potential difference between any two points is 1 V when 1 J of work is required to move a charge of 1 C between these two points.** Potential difference is related to energy in an analogous way to that in which the electric field is related to electrical force. Potential difference is often simply called the voltage.

Potential difference is related to energy in an analogous way to that in which the electric field is related to electrical force.

Physicists find it convenient to define a new unit of work and energy called the electronvolt (eV), which is related to the potential difference in volts. When an electron is placed in an electric field and allowed to accelerate—or "drop"—through a potential difference of 1 V, it gains an amount of kinetic energy equal to $\frac{1}{2}m_e v^2$, expressed in joules, where m_e is the mass of the electron and v is the speed it acquires in the drop. **Let us now define a unit of energy equivalent to the amount of kinetic energy that the electron acquires in accelerating through the potential difference of 1 V, and term it 1 eV.** According to the principle of conservation of energy, the gain in kinetic energy during the potential drop is identical to the loss in potential energy, which, in turn, is equal to the product of the charge of the electron e and the potential difference through which it falls, V. In other words,

$$\tfrac{1}{2}m_e v^2 = \text{charge on electron } (e) \times \text{potential difference } (V) \tag{13.5}$$

Calculation shows that the quantity eV for the acceleration of an electron through 1 V potential difference—that is, the value of the electronvolt we have defined—is 1.6×10^{-19} C times 1 V, which equals 1.6×10^{-19} J.

The electronvolt is an extremely convenient unit in studies of the ultrasmall world of the atom and subatomic particles. Physicists working in those areas often refer to these miniscule units of energy as "volts of energy" rather than electronvolts. A "1000-V electron," for example, is one that has been accelerated through a potential difference of 1000 V. Its energy is really 1000 eV, which is equivalent to 1.6×10^{-16} J ($1000 \times 1.6 \times 10^{-19}$).

In allowing the electron to fall through a potential difference in this discussion of the electronvolt, we were careful not to specify the relative values of the potential at the start and finish of the electron's motion. Positive charges move by themselves, or "flow," from points of high potential to points of low potential, whereas negative charges automatically flow from points of low potential to points of high potential, if left to themselves. Hence, an electron gains kinetic energy when allowed to move from a low potential to a higher potential.

The idea of the flow of charges through a potential difference brings up another significant point—the distinction between the electric potential at a point and the potential difference between two points. In most practical cases we are concerned only with changes in electric potential (V_{AB}) or potential energy (W_{BA}), just as we invariably deal with differences in potential energy in mechanical problems.

Let us calculate the potential energy and the potential in the vicinity of a single positive charge Q_1 that is fixed in a specific position. We know from Coulomb's law that the force between this charge and another charge Q a distance r away is given by

$$F = \frac{kQ_1Q}{r^2}$$

We also know that this equation is analogous to the equation for gravitational attraction between objects. In Chap. 5 we showed that the potential energy (E_p) associated with the force of attraction between two masses m_1, and m_2 is given by the equation

$$E_p = \frac{-Gm_1m_2}{r} \tag{5.6}$$

Because the two force laws have exactly the same mathematical form—an inverse square relationship—we are justified in continuing to apply the analogy to acquire the potential energy in the electrical case. Hence, we have the equation

$$E_p = \frac{kQQ_1}{r} \tag{13.6}$$

We omitted the negative sign that accompanies gravitational mechanical potential energy because of the dual nature of electric charge. If Q is negative, for example, and Q_1 is positive, the force between them is attractive, as is the situation for the gravitational force between any two masses. However, Eq. (13.6) actually makes provision for that fact. By inserting the negative sign into the expression when we introduce the value of Q, we automatically arrive at a negative value of potential energy for this attractive force, just as we should. Note that if Q and Q_1 are both positive or both negative, resulting in a repulsive force, the potential energy will be positive. The potential energy given in Eq. (13.6) represents the work required to move the charge Q from an infinite distance away up to a distance r from the charge Q_1.

This *potential energy* at a point can be converted to the *"potential"* at that same point by dividing the charge Q into Eq. (13.6) (remember that the potential is the potential energy per unit charge). Hence, the potential at point r caused by charge Q_1 is given by

$$V = \frac{E_p}{Q} = \frac{kQQ_1}{rQ} = \frac{kQ_1}{r} \tag{13.7}$$

In this equation for the electric potential at a point in the vicinity of a charge Q_1, we must also insert the sign of the charge because the potential, or energy per unit charge, is a scalar quantity rather than a vector. When Q_1 is positive, the potential increases as r becomes smaller and smaller, that is, the potential rises as one approaches a positive charge. Inclusion of the sign is most important in situations involving more than one point charge. The total potential at any particular point in such cases is calculated by adding the individual potentials due to all charges with their signs included:

$$V = \frac{kQ_1}{r_1} + \frac{kQ_2}{r_2} + \frac{kQ_3}{r_3} + \cdots$$

Electrostatic production of charge imbalance, although important in many experiments, is generally limited in practical application. Not until scientists had learned to maintain potential differences by other means did electricity have any real impact outside the laboratory. But the Short Subjects on pages 387 and 404 indicate two modern uses of electrostatics.

How do we create a potential or a potential difference? The secret is to create an imbalance of charge in some way. An obvious method is that mentioned at the beginning of this chapter: we can produce a potential difference electrostatically by rubbing a rubber rod with fur, a glass rod with plastic wrapping foil, or clean, dry hair with a comb. More practically, we can use a battery or dry cell, which develops a potential difference by chemical means, or a generator, which creates it mechanically.

EXAMPLE 13.8

A charge of $+2\ \mu C$ lies 20 cm away from another charge of $+4\ \mu C$. Calculate the (a) potential energy of this combination of two charges and (b) change in potential energy if the $+2$-μC charge is moved to a distance of 8 cm from the 4-μC charge.

Solution

(a) The potential energy of the system when the two charges are 20 cm apart is given by Eq. (13.6):

$$E_p = \frac{kQQ_1}{r} = \frac{(9 \times 10^9\ \mathrm{N \cdot m^2/C^2})(+4 \times 10^{-6}\ \mathrm{C})(+2 \times 10^{-6}\ \mathrm{C})}{(0.2\ \mathrm{m})}$$

$$= 0.36\ \mathrm{J}$$

(b) The potential energy when the charges are at a distance of 8 cm is again calculated by Eq. (13.6):

$$E_p = \frac{kQQ_1}{r} = \frac{(9 \times 10^9\ \mathrm{N \cdot m^2/C^2})(+4 \times 10^{-6}\ \mathrm{C})(+2 \times 10^{-6}\ \mathrm{C})}{(0.08\ \mathrm{m})}$$

$$= 0.90\ \mathrm{J}$$

The change in potential energy is $0.90\ \mathrm{J} - 0.36\ \mathrm{J} = +0.54\ \mathrm{J}$. This is an increase in potential energy since it takes external work to push the two charges of like sign from 20 cm apart to 8 cm apart. The force between the charges is repulsive.

EXAMPLE 13.9

Calculate the (*a*) potential at a point 20 cm from a $+4\text{-}\mu C$ charge and (*b*) potential energy of a $-2\text{-}\mu C$ charge placed at this point.

Solution

To calculate the potential produced by the $+4\text{-}\mu C$ point charge, we use Eq. (13.7):

$$V_4 = \frac{kQ_1}{r} = \frac{(9 \times 10^9 \text{ N} \cdot \text{m}^2/\text{C}^2)(+4 \times 10^{-6} \text{ C})}{0.2 \text{ m}} = 1.8 \times 10^5 \text{ V}$$

(*b*) The potential energy when a $-2\text{-}\mu C$ charge is placed at this point 0.2 m from the $+4\text{-}\mu C$ charge is calculated using Eq. (13.6):

$$E_p = QV_4 = -2 \times 10^{-6} \text{ C} \times 1.8 \times 10^5 \text{ V} = -0.36 \text{ J}$$

EXAMPLE 13.10

Two $5\text{-}\mu C$ charges, one positive and the other negative, are separated by 20 cm, as shown in Fig. 13.16. Find the electric potential at (*a*) point A, midway between the two charges, and (*b*) point B, which is on the plane midway between the charges but 5 cm above the point A. (*c*) Compute the electric field intensity at both points A and B.

Figure 13.16 Computation of the electric field due to two charges equal in magnitude but opposite in sign at (a) point A midway between the two charges and (b) point B midway between the two charges but 5 cm above the line joining the two charges.

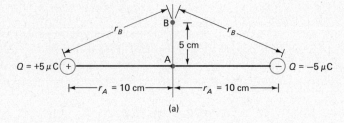

(a)

(b) to determine the electric field intensity at A:

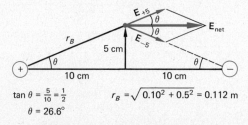

$\tan \theta = \frac{5}{10} = \frac{1}{2}$

$\theta = 26.6°$

$r_B = \sqrt{0.10^2 + 0.5^2} = 0.112$ m

(c) to determine the electric field intensity at B:

Solution

(a) The potential at point A is the sum of the potentials produced by the $+5$- and -5-μC charges as calculated using Eq. (13.7). The potentials are not vectors and so can be added directly if we insert the sign of the charges in the equation:

$$V = \frac{kQ_{+5}}{r_A} + \frac{kQ_{-5}}{r_A} = \frac{(9 \times 10^9 \text{ N} \cdot \text{m}^2/\text{C}^2)(5 \times 10^{-6} \text{ C})}{0.1 \text{ m}}$$

$$- \frac{(9 \times 10^9 \text{ N} \cdot \text{m}^2/\text{C}^2)(5 \times 10^{-6} \text{ C})}{0.1 \text{ m}}$$

$$V = 0 \text{ V}$$

(b) To find the potential at point B, we need the distance r_B as indicated in Fig. 13.16:

$$r_B = \sqrt{10^2 + 5^2} = \sqrt{125} = 11.2 \text{ cm} = 0.112 \text{ m}$$

$$V = \frac{(9 \times 10^9 \text{ N} \cdot \text{m}^2/\text{C}^2)(5 \times 10^{-6} \text{ C})}{0.112 \text{ m}}$$

$$- \frac{(9 \times 10^9 \text{ N} \cdot \text{m}^2/\text{C}^2)(5 \times 10^{-6} \text{ C})}{0.112 \text{ m}} = 0$$

$$V = 0 \text{ V}$$

(c) To compute the intensity of the electric field at points A and B, one must account for the vector nature of the electric field. Figures 13.16(b), (c) are vector diagrams that illustrate how we must add the two vectors. It is clear from these two diagrams that the electric field is not zero; rather, it is finite and directed to the right. To compute the electric field intensity at A, we add the following:

$$E = \frac{kQ_{+5}}{r^2} + \frac{kQ_{-5}}{r^2}$$

$$= \frac{(9 \times 10^9 \text{ N} \cdot \text{m}^2/\text{C}^2)(5 \times 10^{-6} \text{ C})}{(0.1 \text{ m})^2}$$

$$+ \frac{(9 \times 10^9 \text{ N} \cdot \text{m}^2/\text{C}^2)(5 \times 10^{-6} \text{ C})}{(0.1 \text{ m})^2}$$

$$= 2 \times 4.5 \times 10^6 \text{ N/C}$$

9.0×10^6 N/C   to the right

To compute the electric field at point B, we can resolve the vectors $\mathbf{E}_{+5}$ and $\mathbf{E}_{-5}$ into their x and y components. We can simplify the calculation by realizing that the magnitude of $\mathbf{E}_{+5}$ is the same as that of $\mathbf{E}_{-5}$. Since they make the same angle θ with the horizontal, the y components will cancel; the y component of $\mathbf{E}_{+5}$ acting up negates the

The Electrostatic Accelerator

Despite its general image as a prehistoric form of the electricity with which we are familiar today, electrostatics still has plenty to offer the up-to-date world of the 1980s physics laboratory. Even now, the majority of particle accelerators that are used to spark elementary particles to extraordinarily high energies for fundamental studies of the underlying nature of matter are electrostatic devices.

The idea of using electrostatic processes to generate and store large amounts of charge was first put forth in 1931 by Robert J. van de Graaff, a physicist at the Massachusetts Institute of Technology. His device looked rather like half a dumbbell and used a silk belt driven by an electric motor to pick up electrostatic charges generated in the air by discharge of electricity and to transfer the charges to a large, spherical, isolated terminal. This process proved capable of charging the terminal up to hundreds of thousands of volts potential. In 1932, British physicists John Cockcroft and Ernest Walton independently produced their own high-voltage machine that could give protons 700 kV of energy. This is considered small in present-day terms, but at the time the machine was the first capable of giving protons enough energy to induce nuclear reactions.

One major problem of the first generation of electrostatic accelerators was the danger of sparks. To overcome this, modern machines are surrounded by some appropriate insulating gas, such as carbon dioxide or sulfur hexafluoride. The generators have also been improved by the substitution of rubber-coated cotton for the silk in the belts and by the use of wires that actually touch the belt, instead of ionized air, to provide the electrical charges which the belt transports to the terminal.

Modern electrostatic accelerators are, not surprisingly, able to obtain far more voltage than their predecessors, through the use of better materials and better techniques. One high-power variation of van de Graaff's original machine is known as the tandem van de Graaff accelerator. It strips two negative charges from particles that start out with single negative charges and thus sends on a beam of very high-energy, positively charged particles. The most impressive of these machines, given the apt name of the double emperor machine, can produce protons with energies up to 30 MeV (30 million electronvolts). In the 1960s, before the advent of a whole new generation of particle accelerators, the double emperor tandem van de Graaff machine (see the accompanying photograph) at Brookhaven National Laboratory on Long Island, New York, reached the highest voltages of any sort in the world.

Double Emperor Accelerator. (*Brookhaven National Laboratory*)

---

y component of $\mathbf{E}_{-5}$ acting down. The x component of $\mathbf{E}_{+5}$ equals the x component of $\mathbf{E}_{-5}$, and they add together. Hence,

$$E_{\text{net}} = E_{+5} \cos \theta + E_{-5} \cos \theta = 2\,|E_{+5}|\cos \theta$$

$$= \frac{2kQ_5}{r_B^2} \cos\theta$$

$$= \frac{2(9 \times 10^9 \text{ N} \cdot \text{m}^2/\text{C}^2)(5 \times 10^{-6} \text{ C})}{(0.112)^2} \cos 26.6°$$

$$= 6.42 \times 10^{+6} \text{ N/C}$$

13.7 POTENTIAL DIFFERENCE AND POTENTIAL GRADIENT

We already know that force and work (or energy) are related in electrical as well as mechanical terms. Obviously then, the electric field, which is the force per unit charge, must be related in some similar fashion to the potential, which is energy per unit charge. We can obtain this relationship by studying Fig. 13.17, which indicates the use of work to move a positive charge Q from point B to point A in an electric field **E**.

To move charge Q from B to A requires a force $\mathbf{F}_{app}$, as shown in the figure. The electrical force on the charge at B is $F_Q = QE$ acting tangentially to the electric field line at B. Thus $\mathbf{F}_{app}$ must be equal in magnitude to $\mathbf{F}_Q$, although it is oppositely directed. If we assume that F is constant from B to A, then the work done to move the charge from B to A, a distance $\overline{BA}$, is given by

$$W_{BA} = F_{app}(\overline{BA}\cos\theta) = QE(\overline{BA}\cos\theta)$$

It should be pointed out that in our example, to move a positive charge from point B to A does require work by the applied force $\mathbf{F}_{app}$. If we assume that the charge Q does not change its speed in the process, then this work W_{BA} equals the change in potential energy of the charge as it moves from B to A. We let V_A equal the potential at A and V_B equal the potential at B. Then

$$W_{BA} = QV_A - QV_B$$

$$QV_A - QV_B = QE(\overline{BA}\cos\theta)$$

Solving this equation for the component of **E** in the direction $\overline{BA}$ (that is, $-E\cos\theta$), we have

$$-E\cos\theta = \frac{-(V_A - V_B)}{\overline{BA}} = \frac{-\Delta V}{\Delta s} \tag{13.8}$$

The quantity ΔV is the potential difference between A and B, and Δs is the displacement from B to A. This quantity $\Delta V/\Delta s$ represents the rate of change of V in the direction of $\overline{BA}$. If this direction is chosen opposite to the direction of the lines of force ($\theta = 0$), it is called the *potential gradient*. If the displacement is in the direction of the field ($\theta = 180°$), $E = -\Delta V/\Delta s$. The negative sign indicates that a positive

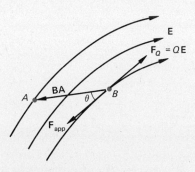

Figure 13.17 Work done in moving a positive charge from B to A.

Figure 13.18 Electric field between two parallel, charged plates. The electric field and the potential difference are related by $E = (V_A - V_B)/d$.

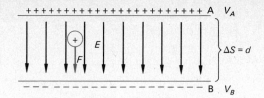

charge moves when let free from a position of higher to one of lower potential, as we noted previously. Hence, E is positive when the value of $\Delta V/\Delta s$ is negative, as this example illustrates.

We can provide a simple and very important example of the use of the potential gradient and Eq. (13.8) by examining the two charged, parallel, conducting plates shown in Fig. 13.18. Here, a charge of $+Q$ is uniformly distributed over plate A and an equal charge $-Q$ is uniformly distributed over plate B. This is called a parallel-plate capacitor, and in this example it carries a charge Q. Essentially an amount of charge is moved from one neutral plate to the other, producing the imbalance. The electric field **E** between charged parallel plates is uniform at all points between the plates away from the edges, at least to a good approximation. A free positive charge would move from the positively charged plate A to the negatively charged plate, that is, from high potential V_A to the lower potential V_B, along the electric field lines. Thus,

Capacitors are important in present-day electronics, as we shall see in Chap. 17.

$$E = \frac{-\Delta V}{\Delta S} = \frac{-(V_B - V_A)}{d} = \frac{V_A - V_B}{d} \quad (13.9)$$

Therefore,

$$\text{Electric field intensity} = \frac{\text{difference in potential}}{\text{plate separation}}$$

The quantity $V_A - V_B$ is a positive number. Note that the units of E as expressed in Eq. (13.9) are volts per meter, which is equivalent to newtons per coulomb. Volts per meter is the more common unit.

EXAMPLE 13.11

The potential difference between two plates 0.5 cm apart is 1000 V. What is the electric field intensity between the plates?

Solution

$$E = \frac{-\Delta V}{\Delta S} = \frac{1000 \text{ V}}{0.005 \text{ m}} = 2 \times 10^5 \text{ V/m}$$

Since no reference is made to direction or the polarity of a charge in this problem, the value of E is taken as positive. We are actually only concerned with the magnitude of the electric field between the plates.

13.8 EQUIPOTENTIALS AND ELECTRIC FIELD LINES

The potential at all points in the vicinity of a charge distribution may be mapped graphically using equipotential surfaces, or lines on a plane. An equipotential surface is a collection of points that are at the same potential. It is similar to a height contour on a map. The equipotential surfaces about a single positive point charge or a single negative point charge, for example, are series of concentric spheres. Figure 13.19 shows the equipotential (dashed lines) and the electric field lines (solid lines) for two charges of opposite sign and for two charges of like sign. Obviously, the electric force does no work on a charge moved along an equipotential line or equipotential surface since the potential energy of the charge remains constant. **The lines of force, or field lines, are perpendicular to the equipotential surfaces.**

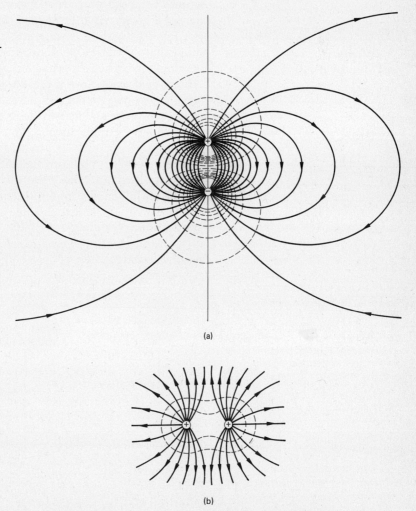

Figure 13.19 (a) Equipotential and lines of force of a dipole (two charges of equal magnitude but of opposite sign). The lines of force are everywhere perpendicular to the equipotentials. (b) Electric field in the neighborhood of a pair of like charges. The equipotential surfaces are shown as dashed curves.

The outer surface of a conductor such as a metal sphere or any oddly shaped conducting object is an equipotential surface. If such an object is placed in an electric field, or if a charge is placed on an isolated conductor, the free electrons in the object quickly redistribute themselves over the metal conducting surface. Since the charges are at rest, the potential must be the same everywhere. The electric field lines intersecting a conductor must always be perpendicular to the surface since any component of electric field along the surface would automatically produce motion of charge and would indicate a potential difference.

In certain cases the distribution of charges on the surface necessary to keep it equipotential may be rather unusual. If a metallic sphere is charged negatively by rubbing a charged rubber rod against it, for example, the excess negative charge on the sphere is uniformly distributed, as shown in Fig. 13.20(a). But if the charged rubber rod is applied to an egg-shaped conductor, tests show that the surface charge density is greater near the pointed portions than near the flatter regions [see Fig. 13.20(b)]. In other words, charges tend to accumulate on the more sharply curved portions of a conductor. At sharp points, the accumulation of negative charge may be so great that the resulting high electric field may cause a breakdown of the air, resulting in charge "leaking off the points." If the excess charge is positive, negative charges from the air "leak to the points." This electrical effect actually has a life-giving application in protecting buildings from the effects of lightning.

The descent of raindrops through the air in a thunderstorm often causes an accumulation of negative charge on the rising moist air, leaving the falling raindrops with a net positive charge, as shown in Fig. 13.21. As a result, the bottom of the cloud becomes negatively charged. With a sufficient buildup of opposite charges, a discharge of lightning in the form of a huge electric spark may flash between the upper positively charged part of the cloud and the lower negatively charged

Charges tend to accumulate on the more sharply curved portions of a conductor.

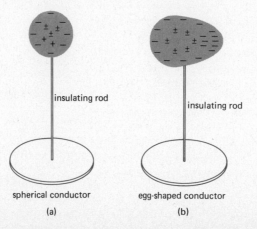

Figure 13.20 Charge distribution on metal surfaces. The surface charge density on a spherical conductor (a) is uniform; but on an egg-shaped conductor (b) it is nonuniform, being greater on the sharpest points.

13.8 Equipotentials and Electric Field Lines 409

Figure 13.21 Rising moist air becomes negatively charged. Falling raindrops become positively charged.

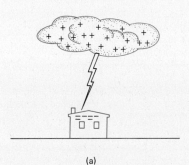

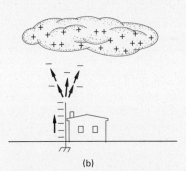

Figure 13.22 Lightning and lightning rods. (a) The induced charge on a building below a positively charged cloud results in lightning. (b) The discharge from a lightning rod neutralizes the charges of a cloud.

part. This is relatively harmless, but on frequent occasions the wind blows away the lower part of the cloud, scattering the negatively charged particles throughout the atmosphere. Then the remaining positively charged upper portion of the cloud attracts negative charges to the top of a building or tree beneath it by induction, as in Fig. 13.22(a). If this buildup of opposite charges reaches a certain limit, a sudden discharge of lightning streaks between the building and the cloud. The building is said to be "struck" by lightning.

Canny technologists use the leakage of charges to protect buildings from lightning. A thin, pointed lightning rod is fastened to the building; the lower end of the rod is buried deeply enough to extend into moist earth. When a positively charged cloud moves over the building, as in Fig. 13.22(b), negative charges move up the rod to the top by induction and quietly leak off the upper point toward the positively charged raindrops. (Remember, the earth is a source of electrons.) As the negative charges reach the raindrops, they neutralize the positive charges on the drops, stopping any dangerous buildup of opposite charges.

On occasion, clouds become negatively charged and hence induce a positive charge on the earth below. If a discharge of lightning takes place in these circumstances, the negative charges move to the earth in the process. Meterologists believe that electrons are the prime movers, both literally and figuratively, in both forms of lightning.

Many of the details regarding lightning and the formation of charges on clouds remain misty and extremely controversial. However, the general picture is fairly well established. It is clear, for example, that electric discharge or lightning usually makes a beeline for points of

Figure 13.23 Bolt of lightning. The electrons are the prime movers. (*Institute of Atmospheric Physics, University of Arizona, Tucson*)

highest electric field above the earth in the region beneath charged clouds. Hence, a person standing in a large, open field during a thunderstorm is taking a dangerous risk. Indeed, a very high proportion of deaths from lightning occur on open golf courses. Furthermore, people are well advised not to seek shelter from thunderstorms under trees, which may be struck and felled by lightning.

SUMMARY

1. Like gravitation, the electric force acts at a distance. Unlike gravitation, the electric force involves both attraction and repulsion.
2. Electric charge can be produced by rubbing together various parts of different materials. Experiments of this sort show that two different types of electric charge exist.
3. Opposite charges attract, and like charges repel each other.
4. Charged objects attract uncharged objects.
5. A charge equal in magnitude to that on the negatively charged electron is known to be the smallest electric charge obtainable.
6. The only true test of whether an object is electrically charged is electrostatic repulsion. Whether an object is charged can be investigated using an instrument known as the electroscope.
7. According to Coulomb's law, the force between two point electric charges is proportional to the product of their magnitudes and inversely proportional to the square of the distance between them.
8. Electric charges produce electric fields in the same way that masses produce gravitational fields.
9. The electrical potential difference is the difference in potential energy per unit charge. Potential difference is usually measured in volts. A potential difference is produced by creating an imbalance of electric charge.
10. Discharge of lightning to earth is a direct result of charge imbalances. It can be restrained by providing a continuous route for removing the imbalance.

QUESTIONS

1. An electroscope has a positive charge. (a) Describe the condition of the electroscope leaves. Describe what happens to the electroscope (b) if a positive charge is brought near the electroscope, (c) if a negative charge is brought near the electroscope.
2. In terms of electrons and their motion, describe what happens when a negatively charged rod is brought near an uncharged small piece of paper.
3. In terms of electrons and their motion, describe what happens when a positively charged rod is (a) brought near an uncharged electroscope and (b) then touched to the electroscope.
4. When you wear leather-soled shoes and walk across a rug in a dry room, you may become charged negatively. In terms of the motion of electrons, describe what happens to you and the rug. What happens in terms of the motion of electrons when you touch a metal door knob?
5. It is not uncommon in dry weather for a long-haired person to comb his or her clean hair and find that the hair becomes electrically charged. Explain what takes place. What is the state of the comb?
6. You are given a charged object. How can you tell the sign of its charge?
7. Describe, in terms of electrons, what will happen when a piece of silk is used to rub a glass rod to charge the rod positively. What happens to both the rod and the silk? If the charged rod were to touch a metal radiator pipe, what would happen? Explain.
8. Describe the steps required to charge a single isolated metal sphere positively using a negatively charged rubber rod. Note how the charges move.
9. Explain why gravitational forces are usually neglected when computing the force between two charged objects. How does the electrostatic force of attraction between two small charged objects change if (a) the separation between the two is doubled? (b) The charge on both is doubled? (c) A positive charge of identical sign and of magnitude equal to that originally on each object is placed on both objects? (d) The mass of the two objects is tripled?
10. List the similarities and differences among the electrostatic force law and the gravitational force law.
11. The accumulation of static electricity in an operating room of a hospital must be avoided because a discharge could produce an explosion owing to the presence of almost any

of the common anesthetics. Should the floor be a conductor or an insulator? Should the humidity in the room be high or low? Explain.

12. In the good old days, trucks carrying gasoline would drive about with a chain dragging on the ground. Nowadays, the truck tires on these vehicles are made to conduct electricity. Why is this important?

13. People inside a steel-framed building are safe from lightning. Why?

14. Explain how a lightning rod protects a building.

15. Explain why lightning rods are pointed, that is, they have hemispheres of very small radius at their end, rather than spheres of 3 or 4 cm in radius. Discuss in terms of the size of the electric field possible at the end of the rods. Note that it is the high electric field producing, in a sense, the large electrostatic force that causes breakdown of the air.

16. Why must the surface of a metallic conductor be an equipotential surface? Does this mean that there are an equal number of excess charges per unit area on an isolated conductor regardless of its shape? Explain.

17. Does an object with twice the electric potential energy of another object have twice the potential? Comment on the reverse statement also.

18. Is it possible for the electric potential to be zero while the electric field intensity at the same point is nonzero? What does this mean in terms of energy?

19. How is the direction of the electric field related to the difference in potential between two points? That is, is it from higher to lower or lower to higher potential? Which way do electrons move, from higher to lower potential or the reverse?

PROBLEMS

Coulomb's Law

1. How many electrons make up 1 C of charge?

2. What is the total charge in coulombs of a mass of protons equal to 1 kg?

3. Two small metal spheres are placed 0.50 m apart. What identical charge must be placed on both spheres so that the force of repulsion between them is equivalent to your weight?

4. A positive and a negative point charge are initially 4 cm apart. When they are moved closer together so that they are only 1 cm apart, how will the force between them change?

5. Suppose we have two small, charged spheres placed a distance D apart. If the charge on each is $+Q$, the force between them will be F. What will be the new force in each of the following cases? Only the parameters mentioned are changed from the original values: (*a*) The charge on one of the spheres is changed to $-Q$. (*b*) The charge on one of the spheres is doubled. (*c*) The charge on both the spheres is doubled. (*d*) The distance between the spheres is doubled. (*e*) The distance between the spheres is reduced to $\frac{3}{4}D$. (*f*) One sphere is moved to the opposite side of the first sphere, but the spheres are still separated by a distance D. (*g*) If one sphere is released, what will happen to it and how will the force between the two spheres change?

6. Two point charges of -3×10^{-9} C and $+6 \times 10^{-9}$ C are placed 0.3 m apart under vacuum. Compute the electrostatic force between them.

7. The force between two pith balls is 0.2 N. If one carries a charge of $+8\,\mu$C and the other a charge of $+24\,\mu$C, what is the separation between the charges?

8. Suppose identical amounts of negative charge were placed upon the earth and moon by a "great giant." What charge would be required so that the resulting electrostatic force would just equal the gravitational attraction between the moon and the earth?

9. Using a positron instead of a proton, it is possible to form "atoms" of positronium where the electron "orbits" about the positron. The positron is the "antimatter" counterpart of the electron. It has a mass equal to the mass of an electron and a positive charge equal in magnitude to the negative charge on the electron. (*a*) Compare the gravitational attraction with the electrostatic attraction for the positronium atom. (*b*) Comment on the adaptability of the Bohr model of the hydrogen atom to positronium. Does the electron "orbit" about the positron?

10. A charge of $+60\,\mu$C is placed on one small metal sphere and a charge of $-30\,\mu$C is placed on another identical sphere. The two spheres are 0.3 m apart. (*a*) What is the force of attraction between them? (*b*) If the two metal spheres are brought in contact with each other and then are moved by their insulated bases to 0.3 m apart, what will be the new force between them?

11. Two point charges, $+20\,\mu$C, $-20\,\mu$C, are 10 cm apart. What is the force on a $+4$-μC point charge placed midway between them?

12. Three point charges are placed in a line. The end charges are each $+20\,\mu$C and separated by 40 cm. The third charge is $+2\,\mu$C and is placed midway between the two charges. What is the force on the $+2$-μC charge?

13. Two point charges, each $+20\ \mu C$, are placed 40 cm apart. A third charge of $+20\ \mu C$ is placed on a line between these two at a point 10 cm from the charge on the left. (a) What is the force on the charge 10 cm from the left end? (b) What is the force on the charge at the left end?

14. Two point charges, one $+20\ \mu C$, the other $-20\ \mu C$, are placed 10 cm apart. Where should a $+4$-μC point charge be placed along the line joining the two charges so that the force on the $+4$-μC charge is zero?

15. Two charges, each $+40\ \mu C$, are placed on 5-g pith balls separated by 2 m. If the pith balls are released and experience no other forces, what will be their acceleration just after release?

16. A 1-, 9-, and 16-μC charge are placed on the vertices of a 3- by 4- by 5-m right triangle, as shown in the figure. Calculate the magnitude of the force on the 1-μC charge.

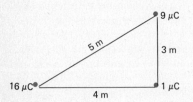

17. Two identical pith balls, each of mass 0.5 g, are attached to the end of fine thread of length 0.5 m. Each pith ball is charged identically. What charge is required to keep the strings at a 20° angle?

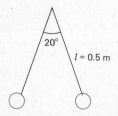

18. Three charges, each $+4\ \mu C$, are placed at the corners of a right triangle of sides measuring 3 by 4 by 5 cm. What is the force on the charge residing on the right angle?

19. Two silk threads are tied together at one end and hung to the ceiling. The threads are each 2 m long, and attached to the lower end of each are identical 0.2-g spheres. When the balls are equally charged, the threads separate by an angle of 10°. What is the charge on each ball?

20. A pair of identical charges of opposite sign is called an electric dipole. Find the force exerted by the dipole shown in the figure on a $+2$-μC charge placed at the point shown.

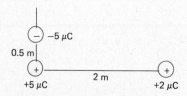

Potential, Potential Difference, Potential Energy, and Work

21. A charge of $+5\ \mu C$ is placed at the origin and a charge of $-45\ \mu C$ is placed at point $(+10\ m, 0)$. (a) Find a point along the x axis where the electric field is 0. (b) What would be the force on a $+25$-μC charge placed at this point? (c) What is the electrostatic potential at this point? (d) What would be the amount of work required to bring a $+25$-μC charge from infinity up to this point?

22. Two charges of $-1.0\ \mu C$ and $+2.0\ \mu C$ are placed on the x axis as shown. (a) What is the magnitude and direction of the electric force acting on the $+2.0$-μC charge? (b) What is the magnitude and direction of the electric field at point A on the x axis? (c) What is the magnitude of the electric potential at point A on the x axis?

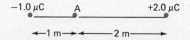

23. (a) Find the resultant electric field at point P in the diagram. (b) Find the force on a -4-μC charge placed at point P. (c) Find the potential of point P if the charges are placed as shown. (d) How much work would be required to move a $+2$-μC charge from infinity up to point P?

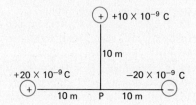

24. Two charges are placed and held at opposite corners of a square as shown in the figure. (a) Find the electric field at point A. Give its direction. (b) What would be the force on a $+2$-μC charge placed at A? (c) What is the potential at point A? (d) How much work is required to bring a $+2$-μC charge from infinity to point A? (e) If a $+2$-μC charge were placed at point A, what would happen to it when it is released?

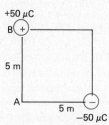

25. A $+5$-μC is placed and held at a distance 20 cm from a -20-μC charge. (a) Find a position a distance x away from the $+5$-μC charge along the line between the two charges where the *potential* is zero. (b) What is the electric field at the position found in (a)? (c) How much work is required to bring a $+2$-μC charge from infinity to the position found in (a)? (d) If a $+2$-μC charge were placed at the position found in (a)

and released, what would it do? Does your answer conflict with the fact that the potential energy is zero at that point? (e) Find the electric potential at point D in the figure.

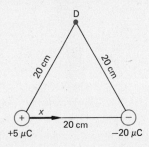

26. Find the potential at points A, B, C, and D in the vicinity of the two charges shown in Fig. 26.
27. How much work is required to move a charge of -4 C from a point at 118 V to a point at 114 V? Must external work be done, or does the electric field do the work?
28. In an x-ray tube, an electron is accelerated through a potential difference of 10 000 V. Find its kinetic energy just before it hits the target.
29. A lightning flash results from a potential difference of 10^9 V, and the flash consists of 20 C of charge moving from the cloud to ground. (a) How much energy is this in electronvolts and joules? If this energy were used to heat water from 20 to 100°C, (b) how much water could one heat with the harnessed lightning bolt?

Figure 26

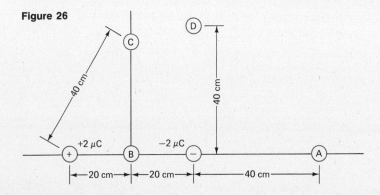

30. (a) What is the difference in potential between two points 20 and 40 cm away from a $+80$-μC charge? (b) How much work is required to move an electron from the 40 point to the 20-cm point? Does the amount of work depend upon the path taken to move the electron?
31. In a hydrogen atom, the electron is about 10^{-10} m from the proton. Compute the electrostatic potential energy associated with the atom. Answer in electronvolts and joules.
32. A large van de Graaff accelerator used in the study of the structure of nuclei accelerates electrons through a potential difference of 5 million V. What is the energy of the electrons released from such an accelerator? Give your answer in electronvolts and joules.
33. A proton approaches the nucleus of a gold atom that contains 79 protons. (a) If the proton is stopped just outside the nucleus of the gold atom a distance of about 10^{-14} m, compute the potential energy in electronvolts of the system. (b) If the proton is now accelerated by the nucleus of 79 protons, compute the ultimate speed of the proton when it is very far from the nucleus. Assume the gold atom remains at rest and does not recoil.
34. An α particle, which is a helium nucleus carrying a charge of $+2\,e$, is 10^{-10} m from a proton. (a) What is the force of repulsion between the two nuclei? (b) What is the potential energy of this system? (c) How fast would the proton be moving if it were allowed to move freely to a distance very far from the fixed α particle?

Electric Field and Electrostatic Force

35. Show that the units of an electric field N/C = V/m.
36. If the force on a -15-μC charge is 5 N to the right at some point in space, what is the electric field at that point?
37. The force on a $+2$-μC charge in an electric field is found to be 6 N, downward. Find the magnitude and direction of the electric field.
38. If an electron is placed between two horizontal plates, how large must the electric field between the plates be if the weight of the electron is just balanced by the electrical force acting on it?
39. A point charge $Q_1 = +8 \times 10^{-4}$ C is 6 m to the right of a second point charge $Q_2 = -3 \times 10^{-4}$ C. What is the electric field at a point midway between the two charges? (b) What would be the force on a $+4$-μC charge placed at the midpoint?
40. Plot the electric field versus r from 1 to 20 cm as you move away from a 5×10^{-9}-C charge.
41. A charge of $+8\,\mu$C is 4 cm away from a charge of $-4\,\mu$C. (a) What is the electrostatic force on each charge? (b) What is the electric field midway between the two charges? (c) What is the electrostatic potential at the midpoint?
42. A charge of $+8\,\mu$C is 4 cm from a charge of $+4\,\mu$C. (a) What is the electric force on each charge? (b) What is the electric field midway between the two charges? (c) What is the electrostatic potential at the midpoint?
43. A $+5$-μC charge is placed on a 0.4-g pith ball. The charged pith ball

is placed in a uniform electric field. The ball starts from rest and in 0.2 s has traveled a distance of 0.2 m. What is the magnitude of the electric field?

44. Identical point charges, each of size $+50\ \mu C$, are placed on the corners of a square. Compute the force on the charge placed at the bottom left-hand corner. (b) What is the force on a charge placed at the center of the square? Does the answer depend on whether the charge at the center of the square is positive or negative? (c) What can be said about the value of the electric field at the center of the square?

45. A pith ball of mass 0.5 g hung from a string 0.25 m long is attached to a vertical wall that is actually a positively charged metal plate. The pith ball becomes charged by contact with the wall and ends up with the string making a 30° angle with the vertical. If the charge on the pith ball is $+80\ \mu C$, what is the electric field due to the wall at the pith ball?

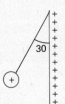

Potential Gradient

46. A pair of parallel charged plates are placed 10 cm apart; the potential difference between them is 1000 V. What is the potential gradient between the plates?

47. The electrical characteristics of a cell membrane can be approximated by a potential of about 0.1 V across two parallel plates of 10^{-7}-m separation. What electric field does this represent?

48. A potential difference of 1000 V exists between two parallel plates separated by 10 cm. An electron is released from the negative plate at the same instant that a proton is released from the positive plate. Compute their velocities just as they strike the opposite plates. How far from the positive plate do they pass each other?

49. What charge must be placed upon a 2-g object so that, when it is placed between two horizontal plates where the electric field is 500 V/m, the charge will remain "suspended" at rest?

50. An electron is placed at the center of two parallel plates. The separation between the plates is 10 cm. The electric field between the plates is 5 V/cm. (a) What is the potential difference between the plates? (b) What is the kinetic energy in joules and electronvolts of the electron just before it hits one of the plates, assuming it is released from rest at the midpoint between the plates? (c) Which plate does it hit? (d) Calculate the velocity of the electron just before it hits the plate.

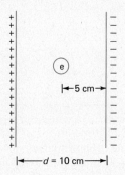

51. Two parallel metal plates are separated by 6 cm. By charging one plate negatively and the other with an equal quantity of positive charge, a uniform electric field of 10^3 V/m is established between the plates. (a) How much work is done against the electric field to move a $+2$-μC charge from the negative to the positive plate? (b) What is the change in potential energy due to moving the charge from one plate to the other? (c) What is the potential difference between the plates? (d) How much work is done by you to move a -2-μC charge from the negative to the positive plate?

Special Topic
How Millikan Measured the Ultimate Electric Charge

One of the most fundamental quantities in all physics is the basic unit of electric charge—the amount of electric charge that cannot be subdivided in any way. This unit is identical to the (negative) electric charge on the electron, the particle whose discovery by Professor J. J. Thomson in 1897 first indicated to physicists that the indivisible atom was not as unsplittable as they had believed all along. Thomson's measurements produced a value of e/m, the ratio of the charge of the electron to its mass, but for more than a decade neither the Cambridge professor nor his scientific colleagues could figure out a way to measure either property individually. Then, in 1909, University of Chicago physicist Robert A. Millikan solved the problem of getting at the charge of the electron with a remarkably ingenious experiment that involved application of physics from three different areas: forces, fluids, and electricity.

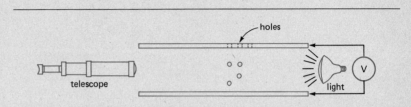

The obvious requirement for any sort of probe of the fundamental electric charge was some kind of vehicle that could carry this charge, for electrons were and still are unseeable in normal terms. Millikan hit on the idea of using small droplets, first of water and later, when he encountered difficulties with evaporation, of oil, that would pick up one, two, or just a few individual electrons or positively charged ions. He sprayed the drops through a number of tiny holes in the upper metal plate of the apparatus shown in the diagram. From there the drops moved down under the influence of gravity toward the lower plate a few centimeters away. Millikan used a battery to charge the two plates, so that each could be electrically charged with respect to the other. Thus, any oil drop that picked up a positive or negative charge—a process helped by beaming x-rays at the air between the plates—would fall under the influence of electrical forces as well as gravitational ones. By changing the relative electric charges on the two plates, Millikan could make his droplets dance up and down like a cowboy under fire. To view this movement, Millikan shone a bright light on the area beween the plates, which picked out the droplets as bright stars against a pitch black sky when viewed through a microscope.

In fact, gravitational and electric attraction were not the only forces that the droplets encountered. They also came under the influence of the buoyancy force of the fluid air and the air resistance, which balanced out the other forces and eventually caused the droplets to move up or down with a constant, terminal velocity. For an uncharged particle, the force equation is simply

$$W - B - kv_g = 0$$

where W is the weight of a droplet, B is the buoyancy force, v_g is the terminal velocity, and k is a constant derived from the Stokes' law expression for the terminal velocity of a droplet of radius r moving through a fluid with viscosity η, $kv = 6\pi\eta rv$.

Now, if m_0 is the mass of the drop, V its volume, and ρ_0 its density, we have the expression for W:

$$W = m_0 g = \rho_0 V g = \tfrac{4}{3}\pi r^3 \rho_0 g$$

According to Archimedes' principle, the buoyancy exerted by the air on the oil drops is equal to the weight of air displaced by the drops. Thus, in the case of the one droplet under consideration, if m_a is the mass of air that the droplet replaces and ρ_a is the density of the air,

$$B = m_a g = \rho_a \tfrac{4}{3}\pi r^3 g$$

Substituting for both B and W in the original equation:

$$\tfrac{4}{3}\pi r^3 g(\rho_0 - \rho_a) - 6\pi\eta r v_g = 0$$

This gives us an expression for the radius of the drop in terms of quantities that are either known or measurable:

$$r = \left[\frac{g\eta v_g}{2(\rho_0 - \rho_a)g}\right]^{1/2}$$

Let us now look at the situation when the droplet picks up an electric charge q and starts to move upward with a terminal velocity v_e. The equation of motion now becomes

$$W - B - qE + 6\pi\eta r v_e = 0$$

where E is the electric field between the two plates—a measure of their difference in potential and the distance between them.

Substituting the expression for $W - B$ from the equation involving no electric charge,

$$qE = kv_g + kv_e$$

or

$$q = \frac{k}{E}(v_g + v_e) = \frac{6\pi\eta r}{E}(v_g + v_e)$$

Thus, using the expression that we derived earlier for the radius of the droplet, we end up with the equation

$$q = 18\pi\eta^{3/2}\left[\frac{1}{2g(\rho_0 - \rho_a)}\right]^{1/2}$$
$$\times \frac{1}{E}v_g^{1/2}(v_g + v_e)$$

This equation produces a value of the charge on the droplet in terms of readily obtainable physical characteristics. As a result of thousands of such experiments in which the main objective was to measure the velocities of the same droplet with the capacitor uncharged and charged, Millikan concluded that the drops always picked up a charge that was a multiple of 1.60×10^{-19} C. This miniscule quantity was thus recognized as the fundamental electric charge of the electron and singly charged positive ions. Thousands of further measurements of the same type in the more than 60 yr since Millikan conducted his historic first measurement have all confirmed this value, and now no serious physicist doubts that this charge is quite unsplittable under all realistic circumstances.

Millikan went on to equally historic achievement in a series of experiments that confirmed Einstein's photoelectric theory—an early application of the quantum theory that was beginning to make its presence felt in the world of physics. For his contributions in both electrostatics and quantum theory, Millikan received the 1923 Nobel prize in physics, becoming the second American citizen to be so honored.

14 Current Electricity

14.1 THE REASON FOR RESISTANCE

14.2 ENERGY AND ELECTRICITY

SHORT SUBJECT: Materials Without Resistance

14.3 OHM'S LAW IN ACTION

SHORT SUBJECT: Electrical Circuit Symbols

14.4 KIRCHHOFF'S LAWS

SHORT SUBJECT: Simultaneous Equations

14.5 MEASURING METHODS

SHORT SUBJECT: Shock Treatment: The Electrical Dangers of Hospitals

Summary
Questions
Problems

SPECIAL TOPIC: How the Nervous System Works

If the study of electricity were limited to the types of phenomena that we can produce by rubbing materials against each other, or even by using such electrostatic generators as van de Graaff machines, it would be a restricted subject at best. Fortunately, there is an electrical world beyond electrostatics. It involves the movement of charges and encompasses phenomena as diverse as the flash of a flashlight bulb, the grinding wheels of industrial processes, and the operation of the nervous system. What these all have in common is current electricity.

The existence of a current of electrons is, in fact, evident in the simplest electrostatic experiments with an electroscope. When a charged rod touches the top of an electroscope, the separation of the instrument's foil leaves occurs because the charges have moved through the device from the top to the leaves. This movement is an electric current, although it is of very short duration—a transient electric current. In the specific case of an electroscope, the current consists entirely of electrons, which move either upward or downward according to the sign of the charge touched to the instrument.

An electric current is defined as the amount of charge per unit time that moves across a cross-sectional slab of conducting material, such as a metal wire. For example, if a charge ΔQ flows past some point in a wire during the brief time interval Δt, then the current I is defined mathematically by

$$I = \frac{\Delta Q}{\Delta t} \tag{14.1}$$

The ampere is named after French physicist André Marie Ampère, who in the early nineteenth century was one of the pioneers in the study of moving electric charges.

The current I is measured in amperes (A), where **1 A of current equals 1 C/s.** Current is a vector quantity whose direction is taken, by perhaps an unfortunate convention, to be that in which positive charges would flow. Of course, we already know that electrons, the most common carriers of electric current, are negative charges. Thus, in the case of metals, through which electrons alone carry current, the actual current or flow of charge in physical terms moves in the opposite direction to that used in mathematical analysis. The situation is less obviously wrong in other types of electrical conductors, such as conducting solutions and the materials known as semiconductors, which lie at the basis of modern electronics. In these cases both negative and positive charges contribute to the current, as they move in opposite directions.

The basic difference between the current in the electroscope and the current through a light bulb is simply a matter of persistence and size.

The charged rod that touches the electroscope produces a transient current. To be useful as a source of power, currents must generally be much larger and last much longer. To obtain an electric current that is large enough to be of value and then maintain it long enough to use, scientists or engineers must, somehow or other, devise ways to produce the electric fields that can exert a continuous "push" on the electrons and/or other charge carriers. To use more technical language, a potential difference must be maintained that, in the simplest case of a wire, means an excess of negative charge at one end and an excess of positive charge at the other. The excess of negative charge means, of course, a larger than normal buildup of electrons, and the excess of positive charge is a deficiency of electrons. Once this condition is established, electrons will move from the negative region to the positive region, in just the same way that gas at high pressure flows into a region of lower pressure. And just as the gas will cease to flow when the pressures are equalized, so will the charges no longer flow if the potential difference no longer exists.

The secret of producing an electric current, therefore, is to maintain an electric potential difference in the wire, which means supplying it with energy. We could, in principle, achieve this electric current by continuing to rub a rubber rod with fur or by using an electrostatic machine that we could continuously crank. But much better is the use of a chemical method of producing electrical energy in what is known as an electric cell or, more familiarly, a combination of cells, which is called a battery.

Cells and batteries also have their limitations as sources of electrical energy. When large amounts of such energy are required, engineers turn to the generating methods outlined in Chap. 16.

It was around 1800 when the Italian physicist Count Alessandro Volta first found that an electric potential could be produced chemically. His setup, now called the voltaic cell, consisted of disks of silver and zinc immersed in a solution of pure water, salty water, or alkali. In its modern manifestation, this has become the dry cell, but the principle is the same. A dry cell consists of a moist paste of ammonium chloride and other chemicals inside a sealed cylinder of zinc, with a rod of carbon in the center. The chemical reactions that occur in the dry cell are much more complex than those in the voltaic cell, but the end result is no different: a fairly large current of electrons moves through any piece of wire that connects the carbon rod and the zinc cylinder outside the cell.

At this point we must take another of our occasional diversions into chemistry, to gain some grasp of the way in which various chemicals produce electric currents. The functional objects are the entities known as ions, which are essentially charged atoms or molecules. Atoms and molecules are ordinarily electrically neutral. However, if an electron is removed by some process, perhaps by close contact of different materials, as in the rubbing analyzed in Chap. 13, the atom or molecule becomes positively charged. In this charged state it is known as a positive ion. Similarly, some neutral atoms or molecules attract extra electrons and become negatively charged. These are known as negative

Figure 14.1 Hydrochloric acid (HCl) diluted in water separates into positive H⁺ ions and negative Cl⁻ ions. If two dissimilar metals are placed in the dilute acid solution, a potential difference will arise between them. If the two plates are connected by a wire, negative charges (electrons) will flow from the negative terminal of the "cell" to the positive terminal.

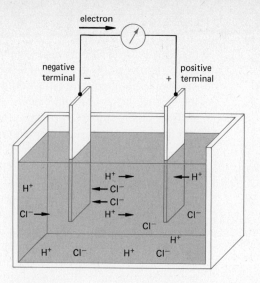

ions. Because the mass of an electron is negligible compared with the mass of the atomic nucleus, ions possess approximately the same mass as the atoms from which they are produced.

When any acid is made into a dilute solution in water, its molecules separate into positive and negative ions, producing the conditions for conduction of electricity. Figure 14.1, for example, shows a solution of hydrochloric acid (HCl) that has separated into positive hydrogen ions (H^+) and negative chloride ions (Cl^-). If we place two dissimilar metal plates in the solution, a potential difference will be created because the ions of the two different materials go into the solution. The process can be approximately described by stating that the negative chloride ions tend to travel through the solution to one of the plates, which we label the negative terminal, where each one releases an electron. The hydrogen ions make their way to the other, called the positive terminal, from which each ion removes an electron. If we link the plates by a metallic wire, the overall result is a two-way current of ions inside the solution, which produces a one-way current of electrons through the wire linking the two terminals. The terminals are called electrodes. In the case of the dry cell, the zinc cylinder is the negative terminal and the carbon rod is the positive terminal.

Ions can also be produced by the effect of high voltages on gases. Neon signs glow, for example, when a high voltage (potential difference) accelerates ions and electrons that in turn collide with the atoms of the gas to produce more ions. The positive ions produce the colorful effects when they recombine with electrons.

These two examples illustrate that electrical conduction in a solution or a gas differs from that in a wire. In the latter case, only electrons move. In ionized gases positive ions, negative ions, and electrons all contribute to the current, whereas in solutions only positive and negative ions act as charge carriers.

Studies of these different types of electrical conductivity have indicated that three separate factors are required to produce an electric current: a potential difference; a closed circuit, or path; and carriers of charge that are free to move through the circuit. The magnitude of the current that results when these three conditions are fulfilled can be expressed mathematically in terms of the velocity and density of the charge carriers.

The individual members of each type of charge carrier—ion or electron—**move with some average velocity when carrying an electric current. We call this average the drift velocity,** v_D. In the case of a metal, through which current is carried solely by electrons, we can say that there are n charge carriers per unit of volume and assume that all possess the drift velocity v_D. Thus, in a time Δt, each charge advances a distance d given by $d = v_D \Delta t$. If the metal has a cross-sectional area A, the number of charges which cross that cross section during the time Δt is plainly the number which occupy the volume Ad, that is, $n(v_D \Delta t)A$. The total charge crossing the cross-sectional area A is therefore given by $\Delta Q = n(v_D \Delta t)Ae$, where e is the charge of the individual charge carriers, the electrons. This transport of electrons results in a current I given by the equation

$$I = \frac{\Delta Q}{\Delta t} = nev_D A \tag{14.2}$$

This equation expresses the current in terms of the characteristics of the charge carrier. Clearly, if more than one charge carrier is involved, as in the case of the glowing neon sign, each type will make a specific contribution to the overall current.

14.1 THE REASON FOR RESISTANCE

Forgetting these complications for the moment and returning to the simple case of electrical conduction via electrons in metal wires, we might backtrack and ask whether we were conned into believing in the idea of drift velocity. More specifically, why don't electrons continue to accelerate when they move along a wire under the influence of a potential difference? This, after all, is what we should expect, because an electrical force acts continually on all the charges.

What really happens, though, is that "frictional" forces, ever present in nature, tend to slow the motion of the electrons under the influence of the electric field. The effect is exactly the same as that of air resistance to motion of an object under the influence of gravity; just as air resistance causes any falling object to move eventually with a constant terminal velocity, so frictional resistance to the motion of charge carriers causes those carriers to settle down to a constant velocity. The effect is known, simply, as resistance.

Resistance is defined as **the ratio of the potential difference across a wire (V) divided by the current in the wire (I).** Experience in the early days of electrical experimentation showed that this ratio V/I is a constant for a metallic wire. This observation that the current in any metallic conductor is directly proportional to the potential difference across the wire, where the proportionality constant is the resistance, is named Ohm's law for its discoverer, the German physicist Georg Simon Ohm. Mathematically, the law is expressed as

$$V = IR \qquad (14.3)$$

where V is the potential difference, in volts; I is the current, in amperes, which results from that potential difference; and R is the resistance, in ohms. The symbol for ohms is the rather appropriate Greek capital omega (Ω). Ohm's law holds invariably for metallic conductors, but for many other materials, the relationship between V and I is more complex. The resistance of any piece of wire is determined by the material of which it is made and by the geometrical properties of the wire, such as its length L and its cross-sectional area A. For simple metallic conductors, the equation for the resistance is

$$R = \frac{L\rho}{A} \qquad (14.4)$$

The quantity ρ is called the resistivity, and its value depends on such properties of the conducting material as its nature and purity, and also on the temperature. Resistivity varies quite spectacularly (see Table 14.1) from a value of about $10^{14}\,\Omega \cdot m$ for good insulators such as mica and glass to between 10^3 and $10^{-5}\,\Omega \cdot m$ for semiconductors and $10^{-8}\,\Omega \cdot m$ for good metallic conductors. Right at the bottom of the

Table 14.1 Resistivity (ρ) and Temperature Coefficients of Resistance (α) at 0°C

Substance	$\rho(\Omega \cdot m)$	α (1/C°)
Silver	1.47×10^{-8}	0.0038
Copper	1.59×10^{-8}	0.0039
Gold	2.27×10^{-8}	0.0034
Aluminum	2.60×10^{-8}	0.0040
Tungsten	5.5×10^{-8}	0.0046
Iron	11.0×10^{-8}	0.0052
Platinum	11.0×10^{-8}	0.00392
Carbon	4×10^{-5}	-0.0005
Germanium	2	
Silicon	3×10^4	
Boron	1×10^6	
Wood	$\sim 3 \times 10^8$	
Glass	$10^{11} - 10^{13}$	
Sulfur	1×10^{15}	
Fused quartz	5×10^{17}	

scale, with values of zero for resistivity, are superconductors—materials that lose all their electrical resistance at ultralow temperatures, as outlined on page 428. It should be emphasized that resistivity, like density, is a property of any material that is independent of its shape.

EXAMPLE 14.1

Compute the electric current through an electric flatiron that has a resistance of 24 Ω when the flatiron is plugged into a house circuit having a voltage of 120 V.

Solution

$$I = \frac{V}{R} = \frac{120 \text{ V}}{24 \text{ Ω}} = 5 \text{ A}$$

We said that the electrical resistance of materials which do possess some resistance is produced by "frictional" force. We can understand how this "frictional" force works by taking a look at the inside of a conductor through which an electric charge is traveling. The skeleton of a metal consists of positive ions, each missing one or more electrons, that are fixed in general position but oscillating about some central spot. The electrons move through this skeleton in a kind of gas, in much the same fashion as that pictured by the kinetic theory of gases which we analyzed in Chap. 11. Occasionally, a single electron encounters an ion that knocks it sideways, backward, or slightly off course. It is this type of "scattering" that is responsible for the "frictional" force on electrons which is translated into electrical resistance. Clearly, it is reasonable to expect that the amount of scattering electrons will suffer, and hence the amount of resistance, will depend on the length of the individual conducting wire. Furthermore, it is not unreasonable that resistance will also vary inversely with the cross-sectional area of the wire.

A down-to-earth analogy for the flow of electrons through a closed circuit is the circulation of water in a community water system pictured in Fig. 14.2. Here, water is pumped out of a river into a storage tank on a hill above the town. The water in the storage tank is at a higher gravitational potential than the homes and thus flows downhill through the pipes to the homes. The water flows onward through the sewer system to the water-treatment plant and then back into the river. The flow of water in the pipes is impeded by friction with the walls of the pipe and its own viscosity. A complete circuit is required, and the pump is necessary to maintain a constant potential difference if the system is to be continuous. Of course, this water analogy is not exactly parallel with the electrical case since some water may be removed in the house, and in the electrical case electrons are not lost from the wires.

In the electrical circuit, the potential difference is developed by a dry

Figure 14.2 Diagram of a water system. The pump lifts the water to the tower, hence increasing the potential energy of the water. The pump and tower together act as a battery in an electrical circuit. The water cycle can be thought of as a closed circuit, although parts of the natural cycle—evaporation, cloud formation, and rain—are omitted here.

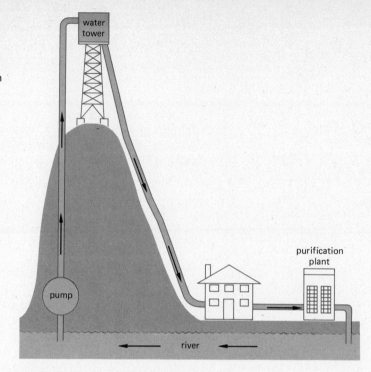

cell battery or a generator, which corresponds to the pump. The closing of a switch in the electrical circuit corresponds to opening of a valve or faucet in the water system.

Some additional electrical terminology also arises from the water analogy. The water pump corresponds to a battery, and obviously work is required to carry a volume of water from the river to the top of the hill. Likewise, work is required to move each charge within the battery. The work per unit charge required to move charges within the battery provides the potential difference across the terminals of the battery. This work per unit charge has been misnamed the electromotive force (emf) $\mathcal{E}$. It is not a force but energy per unit charge. The emf is measured in joules per coulomb, or volts, not in newtons. One might ask why the emf of a battery is not just called the potential difference, since they are both measured in volts. The answer is that the battery has an internal resistance, so when the battery is connected to a circuit to produce a current, the potential difference at the battery terminals will be somewhat less than the emf of the battery. Only in an ideal battery would the emf have the same value as the potential difference.

EXAMPLE 14.2

Compute the resistance of an electric heater coil through which 10-A current flows when the heater is plugged into a house circuit (120 V).

Solution

$$R = \frac{V}{I} = \frac{120 \text{ V}}{10 \text{ A}} = 12 \text{ }\Omega$$

EXAMPLE 14.3

What is the drift velocity v_D of the electrons in copper wire of diameter 0.1 cm carrying a current of 5 A?

Solution

To solve this problem, we use Eq. (14.2),

$$I = nev_D A$$

We need to know the number of free electrons per unit volume (cubic meters). Chemical studies show that the copper atom has one loosely bound electron; thus, we shall assume that each copper atom contributes one free electron. The number of free electrons is hence equal to the number of copper atoms in 1 m³. The atomic mass of copper is 64, and so 64 g of copper contains Avogadro's number, 6×10^{23} atoms. The density of copper is 8.9 g/cm³. Therefore, the number of atoms per cubic centimeter is

$$6 \times 10^{23} \times \frac{8.9}{64} = 8.3 \times 10^{22} \text{ atoms/cm}^3$$

The number of electrons per cubic meters will equal

$$n = (10^6) \text{ cm}^3/\text{m}^3 \times 8.3 \times 10^{22} \text{ atoms/cm}^3 \times 1 \text{ electron/atom}$$
$$= 8.3 \times 10^{28} \text{ electrons/m}^3$$

The area of the wire, in square meters, is

$$A = \pi r^2 = \pi \left(\frac{0.1 \times 10^{-2}}{2}\right)^2 = 7.9 \times 10^{-7} \text{ m}^2$$

The charge on one electron is $e = 1.6 \times 10^{-19}$ C. Using these values and Eq. (14.2), we can solve for v_D:

$$v_D = \frac{I}{neA}$$

$$= \frac{5 \text{ A}}{(8.3 \times 10^{28} \text{ electrons/m}^3)(1.6 \times 10^{-19} \text{ C/electron})(7.9 \times 10^{-7} \text{ m}^2)}$$

$$= 4.8 \times 10^{-4} \text{ m/s} \quad \text{or} \quad 4.8 \times 10^{-2} \text{ cm/s}$$

If the wire were 1 m long, the time taken for an electron to drift the length of 1 m would be

$$\frac{1}{4.8 \times 10^{-4}} = 2.1 \times 10^3 \text{ s} \quad \text{or} \quad 35 \text{ min}$$

This result indicates that an individual electron need not move a long distance to produce a current. In fact, electrons drifting very slowly produce ordinary, sizeable currents. The low drift velocity does not contradict the observation that lights go on almost instantaneously when the switch is closed. The electrons in the light begin to move immediately.

EXAMPLE 14.4

What is the resistance of a 20-m length of copper wire with a diameter of 0.02 cm at $0°C$?

Solution

We use Eq. (14.4):

$$R = \rho \frac{L}{A}$$

$$\rho = 1.59 \times 10^{-8} \, \Omega \cdot m$$

$$L = 20 \, m$$

$$A = \pi r^2 = \frac{\pi D^2}{4} = \frac{\pi (0.02 \times 10^{-2} \, m)^2}{4} = 3.14 \times 10^{-8} \, m^2$$

$$R = \frac{(1.59 \times 10^{-8} \, \Omega \cdot m)(20 \, m)}{3.14 \times 10^{-8} \, m^2} = 10.1 \, \Omega$$

The resistivity, and hence the resistance, of most metallic conductors tends to increase with increasing temperature. This can be understood on a qualitative level by returning to the kinetic theory. As the metal increases in temperature, the ions oscillate more furiously around their basic positions and their oscillations carry them further from the center points. Thus, the ions present more of an impediment to the flow of electrons through the metal. Experiments have shown that the change in resistivity of a particular piece of wire is proportional to the initial resistivity of the wire, ρ_0, and the change in temperature. The equation is

$$\Delta \rho = \alpha \rho_0 \Delta t \tag{14.5}$$

Because resistivity varies directly with resistance, this equation can also be expressed in the form

$$\Delta R = \alpha R_0 \Delta t$$

Both equations are analogous to that for the linear expansion of a solid on page 270. If ρ_0 is the resistivity at $0°C$, then Δt is the temperature in degrees Celsius. The coefficient of resistivity, α, has units of $1/°C$; a few typical values for α measured at $0°C$ appear in Table 14.1.

We mentioned the use of resistance thermometers, in Chap. 10. Equation (14.5) illustrates the rationale for them: the direct relationship between temperature differences and resistivity. In practice, the thermometers are both extremely accurate and capable of monitoring temperatures continually. Platinum is generally the material of choice for recording high temperatures (of a few hundred degrees or more) because its resists corrosion and has a high melting point of 1773 °C.

EXAMPLE 14.5

An iron wire has a resistance of 200 Ω at 20°C. What will be its resistance at 120°C if the temperature coefficient of resistance is 0.0052/°C?

Solution

To determine R at 120°C, we recall that $\Delta R = R - R_0$, where $R_0 = 200\ \Omega$ at $t = 20°C$. Hence,

$$\Delta R = R - R_0 = \alpha R_0\, \Delta t$$
$$R = R_0 + \alpha R_0\, \Delta t = 200\ \Omega + (0.0052/°C)(200\ \Omega)(120°C - 20°C)$$
$$R = 200\ \Omega + 104\ \Omega = 304\ \Omega$$

where we neglect the change of α with reference temperature.

14.2 ENERGY AND ELECTRICITY

Returning once more to the kinetic theory as it pertains to electrical resistance, we see that the collisions between electrons and ions which are responsible for resistance actually represent means of transferring energy from the electrons to the ions, and hence to the wire of which the ions are parts. The loss of the electron's electrical energy is compensated by a gain in the wire's energy in the form of heat. This process is often called Joule heating, after the nineteenth century English physicist who pioneered studies of the conversion of electrical energy into heat. (We first encountered Joule in Chap. 10 in connection with the conversion of *mechanical* energy to heat.)

Joule discovered that the rate at which heat is produced in a metallic conductor is proportional to the square of the current in the conductor. This relationship is known as Joule's law. It can be derived relatively simply.

Imagine a wire with a potential of V_a at one end and a potential of V_b at the other, which is carrying a current I. To move a specified charge Q (where $Q = It$) from V_a to V_b requires a certain amount of work, given by the product of the charge and the potential difference. In other words,

$$\text{Work} = Q(V_a - V_b) = It\, \Delta V$$

This work is equal to the amount of energy given up by the charge in moving along the wire. The power dissipated (P) as the charge moves

Materials Without Resistance

Extreme conditions produce extreme properties. Certainly, few properties of materials are stranger than that exhibited by certain substances at ultralow temperatures. When cooled close to the absolute zero of temperature, these substances lose all their electrical resistance. A current induced in a ring of such material flows literally for days without dissipating, as long as the setup is kept sufficiently cold. The name of this strange characteristic, which was first discovered in 1911 by the Dutch physicist Kamerlingh Onnes, is superconductivity.

The basic limit on superconductivity is temperature. Every superconducting material possesses what is known as a transition temperature. Below this temperature, the material's resistance is effectively zero; above it, the superconductivity disappears. The first superconducting materials to be identified all had transition temperatures within single degrees Kelvin of absolute zero; but in recent years, physicists at the Bell Telephone Laboratories and elsewhere have produced superconductors that can support the flow of current endlessly even above 20 K. Transition temperatures can be reduced by the application of strong magnetic fields.

The exact details of the cause of superconductivity remain sources of debate among physicists, but the broad outline of why superconductors are as they are was formulated in the mid-fifties. The theory is known as the BCS theory, after its authors John Bardeen, Leon Cooper, and John Robert Schrieffer, who received the Nobel physics prize for their efforts in 1972. Basically, their theory proposes that the internal vibrations of the atoms and electrons of superconductors somehow harmonize at very low temperatures, giving the electrons unimpeded paths through their metals.

The potential savings made possible by overcoming electrical resistance have already persuaded engineers to invest in the expensive cooling equipment necessary to maintain superconductors in their nonresistive state in some particle accelerators and electric motors. Experts are even considering using superconducting transmission lines, cooled with liquid helium, to transport electric energy over large distances from power stations to consumers. And in the far distant future, some theoreticians believe that organic materials may one day be developed that are superconductors at normal room temperatures. If such materials can be made, they will undoubtedly cause a revolution in our way of dealing with electricity.

---

from one end of the wire to the other is the amount of work per unit time:

$$P = \frac{\text{work}}{t} = \frac{It\,\Delta V}{t} = I\,\Delta V$$

If we replace ΔV by V, bearing in mind that it represents a *difference* in potential, we come up with the simple expression $P = IV$. This is one form of Joule's law. Joule's law applies equally to the calculation of the amount of useful work obtained from a certain current and a corresponding voltage.

We can extend the expression by substituting for the potential using its value as obtained from Ohm's law, that is, IR. Then, we obtain the equation for the power dissipated in metallic conductors:

$$P = IV = \frac{V^2}{R} = I^2 R \qquad (14.6)$$

The power is thus proportional to the square of the current in the wire and directly proportional to the wire's resistance. The units of electrical power are familiar to anyone who has read the power rating on an electric light bulb, air conditioner, or similar electrical device, namely watts. However, although electrical devices are rated according to their power usage, the consumer who uses electricity to operate them actually pays for **electric energy,** which **is the product of the power consumed by the device and the time during which it is operating.** This is the rate at which energy is delivered to an appliance and the length of time it operates. Most electricity bills charge for electrical energy on the basis of the number of kilowatthours (kWh) consumed, that is, the product of the power expressed in thousands of watts, and the time for which that power is used, expressed in hours.

EXAMPLE 14.6

What is the resistance of a 100-W electric light bulb that draws a current of 0.9 A when it glows at full brightness?

Solution

Since we are given the power and current, we can use the following form of Eq. (14.6):

$$P = I^2 R$$

$$R = \frac{P}{I^2} = \frac{100 \text{ W}}{(0.9 \text{ A})^2} = 123 \text{ }\Omega$$

EXAMPLE 14.7

A 1440-W room air conditioner operates on a 120-V house circuit for a total of 300 h during the month of August. Compute the (*a*) current through the air conditioner, (*b*) equivalent resistance of the air conditioner, and (*c*) cost of using this air conditioner during the month of August if the electrical company rate is 4¢/kWh.

Solution

(*a*) To compute the current in the air conditioner, we use Joule's law, Eq. (14.6):

$$P = VI \quad \text{or} \quad I = \frac{P}{V} = \frac{1440 \text{ W}}{120 \text{ V}} = 12.0 \text{ A}$$

Note that most household circuits are fused for 15 or, at most 20 A; hence, the circuit on which the air conditioner is placed cannot take

New Yorkers who wish these examples to reflect real-life situations can substitute a rate of 9¢/kWh for the electric utility charges in this calculation. In fact, utility charges vary widely from region to region and state to state according to such factors as the sources of energy for electrical generation, the extent of such sources, state laws, and the economic state of the local utilities.

additional appliances while it is operating, or the fuse may blow.

(*b*) To compute the effective resistance, we can use Ohm's law:

$$V = IR \quad \text{or} \quad R = \frac{V}{I} = \frac{120 \text{ V}}{12 \text{ A}} = 10 \, \Omega$$

(*c*) To compute the cost of the air conditioner over the 300 h, we multiply the power it uses times the 300 h times the cost, 4¢/kWh:

$$\text{Cost} = \frac{1440 \text{ W}}{1000 \text{ W/kW}} \times 300 \text{ h} \times \frac{\$0.04}{\text{kWh}} = \$17.28$$

14.3 OHM'S LAW IN ACTION

The first step in applying Ohm's law—or Joule's law, for that matter—is to get a clear understanding of what is really happening in the electrical circuit to which the law is being applied. The simplest way of achieving this is to sketch out the circuit using the electrical symbols, indicating at the location of each symbol the size of the resistance, potential difference, or other appropriate unit represented by the symbol.

In the particular case of electrical resistance, one normally lumps the resistance of connecting wires together with that of a resistance device or neglects the resistance of the connecting wires altogether.

Figure 14.3 shows a typical circuit diagram. The dry cell, or battery (a group of two or more cells), provides an emf that acts as the energy supply for the circuit carrying the current I. When the switch is closed, this current passes through the resistor R_1, whose value of resistance is constant, and resistor R_2, whose resistance can be varied. The diagram indicates both the direction of the convential current (the motion positive charges would take) and the direction in which the actual charge carriers, the negative electrons, move around the circuit. Normally, such diagrams show only the direction of the conventional current.

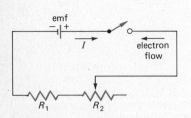

Figure 14.3 A simple circuit containing a dry cell (emf), a switch, a fixed resistor, and a variable resistor all in series.

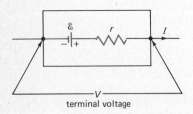

Figure 14.4 Circuit for a cell or battery. The cell has a specific emf, $\mathcal{E}$, determined by the electrodes and internal chemistry. The terminal voltage is different from the emf when a current exists because the cell has an internal resistance.

Ohm's law can be applied to a circuit as a whole, or to individual parts of a circuit. However, to apply the law accurately to a complete circuit, one must first understand how the parts interact. Consider, for example, the case of a cell whose emf is ideally 1.5 V. When the cell takes its place in a circuit and starts providing current, the potential difference across its terminals—called the terminal voltage—is less than 1.5 V because the cell has an internal electrical resistance of its own. This loss of potential energy can be best understood by creating a circuit diagram of the cell itself, as shown in Fig. 14.4. Here, the cell is represented by an emf ($\mathcal{E}$)—which is the voltage drop across the cell when no current is drawn from it and is thus equal to 1.5 V—combined with an internal resistance r. Whenever a current I passes through the cell, that internal resistance causes a loss of voltage equal to Ir. The potential difference across the cell when a current I is passing through

Electrical Circuit Symbols

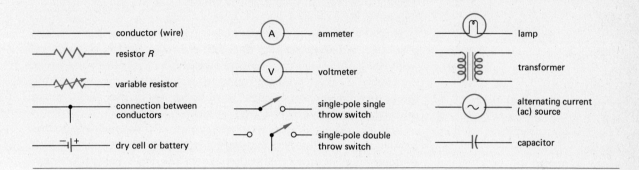

it is equal to the difference between the cell's emf and the voltage lost as a result of the internal resistance. In other words,

$$V = \mathcal{E} - Ir \tag{14.7}$$

Plainly, potential difference across the terminals is identical with the emf of the cell only when no current is drawn from it.

The internal resistance of a cell can vary widely, as a result of such factors as its age and the state of its electrolyte. A dry cell straight off the production line is likely to have an internal resistance of no more than 0.05 Ω, whereas an old cell whose electrolyte has dried out on the shelf might have a massive internal resistance of 100 Ω or more. Therefore, although the two cells have the same emf, determined by the chemistry, the old one wastes most of that energy internally.

Another complication in computing the values of basic electrical quantities in circuits involves the way in which cells are linked together in circuits. Cells are normally combined with the negative terminal of one adjacent to the positive terminal of the next, thus ensuring that the current passes through both in the same direction, producing a net voltage which is the sum of those generated by the individual cells, as shown in Fig. 14.5(a). The cells are said to be linked "in series aiding." But what if the two negative (or positive) terminals are adjacent to each other in the circuit? In this case the voltages of the cells oppose each other, as demonstrated in Fig. 14.5(b). The terminal voltage V_1 across

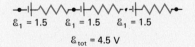

(a) cells in series to form a "battery"

(b) two emf's opposing one another

Figure 14.5 Connection of cells. (a) Three cells (emf's) in series producing a "battery" (a battery of cells). (b) Two cells (emf's) in series opposing each other.

cell 1 in this case is given simply by the equation

$$V_1 = +\mathcal{E}_1 - Ir_1 = (6\text{ V}) - (2\text{ A})(1\ \Omega) = 4\text{ V}$$

The current passes through the second cell in reverse, however, producing a slightly different equation for its terminal voltage V_2:

$$V_2 = -\mathcal{E}_2 - Ir_2 = (-1.5\text{ V}) - (2\text{ A})(0.5\ \Omega) = -2.5\text{ V}$$

The net potential difference across both cells is the sum of V_1 and V_2, that is, $(4 - 2.5)$ V, or 1.5 V.

How should batteries or cells be connected together? Clearly when connected with their polarity opposed, as in Fig. 14.5(b), the cells are working against each other, reducing the net driving voltage and hence the current. Usually cells are connected together in series by hooking the positive of one to the negative of the other. In Fig. 14.5(a) three cells are added in series to form a "battery" of 4.5 V.

Before we apply Ohm's law to real circuits it is convenient to discuss how we can add resistances that appear in real circuits. We consider two configurations: a series connection of resistors and a parallel configuration. Using Ohm's law it is possible to develop a prescription that enables us to replace more complicated collections of resistors by an equivalent resistance and hence to determine the current drawn by various parts of a circuit. Consider first the three resistors R_1, R_2, and R_3 in Fig. 14.6(a). They are connected in *series*. The current I has only one path open to it—that which takes it through all three resistors. The same size of current passes through each resistor. Can we add the resistances of the three resistors in series to obtain a single equivalent

Figure 14.6 Resistors in parallel and series. (a) Three resistors in series. (b) The equivalent circuit for (a). (c) Three resistors in parallel. (d) The equivalent circuit for (c).

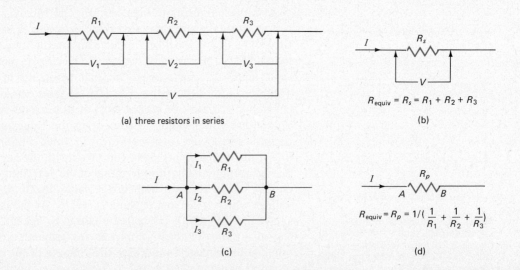

resistance in the same way that we added the terminal voltages of the batteries? We can approach this problem by starting with the sum of the potential differences across each resistor:

$$V = V_1 + V_2 + V_3$$

Applying Ohm's law to each individual voltage,

$$V = IR_1 + IR_2 + IR_3 = I(R_1 + R_2 + R_3)$$

Thus, $V = IR_s$, where R_s is the sum of the three individual resistances:

$$R_s = R_1 + R_2 + R_3 \tag{14.8}$$

Then, we can replace the three resistances in series by a single resistance equal to their sum. Diagrammatically, the circuit in Fig. 14.6(b) can validly be substituted for the more complicated circuit shown in Fig. 14.6(a).

EXAMPLE 14.8

In a simple circuit, a dry cell which has an emf of 1.5 V and an internal resistance of 0.5 Ω is connected to a coil of wire that has a resistance of 19.5 Ω (this includes the resistance of the connecting wires). Compute the current through the coil.

Solution

$$I = \frac{V}{R} = \frac{1.5 \text{ V}}{(19.5 + 0.5) \text{ Ω}} = \frac{1.5 \text{ V}}{20 \text{ Ω}} = 0.075 \text{ A}$$

We might expect that the situation will be different for resistors linked in parallel, as illustrated in Fig. 14.6(c). In this case the potential difference is the same across each of the three resistors, and the current splits into three separate parts, not necessarily equal, to travel through the resistors. Since charge is conserved, the sum of the individual currents in the branches equals the value of the single current before and after it splits; thus,

$$I = I_1 + I_2 + I_3$$

Using Ohm's law, and expressing these currents in terms of voltage (which is the same for all resistors) and resistances, we obtain the equation

$$\frac{V_{AB}}{R_p} = \frac{V_{AB}}{R_1} + \frac{V_{AB}}{R_2} + \frac{V_{AB}}{R_3}$$

where R_p is the overall resistance of the three resistors in parallel.

Canceling out the voltage V_{AB},

$$\frac{1}{R_p} = \frac{1}{R_1} + \frac{1}{R_2} + \frac{1}{R_3} \tag{14.9}$$

Thus, the circuit shown in Fig. 14.6(d) can be substituted for that of Fig. 14.6(c).

EXAMPLE 14.9

Three resistances, 2 Ω, 4 Ω, and 5 Ω, are connected in series with an emf of 6 V, as shown in Fig. 14.7. The internal resistance of the battery is 1 Ω. Compute the (a) equivalent resistance external to the battery, (b) current in the circuit, and (c) terminal voltage of the battery.

Solution

(a) The equivalent resistance of the three series resistances is

$$R_s = R_1 + R_2 + R_3$$

or

$$R_s = 2\,\Omega + 4\,\Omega + 5\,\Omega = 11\,\Omega$$

(b) We can use Ohm's law to determine the current in the circuit. The voltage drop across the equivalent resistance must equal the terminal voltage V_{AB} of the battery. Using Eq. (14.7) and Ohm's law, Eq. (14.3), we can write

$$V_{AB} = IR_s = \mathcal{E} - Ir$$

or

$$\mathcal{E} = I(R_s + r)$$

$$I = \frac{\mathcal{E}}{R_s + r} = \frac{6\text{ V}}{11\,\Omega + 1\,\Omega} = 0.5\text{ A}$$

Note that the current is just the total emf ($\mathcal{E}$) divided by the total resistance of the circuit. We can obtain it by applying Ohm's law to the complete circuit.

(c) The terminal voltage is given by

$$V = \mathcal{E} - Ir = (6\text{ V}) - (0.5\text{ A})(1\,\Omega) = 5.5\text{ V}$$

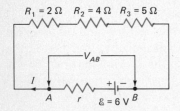

Figure 14.7 A series circuit.

EXAMPLE 14.10

Three resistances are connected as shown in Fig. 14.8, one in series, with a pair in parallel. For the values indicated in Fig. 14.8, determine the (a) equivalent resistance of the parallel group, (b) total resistance external to the battery, and (c) current in all three resistances. Assume the battery has no internal resistance.

Solution

(a) To compute the equivalent resistance for the pair of resistances in parallel, we use Eq. (14.9):

$$\frac{1}{R_p} = \frac{1}{R_1} + \frac{1}{R_2} = \frac{1}{6} + \frac{1}{3} = \frac{1+2}{6} = \frac{1}{2}$$

$$R_p = 2\,\Omega$$

(b) The total resistance external to the battery is the sum of R_3 and R_p which are in series with each other:

$$\text{Total } R_{\text{equiv}} = R_p + R_3 = 2\,\Omega + 4\,\Omega = 6\,\Omega$$

(c) We use Ohm's law to compute the current I in R_3 and R_p:
$$\mathcal{E} = IR_{\text{equiv}} = I \times 6\,\Omega$$

or

$$I = \frac{6\text{ V}}{6\,\Omega} = 1\text{ A}$$

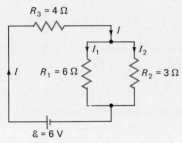

(a) original circuit

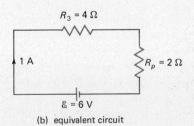

(b) equivalent circuit

Figure 14.8 A combination series circuit with one parallel branch. (a) Original circuit. (b) The equivalent circuit.

This current of 1 A divides into I_1 and I_2 when it enters the parallel branch. To compute I_1 and I_2, recall that the potential differences across R_1 or R_2, and across the equivalent resistor R_p, are all identical. That is,

$$V = IR_p = (1\text{ A})(2\,\Omega) = 2\text{ V}$$
$$V = I_1 R_1 = I_1 \times (6\,\Omega)$$
$$V = I_2 R_2 = I_2 \times (3\,\Omega)$$

Equating these potential differences yields the result

$$2\text{ V} = I_1 \times (6\,\Omega) \quad \text{or} \quad I_1 = \tfrac{2}{6} = \tfrac{1}{3}\text{ A}$$
$$2\text{ V} = I_2 \times (3\,\Omega) \quad \text{or} \quad I_2 = \tfrac{2}{3}\text{ A}$$

Since $I = I_1 + I_2$, we can check this result:

$$1\text{ A} = \tfrac{1}{3}\text{ A} + \tfrac{2}{3}\text{ A}$$

14.4 KIRCHHOFF'S LAWS

Ohm's law provides an excellent means of understanding what is happening in simple electrical circuits, but when one encounters more complicated combinations, such as those that involved more than one source of emf and a large number of resistors, its use becomes cumbersome, to say the least. A straightforward method of analyzing complete circuits, based on Ohm's law, was advanced in the middle of the nineteenth century by the Prussian physicist Gustav Kirchhoff, who was also the first man to show that electrical impulses move at the speed of light. Kirchhoff's two "laws" are really simple statements that outline how to understand electrical circuitry:

1. The sum of the currents entering any junction, that is, any point

where three or more wires intersect, equals the sum of the currents leaving that particular junction.

2. The sum of the emf's around any closed loop equals the sum of the potential drops around that loop; or, in other words, the algebraic sum of the individual differences in potential around any closed loop is zero.

The first law states simply that "what goes in must come out" and represents an expression for the conservation of electric charge. Currents do not pile up at electrical junctions. The second law is just one of many statements of the conservation of energy. If you start at any particular point in a closed electrical loop and add all the potential rises caused by emf's and the potential drops resulting from currents in resistors as you go around the loop, your final sum when you return to the starting point must be zero. If it were not, you would have gained or lost electrical potential, thus violating the principle of conservation of energy.

Kirchhoff's laws are used in closed circuits to determine the currents in individual circuit elements and the potential differences across circuit elements. They can best be applied in a simple stepwise fashion:

1. Assume an arbitrary direction for the current in each individual loop of a circuit. Do not worry about choosing the wrong direction because the mathematics will not suffer. If you have chosen the wrong direction, the current will simply come out with a negative sign in the final calculation.
2. Apply Kirchhoff's first law to all but one of the junctions, to obtain a set of independent simultaneous equations. The current equation for the last junction will serve to check the overall calculation.
3. Apply Kirchhoff's second law to each loop, bearing in mind the correct direction from the emf of each battery or cell and that every resistance produces a drop in the potential if a current passes through it. Recall that a potential drop also occurs when the current passes through an internal resistance of a battery.

EXAMPLE 14.11

Use Kirchhoff's laws to calculate the currents in the circuit shown in Fig. 14.9.

Solution

This is a simple circuit that could be solved using Ohm's law. However, we are instructed to use Kirchhoff's laws. First, we select the current directions; the fact that there is only one emf makes our choice apparent. Applying Kirchhoff's first law for one of the two junctions,

$$I = I_2 + I_4 + I_6$$

The subscripts indicate the resistor through which the currents pass. Next, we apply the second law to more than one loop. Consider the

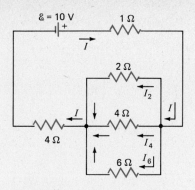

Figure 14.9 A series and parallel combination.

loop to include the main circuit and the 2-Ω resistor in the top branch of the parallel circuit. Beginning at the emf:

$$10 \text{ V} - 1I - 2I_2 - 4I = 0$$

or

$$5I + 2I_2 = +10$$

Now considering the loop to consist of the main circuit and the 4-Ω resistor in the parallel branch, we have

$$10 \text{ V} - 1I - 4I_4 - 4I = 0$$

$$5I + 4I_4 = +10$$

Similarly, for the 6-Ω resistor and the main circuit, one obtains

$$5I + 6I_6 = +10$$

These equations can be solved simultaneously in several ways. For example, if we solve each of the three loop equations for I_2, I_4, and I_6, we obtain

$$I_2 = \tfrac{10}{2} - \tfrac{5}{2}I$$

$$I_4 = \tfrac{10}{4} - \tfrac{5}{4}I$$

$$I_6 = \tfrac{10}{6} - \tfrac{5}{6}I$$

If these values are substituted into the equation relating the currents, we have

$$I = \tfrac{10}{2} - \tfrac{5}{2}I + \tfrac{10}{4} - \tfrac{5}{4}I + \tfrac{10}{6} - \tfrac{5}{6}I$$

or

$$I = 1.64 \text{ A}$$

Substituting the value for I into the three equations for I_2, I_4, and I_6, one obtains the other three currents:

$$I_2 = 0.89 \text{ A}$$
$$I_4 = 0.45 \text{ A}$$
$$I_6 = 0.30 \text{ A}$$

EXAMPLE 14.12

Using Kirchhoff's laws, calculate the unknown currents in the circuit shown in Fig. 14.10.

Solution

The first step, choosing the current directions, has already been taken. Applying Kirchhoff's first law to one of the two junctions yields

$$I_2 = I_1 + I_3$$

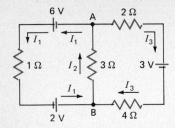

Figure 14.10

Next we write two loop equations. Starting at point A, we can move clockwise around the right-hand loop:

$$-2I_3 - 3 - 4I_3 - 3I_2 = 0$$

or

$$-3 = 6I_3 + 3I_2$$

or

$$-1 = 2I_3 + I_2$$

The negative sign for the voltage arises because the output current of the cell opposes the direction we chose to move about the circuit.

Next we begin at A and move counterclockwise about the left-hand loop:

$$6 - 1I_1 + 2 - 3I_2 = 0 \qquad 8 = I_1 + 3I_2$$

We now have three equations with three unknowns that can be solved simultaneously for I_1, I_2, and I_3:

$$I_2 = I_1 + I_3 \tag{1}$$
$$-1 = I_2 + 2I_3 \tag{2}$$
$$8 = I_1 + 3I_2 \tag{3}$$

Solving Eq. (1) for I_1 and substituting the result into Eq. (3) yields

$$I_1 = I_2 - I_3 \qquad 8 = I_2 - I_3 + 3I_2 \qquad \text{or} \qquad 8 = 4I_2 - I_3$$

This resulting equation multiplied by 2 and added to Eq. (2) will yield a result for I_2:

$$-1 = I_2 + 2I_3$$
$$\underline{16 = 8I_2 - 2I_3}$$
$$15 = 9I_2$$

$$I_2 = \tfrac{15}{9} = \tfrac{5}{3} \qquad \text{or} \qquad 1.67 \text{ A}$$

Substituting for I_2 into Eq. (2) produces the answer for I_3:

$$-1 = 1.67 + 2I_3$$

$$I_3 = -\frac{2.67}{2} = -1.33 \text{ A}$$

The negative sign indicates that this current is in the direction opposite to that originally chosen. Using Eq. (3), we can determine I_1:

$$8 = I_1 + 3I_2$$
$$I_1 = 8 - 3(1.67) = 3 \text{ A}$$

We can check the result by using a third loop equation beginning at A and moving clockwise about a loop made of the extremities of the circuit:

Simultaneous Equations

A single algebraic equation can be used to determine just one unknown quantity. In many instances we are able to write down more than one independent equation in which the same variables appear. For each equation, an additional unknown can be evaluated. If we have a set of N independent equations, then we could obtain values for N unknowns. How do we solve two such independent, simultaneous equations? There are several techniques, some more elegant than others; we shall review briefly the most direct technique. Suppose for simplicity that we have two equations in which the same pair of unknowns appear. We shall use the symbols I_1 and I_2 for our unknowns. To solve the equations, you solve for one variable in terms of the other variable in one equation and then substitute for that first variable into the second equation, resulting in an equation for just one unknown. For example,

$$27 - I_1 - 24I_2 = 0 \quad (1)$$
$$I_2 - I_1 + 2 = 0 \quad (2)$$

If we solve for I_1 in Eq. (1), we have

$$I_1 = 27 - 24I_2$$

Next, we put this value for I_1 into Eq. (2), yielding

$$I_2 - (27 - 24I_2) + 2 = 0$$

or

$$I_2 - 27 + 24I_2 + 2 = 0$$
$$25I_2 = 25$$
$$I_2 = 1$$

To determine I_1, we need only return to Eq. (1) or (2) and substitute 1 for I_2:

$$I_2 - I_1 + 2 = 0$$
$$1 - I_1 + 2 = 0$$
$$I_1 = 3$$

We can check our result by substituting for I_1 and I_2 in Eq. (1):

$$27 - I_1 - 24I_2 = 0$$
$$27 - 3 - 24 = 0$$

or

$$27 = 27$$

This straightforward technique can be used with more than two equations, but things become more confusing as the number of equations increases. Nevertheless, with a little patience, the method will yield the necessary results.

$$-2I_3 - 3 - 4I_3 - 2 + I_1 - 6 = 0$$
$$-11 - 6I_3 + I_1 = 0$$
$$-11 - 6(-1.33) + 3 = 0$$
$$-11 + 8 + 3 = 0$$

One point to note in this calculation: The direction chosen to move around the loop is the opposite of that taken by the current I_1. Thus, the potential difference across the 1-Ω resistor is a rise instead of a drop, and the potential differences across all the cells in the circuit are potential drops.

14.5 MEASURING METHODS

We saw that the internal resistances of batteries and the resistances of wires which link the various devices in an electrical circuit alter such quantities as the terminal voltage of a battery. It is hardly surprising, then, to learn that devices used to measure current and voltage also

Shock Treatment: The Electrical Dangers of Hospitals

Any patient who enters an operating room is taking a positive, if small, risk of dying there from causes unrelated to the ailment. According to current estimates, about 1 person in 2000 dies as a result of the anesthetic, and other fatalities result from occasional mistakes by the operating staff. A minute proportion of patients is at risk from yet another source: electricity. For seriously ill patients an unexpected electric shock or a stray current from a piece of electrical machinery can mean the difference between life and death.

One of the most potentially dangerous situations arises from a combination of anesthetics and electricity. Many modern anesthetics are volatile liquids whose vapors can form extremely explosive mixtures with air. In these circumstances, just a small spark, so small as to be invisible, is sufficient to cause quite a violent explosion. The circumstances for producing such sparks are surprisingly propitious in operating rooms. In low-humidity surroundings, even the routine act of stripping a wool blanket from a rubber mattress can generate a potential difference of several thousand volts between the operating table and the floor. That potential, and static electricity that also builds up on mobile x-ray units with rubber wheels, can easily discharge through the air if an object or person of different potential comes near to it.

Fortunately, these circumstances rarely produce explosions. However, patients who have avoided electrostatic elimination have had their lives put in jeopardy, if not snuffed out, by hazards produced by electrical currents. This type of risk arises largely from the complexity of modern medical technology, particularly in such areas of the hospital as intensive care units. Seriously ill patients can easily receive dangerous electric shocks from defective or improperly maintained devices. Even doctors have on occasion been shocked when using defibrillators to jolt patients' hearts back to action because of poor insulation on these pieces of equipment.

Seriously ill patients who have electrical leads implanted into their hearts or arteries face another risk, that of microshock. In these cases, stray leakages of tiny amounts of current that would normally be harmless might well prove sufficient to stop a patient's heart.

---

tend to alter the very quantities they are recording. The only way to get around this problem, which turns up in a more fundamental form in our studies of the inside of the atom in Chap. 22, is to ensure that the measuring methods create as little disturbance as possible.

The instrument used to measure both current and voltage is known as a galvanometer, although it is referred to as an ammeter when current is its "target" and as a voltmeter when potential difference is being measured. The working principle of the galvanometer must await the next chapter. For the present, we can regard it as a little "black box" that has a certain resistance. To function as an ammeter, the box must be placed directly in the electrical circuit whose current it is measuring. To minimize its disturbance on the circuit, its resistance must, therefore, be as low as possible. On the other hand, when it is operating as a voltmeter to measure the potential drop across a resistor or the terminal voltage of a battery, the black box must be placed in parallel with the main circuit. To minimize the amount of current that it bleeds off the main circuit—the galvanometer's effect on this measurement—its resistance must be as high as possible.

We can illustrate how a single current-sensitive galvanometer can be adapted for both current and voltage measurements. Imagine that we have a galvanometer with a resistance of 20 Ω that can record current up to 0.01 A. If one wishes to adopt the instrument to function as an ammeter that measures currents up to 10 A, one must combine it with what is known as a shunt resistor. The arrangement is shown in Fig. 14.11. Because the galvanometer by itself can accommodate a maximum current of 0.01 A, the circuit must be arranged in such a way that (10.00 − 0.01) A, that is, 9.99 A of the maximum current to be measured, bypasses the meter through a low resistance, called a shunt. To determine the resistance of the shunt resistor that we require, we must recall that the parallel arrangement of this resistor and the galvanometer mandates that the potential differences across the galvanometer and the shunt must be equal. Thus, from Ohm's law, where I_s is the current passing through the shunt, I_{max} is the maximum current allowed through the galvanometer, and R_s and R_g are the resistances of the shunt and the galvanometer,

$$I_s R_s = I_{max} R_g \quad \text{or} \quad R_s = \frac{I_{max} R_g}{I_s} = \frac{0.01 \text{ A}}{9.99 \text{ A}} 20 \text{ Ω} = 0.02 \text{ Ω}$$

Comparison of the two resistances shows that the small shunt resistor will carry most of the current in the circuit. The ammeter consists of both the galvanometer and the shunt, and its resistance can be obtained using Eq. (14.9) for the sum of resistances in parallel:

$$\frac{1}{R_{ammeter}} = \frac{1}{R_g} + \frac{1}{R_s} = \frac{1}{20 \text{ Ω}} + \frac{1}{0.02 \text{ Ω}}$$

$$\frac{1}{R_{ammeter}} = \frac{1}{20} + 50 = 50.05$$

$$R_{ammeter} = 0.02 \text{ Ω}$$

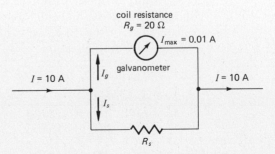

Figure 14.11 An ammeter constructed from a sensitive galvanometer and a shunt resistor.

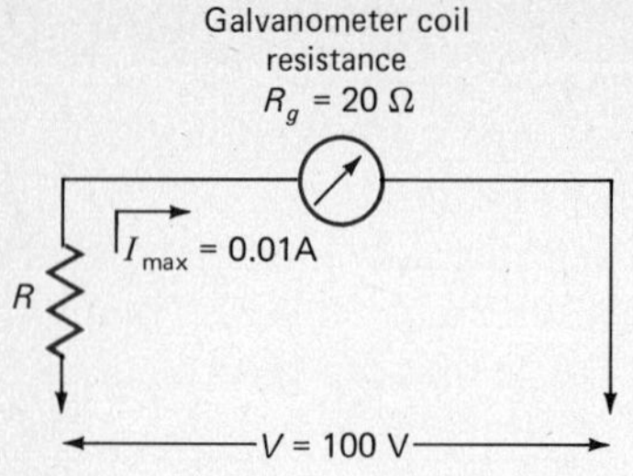

Figure 14.12 A voltmeter, 0 to 100 V. A galvanometer is placed in series with a resistor R to make a voltmeter with a full-scale deflection corresponding to 100 V.

The ammeter is a low-resistance instrument. Hence, it has, as required, a minimum effect on the circuit whose current it measures.

We can also convert the original galvanometer, with its resistance of 20 Ω, into a voltmeter by suitable combination with a resistor. Suppose we need a voltmeter that will read voltages up to a maximum of 100 V. Ohm's law tells us that 100 V would produce a current of 5 A if it were placed directly across the galvanometer's terminals. Since the galvanometer can withstand a maximum of 0.01 A, that action would obviously burn out the instrument. To reduce the current, and thus allow the galvanometer to read the voltage as required, we must place a large resistance in *series* with it, as shown in Fig. 14.12. We can find out the size of this resistance simply by applying Ohm's law to the combination:

$$V = IR_{tot}$$

where R_{tot} represents the combination of the resistance of the galvanometer and the unknown resistance R, and I is the current through both resistors. Since the maximum voltage is 100 V, and we want that to result in the largest current in the galvanometer and thus show a full-scale deflection,

$$V = 100 \text{ V} \qquad R_{tot} = R + 20 \text{ Ω} \qquad \text{and} \qquad I_{max} = 0.01 \text{ A}$$
$$V = IR_{tot}$$
$$100 \text{ V} = (0.01 \text{ A})(R + 20 \text{ Ω})$$
$$R = 10\,000 - 20 = 9980 \text{ Ω}$$

The total resistance of the voltmeter is 9980 + 20, or 10 000 Ω, which is much larger than the resistance of the galvanometer alone. As a result, very little current will pass through the voltmeter. Clearly the current in the original circuit will be disturbed only minimally if the resistances in the circuit are much lower than the 10 000 Ω of the voltmeter.

EXAMPLE 14.13

A milliammeter reads 1 mA full scale and has a resistance of 50 Ω. How would you convert this to a voltmeter reading 10 V full scale?

Solution

You require a single resistor in series with the meter movement, as shown in Fig. 14.12. Since 10^{-3} A produces full-scale deflection and the maximum is to be 10 V, we can use Ohm's law to determine the resistance R of this resistor:

$$V = I_{max} R_{tot}$$
$$V = (I_{max})(R + 50 \text{ Ω})$$
$$10 \text{ V} = (10^{-3} \text{ A})(R + 50 \text{ Ω})$$
$$R = 10^4 - 50 = 9950 \text{ Ω}$$

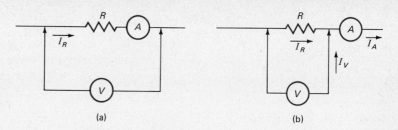

Figure 14.13 Voltmeter-ammeter method of determining resistance. (a) The ammeter measures the correct current; the voltmeter measures the potential difference across the resistor and the ammeter. (b) The voltmeter measures the correct value, but the ammeter measures the current through both the resistor R and the voltmeter.

Since how to measure both current and voltage has been demonstrated, one might think that it is a simple matter to obtain accurate values of unknown resistances just by carrying out both measurements simultaneously. One would be quite wrong because the method of measuring current compromises the means of determining voltage, and vice versa. The two circuits in Fig. 14.13 illustrate the problem.

In Fig. 14.13(a) the ammeter registers the current that passes through the resistance adequately enough, but the voltmeter measures the potential difference across both the resistance and the ammeter. Low as it might be, the ammeter's resistance affects the reading of the voltmeter, causing it to give a higher reading than the voltage across the resistance alone. This particular problem has been solved in Fig. 14.13(b), in which the voltmeter registers only the potential difference across the unknown resistance. But this adjustment interferes with the ammeter's ability to measure the current through the resistance. The instrument measures the sum of the required current and the certain small current that passes through the voltmeter. Obviously, neither method is anywhere near perfect for measuring resistance.

The way to overcome the problem is to use a complicated-looking device known as a bridge, which is designed so that the current through the galvanometer is exactly zero when a measurement is made. In the case of resistance measurement, the instrument is known as a Wheatstone bridge, after the nineteenth century English physicist Sir Charles Wheatstone who publicized the device, although he did not invent it.

Figure 14.14 shows a Wheatstone bridge circuit. The current from the battery splits into two branches, I_1 and I_2, at point A. Before they recombine, at D, the two currents split again, into a branch that joins the two as it passes through a galvanometer. The object of the exercise is to "balance" the bridge, by adjusting the values of the individual resistances until the galvanometer reads zero, indicating that there is no current in its branch of the circuit. In this circumstance, points B and C must be at the same potential. Thus, the potential difference between A and B equals that between A and C:

$$V_{AB} = V_{AC} \quad \text{or} \quad I_1 R_1 = I_2 R_3$$

Similarly, the potential difference between B and D is the same as that between C and D. Hence,

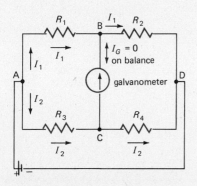

Figure 14.14 Wheatstone bridge circuit. This bridge circuit is used to compare unknown resistance with a known resistance.

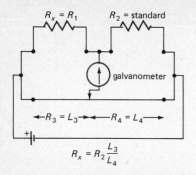

Figure 14.15 Wheatstone bridge circuit—slide-wire bridge. Two of the variable resistances are replaced by a long, straight wire.

$$V_{BD} = V_{CD} \quad \text{or} \quad I_1 R_2 = I_2 R_4$$

If we solve these two equations simultaneously, dividing one by the other, for example, then the currents I_1 and I_2 are eliminated:

$$\frac{I_1 R_1 = I_2 R_3}{I_1 R_2 = I_2 R_4} \quad \text{or} \quad \frac{R_1}{R_2} = \frac{R_3}{R_4}$$

Now, if R_1 is an unknown and R_2 is a standard resistance, we can determine R_1 in terms of R_2, using the ratio of resistors R_3 and R_4:

$$R_1 = R_2 \left(\frac{R_3}{R_4}\right)$$

Resistors R_3 and R_4 can be variable resistance boxes that allow for adjustment to produce the zero reading, otherwise known as the null condition. However, in many cases R_3 and R_4 can be combined in a long slide wire of uniform diameter and resistivity, as shown in Fig. 14.15. Here, the resistances R_3 and R_4 are replaced by the lengths L_3 and L_4 of the wire. One obtains the zero reading on the galvanometer by moving the pointer along the wire. In an elementary laboratory, $L_3 + L_4$ is often a 1-m wire mounted over a meter stick. In this bridge, the emf of the driving cell need not be known, but R_2 must be known accurately.

One can apply the same principle to measure an emf or a potential difference without disturbing the circuit or circuit element being measured; this type of null detector is called a potentiometer. It compares an unknown potential difference with a known standard, as shown in Fig. 14.16. Here, a working battery produces a current through a resistance—perhaps a long wire. There is then a potential difference IR across this wire. A parallel circuit is added which can be switched through either an unknown emf ($\mathcal{E}_x$) (or potential difference) or the known standard cell ($\mathcal{E}_s$). The connecting wire returning to the resistance wire R contains a galvanometer and a protective resistance R'. The latter resistance may be removed by closing the switch K when

Figure 14.16 A potentiometer circuit, used to determine unknown emf's or potential differences by comparison with a standard using a null detection method.

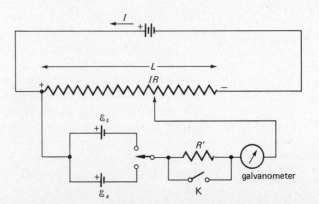

the galvanometer current is almost zero. The principle is first to balance the potential drop along the wire against the standard $\mathcal{E}_s$ and then balance it against the unknown $\mathcal{E}_x$ by obtaining a zero reading on the galvanometer in each case. Suppose the resistance wire is a single long, uniform wire; the resistance required to balance $\mathcal{E}_s$ is L_1; and the resistance required to balance $\mathcal{E}_x$ is L_2. Clearly, $\mathcal{E}_s = L_1 I$ and $\mathcal{E}_x = L_2 I$, and hence,

$$\frac{\mathcal{E}_x}{\mathcal{E}_s} = \frac{L_2}{L_1} \quad \text{or} \quad \mathcal{E}_x = \mathcal{E}_s \frac{L_2}{L_1}$$

With such a device you balance both $\mathcal{E}_x$ and $\mathcal{E}_s$ against L (or R) when no current is drawn from the standard $\mathcal{E}_s$ or the unknown $\mathcal{E}_x$. This enables you to determine the true emf of a cell rather than the terminal voltage because each measurement is made under null conditions. Of course, any potential difference can be measured by using the potentiometer. The standard cell must be protected, and current must not be drawn from it so that it does not deteriorate and vary from its standard value.

SUMMARY

1. An electric current is produced whenever charged particles flow from one place to another. The current is defined as the amount of electric charge that flows across a cross section of conducting material per unit time. The usual unit of current is the ampere, which equals 1 C/s.

2. A lasting current requires maintenance of a potential difference, normally with an electric cell or similar means of generation, a closed circuit, and carriers of charge that are free to move through the circuit. In metals, the charge carriers are electrons.

3. According to Ohm's law, the ratio of potential difference (voltage drop) across a metallic conductor to the current through the conductor is a constant for that piece of conductor. This constant is called the resistance and can be traced to the collisions of the electrons with the ions in the metal as the electrons move through the metal.

4. The resistance of any piece of material is related to its length, its cross-sectional area, and the material's resistivity. The latter is an inherent property related to the nature, purity, and temperature of the material.

5. Electrical resistance leads to loss of electrical energy in the form of heat.

6. Ohm's law accounts for the passage of current through simple circuits and junctions. More complex circuits can be understood using Kirchhoff's laws, which state that no charge accumulates at junctions and the sum of potential differences around any circuit is always zero.

7. The instrument known as a galvanometer can be linked with resistors in varous ways to produce accurate measurements of currents, voltages, and resistances.

QUESTIONS

1. Is the resistance of a 100-W bulb larger than, less than, or the same as that of a 60-W bulb?

2. What is the difference between terminal voltage, open circuit voltage, and the emf of a cell or battery?

3. Describe a way to determine in the laboratory the internal resistance of a battery.

4. To determine whether or not a dry cell is "good," it is really not sufficient to measure the emf of the cell. Why not? What measurement should be made to test a dry cell?

5. As little as 0.05 A of current can be fatal to a human being. Discuss the danger of the following sources: a 12-V automobile battery capable of delivering 35 A; a van de Graaff generator that can develop a potential difference of 1 million V; a 110-V household line; a 220-V circuit or a

PROBLEMS

Resistance and Resistivity

1. A charge of 240 C passes a point in a wire in 1 min. (a) What is the current in the wire, and (b) how many electrons moved through the wire in the minute?
2. In a house circuit, 10 A are drawn by an iron. In a period of 5 min, what charge passes through the iron?
3. A copper wire 0.8 mm in diameter carries a current of 1 A. If there are 8×10^{28} free electrons per cubic meter of copper, what is the drift velocity of the electrons in the copper wire?
4. A silver wire has a resistance of 10 Ω. If it is melted down and made into a new wire of only one-half the original length, what is its new resistance?
5. What is the resistance of 5 m of copper wire of diameter 2 mm? On the basis of this result, is it a good approximation to neglect the resistance of copper wires in circuits?
6. Try to identify the material of which a wire is made by determining its resistivity from the following data. The wire is 10 m long, has a diameter of 10^{-3} m, and has a total resistance of 0.33 Ω.
7. Compare the resistance of a series and parallel connection of three 4-Ω resistors.
8. Determine the equivalent resistance of the arrangement of resistors shown in the top figure.
9. Determine the equivalent resistance of the arrangement of resistors shown in the second figure.

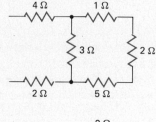

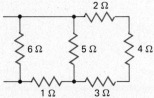

10. A circuit is formed using 6- and 12-Ω resistors connected as shown. What is the equivalent resistance?

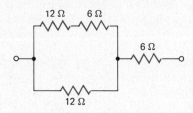

11. If the resistance of a long coil of wire is 100 Ω at 20°C and 116 Ω at 60°C, what is its temperature coefficient of resistivity?
12. The resistance of a lamp is 180 Ω when cold ($t = 20°C$). The resistance rises to 270 Ω when it is hot. If the temperature coefficient of resistance is 0.0005/°C, find the temperature of the hot lamp.
13. The tungsten filament of an ordinary 100-W light bulb when operated in 110 V operates at about 2000°C. Using the temperature coefficient of resistance for tungsten in Table 14.1, compute the resistance of the light bulb filament at room temperature 20°C.
14. A platinum resistance thermometer has a resistance of 10 000 Ω at 20°C. What is the temperature of a furnace in which the thermometer has a resistance of 24 120 Ω. See Table 14.1 for the temperature coefficient.
15. The resistance of a copper wire is 2 Ω at 20°C. What is its resistance at absolute zero? Is this answer suspect?
16. If the resistivity of cell membrane is 1.30×10^7 Ω·m, what is the resistance of 10 μm² of membrane that is 8×10^{-9} m thick? If a potential difference of 100 mV is established across the membrane, what current will be established and how many charges flow across the membrane each second?
17. A block of copper of volume 100 cm³ is to be made into a 2-Ω resistor. The resistivity of copper is 1.59×10^{-8} Ω·m. What is the cross-sectional area of the required wire?

Ohm's Law

18. A flashlight bulb carries 0.8 A when 3 V are used. What is its resistance?
19. A potential difference of 110 V is placed across a lamp. What current will be carried by the lamp if the lamp has a resistance of 200 Ω?
20. What are the largest and smallest resistances obtainable from a combination of ten 10-Ω resistors, each rated at 1 W? What is the maximum

current that can be allowed in each of these two combinations?

21. To determine the resistance of a human being under different conditions, connect a convenient voltage source of less than 10 V with one terminal to each hand. To determine the resistance, you will require a milliammeter and a voltmeter. Indicate where these go in a circuit diagram. If the voltage is 6 V and a current of 0.2 mA is present, what is your resistance?

22. A 1-A fuse is placed in a circuit with a battery having an emf of 6 V. If the internal resistance of the battery is 0.5 Ω, what minimum resistance should be placed in series with the fuse?

23. What is the smallest number of 10-kΩ, 1-W resistors needed to make a 1000-Ω, 10-W resistor? How are they connected?

24. A voltmeter with a resistance of 1000 Ω used to determine the voltage of a worn-out flashlight cell indicates 0.9 V. What is the internal resistance of the cell?

25. Determine the (a) internal resistance and (b) emf of a battery by the results of the following experiment. First, place a 10-Ω resistor in series with the battery, to measure a current of 0.48 A in the circuit. Next, add a second 10-Ω resistor in series, resulting in a current of 0.25 A. Neglect the resistance of the ammeter.

26. An old dry cell of 1.5 V has an internal resistance of 0.3 Ω and is used in a circuit with three lamps in series, each lamp having a resistance of 1 Ω. (a) What is the current in the circuit? (b) Compute the power loss in the dry cell. (c) Compute the terminal voltage of the dry cell.

27. The old-type Christmas tree lights were collections of identical lamps connected in series. Consider 10 identical lamps of resistance R in a string. The string dissipates 5 W when operated on 110 V. Compute the (a) resistance of each lamp, (b) current in each lamp, and (c) potential difference across each lamp. If the fourth lamp from one end burns out, (d) what will be the potential drop across it? (e) What will be the potential drop across the fifth lamp?

28. The more modern Christmas tree lights are connected in parallel. If 10 lamps are connected in parallel in one string and the power dissipated in each is 5 W when the string is connected to 110 V, (a) what is the resistance R of each lamp? (b) What is the potential difference across each lamp? (c) What is the total resistance of the array of 10 lamps? (d) How much power is dissipated by the string of lamps? (e) What is the current in each lamp? (f) If one lamp burns out, do the other lamps go out, burn brighter, burn less bright, or stay the same?

Power

29. What current exists in a 100-W light bulb when it is in a 110-V household circuit?

30. What is the maximum voltage you can safely place across a 10-kΩ resistor if it is rated for 0.50 W?

31. A 12-V automobile battery can deliver 35 A to start the engine. What power does the battery deliver? Compare this power with the power dissipated by the headlights run off the battery when the engine is off. The lights require 2 A.

32. A dry cell that delivers an emf of 1.5 V is connected in series with three bulbs with resistances $R_1 = 2\,\Omega$, $R_2 = 3\,\Omega$, and $R_3 = 5\,\Omega$. Compute the current in each bulb and the power dissipated in each bulb.

33. If a resistance of 6 Ω is connected across the terminals of a 120-V source, how much current flows in it? How much power is developed in it?

34. An electric lamp carries a current of 2.5 A when a voltage of 110 V is placed across it. (a) What is the resistance of the lamp? (b) If the current is on for 60 s, how much energy is delivered to the lamp? (c) What happens to this electrical energy?

35. Normally in a household, each circuit (110 V), which may contain several outlets or lighting fixtures, is fused to prevent more than 15 A being drawn from that single circuit. (a) Why is it unsafe to bypass the fuse so that we can draw, say, 30 or 60 A? (b) Could you plug a 1000-W toaster and a 400-W coffee percolator into this circuit without blowing a fuse? (c) Could you plug a 150-W electric blanket, a 40-W radio, four 100-W lamps, and a 2-W electric clock into this circuit without blowing the fuse?

36. A glass-blowing shop has an electric furnace for annealing glass. Glass is left in the furnace for 4 h while the temperature rises to about 1050°F. Then the electricity is turned off, and the furnace cools gradually for 16 h before the glass is removed. This electric furnace uses 35 A on a 220-V circuit when in use only 4 h a day for only 2 days a week. (a) Compute the rate at which electrical energy is used by this furnace. (b) Compute the resistance of the heating elements (coils of wire) in this furnace. (c) At a rate of 5¢/kWh, compute the cost of operating this electric furnace for an 8-week billing period.

37. A current of 10.0 A passes through a wire that is 3.0 m in length and has a cross-sectional area of 5×10^{-9} m² and a resistance of 15.0 Ω. (a) How much power is dissipated in the wire? (b) What is the voltage drop along 1 m of the wire? (c) If this wire were drawn out so that its cross-sectional area is reduced to

Circuit Problems

38. Consider the three resistors connected in the four ways shown in Fig. 38. Compute the current in each resistor, the current in the 6-V battery, and the power dissipated in each resistor.

39. What is the current in each resistor in the circuit shown?

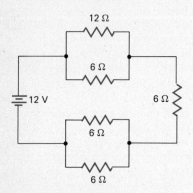

40. Find the power dissipated in one of the 10-Ω resistors.

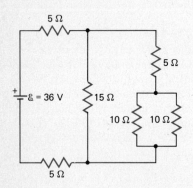

41. Determine the current through the battery and the current through each resistor in the circuit shown in Fig. 41.

42. Determine the current in each resistor and the potential difference across each resistor in Fig. 42.

43. Compute the current in the 3-Ω resistor in the circuit in Fig. 43.

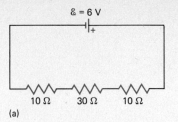

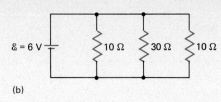

(a)   (b)

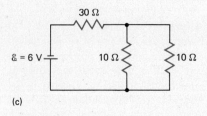

(c)

Figure 38

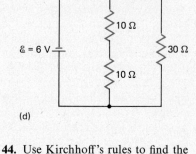

(d)

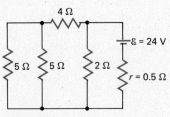

Figure 41

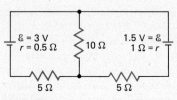

Figure 42

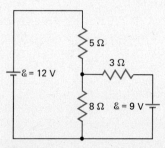

Figure 43

44. Use Kirchhoff's rules to find the current in the 4-Ω resistor in the figure below.

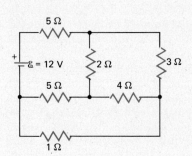

45. Solve for the currents through each battery in the figure below. Use Kirchhoff's laws.

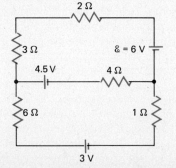

46. A circuit is formed of four resistors of 6, 9, 18, and 2 Ω, as shown below; the internal resistance of the 25-V source is 2 Ω. (*a*) Calculate the effective resistance in the circuit. (*b*) Calculate the current through the 2-Ω resistor. (*c*) Calculate the voltage drop across the 6-Ω resistor. (*d*) Calculate the power loss in the 25-V battery.

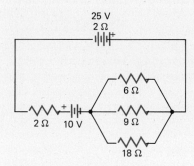

Measuring Devices

47. A voltmeter has a full-scale deflection of 150 V. Its sensitivity is rated as 20 000 Ω/V. What is the resistance of the meter, and what current is required by the voltmeter when it is registering 150 V?

48. You have a meter movement that will give a full-scale deflection when 10 μA pass through it. To make an ammeter with a full-scale deflection corresponding to 10 A, how must you alter the meter?

49. You have a meter movement that will give a full-scale deflection when 10 μA pass through it. To make a voltmeter with a full-scale deflection of 1 V, how must you alter the meter?

50. A milliammeter reads 1 mA full scale and has a resistance of 30 Ω. Draw a circuit diagram to illustrate how to convert this milliammeter to a voltmeter with a full-scale deflection of 5 V. What resistor is required for this alteration?

51. A 10-Ω galvanometer with full-scale sensitivity of 20 μA must be made into a voltmeter reading 100 V, full scale. (*a*) What resistance must be used, and where is it placed? (*b*) What resistance will convert the galvanometer into an ammeter reading 15 A full scale, and where is it placed?

52. A Wheatstone bridge circuit is shown in the figure below. The resistances are 5, 10, and 30 Ω. Find the unknown resistance.

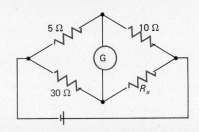

53. In the Wheatstone bridge shown below, R_1 is 20 Ω and R_2 is 40 Ω for balance. If the known standard resistance is 15.0 Ω, what is the unknown resistance R_x?

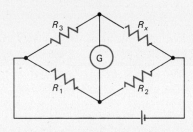

54. In a commercial Wheatstone bridge, the ratio R_2/R_1 can be set between 0.001 and 1000 in integral powers of 10 by a single switch. If the ratio R_2/R_1 is 100 and R_s is 47.0 Ω when the galvanometer reading drops to zero, what is the value of R_x, the unknown?

55. A voltage divider is shown in the figure below. What is the output voltage, that is, the potential difference, across the 1-kΩ resistance?

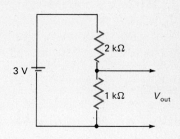

56. In a slide-wire potentiometer (see Fig. 14.16) the standard cell of emf = 1.0179 V gives a null reading when using 0.38 of the length of the wire. The unknown emf produces a null using 0.67 of the length of the wire. What is the emf of the unknown cell?

57. A slide-wire potentiometer like that shown in Fig. 14.16 is used to measure the terminal voltage of a cell while it is connected in series with a 25-Ω resistor. The slide wire balances the terminal voltage of the cell at 65 cm, and it balances the emf of the cell at 80 cm. What is the internal resistance of the cell?

Special Topic
How the Nervous System Works

The system that coordinates all the muscular activities of living creatures is a complicated mixture of electrical and chemical signals. The fundamental vehicles for transmission of messages between the brain and the rest of the body are cablelike cells known as neurons, which pass messages in the form of electrical pulses. Interconnections between the neurons in the body—humans possess typically 10^{10} such cells—and between neurons and muscles are generally made through a chemical link.

A typical neuron consists of a cell body, complete with its cell nucleus, which receives signals via ends that are known as dendrites, and passes the signals on as electrical pulses through a long tail that is called the axon. Some axons in the human body, such as those linking the spine with the fingers and toes, can be as much as a meter in length. Each individual neuron reacts in exactly the same way to an incoming signal. The electrical pulse that passes along the individual axon never varies in strength or duration. The only indication given by the neuron of the strength of any stimulus is the number of pulses produced by that stimulus.

Neurons can be divided into three separate types. Sensory neurons are those that are influenced directly by the body's sensory organs and thus provide information for the brain. Interneurons pass this information from one neuron to another into the central nervous system. And motor-neurons receive the response from the brain and pass it on, again via the interneurons, to the body's muscle cells.

The passage of signals between neurons and from neurons to other types of cells can occur either electrically or chemically. Electrical transfer is the rule in the more primitive organisms, but in higher organisms, the method is strictly chemical. A gap of a few hundred angstroms exists between the nerve ending of an axon and the cell to which it must transfer its message. When the signal reaches this gap, which is known as a synapse, it causes release of a chemical which rapidly moves across the gap and stimulates the adjacent cell, thus ensuring reliable passage of the message.

15 Moving Charges and Magnetic Fields

15.1 THE EARTH'S MAGNETIC FIELD

15.2 ELECTRIC CHARGES IN MAGNETIC FIELDS

SHORT SUBJECT: The Race for the Monopole

15.3 THE MAGNETIC FIELD AND ELECTRIC CURRENTS

SHORT SUBJECT: The Electromagnetic Accelerator

15.4 CURRENT LOOPS AND MAGNETS

15.5 METERS AND MOTORS
Summary
Questions
Problems

SPECIAL TOPIC: Earth's Changing Magnetic Field

Most of us have played with a horseshoe or bar magnet and seen it attract small objects made of iron and steel but fail to attract similar objects constructed of other metals such as aluminum and copper. Later, in more sophisticated experiments, we might have looked at the operation of a magnetic compass or studied the effect of a bar magnet on iron filings scattered over a piece of paper above the magnet. The iron filings line up in the magnetic field produced by the magnet, as shown in Fig. 15.1. These simple experiments—the most obvious examples of the strange phenomenon called magnetism—are the starting point for a most important part of physics. Magnetism's really important applications are those involving the link between magnetism and electricity—the subject termed **electromagnetism.** The generation of electrical energy and the operation of electric motors are just two of the vital everyday events that stem from electromagnetic effects.

We can best introduce electromagnetism by summarizing the basic properties of natural magnetism. Magnetism is named after the city of Magnesia in Asia Minor, where civilized people first unearthed the magnetic material lodestone (or leading stone). This material is an ore

Figure 15.1 (a) Field from a bar magnet. (*Trudy Tuttle*) (b) Field lines around a horseshoe magnet. (*Kodansha Co. Ltd., Tokyo*) Iron filings are oriented along the magnetic field lines.

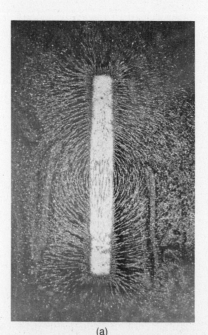
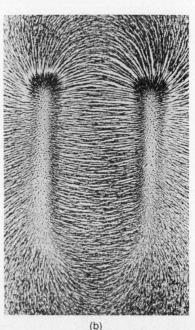

(a) (b)

The city of Magnesia also gave us the laxative Milk of Magnesia because the compound MgO, called magnesia, was first obtained there.

of iron, but scientists today recognize many elements and compounds that possess the property most people call magnetism and to which experts refer more correctly as ferromagnetism. In addition to iron, these elements and compounds include cobalt and nickel and a number of what are known as rare earth elements. Most steels are ferromagnetic because they are alloys of iron. Curiously enough, however, that steel contains iron is not in itself sufficient to make it ferromagnetic, as proved by certain stainless steels that are not ferromagnetic.

What is so special about ferromagnetic substances as opposed to all others? It is not their ability to be magnetized when placed in a magnetic field, because almost every element and compound is magnetized *while* in a magnetic field. However, most materials lose their magnetism when removed from the magnetic field; materials are known as either paramagnetic or diamagnetic, according to whether they are magnetized in the same direction as the magnetic field to which they are exposed or in the opposite direction. Ferromagnetic substances, by contrast, exhibit magnetic effects that are 1000 times larger than paramagnetic or diamagnetic effects and always retain a portion of the magnetization which they experience in a magnetic field and, under appropriate conditions, can be induced to retain so much magnetization that they become magnets—strictly, permanent magnets—themselves. A magnet's permanent magnetism can be destroyed by forcibly hammering the magnet, heating the magnet above a specific temperature known as the Curie temperature, or influencing the magnet with a time-varying electromagnetic field.

In addition to possessing a magnetic field similar to an electric field, permanent magnets resemble electricity in that they have two types of magnetic poles: north and south. These poles are similar in their way to positive and negative charges. Just as unlike electric charges attract each other and like electric charges repel, so magnetic poles have similar interactions. However, magnetic poles do not "attract" or "repel" electric charges. Place two bar magnets side by side, with the opposite poles adjacent to each other, and the two magnets come together with a bang. Place them together, with the like poles contiguous, and the magnets separate.

A typical bar magnet contains a north pole at one end and a south pole at the other. One interesting and vital fact about permanent magnets is that when one splits them carefully, each part becomes a permanent magnet itself, with its own north and south poles, as shown in Fig. 15.2. These poles appear in pairs, unlike electric charges, which can be isolated. Careful scientific investigation has shown that the individual atoms or molecules of a ferromagnetic material are themselves magnets. The difference between permanent magnets and other ferromagnetic materials involves the cooperative lineup of their respective atoms or molecules.

Any piece of iron, steel, or other ferromagnetic material consists of magnetic domains, which are collections of about 10^{15} atoms whose

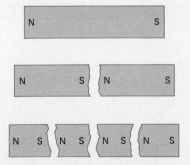

Figure 15.2 Cutting a bar magnet always results in a pair of magnets. The north and south poles always appear as pairs.

Figure 15.3 Magnetic domains. Groups of "atomic magnets" in local regions align with one another, producing magnetic domains. (a) These domains are randomly oriented when the material is demagnetized or in its unmagnetized state. (b) When the material is placed in an external magnetic field, some domains grow at the expense of others, producing a net permanent magnetization.

(a) demagnetized

(b) magnetized—not all domains are united to form a single domain

north and south poles are aligned in exactly the same direction. These domains are relatively small—occupying a space of about 0.01 mm on a side—and in a piece of magnetic material that is not a permanent magnet they are all aligned in different directions, as shown in Fig. 15.3. When the magnetic material is placed in a magnetic field, however, the field forces them to line up together, as shown in Fig. 15.3(b). Domains that point in the same direction as the magnetic field tend to grow at the expense of other domains pointing in different directions: if the magnetic field is strong enough, the object ends up with only a few very large domains, and thus becomes a permanent magnet itself. Heating or beating a permanent magnet sends these nicely aligned, large domains into disarray and thus destroys the magnet's magnetic properties. This effect is an illustration of the introduction of entropy, or disorder, of the type outlined in Chap. 11.

The ultimate permanent magnet would consist of one single magnetic domain; thermal vibrations and faults in materials prevent this perfection in real life.

Figure 15.4 The earth's magnetic field, highly idealized. A diagrammatic sketch of the lines of force of the earth's magnetic field, showing magnetic poles, geographic poles, the magnetic axis, and the axis of the earth's rotation. The magnetic poles, indicated by SMP and NMP, shift positions slowly but independently through the years and are not always opposite each other in the same locations on the same axis. Thus the intersection of the magnetic axis and the plane of the magnetic equator does not coincide with the intersection of the earth's axis of rotation and the plane of the geographic equator. Small arrows on broken lines indicate positions of a magnetic needle free to move in space at various locations on the earth's surface—in short, the inclination, or magnetic dip.

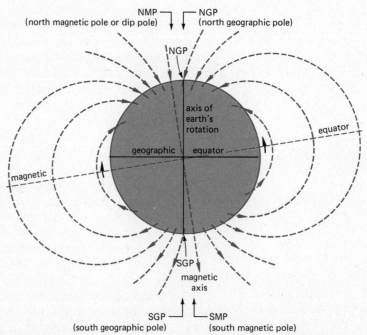

15.1 THE EARTH'S MAGNETIC FIELD

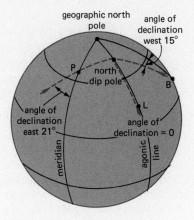

Figure 15.5 Magnetic declination. Because the magnetic or dip poles do not coincide with the geographic poles, the compass needle does not point to true north, except along the agonic line. The angle of divergence of the compass from geographic north is the declination and is measured east, west, of north. B = Boston, Massachusetts, L = Lansing, Michigan, P = Portland, Oregon.

Because of the nature of magnetic attraction, the planet's magnetic poles are the reverse of the geographic ones.

15.2 ELECTRIC CHARGES IN MAGNETIC FIELDS

The largest magnet with which we are all familiar is the earth itself, whose magnetic field lines form the pattern illustrated in Fig. 15.4. Although we regularly use magnetic compasses to locate both north and south, this picture illustrates that our planet's magnetic poles do not coincide with its geographic poles. Unlike the geographic poles, the magnetic poles are not fixed in position. Over the eons the poles have drifted considerably, and according to many geophysicists they have actually switched over from north to south, and vice versa, at certain times in prehistory, as discussed in this chapter's special topic.

The magnetic poles can be traced using a magnetic dip needle, which is a magnetized needle that is free to rotate around a horizontal axis. This type of compass needle always aligns itself with the magnetic lines of force at its location, unlike a normal compass needle that, lying in a horizontal plane, aligns itself with only the horizontal component of the earth's magnetic field. The principal magnetic poles of the earth are located far beneath two of the positions on the earth at which the magnetic lines of force are vertical. At present, the north end of a magnetic dip needle points directly downward at a spot about 1800 mi north of Winnipeg, Canada, at a longitude of 100.5°W and a latitude of 75.5°N, and points vertically upward on the icy coast of Antarctica, at longitude 143°E and latitude 67°S. The angle between the magnetic dip needle and the earth's surface is known as the **magnetic inclination,** or angle of dip. This angle is 90°, just above the two magnetic poles.

Another definition relevant for description of the earth's magnetic field is the angle of variation, or **magnetic declination,** at any point on the surface of the globe. This is the angle between the direction in which a compass needle points and the true north. The magnetic declination is zero only on the single meridian, which passes through both the geographic and magnetic poles, as can be seen in Fig. 15.5. This meridian is termed the agonic line; it passes close to Lansing, Michigan, and Atlanta, Georgia.

A final, rarely noted point regarding the earth's magnetism: because of the nature of magnetic attraction, the planet's magnetic poles are the reverse of the geographic ones. Since unlike magnetic poles attract each other, the north end of a compass needle seeks out a magnetic south pole—the pole located near the north geographic pole—and the south end of the needle points to a magnetic north pole located near the south geographic pole. The confusion arises because historically the end of the compass that points north was called the north-seeking (north) pole. To be consistent with the known fact that unlike poles attract, the magnetic pole near the north geographic pole must be labeled a magnetic south pole.

With this basic background under our belts, we can now approach the major topic of this chapter: the influence of magnetic fields on moving electric charges. The effect of a magnet on a moving charge can be

The Race for the Monopole

Like the effort to split the atom, the search for a single unit of magnetism has consumed many years of scientific effort. But unlike atom splitting, the quest for an entity that is a pure north or south magnetic pole has not yet reached a definite conclusion. Many physicists feel that pure magnetic poles, termed magnetic monopoles, do exist, because single positive and negative electric charges exist. But so far no one has been able to prove convincingly that he or she has spotted a monopole.

Because the magnetic monopole has proved so elusive, scientists are concentrating their search for it in out-of-the-way places they judge likely to attract individual pieces of magnetism, such as magnetic rocks, the sea bed, particle accelerators, and even the soil gathered by astronauts on the surface of the moon. But the greatest stir over magnetic monopoles in recent years resulted from a probe in the earth's upper atmosphere. In August 1975 physicists from the University of California at Berkeley and the University of Houston announced that they had identified a strange particle track in a detector consisting of layers of plastic that had been raised to an altitude of 130 000 ft by a balloon. So unusual was the track, the physicists reported, that only a magnetic monopole could have caused it. The would-be monopole, analysis suggested, was hundreds of times more massive than the proton; if two opposite magnetic monopoles approached each other, the measurements seemed to show, they would be attracted by a force 18 000 times as strong as that between a proton and an electron.

The assertion that a monopole had indeed been detected was quickly questioned. Many physicists argued that the find was actually a heavy, ionized nucleus. Today, the report remains in limbo, with the general opinion among physicists in the field being that the track was not caused by a monopole. Nevertheless, many experts are convinced that a monopole will turn up sooner or later, and some still believe that the balloon "detection" was valid.

---

clearly seen in a simple experiment using the apparatus sketched in Fig. 15.6. Here a high voltage is applied between the two electrodes of an evacuated tube (a cathode ray tube) that contains just a small amount of mercury vapor. As the electrons make their way from the negative electrode to the positive one carrying the electric current, they strike the mercury atoms and cause the atoms to emit bursts of light. These bursts from the individual atoms come together into a beam that outlines the track of the moving, negatively charged electrons.

Having set up apparatus in this fashion, one could bring up a permanent magnet close to the tube; the beam would bend, showing that the magnet is altering the direction of the moving electrons. In Fig. 15.6 the magnet used is actually a bar magnet twisted into a C shape. A little experimentation shows that the bending of the electron beam is most pronounced when the bar magnet is perpendicular to the beam, and the deflection is next to nothing when the beam and the bar magnet are parallel. The **direction in which the beam actually bends is perpendicular to both its own direction** and **the direction of the magnetic field** through which it travels; in the particular case shown in Fig. 15.6,

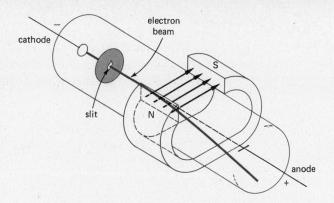

Figure 15.6 Bending of a beam of electrons as they pass through a magnetic field. The electrons emitted from the cathode of a vacuum tube are collimated by a slit to provide a beam of electrons that is accelerated to the anode. In the presence of a magnetic field, the beam is deflected perpendicular to both the magnetic field and the direction of the beam. Electrons colliding with a small amount of gas in the tube enable detection of the beam since the excited atoms give up light when they deexcite.

Here is a convention used to indicate the direction of an electric current in a wire that is perpendicular to the printed page. Imagine an arrow pointing in the direction of the conventional current—the direction in which positive charges would move. A cross inside a circle, Fig. 15.7(b), represents the tail feathers of an arrow and shows that the current is traveling away from the reader, directly into the page. A circle with a dot in its center, Fig. 15.7(c), represents the front of the arrow and indicates that the current is coming out of the page toward the reader.

the electron beam tends to curve down when the magnetic field, directed from the north to the south pole of the magnet, is oriented as shown. If the magnet is turned around, reversing the position of the north and south poles and hence reversing the direction of the magnetic field, the beam of electrons bends upward.

This type of influence is not restricted to the relatively free movement of electrons through a vacuum or mercury vapor, as Fig. 15.7(a) amply demonstrates. A wire, connected to a battery via a switch, is placed in the magnetic field between the two poles of a horseshoe-shaped magnet. When the switch is turned on to allow an electric current through the circuit, the wire moves downward. If either the current or the magnetic field were reversed, the wire would move upward once the charges started to flow.

Experiments show that the maximum force on a current-carrying wire between two magnetic poles occurs when the wire is perpendicular to the magnetic field, whereas the minimum force, zero, occurs when the wire is parallel to the magnetic field. The **force moves** the **wire in a direction perpendicular to both the magnetic field** and the **current that it is carrying;** its strength depends on the intensity of the magnetic field and the magnitude of the current in the wire, in addition to the relative orientations of the wire and the magnetic field.

We can put the observations concerning the force on a charge in a magnetic field into mathematical form. The force is found experimentally to be proportional to the magnitude of the charge moving through the magnetic field, the speed at which it moves, and the magnitude of the component of the magnetic field perpendicular to the velocity of the charge. The complication here arises from the directional character of this interaction. Figures 15.6 and 15.7 show the geometry for a maximum force. For example, in Fig. 15.6 the force on an individual electron is perpendicular to both the direction of the magnetic field and the direction in which the charge was moving (the direction specified by the velocity v). In Chap. 7 we ran into a similar interaction when we found that the maximum torque occurred when the direction of the force applied at a point on an object had the largest lever arm. We expressed the torque and the angular momentum as vector cross products. For a positive charge Q moving with velocity **v** in a magnetic field whose strength and direction are represented by a vector quantity **B**, the so-called magnetic induction, we can write the vector force on the charge as

$$\mathbf{F} = Q\mathbf{v} \times \mathbf{B} \tag{15.1}$$

Of course, the vector cross product can be rewritten for the magnitude of the force as

$$F = QvB \sin \phi \tag{15.2}$$

where ϕ is the angle between the direction of the magnetic field and the direction the positive charge moves. When **B** and **v** are at right angles, $\phi = 90°$ and $\sin \phi = 1$. The force will then be a maximum, in agreement with experiment. Likewise, if **v** and **B** are parallel, that is, when the charge moves parallel to the magnetic field, the force will be zero, since $\phi = 0°$ or $180°$ and $\sin 0° = \sin 180° = 0$. Of course, the force can also be zero if the magnetic field is zero, or, perhaps more important to remember, **the force will be zero if the charge is at rest in the magnetic field. When v = 0, then F must equal zero.** Charges only experience a force in a magnetic field when they move.

To determine the direction of the force on the charge, we can also use the definition (Chap. 7) of the vector cross product. First, we shall stick with convention and consider the direction of the magnetic force on a moving *positive* charge since the direction in which positive charges would move is the direction of a conventional current.

We use *the right-hand screw rule* for our vector equation $\mathbf{F} = Q\mathbf{v} \times \mathbf{B}$. To discover the direction of the force, rotate the vector **v** (the direction positive charges would move) through the smaller angle ϕ toward the vector **B**. The direction of **B** is the direction of the magnetic field lines from magnetic north to south pole. Imagine that you carry out this process by turning, with a screwdriver, a right-hand screw through the angle ϕ. Then, the direction of the force *for a positive charge* is the direction in which the screw advances forward, as shown in Fig. 15.8(a). Our rule is defined in terms of the motion of positive charges, and most often the charge carriers are negative.

To use the cross product and the right-hand screw rule for negative charges, just consider that positive charges would move in exactly the opposite direction to the negative charges. Hence, the direction of **v** used in the right-hand screw rule should be opposite the direction of the moving electrons. The resulting direction for the force will then be correct for the negative electrons.

There are several alternative "rules" that can be used to find the direction of the force on a charge moving through a magnetic field. However, the right-hand screw rule is usually the easiest to use once

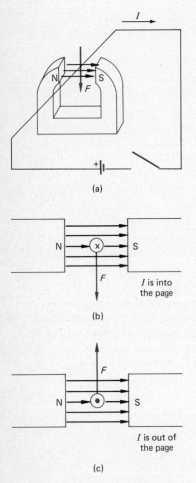

Figure 15.7 Force on a current-carrying wire in a magnetic field. (a) The force is perpendicular to both the current and the magnetic field. (b) Front view of (a). (c) Front view of (a) but with the current reversed. *Note:* the horizontal black lines indicate only the magnetic field due to the permanent magnet.

Figure 15.8 (a) The direction of magnetic force on a moving positive charge is in the direction that an ordinary (right-hand) screw would advance when rotated from the direction of **v** toward the direction of **B**. (b) A right-hand rule that is consistent with the screw rule. If the right thumb is placed in the direction of the moving positive charge and the fingers in the direction of **B**, the force will be in the direction the palm is pushing.

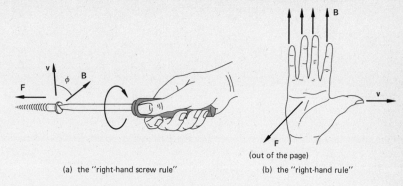

(a) the "right-hand screw rule"

(b) the "right-hand rule"

you learn how to approach it. But you might wish to resort to the following: Open your right hand and place your right thumb along the direction in which a *positive* charge would move, that is, along the direction of the velocity **v**, and place your other fingers in the direction of the magnetic field, which always goes from north pole to south pole. Your palm will then be pushing in the direction of the force on the charge, as shown in Fig. 15.8(b).

We can use Eq. (15.1) to *define* the units of the quantity **B**, which is called the magnetic induction; **B** is a measure of the strength of the magnetic field. In the mks units, **B** is measured in newtons per coulomb times meters per second, or in newtons per ampere-meter (N/A·m). This unit for **B** is called the tesla (T), in honor of an American engineer Nikola Tesla (1857–1943), who around the turn of the century made great strides in developing alternating current and transmission of electrical power by electromagnetic waves. A more commonly used unit of magnetic induction is the gauss (G), named after the German mathematician and physicist Johann Karl Friedrich Gauss (1777–1855), who in the early nineteenth century achieved a great deal of understanding of the earth's magnetism. One tesla is equal to 10^4 G.

EXAMPLE 15.1

Calculate the magnitude of the force on a free electron in a **B** field of magnitude 10 T if **v** is initially perpendicular to **B** and of magnitude 10^7 m/s.

Solution

$$F = QvB \sin \phi = (1.6 \times 10^{-19} \text{ C}) \times (10^7 \text{ m/s}) \times (10 \text{ T})(1)$$
$$= 1.6 \times 10^{-11} \text{ N}$$

EXAMPLE 15.2 Calculate the magnitude of the force on length ΔL of a current-carrying wire if n and v_D and the diameter of the wire are given; see Eq. (14.2). The wire is in a magnetic field **B**, not necessarily perpendicular to $\mathbf{v}_D$.

Solution
$$F = QvB \sin \theta$$
$$= (neA \, \Delta L) v_D B \sin \theta = (I \Delta L) B \sin \theta$$

The effect of magnetic fields on moving electric charges was one of the most important principles involved in advancing the science of atomic physics early in this century because it made possible the design of apparatus capable of determining the ratio of the charge to the mass of electrons (and ions) and of the mass spectrometer. The mass spectrometer, an instrument that can identify individual elements, is an important analytical tool used today by chemists and other scientists.

The experimental setup for measuring e/m, the ratio of the charge to the mass of electrons or other charged particles, is shown in Fig. 15.9. Choosing electrons for simplicity, we fire a beam of those negatively charged particles into a magnetic field at right angles to the beam. The electron beam is contained in a vacuum tube that is not shown in Fig. 15.9. Figure 15.9 shows the electron beam moving from left to right, which means that *positive current* travels from right to left. The magnetic field is directed into the page. Use of the right-hand rule, or right-hand screw rule, indicates that the resulting force on the electrons acts downward initially, after which it continues to act in such a way that the electrons move in a circle. The force is always perpendicular to the velocity! Because the angle between the velocity and the field is 90°, for which angle $\sin \phi$ is 1, the force F is given by the simple expression $F = evB$. Here the charge Q equals the charge on the electron e. All three factors are constant, and thus the magnitude of the force is steady also.

Now, the force is perpendicular to the velocity of the electrons, and hence it cannot change their speed. Therefore this force does not change the energy—it does no work. In fact, this force provides the

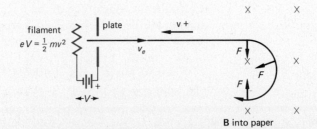

Figure 15.9 Motion of electrons in a magnetic field. The magnetic field is directed into the paper. Negative charges entering from the right are bent in circular orbit.

centripetal force for the circular motion. We encountered in Chap. 4 the equation for the required centripetal force:

$$F = \frac{mv^2}{R}$$

where m is the mass of the electron and R the radius of the circle in which it moves. Equating these two expressions for the center-seeking force, we develop the expression

$$\frac{mv^2}{R} = evB$$

or

$$R = \frac{mv}{eB} \tag{15.3}$$

In other words, **the radius of the circle traveled by the electron beam is directly proportional to the speed of the electron and inversely proportional to the strength of the magnetic field.** A magnetic field bends fast-moving electrons very little, although increasing the magnetic field increases the amount of bending.

We can clearly determine the ratio of the electron's charge to its mass using this equation if we know the strength of the magnetic field and can measure both the speed of the electron beam and the radius of the circle it creates in the field. The only real poser here is the speed of the electrons. We can work this out most easily by using an "electron gun" to produce our beam of electrons.

An electron gun consists of two components. A hot-wire filament acts as the negative terminal from which electrons are "boiled off," and a plate with a hole in its center, charged to a potential V above that of the filament, is the positive terminal. Electrons accelerate from the filament to the plate, and some pass through the hole in the center as a thin beam. This beam continues in a straight line with a speed v until it reaches the region of the magnetic field.

The energy given to the electrons is the product of the charge on the electron and the potential difference between the terminals in the tube. Because the electrons start off from rest at the negative terminal, or at least with a negligibly small speed, this energy equals the kinetic energy of the beam, $\frac{1}{2}mv^2$. Thus,

$$eV = \tfrac{1}{2}mv^2$$

from which we obtain

$$v = \sqrt{\frac{2eV}{m}}$$

Thus, the speed of the electrons is related to both the potential difference between the terminals of the electron gun and the e/m ratio in which we are interested. Substituting this formulation in Eq. (15.3),

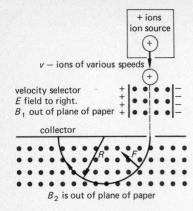

Figure 15.10 A mass spectrometer employing a velocity selector.

$$R = \frac{mv}{Be} = \frac{m}{Be}\sqrt{\frac{2eV}{m}}$$

Using an algebraic move or two, we obtain the ratio of the charge of the electron to its mass in terms of measurable quantities:

$$\frac{e}{m} = \frac{2V}{B^2 R^2}$$

The same principle is used in a slightly different way in the mass spectrometer to separate charged ions of different masses. This instrument, invented early in this century by the English physicist John Aston, was responsible for huge leaps in the understanding of atomic physics during the earliest years of that subject. The mass spectrometer identifies individual ions in mixtures. The instrument, shown in Fig. 15.10, consists of a source of ions, a velocity selector, a region in which the ions are bent by a magnetic field, and a collector.

In action, positive ions leave the ion source in much the same way that a beam of electrons departs from an electron gun. Because of the variety of their masses and charges, however, various ions depart at different speeds. The ions encounter an electric field in the velocity selector, which is applied at right angles to the magnetic field. This means that the force on the moving charges produced by the electric field exactly opposes that produced by the magnetic field B_1. As a result, some ions will pass through both fields without suffering any net deflection from their original path. Let us say that these ions have charge Q and velocity v. Then, balancing the two sets of forces,

$$QE = QB_1 v$$

Hence, the speed of these undeflected ions is given by $v = E/B_1$, independent of Q or m.

Ions moving faster or slower than this velocity v will be deflected away from the mainstream and thus will not reach the region where the second magnetic field, B_2, is applied. In other words, appropriate adjustment of the electric field and the first magnetic field allows the researcher to select ions with specific *velocities* for further study.

The further study involves looking at the effect of the second magnetic field on the ions. In the region of field B_2 the magnetic force provides the centripetal force necessary to bend the beam of charges with velocity v. Solving the equation of the type already dealt with in the e/m example for the radius R, we obtain

$$\frac{mv^2}{R} = QvB_2$$

$$R = \frac{vm}{QB_2}$$

If we use the velocity of the ions that pass through the velocity selector for the value of v, we have a simple expression for the radius of curvature in terms of easily measured quantities:

$$R = \frac{Em}{QB_2 B_1}$$

That is, the radius of curvature is proportional to the mass of the ions, as long as the ions possess the same charge. Early atomic chemistry established that the charges on normal ions are either equal to that on the electron or two or three times as great. Given so few allowed charges, the device can obviously separate and identify ions, and hence the atoms from which they derive.

So far in this chapter we have concentrated on electric charge in our equations. However, we can also derive Eq. (15.1) in terms of electric current by slightly altering our theoretical approach. Consider the force on a segment of wire whose length is Δl and through which electrons drift with a speed v. Because each electron carries a single unit of charge (1.6×10^{-19} C, as we recall from Chap. 13), the total charge Q is given by Ne, where N is the number of electrons. The current is given by the equation

$$I = \frac{Q}{t} = \frac{Ne}{t}$$

In addition, the speed of the electrons along the wire is given by the elementary equation $v = \Delta l/t$. Combining these equations and eliminating t,

$$I = \frac{Nev}{\Delta l}$$

Thus,

$$I \Delta l = Nev$$

Substituting this in Eq. (15.2) after first expressing Q as Ne, $F = QvB \sin \phi = NevB \sin \phi$,

$$F = I \Delta l B \sin \phi$$

or, using vectors,

$$\mathbf{F} = \Delta l \mathbf{I} \times \mathbf{B} \qquad (15.4)$$

The angle ϕ is the angle between $\mathbf{I}$ and $\mathbf{B}$, and we use the direction of the flow of positive charge for the direction of $\mathbf{I}$. We can calculate the direction of the force by way of this equation using exactly the same two rules that we discussed on pages 458 and 459, replacing the velocity by the current. Apply these rules to Fig. 15.7 and see if you agree with the direction of the force indicated there.

EXAMPLE 15.3 An electron is projected horizontally from east to west into a region where the magnetic field is vertically downward. The velocity of the electrons is 2×10^6 m/s, and the magnetic induction 6×10^{-5} T (about the strength of the earth's magnetic field). Find the magnitude and direction of the magnetic force on the electron.

Solution We calculate the force from Eq. (15.1). Because the electron moves at right angles to **B**, $\sin \phi = 1$:

$$F = QvB \sin \phi = QvB = (1.6 \times 10^{-19} \text{ C})(2 \times 10^6 \text{ m/s})(6 \times 10^{-5} \text{ T})$$
$$= 1.92 \times 10^{-17} \text{ N}$$

We find the direction of **F** using the right-hand screw rule. The problem indicates an electron moving east to west, the equivalent to a positive charge's movement from west to east. Rotate a vector pointing east downward toward the vector **B** the way one turns a right-handed screw (clockwise), and the screw advances northward, which is our answer.

EXAMPLE 15.4 A beam of protons is accelerated through a potential difference of 1 million volts and then sent into a magnetic field B. Because the direction of the beam is perpendicular to the magnetic field, the beam is bent into a circle of radius $R = 10$ cm. Determine the strength of the magnetic field.

Solution We could calculate the radius of curvature of a charged particle of mass m moving with speed v perpendicular to the field B using Eq. (15.3):

$$R = \frac{mv}{Be}$$

where the charge is that on the proton, $+1.6 \times 10^{-19}$ C, $m_p = 1.67 \times 10^{-27}$ kg, and R is 0.1 m. We do not know the value of v, but we can calculate it from the energy of the protons. The protons initially at rest are accelerated through a potential difference of 10^6 V. Setting this potential energy equal to the kinetic energy $\frac{1}{2}mv^2$ gained by the protons, we have

$$\tfrac{1}{2}mv^2 = 10^6 \text{ eV} \times 1.6 \times 10^{-19} \text{ J/eV} = 1.6 \times 10^{-13} \text{ J}$$

$$v = \sqrt{\frac{2 \times 1.6 \times 10^{-13} \text{ J}}{1.67 \times 10^{-27} \text{ kg}}} = 1.38 \times 10^7 \text{ m/s}$$

Using this value of speed, one obtains a magnetic induction

$$B = \frac{m_p v}{Re} = \frac{1.67 \times 10^{-27} \text{ kg} \times 1.38 \times 10^7 \text{ m/s}}{0.1 \text{ m} \times 1.6 \times 10^{-19} \text{ C}} = 1.44 \text{ T}$$

EXAMPLE 15.5

A wire is in a magnetic field of intensity $\mathbf{B} = 0.2$ T at an angle of 30°, as shown in Fig. 15.11. If the part of the wire in the magnetic field is 12 cm long and the wire carries a current of 4 A, determine the magnitude of the force and its direction.

Solution

The direction of $\mathbf{I}$ is as indicated in Fig. 15.11. To find the direction of the force, use the right-hand screw rule. Rotating $\mathbf{I}$ into $\mathbf{B}$ through the angle $\phi = 30°$ the way a right-hand screw is turned advances the screw upward. The force $\mathbf{F}$ is upward. Using Eq. (15.4), solve for the force:

$$F = BI \Delta l \sin \phi = (0.2 \text{ T})(4 \text{ A})(0.12 \text{ m})(\sin 30°)$$
$$= 4.8 \times 10^{-2} \text{ N}$$

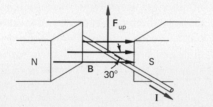

Figure 15.11 Force on a current-carrying wire in a uniform magnetic field.

So far we have examined the path of charged particles that enter a magnetic field moving in a direction perpendicular to the magnetic field. This leads to a circular orbit in which the charge could actually be trapped if the magnitudes of v and B were appropriate and the magnetic field influenced a sufficiently large area. What happens to charged particles that enter a uniform magnetic field with a component of their velocity parallel to the magnetic field? In such a situation only one component of the velocity, the component at right angles to the magnetic field, is influenced by the magnetic force. From Eq. (15.1), the magnitude of the force is given by $F = Q(v \sin \phi)B$. Since ϕ is the angle between $\mathbf{v}$ and $\mathbf{B}$, $v \sin \phi$ is the component of $\mathbf{v}$ perpendicular to $\mathbf{B}$. The particle will undergo circular motion but at the same time will move parallel to the magnetic field at a uniform speed equal to the component of its original velocity parallel to the magnetic field upon entrance

Figure 15.12 Motion of a positive charge in a uniform magnetic field. (a) The positive charge starts its motion in a uniform magnetic field in a direction perpendicular to the field, resulting in a circular path. (b) The positive charge starts its motion in a magnetic field with a component of its velocity parallel to the magnetic field—opposite the field lines, in this case. The resulting motion is a screw-type motion.

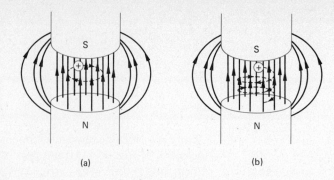

Figure 15.13 Northern Lights. (*Wide World*)

The Van Allen radiation belts are named after James Van Allen, a University of Iowa physicist who suggested their existence based on data obtained in 1958 from the United States satellite Explorer I.

into the field. The motions parallel to the magnetic field and perpendicular to the magnetic field are independent, just as in projectile motion, where a mass moves horizontally at a uniform speed but undergoes free fall in the vertical direction. The result of these two independent motions is a corkscrew (helix)-type motion of the charged particle, as shown in Fig. 15.12, which compares the resulting motions for the initial **v** perpendicular to **B** and the initial **v** not perpendicular to **B**.

Experiments at nuclear particle accelerators have demonstrated a number of important examples of this "screw"- or "coiled"-type motion, including motions in cloud or bubble chambers of charged particles in magnetic fields. The result of this motion in the earth's magnetic field is the spectacular northern or southern lights, or aurora, shown in Fig. 15.13. Streams of charged particles, primarily from the sun, enter the vicinity of the earth. Many of these particles are trapped in the earth's magnetic field and deflected by it into helical paths around magnetic field lines.

Zones of trapped, charged particles are known as the Van Allen belts. The particles work their way toward the poles and, on reaching the atmosphere, collide with atoms of the atmosphere, exchanging energy with them. High-energy atoms subsequently emit light as they return to their normal state. This emitted light produces the aurora. A sketch of this motion is shown in Fig. 15.14.

15.3 THE MAGNETIC FIELD AND ELECTRIC CURRENTS

What is it that causes magnetic fields to influence electron beams in a vacuum and a current in a wire? Since we know that permanent magnets can either repel or attract each other, that is, their magnetic fields interact, it seems reasonable to assume that anything influenced by a magnetic field must have its own magnetic field associated with it. Specifically, one might conclude from observations of such instruments as the mass spectrometer that a current of electrons or ions produces its own magnetic field, which then interacts with the field of the permanent magnet. We can examine this possibility by using a magnetic compass or iron filings as a simple measuring device. The iron filings align with

The Electromagnetic Accelerator

The combination of electric and magnetic fields is used by nuclear physicists in a number of ways other than for identifying elements in the mass spectrometer. The most spectacular application of this combination is in such machines as the cyclotron and the synchrotron, accelerating subatomic particles to the extremely high energies necessary for fundamental investigations of the nature of matter. These accelerators compete with and complement the electrostatic van de Graff devices we encountered in Chap. 13.

For the average person the cyclotron is undoubtedly the best recognized type of accelerator in the nuclear physicist's armamentarium. It consists of an evacuated chamber that carries a homogeneous magnetic field produced by an electromagnet and an alternating electric field made by two hollow semicircular electrodes known as dees to which an oscillating voltage is applied. Ions produced by a probe near the center of the chamber are influenced by the magnetic field as soon as they are shot into the chamber. As a result, the ions start to move in circles inside the chamber. The oscillating voltage driving the dees is adjusted to the same frequency as that of the circling ions, so whenever an ion passes from one electrode to another it is accelerated by the electric field between them. This acceleration increases the ion's speed and thus decreases its deflection, which drives its orbit further from the center of the chamber. During the whole operation the frequency remains constant. Eventually, after many such accelerations, the ion reaches the edge of the chamber and moves out through a special exit into an experimental chamber.

Cyclotrons are generally used to give relatively heavy particles quite low velocities. By contrast, synchrotrons are designed to accelerate very light particles to velocities close to that of light. Although based on the same principle as the cyclotron, the synchrotron's particles remain in a constant orbit. The magnetic field around the particles is increased to maintain a constant orbit as the energy and speed of the particles is increased because of the oscillating electric field.

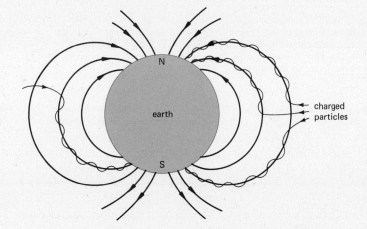

Figure 15.14 Sketch of charged particles entering the earth's magnetic field, being trapped, and entering the earth's atmosphere near the north and south poles, resulting in the aurora.

Figure 15.15 The pattern obtained when iron filings are placed on a paper through which a current-carrying wire passes perpendicularly. The filings form concentric circles around the current-carrying wire, which indicates the magnetic field around the wire. (*Kodansha Co., Ltd., Tokyo*)

the magnetic field as the filings' magnetic domains become oriented in the magnetic field. *When iron filings are placed on a piece of paper or board through which a wire carrying an electric current passes at right angles, the filings line up in concentric circles around the wire,* as shown in Fig. 15.15.

The compass can be applied in either of two ways; the first way, illustrated in Fig. 15.16(a), duplicates the iron filing method. The needle on the piece of board that is penetrated by a current-carrying wire points along the magnetic field lines wherever the compass is placed and traces a series of circles if the compass is moved about with sufficient care. The other technique, which was developed by the Danish physicist Hans Christian Oersted (1777–1851) in the early nineteenth century, is shown in Fig. 15.16(b). A compass needle is suspended close to a wire carrying an electric current. When the switch in the electrical circuit is opened and the current is "turned off," meaning that no charges are flowing, the needle points north. But after the switch is thrown (closed) and electric charges start to flow, the needle swings round as indicated.

How do you predict the direction of this magnetic field? The simplest method is the manual one illustrated in Fig. 15.17. **When you wrap the fingers of your right hand around the current-carrying wire, your thumb pointing in the direction of the conventional (positive) current, the magnetic field that encircles the wire is oriented in the direction in which your fingers are pointing.** This rule lets you predict the direction of the magnetic field consistent with experimental observation.

Having described in qualitative terms the magnetic field produced when current travels through a wire, we must now examine the situation quantitatively. If we do so experimentally, we find that the magnitude of the magnetic induction produced by charges Q moving with speed v decreases inversely as the square of the distance from the moving charges or current-carrying wire and is proportional to the

Figure 15.16 (a) The direction of circular magnetic lines of induction due to an electron current flowing up through a wire. (b) The turning of a compass needle placed in a magnetic field due to current I.

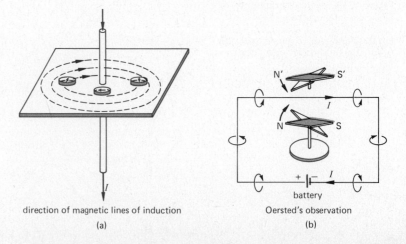

direction of magnetic lines of induction

(a)

Oersted's observation

(b)

magnitude of the charge and the speed of the charge or charges. In addition, we find here another situation where the magnitude of a vector quantity, **B** in this instance, depends upon direction. For example, consider the charge Q moving with a velocity **v**, as shown in Fig. 15.18. When the charge is at point P_1, the magnetic field it produces at some arbitrary point P_2 in space depends upon the angle θ between the velocity of the charge **v** and the vector **R**, from point P_1 to point P_2. In fact, the magnitude of **B** is found to depend specifically upon $\sin \theta$ in such a way that when the charged particle moves toward the point in question (so that **R** and **v** are along the same line), the magnetic field at the point will be zero (when $\theta = 0$, $\sin \theta = 0$). When $\theta = 90°$ and the point P_2 lies on a perpendicular to the velocity **v**, the magnetic field is a maximum. This is consistent with a sine function; $\sin 90° = 1$, the maximum value.

These observations can be written in equation form if we introduce a constant k' whose value depends on the units employed:

$$B = \frac{k'Q}{R^2} v \sin \theta \tag{15.5}$$

Figure 15.17 Right-hand rule to determine the direction of the magnetic field produced by a current-carrying wire. Place the thumb of your right hand along the wire in the direction of the conventional current. Wrap your fingers around the wire; your fingers point in the direction of the magnetic field **B** around the wire.

Equation (15.5) can actually be written as a vector equation that, in fact, amounts to a vector product of $\mathbf{v} \times \mathbf{R}$. Hence the direction of the magnetic field generated by a moving charge at any point P_2 as in Fig. 15.18 can be predicted using the right-hand screw rule, rotating **v** into **R**. In Fig. 15.18 **v** and **R** are in the plane of the paper, and the direction of the magnetic field as specified by **B** is into the paper.

Equation (15.5) deals only with moving *individual* charges. For practical purposes, physicists are really more interested in the magnitudes of magnetic fields produced by electric currents in wires. Figure 15.19 illustrates how to go about extending the single charge equation to the more general situation. A segment of wire of length Δl_1 is carrying a current **I** that produces a particular contribution ΔB_1 to the total magnetic induction induced by the current at a point in space, P_2.

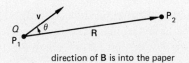

direction of B is into the paper

Figure 15.18 A positive charge moving with velocity **v** when it is at point P_1 produces a magnetic field **B** at point P_2. **B** is directed inward in this example.

Figure 15.19 The current **I** in an element wire Δ / l_1 produces a magnetic field $\Delta \mathbf{B}$ at point P_2. Point P_2 is located by **R** and the angle θ.

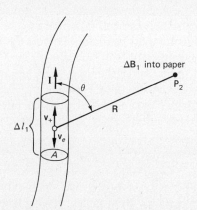

v_+ = direction positive charges would move
I = direction of the conventional current
v_e = direction electrons drift in the wire to produce the current **I**
A = cross-sectional area of the wire

The current is produced by electrons moving with a drift velocity v_D, whose direction we already know is opposite that of the conventional current. The angle between the segment of wire and the line **R** that joins the wire segment and the point P_2 is θ. Finally, we specify that the distance R between the segment of wire and the point P_2 is much larger than Δl_1, the length of the segment.

Now, we can adapt Eq. (15.5) to this situation. Since $n\, \Delta l_1\, Ae$ is the total charge in the wire element, its contribution to the magnetic induction would be

$$\Delta B_1 = \frac{k' n\, \Delta l_1\, Aev_D \sin\theta}{R^2}$$

which, using Eq. (14.2), can be expressed as

$$\Delta B_1 = \frac{k' I\, \Delta l_1}{R^2} \sin\theta \qquad (15.6)$$

This equation is known as the law of Biot and Savart, after Jean Baptiste Biot and Felix Savart, two French physicists who, only a few weeks after Oersted's discovery that a compass needle was deflected by a current-carrying wire, began to examine in a quantitative way the magnetic fields generated by current-carrying wires.

If we measure the current in amperes, the distance in meters, and the field in teslas, then the value of the constant k' turns out to be 10^{-7} N/A². To work out the direction of the magnetic field, we once again turn to the trusty right-hand screw rule. Equation (15.6) for the magnitude of ΔB_1 can be written as a vector cross product involving **I** × **R**. Rotating the vector **I** into the vector **R** that joins the segment of wire with the point P_2 through the angle θ, we see that $\Delta \mathbf{B}_1$ goes into the paper.

Of course, the magnetic induction produced by a small segment of current-carrying wire is hardly a useful quantity by itself. What we really need to know is the total magnetic induction at the same point produced by the current in the entire length of wire. To find this quantity we must sum the contributions to the magnetic induction from all the small segments of the entire current-carrying wire. This is an extremely difficult task in most cases, owing to geometric complexities. It can be determined relatively easily, however, in two uncomplicated examples: a long, straight piece of wire and a circular loop.

The case of the long, straight wire is shown in Fig. 15.20. Here, we wish to work out the magnetic induction B at a point P whose perpendicular distance from the wire is R_0. We know from the right-hand screw rule that the magnetic field's direction is into the paper. The problem is to find the sum of all the $\Delta \mathbf{B}$'s from all the segments of the wire.

One simplifying factor is that all these segmental fields point in the same direction. We can, therefore, add them all arithmetically. To do this addition we must realize that each segment of the wire is at a

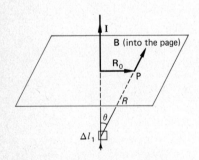

Figure 15.20 Magnetic induction in the vicinity of a long, straight, current-carrying wire. The element of wire Δl_1 produces a contribution to the field at point P, which is into the page.

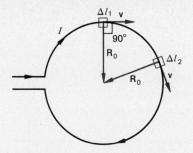

Figure 15.21 The field at the center of a current loop is the sum of the fields produced by the current *I* in each element of wire Δl_1, Δl_2, Δl_3, ... The direction of the current is always perpendicular to the vector **R**, to the center of the loop.

different distance from and angle to point P. This sum can only be performed practically using integral calculus, which yields the result

$$B = k' \frac{2I}{R_0} \tag{15.7}$$

The strength of the magnetic field produced by a long, straight piece of wire that carries a current I depends directly on the size of the current and varies inversely with the distance from the wire.

A wider ranging application of the law of Biot and Savart involves the magnetic field at the very center of a loop of wire carrying an electric current. Here the calculation requires us to add all the equal contributions from equal segments of the wire to the magnetic field in the center spot, as illustrated in Fig. 15.21. Applying Eq. (15.6) to each individual segment,

$$\Delta B = \frac{k' I \, \Delta l}{R_0^2} \sin \theta$$

Since θ is 90° for each and every segment, each contribution to the overall magnetic induction at the center is directed into the paper and has a magnitude $(k'I/R_0^2)(\Delta l)$. The sum of the lengths of all the segments in the loop is the loop's circumference, that is, $2\pi R_0$. Hence, the overall magnitude of the magnetic induction at the center, B, is given by

$$B = \frac{k' I}{R_0^2}(2\pi R_0) = \frac{k' 2\pi I}{R_0} \tag{15.8}$$

This particular expression gives the size of the magnetic induction at, and only at, the very center of the loop. Elsewhere inside the loop of wire the field is somewhat smaller, but its direction is the same, as you can determine using the version of the right-hand rule outlined in Fig. 15.17. The same rule indicates that the magnetic field is in the opposite direction outside the loop.

Figure 15.22 provides an almost complete picture of the magnetic field produced by a current-carrying loop of wire; the arrows indicate the direction of the field at specific points, and the separation of the lines gives an idea of the field strength. The right-hand side of this figure acts as the north pole and the left-hand side as the south pole.

Thus far we have calculated the strength of the magnetic field that results from a single current-carrying wire in two different configurations. Our next step is to look at the extent of magnetic interaction between the fields of two individual wires. Common sense tells us that the magnetic force between any two wires should depend on the relative orientations of the wires.

As a relatively simple case, consider the magnetic interaction between two parallel straight wires in Fig. 15.23. We assume that wire 1, carrying a current I_1, is so long that its length is effectively infinite, whereas the length of wire 2, which supports a current I_2, is Δl_2. Now,

Figure 15.22 Magnetic lines of force due to a current in a circular loop of wire.

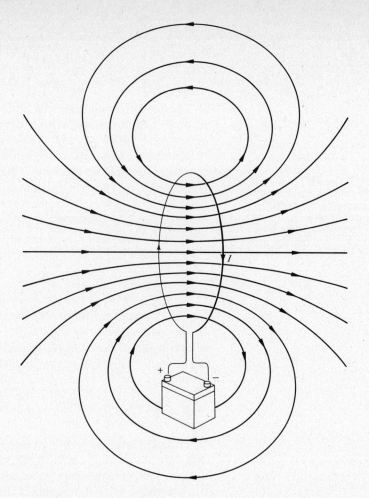

wire 1 produces a magnetic field which circulates around it in such a way that it goes into the page—as indicated by the crosses—on the right-hand side of the wire. We know from Eq. (15.7) that the magnetic induction produced by wire 1 is given by

$$B_1 = \frac{k'2I_1}{R_0}$$

This magnetic induction exerts a force on the current-carrying wire 2, whose strength we can determine from Eq. (15.4):

$$\mathbf{F} = \Delta l_2 \mathbf{I}_2 \times \mathbf{B}_1$$

or in magnitude

$$F = I_2 \Delta l_2 B_1 \sin \phi$$

Since $\mathbf{B}_1$ is perpendicular to $\mathbf{I}_2$, $\sin \phi$ is 1, and the expression simplifies

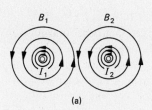
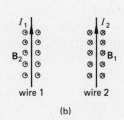

Figure 15.23 (a) and (b) Force of attraction between two current-carrying parallel wires. Each wire resides in the magnetic field of the other wire. View (a) is the top view of (b). (c) To calculate the force on wire 2 due to the magnetic field B_1 from wire 1, we need to know I_2, Δl_2, and R. The force is one of attraction in this case. Note that the field lines indicate the field due to each individual current and not the resultant of the two.

to

$$F = I_2 \, \Delta l_2 \, B_1$$

Substituting for B_1, we have

$$F = I_2 \, \Delta l_2 \, \frac{k' 2 I_1}{R_0} = 2k' \frac{I_1 I_2 \, \Delta l_2}{R_0} \qquad (15.9)$$

What about the direction of this force? Because $\mathbf{I}_2$ is upward and $\mathbf{B}_1$ is directed into the page, rotation of $\mathbf{I}_2$ into $\mathbf{B}_1$ in a right-handed way, using the ever-faithful right-hand screw rule, indicates that the force is an attractive one; it pulls wire 2 toward wire 1. By Newton's third law—remember Chap. 3—the reaction to this attractive force is an equal attractive force by wire 2 on wire 1. This latter force derives directly from the fact that the current in wire 2 creates a magnetic field at wire 1.

The result of the analysis in Eq. (15.9) shows us that two straight current-carrying wires possess a force of mutual attraction between them whose strength is proportional to the magnitude of the two currents, inversely proportional to the separation of the two wires, and proportional to the distance over which they are parallel. In fact, the ampere of current can be defined from Eq. (15.9) in terms of newtons and meters. The constant k' in this equation is given a value of 10^{-7}, which makes the constant used in Coulomb's law about 9×10^9.

If the current in one of the wires is reversed, the force becomes one of repulsion, as students can see for themselves by repeating the above analysis using the right-hand screw rule with $\mathbf{I}_2$ moving in the opposite direction. It is also worth noting that the force between the two wires is a maximum when they are parallel. Any change in orientation will make the magnetic induction experienced by one wire due to the other change from point to point in magnitude and direction. This generally results in torques and a smaller net force.

15.4 CURRENT LOOPS AND MAGNETS

Imagine now that we have two circular current-carrying loops of wire side by side, with their centers on the same axis. If the current is circulating in the same direction around each loop, the north pole of the

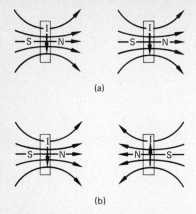

Figure 15.24 A current-carrying loop produces a magnetic field such that one side of the loop is a north pole, and the other side is a south pole. The loops act as magnets that attract or repel each other between the loops. The **I** indicates the direction of the current in the upper part of the loop closest to you. (a) Attraction between the loops when current in the loops is in the same direction. (b) Repulsion between the loops when current in the loops is in the opposite direction.

magnetic field from one loop will obviously be adjacent to the south pole of the other, and the two loops will attract each other (see Fig. 15.24). If the currents are moving in opposite directions around the loops, the like poles will be adjacent and the two loops will repel. These "current loops" behave much like bar magnets. In fact, combinations of many continuous current loops, slightly stretched out, produce an extremely useful and important electromagnetic device known as a solenoid. The similarity between the magnetic field of a solenoid and that of a bar magnet is shown in Fig. 15.25. The major difference between solenoids and bar magnets is that solenoids lose their magnetic field when the current is switched off. We can, in effect, regard solenoids as "temporary magnets," but they are more commonly called electromagnets.

The value of the current loop is not restricted to electromagnetic devices. We can also study current loops to understand two theoretical problems of magnetism: the existence of permanent ferromagnetism and the nature of the earth's magnetic field.

We know from our roundup of magnetic phenomena early in this chapter that the individual atoms of ferromagnetic materials are themselves tiny magnets. Now, the picture of the atom developed by Niels Bohr, which is useful for descriptions of many circumstances, is of a number of negatively charged electrons revolving like tiny planets around a sunlike, positively charged nucleus. The revolution of the electrons around the nucleus is really no different from the movement of electrons around a loop of wire. In other words, each revolving electron produces its own diminutive current loop, which is the basis for a solenoidlike electromagnet.

Figure 15.25 The magnetic field from (a) a single loop of wire and (b) a solenoid. Iron filings are oriented along the magnetic field lines. Compare with Fig. 15.1. (*Kodansha Co., Ltd., Tokyo*)

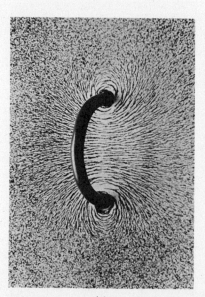

(a)

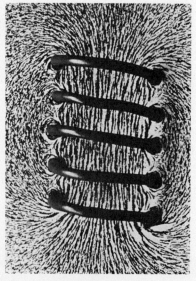

(b)

An alternating current that is decreased gradually will destroy permanent magnetism by creating magnetic fields which rapidly alter direction. Note that this electrical method is far preferable to the hammer technique for demagnetizing watches and other delicate instruments.

Why then are not all atoms of all materials ferromagnetic? The answer comes from atomic chemistry. The electrons in the atoms of some elements form so-called shells of electrons that have no net current loop effect. This can be pictured as a combination of one electron moving in one direction and its partner moving in the other. The atoms of iron, cobalt, and other ferromagnetic materials possess at least one unpaired orbiting electron. It is these little charged particles, acting as current loops, that produce the permanent magnetism we observe. Many other atoms have unpaired electrons but are not ferromagnetic because the atoms do not align cooperatively with one another into domains as do iron ions. The current loops in magnetic domains must also be aligned in some way to produce a permanent magnet. This is why magnets can be demagnetized when they are heated or pounded. These effects all destroy the alignment of the current loops in nearby atoms.

Surprisingly, the atom is not the smallest entity that is inherently magnetic. Electrons, protons, and even the electrically neutral particles known as neutrons also produce magnetic fields, behaving as if they spin on their axes. These phenomena can again be understood in terms of current loops produced by the distribution of the electric charge that make up the spinning particles. The electron, proton, and neutron can all be pictured as spinning on an axis through their respective centers in much the way that the earth spins on its axis.

This intrinsic spin of electrons plays an important role in determining whether atoms exhibit magnetic effects. In most atoms the spin of one electron is canceled by another electron with the opposite spin. The concept of the current loop is also valuable at the other end of the scale—in the attempts by geophysicists to understand the cause of the earth's field.

Explanation of the earth's magnetic field, using satellites, airplanes, and other forms of investigation, shows that its shape is substantially such as that one would expect if a vast bar magnet were embedded in the earth. Because of the high temperature in the core of the earth, a bar magnet cannot be responsible for the earth's magnetic field. However, the fact that the magnetic field of a bar magnet can be reproduced by a current loop suggests that the earth's magnetic field results from currents in the center of our planet. The currents required to produce the magnetic field would have to have a magnitude of about 10^9 A. Such currents are not detected near the earth's surface.

The presently accepted theory is that earth's magnetic field is produced by ions rotating about the earth's axis in the molten core in the vicinity of the center of the earth. The molten core consists of iron and nickel at a temperature of about 4000°C. Presumably, huge current loops and convective currents act simultaneously—behaving like a self-exciting dynamo. One would expect the rotating movement of the ions to be coupled in some way to the overall rotation of the earth, implying that the magnetic axis is close to the axis of rotation; this is indeed the case. According to this scheme, the observed motion of the magnetic poles is explained as resulting from changes in the motion of ions within the planet's molten core.

15.5 METERS AND MOTORS

The practical uses of current-carrying loops of wire range from delicate measuring instruments to mammoth mechanical machines. The magnetic force on wire loops is used to control almost all electrical measurements and to provide the driving force for the electric motor.

The device most commonly used in electrical measurements is the galvanometer, which we encountered in the last chapter when we discussed how to adapt the instrument to different types of electrical measurement. Now we are in a position to understand how the galvanometer actually works.

One of the earliest galvanometers to be developed—by Oersted—was simply a compass needle. However, today practically all galvanometers used are of the so-called D'Arsonval, moving-coil, or pivoted-coil type. Figure 15.26 is a diagram of the D'Arsonval device; it consists of a flat loop of several concentric turns of wire, C, placed between the poles of a permanent magnet. The loop is suspended by a fine piece of wire, F, whose position is fixed; this wire provides a restoring torque whenever the plane of the loop of wire is deflected. Beneath the loop is a weak spring, S, which provides a negligibly small resistance to any twist in the planar loop. A piece of soft iron, A, known as a core, is normally placed in the center of the loop to increase the strength of the magnetic field near the wires of the loop. This increase occurs because the domains in the core A align with the external field, adding to the field. The indicating needle, labeled P, is attached to the restoring wire. This needle indicates the strength of the force on the loop and sweeps across the face of the meter.

Any current that passes through the loop of wire causes forces on the vertical wire, which is located in the horizontal magnetic field of the horseshoe magnet. These forces exert a torque on the coil, which will turn the coil until the elastic restoring torque of the suspension balances the electromagnetic action. The amount of rotation of the loop, and hence the reading produced by the needle, depends on the size of the magnetic force and the stiffness of the suspension wire.

If the current in the loops is reversed, the loop rotates in the opposite direction, sending the needle the other way. One would therefore expect all moving-coil meters to indicate the direction as well as the size of the current passing through them. However, many such meters, particularly those adapted for use as common ammeters and voltmeters, are designed so that the needle moves in only one direction, from a zero marking on the far left of the scale. Putting current through these instruments in the wrong direction can "zap" the meter movement because the needle is mechanically prevented from deflecting to the left of the zero mark.

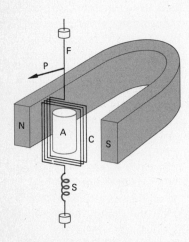

Figure 15.26 Construction of the D'Arsonval galvanometer.

In simplified terms, an electric motor is no more than a galvanometer movement that is free to turn.

The advantage of this type of meter movement is that it can be made very sensitive to minute currents, thus making it a natural device for use as the null detector in potentiometer bridges and Wheatstone bridges. Most other forms of galvanometers resemble the D'Arsonval type in principle.

Another application of the current loop is the electric motor. In simplified terms, an electric motor is no more than a galvanometer movement that is free to turn, rather than being constrained by a suspension wire. Figure 15.27 shows the design of a simple dc motor. In this machine, a current-carrying loop of wire known as an armature is wound on a soft iron core and placed between the appropriately shaped poles of a permanent magnet or electromagnet. The problem to be overcome here is that of somehow connecting the armature to an electrical circuit containing a battery while still allowing the armature to rotate freely. The normal solution is to attach a split ring called a commutator to the end of the armature, which then connects with the rest of the circuit through sliding contacts known as brushes, as shown in Fig. 15.27. When the battery delivers current through the loop, via the brushes and commutator, the magnetic force causes the loop to rotate.

Why is the commutator split, and what is its effect? Plainly, the split reverses the direction of the direct current coming into the loop each half rotation. Just why this reversal is necessary can be understood if we analyze the forces on the armature loop in Fig. 15.28. The loop is a rectangle of sides a and b; its plane makes an angle θ with the direction of the magnetic field **B**. We wish to analyze the total magnetic force on all four wires that make up the loop. In general, the force F on a wire whose length is l is given by the equation

$$F = IlB \sin \theta$$

I being the current in the wire. Now, the top and bottom wires are always perpendicular to the field **B**, no matter what the value of θ happens to be. Thus, applying the equation to these two wires, we find that they each experience a force $F = IaB$. And by using the appropriate rule to work out the direction, we discover that the force on the top wire is directed upward, and on the bottom wire the force is directed downward.

In the case of the two side wires that are perpendicular to the axis of rotation, the forces are directed away from the loop, as shown in Fig. 15.28(b). As with the top and bottom wires, the forces on the two side wires are equal but oppositely directed, have strengths $F = IbB \sin \theta$, and change with the angle of rotation. These forces tend to pull the loop apart; thus, they have no effect on the motion of the rigid loop, so we can ignore them.

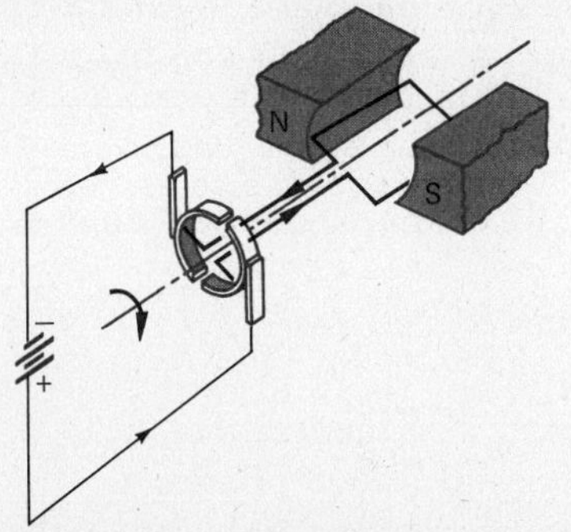

Returning to the upper and lower wires, we see in Fig. 15.28(a) that unless the loop is oriented vertically, with the plane of the coil perpen-

Figure 15.27 A simple motor (schematic; iron core of armature not shown).

Figure 15.28 Forces on the wires of a square loop carrying a current in a uniform magnetic field. Diagram (a) shows the forces on the top and bottom wires, resulting in a torque on the loop. (b) View of the loop as seen from the north pole looking toward the south pole, that is, along the magnetic field lines. (c) Relationship of the lever arm to the width of the loop.

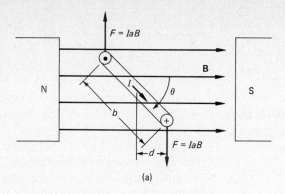

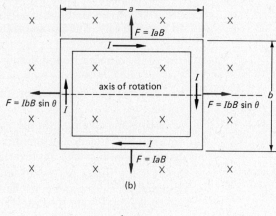

dicular to B, the lines of action of these equal but opposite two forces do not coincide. Therefore, they produce a torque, τ, given by

$$\tau = (d \times F_{\text{bottom}}) + (d \times F_{\text{top}})$$

where $d = (b/2) \cos \theta$ [see Fig. 15.28(c)]. The torque can be written as

$$\tau = IaB \frac{b}{2} \cos \theta + IaB \frac{b}{2} \cos \theta = IabB \cos \theta = IAB \cos \theta \quad (15.10)$$

Here we have used the fact that the product ab is the area A of the loop. This expression actually applies to any shaped loop of area A.

It can be further generalized and simplified using the concept of the magnetic moment **M**. This is defined as a vector perpendicular to the plane of the loop of magnitude IA; that is,

$$\mathbf{M} = IA \quad (15.11)$$

Figure 15.29 Relative orientation of the magnetic moment of the loop, **M**, with respect to the direction of the magnetic induction **B**. (a) **M** at an angle with **B**. (b) **M** ∥ **B**. (c) **M** ⊥ **B**. (d) **M** antiparallel to **B**.

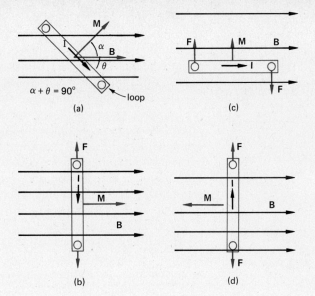

The vector **A** has a magnitude equal to the area of the loop and is assigned a direction that is perpendicular to the plane of the loop.

Equation (15.10) can be rewritten in terms of the angle α made between the magnetic moment, the perpendicular to the plane of the loop, and the direction of the magnetic field as shown in Fig. 15.29(a). The angle α is the complement of angle θ; that is, $\alpha + \theta = 90°$. Thus:

$$\tau = IAB \cos \theta = IAB \sin \alpha$$

In terms of the magnetic moment,

$$\tau = MB \sin \alpha \qquad (15.12)$$

We can use this relationship to examine how the torque τ varies with angle. When $\alpha = 0$, **M** is parallel to **B** and the plane of the loop is perpendicular to the magnetic field, as shown in Fig. 15.29(b). Since $\sin \alpha$ is zero, the torque is also zero. The loop is in a state of equilibrium. In terms of magnetic fields, one might say that the magnetic moment of the current loop represents the direction of the magnetic field from the current loop through its center. Hence the magnetic field of the loop is aligned parallel to the external uniform magnetic field, a condition of stable equilibrium.

If the plane of the loop is along the direction of **B**, $\alpha = 90°$ and $\sin \alpha = 1$; hence a maximum torque occurs, as seen in Fig. 15.29(c). If **M** is antiparallel to **B**, then the torque will tend to reorient the loop. The equilibrium as shown in Fig. 15.29(d) is unstable.

This torque is responsible for the motor action. For the torque to turn the coil in the same direction, the current in the loop must reverse each half rotation. This can be seen by reexamining Fig. 15.28(a).

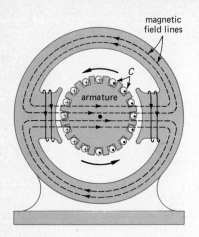

Figure 15.30 Schematic diagram of a dc motor.

When the top wire is to the left of a vertical plane through the center of the loop, the torque produces a clockwise rotation. However, as the momentum moves the loop through its equilibrium point (the vertical plane), the force remaining in the upward direction produces a counterclockwise torque. The loop would stop in the vertical position after a few small oscillations. But if the current is switched just as the loop realizes this vertical position, the force on the top wire acts in a downward direction, maintaining the torque in a clockwise sense. The commutator in Fig. 15.27 provides this periodic reversal of the direction of the current in the loop. Of course, we need not restrict ourselves to a dc motor; we could use an alternating source with no split-ring commutator. If the ac source is synchronized with the rotation speed, we could have a natural means of reversing the current.

Figure 15.30 is a diagram of a practical dc motor. The magnetic field is supplied by the same current that is put through the armature. The armature is a series of coils wound on a soft iron core, to produce a more constant torque than is possible with a single loop.

EXAMPLE 15.6

A rectangular coil consists of 100 turns of wire and is 20 cm wide and 30 cm long. The coil is mounted in a uniform magnetic field of magnetic induction 8×10^{-3} T. A current of 20 A passes through the 100-turn loop. When the plane of the loop makes an angle of 30° with the magnetic field, what is the torque on the loop?

Solution

We can use Eq. (15.10) for one loop $\tau = IAB \cos \theta$. Since we have effectively 100 loops, we multiply this value by 100. The angle θ is 30°, and $A = 20$ cm $\times$ 30 cm, so

$$\tau = 100 \, IAB \cos \theta$$
$$= (100)(20 \text{ A})(0.20 \text{ m})(0.30 \text{ m})(8 \times 10^{-3} \text{ T}) \cos 30°$$
$$= 0.83 \text{ N} \cdot \text{m}$$

SUMMARY

1. Almost all elements and compounds are magnetized when placed in a magnetic field. Only a few substances, known as ferromagnetic materials, are magnetized 100 to 1000 times more and retain a portion of the magnetization when they are removed from the field.
2. Permanent magnets possess two types of magnetic pole: north and south. As with electric charges, like poles repel and unlike poles attract.
3. The earth's magnetic poles tend to wander over periods of eons.
4. Magnetic fields influence moving electric charges, acting on them with a force that is proportional to the size of the moving charge, its speed, the strength of the field, and the direction of motion.
5. The direction of the magnetic force on a moving charge is determined by using the right-hand screw rule, where $\mathbf{v}$ is rotated into $\mathbf{B}$ since $\mathbf{F} = Q\mathbf{v} \times \mathbf{B}$.

6. "Current loops," which consist of almost complete circles of wire through which current is moving, behave magnetically, much like bar magnets. The concept of current loops can explain the nature of ferromagnetism and the earth's magnetic field.

7. Instruments and devices as diverse as galvanometers and electric motors are made possible through the movement of current loops in magnetic fields.

QUESTIONS

1. A proton moves horizontally in a uniform magnetic field that points vertically upward. Describe the motion of the proton you see if you look down into the magnetic field.

2. A wire is placed between the poles of a U-shaped magnet. The wire is vertical, and the U-shaped magnet is in a horizontal plane facing you, with the north pole on the right. In which direction is the wire deflected if a current is established upward in the wire?

3. If the earth's magnetism is considered caused by a circular current loop within the core of the earth, in which direction would the plus and minus charges have to move? In what plane would this motion have to be?

4. Discuss the possibility that the earth's magnetic field is due to the circulation of charge in the upper atmosphere of the earth.

5. Is it important for a navigator to have knowledge of both the angle of dip and the declination?

6. When we are near household wiring, do we experience a magnetic field? If so, describe the field. Consider both single wires carrying alternating current and pairs of wire that are used, for example, as lamp cords.

7. When you look toward an electron, you see it moving in a counterclockwise circular orbit in a vertical plane. What is the direction of the magnetic field responsible for this motion?

8. How can you determine positively whether a bar of metal is magnetized?

9. A circular loop of wire lying in the plane of the paper carries an electric current in a clockwise direction. What is the direction of the magnetic field in the center of the loop and outside the circumference of the loop?

10. For the following cases in Fig. 10, indicate the direction of the force on the electron $\ominus$, and give the value of the magnitude of the force.

11. An electron is shot into a magnetic field with a velocity **v** that is perpendicular to **B**. Indicate the (*a*) trajectory of the electron in the magnetic field, (*b*) magnitude and direction of the force on the electron.

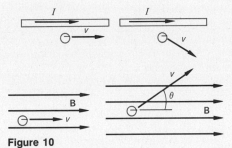

Figure 10

12. Describe the motion of an electron *e* that moves with a velocity **v** in the direction shown below. The electron moves in a uniform magnetic field that is directed into the paper. State clearly the magnitude and direction of the force on the electron.

13. In a moving-coil meter, the pole faces of the permanent magnet are curved to produce a radial magnetic field over the region of the coil. Why is this necessary?

PROBLEMS

Forces on Moving Charges

1. A charge of $+40\ \mu C$ has a velocity of 3×10^5 m/s toward the north as it moves in a horizontal plane. It passes through a uniform magnetic field of 0.2 T, which is directed vertically upward. What are the magnitude and direction of the force on the charge?

2. A charge of $+50\ \mu C$ has a velocity of 3×10^5 m/s toward the north as it moves in a horizontal plane. It passes through a uniform magnetic field of magnetic induction equal to 1.5 T, whose field lines lie in horizontal plane. What are the magnitude and direction of the magnetic force on the charge for the following directions of the magnetic field? *B* is directed toward the (*a*) north, (*b*) east,

(c) northeast, (d) west, (e) 30° south of east.

3. What is the magnitude of the force on an electron moving at a speed of 0.5 the speed of light in a horizontal direction in a uniform magnetic field of the earth having a vertical component of magnetic induction $B = 0.2$ G?

4. An electron enters a magnetic field moving at right angles to the magnetic field. The magnetic field **B** is into the paper. Indicate the path of the electron in the magnetic field, and derive an expression for the radius of the electron's orbit.

5. A particle of mass m and charge $+q$ is accelerated from rest through a potential difference V. The particle then passes into a region where a uniform magnetic field of induction B exists in a direction perpendicular to the velocity of the particle. (a) In a sketch of this apparatus, the direction of **B** is into the paper. Indicate the direction of the force on the particle just as it enters the magnetic field and the path of the particle in the magnetic field. (b) What is the magnitude of the force on the particle while it is in the magnetic field? (c) Derive an expression for the radius R of the orbit of the particle. This expression should be in terms of the quantities V (the potential difference), B, q, and m. (d) Which quantities in the expression you obtained in (c) are easily measured directly?

6. A positive charge is accelerated through a potential difference of 300 V. If the charge of $+60\,\mu C$ moves into a region where there is an electric field of 600 V/cm, as indicated in the diagram, what magnitude of magnetic field would have to be applied, and in what direction, to keep the charge moving straight ahead? The mass of the charge is 90 mg.

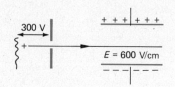

7. Electrons in a cathode ray tube are accelerated to 10 000 V and then sent in a horizontal direction into the deflecting part of the tube. If the earth's vertical component of magnetic field has a magnetic induction of 0.6 G, what would be the radius of the arc into which the beam would be bent?

8. A proton has a velocity of 2×10^7 m/s in a direction at right angles to a uniform magnetic field of magnetic induction equal to 50 T. Describe the motion of the proton relative to the direction of **B** and its initial velocity. What is the magnitude of the force on the charge and the radius of its orbit?

9. An electron moving in a uniform magnetic field of 4×10^{-4} T has a circular orbit of radius 0.8 m. Compute the speed of the electron and its energy in electron volts.

10. Compute the radius of curvature of the path of electrons accelerated through 1 kV and then allowed to move with constant speed at right angles to a uniform magnetic field of magnetic induction equal to 100 G (0.01 T).

11. Compare the radii of the paths of protons, hydrogen nuclei with charge $+e$ and mass 1 u, and α particles, helium nuclei with charge $+2e$ and mass 4 u, when they move into the same magnetic field at identical speeds.

12. Compare the radii of the paths of electrons and protons moving with the same speed in the same magnetic field. The mass of the proton is 1836 times the mass of the electron.

13. Find the radius of the circle in which an oxygen doubly negative ion (O^{2-}) moves when it is injected into a uniform magnetic field of 0.45 T (4500 G) at a speed of 10^5 m/s moving perpendicular to the magnetic field.

14. A deuteron, consisting of one proton and one neutron, is the nucleus of heavy hydrogen. If the deuteron of mass 3.35×10^{-27} kg is projected into a region of magnetic induction $B = 1.5$ T so that its velocity is perpendicular to the magnetic field, the deuteron moves in a circular path of radius equal to 30 cm. Compute the speed of the deuteron.

B from Moving Charges

15. A charge of $+4\,\mu C$ has a velocity of 3×10^4 m/s toward the north moving in a horizontal plane. What are the magnitude and direction of the magnetic field produced by the charge at a point 4 cm directly east of the charge?

16. An α particle, carrying a charge of $+2e$, moves vertically upward. What is the magnetic field it produces at a point 10^{-10} m below and to the right of it? See the diagram.

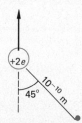

17. Two positive particles A and B are traveling at the same speed to the right. Describe the magnetic force between them. See the diagram.

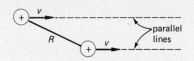

18. Two electrons, each traveling at speeds 2×10^7 m/s, pass each other at a distance of roughly the size of the atom, 10^{-8} cm. They are moving in opposite directions, that is, antiparallel to each other. Compare the electrostatic and magnetic forces between the two electrons at the instant they pass. From your answer, what effect do these forces have on the shape of a beam of electrons originally consisting of many electrons moving parallel to one another closely bunched together?

19. An electron is moving around in a circle of radius R and with a speed v. What is the magnetic moment of this current loop? Relate this to the angular momentum of the electron.

20. An iron ring 3 cm in radius is charged with 10^5 extra electron charges that are spaced by coulombic repulsion equally about the ring. If the ring is rotated about an axis through its center at a frequency of 12 600 r/min, what is the magnitude of the magnetic field at the center of the circle? Describe the direction of this magnetic field in terms of the axis of the ring and the direction you choose for the rotation.

B from Wires

21. A long, straight, vertical wire carries a current upward of 10 A. What are the magnitude and direction of the magnetic field produced by the wire a distance of 15 cm from the wire in a direction perpendicular to the wire?

22. What is the magnitude of the magnetic induction produced by a segment of wire 0.01 m in length that carries 200 A of current at a point 0.5 m from the wire? This point is to the right of the segment but 30° in front of the segment, as seen in the figure.

23. Consider a single current-carrying wire (20 A) in a laboratory. If the wire is 4 ft above your head, what is the magnitude of the magnetic field in the vicinity of your brain due to this long wire? Compare your answer with the magnitude of the earth's magnetic field (about 0.5 G).

24. The magnitude of the magnetic field at a point 4 cm from a long, straight wire is 10^{-4} T (1 G). What is the current in the wire? What is the magnetic field 16 cm away from the wire?

25. A long, straight wire with a resistance of 2 Ω has a potential difference of 12 V placed across its ends. What will be the magnitude of the magnetic field 0.6 m away from the wire in a direction perpendicular to the wire?

26. Two long, parallel wires, each carrying a current of 8 A in the same direction, are held in a vertical plane 16 cm apart. Compute the resultant magnetic field at a point 8 cm directly above the wires and give its magnitude and direction.

27. Two long, straight, parallel wires, each carrying 10 A, are 80 cm apart. What are the magnitude and direction of the magnetic induction midway between the two wires when the currents are (*a*) in the same direction, (*b*) in the opposite direction?

28. A long, straight wire carrying 20 A is placed next to a cathode ray tube parallel to the beam of electrons in the tube. If the electrons were accelerated through 15 kV and the wire is 16 cm away from the beam, compute the force on the individual electrons.

29. A circular coil of wire has 50 turns of wire and a radius of 4 cm. What current in this coil will produce a magnetic induction of 8 G at the center of the loop?

30. Compute the magnetic induction in the center of a circular loop of wire carrying a current of 4 A if the diameter of the loop is 8 cm. If the loop's diameter were doubled, how would the magnetic field change? If the number of turns of wire were increased from 1 to 12, what would be the new magnetic induction at the center of the loop? Compute the magnetic moment in each case.

31. If two loops are concentrically wound so that they resemble the accompanying figure, what is the resultant magnetic field (both direction and magnitude) at the center of the loops?

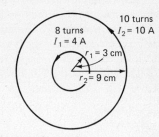

32. A long, straight wire carries a current of 3 A along the y axis, as shown in Fig. 32. A uniform magnetic field of magnitude $B_0 = 4 \times 10^{-4}$ T is directed parallel to the x axis and it covers all space. Find the *resultant* magnetic field at the point $x = 0$, $z = 2$ cm.

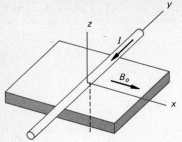

Figure 32

Force on Current-Carrying Wires

33. A long, straight wire lies in a horizontal plane, with 0.3 m of the wire lying in a uniform magnetic field of magnetic induction $B = 1.5$ T, which points upward, making a 30° angle with the vertical. What are the magnitude and direction of the force on the wire if a current in the wire is 7.5 A toward the right?

34. We wish to "levitate" a 0.2-m-long piece of wire in a magnetic field. If the mass of the wire is 2 g and the wire carries a current of 20 A, describe the smallest magnetic field required to support the wire. Give the magnitude and direction the magnetic field required.

35. A long wire hangs in a uniform magnetic field, as shown in the figure. In which direction is the force on the wire in the magnetic field? Compute this force if the resistance of the wire is 0.2 Ω, the battery delivers 6 V, and the magnetic field of 0.1 T intersects a 0.25-m-long piece of wire.

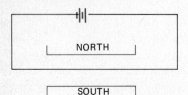

36. If 0.5 m of a long, straight wire is placed in a magnetic field with an orientation such that the wire makes an angle of 37° with the magnetic field of magnitude $B = 0.4$ T, what current carried by the wire will result in a force of 5 N on this segment of wire?

37. A wire with a mass per unit length equal to 0.25 g/cm is placed horizontally in a magnetic field where $B = 0.05$ T. The magnetic field is perpendicular to the wire. What current in the wire will result in a force sufficient to keep the wire from falling?

38. Compute the force between the two parallel wires in Fig. 38, one of which is very long, the other 20 cm long. The currents in the two wires are in opposite directions. What is the direction of this force?

39. Two long, straight, parallel wires are 0.15 m apart. Each wire carries a current of 12 A in the same direction. What is the magnitude of the force per unit length of wire between the two wires? What is the direction of this force?

40. One long, straight wire carries a current of 6 A and is parallel to a short (0.05 m), straight wire carrying a current of 12 A. The two wires are 20 cm apart. (a) What is the force on the short segment of wire? (b) What is the force on the long segment? Give both magnitudes and directions.

41. A rectangular loop of length l, width D, and negligible mass is free to pivot about one of its horizontal sides. A mass M is attached to the center of the free side. A uniform magnetic field B is directed vertically upward, and a current I flows around the loop as shown in Fig. 41. Find the equilibrium angle θ between the wall and the plane of the loop.

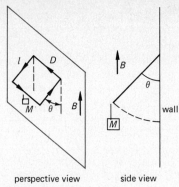

perspective view   side view

Figure 41

42. Compute the net force on the loop shown below.

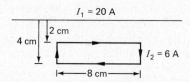

Torques on Current Loops

43. A single rectangular loop of wire 5 × 10 cm and carrying a current of 8 A is placed in a magnetic field where $B = 0.4$ T. If the plane of the loop is parallel to the magnetic field, what is the resultant torque on the loop?

44. A single circular loop of wire of radius 6 cm and carrying a current of 5 A is placed in a uniform magnetic field of 0.2 T. (a) If the plane of the coil is parallel to the magnetic field, what is the torque on the loop? (b) If the loop were made of 50 turns instead of 1, how would the torque change?

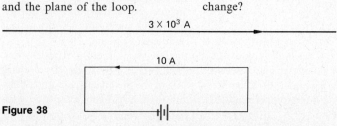

Figure 38

45. A circular coil of 100 turns of wire has a radius of 3 cm and carries a current of 2 A. The plane of the coil makes an angle of 30° with the direction of the magnetic field of magnitude $B = 0.04$ T. (*a*) Compute the torque on the loop. (*b*) What is the maximum torque obtainable in this coil, and at what orientation does this occur?

46. A galvanometer coil consists of 400 turns with dimensions 2×4 cm. The coil is mounted in a magnetic field where $B = 0.15$ T. A current of 1 mA passes through the galvanometer coil. What is the maximum resulting torque on the coil?

47. A rectangular coil with 50 turns and an area of 50 cm^2 is suspended in a magnetic field so that it can rotate. When a current of 10 A passes through the coil, a maximum torque of 1.5 N·m is exerted on the loop. Compute the magnitude of the magnetic induction.

Special Topic
Earth's Changing Magnetic Field

Unlike its geographic poles, the earth's magnetic poles are in a continual state of motion. During any particular century the magnetic poles are likely to be in quite different locations from those they occupied 100 yr before, even though the average positions of the magnetic poles over tens of thousands of years tend to conform to the fixed positions of the geographic poles. These movements provide valuable clues for geologists trying to piece together the past history of our planet.

A major geological effect of the earth's magnetic field is its ability to impress its own direction on any magnetic rock that comes under its influence. Such rocks are those that cool down under the influence of the earth's magnetic field, obtaining a magnetic orientation in the process. The "fossil magnetism" the rocks acquire in this way is extremely stable over hundreds of millions of years. Thus, magnetic rocks retain a measurable memory of the direction of the earth's magnetic field at the time when they last cooled down. The best examples of these rocks are volcanic lavas, which actually cool from molten form.

Geophysicists have used this effect to study one of the most widely publicized geological theories: continental drift. The theory that the earth's continents once occupied the same land mass was first suggested in the Middle Ages; it was then revived in 1912 by the German meteorologist Alfred Wegener. The obvious justification for the theory is the excellent fit between the outlines of Africa and South America when they are placed together on a globe and matches that are almost as good between other continents separated by oceans. However, the theory faced great resistance because geologists could not envision a way in which the continents actually could have drifted apart on the solid rock that makes up the earth's mantle.

One strong piece of evidence in favor of continental drift emerged from studies of the fossil magnetism in rocks on various continents. By studying the directions in which rocks of different ages were magnetized on the different continents, geophysicists were able to plot "polar wandering curves" that indicated the variation in direction of the magnetic field during the past millions of years. When they plotted these polar wandering curves for individual continents on globes in which the continents were placed together, as dictated by continental drift, geophysicists discovered that all the curves coincided when taken back more than 450 million years and that the wandering curves of two or more continents coincided up to 250 million years ago. This suggested strongly that all the earth's landmass was concentrated in a single supercontinent which split into two about 450 million years ago and that the two components started to break up 200 million years later, forming the continents as we know them today.

This type of study, and other circumstantial evidence for continental drift, did not solve the problem of how it could have happened. In the early 1960s, however, a few geologists suggested that the continents were actually carried along on huge solid "plates"—large slabs of rock that floated on the underlying rock. Most experts scoffed at this idea, arguing that such diffusion of solid objects was patently absurd. But then a new possibility arose: perhaps molten rock emerging from the depths of the earth through deep trenches in the middle of the oceans provided the driving force that pushed the plates apart. Proving this contention, however, seemed next to impossible.

The solution came from a surprising finding about the nature of the earth's magnetic field. Every few hundred thousand years the field has quite suddenly changed polarity, reversing from one direction to the other. Geophysicists have no clues as to why this reversal occurs and little evidence on which to forecast the next occurrence. However, they know that the reversals alter the direction of the magnetic field impressed on rocks that are cooling. In 1963, a magnetic study of the bottom of the mid-Atlantic indicated that the region of the ocean floor around the mid-Atlantic trench was magnetically striped—marked by broad bands of rock magnetized in alternately different directions. These bands were exactly what geophysicists would have expected if the molten rock flowing out of the trench were carrying apart the continents of Europe and North America. Thus the study, and similar magnetic measurements in other parts of the oceans, provided yet another vital piece of evidence for the idea that the new world today is further from the old than it was at the time Columbus sailed here.

16 Induced Electric Currents

16.1 INDUCTION IN THE BEGINNING

16.2 THE CONTRARINESS OF NATURE

SHORT SUBJECT: The Father of Electromagnetic Induction

16.3 THE AC GENERATOR

SHORT SUBJECT: Magnetic Levitation: Tomorrow's Transportation System?

16.4 STEPPING UP AND STEPPING DOWN
Summary
Questions
Problems

SPECIAL TOPIC: The Oscilloscope

Turnabout is fair play. If this cliché applies to daily life, why should it not also apply to science? We know that the motion of electric charges through a magnetic field produces a force which can be harnessed in such devices as electric motors. Now we can ask whether this process can be turned about. In other words, is it possible to carry out mechanical work in a magnetic field and thereby produce electric energy?

For the benefit of students who cannot abide mystery tales whose denouement does not occur until the last paragraph, the answer is yes. Mechanical motion of conductors through magnetic fields is a proved means of generating electric power that accounts for many of the creature comforts of our existence. The turbine-driven generator, which we encountered briefly in Chap. 5 in a different connection, is the archetypal example of the use of mechanical force to produce electric energy.

The underlying principle is known as electromagnetic induction. It can be demonstrated on a small scale using a minimum of apparatus: a length of wire, a sensitive galvanometer, and a strong magnet. Attach the two ends of the wire to the two terminals of the galvanometer and then move the wire rapidly up and down in such a way that it travels through the magnetic field. When the wire is stationary, the galvanometer reading is zero, but when the wire moves through the magnetic field, the needle on the instrument deflects, indicating the presence of a small electric current. This is the induced current produced by the movement of the wire through the magnetic field. A little experimentation will reveal more about the phenomenon of electromagnetic induction, notably that *the faster one moves the wire through the field, the greater the magnitude of the current induced.*

We can look at this simple experiment schematically in Fig. 16.1, which shows a segment of wire whose length l is large enough to support a measurable potential difference between the ends when the wire is moved through the magnetic field. The field in this case is directed out of the page toward the reader. We must imagine that the two loose ends of the wire pictured are somehow joined to complete an electrical circuit through which current can be established. This can be accomplished by sliding the wire along a U-shaped frame that contains a galvanometer.

Now the straight length of metal wire contains relatively free electrons. Thus, when we move the wire perpendicular to the magnetic field, we actually move individual charges in the field. We know from

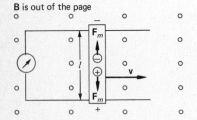

Figure 16.1 Force on charges in a wire moving through a uniform magnetic field. The magnetic induction **B** is directed out of the page; **v** and **B** are perpendicular. The force on the charges, electrons, produces a potential difference between the ends of the moving wire.

Eq. (15.1) that these charges will experience a force F_m given by the equation

$$F_m = QvB \sin \phi$$

Since each charge Q represents the charge of an individual electron, and we have already specified that ϕ is 90°, we can simplify this equation to

$$F_m = evB$$

The effect of the force is to move the electrons. What is a movement of electrons in a wire? Nothing more or less than an electric current. Hence our theory predicts the induction of current that we observed with our simple setup.

What is the direction of the force? Students who thought that they had left the right-hand screw rule behind in Chap. 15 have another thought coming. We press the rule into service once more to solve this problem. We know that the vectors of the magnetic field, the velocity of the electrons, and the force on the electrons resulting from their movement through the field are all perpendicular to one another. Rotating **v** (the direction positive changes would move) into **B** in Fig. 16.1, we find that the force on *positive* charges would be downward.

The negative electrons will move upward in the figure. However, the conventional current of positive charges will be directed downward. If we move the wire parallel to the magnetic field—that is, into or out of the page—the force is zero and no current results. In this case the angle ϕ is equal to either 0° or 180°, and $\sin 0° = \sin 180° = 0$. Hence our theory predicts this as well.

To work out the potential difference V between the ends of the moving wire we need only calculate the amount of mechanical work, W, required to move a single charge from one end of the piece of wire to the other, and then divide this amount by the value of the single charge. Using familiar equations,

$$W = \mathbf{F} \cdot \mathbf{s} = evBl$$

$$V = \frac{W}{q} = \frac{evBl}{e} = lvB \qquad (16.1)$$

The induced voltage is therefore constant if the wire moves through the field at a constant speed, and its value is proportional to both the speed and the strength of the field.

EXAMPLE 16.1

A 0.5-m-long piece of wire moves at a constant velocity of 2 m/s in a direction perpendicular to its own length and perpendicular to a magnetic field whose strength is 0.6 T. Calculate the induced voltage.

Solution Direct substitution into Eq. (16.1) is all that is necessary:

$$V = Blv = (0.6\text{ T})(0.5\text{ m})(2\text{ m/s}) = 0.6\text{ V}$$

16.1 INDUCTION IN THE BEGINNING

This analysis may seem obvious. However, the discovery of induced currents was a major achievement that followed many working years of serious investigation. The finding was made in the early 1830s by two men working independently. One man was Michael Faraday, an English physicist without any formal education who was immensely productive as a scientist developing people's understanding of electricity. The other man was the American Joseph Henry, another poor boy, whose subsequent researches were overshadowed by those of his English colleague.

Faraday's name was immediately associated with the discovery of electromagnetic induction, even though Henry was the first man to actually observe the phenomenon. Henry suffered from geography; he was based in Albany, New York, at a time when most of the world's important scientific work was going on in Europe. Moreover, Faraday conducted exhaustive studies of electromagnetic induction before and after its discovery. And perhaps the best reason for associating Faraday's name with this branch of physics is the answer he reputedly gave a woman attending a lecture by him in London, who asked what possible use the new phenomenon could have. "Madam," Faraday replied, "of what use is a newborn baby?" As for the unfortunate Henry, he and his heirs had the consolation of a unit named after him.

This situation preechoed that which occurred in the field of evolution 25 yr later, when English biologist Charles Darwin and American scientist Alfred Wallace both advanced the theory of natural selection at the same time, only for Darwin to receive all the accolades of the scientific world.

"Madam," Faraday replied, "of what use is a newborn baby?"

Faraday's search for induced currents is one of the most stirring stories in all of scientific endeavor. He knew that a force should exist on a current-carrying wire in a magnetic field, and for seven full years, starting in 1824, he tried to produce detectable currents via magnetism. He finally succeeded by virtue of serendipity. His apparatus, shown in Fig. 16.2, consisted of two separate coils of wire wound around an iron ring, one coil to each side of the ring. One coil was attached to an electrical circuit containing an 1830s version of a battery. Passage of current through this circuit made the iron ring an electromagnet, rather like a solenoid. The other coil was connected in a closed circuit containing a galvanometer that measured any current induced in the coil.

Unfortunately, no detectable current appeared in the second circuit when a constant current was surging through the first circuit, no matter how small or large the constant current was. But one day Faraday noticed that a small current appeared in the second coil at just the moment the current in the first coil was being switched on. This induced current lasted only momentarily and disappeared once the current was established at its constant value in the first coil—only to reappear when the current was switched off. Thorough investigation on this peculiar effect—as a result of which he was able to induce in a number of ways

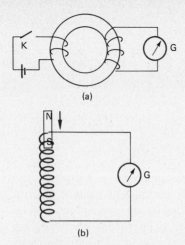

Figure 16.2 Electromagnetic induction. (a) Two coils wound on the same ring of magnetic material. When the key K is closed, the galvanometer G will deflect. When the current in the left-hand loop is constant, there is no induced current in the right-hand loop. When the key K is opened again, the galvanometer deflects. (b) Moving the south pole of a bar magnet into the coil causes the galvanometer to deflect; pulling out the south pole causes a reverse deflection of the galvanometer.

an electric current in a completed circuit without a known source of emf such as a chemical cell—convinced Faraday that every successful example had one ingredient in common: **there was some relative movement between the magnet and the wire in which current was induced.**

In the simple example that opened the chapter, we would find on investigation that we induce current only when the wire is moving through the magnetic field lines. No current flows when the wire is stationary, even if it is situated in the magnetic field. Similarly, if we had used a bar magnet and a coil of wire, a current would have been induced in the coil only when we pushed the magnet into or pulled it out of the coil. No current is induced when the magnet is stationary inside the coil.

In the first example that Faraday probed, the relative movement of magnet and wire is more subtle. The magnetic field is "moving" by increasing quickly in strength, from zero to its stable value, as the current is switched on. Once the current is established, the magnetic field strength becomes constant, and no current is induced in the second wire. Then when one turns off the current in the first coil, the magnetic field in the iron ring drops quickly to zero—moving again and inducing once more a small current in the second coil. As we might expect, the current induced in the second coil in the first case, when the current in the first coil is switched on, is in the opposite direction to that induced in the second example, when the current in the first coil is switched off.

Faraday discovered that the **induced electromotive force, or voltage, was proportional to the rate at which the magnetic lines of force changed in the vicinity of the charges that ultimately constitute the induced current.** To understand this relationship in strict mathematical terms, we must define a new quantity, which we call the magnetic flux Φ (Greek letter phi). According to the definition, the magnetic flux over an area A is given by

$$\Phi = AB \cos \theta \tag{16.2}$$

where B is the magnetic field and θ is the angle between the field and the perpendicular to area A. We can write the equation as the scalar products of two vectors **A** and **B** if we define the vector **A** as having the magnitude of the area and a direction perpendicular to the area. Hence, Eq. (16.2) can be written

$$\Phi = \mathbf{A} \cdot \mathbf{B} = A \cdot B \cos \theta$$

For simplicity, we can regard the magnetic flux as Faraday did, as the number of lines of magnetic force that cross the area A (see Fig. 16.3). The unit of magnetic flux is the weber (Wb), named after Wilhelm E. Weber, a German physicist; 1 Wb/m² is equal to 1 T, our unit of magnetic induction B. The equation simplifies to $\Phi = AB$ when θ is 0°, that is, when the field is perpendicular to the area under consideration because then $\cos \theta$ is 1.

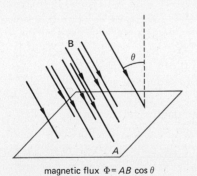

magnetic flux $\Phi = AB \cos \theta$

Figure 16.3 Definition of magnetic flux.

We can now put Faraday's discovery, with which we opened the last paragraph, into mathematical form. The induced emf ($\mathcal{E}$) is proportional to the rate at which the magnetic lines of force change in the vicinity of the charge carriers or, more formally, to the rate of change of the magnetic flux, and is simply expressed in the form

$$\mathcal{E} = -\frac{\Delta \Phi}{\Delta t} \tag{16.3}$$

This formulation is known as Faraday's law.

16.2 THE CONTRARINESS OF NATURE

A striking thing about the mathematical statement of Faraday's law is its negative sign. **This indicates that the current induced electromagnetically moves in a direction such that its magnetic effects oppose the changing of the magnetic field which is responsible for producing that current.** It is an example of a general rule first put into words by Heinrich F. E. Lenz, a Russian who duplicated much of the electromagnetic induction work of Faraday and Henry. In its general form, the principle states that the direction of an induced current is such as to oppose the effect that caused it. This statement is known as Lenz's law.

Lenz's law expresses what might be termed "the contrariness of nature." However, Lenz's law is a necessary consequence of the conservation of energy. If the induced emf did not oppose the effect that caused it but supported that effect, it would be possible to design a perpetual motion machine, that is, get more energy out of an electrical machine than we put into it. The practical meaning of Lenz's law can be best understood by considering a number of examples of the law in action.

Figure 16.4 shows two parallel loops of wire, one in a circuit containing a battery and a switch, the other connected in a closed circuit to a galvanometer. When the switch is closed in loop 1, a current builds up in the direction indicated; the increasing current produces a changing magnetic flux over the second loop, which induces a current in loop 2. Because the cause of this induced current is the increase in the magnetic field through loop 2, the induced current in loop 2 must travel in such a direction that it produces a magnetic field of its own which points in the opposite direction to the original produced by loop 1. Hence the current in loop 2 must move in the opposite direction to the current in loop 1.

Once the current in loop 1 settles down to its constant value, we know that no further emf is induced in loop 2 because the magnetic field produced by loop 1 also remains steady in value. We can alter the situation by switching off the current in loop 1, as shown in Fig. 16.4(b). As the current decays to a magnitude of zero, the magnetic field produced by loop 1 in the vicinity of loop 2 also decreases, inducing a fresh current in loop 2. This time, according to Lenz's law, the induced current in loop 2 must produce a magnetic field that opposes the decay of the magnetic field of loop 1. Hence the magnetic field of loop 2 points in the same direction as the magnetic field produced by the

Figure 16.4 Currents induced by moving the source of the magnet's field. (a) The switch in loop 1 is closed, inducing an opposing magnetic field in loop 2. (b) The switch in loop 1 is opened, inducing a supporting magnetic field in loop 2. (c) Loop 1 with a steady current moves toward loop 2. (d) Loop 1 with a steady current moves away from loop 2. (e) A north pole is pushed toward loop 2. (f) A north pole is pulled away from loop 2.

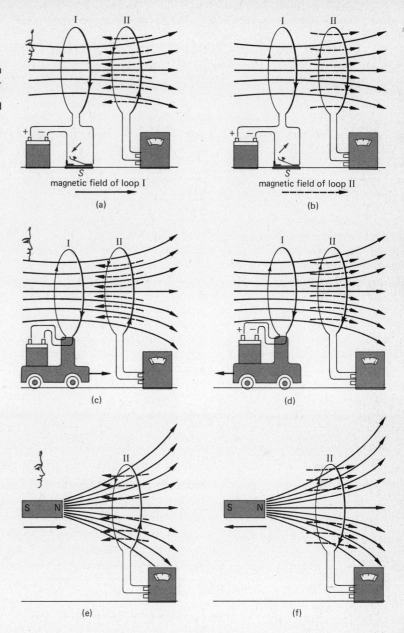

decaying current in loop 1. The current in loop 2 moves in the same direction as that in loop 1.

Figures 16.4(c), (d) show a slightly different situation, involving relative movement of a loop through which a constant current is flowing and another loop is connected to a galvanometer. In this case the physical movement produces the changing magnetic field that

The Father of Electromagnetic Induction

The story of Michael Faraday is reminiscent of the stirring biographies of heroes who overcome poverty, jealousy, and other misfortunes in their rise to glory. Faraday, one of 10 children of a London blacksmith, was fortunate enough to be apprenticed to a bookbinder and to acquire through his apprenticeship a great desire for book reading in general and science in particular. In 1812, when he was 21, Faraday attended a lecture by the famous chemist Sir Humphry Davy at London's renowned Royal Institution. Bitten by the bug of scientific experimentation, Faraday persuaded Davy to take him on as his assistant, even though the new job carried a lower salary than his bookbinder job.

First guided by Davy and then on his own, Faraday became an original thinker and experimenter in many areas of science. In 1823 he liquefied for the first time such gases as carbon dioxide and hydrogen sulfide. Two years later he discovered benzene, which even today is one of the most vital of all organic chemical compounds. He also extended Davy's fundamental studies of electrolysis and, in 1824, embarked on the work that was to gain him most fame: the investigation of electromagnetic induction.

Faraday was probably one of the last of the truly great scientists to lack any mathematical education. He more than made up for this chink in his armor by his extraordinary ability to picture concepts in a nonmathematical way, such as his visualization of magnetic lines of force. Faraday was modest and forebearing almost to a fault. He turned aside with equanimity all efforts of Davy, his one-time master, to discredit him when it became clear that Faraday outshone him, and in his later life he turned down both a knighthood and the presidency of the Royal Society, thus giving up all the prestige that the honors entailed.

---

induces a current in the second loop. According to Lenz's law, the induced current moves in such a direction as to oppose this motion. Hence, when the two loops are moving toward each other, as in Fig. 16.4(c), in a form of motion that increases the strength of the magnetic field of loop 1 as experienced by loop 2, the induced current tends to reduce the total strength of the field. The field of the induced current in loop 2 opposes the field produced by the current in loop 1. This requires the current in loop 2 to travel in the opposite direction to that in loop 1.

When the two loops move away from each other, as shown in Fig. 16.4(d), the strength of the magnetic field of loop 1 as experienced at loop 2 decreases, and the field of the induced current in loop 2 must, therefore, act to increase the magnetic field. Thus the magnetic field of loop 2 points in the same direction as that of loop 1, which means that the current induced in loop 2 travels in the same direction as that already existing in loop 1. Figures 16.4(e), (f) represent the similar but simpler situation that occurs when a bar magnet rather than a current-carrying coil is moved relative to a coil attached to a galvanometer.

As usual, we can analyze both Faraday's and Lenz's laws mathematically by taking the simplest case imaginable—a single strand of

Figure 16.5 Induced emf in a wire moved through a uniform magnetic field. The induced emf is computed by finding the rate of change of flux.

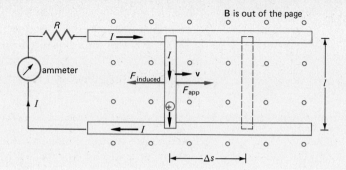

wire that moves perpendicularly through a uniform magnetic field. This is the same example we examined in Sec. 16.1. This situation is illustrated in Fig. 16.5. A segment of wire whose length is l is free to slide along a U-shaped piece of conducting material through which the wire completes a circuit. We imagine that this setup is placed in a magnetic field whose strength is **B**, pointing out of the paper. Let us also postulate that someone pushes the length of wire to the right along the U-shaped conductor with a force F_{app}, which is just sufficient to overcome friction and other retarding forces; as a result, the wire travels with a constant speed v. (The *net* force on the wire is zero.)

Let us now work out the magnitude of the voltage induced in the length of wire using Faraday's law. To do this we must first calculate the rate of change of the magnetic flux. Since the field is perpendicular to the direction of the wire, the value of the change of flux Φ is simply $B \, \Delta A$, where ΔA is the area swept out by the moving wire. If the wire of length l travels a distance Δs in time Δt, it obviously sweeps out an area $l \, \Delta s$ in that time. Thus, the change of flux is $lB \, \Delta s$. The rate of change of the magnetic flux is then given by

$$\frac{\Delta \Phi}{\Delta t} = \frac{lB \, \Delta s}{\Delta t} = lBv$$

substituting v for $\Delta s/\Delta t$. The term lBv is, according to Faraday's law, numerically equal to $\mathcal{E}$, the emf induced in the wire. Note that this result is identical to the one obtained by analyzing this same example from the perspective of the forces on the charges in the wire as they are moved with the wire through the magnetic field. This earlier result was expressed in Eq. (16.1). And if we knew the resistances of the piece of wire and the U-shaped conductor to which it is attached, we could then work out the current in the whole loop, using Ohm's law.

What about the direction of the induced current? Here we apply Lenz's law. To do so we must identify the cause of the induced current. In this case it is clearly the applied force F_{app} that pushes the length of wire to the right. According to Lenz's law, the induced current must oppose this force, which it does by linking up with the magnetic field **B** to produce a force that points to the left in the figure. We can find out the direction of the current by trial and error, applying the right-hand screw rule to the two possible directions of the current. If you try this yourself, you will find that the downward current shown in the figure produces the required opposing force, $F_{induced}$, when rotated into the vector of the magnetic field. Refer to Eq. (15.4).

EXAMPLE 16.2

Two coils are adjacent to one another, as in Fig. 16.6. A current in coil 1 produces a flux of 5×10^{-4} Wb over circuit 2. When circuit 1 is opened, the flux falls to zero in 10^{-4} s. What average emf is induced in circuit 2?

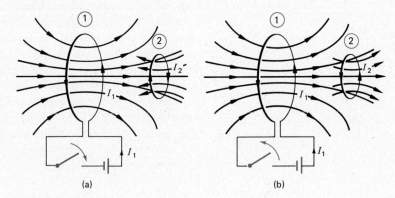

Figure 16.6 (a) When the switch in circuit 1 is closed, the magnetic flux through circuit 2 changes, inducing a current in loop 2 (I_2), which moves in a direction opposite to the current I_1 in loop 1. (b) When the switch in circuit 1 is opened, the magnetic flux through circuit 2 is reduced, inducing a current in loop 2 (I_2), which moves in the same direction as the current in loop 1. Hence, the current induced in loop 2 is attempting to keep the magnetic field from dying out.

Solution

The magnitude of the induced emf is given by Eq. (16.3) [we neglect the negative sign of Eq. (16.3) since it relates to the direction of the induced emf]:

$$\mathcal{E} = \frac{\Delta \Phi}{\Delta t} = \frac{5 \times 10^{-4} \text{ Wb}}{10^{-4} \text{ s}} = 5 \text{ Wb/s} = 5 \text{ V}$$

EXAMPLE 16.3

A coil consisting of 50 turns of wire with an area of 10^{-3} m² is placed in a region of constant flux density equal to 3 T (3 Wb/m²). The flux density is reduced to zero in an interval of 10^{-3} s. If the magnetic flux makes an angle of 30° with the perpendicular to the plane of the coil of wire, what is the induced emf?

Solution

Once more we apply Faraday's law, Eq. (16.3), but we must realize that **B** and **A** are not parallel and that we are working not with 1 but 50 loops, which means that we must multiply the changing flux in one loop by 50 to obtain the induced voltage:

$$\mathcal{E} = 50 \frac{\Delta \Phi}{\Delta t} = 50 \frac{\Delta BA \cos \theta}{\Delta t}$$

$$= 50 \frac{(10^{-3} \text{ m}^2)(3 \text{ Wb/m}^2 - 0)(\cos 30°)}{10^{-3} \text{ s}} = 130 \text{ V}$$

EXAMPLE 16.4

A 0.5-m length of wire moves at a constant velocity of 2 m/s in a direction perpendicular to its own length and at an angle of 60° with respect to the field lines from a magnetic field of flux density equal to 0.6 Wb/m². Calculate the induced emf.

Solution

We can use Eq. (16.1) if we correct for the fact that the wire is not moving at right angles to the magnetic field but at an angle of 60° with respect to the field lines. From Faraday's law we know that the component of the velocity perpendicular to **B** contributes to the changing flux AB, the important quantity in determining the induced emf. Examination of Fig. 16.7 shows that the important part of the velocity is $v \sin \theta$, where $\theta = 60°$. Therefore, the induced emf is given by

$$\mathcal{E} = Blv \sin \theta = (0.6 \text{ Wb/m}^2)(0.5 \text{ m})(2 \text{ m/s})(\sin 60°)$$
$$= 0.3 \text{ V}$$

Figure 16.7

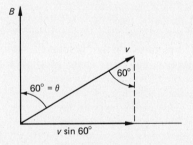

16.3 THE AC GENERATOR

The ac generator produces electric current by moving conducting material physically through a magnetic field.

Probably the most important application of everything we have so far learned about electromagnetic induction is the production of electric energy in the ac generator. In a sense this device is the reverse of the electric motor outlined in Chap. 15. Instead of generating movement by passing an electric current through a conducting material in a magnetic field, as the electric motor does, the ac generator produces electric current by moving conducting material physically through a magnetic field.

Magnetic Levitation: Tomorrow's Transportation System?

Throughout the world, transport between cities is a mess. Automobiles cause pollution, strain the drivers, lead to traffic jams, and waste time as drivers look for parking spaces, and are inherently inefficient. Bus transport overcomes a few of these problems but is still bad. Often buses run on infrequent schedules, are extremely crowded and thus dangerous, and are inconvenient in that riders may have to switch buses several times to get to their destinations. Airplane travel, although fast in the air, involves extreme delays at city centers and airports. Trains are restricted to speeds well below the theoretical maximum speeds they could attain on the right tracks and under optimum conditions. But even at ideal speeds—about 200 mi/h—trains can never rationally compete with planes for distances beyond about 300 mi.

Thus, the search is on for an entirely new method of getting between cities a few hundred—or just a few score—miles apart. One of the more promising possibilities is a class of vehicles that actually fly close to the ground above specially prepared tracks, levitated by electromagnetic repulsion, at speeds of hundreds of miles per hour.

The basic idea behind the system, which was first proposed by the French engineer Emil Bechelet in 1912, is that magnets mounted on vehicles should be repelled by other magnets on a track, thus levitating the vehicle. If electromagnets are used in the vehicle, the levitation effect can be coupled with the vehicle's forward movement. As it travels over a conducting surface, the vehicle-borne electromagnet induces currents in the conductor that, according to Lenz's law, act in such a direction as to repel the electromagnet and thus lift the moving vehicle.

The types of propulsion proposed for this transportation system also rely on electromagnetic induction. One type, known as the linear induction motor, consists of a nonmagnetic cylinder that rotates when eddy currents are induced in it by electromagnets. The other, the linear synchronous motor, is similar, except that the cylinder, or rotor, is made of a magnetic material. The latter is a more efficient type of motor but requires more complex equipment. Indeed, it has been made practically possible only by the development of superconducting magnets.

One of the most promising types of flying magnetic vehicles tested to date is the "magneplane," developed by researchers at MIT. This vehicle contains saddle-shaped superconducting coils on its underside. The coils allow the vehicle to fly along about a foot above an aluminum trough, or guideway. Tests of the system indicate that a full-scale version of the vehicle, which would hold 100 people, could travel at 300 mi/h.

The MIT magneplane. (*Massachusetts Institute of Technology*)

Moreover, control of the system would be sufficiently reliable to allow magneplanes to pass any particular location within a minute of each other. Thus, as MIT physicists Henry Kolm and Richard Thornton wrote in *Scientific American,* "Magneplanes could serve dozens of stations connected by branch loops, with vehicles operating at one-minute intervals and none of them making more than one or two stops between Boston and Washington."

Figure 16.8 Production of an alternating current. The letter R is used as a reference to distinguish one side of the loop from the other.

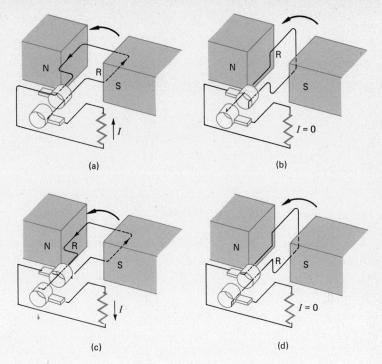

The basic setup of the ac generator is shown in Fig. 16.8. A rectangular coil is rotated between the north and south poles of two magnets, in much the same style as the electric motor. But instead of commutators, the ends of the moving coil are attached to devices called slip rings, which rotate as the coil rotates. The two slip rings are connected through the external circuit in a closed circuit via brushes of the sort we encountered in connection with the electric motor. By rubbing against the slip rings, the brushes maintain continuous electric contact with the rotating loop.

What happens to the coil as it rotates in the uniform magnetic field, whose direction is from the north pole to the south? Only the two parts of the rectangular loop parallel to the axis of rotation, which cross the lines of magnetic force, are effective in producing an electromagnetic emf. Neither the far nor the near end of the coil plays any part in the induction process.

As these two wires move from the position in which the coil as a whole is horizontal to one in which it is vertical, that is, from the position in Fig. 16.8(a) to that in Fig. 16.8(b), they cross the magnetic field lines; therefore an emf is induced in each wire. Since one wire is moving upward as the other wire travels downward, the wires' emf's are in opposite directions. However, instead of canceling each other, these two emf's add to give the induced current indicated in the figure. The direction of this current can be checked using the right-hand screw

Figure 16.9 Calculating the induced emf in a rotating loop. R locates the end of one wire of the loop and serves as a reference. The vector **A** is perpendicular to the plane of the loop.

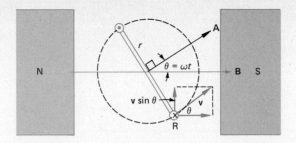

rule, remembering that the coil rotates as a result of a mechanical torque.

What happens when the coil has reached the vertical position of Fig. 16.8(b)? As it "passes over the top," both the upper and lower wires of the coil are traveling horizontally, along the magnetic field lines. They do not cross any field lines, and thus they do not produce any current. This situation immediately alters as the rotation brings the coil back toward the horizontal position of Fig. 16.8(c). Once more, the two parts of the coil in which we are interested start to cross magnetic field lines, and hence current is induced in them once again. But because the right-hand wire that originally moved upward has now become the left-hand wire moving downward, and vice versa, the current is in the opposite direction to that produced in the first quarter of the rotation. As it returns to position (d), the current generated once more drops instantaneously to zero. The net result of all this turning and changing is that the rotating coil generates a current which is always altering in magnitude between zero and a certain maximum during every 90° of rotation and changing in direction every 180° of rotation.

We can understand this apparently peculiar behavior by using Eq. (16.1) as modified in Example 16.4, which expresses the emf induced in a wire in terms of the component of the wire's velocity, which is perpendicular to the magnetic field direction, $v \sin \theta$, and the magnetic field strength B. If the wire's length is l and the angle between **B** and **v** is θ, as shown in Fig. 16.9, then we know that the induced emf $\mathcal{E}$ is given by the equation

$$\mathcal{E} = Blv \sin \theta \tag{16.4}$$

Recalling some of the angular kinematics from Chap. 7, we can substitute the product of the angular velocity of the coil, ω, and the distance of the side of the coil from the axis of rotation, r, for the speed ($v = r\omega$). Remembering also that two wires in the coil are responsible for inducing emf, we come to the overall equation

$$\mathcal{E} = 2Blv \sin \theta = 2Blr\omega \sin \theta$$

We can replace θ by its equal ωt. Now, the area of the coil, A, is equal to the product of the length l and twice the distance r. Thus, the expres-

Figure 16.10 Sinusoidal variation of induced emf with time. R locates the end of one wire of the loop and serves as a reference.

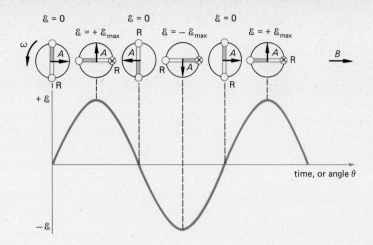

sion simplifies to

$$\mathcal{E} = BA\omega \sin \omega t$$

And if we suppose that the coil contains not just one turn of wire but a larger number, N, then our equation becomes

$$\mathcal{E} = NBA\omega \sin \omega t \tag{16.5}$$

This equation says that when a coil containing N turns of wire with an area A rotates at a constant angular velocity ω in a uniform magnetic field B, then the induced emf, and therefore the induced current, varies with time as a sine function. The actual values of emf and current at any point depend on the magnitudes of B, A, N, and ω as well as on the angle between the coil and the magnetic field. This is shown pictorially in Fig. 16.10.

This diagram shows what we call alternating current; it is in no way a steady passage of electric current. Its value alters cyclically from a value of zero when θ is zero to a maximum corresponding to an induced emf equal to $NBA\omega$ when $\theta = 90°$ and back to zero when $\theta = 180°$, down to a minimum with a numerical value of $NBA\omega$ when θ has the value 270°, and finally up to zero again at the end of the 360° cycle. Strange as the current profile may look, this method is by far the easiest means of generating electric current. The simplicity of this form of electrical generation, in conjunction with the possibility of using transformers (discussed in the next section), is why houses, offices, factories, and other buildings are almost exclusively wired for alternating current. If we had to resort to dragging a wire along a pair of rails at high speed in a configuration as shown in Fig. 16.5, we would find the generation of electricity impractical, to say the least.

We must not forget direct current—the supply of electric current whose direction does not alter. Direct current can be produced by electric generators of the sort we just analyzed if a split-ring commuta-

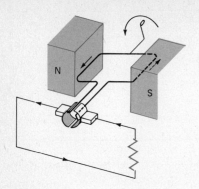

Figure 16.11 A dc generator.

Electronic circuits and other, more advanced, concepts of modern electricity appear in Chap. 17.

tor of the type used in the electric motor is substituted for the two slip rings, as illustrated in Fig. 16.11. As the wires of the coil change direction, between up and down in the magnetic field, the commutator reverses the contacts automatically. The result is a profile of electric current like that sketched in Fig. 16.12—a varying emf and current whose values never drop below zero. This pulsating direct electric current is obviously different from the steady output of direct current from a dry cell or battery, however. Producing direct current of constant magnitude from a generator is a more elaborate process that requires a large number of independent coils orientated in different directions. In this way a drop in the induced emf of one particular coil is compensated for by a rise in the emf induced in another. In addition, an electrical or electronic filtering circuit may be required.

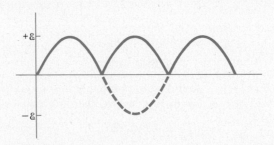

Figure 16.12 Variation of the emf from a single-loop dc generator.

EXAMPLE 16.5

A rectangular coil containing 100 turns of wire with an area 20×30 cm is rotated at a constant speed (frequency) of 600 r/min in a magnetic field whose flux density is 0.2 Wb/m². Find the maximum emf produced.

Solution

We apply Eq. (16.5), being careful to use consistent units:

$$\mathcal{E}_{max} = NBA\omega$$

where $N = 100$ turns, $B = 0.2$ Wb/m², and $A = 0.2$ m $\times$ 0.3 m $= 0.06$ m². Therefore,

$$\omega = (2\pi \text{ rad/r})(600 \text{ r/min} \times \tfrac{1}{60} \text{ min/s}) = 20\pi \text{ rad/s}$$

$$\mathcal{E}_{max} = (100 \text{ turns})(0.2 \text{ Wb/m}^2)(0.06 \text{ m}^2)20\pi \text{ rad/s}) = 75.4 \text{ V}$$

16.4 STEPPING UP AND STEPPING DOWN

Beyond simplicity of production, alternating current has another significant advantage over direct current. Faraday's original experiment on electromagnetic induction, which involved the use of direct current in one loop to induce a current in a second loop, worked only when the current was being switched on and off because those were the only times at which the current, and hence the magnetic field it produced, changed. Alternating current by its very nature continually changes in magnitude. Hence, if we use Faraday's apparatus and drive it with an ac voltage through the first coil, an emf can be induced continually in the second coil. This type of device is known as a transformer. It has a multitude of uses in the electrical industry, from transmitting large amounts of electric power over long distances between the generators and the consumers to stepping down household (110 V) voltages to the different low voltages, around 6 to 12 V, necessary to drive model electric railroad trains, power a door bell, or operate television and radio tubes.

The most common type of transformer is shown schematically in Fig. 16.13; it is modeled roughly after Faraday's original apparatus, providing a remarkable testimonial to the ability of this self-educated Englishman. Instead of direct current, the first coil, known as the primary, is provided with an ac voltage. This coil and the secondary one are both wound around a core of magnetic material. $\mathcal{E}$, varying as a sine wave, is applied to the primary coil and produces a time-varying current in the primary coil. This current, I, induces a time-varying magnetic flux in the magnetic core upon which both the primary and secondary coils are wound. The magnetic flux in the core, in turn, induces an emf in the secondary coil. If we consider that there is no IR drop (power loss) in the primary coil, then the voltage applied to the primary is essentially $\mathcal{E}_1$. The voltage $\mathcal{E}_1$ must be equivalent in magnitude but opposite to the back emf (back emf is defined on page 507). If the primary coil consists of N_1 turns of wire wrapped around the magnetic core, then the flux changes in the magnetic core are related to the magnitude of the voltage $\mathcal{E}_1$ by Faraday's law:

$$\mathcal{E}_1 = N_1 \frac{\Delta \Phi}{\Delta t}$$

We assume that the same flux Φ exists throughout the core, including the vicinity of the secondary coil. If this secondary has N_2 turns of wire around the core, the induced emf will have a magnitude of

Figure 16.13 Transformer.

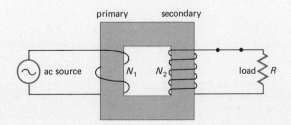

$$\mathcal{E}_2 = N_2 \frac{\Delta \Phi}{\Delta t}$$

Here we have two simultaneous equations that relate the numbers of turns of wire and the voltages of the two circuits. Dividing one by the other, we obtain the simple expression

$$\frac{\mathcal{E}_1}{\mathcal{E}_2} = \frac{N_1}{N_2} \quad \text{or} \quad \frac{\mathcal{E}_1}{N_1} = \frac{\mathcal{E}_2}{N_2} \tag{16.6}$$

In other words, the induced voltage is proportional to the relative number of turns on the two coils—a proportionality that occurs because both coils are coupled with the same varying magnetic flux. Plainly, then, we can use the transformer to increase or decrease voltage by suitably adjusting the numbers of turns of wire in the primary and secondary coils. If the secondary coil has more turns than the primary, the secondary voltage will be larger than that of the primary. This system is known as a step-up transformer. On the other hand, if the secondary has fewer turns than the primary, the voltage in the secondary will be less than that in the primary. This is a step-down transformer.

The gain in voltage that one can obtain with a step-up transformer seems almost too good to be true. The first question that comes to mind is "Where does it really come from?" We can answer this question by using the always-applicable principle of the conservation of energy, which tells us that the amount of power (energy per unit time) transferred between the two loops cannot be greater than the total amount of power available in the primary coil. We know from the early electrical equations in Chap. 14 that the power in any circuit is the product of the voltage and current therein. If we assume that the transformer is 100 percent efficient and dissipates not a single microwatt of power when it operates, we can equate the expressions for the power in the two loops:

$$\mathcal{E}_1 I_1 = \mathcal{E}_2 I_2$$

$$\frac{\mathcal{E}_1}{\mathcal{E}_2} = \frac{I_2}{I_1} = \frac{N_1}{N_2} \tag{16.7}$$

In other words, this means that the currents in the two coils are inversely proportional to the voltages as long as no power is lost in the transfer.

EXAMPLE 16.6

A generator produces 40 A at 600 V. We wish to transmit this power over a high-voltage transmission line, and to do so we must step up the voltage. This is accomplished by connecting this source to the primary of a transformer that consists of 10 turns of wire. If the secondary has

1000 turns, what will be the induced voltage in the secondary and the maximum current permissible in the secondary?

Solution

The induced voltage in the secondary can be computed using Eq. (16.6):

$$\frac{\mathcal{E}_1}{\mathcal{E}_2} = \frac{N_1}{N_2}$$

$$\mathcal{E}_2 = \frac{N_2}{N_1} \mathcal{E}_1$$

$$\mathcal{E}_2 = \tfrac{1000}{10} \times 600 = 60\,000 \text{ V}$$

To compute the maximum current in the secondary, we set the primary power equal to the secondary power—the best possible efficiency:

$$\mathcal{E}_1 I_1 = \mathcal{E}_2 I_2$$

$$I_2 = \frac{\mathcal{E}_1 I_1}{\mathcal{E}_2} = \frac{(600 \text{ V})(40 \text{ A})}{60\,000 \text{ V}} = 0.4 \text{ A}$$

Power losses are now typically 10 to 15 percent of our electrical energy problem.

One other possible means of overcoming power losses in transporting electrical energy over large distances is the use of superconducting power lines of the type outlined in Chap. 14. At present, such schemes are in very preliminary stages.

The use of the step-up transformer described in Example 16.6 raises the question of power losses in electric transmission lines and electrical devices. We know from Chap. 14 that the reduction of power owing to heat losses in any circuit equals I^2R, the product of the square of the current carried by the circuit and the resistance of the circuit. Hence, the lower the current passing through a device, the less the loss of power from that device. By stepping up the household supply of 110 V to 220 V for operating large devices that consume a great deal of electric power, we reduce the operating current and consequently the power losses in the house wiring.

This philosophy is even more evident in the long-distance transmission of electricity. Power losses are now typically 10 to 15 percent of our electrical energy problem. By using high-tension wires, illustrated in Fig. 16.14, which carry electric current with voltages of tens and even hundreds of thousands of volts, electric utility companies reduce the amount of current under transmission and hence minimize their power losses in transit. The application of the high-tension wires requires step-up transformers at the power plants and step-down transformers to provide the normal 220 to 110 V in the consumers' homes and factories. Needless to say, the high-tension lines are extremely dangerous because of the huge voltages.

Transmission lines are not the only potential sources of large power losses. Transformers themselves are hardly models of efficiency, and our

Figure 16.14 High-tension power lines. (*Beckwith Studios*)

previous assumption that these devices lose no power is certainly unrealistic. Transformers and other electric devices lose power mainly as a result of eddy currents. Eddy currents can be minimized in good transformers, but to do so requires engineering design based on knowledge of the mechanism of these losses.

Eddy currents are a result of the same process that gives the transformer its ability to step up or step down voltage. The varying magnetic flux produced by the primary current in the device induces current in any nearby conductor, including the magnetic core of the transformer. Eddy currents can be visualized as circular current loops in the metal. This type of induced current heats up the core at the expense of electrical power, thereby reducing the power efficiency of the transformer. This particular loss of power is normally minimized by building the transformer core with a number of thin plates of iron or other magnetic material; each plate is electrically insulated from the others in such a way that the core contains no large closed circuits through which the eddy currents can travel. This reduction in area of the conductor minimizes the magnitude of the eddy currents.

Another important example of eddy currents in moving machinery can be best understood by examining the situation shown in Fig. 16.15. A rectangular plate of conducting material is allowed to swing freely between the poles of a magnet, acting essentially as a pendulum. As the plate moves into the magnetic field, we know that currents are induced in it in such a direction as to oppose its motion. In other words, eddy currents induced in the plate circulate in a way that produces a pair of electromagnetically induced north and south poles which are opposed to the external magnet as the plate enters the magnetic field. Then, the north pole of the eddy currents opposes the north pole of the magnetic

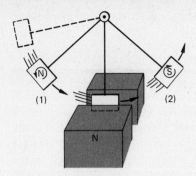

Figure 16.15 Eddy currents induced in the swinging plate are in such a direction as to produce magnetic poles that oppose the motion.

field, and the south pole of the eddy currents opposes the south pole of the magnet. These electromagnetically induced poles tend to keep the plate out of the magnetic field, and their effect is to slow down the motion of the plate. Once the plate has swung through the field and is moving out of it altogether, the circumstances change; now the eddy currents reverse, in the effort to keep the plate in the magnetic field.

The net result of these two opposing tendencies is to slow down severely, or damp, the swinging of the plate. The eddy currents also heat up the plate. This heating can be quite spectacular; in some cases it can actually raise the temperature of a metal plate enough to melt it. And the energy that provides the heat is gained at the expense of useful electrical energy. As a result, engineers must take great care to reduce the extent of eddy currents in transformers and other devices by appropriate methods of construction and design.

Eddy currents are not entirely useless. A damper based on eddy currents is often used in analytical balances to reduce the swinging motion of these laboratory instruments. The brakes on some models of subway cars are also based on the electromagnetic effect that produces eddy currents. These devices involve the attachment of a series of electromagnets to the cars close to the track on which the subway cars run. When the operator wants to stop the cars, the operator energizes the electromagnets in the cars; the electromagnets induce eddy currents in the rails that, through their magnetic fields, oppose the motion of the cars and gradually bring them to a halt.

Another significant consequence of electromagnetic induction, one accompanying electric motors, is termed "back emf." As we know from Chap. 15, in the electric motor a current sent through a loop of wire in a magnetic field produces a torque on the loop. Now, once it starts rotating, the loop also acts as an electromagnetic generator because parts of the loop cross magnetic field lines. Lenz's law implies that the emf induced in this way must oppose the motion of the loop. Therefore, this back emf induced in the motor by the motion of the coil opposes the applied voltage that causes the current making the motor action possible. As a result, the applied voltage minus the back emf reduces the voltage, causing the IR drop in the armature. This reduced voltage equals IR, where I is the current in the armature coil and R is its resistance.

Note that the back emf is zero when the armature coil of the motor is not moving. Since the resistance of the armature coil remains essentially constant whether or not it is moving, the maximum current will exist in the armature when the motor is just turned on. As the armature speeds up, the back emf increases, reducing the voltage applied to the armature coils and thereby causing the current to drop.

In a typical household, starting up a device may draw so much current that it overloads the circuit, causing fuses to blow or circuit breakers to trip. If the motor passes the starting hurdle without blowing a fuse, however, the circuit is unlikely to become overloaded for the induced back emf reduces the net voltage and hence lowers the amount of current drawn by the device.

EXAMPLE 16.7

A series-wound motor, whose field and armature coils are connected in series, has an internal resistance equal to 0.5 Ω. When running at full load on a 110-V line, the motor draws a current of 20 A. What is the (a) back emf in the armature, (b) starting current, (c) power delivered to the motor when running, (d) rate at which heat is generated in the motor, and (e) mechanical power developed?

Solution

a) The back emf in the armature is obtained from the equation

Applied voltage − back emf = net voltage = IR

$110 \text{ V} - \mathcal{E}_B = (20 \text{ A})(0.5 \text{ Ω})$

$\mathcal{E}_B = 110 - 10 = 100$ V opposite to applied voltage

(b) To obtain the starting current, we assume that all the applied voltage of 110 V is applied across the resistance of 0.5 Ω:

$V = IR \qquad 110 \text{ V} = (I)(0.5 \text{ Ω}) \qquad I = 220 \text{ A}$

This is a very large current and would have to be limited in most practical situations.

(c) The power delivered while operating is

$P = IV_{app} = 20 \text{ A} \times 110 \text{ V} = 2200 \text{ W}$

(d) The rate at which heat is generated in the motor is given by

$I^2 R = (20 \text{ A})^2 \times (0.5 \text{ Ω}) = 200 \text{ W}$

(e) The mechanical power developed is just the difference between (c) and (d):

Mechanical power = input power − heat loss
= 2200 W − 200 W = 2000 W

SUMMARY

1. Electric current is induced in a wire whenever it moves through a magnetic field and is part of a closed circuit. The induced voltage is proportional to the speed of movement of the wire and the strength of the field and depends upon the sine of the angle between the direction of the magnetic field and the length of the wire.

2. The same induction effect occurs when a stationary wire is exposed to a changing magnetic field.

3. The induced electric current moves in such a direction that its magnetic field opposes the changes of the magnetic field which produces it.

4. These principles are put to use in the ac generator.

5. Electromagnetic induction is also the working principle of the transformer, which steps up or steps down voltages according to need.

6. Unintentionally induced currents can hinder the operation of transformers and electric motors.

QUESTIONS

1. If a coil is rotated in a uniform magnetic field, will the induced emf be of the same magnitude throughout? Of the same polarity? Describe the time dependence of the emf from the coil.
2. What factors determine the magnitude of the induced emf in a wire?
3. Magnetic storms on the sun produce an eruption of charged particles, some of which enter the earth's upper atmosphere. How could such a deluge of charged particles affect electrical power systems on the earth?
4. Consider a loop of wire in the plane of a piece of paper. Describe what happens in this loop when a bar magnet is dropped into it so that its north pole moves through the loop first.
5. A circular loop of wire rests in a horizontal plane in a magnetic field directed vertically upward. The loop is moved quickly upward. What is the direction of the induced current in the loop when viewed below the coil looking upward in the direction of the loop?
6. What is the direction of the current induced in the resistor R when the swtich S is opened?

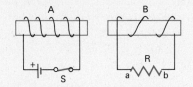

7. Using Lenz's law, determine the direction of the current in the resistor ab when (a) the switch is closed, (b) coil B is brought closer to coil A.
8. When a large motor, such as that in the refrigeration unit of a refrigerator or freezer in a house is started up, the house lights often dim for a very short time. Explain.

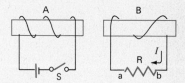

9. The core of a transformer is generally made of thin sheets of iron rather than a solid piece of iron. Why?
10. What are eddy currents, and how do they affect real operation of electrical machines?
11. If the speed of a generator is doubled, how will the voltage output change?

PROBLEMS

The Induced Electromotive Force (emf)

1. A rectangular coil 4 × 6 cm is placed in a uniform magnetic field of 0.60 T, which is directed vertically upward. What is the magnitude of the flux contained in the coil when the plane of the coil (a) is vertical, (b) is horizontal, and (c) makes an angle of 45° with the vertical?
2. A rectangular coil 4 × 6 cm is sitting in a horizontal plane in a magnetic field that is directed vertically upward. The magnitude of the magnetic induction is originally 0.60 T. It is suddenly reduced to 0.20 T in 0.1 s and later turned off so that B goes to zero in 0.025 s. Compute the average induced emf in the coil during both these processes.
3. A magnetic field is produced by turning on a current. This magnetic field is originally zero, and 10 s later it has reached a maximum constant value of 1500 G. If a loop of wire lies in this changing magnetic field perpendicular to the increasing field, calculate the average induced current in this loop. The loop has a resistance of 25 Ω and an area of 200 cm². In what direction does the induced current move?
4. A device that measures the strength of the magnetic induction consists of a "flip coil" of 100 turns of wire 3 cm in diameter. This coil is placed in the magnetic field so that the plane of the coil is perpendicular to the magnetic field. The coil is then flipped out of the magnetic field in 0.02 s. Compute the induced emf in the coil.
5. A uniform magnetic field of 25 T is perpendicular to the plane of a circular loop that has an area of 2 cm². If this magnetic field were reduced to zero in 0.1 s, what would be the induced emf in this loop?
6. Five hundred turns of wire whose resistance is 4 Ω are wound on a plastic tube whose outside diameter is 8 cm. The ends of the wire are connected to an ammeter. If the magnetic field through the coil changes in a uniform rate from 0.05 to 0.50 T in 0.04 s, what maximum current will be indicated on the ammeter?
7. One end of a long rectangular loop of stiff wire is between the pole faces of a large magnet. The uniform field **B** is perpendicular to the plane of the rectangular loop, and **B** is directed inward. The loop is being pulled to the right at constant speed v. The width of the loop is W and its electrical resistance is R. (a) Indicate the direction of the induced current in the loop. (b) Compute the induced

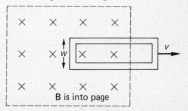

voltage in this loop; $W = 10$ cm, $v = 100$ cm/s, $B = 900$ G. (c) Compute the induced current in the loop if the resistance of the loop is $0.1\,\Omega$.

8. A copper bar 30 cm long is moved through a magnetic field where $B = 0.7$ T, with the length of the bar at right angles to the magnetic field. If the copper bar is moved with a speed of 0.6 m/s at right angles to its length, what is the induced emf in the bar?

9. An automobile moves at 60 mi/h in a northerly direction on a straight horizontal highway. Compute the induced emf in the axle of the car due to the vertical component of the earth's magnetic field $B = 0.6 \times 10^{-4}$ T. The axle of the car is 1.4 m long.

10. A conducting rod AB makes contact with the metal rails. The apparatus is in a uniform magnetic field of 500 G perpendicular to the page and directed into the page. What is the direction of the induced current I in the rod AB when it is moving to the right with a velocity of 40 cm/s? Compute the induced emf in the rod under these conditions. (See Fig. 10.)

11. A 747 jet has a wing span of 60 m, a length of 70 m, and a maximum speed of 250 m/s. If such a plane is flying due north according to the plane's compass at its maximum speed in the presence of the earth's magnetic field ($B_{horiz} = 0.2 \times 10^{-4}$ T, $B_{vert} = 0.5 \times 10^{-4}$ T), calculate (a) the voltage developed from one wing tip to the other wing tip; (b) which wing tip has the positive polarity (east or west); (c) the voltage induced from the front end to the back end of the plane.

12. A horizontal rod 1 m long falls in a horizontally directed magnetic field at a constant rate of 12 m/s. If the magnetic induction is 0.004 T, what is the magnitude of the induced emf in the rod?

13. The speed of an electrically conducting fluid can be measured in the following manner. Assume the liquid moves through a tube 2 cm in diameter and the tube is placed at right angles to a magnetic field of 0.04 T. An emf is set up across the diameter of the tube and is determined to be 1.2 mV. Compute the flow rate of the liquid from these data. Such a system can be used to measure blood flow rates.

14. Consider a small rectangular loop of wire that is near a long, straight wire (Fig. 14). The loop is very narrow (0.1 cm), so the whole rectangular loop can be considered the same distance ($R = 5$ cm) from the long wire. (a) Describe the magnetic field due to the long wire. Give the value of the magnetic field B at the rectangular loop when a current $I = 15$ A is in the long wire. (b) Compute the magnitude of the induced voltage in the rectangular loop when the current is turned on. The current reaches its maximum (and steady) value of $I = 15$ A in 0.01 s. (c) What is the direction of the induced current in the loop?

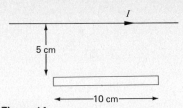

Figure 14

15. Two single-turn coils are positioned as shown below, with the outer coil connected to a cell by a closed double-pole switch. (a) When the switch is closed, what is the direction of the magnetic field inside the coils? (b) If the switch is opened, what is the direction (clockwise or counterclockwise) of the resulting induced current in the small coil?

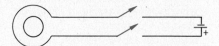

16. In an actual "flip coil" device used to measure magnetic field, the charge that moves through the circuit is measured by a galvanometer. This technique eliminates the measurement of the time to remove the coil from the magnetic field. For example, consider a coil of 50 turns of wire with a diameter of 4 cm. The coil has a resistance of $3\,\Omega$ and is connected to a galvanometer whose resistance is $497\,\Omega$. When the coil is removed by pulling it straight out of the magnetic field with no rotation, a charge of 5×10^{-4} C is sent into the galvanometer. Compute B assuming that the plane of the flip coil was perpendicular to B.

17. In an experiment to determine the flux density of the region between the pole pieces of a magnet, a 30-turn coil of 50-Ω resistance and 2-cm² area is placed with its plane perpendicular to the field and then quickly withdrawn from the field. The charge flowing in the operation is found to be 5×10^{-4} C. Determine the flux density.

Figure 10

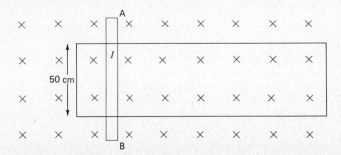

Generators, Transformers, and Back emf

18. The transformer for an electric train has 240 turns in its primary and a variable number of turns in its secondary. If the primary voltage is 120 V ac, (a) what number of turns is required to produce 24 V in the secondary? (b) If a current of 1 A is required to operate the train at 24 V, what is the current in the primary of the transformer?

19. A model-train transformer delivers 12 V to the engine. If the input to the transformer is 110 V and the primary contains 400 turns, how many turns are on the 12-V secondary? If the transformer is ideal and the train draws 4 A, what is the current drawn from the household circuit?

20. A step-up transformer has 40 turns on the primary and 480 turns on the secondary. If the transformer is 92 percent efficient and a current of 20 A is drawn by the 110-V primary, compute the power in the secondary.

21. A rectangular coil with its axis of rotation at right angles to a uniform magnetic field of magnetic induction 0.2 T is used to generate a maximum voltage of 130 V. The coil has an area of 50 cm² and consists of 300 turns. With what frequency must the coil be turned?

22. Design a loop that will produce a maximum emf of $\mathcal{E} = 150$ V when rotated at 60 r/s in a uniform field of magnetic induction of 0.6 T.

23. A circular coil of area 10 cm² made of 150 turns of wire rotates about an axis through its center that is perpendicular to a uniform magnetic field of magnetic induction of 0.05 T. The coil is rotated at a speed of 24 rad/s. What is the maximum induced emf in the coil?

24. A rectangular coil of 500 turns is 10 cm long and 10 cm wide and rotates at a speed of 20 r/s about an axis that lies in the plane of the coil and passes across the center of the coil. The coil rotates in a magnetic field perpendicular to the plane of the coil. The magnetic induction of the field is 0.05 T. If the coil has a resistance of 8 Ω and is connected to an external circuit of 12 Ω, what is the peak current flowing in the external circuit?

25. A series-wound motor designed to operate on 110 V has a resistance of 100 Ω in series with the armature resistance of 10 Ω. When operating at constant full speed, the back emf is 80 V. What is the initial current drawn by the motor and the operating current?

26. The armature of the starting motor in an automobile has a resistance of 0.1 Ω. The motor is driven by a 12-V storage battery. After the generator reaches its operating speed, a back emf of 6 V is generated. What is the starting current and the current at operating speed?

27. The armature of a 220-V dc motor draws a current of 10 A when operating. It has an armature resistance of 0.4 Ω. What is the back emf when operating and the starting current?

Special Topic
The Oscilloscope

The oscilloscope is one of the most widely used electronic devices for measuring electrical quantities. It is an ideal instrument for displaying voltage waveforms, determining the magnitudes of voltages, and working out the phase relationship between voltage signals. The heart of the oscilloscope is the cathode ray tube (CRT), upon which the waveform is displayed. The CRT, the forerunner to the more complex TV tube, is a vacuum tube in which electrons are produced and directed to a phosphorescent coating on the tube face. The electrons striking the coating excite the atoms of the phosphor, which in turn emit visible light photons. Of course, some x-rays are also emitted in this process, and so certain precautions must be designed into the CRT to keep the intensity of x-rays emitted from the tube to a minimum.

A schematic diagram of a CRT is shown in Fig. A. The electrons are emitted from a hot filament and accelerated through a series of electrodes to obtain a sharply focused, single beam. This part of the tube is often called an electron gun, for the electrons are accelerated through a very high potential, of the order of 10 000 V. The high-energy electrons move down through the tube with a constant component of velocity in the direction toward the screen. In the electron gun, electrostatic focusing is usually employed to provide the sharp, collimated beam of high-energy electrons. This beam of electrons is deflected by the potential difference established across either or

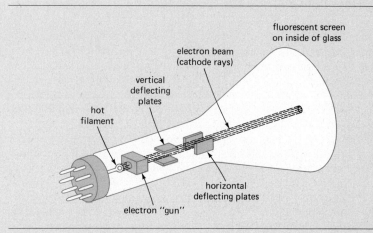

Figure A Cathode ray tube.

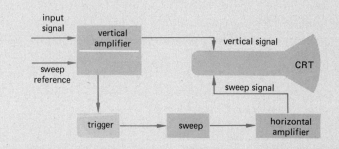

Figure B Block diagram of the major parts of an oscilloscope. The input signal is normally amplified and put on the vertical deflection plates. The sweep signal is normally placed on the horizontal deflection plates and can be triggered by a line voltage signal or an internal or external signal.

both of the horizontal pair and vertical pair of deflecting plates. The deflection plates reflect the qualities of the input waveform to the oscilloscope so that the motion of the electron beam across the face of the tube represents the waveform.

Usually, the signals to the horizontal and vertical deflecting plates are fed to the CRT via amplifiers, to enable the scientist using the instrument to examine small signals. A block diagram of the principal parts of an oscilloscope is shown in Fig. B.

The vertical amplifier attenuates or amplifies the input waveform to produce a magnitude of the vertical deflection that is appropriate for observation on the CRT screen. Generally, a single external signal to be examined is put into the vertical amplifier. The vertical amplifier also supplies a signal related to the input signal that begins the triggering process, eventually leading to the horizontal deflection plate. The triggering allows for the synchronization of the horizontal and vertical deflections. Often the oscilloscope's vertical amplifier has at least two inputs: one for a dc signal and one for an ac signal. The ac input is used to measure a small ac signal of, say, a few volts or even millivolts riding upon a large dc voltage signal. The dc input goes directly to the amplifier, as shown in Fig. C. A constant dc signal produces a constant deflection of the electron beam in a vertical direction. The ac input is made through a capacitor that blocks the dc signal.

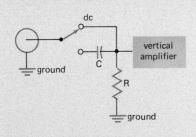

Figure C Input selection to the vertical amplifier. This allows for an ac input through the capacitor or a dc input.

Figure D Sawtooth waveform used to drive the horizontal amplifier. This type of waveform moves the electron beam across the face of the oscilloscope to make a line and then rapidly returns the beam to the starting side of the screen.

The sweep generator feeds the horizontal amplifier, producing the deflection of the electron beam in a horizontal direction. This part of the device can be started up by the input signal to the vertical plate through the trigger; alternatively, this trigger may be initiated by some other signal. The sweep generator, once triggered, produces a sawtooth signal that drives the horizontal deflection plates, creating a time axis, as shown in Fig. D. The linearly increasing voltage moves the electron beam at a constant rate from left to right across the screen. During the rapidly decreasing portion of the sweep, the beam is returned to the left-hand side of the oscilloscope face. Usually the beam is shut off during this return to keep the "fly back signal" from producing confusion. The trigger section is included to make the input signal compatible with the sweep generator, for the input signal may not be of sufficient magnitude or of proper shape to initiate the sweep by itself. This mode of operation of an oscilloscope is often called the time-base mode.

The oscilloscope can be used in other ways. It may be used to compare two external signals directly; this is accomplished by feeding one signal to the horizontal plates and the other signal to the vertical deflection plates, after both signals have passed through the corresponding amplifiers. The triggering circuit and the horizontal sweep circuit are bypassed. Often this direct x-y mode of operation is used to measure the frequency of an unknown signal as compared to a controllable known frequency. The results are called Lissajous figures. If the variable-frequency source is connected to the horizontal amplifier, then the Lissajous figures shown in Fig. E can be obtained. To understand how these patterns result, look at the diagrams in Fig. F, where the patterns resulting from the comparison of two signals of the same frequency but various phases are compared. In this figure the horizontal signal is fixed in frequency and phase, and the vertical deflecting signal of identical frequency is shown for seven different

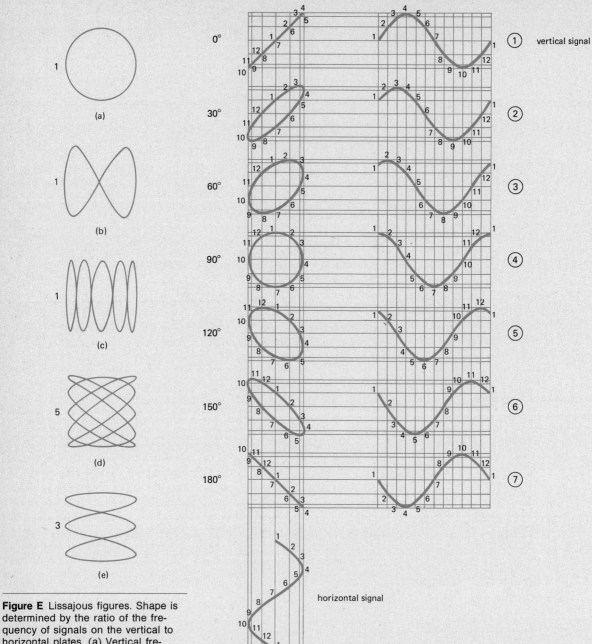

Figure E Lissajous figures. Shape is determined by the ratio of the frequency of signals on the vertical to horizontal plates. (a) Vertical frequency = horizontal; (b) vertical frequency = 2× horizontal: (c) vertical frequency = 5× horizontal; (d) 3× vertical frequency = 5× horizontal; (e) 3× vertical = horizontal.

Figure F Lissajous figures indicating phase difference. (*From "Radar Electronic Fundamentals," U.S. Government Printing Office, Washington, D.C., 1944*)

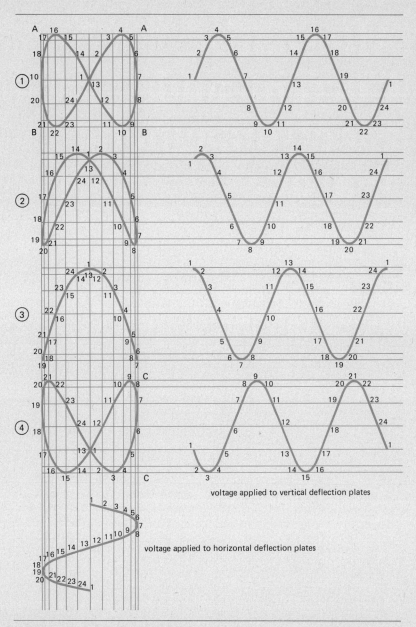

phases. Similarly, see Fig. G for two different frequencies: the vertical signal has twice the frequency of the horizontal frequency.

There are many uses for oscilloscopes, and the machines come in many sizes, shapes, and designs. There are even oscilloscopes that store data for display at a later time. To obtain a permanent record of data on an oscilloscope, short focal-length cameras are often used. The versatility of the oscilloscope makes it an extremely useful measuring instrument.

Figure G Lissajous figure for 1:2 frequency ratio. (*From "Radar Electronic Fundamentals." U.S. Government Printing Office, Washington, D.C., 1944*)

17 Alternating Current and Electromagnetic Waves

- 17.1 CHANGING THE CHARACTER OF CIRCUITS
  Inductance
  Capacitance
- 17.2 AC VERSUS DC POWER
- 17.3 CAPACITIVE AND INDUCTIVE CIRCUITS
  The Combined *LCR* Circuit—AC Ohm's Law
- 17.4 ELECTRICAL OSCILLATIONS
- 17.5 AC TO DC
- 17.6 THE TRANSISTOR AND BEYOND
- 17.7 MAXWELL'S EQUATIONS
- 17.8 ELECTROMAGNETIC WAVES

SHORT SUBJECT: Poor Reception

SHORT SUBJECT: The Fathers of Radio

Summary
Questions
Problems

SPECIAL TOPIC: Is Anybody Out There?

To such pioneers of electrical science as Benjamin Franklin or Michael Faraday, today's electronic devices would have been truly astonishing. Miniature calculators that perform a vast variety of mathematical functions from multiplication to square rooting, tiny chips of silicon no more than a few square millimeters in area that contain hundreds of individual electrical devices, and computers that can store literally millions of pieces of information in volumes no larger than a postage stamp are all part of the modern electronic age. Yet all these devices are made possible by the elementary laws of electricity that we studied in the last four chapters, along with a dash of solid-state physics.

Here we extend our study of electric and magnetic phenomena beyond dc circuits to a series of new topics, including inductance, capacitance, reactance, and impedance, and ac circuits that use a modified version of Ohm's law. We then examine the electric oscillator, which generates electromagnetic waves. We conclude with a description of electrical phenomena in Maxwell's four equations and the prediction of electromagnetic waves. The electromagnetic spectrum is introduced here because it is a vital subject in the remainder of this book.

17.1 CHANGING THE CHARACTER OF CIRCUITS

Most practical and useful electronic circuits employ alternating currents or at least time-varying (nonconstant) currents. In Chap. 14 our discussions of circuitry were limited primarily to direct currents. Figure 17.1 shows the difference between this type of constant current and a pulsating direct current produced by a simple one-coil generator. The most common type of current that we actually use in daily life is neither of these; it is the alternating current whose direction changes every half cycle, as we saw in Chap. 16.

Inductance

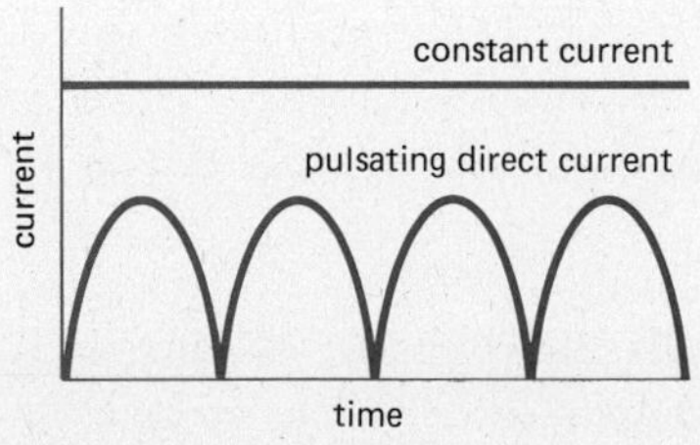

One unavoidable feature of all currents whose strength varies with time, whether they are pulsating direct currents or conventional alternating currents, is the induction of a voltage or emf in any nearby conductor; this is the effect that produces eddy currents and reduces electrical efficiency. When the current increases, the process leads to the production of a back emf, which is a voltage that opposes the original emf in a circuit. **The inductance, given the symbol L, is defined in terms of the back emf induced in a circuit; in fact, the back emf induced in a coil is directly proportional to the rate of change of current, and the proportionality constant is the inductance L.** Expressed as an equation,

$$\text{Induced emf} = L \frac{\Delta I}{\Delta t} \tag{17.1}$$

Figure 17.1 Two types of dc waveforms: one is constant; the other is pulsating.

Henry was the man unfortunate enough to discover a number of the fundamentals of electromagnetic induction at the same time as British physicist Michael Faraday.

The unit of inductance is the henry, H, so named for the nineteenth century American physicist Joseph Henry. A coil with an inductance of 1 H induces 1 V of back emf when the current is changed by 1 A/s. The common values of inductors range from 10^{-6} H (1 μH) into the millihenry range, and coils with inductances of 1 to 10 H are not unknown. When the current is constant ($\Delta I/\Delta t = 0$), there is of course no induced back emf. Therefore, a coil of wire in a dc circuit carrying a constant current experiences no back emf due to its inductance, although the coil still possesses a resistance. The practical inductor, often called a choke, is usually a low-resistance coil to keep the Joule heating to a minimum.

In Chap. 14 we examined and analyzed dc circuits using Ohm's law ($V = IR$), which describes the simple direct proportionality between the current I in a resistor R and the potential difference V impressed across the wire. This is called a *linear relationship,* which adequately describes metal wire conductors under most circumstances. It is possible to generalize Ohm's law to describe the effect of the inductance of a coil of wire on the alternating current in the coil. **A new quantity called the inductive reactance X_L is defined as the proportionality factor between the potential difference V across a coil with an inductance L and the current I in the coil.** That is, $V_L = IX_L$. Clearly, X_L corresponds to R in our original version of Ohm's law because both express the property of the circuit element that opposes the current or charge flow. Resistance produces Joule heating, but the inductive reactance does *not*. The inductive reactance includes *only* the effect of the induced back emf. R and X_L are different although they play analogous roles in the relationship between I and V.

It is clear from Eq. (17.1) that the effectiveness of an inductor depends on the frequency at which the current oscillates. The inductive reactance X_L is directly proportional to frequency and is given by

$$X_L = 2\pi\nu L \tag{17.2}$$

As the frequency ν increases, the inductor offers more and more impedance, or "resistance." Thus it will pass a direct or low-frequency current as it "chokes off" a high-frequency current. Hence, an inductor acts like a frequency filter. An inductor affects a direct current of constant amplitude only when the circuit is first closed or later opened because these are the only times when the current changes. As the current builds to the steady, constant value, an induced back emf is present in such circuits. This buildup in current as the switch is closed is exponential in nature, as shown in Fig. 17.2, and is expressed by

$$I = I_{\max}(1 - e^{-t/\tau})$$

The characteristic time associated with this buildup of current is $\tau = L/R$. In a time τ, the current builds up to within $1/e$ ($e = 2.72$, the base of the natural logarithms) of the maximum value, or 63.2 percent of the maximum current. Energy can be stored in an inductor, which makes it a useful element in an oscillator, as we shall see later.

Figure 17.2 Exponential rise in current in an inductor as a function of time.

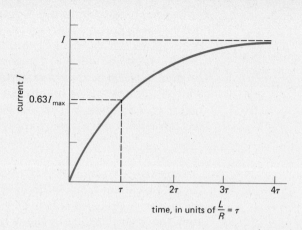

EXAMPLE 17.1

A coil with an inductance of 0.5 H is connected to a 130-V, 60-Hz source. If the resistance of the coil is neglected, compute the current through the coil.

Solution

This can be computed using a generalized version of Ohm's law. If we consider the inductive reactance as equivalent to a resistance, then we can write Ohm's law as $I = V/X_L$, where $X_L = 2\pi \nu L$. Computing the reactance of the coil, we have $X_L = 2\pi \nu L = (2\pi)(60 \text{ Hz})(0.5 \text{ H}) = 188.4 \, \Omega$. Substituting the reactance in Ohm's law yields

$$I = \frac{V}{X_L} = \frac{120 \text{ V}}{188 \, \Omega} = 0.64 \text{ A}$$

Capacitance

A second electrical device that can filter out current of different frequencies is the capacitor. This device usually consists of two thin metal plates separated by a layer of insulating material, which is known as a dielectric. The fundamental characteristic of the capacitor is its ability to store electric charge and energy.

We can understand how the capacitor does this by examining the circuit in Fig. 17.3(a), which shows a capacitor in series with a resistor, a battery, and a switch. When the switch is closed, as in Fig. 17.3(b), the driving potential of the battery starts to move electrons from one plate of the capacitor to the other. The charge cannot move through the insulating material between the plates of the capacitor. Thus, a positive charge $+Q$ accumulates on one plate of the capacitor and a negative charge $-Q$ on the other plate. This buildup lasts for only a short time after the switch is closed for the potential difference across the capacitor soon equals the emf of the battery—a state of equilibrium that halts the current in the circuit.

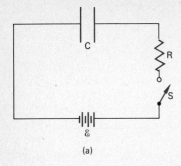

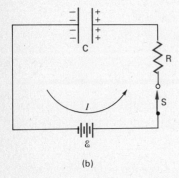

Figure 17.3 Charging of a capacitor in a dc circuit.

Figure 17.4 shows how this transient current varies with time. The decrease in the current and the corresponding buildup of charge on the plates of the capacitor are governed by an exponential law similar to that for radioactive decay. The current decays in such a way that it drops to $1/e$, 36.8 percent of its starting value, in a characteristic time τ, whose value depends on the amount of resistance in the circuit and the nature of the capacitor. The voltage across the capacitor plates when the capacitor is discharging varies as

$$V = V_0 e^{-t/RC}$$

Once the capacitor is fully charged, it acts as an open switch in a dc circuit. No current can flow through the circuit owing to the presence of the insulating properties of the dielectric. If one removes the capacitor from the arrangement and shorts it out by connecting its two plates with a conductor, the capacitor will discharge rapidly, passing the accumulated charge Q from one plate to the other and returning the individual plates to electrically neutral states. The existence of charge on capacitors in circuits even after they have been disconnected from a voltage source has shocked—literally—many who have dabbled in electronics.

The ability of a capacitor to store charge is expressed as the capacitance, C. **Since the amount of charge stored by any individual capacitor is directly proportional to the voltage developed across the plates of the device ($Q \propto V$), the capacitance is defined as the ratio of these quantities:**

$$C = \frac{Q}{V} \tag{17.3}$$

The unit of capacitance is clearly coulombs per volt. One coulomb per volt is generally known as a farad (F). Since a coulomb is a large amount of charge, this capacitance unit is extremely large by conventional electronic standards, so electrical engineers generally use microfarads (1 μF = 10^{-6} F) and even picofarads (1 pF = 10^{-12} F).

The capacitance of any particular capacitor is determined both by its geometry and the nature of the dielectric material. Conceptually, the

Figure 17.4 Decay of current in a dc circuit after the switch is closed as a capacitor charges.

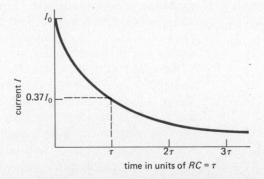

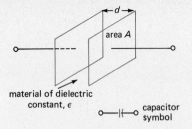

Figure 17.5 A parallel-plate capacitor.

simplest capacitor is the parallel-plate device shown in Fig. 17.5, whose capacitance C is given by the equation

$$C = \frac{\epsilon A}{4\pi k d}$$

where A is the area of the plates, d is their separation, k is a constant whose value is 9.0×10^9 N·m²/C² to ensure the correct units in the equation, and ϵ the dielectric constant of the insulating material between the plates. **This dielectric constant is a dimensionless number characteristic of the insulating material.** Its value is defined as 1 for air, whereas for most solid insulators it ranges between 2 and 6.

As the previous equation indicates, capacitance depends on geometric factors; its value alters with different placement of the two conducting plates of a capacitor with respect to each other. Commercial capacitors come in a wide variety of shapes and sizes. The most typical is a parallel-plate device that consists of two pieces of metal foil with a sheet of plastic dielectric between them. The whole arrangement can be rolled up into a compact package.

Our analysis of how capacitors charge up showed that these devices have little use in dc circuits, where they simply act as open switches. However, if the battery in the circuit drawn in Fig. 17.3 is replaced by an alternating voltage source (emf), the situation becomes more interesting. As the current reverses every half cycle, the capacitor plates alternately charge and discharge; as a result, the potential across the plates keeps reversing, and current flows in the circuit. The capacitor therefore offers only token impedance to the current. This generalized resistance is called an impedance or, more specifically, the capacitive reactance of the capacitor. Its value is given by

$$X_C = \frac{1}{\omega C} = \frac{1}{2\pi \nu C} \tag{17.4}$$

where C is the capacitance and ω is the product of 2π and the frequency ν of the oscillating emf. Plainly, when ω is zero—in the case of a direct current—the impedance of the capacitor after it has charged up is infinite. By contrast, at very high frequencies of alternating current the capacitor has virtually no effect on the passage of the current. Thus, capacitors can be used in electronic circuits to filter out direct currents from alternating currents—the reverse effect of inductors.

Capacitors can be used in electronic circuits to filter out direct currents from alternating currents.

Capacitors can be connected in several ways, and one can determine the equivalent capacitance for the combination in much the same manner as that used to determine the equivalent resistance for combinations of resistors. When the capacitors are connected in series, as shown in Fig. 17.6(a), each capacitor must have the same charge, $+Q$ and $-Q$, on its pair of plates. The battery produces an excess of negative charge on the left-hand plate of C_1. This charge comes ultimately from the right-hand plate of capacitor C_3. The charge $-Q$ on the left-hand plate of C_1 causes charges on the right-hand plate to

Figure 17.6 (a) Capacitors in series. (b) Capacitors in parallel.

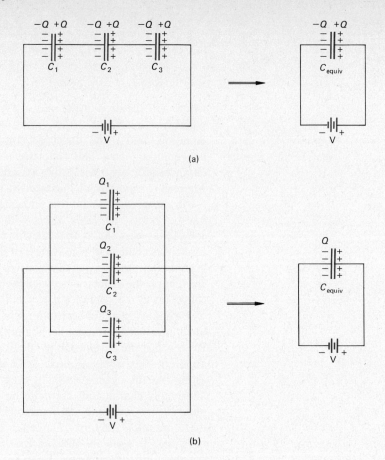

be repelled as far as possible, namely to the left-hand plate of capacitor C_2. A similar process occurs on capacitor C_3. The result is a charge of Q on each capacitor (which means a charge of $-Q$ on one plate and a charge of $+Q$ on the other). The potential difference across the three capacitors is equal to V and must equal the sum of the separate potential differences across the three individual capacitors. Therefore,

$$V = V_1 + V_2 + V_3$$

Using definition of capacitance $V = Q/C$ for each capacitor and for the equivalent capacitor C_{equiv}, we can write

$$\frac{Q}{C_{\text{equiv}}} = \frac{Q}{C_1} + \frac{Q}{C_2} + \frac{Q}{C_3}$$

Since all the capacitors have the same charge, the equivalent capacitance for the series arrangement is given by

$$\frac{1}{C_{\text{equiv}}} = \frac{1}{C_1} + \frac{1}{C_2} + \frac{1}{C_3} \quad \text{for capacitors in series} \quad (17.5)$$

Capacitors may also be connected in parallel, as shown in Fig. 17.6(b). With this arrangement the potential drop across each capacitor is equal to the voltage developed by the battery. All capacitors are charged to the same voltage, the emf of the battery. The charge stored on each capacitor is given by the definition for the capacitance ($Q = CV$) and will be different for each different capacitor. The total charge on the equivalent capacitor must then be made equal to the sum of the charges on the individual capacitors:

$$Q = Q_1 + Q_2 + Q_3$$

Using the definition of capacitance, we can write

$$C_{\text{equiv}} V = C_1 V + C_2 V + C_3 V$$

Canceling the V's, one obtains

$$C_{\text{equiv}} = C_1 + C_2 + C_3 \quad \text{for capacitors in parallel} \quad (17.6)$$

Note that these relationships for adding capacitors in series and parallel are just the reverse of those for resistors.

EXAMPLE 17.2

A capacitor with a capacitance of 4 μF is connected to a 60-V battery. What is the resulting charge on the capacitor?

Solution

The charge on a capacitor refers to the magnitude of the charge on either plate of the capacitor. From the definition of capacitance, Eq. (17.3),

$$Q = CV = 4 \, \mu\text{F} \times 60 \text{ V} = 240 \, \mu\text{C}$$

EXAMPLE 17.3

The plates of a parallel-plate capacitor are 3 mm apart in air. If the area of each plate is 0.2 m², what is the capacitance?

Solution

The capacitance of a parallel-plate capacitor is given by

$$C = \frac{\epsilon A}{4\pi k d}$$

where $k = 9 \times 10^9$ N·m/C², $\epsilon = 1.00$ for air

$$C = \frac{1 \times 0.2 \text{ m}^2}{(4\pi)(9 \times 10^9 \text{ N·m}^2/\text{C}^2)(3 \times 10^{-3} \text{ m})} = 5.89 \times 10^{-10} \text{ F}$$

$$C = 589 \text{ pF}$$

EXAMPLE 17.4 If the air between the plates of the capacitor in Example 17.3 is replaced by mica whose dielectric constant is $\epsilon = 5.0$, what would be the capacitance of the capacitor? Would the maximum charge on this capacitor increase or decrease?

Solution In the expression for a parallel-plate capacitor the capacitance depends linearly on the dielectric constant of the material between the plates. Since the dielectric constant of the mica is five times that of air, the capacitance will increase by five times, or

$$C = (5)(589 \text{ pF}) = 2.95 \times 10^{-3} \, \mu\text{F}$$

Since the capacitance has increased, the amount of charge on the capacitor for a given applied voltage would increase, since by definition

$$Q = CV$$

EXAMPLE 17.5 A 12-V battery with an internal resistance of 1.5 Ω is connected to a 4-μF capacitor in series with a 2-Ω resistor. (a) What is the initial current delivered to the capacitor? (b) How long does it take for the capacitor to reach 0.63 of its maximum charge? (c) What is the maximum charge on the capacitor?

Solution (a) Since there is no initial charge on the capacitor, we can use Ohm's law to determine the initial or maximum current:

$$I = \frac{\mathcal{E}}{R + r} = \frac{12 \text{ V}}{2 \, \Omega + 1.5 \, \Omega} = 3.43 \text{ A}$$

(b) The capacitor will reach 0.63 of its full charge in one time constant or $\tau = R_{tot} C$

$$\tau = R_{tot} C = (3.5 \, \Omega)(4 \, \mu\text{F}) = 14 \, \mu\text{s}$$

(c) The maximum charge occurs when $I = 0$ and the potential difference across the capacitor is equal to the battery emf, or 12 V. Using Eq. (17.1),

$$Q = CV = (4 \, \mu\text{F})(12 \text{ V}) = 48 \, \mu\text{C}$$

EXAMPLE 17.6 In the circuit in Fig. 17.7 a 60-Hz ac voltage is placed across the three capacitors. The maximum value of the voltage, or peak value, is 170 V.

Compute the (a) equivalent capacitance, (b) total charge on the equivalent capacitance, (c) charge on the 3-μF capacitor, and (d) reactance of the circuit.

Solution

(a) First, compute the equivalent capacitance of the 2- and 4-μF capacitors in series. Using Eq. (17.5) for capacitors in series,

$$\frac{1}{C_{\text{equiv}}} = \frac{1}{C_1} + \frac{1}{C_2} = \frac{1}{2\,\mu\text{F}} + \frac{1}{4\,\mu\text{F}} = \frac{3}{4\,\mu\text{F}}$$

$$C_{\text{equiv}}\ (\text{series}) = \frac{4\,\mu\text{F}}{3} = 1.33\,\mu\text{F}$$

Next, this equivalent capacitance must be added to the 3-μF capacitor in parallel with it. For capacitors in parallel, we just add the capacitance as indicated in Eq. (17.6):

$$C_{\text{equiv}}\ (\text{parallel}) = C_s + C_3 = 1.33\,\mu\text{F} + 3\,\mu\text{F} = 4.33\,\mu\text{F}$$

(b) To compute the total charge, use Eq. (17.1),

$$Q = CV = (4.33\,\mu\text{F})(170\text{ V}) = 736\,\mu\text{C}$$

(c) The charge on the 3-μF capacitor is given by Eq. (17.1)

$$Q = CV = (3\,\mu\text{F})(170\text{ V}) = 510\,\mu\text{C}$$

(d) To compute the capacitive reactance, we employ its definition

$$X_c = \frac{1}{2\pi\nu C}$$

$$X_c = \frac{1}{(2\pi)(60\text{ Hz})(4.33\,\mu\text{C})} = 613\,\Omega$$

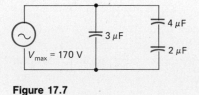

Figure 17.7

17.2 AC VERSUS DC POWER

The ordinary alternating voltage from a generator has a magnitude that varies with time, usually in a manner which can be represented by a sine function, for example, $V = V_0 \sin 2\pi\nu t$, where V_0 is the maximum value of the voltage and ν is the frequency of the oscillation of the voltage. This time-varying voltage produces an alternating current that can be represented by $I = I_0 \sin 2\pi\nu t$. If the circuit under consideration contains only resistance, no capacitance or inductance, both the current and voltage vary with time at the same frequency (ν) and are in phase with one another. That is, when the voltage reaches a maximum value in a circuit containing only a resistance, the current also reaches its maximum value.

An interesting observation concerning the sinusodial variation of current with time is that the average current is zero. After all, over a single period the average value of the sine function is zero. What good is an alternating current with an average value of zero? Recall that in

Chap. 14, when we examined the effects of current, we noted that a current through a resistor produced heat determined by Joule's law, where the power dissipated in the resistor was I^2R. Since I is squared, it makes no difference whether I is positive or negative. From a physical point of view, the $+$ and $-$ on the current just indicate a change in direction, and the heating effects of the current would be independent of the direction of the current in the wire. If we are interested in the heating effects of current, it is I^2 that is important. In fact, the average power dissipated in a resistor (average I^2R) is the really appropriate concern rather than the current alone. Since R is constant, we need to know the average value of $\overline{I^2}$, the so-called mean-square current. If you square $I = I_0 \sin 2\pi\nu t$, it will always be positive. For a sinusodially varying current, the average value of I^2 is just one-half the maximum value I_0^2—that is, the average value of $\overline{I^2} = \frac{1}{2}I_0^2$.

The "effective" current can be obtained by taking the square root of $\overline{I^2}$ to obtain the root-mean-square (rms) current:

$$I_{\rm rms} = \frac{I_0}{\sqrt{2}} = 0.707\, I_0 \qquad (17.7)$$

Since the average value of $\overline{V^2}$ is $\frac{1}{2}V_0^2$, then the rms voltage will be

$$V_{\rm rms} = 0.707\, V$$

These rms values of current and voltage are called the effective values because they are the appropriate ones when dealing with power or Joule heating.

The rms value of a current $I = I_0 \sin 2\pi\nu t$ would produce the same power in a resistor as a constant amplitude direct current of the same value. Figure 17.8 shows a graph of I versus time and indicates I_0 and $I_{\rm rms}$. Most ac voltmeters and ammeters read the effective or rms value. Hence the peak ac voltage is $\frac{1}{0.707} V_{\rm rms} = 1.41\, V_{\rm rms}$, and the peak value of an alternating current is $1.41\, I_{\rm rms}$. Note that the labeling of common electrical devices such as radios, light bulbs, hair dryers, stoves, and so on is given in rms values.

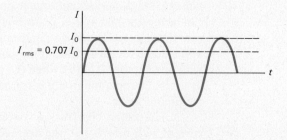

Figure 17.8 Variation of current with time for a current $I = I_0 \sin 2\pi\nu t$. The effective current or rms value $I_{\rm rms} = 0.707 I_o$.

17.3 CAPACITIVE AND INDUCTIVE CIRCUITS

The previous section stated that in a purely resistive circuit the alternating voltage and alternating current are in phase. All power losses in a circuit are due to resistance, as given by $I_{rms}^2 R$. What do a capacitance or an inductance do to the voltage and current relationships in an ac circuit?

First, for the capacitor, if $V = V_0 \sin 2\pi v t$, then since $Q/C = V$ for a capacitor, we see that the charge on a capacitor to which an ac voltage has been applied varies with time. Since C is a constant at a given frequency for a particular capacitor, the charge on a capacitor varies as

$$Q = CV_0 \sin 2\pi v t$$

The charge on the capacitor oscillates from a peak value of $+CV_0$ to a minimum of $-CV_0$, with a frequency identical to the time variation of the voltage. Clearly the charge on the capacitor will be a maximum when $V = V_0$, its maximum value. How does the current vary with time? Since $I = \Delta Q/\Delta t$, then the current varies as the derivative or the slope of a Q-versus-t relationship. This is just a cosine function. The result is that the current is 90° out of phase with the voltage.

The current "leads" the voltage function by one-fourth of a cycle or one-fourth of a period, as seen in Figure 17.9. This is easily understood if we realize that as the voltage on a capacitor increases, the charge is building up at the same time. However, as the capacitor is charging the current in the circuit is decreasing because the buildup of charge on the capacitor opposes the applied voltage. When the capacitor is fully charged, no charge is flowing; the current has been reduced to zero. As the voltage across the capacitor decreases to zero, the current rises to its maximum value. It might be helpful to refer to the discussion of simple harmonic motion in Chap. 12, where the displacement s was represented by a cosine function and the velocity $v = \Delta s/\Delta t$ by a sine function.

For an inductance, a different phase shift occurs. The back emf developed in an inductance is given by emf $= L \Delta I/\Delta t$. If a voltage of the form $V = V_0 \sin 2\pi v t$ is applied to the inductance L, it equals the induced emf in the inductor:

$$V_0 \sin 2\pi v t = L \frac{\Delta I}{\Delta t}$$

This relationship says that the voltage is proportional to the slope of the current-versus-time graph, since L is a constant for a given inductor and a particular frequency. As a result, the voltage leads the current by 90° or, equivalently, by one-fourth of a cycle or one-fourth of a period. One can again explain this phase shift qualitatively by examining how the inductance functions in the circuit. Recall that the inductor induces an emf which opposes the applied voltage resulting in the peak voltage that occurs when the current is zero, as shown in Figure 17.10.

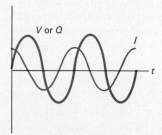

Figure 17.9 Comparison of voltage and current versus time for a capacitive circuit.

Figure 17.10 Comparison of voltage and current versus time for an inductive circuit. The voltage leads the current by 90°, one-fourth of a cycle, or one-fourth of a period.

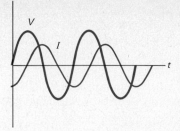

The Combined *LCR* Circuit—AC Ohm's Law

For a resistive circuit Ohm's law tells us that $I = V/R$; for the capacitive circuit $I = V/X_C = V/\tfrac{1}{2\pi\nu C}$; for the inductance $I = V/X_L = V/2\pi\nu L$. What happens when all three elements, L, C, and R, are present in a circuit such as that shown in Figure 17.11? For this type of circuit the equivalent of Ohm's law is given by

$$I = \frac{V}{\sqrt{R^2 + (X_L - X_C)^2}} = \frac{V}{Z} \tag{17.8}$$

where the quantity $Z = \sqrt{R^2 + (X_L - X_C)^2}$ is called the impedance of the circuit. Hence, Ohm's law becomes $V = IZ$ for a series ac circuit. Either the peak values for I and V or the rms values for I and V may be used in this expression for Ohm's law for the ac circuit with R, L, and C in series. However, this relationship does not apply to the instantaneous values of I and V. As we saw in Sec. 17.2, the voltage and current are not in phase with one another. In the general series ac circuit containing various values of L, C, and R, the phase angle θ between the voltage and the current can be obtained by a vector diagram, as shown in Fig. 17.11(b), which gives the proper relationship between the impedance Z, the net reactance $(X_L - X_C)$, and the resistance R of the circuit. Here we have added R to $(X_L - X_C)$ in a vector manner to obtain a resultant Z of the right magnitude by the Pythagorean theorem. Note that the value of θ is zero when $X_L - X_C = 0$, that is, when the circuit either has no capacitance and inductance or if these two reactances are equal in magnitude. In such cases the impedance Z is due only to the resistance R. As we have already learned, when we have a resistive circuit, then V and I are in phase—that is, $\theta = 0$. As the balance between X_L and X_C changes from a predominantly greater inductive reactance to a predominantly greater capacitive reactance, the value of θ changes from positive to negative. This corresponds to the voltages going from leading the current (θ positive) to the current's leading the voltage (θ negative).

It is interesting to examine the effect of the phase shift between voltage and current on the power. One form of Joule's law which is convenient to recall here is that power equals the product of the voltage times the current. If we have a circuit with only a resistance, the

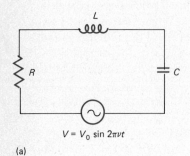

(a)

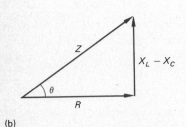

(b)

Figure 17.11 An ac series circuit containing a resistor, an inductor, and a capacitor. (b) The relationship between the impedance $Z = \sqrt{R^2 + (X_L - X_C)^2}$ and the resistance and net reactance $X_L - X_C$.

instantaneous values of V and I rise to a maximum and then decrease together, so that the power is just the product of V and I. However, if the instantaneous values of V and I are 90° out of phase, so that V is a maximum and I is a minimum, the average of the product VI is zero. Clearly this is consistent because we stated that the power dissipated depends only on resistance, not capacitance or inductance. When we have R, C, and L components in an ac circuit, power = $VI \cos \theta$.

The value $\cos \theta$ is called the power factor. When $\theta = 0$, $\cos \theta = 1$, and the power dissipated will be a maximum because we have a resistive circuit. As θ approaches $\pm 90°$, a pure inductive or capacitive circuit, $\cos \theta$ approaches zero. Plainly we can consider the voltage and current relationships in an ac circuit from the perspective of simple right-angle trigonometry and vector addition.

In our discussion we noted that the quantity $X_L - X_C$ could be zero if there were no inductance or capacitance or if $X_L = X_C$. In the next section we shall examine the latter conditions.

EXAMPLE 17.7

A 100-Ω resistor, a 0.5-H inductor, and a 10-μF capacitor are connected in series to 120-V, 60-Hz alternating voltage. (*a*) Find the impedance of the circuit. (*b*) Determine the effective current in the circuit. (*c*) What are the phase angle and power factor? (*d*) What is the power dissipated in this circuit?

Solution

(*a*) To compute the impedance $Z = \sqrt{R^2 + (X_L - X_C)^2}$, we must first compute the reactances X_L and X_C.

$$X_L = 2\pi \nu L = (2\pi)(60 \text{ Hz})(0.5 \text{ H}) = 188 \text{ }\Omega$$

$$X_C = \frac{1}{2\pi \nu C} = \frac{1}{(2\pi)(60 \text{ Hz})(10 \times 10^{-6} \text{ F})} = 265 \text{ }\Omega$$

Therefore, the impedance of the circuit is

$$Z = \sqrt{R^2 + (X_L - X_C)^2} = \sqrt{(100 \text{ }\Omega)^2 + (188 \text{ }\Omega - 265 \text{ }\Omega)^2} = 126 \text{ }\Omega$$

(*b*) To determine the effective current we use the ac version of Ohm's law: $I = V/Z$. We must understand that the voltage 120 V in the problem is the value measured with a normal meter, which is the effective voltage or V_{rms}. Hence, in terms of effective or rms values

$$I_{rms} = \frac{V_{rms}}{Z} = \frac{120 \text{ V}}{126 \text{ }\Omega} = 0.95 \text{ A}$$

(*c*) To compute the phase angle, refer to Figure 17.11(*b*), where it is clear from our vector model that

$$\tan \theta = \frac{X_L - X_C}{R} = \frac{188 \text{ }\Omega - 265 \text{ }\Omega}{100 \text{ }\Omega} = -0.77 \quad \text{or} \quad \theta = -37.6°$$

The negative sign means that the capacitative reactance dominates the inductive reactance and the current leads the voltage.

(d) The power dissipated in the circuit is given by

$$P = I_{rms} V_{rms} \cos \theta = (0.95 \text{ A})(120 \text{ V})(\cos 37.6°) = 90.3 \text{ W}$$

Note that to obtain the power we must make use of the effective or rms values of I and V.

17.4 ELECTRICAL OSCILLATIONS

The generation of electromagnetic waves in the radio-to-microwave region of the spectrum is an extremely important application of electronic circuits. An electric oscillation is produced by an oscillator in which electrical energy is stored alternately in a capacitor and an inductor in an LCR circuit, often called a "tank" circuit. We can provide a simple discussion of the oscillation action by examining an ideal circuit consisting of a capacitor and an inductor. We assume that no energy is lost owing to Joule heating. Hence, there is no resistance (R) in our ideal circuit. Consider a fully charged capacitor connected through a switch in series with an inductor, as shown in Fig. 17.12(a). When the switch is closed, as in Fig. 17.12(b), the electrons begin to move through the inductor toward the positive plate, discharging the capacitor into the inductor. Associated with the current is a growing magnetic field that induces a back emf in the inductor. At the instant the capacitor plates are neutralized (that is, uncharged), the current decreases, causing the magnetic field in the inductor to collapse and inducing an emf in the inductor that opposes the decrease. The current

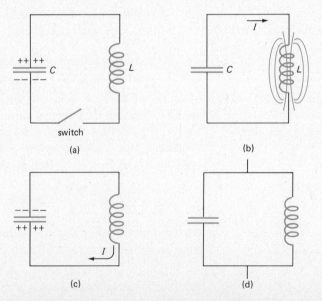

Figure 17.12 Oscillation of electrical energy in an LC tank circuit.
(a) The capacitor is charged but not connected to the inductor L. (b) The capacitor discharges into the inductor L. (c) The inductor recharges the capacitor in the opposite sense. (d) A parallel LC tank circuit.

therefore persists in the same direction until the capacitor has been charged in the opposite polarity as shown in Fig. 17.12(c). After the induced current has dropped to zero and the capacitor is fully charged to the initial magnitude of potential difference, the process repeats itself; the capacitor discharges, forcing a current in the opposite direction through the inductor.

In terms of energy, the electrical oscillation consists of a transfer of energy back and forth between the capacitor and inductor. The total energy remains constant in our ideal isolated circuit. This process is analogous to the transfer of energy back and forth between potential and kinetic energy in a mechanical system with no frictional losses to heat, such as a simple pendulum oscillating from its highest point, of maximum gravitational potential energy, to its lowest, or equilibrium, point, where its energy has been transformed into kinetic energy and then back to maximum potential energy. The fully charged capacitor could be considered equivalent to the state of maximum potential energy, and the current in the inductor represents the kinetic energy.

As one might expect, the frequency of the electrical oscillation is determined by the values of L and C. The "natural" resonant frequency v_0 of the LC oscillation is the frequency at which the capacitive reactance equals the inductive reactance $X_C = X_L$. When this is true, then

$$\frac{1}{2\pi v_0 C} = 2\pi v_0 L$$

resulting in the frequency

$$v_0 = \frac{1}{2\pi \sqrt{LC}} \tag{17.9}$$

One can further develop the comparison between mechanical oscillators and electrical oscillators. For example, the frequency of a mass m on a spring of spring constant k is given by

$$v = \frac{1}{2\pi} \sqrt{\frac{k}{m}} \quad \text{for simple harmonic motion}$$

We can write the frequency for the electrical oscillation as

$$v = \frac{1}{2\pi} \sqrt{\frac{1/C}{L}}$$

and relate $1/C$ to the spring constant and L to inertia. The inductance is, in a sense, a measure of electrical inertia and hence analogous to mass. It is not too difficult to show that the energy stored in the capacitor is equal to $\frac{1}{2}(1/C)Q^2$ and the magnetic energy associated with the inductor is $\frac{1}{2}LI^2$. The latter expression looks much like our expression for kinetic energy $\frac{1}{2}mv^2$ if we realize that I is analogous to v. The potential energy stored in a spring is given by the expression $\frac{1}{2}kx^2$, where x represents the displacement of the object from its equilibrium position. Hence the potential energy of a charged capacitor depends in an analogous way upon a constant of the capacitor $1/C$ and the degree to which it is charged. The amount of charge Q thus corresponds to the stretch of the spring.

By changing the values of L and C we can produce electrical oscillations of various frequencies. For example, for a small value of C, $1/C$ is large, corresponding to a stiff spring that results in a high-frequency oscillation. Increasing the inertia (mass) of the object on the spring (which corresponds to an increase in inductance L in the electrical oscillator) causes the frequency of a mechanical oscillator to decrease. By analogy, then, the frequency of an electrical oscillator decreases if the inductance is increased.

It is clear that a real system loses energy, which means the electrical oscillation goes to zero in much the same way that a mechanical oscillator like a pendulum gradually comes to a stop unless energy is continually fed into it. Resistance in the electrical circuit acts just as friction in the mechanical oscillation (transforming mechanical energy to heat). Hence energy must be fed into the circuit to sustain the oscillation.

The most common way to accomplish this is to use an amplifier circuit that employs a transistor or electron tube. For example, the parallel-tank circuit shown in Fig. 17.12d could be placed into the amplifier circuit in Fig. 17.21, where the small ac signal is indicated in the base-to-emitter circuit. The amplifier circuit is discussed later in this chapter. There are numerous types of oscillators; most are based upon the principles described here.

EXAMPLE 17.8

We have a 600-pF capacitor and wish to make an LC circuit whose resonant frequency is 92 kHz. What size inductance is required in our tank circuit if we assume that it is an ideal circuit with no resistance?

Solution

The resonant frequency of an LC tank circuit is given by Eq. (17.9),

$$\nu_0 = \frac{1}{2\pi\sqrt{LC}}$$

so

$$\nu_0^2 = \frac{1}{4\pi^2 LC}$$

or

$$L = \frac{1}{4\pi^2 C \nu_0^2} = \frac{1}{(4\pi^2)(600 \times 10^{-12} \text{ F})(9.2 \times 10^4 \text{ Hz})^2}$$

$$= 4.99 \times 10^{-3} \text{ H} = 4.99 \text{ mH}$$

17.5 AC TO DC

An electronic device that can alter and in a sense filter current is known as a rectifier. This element conducts current in only one direction. A rectifier is also a nonlinear device—the meaning of this is illustrated in Fig. 17.13, which shows the relationship between current and voltage for a theoretically perfect rectifier (the thick line), a real rectifier (the thin line), and (for simple comparison) a resistor that obeys Ohm's law (dashed line).

A perfect rectifier, or diode, acts exactly like a switch. When this rectifier is **on** its resistance is zero; current passes through the device without any drop in voltage, as shown by the vertical line on the y axis in Fig. 17.13. Real rectifiers actually present a small amount of resistance, and thus the actual effect is a steep rise in current for a very small increase in voltage across the device, as illustrated. This means that only a minute voltage drop exists across the rectifier as the current drawn through it increases by a large amount.

During the **off** stage of a perfect rectifier the resistance should be infinite so that the rectifier lets no current pass through it, no matter how large a voltage of the reverse polarity is applied. This corresponds to the heavy line along the negative x axis, or the negative voltage axis. Actual rectifiers do pass a small current in the **off** cycle, where the voltage is negative. The slopes of the characteristic curves shown in Fig. 17.13 give the reciprocals of resistance. These indicate that the **on** resistance is much lower than the **off** resistance and the ratio of back-to-front resistance R_b/R_c is a useful measure of the effectiveness of a rectifier. A value of at least 25 for this ratio should be expected.

The action of a typical rectifier is best illustrated by the vacuum-tube diode shown in Fig. 17.14. The device consists of a heated filament, which acts as a source of electrons and one of the two electrodes in an

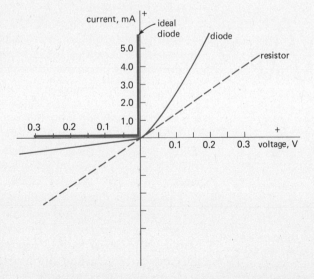

Figure 17.13 Voltage-current relationship (or characteristics) of a real diode and an ideal diode (heavy line) compared to a resistor. The diode is a nonlinear device that does not obey Ohm's laws.

Figure 17.14 (a) Diagram of a vacuum diode. (b) Symbols for a vacuum diode. (c) Symbols for a solid-state diode.

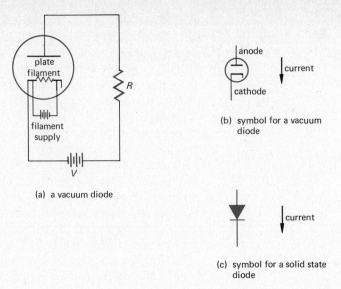

evacuated glass container. If the plate in the diode is maintained at a positive voltage with respect to the cathode or filament—the **on** condition—it will draw electrons across the tube, thus maintaining the current in the circuit of which it is a part. If the potential across the tube is reversed, however, putting the diode in its **off** condition, electrons will no longer be attracted across the tube and the electrical circuit will be shut off.

Now, if the battery in Fig. 17.14 is replaced by an alternating voltage, the diode will conduct only on the positive half cycle, as illustrated in Fig. 17.15, when the plate is at a positive potential above the filament. The diode therefore converts, or rectifies, alternating current to pulsating direct current.

The development of solid-state technology quickly led to a new type of diode to replace the cumbersome vacuum-tube device. This diode uses the electrically "in-between" compounds known as semiconductors. The most common semiconducting diode is a *p-n* junction, which can be put together in a number of ways. All have the end result of placing a piece of *p*-type semiconductor in contact with a piece of *n*-type semiconductor. The junction may even be formed in the same crystal by doping one region to be *p* type and an adjacent region to be *n* type. However it is constructed, a built-in potential difference exists at the transition region at the place where the *p*- and *n*-type regions meet.

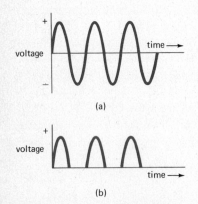

Figure 17.15 Rectified ac signal. (a) An ac voltage applied to a diode. (b) Rectified voltage showing the voltage drop across the resistor in Fig. 17.12.

The symbols *p* for positive and *n* for negative indicate the carrier mainly responsible for the conduction in a semiconducting crystal, such as silicon or germanium. The term *n* type indicates that the majority carrier contributing to the electrical current is the electron; *p* type indicates that conduction is primarily via holes. A hole is the absence of

17.5 AC to DC

an electron and is, in effect, a positive charge. Conduction in semiconductors takes place when negative electrons and positive holes move through the crystal. Whether a particular semiconductor is n or p type is related to the impurities in the crystal.

The p-n junction acts like a diode because it will conduct only when in the forward-biased condition, which is when the p side of the junction is at a positive potential with respect to the n side of the junction. The energies required to turn on the diode, that is, lower the bias potential, are small—about 0.5 V. Even in its unbiased state, the p-n junction has a potential barrier sufficiently high to prevent appreciable conduction of electrons from the p- to the n-type region. A slight reverse bias (reverse potential) results in a very small reverse current.

One major problem posed by use of single diodes to convert alternating to direct current is that they produce not only pulsating dc voltage but a voltage whose value is zero half the time, as shown in Fig. 17.15(b). Electronic engineers who incorporate diodes into their circuits generally need to obtain constant and consistent dc voltages. They can achieve this effect using the full-wave rectifier in Fig. 17.16(a), which contains two diodes. Each diode is **on**, and therefore able to conduct electricity, for half an ac cycle. Each diode therefore produces a pulsed, rectified voltage during the half cycle that it is **on**, and the other diode passes no current. The combined result, as observed across the resistor R, is the pulsed, dc waveform given in Fig. 17.16(e).

Clearly, the rectified voltage drop across the resistor is far from constant. However, electronic circuits consisting of a combination of resistors and capacitors or resistors and inductors can rather easily filter and smooth this waveform. The most common filter is called an RC network; it contains a capacitor in parallel with a resistor, as indicated in Fig. 17.17(a). As the voltage builds up in the resistor during the first cycle, the capacitor charges up to its peak value, which is determined by the peak voltage. Once the peak voltage has passed, the capacitor starts to discharge, thus keeping the current through R from dropping to zero. If the RC time constant of the network is slower than the ac period, the voltage will never get down to zero because the capacitor will never discharge fully. By proper choice of the capacitor, one constructs a voltage such as that shown in Fig. 17.17(b).

In practice, there are a variety of solid-state diodes having a multiplicity of uses in electronic circuits beyond simple rectification. They

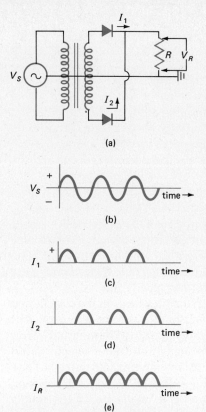

Figure 17.16 A circuit used to rectify an ac signal to a pulsating direct current using two diodes.

Figure 17.17 A rectifier circuit with a smoothing RC circuit.

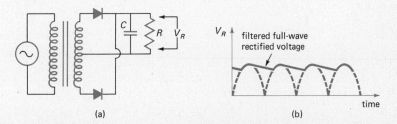

are frequently used to regulate voltages—maintaining them at a fixed value even when such basic electric characteristics of the circuit as the current are changing slightly or even drastically. Diodes are also used in digital circuits to generate pulses of electricity.

17.6 THE TRANSISTOR AND BEYOND

The invention that more than any other led to the modern era of solid-state electronics was that of the transistor, by Bell Laboratories scientists John Bardeen, Walter Brattain, and William Shockley in 1948. The transistor has the dual purpose of rectifying and amplifying electrical currents; it is named for its ability to transfer current across a resistor.

The transistor is essentially a three-terminal device consisting of an emitter, a collector, and a base. An electrical signal is injected into the emitter and taken out of the collector, and the performance of the device is controlled by the potential of the base. The most common type of transistor is the junction transistor, which consists of two *p-n* junctions. These can be arranged in either the *p-n-p* or *n-p-n* fashion. Figure 17.18 shows the operation of the former type.

The potential between the emitter and base controls the current that passes from the emitter to the collector through the transistor. The emitter is biased positively with respect to the collector; this forces positive holes from the left-hand *p* region into the *n* region; if this region is thin enough, it allows the holes to move on into the right-hand *p* region. The emitter is also forward biased with respect to the base, thus allowing electrons from the *n* region to move into the left-hand *p* region. At the same time, this base-to-emitter bias reduces the barrier potential for the holes to move from the left-hand *p* region into the *n* region. In the *n* region, holes are neutralized with electrons. However, if the *n* region is thin, then moving holes migrate into the right-hand *p* region and hence produce a current of positive holes that moves out of the transistor to the collector. The base-to-emitter voltage acts as a switch to control the larger emitter-to-collector current.

Bardeen, now at the University of Illinois, is the only scientist to have been awarded the Nobel Prize twice in the same subject. He shared the 1956 prize for his part in developing the transistor and the 1972 prize for his part in formulating the BCS theory of superconductivity outlined in Chap. 14. His appearance at a press conference following the 1972 award was delayed, incongruously, when his transistorized automatic garage door jammed.

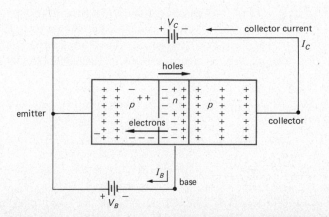

Figure 17.18 A *p-n-p* junction transistor.

Figure 17.19 Current-voltage characteristic curves for a transistor; I = collector current, I_B = base current, and V_{CE} = collector-to-emitter potential.

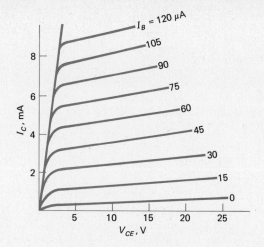

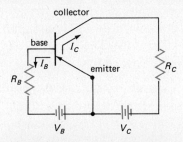

Figure 17.20 A *p-n-p* transistor in the common emitter configuration.

The characteristics of a transistor in a circuit are illustrated in Fig. 17.19, which shows the collector current as a function of V_{CE}, the potential difference between the collector and the emitter, at different values of the base current. The measurements refer to the typical transistor circuit sketched in Fig. 17.20. The potential difference between the emitter and collector is the difference between the potential drop across the resistor R_C (given by the product IR_C) and the applied collector voltage V_C. When the base current I_B is zero, the transistor is essentially off for all values of V_{CE}. However, because I_B is measured in microamperes (μA), whereas the collector current I_C is in milliamperes (mA), if V_{CE} is held at 10 V, a small change in I_B from 15 to 30 μA produces about a 1-mA change in collector current I_C. It is this property of a transistor that enables it to be effective as an amplifier and in control circuits.

Figure 17.21 shows an example of a simple transistor circuit used to amplify an ac signal. A change in the base current I_B produced by the small ac signal from V_{in} causes changes in the collector current in the milliampere range, producing an amplified output voltage V_{out}. Electronic engineers can easily design circuits with 100fold amplifications by proper choice of transistor, voltages V_C, and the resistors. By connecting several transistors in series, they can obtain even larger amplification factors.

Electronics design has now progressed to the point where arrangements of transistors and diodes can be put in a single miniaturized integrated circuit (IC), which is a collection of transistors on a specially processed silicon chip that operates as an entire electronic circuit. A chip of only a few millimeters in diameter can act as the equivalent of a number of capacitors, resistors, and transistors wired together. Because of their miniaturized sizes and low power requirements, ICs have revolutionized the electronic industry of the 1970s in the same way that

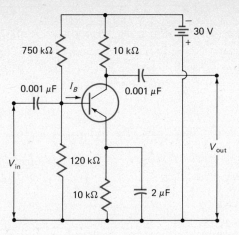

Figure 17.21 A simple transistor amplifier. A small ac input voltage V_{in} alters the base current I_B, producing a change in the collector current. This results in an amplified output voltage V_{out}. The single battery V_C provides proper biasing of the transistor base, collector, and emitter by means of the resistive network.

transistors revolutionized the same industry in the 1960s by replacing vacuum-tube electronics. Microminiature circuits contained on a small chip have made the pocket calculator an everyday item and promise to make miniature components as common in the house of tomorrow as televisions are today.

17.7 MAXWELL'S EQUATIONS

Before leaving the study of electric and magnetic phenomena, we must mention the theoretical work of the Scottish physicist James Clerk Maxwell in the mid-1860s. In a series of four powerful and fundamental mathematical relationships, Maxwell provided a unification of the earlier electric and magnetic laws.

These four equations not only summarize all the electrical and magnetic effects with which we have dealt so far in the past four chapters but predict the existence of the electromagnetic waves that yield all forms of information about objects: beams of light, radio waves, x-rays, and other parts of the electromagnetic spectrum. As an indication of Maxwell's extraordinary ability in theoretical physics, note that the actual existence of the waves his theory predicted was not confirmed by experiment until 20 yr later, in the 1880s. Perhaps even more important than these predictions is that Maxwell's equations pointed the way to the existence, and predicted the value, of one of the most fundamental constants in the whole universe, namely, the speed of electromagnetic waves: the speed of light.

Maxwell's equations basically relate the behavior of electric and magnetic fields at single points in space to changes in the electric and magnetic fields at other points in the immediate vicinity. Before examining the equations themselves, we might review the basic similarities and differences between electric and magnetic fields. In fact, the opportunities for confusing the two types of fields and substituting one for the other are numerous.

Magnetic fields and electric fields are inseparable.

Magnetic fields occur between the poles of permanent magnets and electromagnets and also accompany the passage of electric currents. Electric fields exist wherever there is an electric charge, in just the same way that a gravitational field can be found wherever there is an object with a finite mass. In addition, circular electric fields, whose field lines are closed loops, are produced by changing *magnetic* fields during the course of electromagnetic induction. And careful studies of electrical effects show that changing *electric* fields induce magnetic fields. Magnetic fields and electric fields, in other words, are inseparable. Maxwell's equations account for this togetherness and the way in which magnetic and electric fields move through space.

Maxwell's first equation is basically a restatement of Coulomb's law, although this is not obvious at first glance. The equation states that if we draw a spherical shell in space and find the number of electric field lines entering the sphere is exactly the same as the number leaving it, then the sphere contains no net electric charge. This statement is a sophisticated way of saying that what goes in must come out. Maxwell's first equation goes beyond the no-charge situation, however. In cases where a sphere drawn in space does contain an electric charge, the equation states that the net number of lines of electric force entering or leaving the sphere is proportional to the net electric charge inside the sphere.

Maxwell's second equation deals with magnetic fields. It states that, unlike electric field lines, which can originate on positive charges and terminate on negative charges, magnetic field lines do not suddenly change density. This means that individual magnetic charges—single magnetic poles—do not exist. This equation is perhaps the most controversial of the four. Even today, physicists studying high-energy particle physics are searching for what they call the magnetic monopole, the single isolated magnetic north or south pole that this law seems to preclude. However, classical experiments have shown that magnetic field lines do not end in magnetic (single) poles, and Maxwell's equations as he formulated them have been shown to apply to most electrical and magnetic phenomena. Complexities do arise in the realms of particles with high speeds and high energies.

Maxwell's third equation is related to Faraday's law of electromagnetic induction. It states that changing magnetic fields produce electric fields. As the magnetic field increases or decreases, it brings about an electric field whose field lines all form circles in planes perpendicular to the magnetic field lines.

The fourth Maxwell equation treats the reverse process: the production of a magnetic field by either a current of electricity or a varying electric field.

We can understand the first case, the one considered in the third equation, very simply. Moving charges produce a current, and we know that magnetic field lines always form circles around an electric current.

The second case, the production of a magnetic field by a changing electric field, is analagous to Faraday's law; it was first predicted by Maxwell in his fourth equation and is yet another example of a theoretical forecast, made by the Scottish physicist, that stood up to experimental scrutiny years later.

17.8 ELECTROMAGNETIC WAVES

Beyond giving us a self-contained explanation of all electrical phenomena, Maxwell's equations spell out something totally unexpected, the movement of a wave through space. Unlike the waves we dealt with in Chap. 12, Maxwell's wave motions do not involve any visible alteration of a material medium, such as the lazy up-and-down movement of a clothesline rippled by a light breeze. The physical phenomena that are actually waving, according to the Maxwell equations, are the electric and magnetic fields, whose amplitudes in planes perpendicular to one another alter as shown in Fig. 17.22. Because these oscillations are also perpendicular to the direction in which the wave travels, these "electromagnetic waves" are transverse waves. And although they are not strictly "visible" (except for light), the waves can be detected using simple apparatus.

To detect the electric field component of the waves, place a standard electric charge at some point along the x axis in Fig. 17.22 and measure how the electrical force on it changes with time. The force oscillates as the wave moves by. By using the definition of electric field as the electrical force per unit charge, one could determine the strength of the field at any time and work out the maximum and minimum values of the oscillating field. One could also find the strength of the magnetic field by measuring the torque on a loop of wire that is carrying direct current. Combining these observations, one would discover that the electric and magnetic fields vary together, as shown in the figure. To use the technical term from Chap. 12, the two fields are in phase with each other. This means that the maximum and minimum of both fields occur at the same places perpendicular to the direction of propagation.

What is the medium that supports this wave motion? In common with other scientists of his time, Maxwell believed that the whole of space, including vacuum, was permeated by a massless form of matter known as the **ether**. Maxwell pictured the passage of electromagnetic

This type of ether should not be confused with the very substantive ether gas that was one of the original anesthetics.

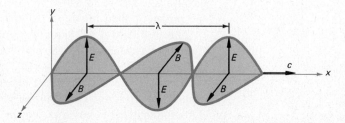

Figure 17.22 An electromagnetic wave consisting of an electric field E and a magnetic field B oscillating in phase but in planes perpendicular to one another. The wave is transverse, propagating at speed c in the x direction.

waves as occurring via stresses and strains in the ether. But in 1887, shortly after Maxwell's untimely death from cancer, a crucial experiment by American physicists Albert Michelson and Edward Morley failed to detect any trace of the mysterious ether. Eighteen years later, in one of the most famous scientific papers, Albert Einstein showed that the theory of relativity could account for both that experiment and the existence of electromagnetic waves without an ether. In effect, we now understand the transmission of such waves to result from oscillations in the strengths of the electric and magnetic fields alone, without support from any form of matter. Electromagnetic waves, in other words, can move through total vacuum.

We shall deal with the reasoning that produced this conclusion in Chap. 20. Meanwhile, note that the lack of an ether did not really alter the general understanding of electromagnetic waves in terms of Maxwell's four equations.

In quantitative terms, the most important contribution of Maxwell's equations was their use to calculate the speed of electromagnetic waves. This speed turned out to be about 3×10^8 m/s in a vacuum, an extremely large but nonetheless finite speed. A wave traveling at that speed takes 10^{-4} s to travel from one hilltop to another 20 mi away and then return, for example, and just 16 min for the round trip between the sun and the earth. Even more significant than the numerical value of the speed was its similarity to the speed of light, which had been measured by the French physicists Armand H. L. Fizeau and Jean Foucault when Maxwell worked out his four famous equations. So close were the two values that physicists became convinced, correctly, that light itself was a type of electromagnetic wave.

This realization that all electromagnetic waves, including light, travel at the same superspeed gave physicists an obvious means of identifying individual waves. Recalling the wave equation of Chap. 12 and substituting for speed the symbol c, which represents the speed of light in vacuum, we have

$$c = \lambda \nu \tag{17.10}$$

where λ is the wavelength of the wave and ν is its frequency. Obviously, there is a whole spectrum of values of λ and ν that give the value of c when multiplied together, ranging from close to $\lambda = 0$ and $\nu = \infty$ to close to $\lambda = \infty$ and $\nu = 0$. In fact, each point on the electromagnetic spectrum is identified by a wavelength and its corresponding frequency.

All electromagnetic waves are generated by the oscillation—or more strictly, acceleration—of electric charges. The exact frequency and wavelength of any wave are determined by the situation of the oscillating charge that produces them—whether they are in the atomic nucleus, the atom, or the molecule.

Figure 17.23 shows the entire electromagnetic spectrum. This spectrum is actually a continuous one, and the names of the different

Figure 17.23 The electromagnetic spectrum. On the left-hand side of the diagram are the frequencies; the corresponding wavelengths are on the right-hand side. The frequencies and wavelengths are related by $c = \nu\lambda$, where $c =$ speed of light in free space (3×10^8 m/s) and is the same for all wavelengths of the electromagnetic spectrum.

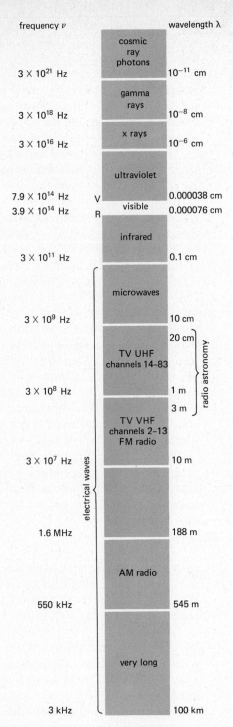

regions indicate primarily how we identify the radiation in those regions. Some regions of the spectrum even overlap. Waves whose wavelengths are between 0.3 and 3.0 cm, for example, can be produced by hot objects that radiate infrared rays (heat) or microwave sources that emit so-called radio or radar waves. Similarly, waves in the region of 10^{-6}-cm wavelength can be produced by either x-ray or ultraviolet sources, and those with wavelengths around 10^{-8} cm can emanate from either x- or gamma (γ)-ray sources.

Taking the spectrum from the top, the known shortest wavelength and highest frequency waves come from outer space in the form of cosmic rays. The cause and even the origin of such rays, which penetrate the earth's atmosphere and much of the earth's crust, remain unknown. Some astronomers believe that the bulk of cosmic rays reaching the earth come from somewhere inside our own galaxy, and most likely the galaxy's center, while others argue that at least a large proportion come from outside our immediate galactic confines.

The next shortest members of the electromagnetic spectrum in terms of wavelength are γ-rays. These intense penetrating rays are emitted from the nuclei of radioactive substances, as we shall see in more detail in Chap. 24. Exposure to radiation can be extremely dangerous to one's health, but in controlled doses this type of radiation is used in the treatment of certain cancers, in goiter, and in other ailments. In the laboratory, γ radiation is used to detect and identify radioactive substances.

x-rays, the next category down the spectrum, need little explanation. They are generally produced by the impact of beams of electrons on metals and are used in the diagnosis of faults in metals and other material objects as well as patients in hospitals and doctors' offices. Like γ-rays, x-rays can be dangerous in large quantities. The x-ray devices used in the 1940s to check whether children's shoes fitted correctly are definitely on the blacklist today, and even dentists and hospital radiologists take great care not to expose their patients to more than the necessary amounts of x radiation.

Ultraviolet radiation, which literally means beyond the violet end of visible light, is next along the spectrum, and it can be equally hazardous in large quantities. This type of radiation is responsible for producing both suntans and skin cancer. The controversy over the possible destruction of ozone in the atmosphere by certain types of aerosol sprays and supersonic airplanes is in the final analysis a debate about ultraviolet radiation. The ozone layer in the upper stratosphere at an altitude of about 10 mi acts as a large-scale screen for the inhabitants of earth, absorbing all but a small percentage of the ultraviolet radiation that reaches earth from the sun. If the effectiveness of this screen is jeopardized by a reduction in the amount of ozone, experts believe that the number of cases of skin cancer in the human population might increase significantly.

The x-ray devices used in the 1940s to check whether children's shoes fitted correctly are definitely on the blacklist today.

On the practical level, the argument revolves around the question of whether chlorine atoms, produced by the breakup of the chlorofluorocarbons that power certain aerosols, linger long enough in the upper stratosphere to cause reduction of the amount of ozone there.

One major problem of exposure to ultraviolet radiation is that we cannot see it. Many sunbathers tend to lose their caution on cloudy days and reckon that they cannot get burned because they cannot see the sun. But some of those ultraviolet rays that manage to penetrate the ozone barrier can also get through cloud layers, as many burned bathers have learned to their regret. One sure way of discovering the presence of ultraviolet rays is to rely on an old aid to our visual senses: photographic film. Ultraviolet rays blacken most photographic films even more readily than do rays of visible light.

Ultraviolet radiation is not all bad. Our bodies need exposure to a certain amount of it to be able to produce vitamin D, an essential chemical for proper functioning. It is also used to produce vitamin D in milk. One can obtain extra ultraviolet light from lamps containing mercury-vapor bulbs, whose output contains a high proportion of ultraviolet rays.

The smallest region of the electromagnetic spectrum is in many ways the most important to people. This is visible light, whose boundaries in the spectrum are very tightly defined. If you direct a beam of sunlight through a prism, you produce a rainbow of colors whose wavelengths vary from 0.000 038 cm at the violet end to 0.000 076 cm at the red end. Of course, not everyone can perceive the full range of the visible spectrum; different eyes interpret what they see in different ways. The two limits of wavelength indicate the maximum range that the human eye can sense.

Humans perceive the next part of the spectrum, the infrared, as heat. It is often called heat radiation. We obtain energy from the sun in the form of infrared or heat radiation. Infrared radiation is far less likely than visible and ultraviolet light to be absorbed by the water vapor in the atmosphere, and photographic films sensitive to the infrared region are used in aerial surveys of the ground on cloudy, hazy days. These films show a thermal profile of the surface of the earth, therefore providing useful data to farmers, geologists, and even archeologists. For example, infrared radiation emitted from infected crops is different from that emitted from healthy crops. Many pollution sources can be identified in this manner because the pollution-containing water is usually at a different (warmer) temperature from that of the lake or river into which it flows.

Beyond the infrared region are the electric waves, so named because they are produced by electrical processes. Microwaves, whose wavelengths range from 0.1 to 10 cm, are the shortest wavelengths of any in this range. They are produced in microwave ovens and radar units and have frequencies that vary from 3×10^5 to 3×10^3 MHz.

Beyond the microwave region is the ultrahigh frequency (UHF) poriton of the spectrum. This area, with wavelengths between 10 cm and 1 m and frequencies between 3000 and 300 MHz, is reserved by the Federal Communications Commission (FCC) for UHF television stations. According to rules of the FCC, Channel 14 in any area

The hertz is named in honor of Heinrich Hertz, who first observed radio waves in 1887. A megahertz is 1 million cycles per second.

Poor Reception

When Apollo astronauts Tom Stafford, Vance Brand, and Deke Slayton were orbiting the earth on their historic mission to link up with the Soviet Soyuz spacecraft, they were amazed to hear a clear instruction to come in for a landing—at Los Angeles International Airport. Somehow the signals had become crossed, and the Los Angeles tower radio beam was reaching the spacecraft at an altitude of more than 100 mi. Unusual as it was under the circumstances, this type of incident is common enough for mortals down on earth. Just consider the number of times that you cannot tune in the radio station you want because nearby stations on the dial seem to overlap. The radio spectrum, and indeed, the electromagnetic spectrum in general, is overextended just as surely as are the earth's resources of land. And just as politicians and administrators seem unable to solve the latter problem, there appears to be no present redress for the former. For the foreseeable future cluttered reception is likely to be a fact of life for most of us who use radio.

operates at a frequency of 475.75 MHz and a wavelength of 0.63 m, and Channel 83 operates at a frequency of 889.76 MHz and a wavelength of 0.337 m. All channels between transmit at appropriate intervals between these two limits.

The UHF band of wavelengths, and a small part of the adjacent very high frequency (VHF) band, also fulfill a vital need for radio astronomers. This portion of the electromagnetic spectrum contains most of the radio frequencies on which stars, intergalactic gas and dust clouds, and other objects in the skies transmit information about themselves. Not surprisingly, radio astronomers go to great pains to locate their instruments in areas where there is little local UHF activity of the self-created variety.

The VHF band stretches from a wavelength of 1 m and a frequency of 300 MHz to 10 m and 30 MHz. Its main function is to host the signals from the familiar television channels between 2 and 13. Channel 2, for example, transmits at a frequency of 59.5 MHz and a wavelength of 5 m, and Channel 13 at a frequency of 215.75 MHz and a wavelength of 1.39 m. Wavelength bands used for certain radio purposes are squeezed in at appropriate points between the wavelengths assigned to the television operators. Police and taxicab radios, for example, broadcast on wavelengths of about 1 m, at the boundary between UHF and VHF waves, and FM radio stations put out their signals in a narrow band from about 80 to 110 MHz, in the midst of the wavelengths of television channels. Finally, the band for AM radio is well beyond this range, varying between 1600- and 500-kHz frequency and wavelengths between 188 and 545 m.

The Fathers of Radio

Like many scientific finds before and since, the discovery of radio was an event that arose from a project with limited aims. In the middle 1880s, the young German physicist Heinrich Hertz was studying Maxwell's equations of electromagnetic radiation in the quest for a prize that had been offered by the Berlin Academy of Science. Hertz set up an oscillating electrical circuit that, according to the Maxwell equations, should produce radiation with extremely long wavelengths. To detect this radiation, Hertz used a broken loop of wire. Any radiation, he reasoned, would create an electric current in the wire that would produce a spark across the broken ends.

In 1886 Hertz hit the jackpot. His detector started to produce small sparks, and when he moved it to various parts of his laboratory, he was able to establish the shape—sinusoidal—of the waves and their wavelength, 66 cm, which was 1 million times as great as that of visible light. Eight years later, the Italian engineer Guglielmo Marconi started to repeat Hertz's experiments and was able to improve on them sufficiently to send the waves over increasing distances. They were named radio waves as an abbreviation for "radiotelegraphy."

By the end of the century, Marconi was sending the radio beams about 20 mi; in December 1901 he made history by transmitting waves from Goonhilly Down, in southwest England, all the way over the Atlantic to Newfoundland. With this event, an industry was born. The world has been listening in ever since.

Heinrich Hertz. (*Culver*)

Guglielmo Marconi. (*Wide World*)

EXAMPLE 17.9

Compute the wavelength in meters of the television wave transmitted by a Channel 2 station telecasting on its assigned frequency of 60 MHz.

Solution

$$\lambda = \frac{c}{\nu} = \frac{3 \times 10^{10} \text{ cm/s}}{60 \text{ MHz}} = \frac{3 \times 10^{10} \text{ cm}}{60 \times 10^6} = 500 \text{ cm} \quad \text{or 5 m}$$

SUMMARY

1. Inductors and capacitors can act as filters for electric currents. Inductors present high impedance to high-frequency alternating current, and capacitors have high impedance for low frequencies.
2. Circuits consisting of capacitors and inductors produce electrical oscillations.
3. Rectifiers are instruments that conduct current in only one direction. They convert alternating current to pulsating, direct current.
4. The ac voltage and current obey an Ohm's law linear relationship if the dc resistance is replaced by the impedance.
5. Transistors based on doped semiconductors can amplify ac signals up to hundreds of times.
6. The movement of waves through space involves the oscillation of the magnitudes of magnetic and electric fields that are perpendicular to each other. Electromagnetic waves require no substance to support them; they can move through complete vacuum.
7. All electromagnetic waves move with the same speed of about 3×10^8 m/s in vacuum.
8. Electromagnetic waves are generated by the oscillation of electric charges. The frequency and wavelength of every wave is determined by the situation of the oscillating charge that produces it.

QUESTIONS

1. What is inductance? Compare it to resistance. How are they alike; how do they differ? What is inductive reactance?
2. Under what conditions does inductance affect current in a dc circuit?
3. How is the inductive reactance dependent upon frequency? Does resistance depend upon frequency of the emf? Comment.
4. Under what conditions does an inductance act as a choke and assist in filtering out various frequencies?
5. Do inductors in parallel add like resistors or like capacitors?
6. When five capacitors are in parallel, what can be said about the charge on each and the potential drop across each?
7. When a dielectric is placed between the plates of a capacitor, does the capacitance go up or down or does it remain the same? How is the stored charge affected?
8. How can a capacitor act as an ac filter? For what frequencies will the capacitance appear as a short circuit (a very low impedance)?
9. Explain how an electric oscillation transfers from an inductance to capacitance and back again.
10. Since any oscillation due to energy losses will eventually become less in amplitude, how can an electrical oscillation be maintained?
11. What is meant by a nonlinear circuit element?
12. An incandescent lamp is connected to a variable capacitor in series with a 110-V ac source. Will the lamp glow more or less brightly if the capacitance is increased?
13. What happens to the resonance frequency of a tank circuit as the capacitance is increased? As the inductance is increased?
14. When a circuit is tuned to its resonance frequency, what is the power factor?
15. How do you expect that the inductive reactance of a coil changes when an iron bar is inserted into the coil?
16. What is the source of electromagnetic waves? For each region of the electromagnetic spectrum, describe a typical source of the waves.
17. How do electromagnetic waves differ from sound waves in air?
18. An infrared sensing device developed for the military allowed soldiers to see in the dark. What is meant by "seeing in the dark?" What is "detected" by the infrared detector? Could the device be used to "see" nonliving objects in a dark room?
19. If you charge a rubber rod by rubbing it with fur and then you wave the rod back and forth, will you generate electromagnetic waves? At what rate would you have to move the rod to generate light waves?
20. x- and γ-rays and light waves are all emitted by a distant star. Which of these types of electromagnetic radiation will reach the earth first if they are all emitted simultaneously toward the earth?
21. What sort of magnetic fields are produced by changing electric fields? How are the directions of such magnetic fields related to the direction free positive charges would move in the changing electric fields?
22. An electromagnetic wave is moving horizontally in a direction toward you. If at a specific instant of time the electric field points down, in which direction does the magnetic field point?
23. Place the following electromagnetic waves in order, from lowest frequency to highest frequency: ultraviolet, infrared, microwaves, blue light, x-rays, radio waves, γ-rays, yellow light, red light.

PROBLEMS

Capacitance

1. How much charge can be stored on a 20-μF capacitor if it is connected to a 100-V battery?
2. A 100-μF capacitor has a charge of 200 μC on its plates. What is the potential difference across the capacitor?
3. (a) What is the capacitance of two parallel circular plates 4 cm in radius and separated by 0.05 cm of air? If 100 V are placed across these plates, (b) what charge is on each plate?
4. A parallel-plate air capacitor has a gap of 0.1 cm between plates and the electric field between them is 450 V/cm. (a) What is the potential difference across the plates? If the area of the plates is 15 cm^2, (b) what is the capacitance of the capacitor and (c) the charge on its plates?
5. A capacitor has a capacitance of 12 μF when it contains air. It is charged to 200 V and then the battery removed. If the 0.4-mm air space between the plates is replaced by a 0.4-mm sheet of mica whose dielectric constant is 5, what will be the (a) potential across the plates and (b) capacitance of the system?
6. A capacitor with an air dielectric has a capacitance of 24 μF. The capacitor is connected to a power supply of 1000 V. What is the charge on each plate? If plastic with a dielectric of 8 is inserted between the plates, compute the (a) new capacitance, (b) charge on each plate, and (c) potential difference across the capacitor.
7. A capacitor has a capacitance of 12 μF when its plates are separated by 0.4 mm and contain air between them. The capacitor is charged to 200 V. (a) What is the charge on the capacitor? If the plates are brought closer together so that they are separated by 0.2 mm after the battery is disconnected, (b) what is the charge on the capacitor? (c) What is the new capacitance of the capacitor? (d) By what factor does the potential difference of the capacitor change?
8. Three capacitors are connected as shown in the figure. Compute the potential difference across each and the charge on each.

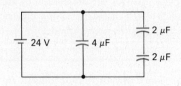

9. Compute the equivalent capacitance of all the capacitors and the charge and potential difference on the 2-μF capacitors.

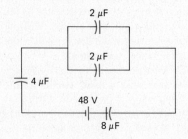

10. At what frequency does a 0.1-μF capacitor have a 2-kΩ impedance?
11. A 25-μF capacitor is connected with a resistor in series with a 24-V battery. A potential difference of 18 V develops across the capacitor in 10 ms. What is the total resistance in the circuit?
12. A 4-μF capacitor and a 10-kΩ resistor are connected to a 6-V battery. What is the time constant for charging the capacitor?
13. A 600-μF capacitor is placed in series with a battery and a resistor. The maximum charge on the capacitor is 0.12 C, and the time constant of the circuit is 0.5 s. Find the resistance and the emf in the circuit.
14. A 2-μF capacitor is in series with a 5-kΩ resistor and a 24-V battery. What is the time constant for this circuit, and what was the initial current in the circuit?

Inductance and Resonance

15. An 8-mH inductor in series with a 120-Ω resistor is connected to 90 V. What is the time constant and the maximum current in this dc circuit?
16. What is the inductive reactance of a 50-mH coil in a 120-V, 60-Hz household circuit?
17. What is the reactance of a 15-mH coil at a frequency of 20 kHz?
18. If the current in a 200-mH inductor changes from 5 A to 0 in 2 ms, what is the induced emf?
19. A 12-V battery in series with a circuit containing two circuit elements results in a current of 1.5 A. If the 12-V battery is replaced by a 12-V, 60-Hz ac source, the current reduces to 1.0 A. What are the values of the two circuit elements?
20. At what frequencies will a 0.1-H inductor and a 10-μF capacitor have reactances of 500 Ω?
21. A 10-mH inductance and a variable capacitor are connected in series to a 20-V, 1-kHz ac source and the capacitor adjusted for resonance. What is the value of the capacitance required?
22. What is the resonance frequency of a circuit containing a 600-μH inductance and a 120-pF capacitor?
23. In making a radio, the resonance condition is obtained with a 10-mH inductor and a variable capacitor. To tune in resonance from 500 kHz to 1.5 MHz, what must be the (a) maximum and (b) minimum values of capacitance obtainable with the variable capacitor?

AC Circuits

24. A 2-H inductor with negligible resistance is connected to a 50-V, 20-Hz ac line. Compute the (a) reactance and (b) current in the circuit.

25. A 6-μF capacitor is connected to a 110-V, 60-Hz power line. What is the current in the circuits?

26. A 50-μF capacitor and a 80-Ω resistor are connected in series with a 120-V, 60-Hz power line. Determine the (a) current in the circuit, (b) power dissipated in the circuit, and (c) power factor.

27. A 0.10-H inductor that has a resistance of 40 Ω is connected to a 120-V, 60-Hz power line. What current exists in the circuit?

28. A 0.11-H inductor, a 1-μF capacitor, and 20-Ω resistor are connected in series with a 120-V, 60-Hz source. (a) Find the current in the circuit. (b) Repeat for a frequency of 6 kHz.

29. An inductor draws a current of 2 A when connected to a 12-V battery. When it is connected to a 110-V, 60-Hz source, it draws 3 A. Compute the (a) power dissipated in the ac circuit and (b) inductance of the coil.

30. A 10-Ω resistor, a 2.5-H inductor, and a 2.5-μF capacitor are connected in series with a 90-V, 60-Hz source. Compute the (a) current in the circuit, (b) power dissipated in the circuit, (c) power factor, and (d) phase angle. (e) Does the current lead or lag the voltage?

31. A series ac circuit contains a 100-Ω resistor, a 0.2 H-inductor, and a 3-μF capacitor connected to a 110-V, 60-Hz power line. (a) What is the inductive reactance? (b) Compute the capacitive reactance. (c) What is the impedance of the circuit? Calculate the (d) current in the circuit, (e) power factor, and (f) phase angle.

Transistors

32. What base current is required with a collector voltage of 15 V to produce a collector current of 3 mA? See Fig. 17.19.

33. Using Fig. 17.19, determine the collector current when the base current is 45 μA and the collector voltage is 10 V.

34. Using Fig. 17.19, estimate the amplification of this transistor if the collector voltage is kept at 10 V while the base current is varied from 15 to 45 μA. What is the resulting variation in the collector current?

Electromagnetic Waves

35. Compute the wavelength of the radio waves at either end (the low- and high-frequency extremes) of the FM and AM bands on your radio.

36. How long are the wavelengths in Prob. 35 in terms of some "everyday" physical object?

37. Express 5.45×10^{-7} m in nanometers (nm) and micrometers (μm).

38. What is the frequency of the extreme blue light in the visible spectrum if the wavelength is 400 nm?

39. What is the frequency of x-rays of wavelength 0.1 nm or 10^{-8} cm?

40. Green light has a frequency of 5.6×10^{14} Hz. What is the wavelength of green light?

41. What type of electromagnetic radiation has wavelengths similar to the diameter of a softball?

Special Topic
Is Anybody Out There?

The Arecibo 1000-ft radiotelescope. (*Cornell University*)

Does intelligent life exist beyond earth? And if it does, can we get in touch with it? During the first 60 yr of this century these questions were strictly in the realm of science fiction. But then, the dawning of the space age, the onset of radio astronomy, and the efforts of a few dedicated astronomers began to convince many scientists not only that the chances are favorable that many intelligent civilizations exist elsewhere in the universe but that getting in contact with them might not be as difficult as it first appeared. Face-to-face contact still seems next to impossible, owing to the vast distances over which people or other intelligent beings must travel. But conversation via radio waves, in a kind of cosmic communications network, seems not unlikely.

Neglecting, on fairly firm grounds, the notion that other intelligent life exists within our own solar system, astronomers face the challenge of looking for life in the stars, or at least in other solar systems connected with stars, the closest of which are a few light-years distant. They can set about the task in one of two ways. The active method is to transmit information into space via radio waves, in the hope that some advanced civilization will intercept the messages and reply to them. The passive means is to "listen in" with radio telescopes for any type of signal that appears to be created by a superior intelligence.

To some extent people started sending messages from earth as soon as the radio was invented. A small fraction of the beams carrying the word that people had reached that technological plateau managed to escape into space. But effective radio beams, powerful enough to outshine the sun at their particular frequencies, have been on the air only since the mid-fifties.

Astronomers at Cornell University's Arecibo radio telescope in Puerto Rico are trying a more venturesome form of interstellar communication. In late 1974, the 1000-ft radio telescope built in a natural valley transmitted a radio signal in binary computer code. The message, aimed at the Great Cluster in the constellation Hercules, indicates such basic parameters of human life as the chemical formula for the DNA molecule, the atomic numbers of various elements, the position of earth in the solar system, and the typical shape and height of human beings. Since the target is about 2000 light-years away, a reply is hardly expected tomorrow.

Listening in for signals sent intentionally or unintentionally in the direction of earth is an even more popular occupation among radio astronomers, and at present at least five different radio astronomical teams in three nations—Russia, Canada, and the United States—are

Message beamed to the universe from Arecibo in 1974. (*Cornell University*)

doing just that. In their efforts to eavesdrop in a stellar sense, radio astronomers tune their telescopes to specific frequencies that seem likely to carry the intergalactic traffic, such as the well-known hydrogen transition wavelength of 1420 MHz. This was the wavelength selected by Cornell University astronomer Frank Drake in 1960 when he undertook Project Ozma, the first search of this nature ever tried. Like all the others that followed him, Drake was unsuccessful in detecting any messages that seemed to emanate from intelligent beings.

There has been one close call. In 1967, British radio astronomers at Cambridge University detected signals so regular and precise that an intelligent source seemed to be the only explanation. The excited scientists dubbed the source LGM-1 (for little green men), but their sense of scientific thrill took on a different quality a few months later when they detected another source emitting signals just as precise, then another, and yet another. The radio sources turned out to be pulsars—awesome objects scientifically because of their nature of operation rather than because of any connection with intelligent creatures. Thus, the search for other forms of intelligent life continues in expectation, and in the firm belief among a small coterie of radio astronomers that sooner or later we shall indeed contact a world of intelligence well beyond our own.

18 The Nature of Light

18.1 HUYGENS SEES THE LIGHT

18.2 THE INTERFERENCE OF LIGHT

18.3 OTHER FORMS OF INTERFERENCE

18.4 DIFFRACTION: INTERFERENCE AT THE EDGES

18.5 POLARIZATION

18.6 HOLOGRAPHY: PHOTOGRAPHS WITHOUT LENSES

SHORT SUBJECT: Holography in Action

Summary
Questions
Problems

SPECIAL TOPIC: The Search for c

What is light? Essays, theses, and books have been written in answer to this question. Philosophers have spent long nights arguing over subtleties in the definition of light. Artists have gloried in it, poets have praised it, astronomers have appreciated it, and scientists have been intrigued by it. It is the basic form of energy that makes life on this earth possible, and it is the means by which we transmit messages to and receive information from objects as close as these pages and as distant as the farthest reaches of the universe. Light is, in short, an essential ingredient of our physical, mental, and aesthetic health on planet earth.

The history of people's perception of the nature of light reflects in a number of important ways the development of science itself. The Greeks had a word for light, or at least a basic description. They believed that light consisted of little particles, generally known as corpuscles, which were emitted from the light source and "seen" when they stimulated visual perception by hitting the observer's eye. The Greeks knew from their observations of shadows that light traveled in straight lines, and the corpuscular theory appeared to account for that particular effect in a satisfying fashion.

The corpuscular theory was sensible enough to appeal to Isaac Newton, who used it to explain both **reflection—the mirror effect—**and **refraction—the bending of light when it passes from one medium to another,** such as from water to air. But in 1678, at which time the 36-year-old Newton had already been a Professor at Cambridge University for more than 12 yr, a powerful new theory of light arose. The Dutch scientist Christiaan Huygens proposed formally what a number of predecessors had suspected on the basis of hunch—that light was actually a wave motion. There was motivation for this different view, for the high speed of light, now known to be 3×10^8 m/s in vacuum, seemed more consistent with the motion of waves than with the motion of particles. Huygens' model proved able to account for reflection and refraction as satisfactorily as the corpuscular theory, but Newton's support for the latter hypothesis meant that the wave theory had little chance of acceptance at the time it was proposed and for about one-and-a-half centuries afterward. We now know that Huygens and Newton were both right, as we shall see when we discuss the wave-particle duality of electromagnetic waves.

Newton's objections to the wave theory were based on apparently firm grounds. He felt that light could not be a form of wave motion because it did not bend around corners the way sound waves did. This

This was just one example of the misdirected influence of a great scientist. Unfortunately, such influence by scholars of high repute tends to lead their less eminent colleagues astray or to suppress new ideas.

bending effect is known as diffraction. The famous physicist was right in his interpretation—if light did not diffract, it certainly could not be an example of wave motion—but wrong in his facts. Light actually does undergo diffraction, and also interference, another property of wave motion, but because the wavelength of visible light is so small, these effects were too slight to be observed by the primitive instruments of the seventeenth century.

Once the effects had been detected, in the first quarter of the nineteenth century, everything seemed to be sweetness and light. But then at the end of the nineteenth century and at the beginning of the twentieth, Max Planck and Albert Einstein returned to a corpuscular theory to explain the emission of radiation from hot objects and the photoelectric effect. In this latter process, light causes a metal on which it shines to emit electrons. Einstein won the Nobel Prize for this contribution to physics (his special and general theories of relativity went unrecognized by the Nobel committee, probably because the committee has been programmed toward applications) and once more put the physics of light into a turmoil.

The scientific revolution wrought by Planck and Einstein is the subject of Chap. 20.

Today, scientists take what one can view as either the ultimate copout or the ultimate compromise on the nature of light. They regard light as a phenomenon with a dual nature, acting sometimes as if it were particulate, sometimes as if it were wavelike. We have not said what light is but *how it behaves*. This concept of the dual nature of light does not keep theoretical physicists awake at nights. They accept it quite readily, in just the same way that they accept the existence of elementary particles which they will never be able to see directly and the existence of stars and galaxies that emitted, billions of years ago, the light we see tonight.

Today, scientists take what one can view as either the ultimate copout or the ultimate compromise on the nature of light.

18.1 HUYGENS SEES THE LIGHT

Christiaan Huygens' basic contribution to the theory of light was to develop a geometric method of finding at any time the shape of a "wave front," the line that joins points of equal phase, from knowledge of the shape of the wave front at some earlier time. Light spreads out uniformly in all directions from a point source, such as a miniature light bulb. Such a source is said to produce spherical waves. **Huygens' principle states that every single point in a wave front can be regarded as a new source of spherical waves which spread out in all directions with the same speed and frequency as the original waves.** This is illustrated in Fig. 18.1 for the case of both a spherical and a plane wave front.

Figure 18.1(a) shows wave fronts moving out in all directions from a point source of light—perhaps a firefly—or ripples expanding from a point on a smooth water surface after a pebble is dropped into the water. The distance between succeeding wave fronts, λ, is the wavelength of the waves. If we examine, say, the fifth wave front out from the source, Huygens' principle tells us that each point on this spherical front represents a source of new waves. Thus, in the same time that the main wave front moving out from the firefly would travel a single

Figure 18.1 Huygens' principle. (a) Spherical waves travel out from a point source. The wave fronts are spherical (circular on a plane) and separated by 1 wavelength. A wave front represents adjacent points of equal phase. (b) Any point on a wave front acts as a new source of waves. A series of points on the fifth wave front are treated as point sources, and one period later, if these wavelets are connected by a tangent, one obtains the new wave front. (c) A plane wave front. On the third wave front new point sources were arbitrarily selected, and the next two plane wave fronts were generated using Huygens' principle.

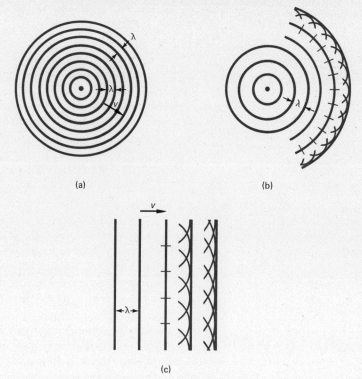

wavelength, each new spherical wave from these imagined sources will also move a distance of 1 wavelength. By putting together a surface at a tangent to all these "wavelets," which connects all points of equal phase at a given time, we can reconstruct a new wave front.

Figure 18.1(c) shows the use of Huygens' principle to reconstruct a plane wave front. Here we can imagine that the source of light is far away so that the wave fronts become approximately parallel planes. If we arbitrarily choose a number of points at regular intervals along the third wave front, draw circles in diameter λ with these points as their centers, and join the tangents to all the circles, then we end up with another plane wave front, as the figure shows.

What does all this work achieve? Apparently nothing beyond proving that we can reconstruct a fresh wave front from an existing one. So far, however, we have worked our wave fronts in a single medium, air. The value of Huygens' approach becomes clear when we study the behavior of wave fronts that move from one medium to another, such as from air to glass or from water to air. This is the physical phenomenon known as **refraction—the bending of light as it passes from one medium to another.** Refraction is responsible for the apparent distortions of our limbs in the bath and in swimming pools, for the larger-than-life appearance of printed words when they are placed under transparent glass paperweights, and for mirages on hot days.

Figure 18.2 Huygens' principle and the law of refraction. The waves travel more slowly in water (2) than in air (1). This results in bending of the wave front at the interface.

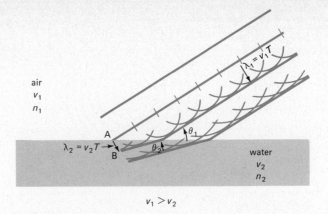

We can analyze the situation in terms of Huygens' principle using the sketch in Fig. 18.2, which shows the transfer of a plane wave from air above, in which it travels with a speed v_1, to water below, through which it passes with a different speed v_2. We know that the speed v, frequency ν, and wavelength λ of a wave are related by the expression $v = \lambda\nu$. Now, the frequency of any wave is an inherent property that is determined by its source; it does not alter with the medium through which the wave travels. Hence, a change in the speed of a wave as it passes from one medium to another mandates also a change in its wavelength. The higher the speed of the wave, the greater its wavelength. In effect, waves stretch out to move faster when they travel from a slow medium to a fast one, and compress to move more slowly when they make the passage in the opposite direction.

In our example of the transfer of a parallel beam of light from air to water, we have an advantage over Huygens: we know from differential measurements of the speed of light that it travels faster in air than in water. Hence, the wavelength of the light in air, λ_1, is larger than its wavelength in water, λ_2. If T is the period of the wave motion, that is, the reciprocal of the frequency, then $\lambda_1 = v_1 T$ and $\lambda_2 = v_2 T$. In light of this knowledge, we can look at the basic geometry of Fig. 18.2.

Huygens' principle states that points that are along a wave front can be considered as sources for new spherical wavelets (or circular wavelets, if we look at the effect in a two-dimensional diagram). We apply the rule to the whole of the wave front. The point A in Fig. 18.2, the first on the wave front to reach the water, sends wavelets into the water with speed v_2. After a time T, these wavelets have traveled a distance λ_2 equal to $v_2 T$.

Next, we consider the source of wavelets adjacent to that at point A in Fig. 18.2. During the time interval T, waves from this source travel first at high speed through the air and then at lower speed in the water. The next source out from this point produces waves that travel all the

Another part of the wavelet reaching the water travels backward and sideways into the air, creating a reflection from the surface of the water; we can neglect this part of the problem without undermining any understanding of refraction.

Figure 18.3 Expanded view of the geometry used to derive Snell's law of refraction.

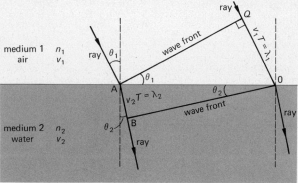

θ_1 = angle of incidence
θ_2 = angle of refraction

way in air during the same period of time, as shown in the figure. Thus, when we connect the points of equal phase after time T, by linking the tangents to the circular wavelets, we reconstruct a complete wave front that is bent. The wave front that comes in through the air at an angle θ_1 to the surface of the water moves on into the water at a smaller angle, θ_2, to the surface. These two angles are called, respectively, the angle of incidence and the angle of refraction.

To relate the two angles to each other, consider Fig. 18.3, which is an enlarged view of events at the interface between air and water. This figure contains two right triangles, AQO and ABO, which between them contain all the action from the point at which the left-hand edge of the wave front, A, first touches the water to the point at which the right-hand edge, Q, reaches the surface of the water. Applying simple geometry to the two triangles, we obtain the equations

$$\sin \theta_1 = \frac{v_1 T}{AO}$$

$$\sin \theta_2 = \frac{v_2 T}{AO}$$

Eliminating the distance AO from the equations,

$$\frac{\sin \theta_1}{\sin \theta_2} = \frac{v_1}{v_2}$$

In other words, the sine of the angle between a wave front moving through any medium and the boundary of the medium is proportional to the speed with which the wave front moves through that medium.

Scientists find it convenient to express the velocity of light in different media in terms of its velocity in vacuum and what is called the refractive index of the medium. **The refractive index of any material, denoted by the symbol n, is the ratio of the speed of light c in a vac-**

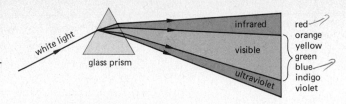

Figure 18.4 Dispersion. A beam of white light is "broken up" after passing through a prism into the various colors forming the visible spectrum, which is flanked by the invisible infrared on the long-wavelength side and by the invisible ultraviolet on the short-wavelength side.

Table 18.1 Index of Refraction for Yellow Sodium Light[a]

Substance	Index of Refraction
Air[b]	1.000 29
Hydrogen[b]	1.000 13
Water	1.33
Ice	1.31
Ethyl alcohol	1.36
Fluorite (CaF$_2$)	1.43
Carbon tetrachloride	1.46
Benzene	1.50
Rock salt (NaCl)	1.54
Quartz (SiO$_2$)	1.54
Crown glass	1.52
Flint glass	1.66
Diamond (C)	2.42
Rutile (TiO$_2$)	2.7

[a] $\lambda = 589 \times 10^{-9}$ m
[b] At standard temperature and pressure

uum to the speed of light in the material. If the latter speed is v, then

$$n = \frac{c}{v} \tag{18.1}$$

The refractive index of light in a vacuum is obviously 1, and the index in air is very close to that, varying from about 1.000 296 for violet light to 1.000 291 for red light.

A vivid result of different wavelengths of light traveling at different speeds in the same medium (or having different refractive indices) is the separation of ordinary white light into its component colors (different wavelengths) when light passes through a piece of glass such as a prism. This property of waves is called *dispersion* and is shown schematically in Fig. 18.4.

For many purposes, however, one can justifiably regard the refractive index of all wavelengths of light in air as simply 1. Table 18.1 gives the refractive indices for the yellow light of wavelength 589×10^{-9} m, known as the sodium D light, in a variety of materials. As the table shows, the refractive index of water is about 1.33 and that of glass anywhere between 1.4 and 1.9, according to the type of glass. Only a few substances, such as diamonds, have a refractive index greater than 2.

Returning to our analysis of refraction, we can substitute the expression c/n for each of the two velocities. If n_1 is the refractive index of light in air and n_2 is its index in water, then the basic equation becomes

$$\frac{\sin \theta_1}{\sin \theta_2} = \frac{c/n_1}{c/n_2} = \frac{n_2}{n_1}$$

or, alternatively,

$$n_1 \sin \theta_1 = n_2 \sin \theta_2 \tag{18.2}$$

This equation is known as Snell's law, after the Dutch scientist Willebrod Snell, who discovered the relationship experimentally in 1621. It enables one to predict the angle of refraction of a wave front in terms of the angle of incidence and the refractive indices of the two media between which the wave front passes.

One point of nomenclature is worthy of mention. Figure 18.3 contains the terms wave fronts and rays. **A ray is defined as the**

perpendicular to a wave front and indicates the direction in which the wave front actually travels. Because of their relative simplicity, rays are often used in geometrical optics—problems of reflection and refraction—in place of wave fronts. In these cases, the angles of incidence, reflection, and refraction are the angles between the ray and the perpendicular to the surface that the ray strikes, as is illustrated in Fig. 18.3.

The use of Huygens' principle to explain the experimentally observed behavior of light is not, of course, restricted to refraction. By using the same type of approach which we have just used for refraction, students can prove the experimentally tested fact that **the angle of incidence of a wave front at a surface equals the angle at which the light is reflected from that surface.**

EXAMPLE 18.1

Calculate the speed of light in quartz.

Solution

From Table 18.1, the index of refraction for quartz is $n = 1.54$. From the definition of the index of refraction and the value for the speed of light in vacuum, $c = 3 \times 10^8$ m/s, we have

$$n = \frac{c}{v_n}$$

or

$$v_n = \frac{c}{n} = \frac{3 \times 10^8 \text{ m/s}}{1.54} = 1.95 \times 10^8 \text{ m/s}$$

EXAMPLE 18.2

A beam of red light of wavelength 650 nm passes into a pool of water. What is the speed of the light in the water, and what is the wavelength of the light in water?

Solution

We require the index of refraction in water, which is listed in Table 18.1 as $n = 1.33$. The speed in water is given by

$$n = \frac{c}{v_n}$$

or

$$v_n = \frac{c}{n} = \frac{3.0 \times 10^8 \text{ m/s}}{1.33} = 2.26 \times 10^8 \text{ m/s}$$

To find the wavelength in water, we use the wave equation $c = \lambda \nu$ in air, or $v_n = \lambda_n \nu$ for the case of water. Since ν remains constant, we can solve these two equations simultaneously by removing ν:

$$\frac{c}{\lambda} = \frac{v_n}{\lambda_n}$$

$$\lambda_n = \lambda \frac{v_n}{c} = \frac{\lambda}{n}$$

The original wavelength was 650 nm, so

$$\lambda_n = \frac{650 \text{ nm}}{1.33} = 489 \text{ nm} = 4.89 \times 10^{-7} \text{ m} = 4890 \text{ Å}$$

One ångstrom (Å) is defined as 10^{-10} m. This unit is frequently used for ultrasmall distances, such as those involved in atomic processes.

Note that the wavelength in air and the wavelength in a material medium having a refractive index n are related by

$$\lambda_n = \frac{\lambda}{n} \tag{18.3}$$

as derived in this problem.

EXAMPLE 18.3

Light passes from air into water with an angle of incidence of 35°. Calculate the angle of refraction in the water if the index of refraction is 1.33.

Solution

We can obtain the angle of refraction from Snell's law:

$$n_1 \sin \theta_1 = n_2 \sin \theta_2$$

$$\sin \theta_2 = \frac{n_1}{n_2} \sin \theta_1 = \frac{1}{1.33} \sin 35° = 0.43$$

$$\theta_2 = 25.5°$$

EXAMPLE 18.4

A glass beer stein with a thick bottom contains a beer with an index of refraction close to that of water, $n = 1.33$, as in Fig. 18.5. The index of refraction of the glass is 1.50. What is the angle of refraction in the glass if light is incident upon the glass-beer interface at an angle of 38°?

Solution

Again, application of Snell's law is called for. The beer is the first medium, so $n_1 = 1.33$ and θ_1, which is the angle of incidence, is 38°.

Figure 18.5

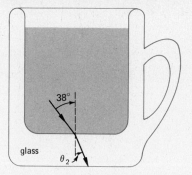

Since $n_2 = 1.5$, we can obtain θ_2 by the following calculation:

$$n_1 \sin \theta_1 = n_2 \sin \theta_2$$

$$1.33 \sin 38° = 1.50 \sin \theta_2$$

$$\sin \theta_2 = \frac{1.33}{1.50} \sin 38° = 0.55$$

$$\theta_2 = 33.1°$$

18.2 THE INTERFERENCE OF LIGHT

Young, an infant prodigy who could read at the age of 2, also became an expert Egyptologist, making remarkable contributions to the understanding of hieroglyphics.

Over the years, accurate measurement of the speed of light in different media, taken together with the type of analysis of refraction that we have just carried out, raised serious doubts as to the validity of the old corpuscular theory of light, at least in the form in which Newton understood it. By the time Foucault measured those speeds, in 1850, in a classic experiment outlined in this chapter's Special Topic, the theory had already been under continuous and all but irresistible assault for almost half a century. This attack started in high gear in 1802 when Thomas Young, an English physician of extraordinarily wide-ranging talent, whom we encountered in connection with Young's modulus in Chap. 8, first observed the interference of light beams and realized what he had seen. Interestingly, Newton had observed a type of interference during his experiments with prisms; it consisted of fringes of darker and brighter light, and is now referred to as Newton's rings. However, the consummate scientist did not try to puzzle out this particular observation.

The basis for any interference between waves, as we saw in Chap. 12, is the addition of two separate waves or wave trains. We obtain the resulting wave by the *principle of superposition,* adding the instantaneous displacements that would be produced at every point by the separate waves if they were present alone. The use of this principle is restricted to so-called linear waves. This restriction hardly concerns interference experiments with light because few sources of light are

562 The Nature of Light

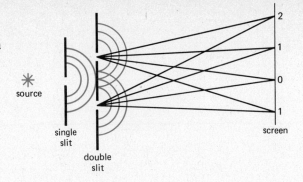

Figure 18.6 Young's double-slit experiment. Diagram of light paths for constructive interference. The figure represents a cross-sectional view or a view from above a set of vertical slits.

Although three slits are actually involved in the experiment, physicists refer to it as the "double-slit experiment" because the use of a double slit to split the beam into two interfering parts is the critical operation. You could just as well use a narrow-filament showcase lamp in place of the first slit.

intense enough to act in a nonlinear way. However, the laser is capable of sufficiently large light intensities so that it produces nonlinear effects. Nonlinear effects for sound waves were discussed briefly in Chap. 12 in respect to shock waves and sonic booms.

Laser light does possess one property that guarantees interference of a type that can be readily observed. This property is known as *coherence*. All parts of a coherent source of light are in phase with one another. The normal sources of light with which we are familiar, such as electric light bulbs, fires, and even the sun, emit light in a very haphazard manner because different parts of the source have no correlation with other parts. The atoms of these sources act as a multitude of independent sources, each source emitting light of varying phase, resulting in a jumble of different phases, to say nothing of the different colors (or wavelengths).

The simplest way to give two beams of light some kind of coherence when emitted from an incoherent source of light is first to extract a thin beam of light using a narrow slit or hole. This opening produces a narrow beam of light. This is just what Young did in his now famous experiment, which is shown in Fig. 18.6. He sent sunlight through a single small hole about the thickness of the cutting edge of a razor blade, directed the resulting beam through two narrow slits, and looked at the result on a screen beyond the double slits. What he saw on the screen was the series of equally spaced bright and dark bands, produced as the two halves of the original beam of light that had passed through the slits and interfered with each other. You can repeat Young's experiment by blackening a couple of plates of glass with soot from a match or candle and then drawing the edge of a razor blade across the blackened areas to make the requisite three slits.

Young explained the effect he observed with the intricate set of drawings of wave fronts shown in Fig. 18.7. We can understand the experiment just as efficiently by examining the simplified sketch in Fig. 18.8. If we neglect the actual size of the double slits for the time being, we can regard each slit as the start of a new series of wave fronts that spread out in all directions according to Huygens' principle. The wave

Figure 18.7 Thomas Young's original drawing showing interference effects in overlapping waves. The alternate regions of reinforcement and cancellation in the drawing can best be seen by placing your eye near the right edge and sighting at a grazing angle along the diagram.

Figure 18.8 Young's double-slit experiment. Geometric diagram.

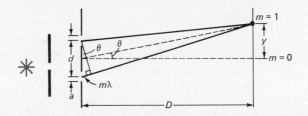

fronts, or beams of light, from the two slits interfere just as two waves on a clothesline interfere. At points where the beams (or wave fronts) are in phase, they add constructively; at points where they are 180° out of phase, they cancel out destructively. Thus, at the center point of the screen, marked 0, the light is extra bright because both beams have traveled the same distance to that point and are therefore in phase. This patch of light is known as the zeroth-order bright interference fringe.

Moving away from this point, either up or down in the diagram, one passes points at which the two beams are increasingly out of phase, until a certain spot is reached at which the light from one slit has traveled a distance equal to half a wavelength further than the light from the other slit. Here, the light from the two slits is 180° out of phase, and the two waves cancel out each other, leaving darkness on the screen. This is the first-order dark interference fringe.

Moving further along the screen in the same direction, one reaches a point at which the difference between the distance traveled by the two beams is exactly 1 wavelength. Here the two beams are once more in phase, and another bright fringe results, this one the first-order bright fringe. Moving further from the center, one passes another dark area (the second-order dark fringe), then another bright one, and so on.

Looking at the situation in Fig. 18.8 from the strictly mathematical point of view, we must note first that the figure is out of proportion. The separation of the slits d and the widths of the slits, which we can regard as infinitesimal for now, are actually far smaller in comparison with the distance from the double slits to the screen than the diagram shows, as is the distance y between the zeroth- and first-order brightness fringes. More concisely, the angle θ in the figure is much smaller than indicated.

Figure 18.8 contains all the geometry we need to put interference of light into a mathematical context. The lines from the two slits to the screen mark the paths of the rays that are interfering. These produce bright, constructive interference fringes on the screen whenever the distance y between the center of the screen and the meeting point of the two rays is such that the difference in the distance from that point on the screen to the two slits is equal to $m\lambda$, λ being the wavelength of the light and m an integer (0, 1, 2, 3, and so on). In the case of the first bright interference fringe shown in Fig. 18.8, m has the value of zero at the center of the screen. The next bright interference fringe, called the first-order bright fringe, occurs when $m = 1$. We now join, with a line, the top slit to a point on the ray from the bottom slit, choosing the point so that the distance from the point to the screen along the ray is equal to the distance the top ray travels between the slit and the screen. Since the angle θ is very small, we construct the line we drew to make a right angle with the lower ray. Simple geometry, then, shows that this line forms a triangle with the lower ray and the line between the two slits; this triangle has an angle θ, an opposite side $m\lambda$, and hypotenuse d, as shown in the figure. Hence,

$$\sin \theta = \frac{m\lambda}{d} \quad \text{or} \quad m\lambda = d \sin \theta \tag{18.4}$$

This is the basic condition for constructive interference and is known as Young's double-slit relationship.

We can next eliminate the angle θ from Eq. (18.4). Viewing the large triangle, we see that the tangent of angle θ equals y/D. Since the angle θ is very small, this tangent can be substituted for the sine of the angle. Substituting into Eq. (18.4), then, we obtain the relationship

$$\frac{m\lambda}{d} = \frac{y}{D} \tag{18.5}$$

This relation actually applies with an error less than 10 percent for all values of θ up to 20°.

Note in this equation that the distance y from the center of the screen to the bright fringe is proportional to the wavelength of the light. Fringes, therefore, occur at different points for the different color components of white light. As it turns out, the fringes of red light spread out the greatest distance, providing evidence that red light has the longest wavelength of all visible light. This spread does not apply, however, to the zeroth-order fringe, right at the center of the screen. Here, every color has a constructive interference fringe.

An extremely practical outcome of Eq. (18.5) is that it allows any experimenter with a minimum of apparatus and a certain amount of care to measure that extremely small quantity, the wavelength of light. Cross multiplying Eq. (18.5) to obtain an expression for λ alone, we obtain

$$\lambda = \frac{dy}{mD}$$

Both y and D can be measured with a ruler, and d, the separation of the two slits, can be obtained using a microscope; m, of course, is an integer. This is a direct and absolute means of measuring the wavelength of light that requires no standard or reference. And since the wavelength of light is around the value 5×10^{-4} mm, the method measures a quantity well beyond the resolution limit of any optical microscope.

Actually, ordinary light tends to produce somewhat dim and difficult-to-observe interference fringes when used in Young's double-slit experiment. Laser light of low intensity such as that from an inexpensive helium-neon gas laser is a preferable source of light for the experiment. The principle of the experiment can be demonstrated on a large scale with longer-wavelength sound waves or water waves. For the sound waves, two small speakers should be driven from the same amplifier with a note of a single frequency; for the water waves, two coupled ripplers must be used in a ripple tank to produce an excellent demonstration of two-source interference. (In both cases reflections from walls must be suppressed.)

EXAMPLE 18.5

In a double-slit experiment, the two slits are placed 0.08 mm apart with the screen 1 m from the slits. The third-order bright fringe is displaced 2 cm from the central fringe. Find the (*a*) wavelength of the incident light and (*b*) point at which the third dark fringe appears.

Solution

(*a*) We can use Eq. (18.5) to determine λ; in this case, $y = 2.0$ cm, $D = 1$ m, $d = 0.08$ mm, $m = 3$:

$$\frac{m\lambda}{d} = \frac{y}{D}$$

$$\lambda = \frac{dy}{Dm} = \frac{(8 \times 10^{-5} \text{ m})(2 \times 10^{-2} \text{ m})}{1 \text{ m} \times 3}$$

$$= 5.33 \times 10^{-7} \text{ m} \quad \text{or } 5.33 \times 10^{-5} \text{ cm} \quad \text{or } 533 \text{ nm}$$

(b) We can find the displacement of the third dark fringe by using the following argument. The fringes are spaced equidistantly from one another beginning at a bright fringe in the center of the screen, in the sequence $m = 0$ bright, $m = 0$ dark, $m = 1$ bright, $m = 1$ dark, $m = 2$ bright, and $m = 2$ dark; $m = 2$ dark is the third dark fringe. The distance of the third-order bright fringe is 2.0 cm, which is three times the separation between bright fringes or six times the separation between bright and dark fringes. Dividing this distance by 6 yields a value of 0.33 cm between bright and dark fringes. To reach the third dark fringe, we pass through five intervals of bright and dark fringes. The third dark fringe is then located 5×0.33 cm $= 1.65$ cm from the center of the screen.

An alternate way of determining this distance is to alter Eq. (18.5) by replacing $m\lambda$, the condition for constructive interference, with $(m - \frac{1}{2})\lambda$ to get the condition for destructive interference. Note that $m = 2$ yields the second dark fringe because the first occurs at $m = 1$:

$$\frac{(m - \frac{1}{2})\lambda}{d} = \frac{y}{D}$$

$$y = \frac{D}{d}\left(m - \frac{1}{2}\right)\lambda = \frac{(1 \text{ m})(2.5)(5.33 \times 10^{-7} \text{ m})}{8 \times 10^{-5} \text{ m}}$$

$$= 1.67 \times 10^{-2} \text{ m} = 1.67 \text{ cm}$$

18.3 OTHER FORMS OF INTERFERENCE

Interference of light does not inevitably require the intricate experimental setup of double slits. In fact, interference fringes can occasionally be seen in such unlaboratory-like phenomena as oil films and soap bubbles. The basic condition for all forms of interference of light is the presence of two coherent wave trains; in most cases these are obtained by accident or design when light from a single source is divided into two separate beams, which then interfere with each other.

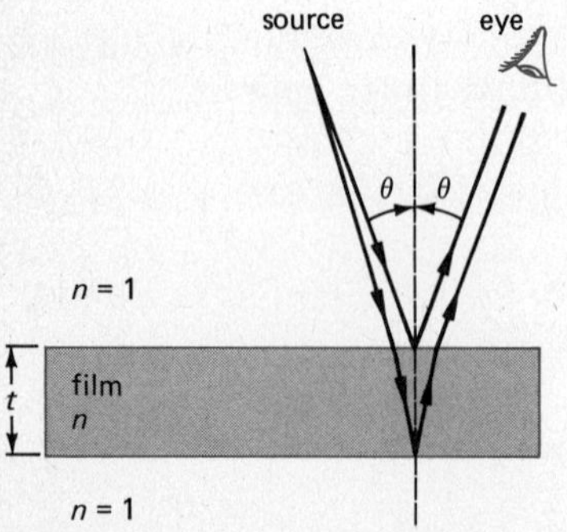

Figure 18.9 shows a magnified portion of a very thin film, such as a portion of a soap bubble or an ultrathin piece of transparent plastic or mica. The film has a refractive index n and is surrounded by air whose refractive index is 1. As we might expect, n is greater than 1; to use technical terminology, the film is *optically denser* than the air around it.

Imagine that the light which strikes the film comes from just a *small* part of an extended monochromatic (single wavelength) light source and is thus relatively coherent. When the light reaches the film, part of it is reflected back into the air, and another part continues on into the film, being bent in the process by refraction. This part of the light then reaches the back surface of the film, where it again splits up. We are interested in just the reflected waves here, which return to emerge from the front surface and mix with the part of the light that was originally reflected from that surface, as shown in Fig. 18.9. The mixing inevitably

Figure 18.9 Interference of reflected light from a thin film bounded by air.

results in interference between the two wave trains; if the film is sufficiently thin, the interference fringes will be clearly visible.

What are the conditions for constructive interference in this case? If we imagine that the angle θ in Fig. 18.9 is close to zero, which means that the light strikes the film at essentially normal incidence, then the requirement for constructive interference seems to be a remarkably simple one—the path difference between the two wave trains ($2t$, if t is the thickness of the film) should equal an integral number of wavelengths. Unfortunately, the situation is not quite that simple. First, we must remember that we are dealing with two different wavelengths of light: the normal wavelength λ in air and the slightly different wavelength in the medium, which we can label λ_n. Furthermore, we must look back to Chap. 12 to recall what actually happens to the phase of waves that are reflected and refracted, because these two effects have a critical influence on the nature of the interference.

Page 349, Chap. 12, reveals that a wave reflected at a boundary with a medium denser than the one in which it is traveling changes phase by 180° as a result of the reflection. This phenomenon is as valid in the case of light waves as it is for mechanical waves; thus, the part of the beam in Fig. 18.9 that is reflected at the upper surface of the thin film is exactly out of phase with the incoming beam. The phase of the other beam—the part of the original that is refracted into the film, reflected by the bottom surface, and then refracted out to the air—does not change. Refraction has no effect on the phase of a wave train, and reflection inside the thin film, at the boundary between the more optically dense film and the less optically dense air, is equivalent to a wave train reflecting at the free end of a rope—an operation that also leaves the phase of the wave train undisturbed. What is the net effect of the change and lack of change in phases of the two parts of the beam? Plainly, if we assume that the path difference, that is, twice the width of the film, or $2t$, is an integral number of wavelengths of the light *in the film,* then the refracted and reflected beams will be exactly out of phase, rather than in phase as would be the case if reflection did not produce a phase change. Hence, a path difference of an integral number of wavelengths causes destructive rather than constructive interference. To compensate for the phase change, the condition for constructive interference becomes

$$2t = (m + \tfrac{1}{2})\lambda_n$$

where $m = 0, 1, 2, 3$, and so on, and, of course, λ_n is the wavelength in the film. The simple formula that applies to destructive interference in this case is

$$2t = m\lambda_n$$

We generally refer to the wavelengths of light in air rather than in other media such as glass. It is therefore convenient to write the equation above in terms of the wavelength of the light in air. Using

> Normal incidence means arrival perpendicular to the reflecting or refracting surface.

Eq. (18.3), $\lambda_n = \lambda/n$, our conditions for constructive and destructive interference of light shone on the thin film whose thickness is t become

$$2t = \frac{\lambda}{n}(m + \tfrac{1}{2}) \quad \text{constructive interference}$$

$$2t = \frac{m\lambda}{n} \quad \text{destructive interference}$$

In practice, the interference fringes are not spaced equally in accordance with these equations because these conditions apply strictly only to light that impinges on the thin film at angles of about 90°. When the angle is different from 90°, the path difference, that is, the distance traveled in the film by the refracted beam, becomes rather more than $2t$. The overall result, as one can see in a very thin layer of gasoline spilled at a gas station or in an ultrathin piece of plastic foil, is a series of interference fringes alternately bright and dark. And since the position of fringes depends critically on the wavelength of the light involved, it is clear that the positions of the bright and dark circles will be different for different colors. Hence white light that is reflected and refracted from a thin film produces colored fringes, owing to the difference in the wavelengths of the colors that make it up.

A practical example of the use of interference between beams of light reflected from the front and back surfaces of a thin film is termed a "nonreflecting coating," which is used to prevent reflection of a single wavelength of light from the front surface of a lens. This situation, shown in Fig. 18.10, differs from that of our previous example insofar as the light passes from air to a film whose index of refraction is greater than that of air but less than that of the glass which completes the sandwich. Since the index of refraction of lens glass is normally about 1.5, that of the film must be between 1 and 1.5.

The object of the exercise is to prevent reflection from the coating and the lens by causing the waves reflected from the top and bottom of the coating to interfere destructively. And since both rays reflect at a surface between a less dense and a denser medium, the effects of phase change on the reflected rays cancel out. If we consider wave trains that come in at normal incidence, then the path difference between the two waves must be just $\tfrac{1}{2}$ wavelength. Thus, if t is the thickness of the film, the condition for the nonreflectivity that we seek is

$$2t = \frac{\lambda_n}{2} \quad \text{or} \quad t = \frac{\lambda_n}{4}$$

A film whose thickness is $\tfrac{1}{4}$ wavelength therefore prevents any reflection of light of that wavelength, resulting in high transmission of that particular wavelength. Binocular lenses are often coated this way.

Another type of interference effect is that observed but not pursued by Newton, namely, Newton's rings. The interference occurs when light

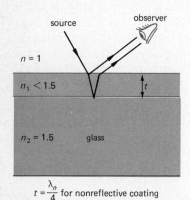

Figure 18.10 A nonreflective coating. Light reflects from a film placed upon a more optically dense medium.

Figure 18.11 Newton's rings.

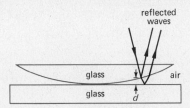

passes between a wedge of air that is sandwiched between two plates of glass. Figure 18.11 illustrates one appropriate arrangement; the air is sandwiched between a glass lens and a flat glass plate. This air wedge, which is extremely thin, provides top and bottom surfaces from which beams of light can reflect and interfere with each other. The condition for constructive interference is exactly the same as that involving a thin film bounded on both sides by air. In this case, however, the light reflected at the top of the air wedge undergoes no phase change because the boundary links a denser medium with a less dense one, and the waves reflected from the bottom of the wedge are changed in phase by 180°. Thus, if the thickness of the wedge is d and the wavelength of the light is λ, the condition for constructive interference of light that comes into the system at roughly normal incidence is

$$2d = (m + \tfrac{1}{2})\lambda$$

Destructive interference occurs between the center of the lens and the point at which the thickness of the air wedge is less than $\tfrac{1}{2}$ wavelength, producing a dark spot in the center of the lens. Then there is a bright ring, another dark ring, and so on. The overall effect is shown in Fig. 18.12. The distance between adjacent dark or bright fringes decreases with distance from the center of the lens.

Newton's rings are used in the laboratory to examine lenses and to find out how flat an unknown surface really is. An experimenter places an optical flat—a piece of glass that has been carefully ground and polished to be really flat—over the unknown surface of metal, glass, or other solid material and then observes the interference fringes. If the fringes are uniform or nonexistent, the upper surface of the object under investigation is completely plane. If the fringes are wavy, the surface is not properly flat. The interference pattern indicates how the thickness of the air wedge between the optical flat and the surface under study varies across the surface. A pit in the surface yields a series of fringes that follows the contour of the pit. This type of technique enables experts to grind surfaces accurately to within fractions of a wavelength of light.

Another illustration of the use of interference occurs in the device known as the Michelson interferometer. The device, named after the nineteenth century American physicist Albert A. Michelson, is useful for the calibration of a centimeter or millimeter scale in terms of the

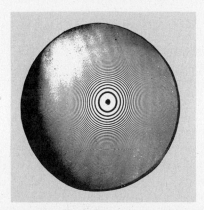

Figure 18.12 Newton's rings formed by interference in the air film between a convex and a plane surface. (*Bausch and Lomb*)

570 The Nature of Light

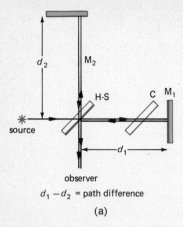

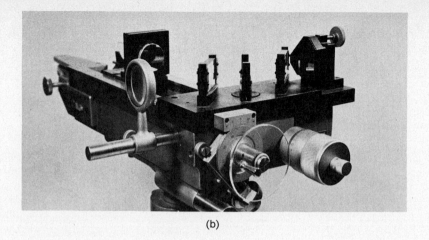

Figure 18.13 Michelson interferometer. (a) Schematic diagram; M_1 and M_2 are plane mirrors; C is a compensating plate; H-S is a half silvered mirror. (b) Photograph of a modern Michelson interferometer. (*Trudy Tuttle*)

Monochromatic means literally single-colored, and thus of a precise, single wavelength.

wavelength of light, as opposed to some less accurate length standard. It is sketched schematically in Fig. 18.13(a).

In action, light from a monochromatic source, such as a sodium vapor or mercury lamp, reaches a half-silvered mirror (H-S), which splits it into two separate beams. The first beam continues on through another glass plate (C) to a plane mirror M_1, from which it is reflected back through the glass plate to the back side of the half-silvered mirror, which reflects it to the observer's eye. The second half of the original beam is reflected from the half-silvered glass plate to the mirror M_2 and returns to be recombined with the first beam in the eye of the observer. The glass plate on the route to mirror M_1 has the same width as that of the half-silvered mirror and is placed parallel to it; it therefore compensates for the distance that the second beam travels inside the latter piece of glass.

Obviously the two halves of the beam interfere; the type of interference depends on the path difference between their routes. If the distances from the silvering of the half-silvered mirror to the two mirrors M_1 and M_2 are exactly equal, the two parts will interfere constructively because the beams will be exactly in phase after their separate round trips. If either mirror is moved $\frac{1}{2}$ wavelength, constructive interference will occur once more because the movement adds a full wavelength to the path of the beam that reflects from it. Thus, if one fixes the position of mirror M_2 and moves M_1 slowly on a screw drive, one can count the number of half wavelengths it moves simply by counting the number of fringes passing by.

Michelson invented this device, a modern version of which is shown in Fig. 18.13(b), to search for something more esoteric than interference fringes. He and his compatriot Edward Morley tried to detect the ether, the mysterious medium that Victorian scientists regarded as permeating all of space. Their failure to find it set the stage for Einstein's special theory of relativity, as we shall see in Chap. 20.

EXAMPLE 18.6

Two rectangular pieces of plate glass are placed, one upon the other, with a thin strip of foil put between them at one edge, forming a very thin, triangular air wedge. Light of wavelength $\lambda = 580$ nm illuminates the plates at normal incidence. Bright and dark bands or fringes are formed, with 10 of each per centimeter of length of wedge measured normal to the edges in contact. Find the angle of the wedge.

Solution

The situation is illustrated in the diagram in Fig. 18.14. A dark band occurs near the point at which the two glass plates touch because the light reflecting from the top of the air wedge experiences no phase shift from the incident light and light at the bottom of the air wedge does experience a phase shift of 180°. In the former case, the light reflects from a boundary between material of higher index of refraction and of low index; in the latter the situation is the reverse. There is no physical path difference at the edge where the two plates meet because the air wedge has no thickness. In Fig. 18.14 the first 10 dark bands are drawn; actually there are 11 if you count the $m = 0$ dark band at the point where the two pieces of glass meet. These dark bands out to $m = 10$ include the first 10 bright bands, and according to the statement of the problem, cover a distance of 1 cm. Since the reflected waves from top and bottom of the air wedge are out of phase by 180°, the condition for a dark band is that the thickness of the air wedge t must be such that $2t = m\lambda$, where $m = 0, 1, 2, 3, \ldots$. For the tenth band, $m = 10$, and the thickness at the tenth dark band is given by

$$t = \frac{m\lambda}{2} = \frac{(10)(580 \text{ nm})(10^{-9} \text{ m/nm})}{2}$$

$$= 2.90 \times 10^{-6} \text{ m} = 2.90 \times 10^{-4} \text{ cm}$$

The angle of the wedge is given by

$$\tan \theta = \frac{t}{1 \text{ cm}} = \frac{2.9 \times 10^{-4} \text{ cm}}{1 \text{ cm}} = 2.9 \times 10^{-4}$$

At such a small angle the tangent is approximately equal to the angle itself in radians.

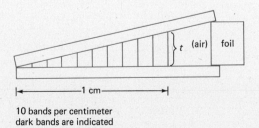

Figure 18.14 An "air wedge."

10 bands per centimeter
dark bands are indicated

EXAMPLE 18.7

An oil film 0.40 μm thick is illuminated with white light normal to the film. The index of refraction of the oil is 1.50. What wavelengths within the limits of the visible spectrum ($\lambda = 4.0 \times 10^{-7}$ m to $\lambda = 7.0 \times 10^{-7}$ m) will be strongly reflected?

Solution

From our discussion of reflection of light from a soap film, we found that the condition for constructive interference in the reflected light is

$$2t = (m + \tfrac{1}{2})\lambda_n$$

or

$$2t = \frac{\lambda}{n}(m + \tfrac{1}{2})$$

We can determine which wavelengths are most strongly reflected by solving the equation for wavelength as m varies from 0 to 1 to 2, and so on:

$$\lambda = \frac{2tn}{m + \tfrac{1}{2}} = \frac{(2)(0.4\,\mu\text{m})(10^{-6}\,\text{m}/\mu\text{m})(1.50)}{m + \tfrac{1}{2}}$$

$$= \frac{1.2 \times 10^{-6}\,\text{m}}{m + \tfrac{1}{2}}$$

For $m = 0$, $\lambda = (1.2/\tfrac{1}{2}) \times 10^{-6} = 2.4 \times 10^{-6}$, or 24×10^{-7} m. Since this wavelength is beyond our range of interest, m will have to be larger. For $m = 1$, $\lambda = (1.2/\tfrac{3}{2}) \times 10^{-6} = 8 \times 10^{-7}$ m, which is in the infrared region, still too large. For $m = 2$; $\lambda = (1.2/\tfrac{5}{2}) \times 10^{-6}$, or 4.80×10^{-7} m. This will, in fact, be the only wavelength in the visible region strongly reflected, since for $m = 3$, $\lambda = 3.4 \times 10^{-7}$ m, which is in the ultraviolet region of the specrum. Hence, the film will be predominantly blue in color.

18.4 DIFFRACTION: INTERFERENCE AT THE EDGES

When light or any other type of wave motion passes through an opening or simply goes by the edge of any obstacle, the wave bends into the region not directly exposed to the wave front. This phenomenon is called **diffraction**. We already encountered diffraction effects. Young's double-slit experiment would not have worked if the light passing through the small slits in the experiment had not been bent around the edges of the slits. Once the light was bent by diffraction it was possible for the beams from the two slits to interfere. For the moment, consider light or some other type of wave that passes through a single slit.

Figure 18.15 shows the effect for surface water waves passing through a slit in a ripple tank. It is difficult but not impossible to observe diffraction with light in normal life. The simplest means of

Interference and diffraction of this sort can also be observed when viewing a distant street light through a window screen or a fabric curtain.

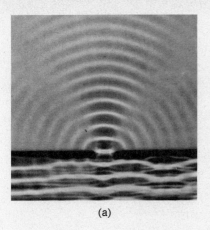
(a)

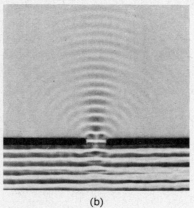
(b)

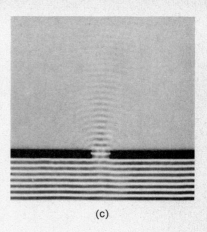
(c)

Figure 18.15 Three views of waves passing through the same opening. Note the decrease in bending of the shorter wavelengths. The wavelength is longest in the left picture and shortest in the right picture. (*From PSSC Physics, D. C. Heath & Company, Lexington MA, 1965*)

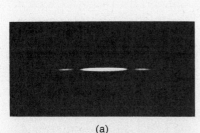
(a)

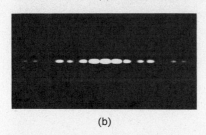
(b)

Figure 18.16 (a) Diffraction pattern of a single slit produced by a helium-neon laser. (b) Interference bands or fringes produced by the same laser source passing through two slits closely spaced. These slits each have a width similar to that of the single slit used in (a). Note that the double-slit pattern is superimposed on the single-slit pattern. (*Trudy Tuttle*)

seeing it is to look at a distant street lamp through the narrow opening between two adjacent fingers. The light appears somewhat blurred, with a faint pattern of light and dark fringes, as shown in Fig. 18.16(a). The pattern alters as you change the distance between your fingers. If this "slit" becomes too wide, the diffraction effect seems to disappear; if you bring your fingers so close together as to make the opening between them all but nonexistent, the central band of light in the pattern of the figure becomes very wide. Figure 18.16(a) actually shows the diffraction pattern of light of a single wavelength; if you look at white light through the finger slit, you will see that light splits up into its constituent colors.

Diffraction actually involves two separate effects: the bending of the light as it passes an obstacle or goes through a slit and the interference between separate waves after they have been bent. We can use Huygens' principle to understand the diffraction pattern of Fig. 18.17. This type of diffraction was named after the German scientist Joseph Fraunhofer, an optical expert of the early nineteenth century. It involves a plane wave front that encounters a slit of width w and is then displayed on a screen a relatively long distance from the single slit. We can analyze the situation mathematically by setting up a Huygens construction similar to the one we used in piecing together Young's double-slit experiment. As the wave front passes through the slit—a single one this time—we regard points along the wave front as acting as the sources for a number of new wavelets, all in phase, which spread out in all possible directions. Figure 18.17 shows a greatly simplified view of this effect, including only those portions of wave fronts that travel away at an angle θ to the original direction of the wave front.

The slit includes nine separate new rays, and the angle θ is chosen to ensure that the middle ray, number 5, is exactly half a wavelength behind the one that emerges from the bottom of the slit, number 1, and half a wavelength ahead of the ray whose source is at the top of the slit, ray 9. Plainly, rays 1 and 5 are 180° out of phase in this direction, and

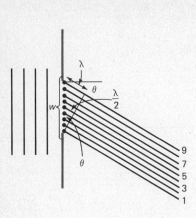
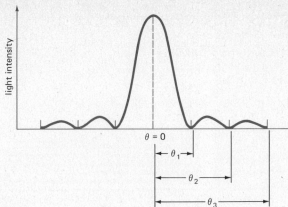

(a) geometric construction

(b) intensity pattern on a screen from a single slit
$\sin \theta_1 = \frac{\lambda}{w}$, $\sin \theta_2 = \frac{2\lambda}{w}$, $\sin \theta_3 = \frac{3\lambda}{w}$

Figure 18.17 Diffraction by a single slit.

therefore cancel out each other. Similarly, ray 2 cancels out ray 6, ray 3 cancels out ray 7, and ray 4 cancels out ray 8, and what remains is only ray 9. If we consider that the rays are actually infinitesimal in intensity and infinite in number, then including more rays in a similar analysis will still leave us with a single ray like number 9 in our finite example. Hence what results at angle θ is minimum brightness or "darkness." The angle θ thus defines the direction in which a dark band occurs in the diffraction pattern. This angle is defined by the equation

$$\sin \theta = \frac{\lambda}{w} \tag{18.6}$$

This is the smallest angle at which destructive interference occurs. The values of θ on either side of the normal to the slit define the angular width of the bright band of constructive interference that occurs at the center of any diffraction pattern, as shown in Fig. 18.17. The other points of total darkness in this interference pattern, which occur symmetrically around the central bright band, are defined by the general equation

$$\sin \theta = \frac{m\lambda}{w}$$

where m is any integer.

One important point to note about the series of equations which this expression represents is that the angular width of the central bright band in a diffraction pattern depends critically on the ratio λ/w. When the width of the slit is much greater than the wavelength of light that passes through it, θ is very small. When w is made smaller, however, the angle increases until the point at which the two quantities are identical is reached. At this stage, $\sin \theta$ is 1, and θ is 90°; this means that the central bright band extends across the screen from one side to the other.

Figure 18.18 Resolution of a diffraction grating increases with N, the number of slits.

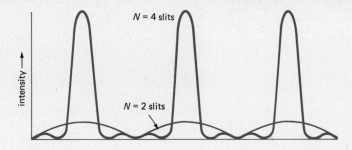

Recall that we neglected the widths of the slits in our original discussion of Young's double-slit experiment. In fact, this would have been justified only for slits which are narrow compared to the wavelength; in general the width of the slits can and does affect the interference pattern that results from the two slits. The rays of light from the two slits that interfere on the screen are not what we might term "pure" rays, but diffracted rays, which produce their own dark and bright fringes. Thus, the overall interference pattern, shown in Fig. 18.16(b), is a mixture of the diffraction patterns of two separate rays and the interference pattern between two rays. There might, for example, be certain points at which one would expect a bright interference fringe, but if those points coincide with dark diffraction bands from both single slits, no interference fringes will be visible because there will be no light at those points to form it.

Another look at Fig. 18.16(b) shows that a double-slit arrangement produces clearly defined Young's double-slit fringes superimposed upon a single-slit diffraction pattern. This definition of the double-slit fringes improves even further with the addition of more slits. This fact is applied in a piece of laboratory equipment known as the diffraction grating, which is used for a variety of optical measurements. Diffraction gratings consist of many hundreds, or even thousands, of identical slits separated from each other by the same distance. They are generally made by a special machine that mechanically rules (cuts) grooves in a glass or metal surface rather than cutting physical slits of the type used by Young. The grooves are generally referred to as "lines." The multiple slit accomplishes two purposes: it increases the intensity of the light that reaches the screen, in comparison with that emerging from a simple double-slit arrangement, and it greatly improves the resolution of the interference pattern, bringing both dark and bright fringes into sharper perspective, as shown in Fig. 18.18. Note that light reflected from a diffraction grating also produces a diffraction pattern.

The condition for constructive interference involving a diffraction grating (see Fig. 18.19) is just the same as that for a double slit, that is,

$$\sin \theta = \frac{m\lambda}{d}$$

Figure 18.19 A helium-neon laser is used here to show the pattern resulting after the beam of light passes through a diffraction grating. (*Trudy Tuttle*)

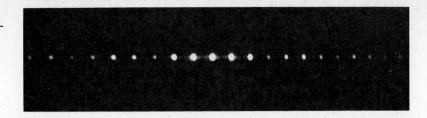

where d is the separation of adjacent slits on the grating. The first-order bright fringe occurs when m is 1, and the angle θ varies with the wavelength of the light passing through the grating. Thus, red light, which has a longer wavelength, forms a bright fringe at a larger angle than blue light for any particular value of the integer m. This separation is so obvious that one can use it to measure the wavelengths of the individual components of white light with remarkable accuracy.

Of course, the spectrum of white light repeats for each order in the interference pattern, that is, for increasing values of m. The dispersion of the different wavelengths actually increases as m increases; the second-order pattern, for example, is twice as wide as the first-order one. But the higher-order spectra produced by diffraction gratings also tend to overlap, producing some confusing effects. For example, light of two different wavelengths from different orders of the pattern may arrive at the same point on the screen and produce constructive interference. A little thought and mathematics will show that this situation must occur when light of wavelength λ in the second-order spectrum is reinforced at the same angle as light with a wavelength $\frac{2}{3}\lambda$ in the third order.

EXAMPLE 18.8

A diffraction grating with 10 000 lines per centimeter is illuminated by yellow light of wavelength 589 nm (5.89×10^{-9} m). At what angles are the first- and second-order bright fringes seen?

Solution

We can use the grating equation $m\lambda = d \sin \theta$ to find the angles if we know the slit separation d. Since there are 10 000 slits per centimeter, the separation between pairs of slits is $\frac{1}{10\,000}$ cm. For first order,

$$\sin \theta = \frac{m\lambda}{d} = \frac{(1)(589 \times 10^{-9} \text{ m})}{\frac{1}{10\,000} \text{ cm} \times \frac{1}{100} \text{ m/cm}} = 0.589$$

$$\theta = 36.1°$$

For second order,

$$\sin \theta = \frac{m\lambda}{d} = (2)(0.589)$$

This number is larger than 1, and so θ does not represent any rational angle. Hence, only the first order appears for this wavelength.

EXAMPLE 18.9

Show that the violet ($\lambda = 400$ nm) of the third-order spectrum overlaps the red ($\lambda = 700$ nm) of the second-order spectrum.

Solution

Using the grating equation, we solve for the angle θ for the violet in the $m = 3$ order and for the red in the $m = 2$ order:

$$\sin \theta_{viol} = \frac{m\lambda}{d} = \frac{(3)(4 \times 10^{-7} \text{ m})}{d} = \frac{1.2 \times 10^{-6}}{d}$$

$$\sin \theta_{red} = \frac{m\lambda}{d} = \frac{(2)(7 \times 10^{-7} \text{ m})}{d} = \frac{1.4 \times 10^{-6}}{d}$$

The sine of the angle of the second-order red is larger than that of the third-order violet, regardless of the value of d.

18.5 POLARIZATION

Polarization was discovered in Newton's time by the Danish scientist Erasmus Bartholinus. He noted that crystals of calcite (Iceland spar) would split a ray of light into two rays as the light passed through the crystal. Not until nearly 1820 did Thomas Young and Augustin Fresnel show that this phenomenon, called polarization, could readily be accounted for if light behaved like transverse waves.

A simple mechanical demonstration of polarization is shown in Fig. 18.20. An experimenter sets up a wave in a clothesline that passes through a "vertical grate"—the back of a common chair. If the experimenter waves the rope in a horizontal plane, as shown in Fig. 18.20, or in a diagonal plane, the wave motion will change abruptly as it reaches the chair back. Only one form of wave motion will pass through the chair unscathed and unaltered, namely, the vertical wave shown in Fig. 18.20. This wave is said to be polarized in the vertical plane. If the expert somehow or other managed to induce a number of waves in different planes in the rope, all but vertical vibrations would be stopped by the chair back (see Fig. 18.21). This barrier would act as a polarizing device for the waves. Obviously, polarization can apply only to transverse waves for longitudinal waves do not have anything oscillating at right angles to their directions of travel and no directional preference or property other than the direction of propagation.

Light is a form of wave motion whose constituents wave in many different planes, and certain substances, notably transparent crystalline materials whose molecular structure is not symmetric, have the same

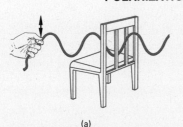

(a)

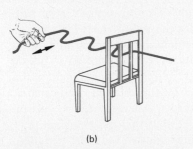

(b)

Figure 18.20 In (a), the wave is polarized in a vertical plane. In (b), the wave is polarized in a horizontal plane. Only wave oscillations in a vertical plane (a) can pass through the rungs of the chair back.

Figure 18.21 (a) Unpolarized wave on a rope oscillating in any plane since the opening allows any oscillation to pass. (b) A plane-polarized wave on a rope; this occurs because only vertical oscillations can pass through the slit.

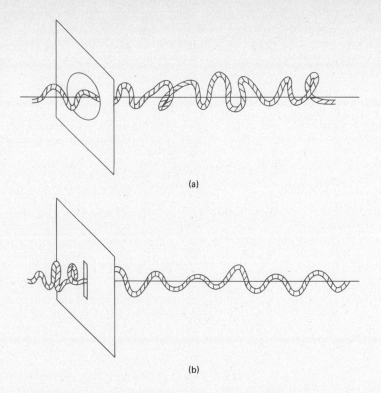

effect on light waves as does the chair back on a waving clothesline. When unpolarized, ordinary light passes through a polarizer such as a Polaroid sheet, only the components oscillating parallel to the plane of oscillation of the sheet are transmitted.

Iceland spar, the earliest material that was identified as having this curious property, enabled scientists to study polarization. However, the practical importance of polarization effects was not realized fully until the invention in 1928 by Edwin H. Land of an inexpensive polarizing plastic sheet called Polaroid. It was not the plastic sheet that accomplished the polarization of light but needlelike crystals of herapathite, a sulfate of iodoquinine. These crystals were imbedded in a plastic material, after which the plastic was stressed to cause alignment of the crystals. Land, who is now better known for the Polaroid Land camera, later improved his process and replaced the herapathite crystals by other assymmetric microcrystalline materials that polarize light.

Such materials are used in polarizing sunglasses that are designed, according to their advertisements, to "reduce glare." That such glasses do contain a polarizing material can easily be checked with a piece of guaranteed polarizing material. Put on the glasses and then hold the polarizing material in front of the glasses. Rotate the polarizing sheet in front of the glasses. As you do so, you will notice that the light reaching your eye alternately brightens and dims with maximums of brightness and darkness occurring every 90°. You could, of course, use another pair of polarizing sunglasses as your test sheet in front of your glasses. Or you could even demonstrate the effect with a broken pair of glasses, holding one lens in front of the other.

What is happening in this experiment? Normal light is unpolarized, with the light waves oscillating in any and all planes. When the light passes through the polarizing sheet, only the components of the electric field vector oscillating in a specific plane are allowed by the crystals to pass through. The transmitted light is plane polarized. Now, when the plane polarized light falls upon the second polarizing sheet—your glasses, say—the amount of light transmitted will depend upon the orientation of the second polarizing sheet. When the two sheets of polarizing crystals are oriented the same way, maximum transmission occurs. However, when one sheet is rotated by 90° from this situation of maximum transmission, the polarizing plane (plane of oscillation) of one sheet is at right angles to the other and essentially no light passes through the second sheet. In this latter orientation, the polarizing sheets are said to be "crossed."

Now that we have some feel for the polarizing phenomenon, let us consider the questions "Why do sunglasses containing a polarizing material reduce glare?" and "How are they different from tinted, darkened glasses that do no more than reduce the intensity of any light as it passes through the glasses?" Reflected light from the road or from puddles of water in the road is generally polarized with the plane of oscillation parallel to the reflecting surface. The polarizing sunglasses are thus designed to allow maximum transmission for light polarized perpendicular to that plane. The effect is that the reflections from the road, which are largely responsible for glare, are greatly reduced, and light reflected from other objects is only slightly reduced.

Blue sky light consists of sunlight that has been reflected, or "scattered," by gas molecules in the upper reaches of the atmosphere, and because of this reflection it is partially polarized. The magnitude and orientation of the polarization of the sky's light depend upon the position of particular points in the sky relative to the sun. The minimum polarization occurs near the sun. That sky light is polarized is apparently used by bees for navigational purposes. Austrian Nobel laureate Karl Von Frisch and his collaborators concluded after many years of experiments that bees are able to compensate for the sun moving across the sky during the day and measure angles based upon the detection of the polarized scattered light from the blue sky. If a polarizing sheet is placed between the blue sky and the bees, the bees will orient themselves with respect to the polarized light they receive from the sheet.

There are many practical applications in science and everyday life for polarized light. For example, placing crystals between a pair of polarizing sheets often allows you to detect whether you have a single crystal or several crystals, that is, a polycrystalline material. An inter-

Figure 18.22 Rotation of the plane of polarization by an optically active material. The angle θ indicates the degree of rotation of the plane of polarization of the light.

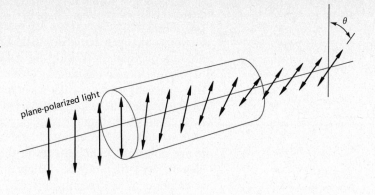

esting children's toy consists of a light bulb in a box with a top made of polarizing material. Crumpled sheets of cellophane or various layers of cellophane tape are placed over the polarizing material, and the light transmitted is polarized depending on the thickness of the cellophane tape. If a second polarizing sheet is placed over the tape and rotated, a spectacular array of changing colors, produced by rotation of the plane of polarization of the light as it passes through the cellophane, is clearly visible.

Polarization is also used to measure the concentrations of solutions of a group of compounds called "optically active chemicals." If a beam of plane-polarized light is passed through a tube containing an optically active compound in solution, the direction of the plane of oscillaton rotates as the light passes through the solution, as shown in Fig. 18.22. The rotation may be right- or left-handed, that is, clockwise or counterclockwise, as viewed looking toward the source. The amount of rotation depends on the concentration and distance traveled by the light through the solution. Sugar is optically active, and sugar solutions are often analyzed using this technique in an instrument called a polarimeter. For a fixed-length sample cell, the angle of rotation, θ in Fig. 18.22, is calibrated for known sugar concentrations; then unknown concentrations can be determined. Two sugars, *dextrose,* which produces a right-hand rotation, and *levulose,* which produces a left-hand rotation, drive part of their names from their optical activity.

There are many other important uses for polarized light and polarizing materials, including the identification of mechanical strains in glass and plastics and three-D glasses. Perhaps the student might look into these most interesting phenomena on his or her own.

18.6 HOLOGRAPHY: PHOTOGRAPHS WITHOUT LENSES

One of the most dramatic and exciting examples of the interference of light is the type of photography known as holography. Developed in the early 1960s, this field of physics seems to the uninitiated to be almost beyond belief. It allows one to produce a three-dimensional

Figure 18.23 Construction of a hologram.

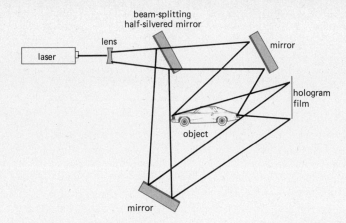

image using a two-dimensional film without using the lenses that are the bases of any normal camera. The three-dimensional result is so effective that one can peek around individual objects in the image and see other objects behind them, objects that are hidden from view at first glance. And every small part of a holographic film contains all the information necessary to reproduce the whole three-dimensional scene that has been recorded.

The growth of holography followed the development of the laser in 1960, but the concept was actually thought out in 1948 by the incredibly inventive physicist Dennis Gabor, who fled from his native Hungary to Great Britain in the 1930s. Gabor's basic system for producing holograms is shown in Fig. 18.23.

The key to the system is a strongly coherent light source, which is nearly always a laser beam. (Gabor's original experiment with mercury lamps produced extremely fuzzy images.) The thin beam is spread out when it passes through a lens—a necessary step because the whole of the object being photographed must be illuminated. Then the beam is split into two components. One component is directed toward the photographic film by means of a plane mirror; the other is used to illuminate the object. Some of the reflected light from the object hits the photographic film, there to interfere with the first part of the beam, which is known as the reference beam. The result is not in any way a normal photograph. Instead, development produces a series of whorls that appear like fingerprints. They represent the interference fringes produced by the interference of the two parts of the beam. The film is known as a hologram.

To produce a three-dimensional image of the original scene, one need only illuminate this hologram with coherent monochromatic light, that is, a coherent beam of a single wavelength as shown in Fig. 18.24. The image produced is known as a virtual image. As shown in the figure, the virtual image cannot be captured on a screen but can be viewed by peering through the hologram. The three-dimensional effect

Figure 18.24 Re-creation of a holographic image.

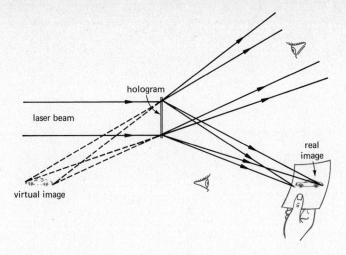

that the image gives the first-time observer is truly breathtaking.

It is the absence of an image-forming lens to focus the light from the object onto the film that produces the all-or-nothing effect of holography. Each point on the object reflects light to the entire photographic plate, and thus every part of the plate is exposed to light—and hence to the resulting interference pattern—from every piece of the object. Thus, when one cuts a hologram into pieces, each piece contains all the information necessary to produce the three-dimensional picture, although the picture from very small pieces of hologram may be somewhat blurry. Blurriness is the penalty of another avoidable error in making holograms. The holographer must take extreme care to avoid any vibration of the object or the apparatus for even motions as small as a fraction of a wavelength can blur the recorded interference pattern beyond recognition.

As a practical matter, holography is currently used to store information, which is recorded in a coded form in different positions on the holographic image and in different wavelengths. A typical hologram the size of a postcard can potentially store hundreds of thousands of bits of information, a bit being the basic unit of information used in computer programming.

In a more futuristic vein, inventors such as Dennis Gabor—who died in 1978—have suggested the use of holography to produce decorative picture windows for apartments in grimy cities, three-dimensional movies, and even three-dimensional television. Physicists are also intrigued by the possibility of using x-rays to produce holograms. If such holograms were viewed in visible light, whose wavelength is much greater than that of x-rays, their contents would be greatly magnified. Thus, scientists could view such normally unseeable objects as atoms in crystals or molecules making up biologic material.

Holography in Action

When the science of holography was first perfected in the early 1960s, it was viewed as little more than a laboratory curiosity—a form of three-dimensional photography without lenses that amazed spectators at physics exhibitions and illustrated basic physical principles but offered little to the practical technologist. But as engineers began to realize the scope of the technology, this attitude started to change. Today, holograms play vital roles in many materials testing and quality control laboratories, and in time they promise to make unique contributions to the world of the arts.

A major advantage of holography is the speed with which it can capture events. Holographers using ultrafast pulses of laser light have been able to photograph in minute detail exploding rocket fuel, fog particles, and similar elusive phenomena. Holograms are also capable of storing great quantities of information, particularly if various frequencies of light are used to produce a single holographic plate. Dennis Gabor, who invented the process of holography, estimated that the contents of an entire encyclopedia can be printed on a single hologram that measures no more than 8 × 11 in.

The most widespread application of holograms, however, has been in component testing. If one takes two holograms of any particular object at different times and superimposes the two, any changes in the object between the exposures show up as interference fringes. Thus, investigators are able to check such hard-to-spot effects as the bonding between the layers of rubber in automobile tires by viewing superimposed holograms of the tire surface taken before and after the tire is inflated.

The future is even more promising for holography. Researchers at the University of Michigan have developed holographic reels, rather like continuous film strips, that can show art objects, old musical instruments, and similarly rare pieces to interested students. In time, experts believe that three-dimensional holographic movies may become possible. Other physicists are trying to develop microwave holograms that could, in theory, photograph objects in complete darkness, and medical researchers have used holographic interfering sound waves to study heartbeats and skeletal structures of humans and animals.

Hologram shows interference fringes produced during tire tests. (*TRW*)

SUMMARY

1. Two theories have been advanced to account for the nature of light: the corpuscular theory, which views light as consisting of tiny particles, and the wave theory, which regards it as a form of wave motion. Physicists today envision light as consisting of both waves and particles.
2. Reflection is the mirror effect. Refraction is the bending of light when it passes from one medium to another.
3. Light travels with the very fast but finite speed of about 3×10^8 m/s in vacuo, as do all forms of electromagnetic radiation. It moves more slowly in other media.
4. According to Huygens' principle, every single point on a wave front

can be regarded as a new source of spherical waves.
5. The refractive index of any material is the ratio of the speed of light in a vacuum to its speed in the material.
6. Because of their wavelike nature, light beams undergo interference. This is shown in the form of dark and bright fringes of light produced by Young's double-slit experiment.
7. Interference of light is also observed in oil films, soap bubbles, and thin wedges of air sandwiched between glass.
8. Individual light beams can interfere with themselves in the process known as diffraction. The intensity distribution in a light beam passing an obstacle involves two effects: the bending of light (diffraction) as it passes an obstacle and the interference between separate waves after they have been bent.
9. A fundamental characteristic of transverse waves is their ability to be plane-polarized; the oscillations are restricted to a single plane or oscillation.
10. A practical example of the interference of light is holography—photography without lenses. This involves interference between a beam reflected from a scene being photographed and an unreflected "reference" beam.

QUESTIONS

1. What conditions are required of two beams of light so that they continue to produce an interference minimum?
2. Under what conditions is it possible to produce interference fringes with white light using a Michelson interferometer?
3. In a single-slit diffraction pattern, on what does the width of the central bright maximum depend?
4. Consider Young's double-slit experiment and explain what the following parameters have to do with the light distribution on the screen: (*a*) distance between the slits, (*b*) width of slits, (*c*) wavelength of the incident light, (*d*) increasing the number of slits.
5. If Young's double-slit experiment were performed with white light containing all possible colors, what would the interference fringes look like? Where on the screen would all the light be in phase?
6. How did Newton's picture of the nature of light differ from that developed by Huygens?
7. When a beam of light travels from air to water, is it bent toward or away from the normal? What happens when light travels from water to glass?
8. Which of the following change when light is refracted: wavelength, frequency, speed?
9. When light is diffracted, which of the following change: the wavelength, frequency, speed?
10. Why was holography a rather insignificant, unimportant phenomenon until the discovery of the laser?
11. In Young's double-slit pattern, which aspects of the resulting pattern are due to diffraction? Which are due to interference effects?
12. How does the appearance of the diffraction pattern from a narrow slitlike opening change as the wavelength increases?
13. Distinguish between diffraction and interference. Can there be diffraction without interference; interference without diffraction?
14. If a soap film is made by dipping a wire loop into a soap solution and then the loop is held up vertically as the film flows downward, becoming thinner and thinner at the top, the film becomes black when viewed with reflected white light just before it breaks. Explain.
15. What does the fact that light can be polarized tell you about the nature of light?
16. What is meant by optical activity?
17. It is possible in an apartment complex to keep people in apartments across a court yard from seeing into your apartment by placing polarizing sheets over windows of the apartments. How does this work? At what angles should the sheets be oriented relative to each other? How would these sheets alter the view of the sky and courtyard? Could people outside the apartment see into an apartment window?
18. Unpolarized light is incident upon a pair of sheets of polarizing material oriented so that no light is transmitted. If a third polarizing sheet is placed between the two, will any light be transmitted?
19. With only one polarizing sheet, how could you determine the polarizing direction of the sheet?

PROBLEMS

Speed of Light

1. In measuring the speed of light by the method attempted by Galileo, (*a*) how far apart would two mountains have to be so that the light from one mountain would take 1 s to reach the other mountain peak? (*b*) How many times around the earth is this?
2. The light reaching the earth from the nearest star (Alpha Centauri) requires 4.3 yr to travel from the star to the earth. How far away is Alpha Centauri in kilometers?

3. The French physicist Cornu used a method similar to that of Fizeau to measure the speed of light. However, he replaced the toothed wheel by a disk with 200 slots in it. He found that when he increased the speed of the slotted wheel from zero to 54 000 r/min, he measured 28 eclipses and brightenings of the light. The mirror was 23 km away. Using these data, determine the speed of light. See Special Topic, p. 587.

4. We wish to measure very accurately the distance to the moon by reflecting a laser beam from a mirror on the moon's surface back to the earth. Before we place the required mirror on the moon, we should look carefully at the technical problems. For example, how far will we on the earth move owing to the earth's rotation on its axis while the laser beam we are using goes from our laser to the moon and back? The moon is 3.8×10^8 m away. Radius of the earth is 6.4×10^6 m, latitude is $30°$.

The Refractive Index and Snell's Law

5. Compute the speed of light in ethyl alcohol.

6. The speed of light in a medium is 1.7×10^8 m/s. Compute the index of refraction of the medium.

7. A ray of light in air makes an angle of incidence of $45°$ at the surface of a substance. The ray is refracted at an angle of $30°$. What is the velocity of light in the substance?

8. A piece of glass of index of refraction of 1.4 is placed on top of and in contact with a piece of glass with an index of refraction of 1.6. If a wave front of light strikes the surface between the two pieces of glass at an angle of $30°$, what angle does the refracted wave front make with the surface?

9. If infrared radiation with a frequency of 2.5×10^{14} Hz is passing through a material with a refractive index of 1.5, what is the wavelength of the radiation in this material?

10. A piece of glass 1.0 cm thick has an index of refraction of 1.54. If light of wavelength 630 nm passes through this glass, how many wavelengths of this light correspond to this thickness of glass? How does this differ from the number of waves in 1.0 cm of air?

Young's Double Slit

11. Two slits spaced 0.3 mm apart are placed 50 cm from a screen. What is the distance between the second and third dark lines of the interference pattern when the slits are illuminated with light of wavelength 600 nm (1 nm $= 10^{-9}$ m)?

12. In a Young's double-slit experiment the light has a wavelength of 600 nm. The distance between the slits is 0.05 cm and the screen is 2 m from the slits. What is the distance between the fringes?

13. A green light of wavelength 550 nm passes through two small slits that are separated by 2×10^{-2} cm. Interference fringes are formed on a screen 1 m from the double slits. (a) Compute the angle at which the fourth bright fringe (line) will be located on the screen. Also (b) compute the position up the screen of this fringe as measured from the central bright fringe.

14. Green light of wavelength 540 nm is incident upon a pair of slits separated by 0.1 mm. What is the angular separation of the first dark fringe and the third dark fringe?

15. In an experiment to measure the wavelength of monochromatic light, the light is sent through a Young's double slit, where the slits are 0.2 mm apart. The interference fringes are viewed on a screen 0.8 m from the double slit, and it is found that the first-order bright fringes are each 2.4 mm on either side of the central bright fringe. From these data, determine the wavelength of the incident light.

16. Using blue light with a wavelength of 450 nm, the pattern from a double slit is projected across a small lecture hall a distance of 4 m. If we wish the fringes to be separated by 2.0 m so they are easily visible, what is the separation between the slits?

Diffraction Gratings

17. If yellow light of 580 nm is incident on a diffraction grating of 30 000 lines per inch, what is the highest-order pattern that can theoretically be seen with this grating?

18. Light of frequency 5×10^{14} Hz is incident on a diffraction grating consisting of 15 000 lines per inch. Calculate the angular separation of the first- and second-order diffraction pattern.

19. White light spans the wavelength region from roughly 400 to 700 nm. Using a diffraction grating with 7500 lines per centimeter, (a) what is the angular separation of the spectrum in first order? In second order? Also, (b) by how much (what angle) do the first- and second-order spectra overlap?

20. Light from a helium-neon gas laser of wavelength 630 nm incident on a diffraction grating produces a diffraction pattern with a separation of the bright fringes of 0.50 cm on a screen. A second wavelength from another monochromatic light source is incident upon the same grating arrangement, and the resulting fringes are separated by 0.4 cm. What is the wavelength of the second light source?

Diffraction

21. Light of 500 nm passes through a single slit 0.05 cm in width and is viewed on a screen 1.5 m away.

(a) What is the width of the central bright band? If the experiment is repeated under water, where the refractive index is 1.33, (b) what would be the new width of this central maximum?

22. Sodium light of wavelength 589 nm is incident upon a pair of slits separated by 0.15 mm and the pattern is viewed on a screen 1 m away from the slits. If the width of each slit is $\frac{1}{500}$ mm, how many bright fringes of the double-slit pattern are contained within the central single-slit diffraction band?

23. In a double slit with a slit separation of d, what is the maximum opening w of the slit allowed if the first-order interference fringes are just to appear in the outer edges of the central single-slit diffraction pattern? Does using slit openings this large place any restrictions on what is seen in the higher-order double-slit pattern?

Interference and Thin Films

24. Light from a helium-neon gas laser of wavelength 630 nm is incident normally from air on a 10^{-3}-mm-thick film of ice of refractive index 1.3. The film of ice is on a piece of glass of refractive index 1.5. Part of the light is reflected at the interface, and part passes into the ice and is reflected back at the ice-glass interface. (a) How many waves are contained in the light path in the ice before the reflected light reenters the air? (b) What is the phase relationship between the waves reflected at the air-ice interface and those waves that are reflected at the ice-glass interface when the light reenters the air?

25. Find the thickness of a coating required to make a nonreflecting surface for glass of $n = 1.5$. The index of refraction of the coating material is 1.2 and $\lambda = 500$ nm.

26. A prism made of quartz is coated with a thin film of material whose index of refraction is 1.35. The coating is designed to be nonreflecting for light of wavelength equal to 600 nm when incident normally. What is the appropriate minimum thickness of such a layer?

27. At the edge of a film of oil of refractive index 1.45 on a calm pool of water of refractive index 1.33, a blue color appears when white light shines on the film. Taking the wavelength of the blue light as 450 nm, find the thickness of the oil film.

28. A pair of glass microscope slides is placed one on top of the other. They are 6 cm long and in contact at one end but are separated by metal 0.09 mm thick. If light from a sodium light of wavelength 589 nm is incident normally on the glass plates, what is the number of dark interference fringes per centimeter one observes owing to the trapped air wedge?

29. What is the thinnest film of refractive index $n = 1.40$ in which destructive interference of the violet light (400 nm) from an incident beam of white light in air can take place by reflection? What is then the residual color of the reflected and transmitted beams?

30. A Young's double-slit experiment can be used to determine the thickness of very thin pieces of transparent material. For example, a double-slit experiment is performed with a pair of slits separated by 0.9 mm and the pattern is viewed 1.5 m away from the slits. Inserting a thin film in front of only one slit causes a shift in the fringe pattern, with the center of the pattern moving away from the geometric center of the two slits. If a piece of plastic with refractive index of 1.2 and thickness of 1.5×10^{-2} mm is placed in front of one slit, by how much will the central fringe move? The wavelength of light used is 500 nm.

31. To determine the thickness of a thin piece of cellophane of index of refraction equal to 1.60, the cellophane can be placed in the light path of one arm of a Michelson interferometer. This produces a shift of 20 fringes when the mercury green line of wavelength 546.1 nm is used as the light source. What is the thickness of the film?

Special Topic
The Search for c

The Greeks were remarkably ignorant about many scientific effects, but when it came to light, they were indeed on the beam. In addition to observing that light traveled in straight lines, they believed that it moved with a high but finite speed. The first scientist to suggest an actual experiment for measuring that enormous speed was Galileo. In 1638, the Italian physicist published in his book *Two New Sciences* a simple technique for measuring the speed of light. Two people with lanterns would be stationed on two hill tops about 1 km apart. One person was to uncover his lantern at any time he wanted, and the other person was to uncover his source of light the exact moment he saw the light from the other experimenter's lantern. By timing the interval between his action in uncovering his own lantern and his first sight of light from the other lantern, the first man should, in theory, have captured the essential factor for calculating the speed of light, assuming that he knew the distance between the hill tops. The experiment was impractical because light took much less time to travel over that short distance than the subjects did to react to it. Nevertheless, Galileo remained convinced that the speed of light was finite and could be measured.

His conviction was justified 34 yr after his death. In 1676, the Danish astronomer Olaf Roemer obtained, in a very indirect experiment, evidence that the velocity of light was finite. Roemer was measuring the time that Io, one of Jupiter's many

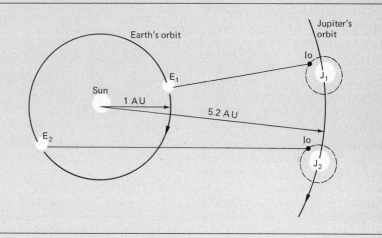

(1) Roemer's method for the determination of the speed of light. He determined the period of Io, a moon of Jupiter (J), in its orbit about Jupiter. This was done over a 6-month period, during which the earth (E) moved a distance equal to the diameter of the earth's orbit about the sun farther away from Jupiter. At this time the period of Io "appeared" to increase, owing to the additional distance the light traveled to cross the diameter of the earth's orbit. The speed of light was calculated using these data.

moons, took to complete a single revolution around that giant planet. He discovered to his surprise that after 6 months the eclipse of Io by Jupiter was behind the predicted time. The difference was quite substantial, namely, about 22 min relative to Io's period of 42 h. Roemer concluded that the differences arose from the motion of earth in its orbit. The light from the Jovian moon, he decided, had to travel a shorter distance to earth when the earth was on the side of the sun nearer to Jupiter than it did when the earth was on the opposite side of the large planet about 6 months later. During these 6 months, Jupiter itself moved only a short distance in its outsize orbit, as shown in the figure. Hence, the extra 22 min represented the time that light took to travel the diameter of the earth's orbit. Soon after, Christiaan Huygens used Roemer's data and his estimate of the diameter of the earth's orbit to determine the speed of light c. His result, $c = 2 \times 10^8$ m/s, is two-thirds of the accepted value today and in error due to Roemer's overestimation of the time light took to travel across the diameter of the earth's orbit. We now know that the trip takes about 16 min.

A direct method of measuring the velocity of light did not come about until 1849, when the French physicist Armand Hippolyte Louis Fizeau developed the piece of apparatus sketched. Light from an appropriate source is focused by an arrangement of lenses to a thin beam that meets a rotating, toothed wheel. Any light that passes through one of the open teeth of the wheel travels on to a mirror, from which it is reflected back along the same path until it reaches the half-silvered glass plate that allows part of the light to pass through while reflecting the rest. Light is reflected from the plate to meet the experimenter's eye. The experimenter will see only the returning beam, however, if the light reflected from the mirror arrives back at the toothed wheel at the exact moment an open tooth appears. The rotating toothed wheel is what we might call a "light chopper."

Fizeau conducted the experiment by rotating the wheel at increasing speed until he reached the speed at which he could see no light emerging from the half-silvered plate. At this point the light was obviously taking the same time to travel from the wheel to the mirror and back as the wheel was taking to move the distance between an open tooth and the adjacent closed tooth. The speed of

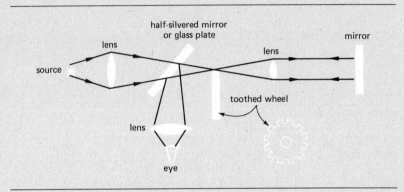

Fizeau's method for determining the speed of light.

the wheel gave an indication of this time, and simple measurement produced the value of the distance; thus, calculating the velocity of light was a matter of simple mathematics. Fizeau's result was the rather high value of 3.133×10^8 m/s. Later, Fizeau's collaborator Jean Bernard Leon Foucault replaced the toothed wheel with a rotating mirror and obtained the much more accurate value of 2.98×10^8 m/s. Foucault also amended the experiment in a more significant way by placing a tube of water between the rotating mirror and the fixed one. This modification allowed him to measure the speed of light in water, and when his results showed that light traveled appreciably more slowly in the water than in air, rather than faster in water than in air as Newton predicted on the basis of corpuscular theory, physicists were temporarily convinced that the corpuscular model of light was dead. That model did not come back for more than 40 yr, when the concept of the speed of light took on an almost mystical significance, as we shall see in Chap. 20.

19 Geometrical Optics

19.1 THROUGH A GLASS DARKLY

19.2 LENSES

SHORT SUBJECT: Communicating with Light

19.3 PICTURING THE IMAGE

SHORT SUBJECT: Over the Rainbow

19.4 CALCULATING THE IMAGE

19.5 THE LENS-MAKERS' EQUATION

19.6 LENSES IN COMBINATION
The Compound Microscope
The Astronomical Telescope

19.7 ABERRATIONS OF LENSES

19.8 REFLECTIONS ON MIRRORS
Summary
Questions
Problems

SPECIAL TOPIC: The Eye: The Window of the Brain

As the saying goes, a little knowledge is a dangerous thing. But on many occasions just a little knowledge is quite sufficient for the job. Millions of people drive cars skillfully and safely, even though they have no more than a passing knowledge of what actually happens under the hood to convert petroleum into driving power. Amateur photographers snap pictures with a passion, although they may not even realize that the critical event in producing a photograph is the chemical reaction which occurs on the photographic emulsion. And in the world of geometrical optics, or ray optics, students can understand certain principles and problems involving reflection and refraction without concerning themselves with the complexities of the true nature of light. Based on the simple laws of reflection and Snell's law, geometrical optics enable us to understand how lenses, prisms, mirrors, and similar optical equipment affect light rays and thus provide the basis for designing telescopes, microscopes, eyeglasses, and other optical instruments.

19.1 THROUGH A GLASS DARKLY

Both reflection and transmission occur when light impinges on—or is incident on, to use the technical phrase—**a boundary between two media in which light travels at different speeds,** that is, **whose indices of refraction are different.** If the light is not incident perpendicular to the boundary, the transmitted ray will be deviated or refracted, as illustrated in Fig. 19.1. Here the ray of light, which, as we recall from Chap. 18 is perpendicular to the wave front and marks the path in which the light waves actually travel, splits into two separate rays when it encounters the boundary between, say, air and glass. The path of the reflected ray is related to that of the incident ray by the law of reflection, which states that the angle of incidence, θ_1, equals the angle of

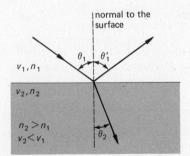

Figure 19.1 Regular reflection. The angle of incidence θ_1 equals the angle of reflection θ_1'. The light passing into the second medium is refracted at an angle θ_2. All angles are measured with respect to the normal to the surface.

reflection, θ'_1. It is important to remember that by convention these two angles are measured with respect to a line which is perpendicular to the reflecting surface, that is, the normal to the surface.

Not unexpectedly, the roughness or smoothness of the surface plays an important role in the reflection of beams of light, which consist of large numbers of individual rays. Every ray is reflected from completely smooth surfaces at exactly the same angle, and the overall reflected beam has the same appearance as the incident beam. This is specular (regular) reflection. Rough surfaces, by contrast, reflect different rays at different angles, as shown in Fig. 19.2, because individual sections of the surface are at different angles to the light beam. This is called diffuse reflection. The effect is similar to that obtained by dropping a dozen Ping-Pong balls on a cobbled street. The individual light rays, like the Ping-Pong balls, scatter in all directions. Hence the light beam reflected from a rough surface appears quite different from the original beam, even though every individual light ray obeys the law of reflection.

Rough surfaces reflect different rays at different angles.

Returning to Fig. 19.1, the second component of the incident ray is the refracted portion, which travels into the glass at an angle θ_2 to the *perpendicular*. This angle is readily calculated using Snell's law:

$$n_1 \sin \theta_1 = n_2 \sin \theta_2$$

where n_1 and n_2 are the refractive indices of the air and glass, or of the two media involved, to make the case completely general.

What of the relative intensities of the reflected and transmitted rays? Observation (and theory) show that the intensity of the transmitted beam is greatest when the angle of incidence is 0° and decreases steadily as the angle of incidence increases. Since the sum of the two must equal the intensity of the incident ray, the intensity of the reflected ray increases as the angle of incidence increases from 0° to 90°. In fact, when the incident ray comes in virtually parallel to the surface, almost all the light is reflected and the refracted ray is so weak that it is almost invisible.

When the situation is reversed—that is, when the incident ray starts in the medium with a higher refractive index and passes into one with a lower index—the refracted ray disappears completely beyond a certain angle for a different reason. This situation is shown in Fig. 19.3, which illustrates, say, the light that passes from a stone at the bottom of a Venetian canal to a gondolier on his gondola. The stone stays still while the gondola drifts slowly from point A to point D. The angle of refraction θ_2 of a ray of light from the stone increases along this path from 0° at A, to about 45° at B, to almost 90° at point C. By the time the boat reaches point D, there is no refracted ray for the gondolier to see. The stone is invisible from this angle because all the light from it is reflected back into the water. This effect is called total internal reflection.

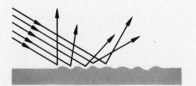

Figure 19.2 Diffuse reflection from a rough surface.

Figure 19.3 Reflection and refraction of light originating from a source in a high-index material that strikes a boundary between a high- and low-index material. Both reflection and refraction occur from angles of incidence from 0° to the critical angle, ray c, beyond which only total "internal reflection" occurs.

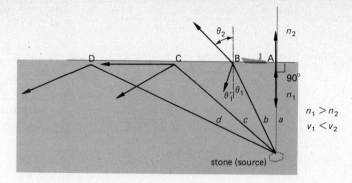

The refracted ray actually becomes invisible when the angle of refraction is exactly 90°. We can work out the corresponding angle of incidence θ_c using Snell's law:

$$n_1 \sin \theta_c = n_2 \sin 90° = n_2$$

Thus,

$$\sin \theta_c = \frac{n_2}{n_1}$$

This angle (θ_c) is known as the critical angle for the two media. Whenever air is one of the media involved, the value of n_2 can be taken as 1, and the sine of critical angle is therefore $1/n_1$, the reciprocal of the refractive index of the other medium. **No ray that has an angle of incidence greater than this critical angle in the medium can be refracted from the medium of higher refractive index into that of lower index. All the light undergoes total internal reflection.**

This effect finds a number of practical applications in everyday life. At a fairly superficial level, opera glasses and prism binoculars contain triangular pieces of glass with angles of 45°, 45°, and 90°. Light coming into the prisms from the stage or from the arena strikes the short legs of the right triangle at an angle of incidence of 45°, as illustrated in Fig. 19.4(a). Since this is greater than the critical angle between glass and air— 42°—*all* this light is reflected with a 90° change in direction, providing a clear picture of the proceedings under observation. On a more technical level, the glass fibers (fiber optics) used to transmit powerful beams of light over long distances and twisting paths with a minimum loss in brightness also apply the principle of total internal reflection [see Fig. 19.4(c)]. These devices are among the most efficient means of getting light from one point to another and are increasingly being used in communications systems and medical devices for such things as internally examining the human body.

Refraction is not always avoided so easily, however. Anyone who fishes, studies the stars, or even looks through a window experiences

Figure 19.4 Total internal reflection (a) in a glass prism used to bend the light 90° with (ideally) no intensity loss. The critical angle is 42°, so the angle of incidence of $\theta_1 = 45°$ exceeds this critical angle. (b) Use of such prisms in binoculars. (c) Use of total internal reflection in a "light pipe."

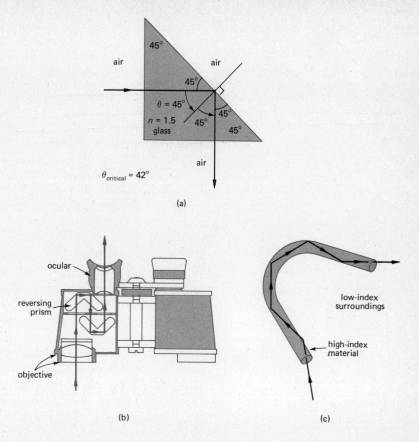

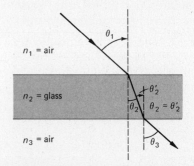

Figure 19.5 Light ray passing through a plate of glass. The beam is displaced parallel to the incident ray.

this type of alteration of light rays. The effects in these individual cases can be analyzed using Snell's law.

One obvious question is, "Does refraction distort light that passes through a window, or any similar plane plate of glass?" We can examine this problem using Fig. 19.5. A ray of light, from a tree perhaps, passes through the air, whose index of refraction is n_1, and reaches the outside of a piece of window glass at an angle of incidence θ_1. It is refracted into the glass, whose index of refraction is n_2, at an angle of refraction θ_2, and continues on through the window until it strikes the inner surface at an angle of incidence θ_2'. The light is then refracted from the inner surface of the glass into the air in the house, at an angle of refraction θ_3. As long as the two surfaces of the windowpane are parallel, then θ_2' is identical with θ_2. We can also assume that the refractive index of the air inside the house, n_3, is the same as that outside, n_1. In fact, these two quantities may well be slightly different, especially on cold days, because the refractive index of any material varies with the temperature. However, the difference would be too slight to affect the calculation significantly.

Applying Snell's law, we obtain the following equations:

At the top surface:

$$n_1 \sin \theta_1 = n_2 \sin \theta_2$$

At the bottom surface:

$$n_2 \sin \theta_2' = n_1 \sin \theta_3$$

since

$$\theta_2 = \theta_2'$$

$$n_1 \sin \theta_1 = n_2 \sin \theta_2 = n_1 \sin \theta_3$$

or

$$\theta_1 = \theta_3$$

This tells us that the parallel windowpane does not cause any overall bending of light that passes through it. Rather, each ray of light that passes through the window is displaced sideways before continuing in a direction parallel to that of the original ray. The amount of displacement is related to the angle of incidence of the original ray. When the angle of incidence is zero, there is no displacement, but as the angle increases, so does the amount of sideways displacement. In real life, many windows do distort light because their surfaces are not parallel. Glass is an example of the state of matter known technically as a supercooled liquid. It behaves as a liquid in some ways, and over periods of many decades it tends to flow slightly. Windowpanes made in the nineteenth century are measurably thicker at the bottom than at the top as a result of this flowing, and they produce marked distortions of scenes viewed through them. Modern glass is more stable than the early glass, which usually had a lower melting point and a higher lead content.

Even modern window glass often contains distortions that result from its production process.

In the case of normal windows, though, refraction plays a relatively minor part in controlling the apparent position of an image viewed through them. The phenomenon is more important in cases where refraction occurs at just one boundary. Any spear fisherman realizes that he must aim below the point at which he sees a fish if he is to have a chance of striking it. The reason is illustrated in Fig. 19.6. Because the light from the fish is bent as it passes out of the water on the way to the fisherman, the fish appears to be swimming a distance q beneath the surface, whereas it is actually a distance p beneath the top of the water. We can calculate this "apparent depth" of the fish's image by applying Snell's law. In this example, the law is

$$n_1 \sin \theta_1 = n_2 \sin \theta_2$$

Applying simple geometry to the figure, we see that

$$\sin \theta_1 = \frac{d}{\text{AO}} \quad \text{and} \quad \sin \theta_2 = \frac{d}{\text{AI}}$$

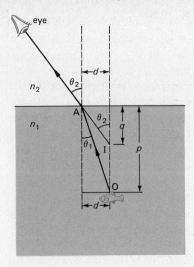

Figure 19.6 Apparent depth. A fish viewed from above the water appears closer to the surface of the water.

$$\sin \theta_1 = \frac{n_2 \sin \theta_2}{n_1}$$

$$\sin \theta_1 = \frac{d}{AO} = \frac{n_2}{n_1} \frac{d}{AI}$$

If we restrict ourselves to cases in which the angles θ_1 and θ_2 are so small that we are effectively looking at the fish from vertically above it, then AI and AO are approximately equal to the depths q and p. Since

$$n_1 \frac{d}{AO} = n_2 \frac{d}{AI}$$

then

$$\frac{n_1}{p} \cong \frac{n_2}{q}$$

The apparent depth of the fish is therefore approximately equal to the product of its actual depth and the ratio of the refractive indices of the water and the air,

$$q = \frac{n_2}{n_1} p$$

If the pond consists of clean water, whose refractive index is roughly four-thirds that of air, the fish will appear to be swimming at three-fourths of its actual depth, and the spear fisher can aim accordingly. The apparent depth will decrease for larger angles.

Some of the most colorful refractive effects are those produced when light passes through transparent materials whose sides are not parallel. The archetype of this phenomenon is the behavior of light that passes through a prism. Prisms are important to physicists mainly because they split, or disperse, light that passes through them into its component colors. This dispersion occurs because the different wavelengths of light travel at slightly different speeds in a material such as glass. The differences in speed translate into dissimilar refractive indices for the various wavelengths of light, as we saw in the previous chapter. Hence some wavelengths are bent more than others when they travel through a prism. White light in particular is split up into its rainbowlike components on passage through any transparent prism, as shown in Fig. 18.4.

The effect is illustrated in more detail in Fig. 19.7, which represents a prism spectrometer. This instrument identifies chemicals by analyzing the wavelengths (colors) of light they emit when heated. Chemists who use prism spectrometers aim to measure what they call the angle of deviation of every specific wavelength of light produced by the unknown chemical. This angle, labeled with the Greek letter δ (delta) in Fig. 19.7(a), is the angle through which the ray of light of the particular wavelength is bent by the prism. Thus, by determining the angle of

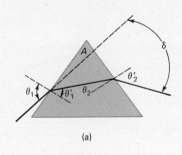

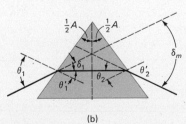

Figure 19.7 (a) Refraction of light by a prism. (b) Condition of angle of minimum deviation.

deviation the chemist can infer the wavelength. Because no two prisms are exactly alike, every instrument of this sort must be calibrated at the outset using light whose wavelength is known.

In practice, the expression for the angle of deviation is a complex one, depending on the angle of the prism, the refractive index of the prism, and the angle of incidence of the ray. The relationship between δ and the angle of incidence is rather unusual. As the angle of incidence decreases steadily from a large value, δ initially decreases, then passes through a minimum value, and starts to increase. The minimum, known as the angle of minimum deviation, δ_m, occurs when the light ray passes symmetrically through the prism, as shown in Fig. 19.7(b). As a result of the symmerical arrangement, the angle δ_m is exactly twice the deviation that occurs at each surface of the prism. Knowing this, observing that $\theta_1' + \theta_2 = A$, and applying Snell's law, we can readily show that

$$n = \frac{\sin\left(\frac{A + \delta_m}{2}\right)}{\sin\left(\frac{A}{2}\right)}$$

19.2 LENSES

The most important devices made possible by refraction are lenses. If two identical glass prisms are placed base to base as in Fig. 19.8, any parallel rays incident on the combination will come to a focus—a bright area of light—on the far side of the arrangement. This particular focus will be somewhat fuzzy because rays near the bases of the prisms will be deviated more than those farther away and hence will come together closer to the prisms. But **if the two prisms are permanently glued together and their surfaces ground to the spherical shapes** shown in Fig. 19.8(b), **the focus will sharpen up markedly, becoming a bright clear**

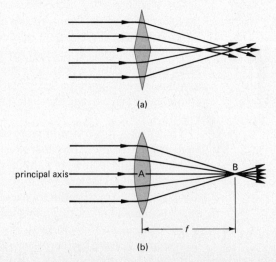

Figure 19.8 Convex lens. (a) Representation of a convex lens by two prisms. (b) Convex, thin spherical lens.

Communicating with Light

One of the most exciting developments of recent years in the communications industry has been that of optical fibers. The idea of using ultrathin strands of plastic with diameters no more than a few thousandths of an inch to carry coded light waves over distances of many miles was first advanced in a very speculative way by Dr. Charles Kao of International Telephone and Telegraph Corporation. Within the short space of a decade, Kao's speculation was firmly grounded in fact and had become the basis for a growing industry.

In late 1975, researchers in England used fibers to link a local police station with its computers a few hundred yards away, and the following year Bell Telephone Laboratories experts started to test the feasibility of using fibers to carry telephone messages over distances of a few miles, using miniature lasers to convert the electronic signals from the phones into light that traveled through the optical fibers.

The fundamental principle on which the technology is based is total internal reflection. Light passing along the fibers strikes the internal walls at angles of incidence that are

Fiber optics delivers the light. (*Bell Labs*)

almost always greater than the critical angle. Hence, virtually no light is lost to the outside of the cable. Certainly, the strength of light beams tends to weaken after they have traveled a certain distance, but this effect can be overcome by installation of "repeaters," devices that sense and regenerate the signals at various points along the fibers.

The major advantage of optical fibers is that they take up far less space than the conventional cables which carry telephone messages. Furthermore, they have such a capacity for messages that they are unlikely to encounter the chronic interference problems which assail messages sent over the air waves. Even today, experts estimate that a single bundle of optical fibers can carry thousands of telephone messages or up to a dozen television channels without difficulty.

spot of light through which every ray passes after refraction. This arrangement is called a lens. Both surfaces are convex in shape, so the arrangement is known as a double convex lens. In the figure, the focus of the lens is the point B, and the line AB, which joins the focus and the center of the lens, is known as the principal axis. The distance AB is

A virtual image, as we shall see later in this chapter, cannot be projected onto a screen.

called the focal length of the lens. The focal length for a lens with a given refractive index is determined by the roundnesses (radii of curvature) of the two surfaces. **All lenses have two focal points (foci), which for thin lenses are equidistant from the center of the lens.** Because this type of lens brings light that passes through it to a focus, it is known as a converging, or positive, lens.

The double convex lens is probably the simplest example of a lens, but lenses come in all shapes and sizes. Another relatively simple type can be constructed by placing two prisms vertex to vertex, as sketched in Fig. 19.9(a), and once more grinding the surfaces into spheres, as illustrated in Fig. 19.9(b). This type of lens, which is thinner at its center than at its edges, is known as a concave, or diverging, lens. The surfaces "cave in," and the lens causes parallel incident rays of light to diverge. Instead of passing through a single point, as they do in the case of a converging lens, the rays of light appear to emanate from one spot, point B in Fig. 19.9(b). This point is called a virtual focus. The word "virtual" is used because the rays of light do not actually pass through it. The line AB is the principal axis of the lens, and the distance AB is referred to as the focal length of the lens, just as in the case of converging lenses. Lenses such as the double concave lens, which cause light to spread out, are referred to as diverging, or negative, lenses.

One factor of crucial importance in dealing with lenses is that they are symmetrical in a particular way. This is illustrated in Fig. 19.10 for the case of a thin converging lens. In Fig. 19.10(a), a beam of light parallel to the principal axis of the lens comes to a focus at F_2, one of

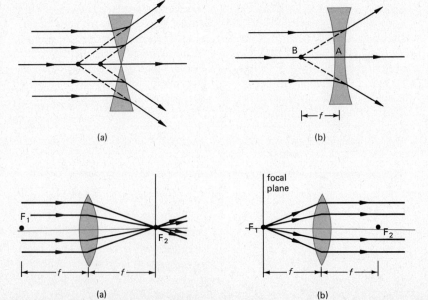

Figure 19.9 Concave, or diverging, lens. (a) Representation of the lens by two prisms. (b) A concave, thin spherical lens.

Figure 19.10 Focal points and planes of a convex lens. (a) Second focal point, where parallel light is focused. (b) First focal point, where a point source will produce a parallel beam.

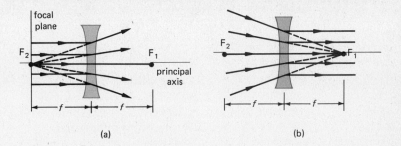

Figure 19.11 Focal points and planes of a diverging (concave) lens. (a) Second focal point; after passing through the lens, parallel light appears to come from the second focal point. (b) First focal point; light directed toward the first focal point leaves the lens parallel to the principal axis.

the two focal point of the lens. In fact, one can locate the focus of a lens in this way, using light from some distant source to ensure that the incoming beam is parallel. When a beam of light that is not parallel to this axis is incident on a lens, the beam comes to a focus off the axis but somewhere on the focal plane—the plane parallel to the lens that passes through the focus.

In Fig. 19.10(b) the symmetry of the lens is illustrated by a point source, such as a light shining through a pinhole at the first focal point F_1. The rays from this focal point produce a parallel beam of light when they are refracted by the same lens. Similarly, a point source of light anywhere else on the focal plane produces, after refraction by the lens, a parallel beam of light that is at an angle to the principal axis. This means that if we reverse the direction of light along any ray, the reversed light will follow back along the path of the original ray.

The symmetrical situation is just the same for diverging lenses, as illustrated in Fig. 19.11. The parallel beam of light in Fig. 19.11(a) leaves the double concave lens as if it originated from the virtual focal point F_2. In Fig. 19.11(b), the rays converging initially toward the first focal point F_1 leave the lens after refraction parallel to the principal axis.

19.3 PICTURING THE IMAGE

Practical uses of lenses stem from their ability to form images—sometimes magnified, sometimes reduced, and occasionally unchanged in size—of objects from which light has originated. The following three rules provide a ready guide for locating the image of objects after light from the object has been refracted by thin spherical lenses:

1. Rays from the object that are parallel to the principal axis of the lens pass through the second focal point of the lens, in the case of converging lenses, or appear to come from the second focal point, in the case of diverging lenses. The former example is illustrated by ray 1 in Fig. 19.12.
2. Rays that pass through the center of a lens are not deviated at all. They pass through in much the same way that light passes through a plane parallel sheet of glass. And since we define our lenses as thin, these rays are displaced negligibly from their path. Ray 2 in Fig. 19.12 illustrates this rule.

Over the Rainbow

"And God said, 'This is the sign of the covenant that I make between me and you and every living creature that is with you, for all future generations: I set my bow in the cloud, and it shall be a sign of the covenant between me and the earth.'"

This passage from the account of the flood in the book of *Genesis* contains perhaps the first mention of the rainbow in written history. But undoubtedly the semicircular apparition in the sky intrigued people long before the time of Noah. The effect occurs during periods of sunshine and showers and stems directly from the fact that different colors in the visible spectrum of light have slightly different indices of refraction.

Whenever sunlight encounters raindrops, part of the light undergoes the refraction-reflection-refraction sequence shown in Fig. A. Because the refractive indices of the components of the "white" sunlight differ, the raindrops split up the light in the same way that a prism does. Owing to the angle of observation, a person on the ground sees only one component of the light from any individual raindrop. But when that person surveys the many thousands of drops in a shower, the net effect is sufficient to produce all the colors and the unforgettable view of the bow-shaped rainbow.

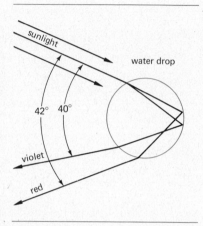

Figure A

Over the rainbow. (*Luoma, Monkmeyer*)

Figure 19.12 Location of an image using three rays. A converging lens with the object outside the focal point. Ray 1, parallel to the principal axis, is directed through the second focal point by the lens. Ray 2, passing through the center of the lens, is undeflected. Ray 3 passes through the first focal point, strikes the lens, and comes out of the lens parallel to the principal axis.

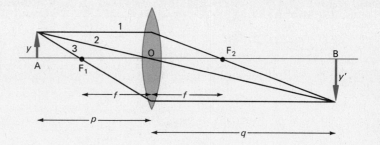

3. A ray that passes through the first focal point (or the case of an object between the first focal point and the lens when extended would have passed through the first focal point), in the case of a converging lens, or that is traveling toward the first virtual focus, in the case of a diverging lens, emerges from the lens parallel to the principal axis. This behavior is shown by ray 3 in Fig. 19.12.

Taken together, the three rays in Fig. 19.12 are sufficient to determine the location and orientation of the image B of an object A whose light has passed through the thin convex lens O. The size of the image is y', which in this case is larger than the size of the original object, y. The image is also farther away from the lens than the object. And most important, while the bottom of both the image and the object are on the principal axis, the top of the image, located by rays 1, 2, and 3, is on the other side of that axis from the top of the object. This means that the image is upside down, or inverted, to use the optics term. Our graphic analysis, therefore, shows us that the image is inverted and magnified. The image is also what optical scientists call "real." Unlike the virtual images produced by diverging lenses, the image can be captured on a screen placed at point B. One further point: a close look at the figure tells us that the location of the top of the image can just as easily be found by constructing two rays rather than three. Remember that the object is what one looks at, and the image is what one actually sees.

Figure 19.13 shows a graphic construction for another circumstance: the object is closer to the lens than the focal point. The paths of the three rays in the sketch diverge after passing through the lens; hence the rays cannot combine to form a real image of the type just encountered. Instead, they can be traced backward to form the virtual image indicated in the diagram. A viewer looking through the lens from the side opposite the object would see this virtual image, which forms a magnified version of the object the right side up. The magnified image appears behind the object. A converging lens used in this manner is called a magnifying glass or simple microscope. Stamp collectors looking at fine details of rare stamps and gem experts studying precious stones can all see an enlarged image when they use a converging lens in this way.

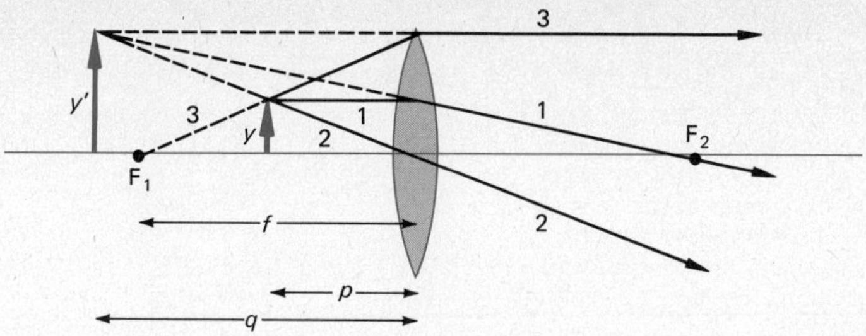

Figure 19.13 A converging lens. Location of the image when the object is inside the focal point; p = object distance, q = image distance, f = focal length, y = size of object, y' = size of image.

Figure 19.14 A diverging lens. Location of the image; p = object distance, q = image distance, f = focal length, y = object, y' = image.

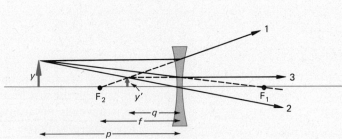

Plainly, the type of image—real or virtual—formed by a converging lens depends on whether the object is farther from or closer to the lens than the focus. Single diverging lenses, by contrast, always form virtual images wherever the object is placed. The geometry of diverging lenses is shown in Fig. 19.14. Ray 1 starts out parallel to the principal axis and is bent by the lens in such a way that it appears to come from the second focus, F_2. Ray 2 passes through the center of the lens undeflected. Ray 3, which sets out from the top of the object en route to the first focus, F_1, ends up parallel to the principal axis after refraction by the lens. The divergence of the rays from the lens is sufficient to tell us that no real image is possible. But anyone who looks back into the lens from the right-hand side will see a virtual image that is upright but smaller than the original object. In fact, virtual images always appear upright when they are produced by single lenses. The converse is also true: real images produced by single lenses are always inverted. The symmetrical effect of lenses insofar as the passage of light through them is concerned can be illustrated very easily with Figs. 19.13 and 19.14 by interchanging the object and the image. In both cases the new object produces an image where the old object was, and the rays of light are simply reversed.

19.4 CALCULATING THE IMAGE

So far we have just looked at the ways in which different types of lenses form images from a qualitative point of view. But a modicum of geometry and common sense allows us to compute the sizes and distance of images, real or virtual, upright or inverted, magnified or reduced. The basic principles for the calculation are shown in Fig. 19.15.

Figure 19.15 Geometry for derivation of the thin-lens formula.

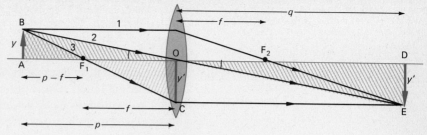

The figure shows an object, AB, which produces a real, magnified inverted image DE on refraction through the double convex lens whose center is at point O. The object's distance from the lens is p; its height is y. The image is also on the principal axis, at a distance q from the lens, and its height is y'. The focal length of the lens is f. The figure as a whole is a series of triangles that can be used to obtain the characteristics of the image in terms of those of the object.

The two similar triangles OAB and ODE (similar in the geometrical sense) yield the equation

$$\frac{y}{p} = \frac{y'}{q} \quad \text{or} \quad \frac{y}{y'} = \frac{p}{q}$$

From the two similar triangles F_1AB and F_1OC,

$$\frac{y}{p-f} = \frac{y'}{f} \quad \text{or} \quad \frac{y}{y'} = \frac{p-f}{f}$$

Setting these two expressions for y/y' equal to one another gives the equation

$$\frac{p-f}{f} = \frac{p}{q} \quad \text{or} \quad pq - fq = fp$$

A couple of algebraic steps (dividing each term by pqf and rearranging) produce the following form for the equation:

$$\frac{1}{p} + \frac{1}{q} = \frac{1}{f} \tag{19.1}$$

This expression is known as the thin lens equation. It relates the distance of the image from the lens to the distance of the object and the focal length of the lens. (All distances are measured from the center of the *thin* lens.)

Another quantity in which lens users are interested is the magnification M produced by lenses. This is the ratio of the sizes of the image and the object, y'/y. Simple geometry shows that in Fig. 19.15 this is the ratio of the distances of the image and of the object from the lens:

$$M = \frac{y'}{y} = -\frac{q}{p} \tag{19.2}$$

The negative sign is inserted in the equation to indicate that the image is inverted. In fact, this is just one of a series of sign conventions associated with the lens equation and similar expressions. The focal lengths of all converging lenses are treated as positive quantities, and those of diverging lenses are given negative signs. In addition, the distances from the lenses of objects and images are considered positive for real objects and images and negative for virtual objects and images. The term **virtual object** sounds contradictory; in fact it appears only when two or more lenses are employed. The virtual object occurs when the light that would have produced an image after passing through one lens falls on a second lens before forming an image. Then the image from the first lens becomes a virtual object of the second lens.

EXAMPLE 19.1

A chess piece 4 cm high is located 10 cm from a spherical, thin converging lens whose focal length is 20 cm. What are the nature, size, and location of the image?

Solution

See Fig. 19.13. The object is between the first focal point and the lens. The image distance is found by using the lens Eq. (19.1):

$$\frac{1}{f} = \frac{1}{p} + \frac{1}{q}$$

$$\frac{1}{+20 \text{ cm}} = \frac{1}{+10 \text{ cm}} + \frac{1}{q}$$

$$\frac{1}{q} = \frac{1}{20} - \frac{1}{10} = \frac{1-2}{20} = -\frac{1}{20}$$

$$q = -20 \text{ cm}$$

Since q is negative, the image is virtual and therefore erect.

The magnification is given by

$$M = \frac{y'}{y} = -\frac{q}{p} = -\frac{-20 \text{ cm}}{10 \text{ cm}} = +2$$

so

$$y' = My = 2 \times 4 \text{ cm} = 8 \text{ cm}$$

The positive value of the image size means that the image is erect. The magnification is +2 in this case, with the final image size being 8 cm.

EXAMPLE 19.2

How far from a converging lens whose focal length is $+20$ cm must a specimen of a Red Admiral butterfly be placed if its image is to be a real one, three times as large as the object?

Solution

Although we know the distance of neither the object nor the image, we can relate p and q from the magnifications. Since the real image is inverted $y' = -3y$, the magnification is given by

$$M = \frac{y'}{y} = \frac{-3y}{y} = -3$$

Hence,

$$M = -\frac{q}{p} \quad \text{or} \quad q = +3p$$

The negative sign is necessary since the image is to be real and therefore inverted. Now, using the thin lens equation,

$$\frac{1}{p} + \frac{1}{q} = \frac{1}{f}$$

$$\frac{1}{p} + \frac{1}{3p} = \frac{1}{20 \text{ cm}}$$

$$\frac{3+1}{3p} = \frac{1}{20 \text{ cm}}$$

$$3p = (4)(20) \text{ cm}$$
$$p = +26.7 \text{ cm}$$

The object is on the other side of the lens than the image and farther from the lens than the first focal point, a situation similar to Fig. 19.12.

EXAMPLE 19.3

How far from a converging lens of focal length $+20$ cm must a butterfly be placed if its image is to be virtual, three times as large as the object?

Solution

This example differs from Example 19.2 only in the requirement that the image be virtual. A sketch of the situation is shown in Fig. 19.13. The object must be placed between the first focal point and the lens; that is, p must be less than f. Again, we use the magnification to determine p in terms of q. Because the image is virtual, it will also be erect; hence, $y' = +3y$:

$$M = \frac{y'}{y} = \frac{-q}{p} \quad \text{or} \quad \frac{+3y}{y} = \frac{-q}{p}$$

so

$$q = -3p$$

Applying the lens equation yields

$$\frac{1}{p} + \frac{1}{q} = \frac{1}{f}$$

$$\frac{1}{p} - \frac{1}{3p} = \frac{1}{20 \text{ cm}}$$

$$\frac{3-1}{3p} = \frac{1}{20 \text{ cm}}$$

$$3p = (2)(20) \text{ cm}$$
$$p = 13.3 \text{ cm}$$

The object is to the left, between the focal point and the lens.

EXAMPLE 19.4

A diverging lens of focal length $f = -16$ cm is held 8 cm from a rare stamp. Where is the image of this stamp located, and what is the magnification of the lens?

Solution

The focal length $f = -16$ cm, $p = 8$ cm, and q is determined using the lens equation

$$\frac{1}{p} + \frac{1}{q} = \frac{1}{f}$$

$$\frac{1}{q} = \frac{1}{f} - \frac{1}{p} = -\frac{1}{16 \text{ cm}} - \frac{1}{8 \text{ cm}} = -\frac{3}{16 \text{ cm}}$$

$$q = \frac{-16 \text{ cm}}{3} = -5.33 \text{ cm}$$

The negative sign indicates that the image is virtual, hence erect, and to the left of the lens. The magnification is

$$M = \frac{-q}{p} = -\frac{(-5.33 \text{ cm})}{8 \text{ cm}} = \frac{+2}{3} \quad \text{or} \quad +0.667$$

The positive value of magnification indicates that the image is erect. Note that the magnification is less than 1, indicating that the image is actually smaller than the object. The image size is two-thirds the object size. The use of the word "magnification" may be misleading since values less than 1 indicate a reduction in size. Magnification as used here in its technical sense indicates any alteration in size, as well as the situation $M = 1$, which means that the image and the object have the same size.

19.5 THE LENS-MAKERS' EQUATION

The people who grind lenses are just as concerned with the mathematics of individual refracting situations as with the overall behavior of their products. They need to produce lenses with specific focal lengths. The focal length of a lens is related to a number of factors, most important being the refractive indices of the lens material and the medium from which the light is to travel into the lens (generally air), and the radii of curvature of the two sides of the lens. The latter require yet another sign convention: the radii of convex lens surfaces are regarded as positive, and those of concave lens surfaces are taken as negative. If the index of refraction of the lens is n_2, that of the medium around it n_1, and the radii of curvature of the two lens surfaces R_1 and R_2, application of Snell's law leads to the following formula for f, the focal length of the lens:

$$\frac{1}{f} = \frac{(n_2 - n_1)}{n_1}\left(\frac{1}{R_1} + \frac{1}{R_2}\right) \tag{19.3}$$

This is known as the lens-makers' equation and applies to thin spherical lenses. In normal situations involving air as the surrounding medium, the value of n_1 is 1, and the formula simplifies to

$$\frac{1}{f} = (n_2 - 1)\left(\frac{1}{R_1} + \frac{1}{R_2}\right)$$

Using the sign convention for the radii, which is illustrated in Fig. 19.16(b), the equation can be applied to both convex and concave lenses and to the lenses that contain surfaces of both curvatures. The

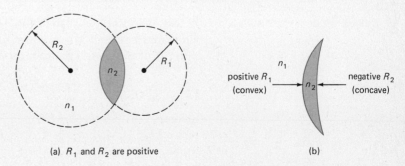

Figure 19.16 (a) The focal point of a lens is determined by the radii of its surfaces and the index of refraction. (b) The sign convention for the radius of a lens surface.

(a) R_1 and R_2 are positive

(b)

equation even accommodates plane lens surfaces; for these, the radius is infinite, and the term $1/R$ is therefore zero. The equation yields both the magnitude and the sign of the particular lens's focal length. If the sign is positive, the lens is a converging one; if the sign is negative, it is a diverging lens.

Note that the above equation produces not the focal length itself but the reciprocal of that value. Lens-makers use this term, which they call the power of a lens, in preference to the actual focal length. Lens powers are measured in units known as diopters. **The power of any lens in diopters is equal to the reciprocal of the focal length of the lens in meters.**

EXAMPLE 19.5

A lens-maker plans to construct a planoconvex lens (a lens with one surface plane and the other convex) from glass of index of refraction 1.5. If the radius of curvature of the convex surface is 30 cm, what will be the focal length of the lens and its power in diopters?

Solution

Both the focal length and power can be calculated from the lens-makers' Eq. (19.3). Assuming that the lens is to be used in air, $n_1 = 1.0$. We know that $n_2 = 1.5$, that R_1 for the plane surface is infinity, and that $R_2 = +30$ cm. Thus,

$$\frac{1}{f} = (n-1)\left(\frac{1}{\infty} + \frac{1}{R_2}\right) = (n-1)\frac{1}{R_2}$$

$$f = \frac{R_2}{n-1} = \frac{30 \text{ cm}}{1.5 - 1} = 60 \text{ cm}$$

The power in diopters is $1/f$ when f is in meters:

$$\text{Power} = \frac{1}{f} = \frac{1}{0.6 \text{ m}} = 1.66 \text{ diopters}$$

EXAMPLE 19.6

A thin spherical lens made of flint glass whose index of refraction is 1.66 has a power of +6 diopters in air. What is the power of the lens when it is immersed in water?

Solution

The focal length of the lens will change in water because it is the change in index of refraction from the outside medium to the lens medium that produces the bending of the light. Air has an index of refraction of 1; water has a larger index of refraction, $n = 1.33$. To solve the problem, we write down the lens-makers' formula for the lens in both air and water using the fact that $1/f$ is the power:

$$\frac{1}{f} = \left(\frac{n_2 - n_1}{n_1}\right)\left(\frac{1}{R_1} + \frac{1}{R_2}\right)$$

For air:

$$\frac{1}{f_{air}} = \left(\frac{1.66 - 1}{1}\right)\left(\frac{1}{R_1} + \frac{1}{R_2}\right) = +6 \text{ diopters} = P_{air}$$

For water:

$$\frac{1}{f_{water}} = \left(\frac{1.66 - 1.33}{1.33}\right)\left(\frac{1}{R_1} + \frac{1}{R_2}\right) = \text{power in diopters} = P_{water}$$

To solve for the power of the lens in water, the quantity $(1/R_1) + (1/R_2)$ can be evaluated from the information given for the lens in air:

$$\left(\frac{1}{R_1} + \frac{1}{R_2}\right) = \frac{6}{0.66}$$

$$\text{Power in diopters in water} = \left(\frac{0.33}{1.33}\right)\left(\frac{6}{0.66}\right) = 2.26 \text{ diopters.}$$

19.6 LENSES IN COMBINATION

Most optical instruments consist of a series of lenses rather than just single ones. Fortunately, the net effect of any instrument on light that moves through it can be pieced together by dealing with one lens at a time. One first works out the characteristics of the image formed by the first lens, and then uses this image as the object for the second lens. Next, having worked out the details of the image produced by the second lens, one uses *this* image as the object for the third lens, and so on. The total magnification produced by a system of lenses is the product of the individual magnifications that result from each single lens, a fact which follows directly from the definition of magnification. This is best illustrated by Example 19.7.

EXAMPLE 19.7

A fly 5 mm high is placed 15 cm in front of a lens of focal length $f = +10$ cm. A second lens of focal length $+5$ cm is placed 40 cm away from the first. Compute the position, nature, and size of the final image and the magnification produced by the system.

Solution

The problem has been solved graphically in Fig. 19.17 to help with the quantitative solution.

For the first lens:

$$\frac{1}{p_1} + \frac{1}{q_1} = \frac{1}{f_1}$$

Figure 19.17 A lens combination.

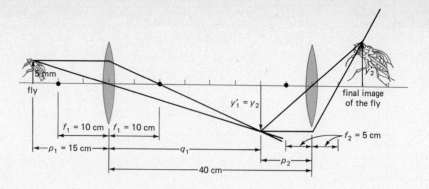

$$\frac{1}{q_1} = \frac{1}{f_1} - \frac{1}{p_1} = \frac{1}{10 \text{ cm}} - \frac{1}{15 \text{ cm}} = \frac{15 - 10}{150 \text{ cm}} = \frac{5}{150 \text{ cm}}$$

$$q_1 = \frac{150 \text{ cm}}{5} = +30 \text{ cm}$$

For the second lens:

If q_1 is 30 cm and the separation between the two lenses is 40 cm, then $p_2 = 10$ cm, as can be seen in Fig. 19.17:

$$\frac{1}{p_2} + \frac{1}{q_2} = \frac{1}{f_2}$$

$$\frac{1}{q_2} = \frac{1}{f_2} - \frac{1}{p_2} = \frac{1}{+5 \text{ cm}} - \frac{1}{10 \text{ cm}} = \frac{2-1}{10 \text{ cm}} = \frac{1}{10 \text{ cm}}$$

$$q_2 = +10 \text{ cm}$$

The final image is a distance 10 cm to the right of lens 2. The image is erect with respect to the original object, and it is real. To compute the size of the object and the magnification, apply the relationship for magnification twice.

For the first lens:

$$M_1 = -\frac{q_1}{p_1} = \frac{-30}{15} = -2$$

The negative sign indicates that the image is inverted. The size of the first image is

$$y'_1 = (-2)(y_1) = -2(5 \text{ mm}) = -10 \text{ mm}$$

The magnification at the second lens is given by

$$M_2 = \frac{-q_2}{p_2} = -\frac{10 \text{ cm}}{10 \text{ cm}} = -1$$

Again, the negative sign indicates that the image is also inverted by the second lens. The size of the second image is

$$y'_2 = (-1)(y_2) = (-1)(-10 \text{ mm}) = +10 \text{ mm}$$

The overall image is upright and 10 mm high. *This could be computed directly from the product of the two magnifications of the individual lenses:*

$$M = M_1 M_2 = (-2)(-1) = +2$$

Hence, the final image size, y_2, can be determined by

$$M = \frac{y_2}{y} \quad \text{therefore} \quad y_2 = 2y = (2)(5 \text{ mm}) = 10 \text{ mm}$$

In this combination, the two lenses perform specific functions. The first magnifies and the second produces an image that is erect with respect to the original object.

The Compound Microscope

A practical example of a simple combination of lenses is the compound microscope, which is more powerful than the simple magnifying glass. The construction of the compound microscope is shown in Fig. 19.18. It consists of two converging lenses, known as the objective lens and the eyepiece. The objective lens forms a real, inverted, and greatly magnified image of the object under the microscope when the object is placed beyond the focus of the lens. The eyepiece acts as a magnifying glass when the real image is being examined. It enlarges this image further, producing a virtual image. The total magnification is the product of the separate magnifications produced by the two lenses. Users of compound microscopes normally adjust them for minimum eyestrain by moving the eyepiece in such a way that the final virtual image is located at infinity. This happens when the image from the objective lens lies exactly on the focal plane of the eyepiece.

The eye is in its most relaxed state when viewing parallel rays of light that appear to come from an infinite distance away.

Figure 19.18 Optics of a compound microscope.

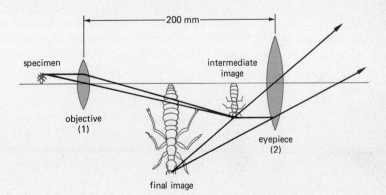

Can we make the final image larger and hence increase the magnification of the system? We can if we substitute another lens system, such as a compound microscope, for the eyepiece. Such changes as these have been applied quite successfully, so much so that the present-day microscopes have reached the theoretical limit of resolution for visible light—the limit at which the image becomes too fuzzy to be useful.

The cause of this fuzziness is diffraction—the formation of multiple images that results from the wavelike nature of light. Thus, to calculate the ultimate resolving power of optical microscopes, we must leave the relatively simple mathematics of geometrical optics and embark on the more complex considerations of wave theory.

Imagine that a lens forms the image of a point source of light produced by a pinhole device. Now, the lens actually serves as an opening for the passage of light that is similar to the slits we met in the last chapter in connection with diffraction effects. Chapter 18 showed that the angular size of the central maximum of a diffraction pattern is related to both the wavelength λ of the light and the width w of the slit. In the case of lenses, the equivalent slit width for a lens whose diameter is D is $0.82D$. Hence, the angular radius of the central bright part of the circular diffraction pattern produced by a lens of diameter D is $\lambda/0.82D$. What this means is that a parallel beam of light refracted by the lens focuses not to a point image but to a diffuse diffraction pattern whose angular radius is $\lambda/0.82D$.

Of course, the wavelengths of visible light are extremely small in comparison with the diameters of typical lenses. Thus, in most cases the factor $\lambda/0.82D$ is so small that the image appears quite sharp. However, as the magnification produced by a lens system increases, the diffraction patterns of individual points on the object start to overlap, rendering the image unclear, much like a picture on a television whose contrast button is incorrectly set. We say that the resolution, or resolving power, of the microscope has reached its limit.

The formula for the angular size of the diffraction maximum, $\lambda/0.82D$, tells us that we can reduce the size of this maximum, and hence improve the resolution of the microscope, in two ways: by reducing the wavelengh of the light involved or by increasing the diameter of the lens. In some cases researchers use ultraviolet light in place of the longer-wavelength visible light to improve the resolution of their microscopes. This is very expensive since quartz lenses are required. Also, a photoplate rather than the eye is required to detect the ultraviolet light. Increasing the sizes of the lenses is a rather less reliable means of achieving that end because the large-diameter lenses tend to contain defects that themselves reduce the resolution of microscopes in which they are incorporated. Furthermore, it is difficult to make large lenses with short focal lengths. Microscope makers have to compromise, by building microscopes that contain as many as 10 separate lenses in the objective lens whose focal length is only a few millimeters. The

Quartz is both more expensive and more difficult to grind than glass.

purpose of many of those lenses is to compensate for the aberrations introduced into the image by the other lenses. Such complex microscopes as these can obtain the theoretical limit of resolution, allowing scientists to see objects just 2×10^{-7} m across.

The Astronomical Telescope

Microscopes are used to magnify artifacts placed just a few millimeters in front of them. Telescopes, by contrast, must provide magnified images of distant objects, for example, planets and other heavenly bodies whose distances range from hundreds of thousands of miles to hundreds of thousands of light-years.

Like the microscope, a telescope consists of an object lens and an eyepiece. Figure 19.19 shows the operation of a simple telescope as it is used to gaze at a star at a vast distance from its objective lens. Because of the vast distance, the light from the star is essentially parallel as it reaches the objective lens. It therefore forms a real image at the focal plane of that lens. The eyepiece then magnifies the intermediate image to produce a final virtual image. Not surprisingly, telescopic images are always smaller than the objects they represent; Neptune is certainly larger than a nickel, and Alpha Centauri is larger than a penny. Plain magnification is therefore not the primary purpose of astronomical telescopes. In fact, all stars appear as points through even our most powerful telescopes. The main advantage of telescopes for star observations is that they gather light. What telescopes do, with singular success, is bring the images of celestial objects much closer to astronomers. Telescopes magnify not the actual size of planets, comets, and natural satellites but their angular size. The **magnifying power** of telescopes, sometimes known as the **angular magnification M^***, is defined by the equation

Plain magnification is not the primary purpose of astronomical telescopes.

$$M^* = \frac{\text{angular size of image}}{\text{angular size of object}} \tag{19.4}$$

We can put this in mathematical terms by considering the setup in Fig. 19.19, which shows light from a planet a distance p_1 from the instrument, whose size is h, coming to a focus at the focal point of the objective lens. This image, whose size is h_1, becomes the object for the eyepiece, which is so placed that the image from the objective lens is at its focal point, a distance p_2 from the eyepiece. The final, virtual image,

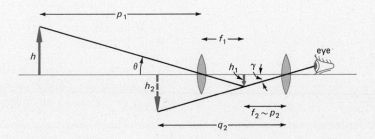

Figure 19.19 An astronomical telescope. Usually $f_1 > f_2$.

whose size is h_2, is formed a distance q_2 behind the eyepiece. The angle θ subtended by the planet when viewed by the unaided eye is h/p_1, in radians, if we assume that p_1 is much larger than the length of the telescope. The angle γ subtended by the final image is h_2/q_2, and from the geometry in the figure this is obviously equal to h_1/f_2. From our definition of magnifying power, then,

$$M^* = \frac{\gamma}{\theta} = \frac{h_1/f_2}{h/p_1}$$

By comparing similar triangles in Fig. 19.19, we see that $h_1/f_1 = h/p_1 = \theta$. Therefore,

$$M^* = \frac{\gamma}{\theta} = \frac{h_1/f_2}{h_1/f_1} = \frac{f_1}{f_2} \tag{19.5}$$

Thus, when the telescope is adjusted to produce a minimum of eyestrain, its magnifying power is the ratio of the focal lengths of the objective lens and the eyepiece.

EXAMPLE 19.8

An astronomical telescope has an objective lens whose power is 2 diopters. This lens is placed 60 cm from the eyepiece when the telescope is adjusted for minimum eyestrain. Calculate the angular magnification of the telescope.

Solution

Our first step is to determine the focal lengths of both lenses. For the objective, we are given a power in diopters. Hence,

$$f_1 \text{ (in meters)} = \frac{1}{\text{power (in diopters)}} = \frac{1}{2} \text{ m} = 50 \text{ cm}$$

The focal length of the eyepiece under conditions of minimum eyestrain can be determined from a glance at Fig. 19.19. Clearly, the sum of the focal distances $f_1 + f_2$ equals the separation of the two lenses. Therefore,

$$f_1 + f_2 = 60 \text{ cm} \quad \text{and} \quad f_2 = 60 - f_2 = 60 - 50 = 10 \text{ cm}$$

The magnifying power of this telescope is calculated from Eq. (19.5):

$$M^* = \frac{f_1}{f_2} = \frac{50 \text{ cm}}{10 \text{ cm}} = 5$$

This is a low-power telescope that is used to examine a large area of the sky. We would require a higher-power telescope to examine individual bodies, such as the moon and planets, in close detail.

Telescopes do not always involve converging lenses. The Galilean telescope, named after the Italian scientist who designed it, eliminates two problems associated with the astronomical telescope: it provides an upright image and is shorter in length. This telescope uses a diverging lens as its eyepiece. The light from the objective is not allowed to reach a focus; instead, it is interrupted with the diverging lens, which then produces an upright image. This arrangement is inherently less troubling, at least to nonastronomers, than one that produces an inverted image.

19.7 ABERRATIONS OF LENSES

In the real world, no lens is perfect.

In the real world, no lens is perfect. For example, there may be small faults produced in the grinding of the lens. However, beyond such correctable imperfections there are inherent limitations caused by the nature of spherical surfaces or because not all wavelengths of light travel at the same speed through the material the lens is constructed of. These inherent limitations are called *aberrations*. Aberration is a result of no system of reflecting and refracting surfaces being able to form perfect images of objects that are at different distances and off-axis to a different extent. These defects require scientists who design optical instruments to use a maximum of ingenuity in an effort to compensate for them.

One common lens limitation is known as spherical aberration. This type of defect causes light rays near the edges of the spherical lens to bend more than those that pass through the center of the lens. The result is a fuzzy focus of the sort sketched in Fig. 19.20; to the viewer, the image lacks clarity. This type of aberration can be minimized, if not overcome entirely, by placing a ring-shaped diaphragm in front of the lens to cut out all rays of light except those that pass through its center. In addition, lenses are often ground parabolic in shape to reduce spherical aberration.

Another common type of fault or limitation is known as chromatic aberration, caused by the index of refraction of the lens varying with the wavelength of the light passing through it. As a result of this variation, the focal length of the lens alters according to wavelength, being greatest for red light and smallest for blue, as shown in Fig. 19.21.

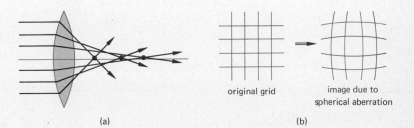

Figure 19.20 Spherical aberration. (a) Light passing through different parts of the lens is brought to focus at different points. (b) Distortion of a grid due to spherical aberration.

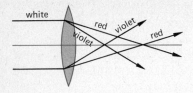

Figure 19.21 Chromatic aberration.

A lens with chromatic aberration tends to split white light slightly into its constituent colors, just like a prism. The image from such lenses appears to be fringed with a halo of red, blue, and other colors. What is most disturbing about this type of fault is that it is impossible to overcome simply and cheaply. By selecting the right type of glass and curvature of the surfaces, one can design lenses that are "achromatic," that is, they do not separate colors, for any two specific wavelengths. But preventing the color dispersion of the whole range of wavelengths that make up white light is a far more complicated matter. Good cameras, for example, normally contain an "achromatic doublet" of two lenses that provides perfect correction for yellow and blue light. This gives excellent results with green light but normally fails to correct fully for violet and red. Really expensive optical instruments contain large numbers of compensating and correcting lenses to overcome spherical, chromatic, and other aberrations. In fact, much of the cost of high-quality cameras pays for these compensatory systems of lenses.

19.8 REFLECTIONS ON MIRRORS

Addition of compensating lenses is not the only approach to overcoming optical aberrations. Astronomers in particular can improve the performance of their instruments remarkably by making simple substitutions of mirrors for lenses. Mirrors of various shapes and sizes are just as valid optical devices as lenses. They have focal points, produce both real and virtual images, and never suffer chromatic aberration because they do not refract light.

To start our analysis of mirrors on the right note, we can consider the plane mirror shown in Fig. 19.22. Because the mirror is plane, it does not have an optical center. However, for convenience we can draw a principal axis perpendicular to O, the geometric center of the mirror. Imagine that a man tying a tie of length y stands a distance p in front of his bathroom mirror to view the knot. The bottom of the tie lies exactly on the principal axis of the mirror as we have defined it. To locate the image of the top of the object—in this case, the knot on the man's tie—we need only construct two rays that obey the law of regular reflection. The first ray, moving parallel to the principal axis, meets the

Figure 19.22 Plane mirror.

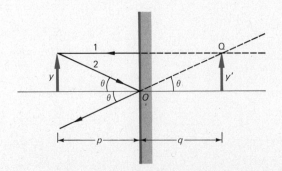

mirror with a zero angle of incidence and is reflected back in the same direction. The second ray, joining the top of the image with the center of the mirror, comes in with an angle of incidence θ and is reflected at the same angle. The two reflected rays diverge on the left-hand side of the mirror, but if they are extended backward, they appear to have come from the point Q. This represents the image of the knot. Geometry shows that it is as far behind the mirror as the tie is in front of it. Furthermore, the image is upright, of the same size as the object, and, because it appears to be behind the mirror where the light can never actually go, virtual. Beyond that, it is also reversed from left to right.

EXAMPLE 19.9

Calculate the minimum height of a plane mirror that enables one to see one's entire reflection. That is, what is the minimum length of a full-length mirror?

Solution

Consult Fig. 19.23. For simplicity, consider your eyes at the very top of your head. Then, to see your head, you look horizontally into the mirror as indicated. To see your feet, the light from your feet must hit the mirror and be reflected into your eyes. Such a ray is indicated in the figure. Clearly, since the angle of incidence equals the angle of reflection, the mirror need be only half your height to allow you to see yourself from head to toe.

You can redraw Fig. 19.23 for the real case of your eyes about 10 cm below the top of your head and arrive at the same conclusion.

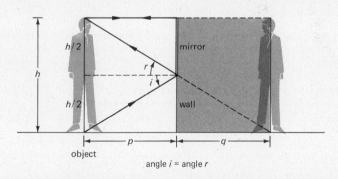

Figure 19.23 Determination of the required length of a full-length mirror.

Plane mirrors are generally used in optical instruments to change the directions of light rays. On the other hand, curved mirrors act more like lenses, insofar as they can focus light. Most curved mirrors are constructed as portions of spheres. Just like lens surfaces, curved mirrors can be either concave or convex.

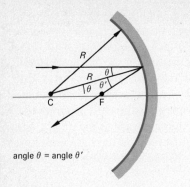

angle θ = angle θ'

Figure 19.24 Concave mirror.

Figure 19.25 Focus of (a) concave and (b) convex mirror.

Figure 19.24 shows a concave mirror whose center of curvature is indicated by the letter C on the principal axis, which we again take for convenience through the geometric center of the mirror. Any ray of light that strikes the mirror is reflected according to the law of regular reflection, as shown for the rays in Fig. 19.24. Geometry indicates that all rays incident on a concave mirror that are parallel to and close to the principal axis pass through the point F on the axis, which is a distance $R/2$ from the mirror, where R is the radius of curvature; $R/2$ is therefore the focal length of the mirror, as indicated in Figs. 19.25(a) and 19.26. The type of reflection produced by a convex mirror is shown in Fig. 19.25(b). The reflected light does not pass through the focus; it appears to emanate from the focal point, as does light refracted by a diverging lens. These two figures indicate the basic rules for locating the image of any object reflected by a spherical mirror. The rules, similar to those for lenses, are:

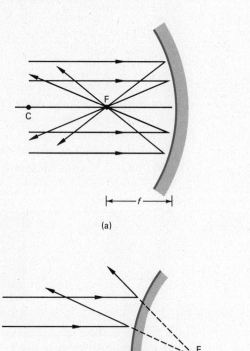

1. A ray parallel to the principal axis passes through, or appears to have come from, the focal point of the mirror after it has been reflected.
2. Rays incident upon the mirror traveling toward the focal point, in the case of a convex mirror, or proceeding from it, in the case of a concave mirror, emerge parallel to the principal axis after they have been reflected.
3. A ray that passes along a radius of a mirror hits the surface with zero angle of incidence and is reflected back along the same path.

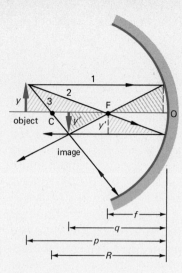

Figure 19.26 Image formation in a concave mirror.

Figure 19.26 illustrates the use of these rules to locate an image reflected in a concave mirror. The object, whose height is y, is placed a distance p from the mirror and beyond the mirror's center of curvature. It produces an image whose height is y' at a distance q from the mirror. The geometrical analysis given here applies only to small apertures, and the constructions should be on the plane through the mirror vertex. Such constructions would only be approximate for the strongly curved mirrors and large apertures used in the accompanying diagrams. These restrictions are equivalent to our previous restrictions with thin lenses.

Examination of the diagram reveals two sets of similar triangles, which are shaded with different tints. These triangles give the relationships

$$\frac{y}{p-f} = \frac{y'}{f} \quad \text{and} \quad \frac{y'}{q-f} = \frac{y}{f}$$

If we solve these two equations simultaneously by eliminating y/y', then we obtain the following expression relating f, p, and q:

$$\frac{y}{y'} = \left(\frac{p-f}{f}\right) \quad \text{and} \quad \left(\frac{y}{y'}\right) = \frac{f}{q-f}$$

Therefore,

$$\frac{p-f}{f} = \frac{f}{q-f} \quad \text{or} \quad f^2 = (p-f)(q-f)$$

Multiplying this equation out and dividing each term by pqf yields

$$\frac{1}{f} = \frac{1}{p} + \frac{1}{q}$$

This is a familiar equation because it is identical to the lens equation [Eq. (19.1)]. Since the focal length of a spherical mirror equals half the radius of curvature ($f = R/2$), this equation can be written

$$\frac{1}{f} = \frac{2}{R} = \frac{1}{p} + \frac{1}{q} \tag{19.6}$$

This equation is called the mirror equation. It applies to both concave and convex mirrors when one uses a sign convention similar to

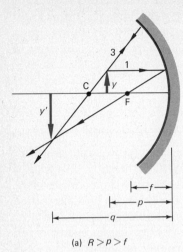

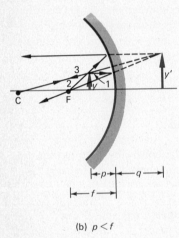

Figure 19.27 Image formation for concave mirror. (a) $R > p > f$. (b) $p < f$.

that applied to lenses. According to the convention, the distances of real objects and images are taken as positive and the distances of virtual objects and images as negative. The radius of curvature, and hence the focal length, is regarded as positive for converging concave mirrors and negative for diverging convex mirrors.

Basic geometry also yields the magnification of the object reflected in a mirror in terms of the distances of the object and image. Recalling that magnification is defined as the ratio of the sizes of the image and the object, and comparing similar triangles, we obtain the relationship

$$M = \frac{y'}{y} = -\frac{q}{p}$$

This is again identical to the formula for lenses.

The arrangement of Fig. 19.26, in which the object is placed beyond the mirror's center of curvature, is frequently used in powerful astronomical telescopes. Mirrors are preferable to lenses in these instruments for two reasons: they do not suffer chromatic aberration, and they can be constructed accurately in much larger sizes than lenses. Such construction is possible because mirrors have only single surfaces and can be shored up from the back to overcome the effects of gravity; lenses, with two surfaces, offer no such opportunity.

Figure 19.27 shows two other forms of reflection from a concave mirror. The object placed between the mirror's center of curvature and focal point in Fig. 19.27(a) produces a real, magnified, but inverted image. The object that stands closer to the mirror than the focus in Fig. 19.27(b) yields a virtual, magnified, and erect image. Women often use this arrangement with a make-up mirror.

Figure 19.28 illustrates the way in which convex mirrors form images. The image of real objects produced by such a mirror is always virtual and erect.

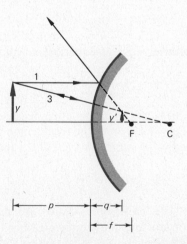

Figure 19.28 A diverging (convex) mirror.

EXAMPLE 19.10

The radius of curvature of a concave mirror is +20 cm. What is the nature of the image formed of an object placed 15 cm from the mirror? Compute the focal length of the mirror and the magnification.

Solution

We can use the mirror equation to solve this problem. First, it is helpful to draw a diagram of the problem, which would be similar to that shown in Fig. 19.27(a) since the object is placed between the center of curvature and the focal point. The focal point is $\frac{1}{2}R$, or 10 cm. Solving the mirror equation, realizing that all quantities are positive, we have

$$\frac{1}{p} + \frac{1}{q} = \frac{1}{f}$$

$$\frac{1}{q} = \frac{1}{f} - \frac{1}{p} = \frac{1}{10 \text{ cm}} - \frac{1}{15 \text{ cm}} = \frac{15 - 10}{(15)(10) \text{ cm}} = \frac{5}{(15)(10) \text{ cm}}$$

$$q = \frac{(15)(10)}{5} = 30 \text{ cm}$$

The image is real. The magnification is given by

$$M = -\frac{q}{p} = -\frac{30 \text{ cm}}{10 \text{ cm}} = -3$$

The negative sign indicates that the image is inverted.

SUMMARY

1. Both reflection and refraction can occur when light strikes a boundary between two media in which the speed of light is different. This difference is expressed by different refractive indices.
2. No refraction can occur when the angle of incidence is greater than the critical angle between the two media involved.
3. The apparent position of an object whose light is refracted is determined by the refractive indices of the two media.
4. Converging lenses are devices that bring parallel beams of light to a sharp focus. Such lenses collimate a point source of light placed at a focal point into a parallel beam. Diverging lenses always form virtual images that cannot be projected onto a screen.
5. The image of any object formed by a lens can be determined by geometrically plotting the paths of two rays.
6. The magnification of any lens is given by the negative ratio of the distances of the image and the object from the lens.
7. The focal length of a thin spherical lens is related to the refractive indices of the lens and the material around it and the radii of curvature of the lens' surface, according to the lens-makers' equation.
8. The positions and magnifications of images of objects whose light passes through more than one lens can be calculated by treating the effect of one lens at a time.
9. The compound microscope forms greatly magnified images of small objects. The astronomical telescope gives highly magnified angular images of heavenly objects.
10. The two most common defects in lenses are spherical aberration and chromatic aberration.
11. Images in mirrors can be determined by tracing the paths of two rays, just as in the case of lenses.
12. Mirrors do not cause chromatic aberrations because they do not refract light.

QUESTIONS

1. Why is blue light bent more than red light when passing obliquely from air into water?
2. A hand magnifier is used as a reading glass. Describe the image formed by the lens.
3. A spectrum may be observed when white light passes through a glass prism or when light passes through a series of narrow slits. Discuss these two phenomena, describing how they differ, how the spectrum is formed in each case, and how the two spectra differ. Which method can be used to determine wavelengths of light using only length measurements?
4. The boundary between two media whose indices of refraction are the same would not be observable. Why?
5. How can you easily find the focal length of a positive lens?
6. What is the advantage of using prisms rather than mirrors to bend light in binoculars?
7. What effect does the distance from a plane mirror have upon the size of the image? The character of the image? The extent of the object contained in the image?
8. Why doesn't light passing obliquely through a plane-parallel plate of glass disperse into a spectrum?
9. What property of diamond makes it sparkle? What internal processes make diamond so attractive?
10. If you are under water and look upward at an angle with the normal to the surface greater than 45°, what would you see?
11. In frequent unidentified flying object reports, headlights from cars on one highway are seen hundreds of feet away on the other side of a swamp shining up into fog. How is this possible?
12. Sunlight from behind you shining on waterdrops far in front of you can produce a rainbow. What causes the spectrum?
13. Why do olives appear to be larger when viewed through the curved walls of a glass jar?
14. How does the focal length of a lens in air change when placed under water?
15. Describe how the image distance and magnification change as a converging lens is moved, from infinity, closer and closer to an object? Is the situation different for a diverging lens?
16. How would you determine the focal length of a diverging lens using a converging lens to assist you?
17. Describe two lens aberrations and how these can be corrected.
18. What are the disadvantages of using lenses in an astronomical telescope?

PROBLEMS

Snell's Law and Internal Reflection

1. A light ray enters a prism with an angle of 30° to the normal. If the refractive index of the glass is 1.5, what is the angle of the ray inside the glass?
2. A wave front is incident at 20° upon a surface between air and a diamond. Calculate the angle of refraction in the diamond.
3. In optical fibers, the fiber of index 1.72 is often covered by a protective glass with index 1.52. What is the critical angle?
4. A light pipe is made of material of index of refraction of 1.45. The light pipe is in air. What is the minimum angle of incidence for which the light will be totally internally reflected down the pipe?
5. If the speed of sound is 380 m/s in warm air high above a lake and 340 m/s in cooler air near the surface, what would be the critical angle for sound in the cool air?
6. The speed of sound is 340 m/s in air and 1496 m/s in water. Compute the index of refraction of sound in water. What is the critical angle in water for sound? What does this mean?
7. A ray of light in air makes an angle of incidence of 45° at the surface of a sheet of ice. The ray is refracted within the ice at an angle of 33°. (a) What is the speed of light in ice? (b) What is the critical angle for the ice? (c) A speck of dirt is embedded 0.75 in. below the surface of the ice. What is its apparent depth when viewed at near normal incidence?
8. A fish looks from below the surface of a lake upward into air at a black fly above the water. The fly is 12 cm above the water. How far above the water does the fly appear?
9. If the bottom of a pond appears to be 2.5 ft deep, how deep is the pond in fact?
10. What is the angle of minimum deviation for an equilateral prism made of crown glass ($n = 1.52$)?
11. Calculate the angle of minimum deviation of a ray passing through a 60° prism if the glass has a refractive index of 1.65.
12. If light is incident at 55° on a parallel plate of window glass 5 mm thick, by how much is the transmitted ray shifted? $n = 1.52$.
13. A ray of light is incident normally on the face AB of a glass prism ($n = 1.52$), as shown in the figure. Find the largest value for the angle ϕ so that the ray is totally reflected at face AC. What is the angle in the glass (measured with the normal to AC) at which the light is reflected from the surface at AC under these

conditions? Consider this when the prism is (a) in air and (b) immersed in water, in which the speed of light is 0.75 c.

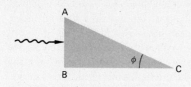

14. Suppose an object were covered with 2 cm of oil of refractive index 1.4. Where does the object really appear to be as viewed by you from directly above?

15. A ray is incident normally on the surface of a prism (45°, 45°, 90°) that is under water. The index of refraction for water is 1.33; for the glass the index is 1.50. (a) Explain (compute angles to back up your explanation) what happens when the ray hits the long side of the prism. (b) What is the critical angle for the glass-water interface? (c) What is the speed of light in the glass? (d) If the top surface of the glass is 10 cm below the surface of the water, where does the surface of the glass appear when viewed in air near normal incidence?

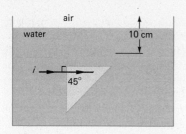

Lenses

16. An object O is placed 5 in. in front of a convex lens of focal length 7.5 in. Sketch clearly on a diagram how to locate the image graphically. Calculate the location of the image.

17. An object is placed 30 cm from a converging lens with focal length 60 cm. Describe the image.

18. An object is 20 cm from a converging lens and a real image appears 40 cm on the other side of the lens. (a) What is the focal length of the lens? If the image is 4 cm tall, (b) what was the actual size of the object?

19. An object 1 mm long is examined with a single lens of focal length $f = 2.5$ cm. Compute the image size and location if the object is (a) 2.0 mm from the lens, (b) 2.4 cm from the lens, (c) 2.6 cm from the lens.

20. A bird is 1.6 m away from you. You attempt to photograph the bird, whose size is 15 cm, and obtain an image 1 cm long that will fit onto a 35-mm slide (35 × 23 mm). What must be the focal length of your lens?

21. How far are the image and object separated if the image is four times larger than the object and the lens has a power of $+20$ diopters?

22. An object is placed two-thirds of the focal length in front of a converging lens. Locate the image and find the magnification.

23. An object is placed 40 cm from a screen. At what distances from the screen must a lens of focal length $f = +8$ cm be placed to obtain an image on the screen? What is the magnification in each case?

24. A glass lens of $n = 1.54$ is constructed with one spherical convex surface of $R = 20$ cm and a plane surface. (a) What is the focal length of the lens? If the lens were placed under water, (b) what would be the focal length of the lens?

25. A lens is constructed of glass of refractive index $n = 1.5$. The lens consists of a concave surface of radius 30 cm and a convex surface of radius 40 cm. What is the focal length of the lens?

26. What is the focal length of a lens, of power 45 diopters?

27. A microscope objective lens has a focal length $f = 16$ mm. What is the magnification for an image distance of 25 cm?

28. A lens of power $+3.0$ diopters is placed 25 cm above a flea 1 mm wide. Describe the image of the flea.

29. A person is able to vary the shape of the lens in her eyes from 60 to 68 diopters. What is the variation in focal length of these eyes?

30. (a) If the distance from the lens in the eye to the retina is 17 mm and the lens has a power of 60 diopters, what is the reading distance? (b) To read at a distance of 40 cm, what must be the power of the eyeglasses required with this lens?

31. A slide projector has a lens with a focal length of 10 cm. The image is projected on a screen 4 m from the lens. (a) What is the distance between the lens and the slide? (b) Is the slide upside down or right side up? (c) What is the magnification?

32. (a) What is the focal length of the lens required in a slide projector if a 35-mm slide is to be projected on a screen so that it is 6 ft across? The projector lens is 25 ft from the screen. (b) What is the magnification?

33. A camera with a lens of focal length 50 mm is used to take the picture of a tree 20 m tall. How far from the tree must the camera be placed to keep the image of the tree 23 mm high, thus allowing it just to fit on a 35-mm slide (35 × 23 mm)?

34. An object is 30 cm from a diverging lens with focal length of -50 cm. Calculate the position of the image.

35. An object is 4 cm from a diverging lens of focal length -12 cm. Locate the image and describe the image.

36. What is the size and image distance of a 1.5-cm object placed 40 cm in front of a diverging lens of focal length -20 cm?

Lens Combinations

37. Two thin lenses, both of focal length 10 cm, the first converging, the second diverging, are placed 5 cm apart. An object is placed 20 cm in front of the first (converging) lens. (a) How far from the first lens will the image be formed? (b) Is the image real or virtual?

38. A 500-mm lens is mounted in contact with a 50-mm lens of a camera. If the lens-to-film distance is 5 cm, for what object distance will a sharp image be formed?

39. If the magnification of the objective of a microscope is $15\times$, what magnification is required of the eyepiece if the overall magnification is to be $300\times$?

40. A microscope has an objective of $f = +0.3$ cm and an eyepiece of $f = 2.5$ cm. Where is the image of the objective if a virtual image for the eyepiece is 30 cm in front of the eyepiece? If the lenses are 20 cm apart, what is the distance from the object to the objective?

41. In a compound microscope the objective lens has a focal length of $f_1 = +2$ cm. The second lens has a focal length of $f_2 = +12$ cm and is placed 17.5 cm from lens 1. An object is located 3 cm from lens 1. (a) Locate the final image and describe its nature. (b) What is the magnification of the system? (c) What is the angular magnification?

42. A telescope has an objective lens of $f = 1$ m and an eyepiece of $f = 2$ cm. A target on a rifle range 6 in. high is viewed at a distance of 75 m with this telescope. (a) What is the distance between the two lenses if the final image is 10 cm in front of the eyepiece? (b) What is the size of the image of the target?

43. A Galilean telescope has a diverging lens of $f = -2.5$ cm as an eyepiece and a positive objective of $f = +30$ cm. An object 80 m away from the objective forms an image 40 cm in front of the diverging lens. (a) How far apart are the lenses? (b) What is the total magnification? (c) What is the angular magnification?

44. If the focal length of the first lens of an astronomical refracting telescope is 11 m and the overall magnification of the telescope is to be 80, calculate the focal length of the eyepiece and the length of the telescope.

45. The radii of curvature of the surfaces of a thin lens are $+10$ cm and $+30$ cm. The refractive index is 1.50. (a) Compute the position and size of the image of an object in the form of an arrow 1 cm high, perpendicular to the lens axis, and 40 cm to the left of the lens. (b) A second lens (diverging) with a focal length of -40 cm is placed 160 cm to the right of the first lens. Where is the final image of this lens system?

46. An image is formed on a screen 20 cm from a converging lens. A diverging lens is inserted midway between the screen and positive lens. It is then necessary to move the screen 20 cm farther away from the converging lens to obtain a sharp image. What is the focal length of the diverging lens?

Mirrors

47. A concave mirror has a focal length of 20 cm. Calculate and determine graphically the image of an object placed at 15 cm, 25 cm, and 75 cm.

48. A concave spherical mirror has a radius of curvature of 4 in. An object in the form of an arrow is placed 8 in. from the mirror. By a ray diagram locate the image indicating clearly how you locate this image. (a) Where would you place an object in front of this mirror so that a virtual image is formed? (b) Calculate the image distance.

49. I wish to find the radius of curvature and focal length of a concave spherical mirror (a positive mirror). When I place an object at 60 cm in front of the mirror, I find an image that has a magnification of -0.5. (a) Where is the image, and is it a real image? Is it erect? (b) What is the focal length and (c) radius of curvature of the mirror?

50. Describe the image of an object placed 20 cm from a concave spherical mirror of radius 60 cm. Indicate on a diagram where the image is located. Determine the image location mathematically.

51. An object in the form of an arrow 10 cm high is placed 30 cm in front (to the left) of a converging lens of focal length 20 cm. If a concave mirror is placed 10 cm to the right of the lens, where is the final image of this system formed? The radius of curvature of this concave mirror is 40 cm. Indicate on a diagram, using two rays, how the image is formed.

52. A 15-cm-long candy cane is 8 cm from the surface of a silvered ball 8 cm in diameter. Where is the image of the candy cane, and how large is it?

53. A convex mirror has a radius of curvature of 10 cm. An arrow 2 cm tall is 5 cm from the mirror. Locate its image using a ray diagram. Is the image real or virtual, inverted or erect, larger or smaller? Note how you draw your rays. Calculate the image distance.

54. If you view yourself in a large convex mirror 0.3 m in radius at a distance of 0.6 m from the mirror, (a) where is the image, and (b) what is its magnification?

55. What is the radius of curvature of a shaving mirror if the magnification is to be three times when one's face is

25 cm from the mirror?

56. An erect object is placed a distance in front of a converging lens equal to twice the focal length F_1 of the lens. On the other side of the converging lens is a converging (concave) mirror of focal length F_2 separated from the lens by a distance $2(F_1 + F_2)$. (a) Find the location of the image formed by the lens alone. (b) Find the location and nature of the final image of the system. (c) Sketch the appropriate ray diagram, noting how you draw the rays used to locate the image.

57. If the 6-in. mirror in an astronomical telescope has a radius of curvature of 1 m, what is the angular magnification of the telescope if it is used with an eyepiece of focal length $f = 8$ mm?

Special Topic
The Eye: The Window of the Brain

Imagine an optical instrument that is guaranteed to continue functioning for the lifetime of its owner, is readily repairable (in most cases) if anything goes wrong with it, requires a minimum of maintenance, and can distinguish such diverse objects as fine print at a distance of little more than a foot and geographic features many miles away with equal facility. In the world of high technology, such a device would represent a major revolution, but we all accept possession of a pair of such objects with equanimity. The instrument is our eyes.

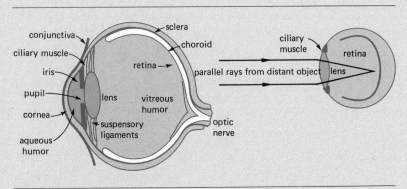

Major parts of the eye.

The eye has often been compared with a camera, and in many ways the comparison is a legitimate one. However, in other vital ways, the eye is both different from and superior to the camera. The diffrences occur in both the makeup and the modi operandi of the two optical devices.

In a simple camera a convex lens focuses light from an object onto film behind the lens. The camera's focus can be adjusted by moving the lens back and forth in relation to the film, and the amount of light reaching the film can be changed by altering the size of the aperture in front of the lens.

In terms of construction, the eye is different right from the start, as shown in the first figure. Instead of being filled with air, as is the case for the camera, the inner eye contains two different types of material—the so-called vitreous humor behind the lens and the aqueous humor in front of it. Both types of humor have the relatively high refractive index of 1.33 and thus refract incoming light by their own existence. In addition, the eye's cornea, which acts as a first line of protection not unlike a transparent lens hood on a camera, has about twice the refractive power of the eye's actual lens.

Plainly, then, the function of the eye's lens is not simply to refract the light coming into the organ. Instead, it sharpens up the image of the light produced on the retina, the coating on the back of the eye that is equivalent to the film in a camera. The lens consists of a flexible, transparent tissue that has a refractive index of about 1.4. The owner of the lens uses it to focus the eye, not by moving it back and forth, as in a camera, but by changing the curvature of its surfaces via the small muscles that actually anchor it in the eye. These changes, which are known as accommodations, can alter the focal length of the human eye from about 17 mm in the relaxed state to a maximum accommodation of about 15 mm. As a result, the power of the human eye can vary from about 59 to 67 diopters. At first glance this may seem a very small range; however, it is sufficient to focus most objects from a "near point" of about 25 cm to distances of many miles on a clear day.

One other most important difference between the camera and the eye concerns the effect of removing the lens; whereas such action puts a camera out of operation, it does not destroy the sight of the eye. If the eye's lens is replaced by a plastic implant as a result of, say, cataracts, its owner can preserve normal vision with an eyeglass lens of about 20-diopters power, although the owner does lose accommodation.

Yet another major difference between camera and eye concerns the image-response mechanism of the film, in cameras, and of the retina, in eyes. Instead of silver compounds, the retina contains more than 100 million light-receptive cells, called rods and cones because of their dis-

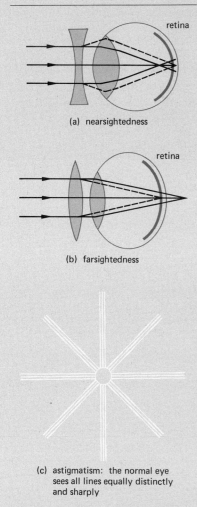

(a) nearsightedness

(b) farsightedness

(c) astigmatism: the normal eye sees all lines equally distinctly and sharply

The three main reasons for poor eyesight.

tinctive shapes. The cones, which tend to be clustered about the exact center of the retina, send clear, sharp images in full color to the brain via the optic nerve. By contrast, the rods are concentrated outside the central area of the retina and record much fuzzier images in black and white. Their role is night vision, which not surprisingly is far less acute than the normal colorful day vision. During both night and day the actual amount of light that reaches the retina is controlled by the eye's iris, which opens and closes in much the same way that a camera's lens stop does.

Despite the lifetime guarantee—in most cases—eyes can develop many defects. Indeed, it is the rare person who goes through life without ever requiring spectacles at some point. Even those people who do not develop problems relatively early in life often fall foul of difficulties that require them to wear eyeglasses in middle age. One typical difficulty of this type is the hardening of the lenses, as a result of which the ciliary muscles that alter their shape can no longer produce the full range of accommodation.

In functional terms, there are three main types of eye defect. The first is nearsightedness, or myopia. This results when the lens of the eye is too thick or the diameter of the eye too long. A myopic person sees such nearby objects as the print in books and newspapers without difficulty but can only distinguish distant objects as blurred images because the rays from such objects cross in front of the retina and thus form fuzzy images. If these rays are to focus squarely on the retina, they must be diverted slightly before they enter the eye. This is accomplished quite simply by use of a concave lens, as shown in the figure.

The opposite type of eye defect is farsightedness, otherwise known as hypermetropia. It occurs when the lens is too thin in the center or the eyeball too short. Rays from nearby objects reach the retina too soon, before they can be brought to a focus. Hypermetropic people have difficulty reading and generally seeing objects right in front of their noses, although they have hardly any problem viewing distant objects. The solution in this case is a corrective converging convex lens, as shown in the figure.

The third common defect, astigmatism, often accompanies the other two and results when the lens of the eye is slightly distorted and causes objects in a particular plane to be seen more or less distinctly than those in other planes. The accompanying figure, part (c), represents a do-it-yourself test for astigmatism. If your eyes are normal, all the lines in the figure should appear equally distinct and sharp to you. If any set of lines appears darker than the others, turn the book through 90°, and see if the perpendicular set, now in the position occupied by the darker lines at first, appears darker than the rest. If it does, you might well be astigmatic and should see an optician or ophthalmologist. Fortunately, the remedy for astigmatism is as painless as that for myopia and hypermetropia: eyeglasses containing lenses ground in such a way that they exactly compensate for the distortions in the lenses of the eyes.

Relativity

20.1 RELATIVE MOTION

20.2 THE SPECIAL THEORY OF RELATIVITY

20.3 SPACE AND TIME

SHORT SUBJECT: The Physicist of the People

SHORT SUBJECT: Faster Than Light?

20.4 THE TRANSFORMATIONS OF LORENTZ

20.5 TIME DILATION

SHORT SUBJECT: Time Flies: The Proving of the Special Theory of Relativity

20.6 VELOCITY TRANSFORMATIONS

20.7 MASS AND ENERGY

20.8 THE GENERAL THEORY OF RELATIVITY
Summary
Questions
Problems

SPECIAL TOPIC: The General Theory of Relativity: Into the Fourth Dimension

Nature and Nature's laws lay hid in light:
God said, Let Newton be! And all was light.

This couplet by the poet Alexander Pope bore testimony to the reverence with which Isaac Newton was held by his peers. Two centuries later, another British poet, J. C. Squire, penned an equally reverential although tongue-in-cheek tribute to the man who inherited Newton's mantle:

It did not last: the devil howling "Ho,
Let Einstein be!" restored the status quo.

This rhyme admirably captures the mind-boggling effect that Einstein had on his colleagues in the scientific world, and on educated people everywhere. His special and general theories of relativity, developed to explain a number of physical effects that turn-of-the-century scientists could not understand according to Newton's laws, seemed to turn the world of the commonplace upside down. They introduced concepts and predicted effects totally alien to everyday experience—the bending of light beams by the sun and other massive cosmic objects, the equivalency of mass and energy, the gradual loss of time by a clock placed at earth's equator with respect to an identical clock kept under exactly the same conditions at the North or South Poles, and the constancy of the speed of light irrespective of whether its source is moving. All these concepts have subsequently been verified by experiment. Einstein's ideas also pointed the way to two of the most far-reaching technical developments, for good or evil, that the world has ever known: nuclear fission and nuclear fusion.

From a philosophical point of view, the real message of Einstein's work is that we on planet earth have a very narrow view of the workings of the universe as a whole. We do not ourselves travel at speeds close to that of light. We do not experience situations in which the concept of time as we understand it becomes meaningless. We do not deal with collisions in which a measurable amount of rest mass is converted into energy. For us, and for physicists dealing with the everyday phenomena that form the bulk of the subject matter of this book, Newton's "common sense" views of how the world works are quite sufficient to explain everything that we observe. But Newton's laws are sufficient only because they hardly differ from Einstein's laws in these restricted circumstances. Newton's theory was just incomplete.

In the more extreme conditions with which we are not familiar, such as the interiors of stars and atoms, the theories of Newton and Einstein predict quite different effects; here, it is Einstein's theory that accounts for the observed facts more accurately.

20.1 RELATIVE MOTION

To start our initiation into Einsteinian physics, we must return to one of the first subjects that we treated in this book: relative motion. In Chap. 2 we learned that all motion is really relative. To talk about any motion in a meaningful sense we must measure distances and displacements with respect to some specific reference point or frame of reference.

If we travel across the country in a jumbo jet and at some point leave our seat to stroll briskly down the aisle, we can relate our movement to a number of different frames of reference. Our speed relative to the seat, which is fixed to the airplane, is perhaps 4.0 mi/h (1.8 m/s). If the airplane's speed with respect to the earth is 600.0 mi/h (about 268.2 m/s) and we move toward the front of the plane, then our speed relative to the surface of the earth is 604.0 mi/h (270 m/s).

We could complicate matters no end by insisting on the center of the earth as our reference point. The speed of different points on the earth's surface can vary by as much as 1000 mi/h (0.50 km/s) with respect to this frame of reference, owing to the earth's rotation. The earth is moving with respect to a frame of reference fixed on the sun. The sun is moving with respect to the center of our galaxy. The galaxy is, in turn, moving relative to a cluster of galaxies of which it is but one.

Is there any end to this series of frames of reference—a point somewhere in the universe that can be used as an absolute frame of reference against which the velocities of all objects in the cosmos should be measured? And if such a frame existed, how could physicists recognize it?

The second question is answered much more simply than the first. An absolute frame of reference would also be an inertial frame of reference. **An inertial frame of reference is one in which the law of inertia (Newton's first law) works exactly without any need of correction.** The earth only approximates an inertial frame of reference because every point on its surface is accelerating toward its axis, and it is itself accelerating in its orbit around the sun; consequently, we should include the acceleration of the earth when we measure the motion of objects on our planet. Since Newton's laws do adequately describe motions on earth, the earth must be a very good approximation to an inertial frame. But if we could somehow find a frame of reference that did not require these corrections, then we would have our absolute cosmic reference point.

The search for such a point was just one of a series of related topics that had scientists and philosophers in an intellectual furor during the last quarter of the nineteenth century. Maxwell's equations for electrodynamics raised some vexing questions related to the whole problem of an absolute point of reference. How, for example, was time involved

Figure 20.1 Motion of the focus of a telescope in the ether.

in electrostatic attraction and repulsion between charged particles? Could the forces between charges "transfer" instantaneously through space? What was the nature of the medium that supported the wave movements of electromagnetic radiation as it propagated at the speed of light? And relative to what frame of reference did one measure the speed of light?

One catch-all solution to these problems seemed to be the existence of what physicists called "the luminiferous ether," which was assumed to be an invisible medium distributed with equal density throughout the universe. It would support the waves of electromagnetic radiation and provide a ready and absolute frame of reference. The concept of the ether did not solve all the questions of time, but it did seem to account for the observed physical facts better than any other idea. One problem, however, was that no one seemed able to detect the ether.

Because ether was supposed to be the medium supporting electromagnetic radiation, one obvious means to detect the ether was to study such forms of radiation, the most obvious of which was light. A potential method of using light this way to observe the existence of the ether is illustrated in Fig. 20.1. If the telescope shown is at rest, then light from a distant star will come to a focus on the cross hairs at P, the focus of the objective. We assume that the light is a well-behaved wave whose speed c is determined by the nature of the ether, just as the elastic properties of a solid determine the speed of sound in a solid. Now imagine that the telescope is moving toward the star. While the wave is traveling toward point P, the moving telescope will carry the cross hairs to point P'; hence, the eyepiece would have to be refocused to view the image of the star. In other words, the focal length of the eyepiece would change according to whether the telescope were moving toward or away from the star through the ether. But when this experiment was tried, no such change in focal length emerged within the limitations of the experimental error.

Scientists were ready with explanations for this apparent experimental failure. Fresnel suggested that moving, transparent objects drag the ether and hence light along with them through the ether. The Dutch physicist Hendrik Lorentz, who applied electromagnetic theory to the problem, argued that the experiment was simply too insensitive to capture the effect. Using the movement of the earth in its orbit to provide the approaching and retreating halves of the experiment, Lorentz calculated that the alteration in the relative speed of light would be only 4 parts in 10^8.

This calculation prompted two American physicists, Albert Michelson and E. W. Morley, to devise an ultrasensitive instrument to seek such tiny changes in the speed of light. The Michelson interferometer, pictured in Fig. 18.13, was used to detect evidence of the relative motion of the earth with respect to the ether. The instrument is sketched in detail, omitting the compensating glass, in Fig. 20.2. It is set up in such a way that the optical path length, l, is exactly the same for

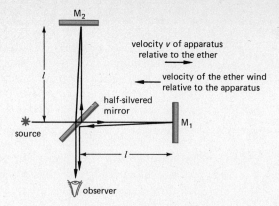

Figure 20.2 Diagram of the light paths in the Michelson interferometer as seen by an observer at rest with respect to the apparatus.

the two beams of light that travel at right angles after striking the half-silvered mirror. If we assume that the whole instrument is moving to the right, relative to the ether with a speed v, or, alternatively, that the ether "wind" is moving from right to left through the laboratory with the same speed, then we would expect the travel times of the two beams to differ.

To understand why this is so, we must delve a little further. A wave of light moving toward mirror M_1 has a speed relative to the apparatus of $c - v$ because the wave travels with a speed c relative to the ether and the apparatus is moving at a speed v relative to the same ether. If the instrument were moving with the speed of light, the speed of the wave relative to it would be zero, for example. For the return trip, the speed of light reflected from the mirror M_1 is plainly $c + v$. Hence, the time that the light takes to make the round trip between the half-silvered mirror and M_1 is

$$t_1 = \frac{l}{c - v} + \frac{l}{c + v} = \frac{2lc}{c^2 - v^2} = \frac{2lc}{c^2\left(1 - \frac{v^2}{c^2}\right)} = \frac{2l}{c\left(1 - \frac{v^2}{c^2}\right)}$$

During the same time, the other ray of light, which moves between the half-silvered mirror and mirror M_2, takes the strange path outlined in Fig. 20.3 owing to the relative motion of the ether and the instrument. If t is the time the light takes to travel from the half-silvered mirror to M_2, we know that the interferometer moves a distance vt and the light, a distance ct, as shown in Fig. 20.3. To return to the half-silvered mirror, the light again takes a time t, moving a distance ct while the device moves a distance vt. The distances vt, ct, and l make up a right triangle, and hence, from geometry, we know that

$$c^2 t^2 = l^2 + v^2 t^2$$

Solving for the time the light takes to reach the mirror and multiplying by 2 to obtain the total time for the journey to M_2 and return, namely, t_2, we obtain

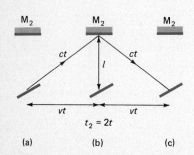

Figure 20.3 Light path to M_2 in the Michelson interferometer if the interferometer moves relative to the ether.

$$t^2(c^2 - v^2) = l^2$$

$$t^2 = \frac{l^2}{c^2 - v^2}$$

$$t = \frac{l}{\sqrt{c^2 - v^2}}$$

$$t_2 = 2t = \frac{2l}{\sqrt{c^2 - v^2}} = \frac{2l}{c\sqrt{1 - \frac{v^2}{c^2}}}$$

The binomial expansions known as Maclauvier's series, which are valid for values of $|x|$ less than 1, are:

$$\frac{1}{1-x} = 1 + x + x^2 + x^3 + \cdots$$

$$\frac{1}{(1-x)^{\frac{1}{2}}} = 1 + \tfrac{1}{2}x + \tfrac{3}{8}x^2 + \cdots$$

Clearly, t_1 and t_2 are not the same. To compare them, we can employ a mathematical approximation by using what mathematicians call a binomial expansion. If we let $x = v^2/c^2$ and expand our expressions for t_1 and t_2 but keep only the first power of v^2/c^2, we obtain

$$t_1 = \frac{2l}{c\left(1 - \frac{v^2}{c^2}\right)} = \left(\frac{2l}{c}\right)\left(1 + \frac{v^2}{c^2}\right)$$

$$t_2 = \frac{2l}{c\left(1 - \frac{v^2}{c^2}\right)^{\frac{1}{2}}} = \left(\frac{2l}{c}\right)\left(1 + \frac{1}{2}\frac{v^2}{c^2}\right)$$

Hence, the time difference $t_1 - t_2$ is

$$t_1 - t_2 = \left(\frac{2l}{c}\right)\left(1 + \frac{v^2}{c^2} - 1 - \frac{v^2}{2c^2}\right) = \left(\frac{l}{c^3}\right)(v^2)$$

Our use of just the first power of v^2/c^2 in the expansion is justifiable because v is normally much smaller than c, which means that v^2/c^2 is small and subsequent powers of the expansion are even smaller. This difference is observed as a shift in the pattern of interference fringes, each shift of one fringe corresponding to a path difference of a single wavelength of the light. The time difference can be converted into the difference in the distance traveled by the two rays, D, using the expression

$$D = c(t_1 - t_2)$$

The number of wavelengths shifted is the actual difference in distance divided by the wavelength of the light,

$$\frac{D}{\lambda} = \frac{c(t_1 - t_2)}{\lambda} = \frac{lv^2}{\lambda c^2}$$

Michelson and Morley made their interferometer large, with the distances between the half-silvered mirror and the two reflecting mirrors both the equivalent of 11 m, to maximize this shift. This was accomplished by mounting the interferometer on a 1.5-m² stone and adding mirrors to reflect each beam back upon itself several times, thus increasing the light paths. To measure the effect they rotated the apparatus 90°, which resulted in doubling the shift. It was necessary for them to rotate the apparatus so they could compare the two results made at 90°. The first measurement served as a baseline and by itself did not provide sufficient information as to whether a shift did exist. The stone mount of the apparatus was rested upon an annular wooden float that in turn floated on mercury (Fig. 20.4). The two physicists used sodium light with a wavelength of 5.9×10^{-7} m, and for the term v in the equation they substituted earth's speed in its orbit around the sun -3×10^4 m/s. Given that the speed of light was 3×10^8 m/s, the equation readily predicted the anticipated shift:

$$\frac{2lv^2}{\lambda c^2} = \frac{(2)(11 \text{ m})}{5.9 \times 10^{-7}} \times \frac{(3 \times 10^4)^2}{(3 \times 10^8)^2} = 0.37 \quad \text{or roughly one-third fringe}$$

The apparatus was capable of measuring a shift of just one-hundredth of a fringe, that is, one-hundredth of the wavelength of light. Yet Michelson and Morley detected no shift whatever.

This negative result raised serious questions about the reality of the ether. Not surprisingly, theorists advanced a number of explanations to account for the unexpected finding. One explanation involved the idea that the earth drags the ether along with it as it moves through space. This meant that the earth had no relative velocity with respect to the ether; the term v is zero, and thus the observed shift is also nothing. Models of this situation indicated, however, that there should be a small shift if the experiment were carried out on a mountain top. It was, but no shift was observed, and the theory bit the dust.

Another proposal contradicted an important cornerstone of wave theory by supposing that the velocity of light assumed the velocity

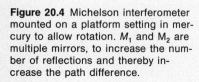

Figure 20.4 Michelson interferometer mounted on a platform setting in mercury to allow rotation. M_1 and M_2 are multiple mirrors, to increase the number of reflections and thereby increase the path difference.

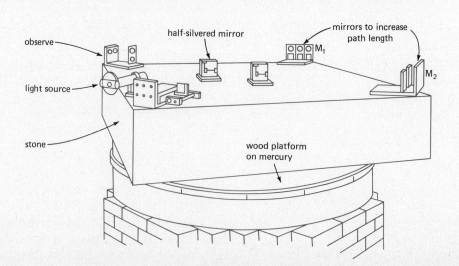

of its source. Physicists discounted this theory by carrying out the Michelson-Morley experiment with light from the sun and the stars as the basic sources rather than the sodium lamp in the laboratory. The fringe shift still failed to show.

In 1895, the Irish physicist George Francis Fitzgerald and the Dutchman Lorentz devised an even more bizarre theory. They argued that the null effect was caused by the interferometer actually contracting in the direction parallel to its movement relative to the ether by a factor $\sqrt{[1 - (v^2/c^2)]}$. Physicists would not be able to measure this contraction, they argued, because their rulers would also shorten when placed next to the interferometer. Scientists tried to get a reading on the validity of that theory by carrying out the experiment with one arm of the interferometer longer than the other. These efforts again produced no evidence for the fringe shift, and the theory was discounted. Ten years after it was first advanced, however, it turned out to be correct, but only part of the reason for the zero shift.

As theory after theory foundered on the rocks of experiment, the truth slowly dawned on physicists: the only satisfactory conclusion of the Michelson-Morley experiment and those that followed was that the ether was nonexistent. Thus, the way was open for Albert Einstein to present his revolutionary view of how the universe could function without an ever-present ether.

20.2 THE SPECIAL THEORY OF RELATIVITY

Einstein was apparently unaware of the Michelson-Morley experiment when he developed his special theory of relativity.

When a local train is passed by an express train traveling in the same direction, passengers in the local train cannot be sure when they look at the express whether their own train is going backward.

Einstein rejected the necessity of an ether that filled all of empty space. He also threw out the idea of an absolute frame of reference. As he saw things, only relative motion was important. The absolute motion that was so valued by Newtonian physicists was, in Einstein's view, of no consequence. You can glimpse some of the thrust of Einstein's thinking by imagining that you are at the wheel of a car that drives up behind another vehicle at a stoplight near the top of a hill. While you wait for the light to change to green, you notice that the space between your car and the other one is rapidly decreasing. For purposes of avoiding an accident, it is hardly important whether your car is creeping forward or the other one is slipping back. The significant fact is that there is relative motion between the two cars. That one car is at rest with respect to a third object, the earth, is inconsequential. If you want to avoid crunching your front end, you will quickly go into reverse. Another common example of the effect is when a local train is passed by an express train traveling in the same direction. Passengers in the local train cannot be sure when they look at the express whether their own train is going backward.

Einstein's special theory of relativity is restricted to inertial frames of reference, that is, reference frames which are not accelerating. It is this restriction that makes it a "special" theory in the terminology of physics. As a result of the limitation, we know that any two reference frames in which we can apply the theory move with a constant velocity with respect to each other.

The first postulate of Einstein's theory states simply that the laws of physical phenomena are the same when stated in all inertial reference frames. This seems to be a statement of the obvious, but it is in fact a profound concept without which the world and universe would have no consistency. It means that the laws of physics are exactly the same in all uniformly moving frames of reference. If, for example, we travel at a uniform speed in a straight line with respect to the earth—in, say, a closed box car in a train, or in a plane—then we know from the first postulate that the laws of physics on the train or plane will be just the same as those on earth (which we can treat approximately as an inertial reference frame). There is no experiment you can perform in your closed surroundings, according to Einstein, that will reveal your frame of reference to be any different from the frame of reference which is the earth. You can try timing a pendulum, but you will find the same relationship for its period to other parameters that you find on earth. You can toss a ball and it will experience the same change in its momentum and kinetic energy as it does on earth. You can, of course, devise ways of detecting whether the closed train or airplane is *accelerating* horizontally with respect to a reference frame. However, we already specified that we are dealing only with frames of reference that are not accelerating with respect to one another.

The requirement that we deal with frames of reference that are not accelerating with respect to each other does not prevent us from examining accelerating objects from the perspective of two inertial reference frames which move relative to each other at constant velocity.

Simple as it may seem, Einstein's first postulate led him directly to the second, more disturbing, one: **the speed of light is the same in all directions and is a universal constant. It is independent of the motion of its source and of the observer,** and it is always exactly the distance traveled by the light divided by the time it takes to travel that distance as measured by any observer in the observer's own frame of reference.

This statement has enough hidden meaning to require closer examination. If, say, a rocket ship moving past the earth at a constant velocity, regardless of its actual speed, sends out a flash of light to communicate with earthlings, we on our own planet would measure the speed of that flash as c. The beings on the rocket would similarly measure speed of the flash as c. If we send a return signal in the form of a light beam from earth to the rocket ship, then both we and they would again measure the speed of the signal as c, no matter how fast the rocket is traveling. This is obviously strange and quite out of line with our intuitive feelings about the way the world works—feelings that are more in tune with the conventional, Galilean, relativity with which we tried to analyze the Michelson-Morley experiment.

The explanation of this peculiar effect is bound up in the concepts of space and time. The observers moving relative to each other actually measure different distances and different times that, when divided, yield the same value of c for the velocity of light. The differences in space and time as seen by observers moving at different speeds in different reference frames are real in each case. However, the differences become large enough to be observable only when the relative speed between different frames of reference starts to approach the speed of light. Under the normal conditions to which we are accustomed, space and time appear to behave in just the way that Newton assumed they behave.

20.3 SPACE AND TIME

This is a convenient point at which to examine how Newtonian mechanics deals with relative motion, the subject that we can best refer to as Galilean relativity, as opposed to Einsteinian relativity. A simple way to define a point in space according to Newtonian mechanics is to refer to it with reference to a set of rectilinear coordinate axes. We locate our point by pinpointing the center (O) of the coordinate axes and stating the x, y, and z coordinates of the point. Then if we further specify the time, we have a complete description of the point in the four dimensions of space and time. According to Galilean relativity, we resolve all physical phenomena into a sequence of events marked by the passage of time. Time itself is absolute (unchanging) and flows continuously—a necessary assumption that allows us to make meaningful use of such concepts as velocity and acceleration, which are intimately tied up with time.

Figure 20.5 shows two inertial frames of reference F and F'. Frame F' moves to the right along the x axis of frame F at a constant relative speed v. In our classical treatment of the physical world, time intervals such as those measured by the period of a swinging pendulum would be the same whether measured in the F or F' frame of reference. For simplicity, we can define the value of time t as zero at the instant that the origins O and O' of the two frames of reference coincide. Time starts flowing onward, and the F' frame moves at a constant speed v to the right, away from the origin of the F frame. After an interval of time t, the origins of the two frames of reference are obviously separated along the x axis by a distance vt. (Of course, the y and z coordinates are exactly the same in both reference frames because the relative movement is taking place just along the x axis.)

Suppose that an explosion booms out at the point P in Fig. 20.5. Observers in the two frames of reference can locate the point of the explosion using their own coordinates. The positions of the point in space and time as measured in the two systems are related by the equations

$$\begin{aligned} x' &= x - vt \\ y &= y' \quad z = z' \quad t = t' \end{aligned} \quad (20.1)$$

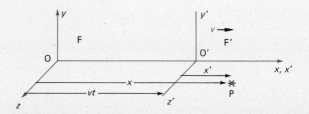

Figure 20.5 Two inertial reference frames. F' reference frame moves to the right with speed v relative to the "at rest" frame of reference F.

The Physicist of the People

The world has never known a scientist comparable to Albert Einstein. He was a veritable legend in his own time, the first man of science since Newton to become the subject of major public adulation. The reverence came about despite the fact that few members of the public had more than an inkling of what Einstein had wrought as a physicist. All people recognized him as a great humanitarian as well as a great scientist.

Einstein's standing in the scientific community was equally high. He was the man selected by his peers in 1942 to acquaint United States President Franklin D. Roosevelt with the scientific potential for producing an atom bomb and to urge Roosevelt to sponsor an American project to develop such a weapon before the Nazis could.

In his early years in the German town of Ulm Einstein gave little indication of the greatness to come. He was such a slow learner as a child that many of his family's neighbors assumed he was retarded; in fact, he could not speak clearly until he was 9 yr old. In later life Einstein was to credit this slow development as a source of his ability to think the unthinkable because, as he claimed, he never asked the normal childhood questions about the universe and so was not taken in by the normal, naïve answers.

Even in high school Einstein had far from a meteoric career. His only strong subject was mathematics, and he failed to graduate. Nevertheless, he managed to win, by virtue of an examination, a college entry to the Swiss Federal Polytechnic School. Graduating from there in 1901 at the age of 22, he could not find the academic post he wanted and thus had to settle for a minor, low-paying job at the Swiss Patent Office in Berne. The advantage of the job was that it gave him free time to consider the nature of the physical universe, and

The physicist of the people: Albert Einstein. (*Wide World*)

These equations are called the Galilean space-time transformations. If we decided to analyze the laws of physics from the points of view of both frames of reference, we would need to know how the quantities such as length, time intervals, and velocity transform from one frame to the other. Since the y and z coordinates and time intervals are the same in both systems, then it is clear that the y and z components of velocity, acceleration, momentum, and similar quantities will also be the same in both frames. The difference between the two systems occurs because of the relative motion along the x axes. The magnitude of any displacement along the x axis is given by the following two expressions for the two frames of reference:

$$\Delta x' = x'_2 - x'_1$$
$$\Delta x = x_2 - x_1$$

Einstein used this time to maximum advantage.

In 1905, Einstein's philosophizing came to magnificent fruition: he published four scientific papers that revolutionized physics. One paper concerned Brownian motion—the jittery movements of tiny pieces of pollen in water. Another paper dealt with the photoelectric effect, which we shall meet in the next chapter. The remaining two papers outlined the most important contribution of all: the special theory of relativity.

Given Einstein's obscurity, the scientific world took its time to digest his thoughts. But in 1913, the Kaiser Wilhelm Institute of Physics in Berlin created a post for the budding genius, even though he had become a Swiss citizen by then. At that time Einstein was working on a grand extension of the special theory, which was published in 1916 as the general theory of relativity. One major prediction of the theory was that gravitational objects such as the sun would markedly deflect light. On March 29, 1919, the scientific world had a literally heaven-sent opportunity to put the prediction to the test during an eclipse of the sun in a part of the sky that contained a number of very bright stars. The British dispatched two eclipse expeditions to the tropics to measure the apparent positions of the stars and compare them with their positions when the sun was not near the paths of their light to earth. A few months later, in one of the most dramatic moments in scientific history, astronomer Sir Arthur Eddington announced that the sun had bent the starlight, just as Einstein had predicted. (The report of an American eclipse expedition that had failed to spot the bending the previous year was virtually ignored; the center of gravity of science at the time was fixed firmly in Europe).

With Eddington's announcement Einstein's fame was established worldwide. It was not sufficient, though, to insulate him from the political currents of the times. He was lecturing in the United States when Hitler and the Nazis came to power in Germany with their antisemitic creed, and his Jewish ancestry made it impossible for him to return. In fact, fervent Nazi physicists refused to teach relativity theory because of Einstein's association with the subject. Thus, Einstein settled at the Institute for Advanced Study in Princeton, and for the rest of his life he undertook the ultimately fruitless effort to develop a unified field theory linking the gravitational and electromagnetic forces and the totally successful task of inspiring all who came in contact with him.

These two equations can be related using Eq. (20.1):

$$x'_2 - x'_1 = (x_2 - x_1) - v(t_2 - t_1)$$

or

$$\Delta x' = \Delta x - v\,\Delta t \tag{20.2}$$

What is the component of velocity along the x direction of an object seen in the two reference frames? Imagine that someone in the F' frame throws a ball with velocity

$$\frac{x'_2 - x'_1}{t'_2 - t'_1} = \frac{\Delta x'}{\Delta t'} = \frac{\Delta x'}{\Delta t}$$

to the right. Note that $\Delta t' \doteq \Delta t$. Then the velocity of the ball as seen by an observer in the F reference frame will be

Faster Than Light?

No scientific theory is immune from modification, and even Einstein's special theory of relativity has been critically examined by physicists who, although accepting its general precepts, argue that it might be improved by minor changes. One of the most unusual modifications suggested by theorists appears to strike at the heart of Einstein's second postulate. The modification suggests that particles can actually travel at speeds beyond that of light. However, theoretical studies of the idea indicate that it can be incorporated into Einstein's theory without destroying its whole fabric.

The assumption which allows these particles to coexist with the theory is that the particles *always* travel faster than light. As long as particles do not have to break the light barrier, by going from speeds below that of light to speeds above it, there is no fundamental objection to

Shape of corona at time of minimum sunspot activity—total eclipse of June 9, 1918. (*Hale Observatories*)

their existence. To add an extra bizzare touch, theory predicts that the particles' speed increases as they lose energy! But until physicists come up with some means of detecting the particles, which are known as tachyons, from the Greek word for speed, it seems that the particles will remain no more than interesting theoretical possibilities.

$$\frac{x_2 - x_1}{t_2 - t_1} = \frac{\Delta x}{\Delta t}$$

Using the transformation relations of Eq. (20.2), we can relate these two expressions as follows:

$$\frac{\Delta x'}{\Delta t} = \frac{\Delta x}{\Delta t} - v \quad \text{or} \quad \frac{\Delta x}{\Delta t} = \frac{\Delta x'}{\Delta t} + v \qquad (20.3)$$

We can best justify these equations by considering a real situation, say a boy standing on a flatbed truck that is moving at 55 mi/h along a straight superhighway relative to an onlooker by the roadside. If the boy throws a ball at 5 mi/h in the direction of the truck's motion, then $\Delta x'/\Delta t = 5$ mi/h is the speed that he measures in his frame of reference.

But as seen by the onlooker on the roadside, the speed $\Delta x/\Delta t$ of the ball is the sum of the speeds of the truck and the ball, that is, 60 mi/h.

Conversely, if the boy on the flatbed truck throws the ball toward the rear of the truck at a speed of 10 mi/h, then $\Delta x'/\Delta t = -10$ mi/h. The speed of the ball along the x direction as seen by the observer on the roadside is determined by

$$\frac{\Delta x}{\Delta t} = \frac{\Delta x'}{\Delta t} + v = (-10 + 55) \text{ mi/h}$$

$$= +45 \text{ mi/h in direction truck is moving}$$

If the boy on the speeding truck were to throw the ball at a speed of 55 mi/h toward the back of the vehicle, then the onlooker by the side of the road would observe a speed of zero for the ball. The ball would be dropping to the ground in free fall, and its path in the view of the onlooker would be straight down to the road (assuming that the flatbed of the truck is short). But the boy who threw the ball would observe the ball moving in a parabolic path away from him. Despite the differences in observation, the ball obeys Newton's laws in both frames of references.

EXAMPLE 20.1

A passenger walks *forward* along the aisle of a train at a speed of 2 mi/h as the train moves along a straight track at a constant speed of 55 mi/h with respect to the ground. What is the passenger's speed with respect to the ground?

Solution

If we choose the train to be the moving or F' system and the ground to be the F system, then $v = 55$ mi/h. Knowing the speed of the passenger in the moving system $\Delta x'/\Delta t = +2$ mi/h, we need to find the corresponding speed $\Delta x/\Delta t$ in the frame of reference fixed on the ground. Using Eq. (20.2),

$$\frac{\Delta x}{\Delta t} = \frac{\Delta x'}{\Delta t} + v = (2 + 55) \text{ mi/h} = 57 \text{ mi/h}$$

EXAMPLE 20.2

Consider dropping a ball in the following manner. Event 1 is the release of a ball from height h, as shown in Fig. 20.6. Event 2 is the collision of the ball with the floor. What are the spatial and temporal separations of these two events as seen by each observer, one in the moving car and the other on the ground?

Solution

The vertical displacement $y_2 - y_1$ or $y'_2 - y'_1$ are equal to $-h$, and the

Figure 20.6 Dropping a ball as seen in two fixed frames of reference: (a) in the train, and (b) on the ground.

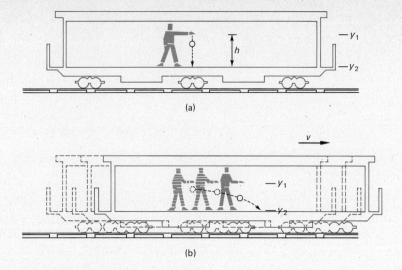

transverse displacement $\Delta z = \Delta z' = 0$. The time of fall in both frames of reference is given by $h = \frac{1}{2}g(\Delta t)^2$, the relation for free fall, so

$$\Delta t = \Delta t' = \sqrt{\frac{2h}{g}}$$

For the displacement in the direction of relative motion—the x direction—there is no displacement in the moving system:

$$\Delta x' = 0$$

This means that to the observer moving in the car, the object falls, landing directly beneath the point of release [Fig. 20.6(a)]. However, for the observer on the ground, Fig. 20.6(b), the displacement Δx is given by Eq. (20.2):

$$\Delta x = \Delta x' + v\,\Delta t$$

or

$$\Delta x = 0 + v\sqrt{\frac{2h}{g}}$$

The purpose of the preceding example is to help you become familiar with the way in which one may think about events and how they are observed. The next obvious step is to extend this Galilean view of relativity to ultrahigh speeds close to the speed of light. Imagine two electrons ejected in opposite directions from the filament of an electron tube that is at rest in a laboratory, as shown in Fig. 20.7. A physicist measures the speed of each electron as two-thirds of the speed of light,

Figure 20.7 Two electrons ejected from a filament. The intuitive Galilean approach to addition of speeds or velocities does not work.

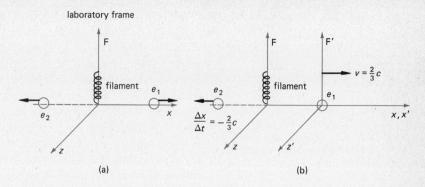

that is, $\tfrac{2}{3}c$. She then wishes to calculate their relative speed using Galilean relativity.

To work this out, assume that the laboratory itself is the F frame of reference, and the electron e_1 that is moving to the right relative to the laboratory with a speed of $+\tfrac{2}{3}c$ is the F' frame. In the laboratory frame of reference, the other electron, e_2, rushes to the left with a speed $\Delta x/\Delta t = -\tfrac{2}{3}c$. Now, the quantity of interest is the speed with which e_2 moves as seen by an observer in the F' frame of reference, that is, $\Delta x'/\Delta t'$. From Eq. (20.3),

$$\frac{\Delta x'}{\Delta t'} = \frac{\Delta x}{\Delta t} - v = -\frac{2}{3}c - \frac{2}{3}c = -\frac{4}{3}c$$

This answer seems logical enough. The electron e_2 is moving to the left (hence the negative sign), away from electron e_1, with a speed equal to the sum of the speeds of the two electrons as they shoot out of the filament. It is the commonsense answer that we also would have obtained by making the F' frame ride along with electron e_2. Galilean relativity predicts that the relative speed of separation of the two electrons in this case is $\tfrac{4}{3}c$. But literally thousands of experiments in physics laboratories have shown that this commonsense answer is wrong. This is a starting point for our excursion into modern relativity theory.

20.4 THE TRANSFORMATIONS OF LORENTZ

The equations that relate space and time as derived from Einstein's two postulates are called the Lorentz transformations, after the man who developed them to explain the null result of the much-touted Michelson-Morley experiment. We do not intend to weave through the complex mathematical formulations whereby the Dutch physicist derived his formulations, but we can explain how such transformations are justified. Once more we consider two frames of reference, F and F', which move with a speed v relative to each other along the x axis alone, as shown in Fig. 20.8. Since no motion takes place in the y or z directions, we have no reason to expect that the y and z coordinates of any point in space will have different values in the two coordinate

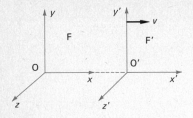

Figure 20.8 Two inertial reference frames moving at speed v relative to one another.

systems. Once more we can take the starting conditions as the instant at which the origins of the two frames pass each other; in other words, $x' = x = 0$ when $t' = t = 0$. An event will be measured by values (x, t) in one reference frame and by (x', t') in the other reference frame. If the unexpected happens and t' is different from t, we expect that at least these differences will increase in direct proportion to the values of x and t because of the homogeneity of space and the uniformity of the natural laws in variable time. In other words, any discrepancy in the values of $(x' - x)$ and $(t' - t)$ should increase over the second kilometer in just the same way that it does in the first kilometer from the origin of the frames of reference at a given time t, and discrepancy in time at the position x should increase during the second second in just the same way that it does over the first. Put mathematically, the relationship between space and time should be linear.

The transformations are given by

$$x' = \gamma(x - vt) \qquad y' = y$$

$$t' = \gamma\left(t - \frac{vx}{c^2}\right) \qquad z' = z$$

$$\gamma = \frac{1}{\sqrt{1 - \dfrac{v^2}{c^2}}} \tag{20.4}$$

Neglecting their actual derivation, we can set out to check these transformations by using them to predict the relationships between lengths, periods of time, and speeds as measured by observers in the two reference frames, and then carry out experiments to check whether these predictions actually agree with the facts.

Imagine a length interval $(x_2 - x_1)$ as measured in frame F, or $(x_2' - x_1')$ as seen in frame F', which moves to the right at speed v. This interval of length might be that of a ruler whose value is L_0 when measured by an observer who is at rest with respect to the ruler. Since the laws of physics are the same in any inertial frame of reference according to Einstein's first postulate, the ruler will always have a length L_0 in any reference frame in which it is at rest. Suppose that we put the ruler into the reference frame F' that is moving with speed v relative to F. To an observer in this frame (F'), the length of the ruler is L_0. Hence, $L_0 = x_2' - x_1'$. To an observer in the F frame, who is moving relative to the ruler, the length of the ruler is $L = x_2 - x_1$. We can now apply the Lorentz transformations to determine how the length of the ruler L, as seen in a frame of reference that is moving with respect to the rule, compares with its basic length L_0.

$$L_0 = x_2' - x_1' = \gamma(x_2 - vt) - \gamma(x_1 - vt)$$

$$L_0 = \gamma(x_2 - x_1) = \gamma L$$

or

$$L = \frac{L_0}{\gamma} = L_0 \sqrt{1 - \frac{v^2}{c^2}} \tag{20.5}$$

Note that the two γvt terms drop out since x_2 and x_1 are measured at the same time.

The faster an object moves, the shorter it appears to become.

The Lorentz transformations predict that the length of the ruler as seen from a moving frame of reference is shorter than the length L_0 measured by an observer at rest with respect to the ruler. In other words, the faster an object moves, the shorter it appears to become to the observer at rest. This effect is called the Lorentz-Fitzgerald contraction. The amount of contraction is actually determined by the ratio of the relative speed of the object to the speed of light, as the equation indicates. At very slow speeds, which include those of people running, cars driving, planes flying, and even rockets soaring into space, this ratio is so small that the length contraction is unmeasurable. However, when the ruler or any other object moves with a speed comparable with that of light, then the contraction appears in plain view. Let us calculate the contraction in length for a ruler traveling at 0.866 of the speed of light:

$$L = L_0 \sqrt{1 - \frac{(0.866\,c)^2}{c^2}} = L_0 \sqrt{1 - (0.866)^2} \cong L_0 \sqrt{1 - 0.75}$$

$$L = L_0 \sqrt{0.25} = L_0 \sqrt{\tfrac{1}{4}} = \tfrac{1}{2} L_0$$

Obviously, the length of the ruler as measured by an observer moving with the same speed as the ruler is just twice its length measured by an observer at rest. As the speed of the ruler approaches the speed of light, the contraction of its length becomes greater and greater, leading to an increasingly smaller measured size for the ruler. As Fig. 20.9 shows, this size for the case of a circular object ultimately ends up as zero, when the object is traveling with the speed of light. This fact provides a reason why we can rationally believe that the speed of light is the upper limit on the speed which any material object can attempt to attain. But it is impossible for a material object to achieve the speed of light because it would have no dimension in the direction of motion.

For some unknown reason, physicists who work on relativity seem to have an addiction to rhymes. The contraction effect that we just discussed has been memorialized with this limerick (Fig. 20.10):

> There was a young fencer named Fisk,
> Whose thrust was exceedingly brisk
> So fast was his action
> The Lorentz-Fitzgerald contraction
> Reduced his rapier to a disk.

The ruler, rapier, or other object does not actually contract as observed by a person holding it; rather, the contraction results when a measurement of the ruler's length is made by an observer in an inertial frame moving at some speed relative to the ruler.

Figure 20.9 The Lorenz-Fitzgerald contraction as seen by an observer at rest with respect to the moving circle for four different relative speeds. Note that the shape of a real object such as a circle as seen by you as the object goes past you would appear very strange because it would in fact seem to bend out of shape as you watched it.

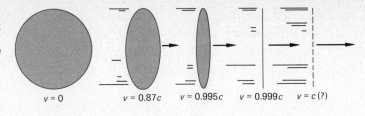

$v = 0 \quad v = 0.87c \quad v = 0.995c \quad v = 0.999c \quad v = c\,(?)$

Figure 20.10 The swordsman, Mr. Fisk.

20.5 TIME DILATION

The Lorentz transformations boil down to the statement that time as observed in a moving frame of reference is different from time measured in a frame of reference which is stationary with respect to any particular event. This is the fundamental point of departure between Galilean relativity and modern relativity as we now understand it. To examine in a mathematical way how time transforms from one frame to another, we can consider an easily reproducible interval, such as two ticks of a wristwatch or the period of a crystal clock or atomic clock.

According to Einstein's first postulate, the laws of physics never vary from one inertial frame of reference to another. Hence, the time between the two ticks, or the period of the clock, is the same whenever it is measured by an observer who is at rest with respect to the timepiece. Let us imagine that the frequency of the vibration measured in the at rest frame of reference (F) in this way is ν_0, the corresponding period is T_0 (where $T_0 = 1/\nu_0$), and the period and frequency measured by an observer who is moving with a constant speed (frame F′ in Fig. 20.8) with respect to the device are T and ν.

The period is the reciprocal of the frequency of any vibration, as we saw in Chap. 12.

According to Eq. (20.4) the period of the instrument measured by the moving observer in F′ is given by

$$t_2 - t_1 = T_0 \qquad t'_2 - t'_1 = T$$

$$t'_2 = \gamma \left(t_2 - \frac{vx}{c^2} \right) \qquad t'_1 = \gamma \left(t_1 - \frac{vx}{c^2} \right)$$

$$T = t'_2 - t'_1 = \gamma \left(t_2 - \frac{vx}{c^2} - t_1 + \frac{vx}{c^2} \right) = \gamma(t_2 - t_1)$$

$$T = \frac{T_0}{\sqrt{1 - \frac{v^2}{c^2}}} \tag{20.6}$$

(The value of x is the same at times t'_2 and t'_1 because the clock is not moving in the F frame of reference.)

Because the quantity $(1 - v^2/c^2)$ is obviously less than 1 as long as v is less than c, the equation predicts that an observer who is moving relative to a timepiece will measure a longer period for that timepiece than one who is stationary in relation to it. Turning this statement around, the equation means that moving clocks appear to slow down, the effect being more marked the faster they travel. This phenomenon is known as time dilation.

Moving clocks appear to slow down.

A fundamental point arises from the fact that the relative speed v between the two reference frames F and F′ is squared in Eq. (20.6). Thus the sign of v, which represents the direction in which the frames of reference are moving relative to each other, has no importance in the equation. Since the frequency of any oscillation is the reciprocal of its period, Eq. (20.6) tells us that the frequency of the oscillation measured by someone in a reference frame moving with respect to the oscillating

Time Flies: The Proving of the Special Theory of Relativity

One of the most unusual predictions of Einstein's special theory of relativity, at least as far as the lay person is concerned, is the so-called twin paradox. If an astronaut travels extremely fast in a spacecraft, according to the paradox, he will age less quickly than his twin brother who remains on earth. The farther and faster the astronaut travels, the greater the difference between his age and that of his less venturesome twin when he gets back to his home planet. If a spaceship could be designed that travels with almost the speed of light, the astronaut returning from a trip whose duration he measures at 1 yr might return to find that his twin had gained thirty or more years in the interim.

To date, space technologists have not managed to perfect a craft that can prove this scenario directly. But in October 1971, two American physicists set out on their own long-range venture to check Einstein's amazing prediction. Joseph Hafele of Washington University, St. Louis (who has since joined Caterpillar Tractor Co.), and Richard Keating of the U.S. Naval Observatory in Washington undertook their experiment in the simplest way possible. They booked space on two round-the-world commercial airline trips—one eastward, the other westward—for themselves and four cesium atomic clocks. Their concept was simple. As the clocks moved round the world in an easterly direction, the same as that of the earth's spin, then according to the special theory they should lose time when compared with a clock that stayed at the U.S. Naval Observatory because they would be moving relatively faster than the "fixed" clock. By contrast, as the traveling clocks circled the world in the westerly direction, in the opposite direction to the planet's rotation, they would be moving relatively more slowly than the earthbound clock and should thus gain a little time.

Because the speed of commercial airliners is a very small percentage of the speed of light, the anticipated effects were minute. After correcting for the planes' speeds, flight paths, altitudes, and sundry other factors, Hafele and Keating concluded that the clocks should lose 40 ± 23 ns (a nanosecond is 10^{-9} s) with respect to the fixed clock on the eastbound leg and should gain 275 ± 21 ns on the westward trip. After months of analysis, the two physicists concluded that the clocks had actually lost 59 ± 10 ns when flying east, while the westbound timepieces gained 273 ± 7 ns. Both figures were well within the range of error. Hence, the two physicists concluded in the magazine *Science*,

> We have shown that the effects of travel on the time-recording behavior of macroscopic clocks are in reasonable accord with predictions of the conventional theory of relativity, and that they can be observed in a straightforward and unambiguous manner with relatively inexpensive commercial jet flights and commercially available cesium beam atomic clocks. In fact, the experiments were so successful that it is not unrealistic to consider improved designs to investigate aspects of the theory that were ignored in the predicted relativistic time differences. In any event, there seems to be little basis for further arguments about whether clocks will indicate the same time after a round trip, for we find that they do not.

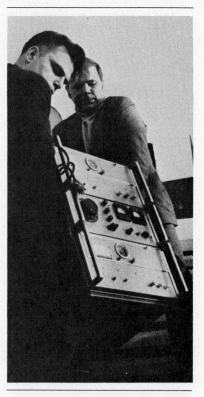

Hafele and Keating deplane with one of their atomic clocks. (*Wide World*)

source will be smaller than that measured by an observer at rest relative to it. This also indicates that the effect of time dilation is not restricted to oscillating clocks and pendulums. Electromagnetic waves such as light arise from the oscillations of electrons in atoms; they obviously will be subject to the effects of time dilation also. Moving observers approaching a stationary source of light will see smaller frequencies than stationary observers. Because the frequency of light is inversely related to its wavelength, through the equation $\lambda = c/\nu$, light will appear to shift toward the longer (red) end of the spectrum as the relative speed of the light source and the observer increases.

Note that this effect with electromagnetic waves alters the classical Doppler effect we encountered in Chap. 12. We found in the classical Doppler effect, applied to sound waves, that a lack of symmetry existed in the relationship. That is, the classical shift depended not only on the speed of the source relative to the observer but on which of the two moved relative to the earth. Relativistic corrections to the classical effect result in a completely symmetric relationship for the Doppler shift as applied to electromagnetic waves. This relationship depends only on the velocity of source relative to the observer.

A brief glance at Eq. (20.6) tells us that the amount of time dilation becomes quite appreciable as the speed v approaches the speed of light, c. If v could reach the value of c, then a zero would appear in the denominator of Eq. (20.6), giving a value of infinity for T. Time T under this circumstance would really stand still! And if v were to exceed c, then we would be faced with an imaginary value of T—a possibility so distressing on philosophical grounds. We are led to conclude that the speed of light is the maximum speed attainable.

How can physicists look for evidence of time dilation? The most obvious means is to find some type of object(s) that moves with speeds close to that of light. Fortunately, these are readily available in the form of particles produced in the powerful machines known as particle accelerators, or atom smashers in the vernacular. Cosmic ray particles from outer space strike atoms and molecules at the top of the earth's atmosphere, producing secondary cosmic ray particles that also fill the bill as objects which move with speeds approaching that of light. Sure enough, experiments on these two types of high-speed objects have confirmed time dilation. Particles that break into pieces after a fleeting existence are found to survive 10 times as long when traveling at 99.5 percent of the speed of light as they do when at rest.

Time dilation is observed directly by high-energy particle physicists who measure the lifetimes of unstable particles created by cosmic rays or particle accelerators. For example, a pion, an elementary particle, has an average lifetime of about 10^{-8} s when measured at rest. However, a pion emerging from an accelerator at a speed of 99 percent that of the speed of light has an average lifetime about seven times longer when measured by an observer on the earth.

One interesting experiment on the lifetimes of unstable particles was the measurement of the decay time of muons produced by cosmic rays when they enter the earth's upper atmosphere. The muon is a particle with a charge of the electron, a mass of about 207 times the mass of the electron, and a half-life of 1.5×10^{-6} s. By half-life we mean the time that elapses during which one-half of the muons disintegrate; after 1.5×10^{-6} s, one-half the muons are gone, after 2 (1.5×10^{-6}) s, three-fourths of the muons are gone, and after 3 (1.5×10^{-6}) s, seven-eighths of the muons are gone. To measure the time dilation, physicists first measured the number of muons striking the top of Mount Washington in New Hampshire, about 1 mi above sea level, during a fixed time interval. They measured the number again with the same equipment at about sea level in a laboratory at M.I.T.

What would we predict would be the change in the number of muons per unit time measured at sea level with respect to the mountain top? From a measurement of the energy of the muons at the mountain top, it is clear that the muons travel very close to the speed of light. Therefore, during the time of one half-life the muons will travel, according to classical computations, a distance

$$D = (1.5 \times 10^{-6} \text{ s})(3 \times 10^8 \text{ m/s}) = 450 \text{ m}$$

During two half-lives the muons would travel twice as far, or 900 m, with a reduction in number of muons to one-fourth of the original number measured at the mountain top. The mountain to sea level distance is about 1 mi or 1610 m. One predicts that the number of muons to reach sea level would be fewer than one-eighth the number striking the top of the mountain during a given period. However, the scientists found that the number of muons reaching sea level was very large, many more than that predicted classically.

This result can readily be explained by time dilation. We can consider the lifetime of a muon as a clock and then realize that the moving muon will have a longer lifetime as measured on the earth. The moving muon appears to have a longer lifetime equal to $(1.5 \times 10^{-6} \text{ s})/(\sqrt{1 - v^2/c^2})$. Hence the muons will travel further before decaying, and many of them will reach sea level.

Interestingly enough from the point of view of the muon whose lifetime is 1.5×10^{-6} s, the earth-distance of 1 mi will appear contracted by the factor $\sqrt{1 - v^2/c^2}$. Hence, the muon in its moving reference frame has a proper lifetime and can travel the contracted distance of 1 earth-mile well within its lifetime. The muon has no trouble reaching sea level during its lifetime.

Time dilation has also been observed, and confirmed in quantitative fashion, in an experiment more closely related to daily life. In 1971, two scientists, J. C. Hafele from Washington University and now with the Caterpillar Corporation, Peoria, Illinois, and R. E. Keating from the U.S. Naval Observatory, flew four extremely accurate cesium atomic clocks around the world on commercial jet flights. When they returned to their laboratory and compared the time shown by the clocks with that of identical clocks which had stayed in the laboratory throughout the journey, they found that the traveling clocks had indeed slowed down slightly as compared with the stationary clocks.

EXAMPLE 20.3

A pilot in a supersonic transport moves at Mach 1.8 ($v = 600$ m/s) with respect to the earth. How long must he fly until his watch has fallen 1 s behind earth-based clocks, assuming that the earth is an inertial frame?

Solution

Consider the earth-based clock to indicate a time interval T_0 and the pilot moving with $v = 600$ m/s will have a clock that has a time interval T. The problem is to find how long T_0 must be so that $T - T_0 = 1$ s. The time transforms by Eq. (20.6),

$$T = \frac{T_0}{\sqrt{1 - \frac{v^2}{c^2}}}$$

The Mach number, named for the nineteenth century Austrian physicist who was one influence on the young Einstein when he was formulating relativity theory, is a measure of speed in relation to the local speed of sound. Mach 1 indicates the speed of sound; Mach 2 represents twice the speed of sound.

Hence $T - T_0 = 1$ s can be written

$$\frac{T_0}{\sqrt{1 - \frac{v^2}{c^2}}} - T_0 = 1 \text{ s} \quad \text{or} \quad T_0 \left(\frac{1}{\sqrt{1 - \frac{v^2}{c^2}}} - 1 \right) = 1 \text{ s}$$

To simplify this problem it is convenient to use the binomial expansion

$$\frac{1}{\sqrt{1 - v^2/c^2}} \cong 1 + \frac{1}{2}\left(\frac{v^2}{c^2}\right) \cdots$$

keeping only the first term, which is applicable when v is much less than c, as is the case in this problem. Making this simplification changes the previous equation to

$$T_0 \left(\frac{1}{\sqrt{1 - \frac{v^2}{c^2}}} - 1 \right) = T_0 \left(1 + \frac{1}{2}\frac{v^2}{c^2} - 1 \right) = 1$$

or

$$T_0 = \frac{1 \text{ s}}{\frac{1}{2}\left(\frac{v^2}{c^2}\right)}$$

The time T_0, which is essentially T because they differ by only 1 s, is given by

$$T_0 = \frac{2c^2}{v^2} = (2)\frac{(3 \times 10^8 \text{ m/s})^2}{(600 \text{ m/s})^2} = 0.5 \times 10^{12} \text{ s}$$

This is about 16 000 yr. Note that even if we consider a rocket moving at 10 times the plane's speed (6000 m/s), this time is very long. However, the accuracy of atomic clocks allows scientists to determine time dilation over relatively short journeys, as was done by Hafele and Keating in 1971 (*Science 177,* pp. 166–170, July 1972).

20.6 VELOCITY TRANSFORMATIONS

Having looked separately at time and space, we can now look at them together, using the Lorentz transformations to examine how speeds transform. We write Eqs. (20.4) in terms of intervals:

$$\Delta x = \gamma(\Delta x' + v\,\Delta t')$$

$$\Delta t = \gamma\left(\Delta t' + \frac{v}{c^2}\Delta x'\right)$$

$$\gamma = \frac{1}{\sqrt{1 - \frac{v^2}{c^2}}}$$

To find the speed $\Delta x/\Delta t$ as measured by an observer in the frame F in terms of the speed $\Delta x'/\Delta t'$ measured by an observer in the F' frame moving to the right with a speed v, we need only divide these two equations:

$$\frac{\Delta x}{\Delta t} = \frac{\gamma(\Delta x' + v\,\Delta t')}{\gamma\left(\Delta t' + \frac{v}{c^2}\Delta x'\right)} = \frac{\Delta x' + v\,\Delta t'}{\Delta t' + \frac{v}{c^2}\Delta x'}$$

If we divide each term on the right-hand side of the equation by $\Delta t'$, then we have what we want:

$$\frac{\Delta x}{\Delta t} = \frac{\frac{\Delta x'}{\Delta t'} + v}{1 + \frac{v}{c^2}\frac{\Delta x'}{\Delta t'}} \tag{20.7}$$

This can be transformed to solve for $\Delta x'/\Delta t'$, yielding

$$\frac{\Delta x'}{\Delta t'} = \frac{\frac{\Delta x}{\Delta t} - v}{1 - \frac{v}{c^2}\frac{\Delta x}{\Delta t}}$$

This equation is sometimes called the velocity-addition law and is used in an analogous way to the Newtonian velocity transformation equation described in Eq. (20.3). Consider a man running down the aisle of a train at a speed u' relative to the train as shown in Fig. 20.11. The

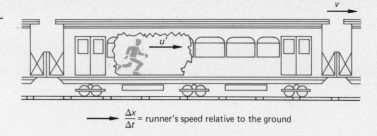

Figure 20.11 Addition of velocities. The passenger runs at a speed u' relative to the train. The train moves at a speed v relative to the ground.

$\dfrac{\Delta x}{\Delta t}$ = runner's speed relative to the ground

speed of the runner as seen by an observer on the ground is $\Delta x/\Delta t$ and is given by Eq. (20.7) when we substitute u' for $\Delta x'/\Delta t'$:

$$\frac{\Delta x}{\Delta t} = \frac{u' + v}{1 + \dfrac{vu'}{c^2}}$$

Here we are adding the speed of the man relative to the train, u', to that of the train which moves relative to the ground at speed v in a relativistic way. With the Newtonian or Galilean addition of velocities, we just add in a commonsense manner and get a value of 1 for the denominator of Eq. (20.7). The vector nature of velocities must be considered and given appropriate signs in both cases. To illustrate the power of velocity transformation, suppose that we substitute for the runner a light wave moving at a speed $u' = c$ and that the source of this light were on a mythical train which itself is moving at a speed $v = c$ relative to the ground. We can find out how fast the light would be with respect to the ground by substituting $u' = c$ and $v = c$:

$$\frac{\Delta x}{\Delta t} = \frac{c + c}{1 + c\dfrac{c}{c^2}} = \frac{2c}{1 + 1} = c$$

This is a direct consequence of Einstein's postulate that the velocity of light is c, no matter from what inertial reference frame you measure the speed of light. Clearly, in the Newtonian analysis, $\Delta x/\Delta t$ would equal $2c$. To obtain a better feel for the relativistic way of handling velocities, we can redo the example we considered on page 643.

EXAMPLE 20.4

Two electrons are ejected in opposite directions from a filament of an electron tube that is at rest in the laboratory. Each electron has a speed of $\tfrac{2}{3}c$, as measured by a scientist in the laboratory. What is the speed with which one electron moves away from the other, according to Einstein's special theory of relativity?

Solution

The statement and setup of the problem are as illustrated in Fig. 20.7. However, we now apply Eq. (20.7) rearranged to solve for $\Delta x'/\Delta t'$:

$$\frac{\Delta x'}{\Delta t'} = \frac{\frac{\Delta x}{\Delta t} - v}{1 - \frac{v}{c^2}\frac{\Delta x}{\Delta t}} = \frac{\frac{2}{3}c - (-\frac{2}{3}c)}{1 - \frac{(-\frac{2}{3}c)(\frac{2}{3}c)}{c^2}}$$

$$= \frac{\frac{4}{3}c}{1 + \frac{4}{9}} = \frac{1.33c}{1.44} = 0.92c$$

Now the result is less than the speed of light. Instead of having the electrons move at $\frac{2}{3}c$, try the calculation at $0.9c$; you will see that the speed of separation $\Delta x'/\Delta t'$ is always less than c. Again the speed of light appears as an upper limit on the speed at which one object can separate from another.

EXAMPLE 20.5

Consider the spaceship shown in Fig. 20.12. As it moves at 90 percent of the speed of light relative to the earth, it is overtaken and passed by Superman zipping along at 95 percent of the speed of light as measured relative to the earth. What do the pilot of the spaceship and Superman measure as their speeds relative to one another?

Solution

Let the spaceship act as the train in Fig. 20.11 and the axis of the F' coordinate system move with the observer in the spaceship. Hence, the speed with which the F' and F reference frames separate is $+v = 0.90c$. Now Superman, traveling relative to the earth at speed $\Delta x/\Delta t = 0.95c$, corresponds to the runner in the train in Fig. 20.11. The problem is to find $\Delta x'/\Delta t'$ (or u' in the train problem in Fig. 20.11), the speed with which Superman moves relative to the moving frame of reference on the spaceship. We use Eq. (20.7) to solve the problem:

$$\frac{\Delta x}{\Delta t} = \frac{\frac{\Delta x'}{\Delta t'} + v}{1 + \frac{v}{c^2}\frac{\Delta x'}{\Delta t'}}$$

or

$$\frac{\Delta x'}{\Delta t'} = \frac{\frac{\Delta x}{\Delta t} - v}{1 - \frac{v}{c^2}\frac{\Delta x}{\Delta t}} = \frac{0.95c - 0.90c}{1 - \frac{(0.95c)(0.90c)}{c^2}} = \frac{0.05c}{1 - 0.855} = 0.345c$$

Hence the two separate at a speed just over one-third the speed of light, which is much greater than that predicted by the "commonsense" Galilean result of $0.95c - 0.90c = 0.05c$.

Figure 20.12 Superman passing a spaceship.

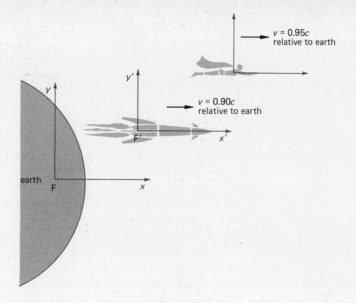

20.7 MASS AND ENERGY

For nonscientists the most awesome outcome of Einstein's theorizing is the nuclear bomb, whose incredible explosive power derives from a relationship between mass and energy predicted by the towering sage. We can start out on the path that led Einstein to this incredible relationship, characterized eternally by the equation $E = mc^2$, by considering just what the transformations of special relativity do to our understanding of the concept of mass. Imagine, for example, the collision in space of two balls thrown by baseball players who are moving relative to each other.

If we assume that the conservation of momentum is a "good law" of physics, then we expect it to be retained in the relativistic world. In Newtonian mechanics, momentum is defined as $\mathbf{p} = m_0\mathbf{v}$ and is conserved as long as no net external forces act. If we analyze a collision using Einstein's special theory of relativity, we find that the quantity $\mathbf{p} = m_0\mathbf{v}$, the classical momentum, is not conserved. However, if we redefine the momentum relativistically as

$$\mathbf{p} = -\frac{m_0\mathbf{v}}{\sqrt{1 - v^2/c^2}} = \gamma m_0 \mathbf{v} \tag{20.8}$$

then we can retain conservation of momentum. Hence we define $\mathbf{p} = \gamma m_0 \mathbf{v}$ as the relativistic momentum. What does this do to classical Newtonian definition of momentum? There is really no conflict if we realize that when v is much less than c, then the relativistic momentum will reduce to the classical momentum $p = m_0 v$.

One interpretation of Eq. (20.8) is that the mass of an object increases with speed. Furthermore, this leads to a definition of a *relativistic mass* given by

$$m = \frac{m_0}{\sqrt{1 - v^2/c^2}} \quad (20.9)$$

The mass of the body when measured by an observer at rest with respect to the body is m_0 and is called the *rest mass*. By interpreting mass as a quantity that raises with speed, as given by Eq. (20.9), we can symbolically retain our classical definition of momentum as $m\mathbf{v}$, where $m = \gamma m_0$. Of course, we realize that to determine the value of inertial mass we must measure how forces change the momentum of an object or measure momentum—not actually mass as an independent quantity.

If we do accept Eq. (20.9) as our interpretation of mass, we see that the mass of an object as measured by an observer moving relative to the object increases as the relative velocity between the mass and observer increases. As the relative velocity v approaches the speed of light c, we see from Eq. (20.9) that the mass appears to increase to infinity as shown in Fig. 20.13 and Table 20.1. As it does, it requires more and more force to increase its speed further. If the mass were actually to reach infinity, it would require an infinite force. Since this seems to be an impossibility, we once more arrive at the conclusion that *no object can be accelerated to the speed of light*.

Just like time dilation, this increase in mass (actually momentum) has been measured by particle physicists, generally by analyzing the deflections of beams of charged particles by magnetic fields. The table clearly shows that mass does not vary significantly from the rest mass defined by Newton as long as the object is traveling at less than one-tenth the speed of light. Above this speed the mass is appreciably greater than the rest mass.

To determine the relationship between mass and energy, we must work out the relativistic definition of kinetic energy. In Chap. 5 we defined energy as that attribute of an object that changed when work is done on the object. The change in kinetic energy, ΔE_k, is the product of the force and the displacement when the two act along the same line. Now using relativistic quantities, we have

An approximation for m/m_0 calculated using the approximation

$$\frac{1}{\sqrt{1 - \frac{v^2}{c^2}}} \simeq 1 + \frac{1}{2}\left(\frac{v^2}{c^2}\right)$$

gives the value of mass quite accurately at speeds up to $0.9c$.

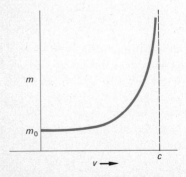

Figure 20.13 Graph of mass as a function of relative speed.

Table 20.1 Change in Mass as a Function of Speed

v/c	m/m_0	Calculated by Approximation m/m_0
0.01	1.000 05	$1 + \frac{1}{2} \times 10^{-4} = 1.000\,05$
0.1	1.005	$1 + \frac{1}{2} \times 10^{-2} = 1.005$
0.5	1.15	$1 + \frac{1}{2} \times (0.25) = 1.125$
0.90	2.3	$1 + \frac{1}{2} \times (0.81) = 1.41 \rightarrow$ the approximation is no longer valid
0.98	5.0	
0.99	7.1	

$$\Delta E_k = \mathbf{F} \cdot \Delta \mathbf{s} = \frac{\Delta(m\mathbf{v})}{\Delta t} \cdot \Delta \mathbf{s}$$

Employing the mathematics of calculus, we reach the conclusion that the kinetic energy of an object in relativistic terms is given by

$$E_k = mc^2 - m_0 c^2 \tag{20.10}$$

rather than the simple equation $E_k = \tfrac{1}{2} mv^2$, which applies only at speeds well below that of light. **The quantity $m_0 c^2$ is called the rest mass energy; it exists because the object has a mass m_0.** The total energy at any time is defined as mc^2. The difference between the total energy and the rest mass energy is the energy of motion, or more familiarly the kinetic energy.

$$E_k = E_{\text{tot}} - E_{\text{rest mass}}$$

$$E_k = \frac{m_0 c^2}{\sqrt{1 - \dfrac{v^2}{c^2}}} - m_0 c^2 = m_0 c^2 (\gamma - 1) \tag{20.11}$$

We can derive our familiar equation for kinetic energy from this expression for the particular case in which speed is relatively small compared with that of light by using the approximate relationship, applicable when v/c is less than one-tenth:

$$\frac{1}{\sqrt{1 - \dfrac{v^2}{c^2}}} = 1 + \frac{1}{2} \frac{v^2}{c^2}$$

$$E_k = m_0 c^2 \left(\frac{1}{\sqrt{1 - \dfrac{v^2}{c^2}}} - 1 \right) = m_0 c^2 \left[1 + \frac{1}{2} \frac{v^2}{c^2} - 1 \right] = \frac{1}{2} m_0 v^2$$

When we talked about kinetic energy in Chap. 5 we neglected the rest mass energy $m_0 c^2$ because it did not change; the problems we dealt with in that chapter involved changes in mechanical energy rather than absolute values of objects' energies. And because mass does not change to any noticeable extent at low speeds, the term $\tfrac{1}{2} mv^2$ was completely satisfactory for kinetic energy.

A positron is a particle identical in every way to the electron except in the sign of its charge, which is positive. It is an example of antimatter—the antiparticle of the electron. Whenever a positron comes close to an electron, the two annihilate each other and produce two or more highly energetic γ-rays. In this case, rest mass is being converted into energy.

The first direct evidence that mass and energy are really different forms of the same thing came in 1932, when physicists at California Institute of Technology observed the conversion of a γ-ray, high-energy electromagnetic radiation, into two particles in a photographic emulsion. The two particles produced by this conversion of energy into rest mass were an electron and a positron.

Experiments of this type have not only proved that rest mass and energy interconvert, they have also proved the accuracy of Einstein's equation. As an illustration of the quantities involved, we might consider the energy required to produce out of nowhere an electron,

Although 0.511 MeV are required to create a single electron, both an electron and a positron must be created together to keep charge conserved. Thus a γ-ray will only be able to produce both an electron and a positron if its energy equals or exceeds twice this figure: 1.02 MeV.

whose rest mass is 9.11×10^{31} kg. According to Einstein's equation, this corresponds to an amount of energy E given by $9.11 \times 10^{-31} \times c^2$, that is, (9.11×10^{-31}) kg $\times (3 \times 10^8 \text{ m/s})^2$, or 8.2×10^{-14} J. Physicists normally express energies this size in terms of electronvolts (eV), where 1 eV = 1.6×10^{-19} J. Thus,

$$E = \frac{8.2 \times 10^{-14}}{1.6 \times 10^{-19}} = 0.511 \times 10^6 \text{ eV} \quad \text{or} \quad 0.511 \text{ MeV}$$

If a rest mass of just 1 g were converted entirely to energy, it would be enough energy to keep a number of large cities lighted, heated, and otherwise satisfied in terms of their energy needs for hours. The energy is actually sufficient to supply the United States for about 5 min. Indeed, this conversion of rest mass becomes about 0.1 percent of the "fuel" used in nuclear reactors that produce great amounts of heat from the small amounts of mass lost when large, unstable atoms split up into smaller fragments. A similar loss of rest mass is responsible for the awesome energy of the hydrogen bomb, the sun, and the futuristic fusion reactors that might, by early next century, be taking over the energy load from the current types of nuclear fission power plant.

Note that any power station converts matter into energy, even if it burns coal. In such nonnuclear processes the amount of mass converted into energy is very small in comparison to the total amount of fuel consumed.

EXAMPLE 20.6

Compute the mass of an electron traveling at half the speed of light.

Solution

$$m = \frac{m_0}{\sqrt{1 - (v^2/c^2)}} = \frac{m_0}{\sqrt{1 - [(\tfrac{1}{2}c)^2/c^2]}} = \frac{m_0}{\sqrt{1 - \tfrac{1}{4}}}$$

$$= 1.15 m_0 = (1.15)(9.11 \times 10^{-31} \text{ kg}) = 1.05 \times 10^{-30} \text{ kg}$$

EXAMPLE 20.7

If 1 g of matter were converted into energy, what would be its value? Assume that energy commands 7¢/kWh (which is roughly the cost of electric energy).

Solution

The total energy $E = mc^2$ can be computed and the units converted:

$$E = mc^2 = (10^{-3} \text{ kg})(3 \times 10^8 \text{ m/s})^2 = 9 \times 10^{13} \text{ J}$$

$$\text{Value of energy} = \frac{9 \times 10^{13} \text{ J}}{3.6 \times 10^6 \text{ J/kWh}} \times \frac{7.0¢}{1 \text{ kWh}} = \$1.75 \times 10^6$$

20.8 THE GENERAL THEORY OF RELATIVITY

The ideas of Einstein's special theory will crop up again in the next three chapters as we investigate the structure of the atom and the atomic nucleus. Indeed, the theory laid the foundation for much of the nuclear physics that transpired during the quarter-century which started in 1905. But Einstein himself was not content to start just one revolution. Having effectively solved the problems of inertial frames of reference moving with constant speeds relative to each other, he set about a more universal problem: accelerating reference frames, such as the sun and the earth that revolves around it. Einstein's theoretical treatment of this type of situation was published in 1916; because it dealt with a much less restricted situation than the special theory, it soon became known as Einstein's general theory of relativity. He dealt with the case of *general* motion of one reference frame relative to another.

The general theory is basically stated in two postulates:

1. The correct laws of physics can be formulated in such a way that they are valid for any observer, however complicated the observer's motion.

2. No experiment performed inside a close laboratory can distinguish between the effects of a gravitational field and the effects of an acceleration of the laboratory relative to the stars.

The second statement, known as the principle of equivalence, connects gravity and relativity and opens up the whole area of cosmology—theories of the origin and nature of the universe—to the techniques of relativity. The principle of equivalence implies the equality of gravitational and inertial mass. Einstein's formulation of the principle produced a huge leap in understanding of the type of universe in which we exist, but even today the field of general relativity remains complex, fascinating, and still incomplete.

SUMMARY

1. The nineteenth century search for the ether—the substance believed to be the basic reference frame for the whole universe—was completely unsuccessful.

2. The notion of the ether was rejected in Einstein's special theory of relativity. This theory applies only to frames of reference that are not accelerating.

3. The first postulate of the special theory states that the laws of physics are the same in all inertial reference frames.

4. The second postulate of the theory is that the speed of light is a universal constant, whatever the speed of the source of the light and the observer.

5. In Galilean dynamics, the relative motion of objects in two reference frames is obtained simply from the velocities of the objects and those of the reference frames.

6. The simple relationship of the previous statement does not apply in relativistic dynamics. The relative motion of objects in two moving reference frames is related to the speed of light.

7. According to Einstein's theory, the faster an object moves, the shorter it appears to become when viewed by an observer with respect to whom the object is moving.

8. The same principle indicates that moving clocks slow down. This phenomenon is known as time dilation.

9. Rest mass and energy are equivalent.

10. Einstein's general theory of relativity, which applies to all frames of reference, connects gravitation and relativity. It produces new understanding of the behavior of the universe on a large scale.

QUESTIONS

1. When riding in a closed truck, could you determine whether you were at rest, moving at a uniform speed in a straight line, accelerating, decelerating, or turning? Discuss.
2. What are the postulates upon which Einstein based his special theory? How are these different from those upon which Galilean relativity is based?
3. Explain why physicists felt the ether should exist, that is, why it was necessary.
4. What statements were made by Einstein to eliminate both the necessity of the existence of the ether and the necessity of finding a fixed (unique) inertial reference frame.
5. What were Michelson and Morley trying to find with their experiment? Was their null result due to inaccurate or insensitive equipment? Explain.
6. One explanation for the null result of the Michelson and Morley experiment when it was carried out totally in their lab was that the velocity of the light might be determined by the velocity of the source. How was this explanation shown to be wrong?
7. What hypothesis of Einstein's clearly explained the null result of the Michelson and Morley experiment? How does this hypothesis explain the null effect?
8. Explain carefully why a laboratory on the surface of the earth is not really in an inertial frame of reference. Can you suggest why it is frequently permissible to assume that it is an inertial frame of reference?
9. How is the frequency of light altered when you are moving at relativistic speeds relative to the source?
10. How do measurements of mass, period, and length of moving objects compare to the same measurements when the objects are at rest?
11. List and discuss several types of measurements that indicate the speed of light is the upper limit of speeds in our universe.
12. You are selected as a space traveler and asked to test the theory of relativity by measuring your pulse, mass, width, and height while you are traveling in a rocket of $v = 0.6c$. When you return to earth 5 yr later, would any of your data indicate changes from your earth-based measurements before you went on the trip?
13. A son of a great hunter is riding in the back of a jeep through the jungle. The jeep is traveling at 45 km/h in a straight line when the boy throws a ball directly upward. Simultaneously, with the ball reaching the top of its path, a monkey drops from an adjacent tree branch from the same height. Describe the motion of the ball relative to the monkey.
14. From your knowledge of physics, give supporting reasons why the earth rather than the stars rotates.
15. How would the predictions of the theory of relativity change if the speed of light were infinitely large?
16. Describe an experimental test of the general theory that has supported the theory of relativity.
17. State two principles upon which Einstein's general theory of relativity is based.

PROBLEMS

Time Dilation and Length Contraction

1. If a modified Michelson and Morley experiment were carried out with the "arms" of the apparatus being 30 m long, what would be the minimum possible value of the velocity of the apparatus through the ether detectable assuming a shift of one-hundredth of a fringe is the limit of resolution? The wavelength used is 630 nm.
2. Suppose the speed of light were 30 mi/h and you were moving along a street at 25 mi/h. You notice that the dimensions of buildings along the side of the street appear different as compared to the dimensions you would measure if you were walking very slowly down the street. (*a*) How do things change? Calculate the change in dimensions of a building 150 ft long and 300 ft tall. (*b*) A woman standing on the side of the street watches you pass by and notices that your car is 5 ft long. How long does *your* car look to *you*? (*c*) Do your answers depend upon whether you are approaching or going away from the woman standing on the roadside? Explain.
3. Suppose the speed of light were 60 mi/h and you are driving along the street at 40 mi/h. You look out the window and notice a building with a square front 200 ft tall and 200 ft long. (*a*) What is the shape of the building viewed by a stationary woman on the street? Give dimensions. (*b*) As you pass the building, you notice an oscillating mass on the end of a spring. What is the period of the oscillator as measured by you in the car if the period of the oscillator was 0.50 s as measured by the woman on the street? (*c*) If you were to have an identical spring (with a mass on it) to that which is on the side of the street, what would its period be as

measured by you? And by the woman on the street? You are still traveling at 40 mi/h.

4. A meter rule moves past with a velocity of 2.5×10^{10} cm/s in the direction of its length. What is its apparent length?

5. How does the volume of a cube 10 cm on a side appear to change if the cube is moving by you at a speed of 2×10^8 cm/s? One edge of the cube is parallel to the direction of motion.

6. A particle called a muon has a half-life of 1.50×10^{-6} s when at rest. What is its apparent half-life when it has a velocity of $0.9c$?

7. A beam of pions whose half-life is 1.8×10^{-8} s when measured essentially at rest has a half-life of 3.2×10^{-7} s. Calculate the speed of the pions.

Relative Velocity

8. A gale is blowing with a velocity of 4×10^3 cm/s. How long will it take for the sound of a siren to travel to a point 500 m away in the direction toward (into the wind) which the wind is blowing? The velocity of sound in air is 3.3×10^4 cm/s. How is this different for light?

9. A person riding on the roof of a speeding train that moves at 140 km/h shoots a bullet horizontally from a rifle at a speed of 490 m/s relative to the gun toward the front of the train. What is the speed of the bullet as seen by an observer on the ground?

10. A spaceship is launched from earth with a velocity of $0.5c$. The ship then launches a small rocket in the forward direction with a velocity of $0.5c$ relative to itself. To an observer on earth, what is the rocket's velocity?

11. A projectile is launched from a spaceship in the direction opposite to that in which the spaceship is moving. The projectile leaves the spaceship at 0.5 the speed of light ($\frac{1}{2}c$) relative to the ship. The ship itself moves at $0.5c$ relative to the earth. What is the projectile's speed as measured by a person on earth?

12. A rocket moves past and away from the earth at a speed of $0.6c$. At a later time this "parent" rocket sends back a "scout" rocket that travels at a speed of $0.9c$ with respect to the "parent" directly back toward the earth. With what speed does the "scout" rocket travel with respect to the earth?

13. Observer A is in a spaceship moving at a speed of $0.5c$ relative to the earth. Observer B is in a spaceship that travels at a speed of $0.9c$ relative to the earth but in a direction exactly opposite to observer A. (a) How fast are observers A and B separating from one another? (b) Observer B sees a vibrator (clock) on spaceship A. The period of this vibrator is 10^{-4} s as measured in spaceship A. What is the period of this clock as seen by observer B?

Relativistic Mass and Energy

14. With respect to the laboratory, the electrons in accelerator A are projected toward the east with a speed of 2×10^8 m/s, and electrons in accelerator B are projected toward the west with a speed of 1.5×10^8 m/s. (a) What is the relative velocity of the A electrons with respect to the B electrons? (b) What is the mass of the A electrons as measured in the laboratory?

15. Find the speed of an electron that has been accelerated to an energy of 5 MeV.

16. At what fraction of the speed of light would the mass of a particle be double its rest mass?

17. If your mass is 70 kg and you move at a speed $0.67c$ relative to the earth, how would your mass change as far as you are concerned? As seen from the earth?

18. At what velocity is the mass of a particle four times its rest mass?

19. What is the kinetic energy in electron volts of a proton when its mass is twice its rest mass? The rest mass of the proton is 1.673×10^{-24} g.

20. (a) A spaceship passes by the earth with a velocity of $0.3c$. What is the pilot's estimate of the mass of the standard pound on earth? (b) This same pilot also notices a pendulum swinging on the earth, and she measures the period of the pendulum. What is the value of the period she measures if the natural period of the pendulum is 2 s when measured by a woman on earth? To do this problem, do you need to know the direction of the swing of the pendulum relative to the motion of the pilot? Explain.

21. An electron in a certain x-ray tube is accelerated from rest through a potential difference of 180 000 V in going from the cathode (negative terminal) to the anode (positive plate). When it arrives at the anode, what is its (a) kinetic energy in electronvolts, (b) relativistic mass, (c) relativistic velocity?

22. What magnetic field strength would confine 100-MeV protons into a path of radius 2 m?

23. An operating nuclear power plant in Wiscasset, Maine, generates 850 MW of power. How much rest mass would have to be changed to energy in this power plant if the plant is only 30 percent efficient?

Special Topic
The General Theory of Relativity: Into the Fourth Dimension

The crowning achievement of Albert Einstein—and indeed of twentieth century science—was the general theory of relativity, which is based on a foundation so powerful that for more than six decades it has survived all attempts to topple or modify it. The theory enjoyed a satisfying surge of fresh interest among physicists around the time of the centennial of Einstein's birth, in 1979; it was then that physicists began to acquire solid evidence of such phenomena predicted by the general theory as black holes and gravity waves.

Before Einstein published his theory in 1916, the world of science had been fully content with Newton's view of gravitational force: a force acting throughout the universe and tugging every pair of objects toward each other. Einstein decided that Newton had not perceived it quite right. Rather than a force, Einstein said, gravity is a field, like a magnetic field. The gravitational equivalent of the magnet is matter itself, which creates a field by distorting the space around it.

The effect of matter on space according to Einstein's schema can best be visualized by imagining a billiard ball on a flexible rubber sheet: the ball creates a dimple in the sheet. If an insect comes along and wants to travel from one point on the sheet to another along the shortest path, it does not take a straight line; instead, it follows the curved contour of the sheet produced by the billiard ball.

The general theory also states that the amount of curvature of space depends on the amount of matter causing it to curve. Thus a cannonball on the rubber sheet would produce a huge dimple, whereas a pea would create a hardly noticeable curvature. And of course the insect's journeys in the vicinity of the two masses would cover quite different ground.

The idea that gravitational fields curve space led directly to the first test of the general theory because it indicated that light would travel along curved paths rather than the straight lines which Newtonian mechanics demanded. In 1917, across the lines of World War I, British astronomers who had received from Dutch theorist Willem de Sitter copies of the German journal containing the general theory decided that the prediction could be quickly tested. The necessary condition was a total eclipse of the sun, which would darken the skies sufficiently to allow observers to spot stars whose light passed relatively close to the solar disk. If the general theory were correct, the British astronomers reasoned, the huge gravitational field of the sun would bend that light, making the stars seem farther from the sun's disk than they should be. If Einstein were wrong, the observers would not notice any displacement in the stars' positions.

On May 29, 1919, British teams at Sobral, Brazil, and the West African island of Principe photographed a total eclipse and the rich field of stars behind the sun. Neither group produced particularly good images, but the results were accurate enough for Sir Arthur Eddington to announce in November 1919 that the general theory was indeed upheld. Einstein became a celebrity overnight, and he lived the rest of his life with the type of instant recognition that only such modern personalities as Muhammad Ali receive.

Einstein certainly deserved the acclaim, although he did not always appreciate it. He was one of many physicists working along the lines that eventually led to his special theory of relativity, but physicists concede that the general theory was a leap of such incredible intuition that only Einstein could have made it.

But no theory—or personality—is sacrosanct in science. Thus it is not surprising that physicists have continually challenged the general theory. One of the most serious attacks was mounted in 1967 by Robert Dicke of Princeton University and Carl Brans of Loyola University. They suggested that Einstein's view of universal gravitational force was basically correct, but incomplete. Gravitational force, the two physicists argued, actually consisted of two components.

One problem facing Dicke and Brans was that there seemed to be little reason at the time for an alternative to general relativity. After all, the theory predicted the bending of light by the sun and also accounted superbly for the variations in the orbit of the planet Mercury that had mystified the astronomical fraternity throughout the nineteenth century and up to 1916. Dicke argued, however, that the general theory accounted too well for the unusual orbit of Mercury; in fact, he claimed, part of the Mercury mystery derived from a lack of perfection in the sun. The solar orb, and hence its gravitational field, were not perfectly spherical, Dicke said.

Dicke's readings designed to prove that contention were controversial, and physicists were hard put to differentiate between the predictions of general relativity and the scalar-tensor theory, as the Brans-Dicke concept was known, because their predictions differed by minuscule amounts. In fact, the scalar-tensor theory has the built-in advantage that the relative proportions of Einsteinian and non-Einsteinian terms can be varied at will. Studies of radio stars by astronomers at the National Radio Astronomy Observatory in Green Bank, West Virginia, in 1974, partly settled the issue. By measuring the bending of radio beams from distant stars, the studies showed that the non-Einsteinian terms in the Brans-Dicke theory cannot contribute more than a few percent to the gravitational fields of matter. The studies did not settle the matter entirely, but they did show that general theory is accurate to within the few percent margin of error that astronomers can obtain. Within the coming years, other physicists plan more stringent tests of the general theory, but most physicists are betting that the theory will survive.

21 Approaching the Atom

21.1   THE RADIATION RIDDLE

        SHORT SUBJECT: Blackbodies and Flaming Fires

        SHORT SUBJECT: The Quantum Man

21.2   THE PHOTOELECTRIC EFFECT

21.3   THE DISCOVERY OF THE NUCLEUS

21.4   THE HYDROGEN SPECTRUM

21.5   THE ATOM ACCORDING TO BOHR

        SHORT SUBJECT: The Great Dane

21.6   CONFIRMING BOHR'S THEORY
       Summary
       Questions
       Problems

        SPECIAL TOPIC: The Solution Without a Problem

Physics is the science that deals with the properties of matter. A major part of the pursuit of this science has involved—and still does involve—the search for what matter is really like in its most fundamental form. By the end of the nineteenth century, physicists were aware from the work of their colleagues in the chemistry laboratories that all elements and compounds consisted of discrete atoms whose atomic weights gave some indication of their chemical properties. They had little idea of the true physical nature of atoms, however. But starting in the 1890s, a series of astonishing experimental discoveries and theoretical concepts started to open up the inside of the atom to the figurative gaze of the wide-eyed physicists. Instead of simple, hard, spherical particles that resembled billiard balls, atoms turned out to be arrangements of more fundamental entities, with quite complicated internal structures. The experimental and theoretical approaches devised to study these complexities had a far-reaching effect on physicists' understanding of the world in general, permanently altering their view of such natural phenomena as energy and radiation. The key to the advances of what the late cosmologist George Gamow described as "thirty years that shook physics" was the development of what we term the **quantum theory of radiation**—the idea that electromagnetic energy, light, and other forms of radiation come in the form of discrete packets rather than, or even in addition to, continuous waves.

The contrast between discrete quanta and continuous phenomena is readily understandable in the case of matter itself. If you slice any piece of matter in two, then slice the halves in two, and so on, you either arrive at pieces of matter that cannot be divided further or fail to arrive at that point, however much you slice. The concept of quanta of other properties is rather less obvious. One simple example illustrates the difference between the quantum and continuous ways of thinking insofar as they refer to time. Imagine that you wish to go to a movie at some distance from your house and have the alternatives of taking a bus or driving by car. If you decide to travel by bus, you can leave the bus stop and arrive at the movie theater only at certain times, determined by the bus schedule; your time of arrival at the theater is therefore quantized into discrete times. By contrast, if you choose to drive your own car, you can leave your house and arrive at the theater at any time you wish; your time of arrival is, in this case, continuous.

21.1 THE RADIATION RIDDLE

Quantum theory actually had its origin in a rather obscure area of physics—the emission of radiation by hot objects. You see this phenomenon whenever you turn on a ring on an electric stove. When the burner is at a low setting, the ring has no obvious color beyond its normal metallic tint, but if you place your hand above the ring (not too close!), you can feel the heat that it emits. At this low temperature almost all the radiated energy takes the form of infrared radiation. As you turn up the electric current, thereby increasing the temperature, more and more energy is radiated as light. First, the burner glows deep or dull red, then bright red, and finally, if the burner has a very high temperature setting, it becomes "white hot." In fact, all the colors of the visible spectrum are present at every temperature; different colors predominate as the temperature changes because they are present in the radiation in greater amounts than the rest. Of course, during this process of increasing temperature, ever-increasing amounts of infrared radiation are emitted.

This can be best understood graphically by plotting an appropriate measure of the intensity of radiation against wavelength. The criterion used by physicists for this type of intensity is the amount of energy radiated per unit time (that is, the power) from a unit area of the surface under consideration. This quantity is indicated by the Greek letter phi (Φ). Figure 21.1 shows the curves of Φ plotted against wavelength λ for a small number of materials at the same temperature.

The exact form of these curves obviously depends on the nature of the material that is radiating, as illustrated in Fig. 21.1. However, studies over the years by physicists revealed that the value of Φ at any particular wavelength and temperature never exceeded the value of the same quantity on a characteristic curve. This is the curve labeled C in Fig. 21.1; it is known as the blackbody curve.

Physicists use the term **blackbody** to indicate the object that emits radiant energy most efficiently at any particular temperature. Since the amount of radiant energy emitted at any temperature depends on the amount of radiant energy the object can absorb, blackbodies are also the most efficient absorbers of heat. In practice, experimentalists obtain the closest approach to the theoretical concept of the blackbody using a small hole in a heated cavity of some opaque material, such as a large hollow iron ball.

Figure 21.2 shows how the form of the curve that relates the power density (Φ) of a blackbody to the wavelength of radiation it emits alters with changing temperature. The wavelength at which the maximum amount of energy is emitted drops as the temperature is increased. The peak of the curve moves to shorter wavelengths. This represents a shift with increasing temperature from infrared, to red, to white, and eventually to blue light. In addition, the total amount of energy emitted, which is represented by the area underneath the curve, increases markedly with temperature, as indicated.

Careful studies in the late nineteenth century showed that two simple relationships applied to the power density-wavelength curves of

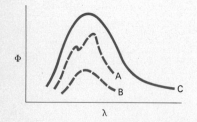

Figure 21.1 Objects A, B, and C are all at the same temperature. Plot of the amount of radiated power per unit area in a wavelength interval λ to $(\lambda + \Delta\lambda)$; C is the curve for the blackbody.

Figure 21.2 Graph of total energy ϕ radiated per unit area per unit time in the wavelength region λ to $(\lambda + \Delta\lambda)$ versus wavelength λ for blackbodies at three different temperatures.

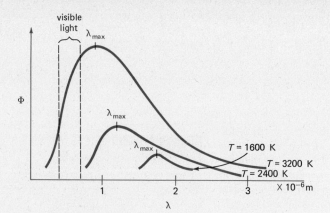

blackbodies. The first, known as the Wien displacement law after German physicist Wilhelm Wien, indicates that the product of absolute temperature and the wavelength of the peak of the curve for that temperature is constant:

$$\lambda_{\max_1} T_1 = \lambda_{\max_2} T_2 = \text{constant}$$

The second equation, the Stefan-Boltzmann law, shows that the total amount of energy radiated per unit time by a unit area of a blackbody is proportional to the fourth power of the absolute temperature:

$$\Phi = \sigma T^4$$

where σ, the Greek letter sigma, is a constant. The law is named after physicists Josef Stefan and Ludwig Boltzmann. Stefan discovered the relationship by careful experimentation in 1879. Five years later Boltzmann showed that the law would be derived directly from thermodynamic principles, thus fitting into the structure of classical thermodynamics that had been built up in the latter part of the last century as Victorian physicists tried to explain the world around them in a strictly mechanistic way.

However, no theoretical physicist could derive an expression to predict the actual distribution of wavelengths in the individual curves of emitted radiation at individual temperatures. The problem was understanding the relationship between radiation and temperature. Some physicists tried to view the problem in much the same terms as those used by Maxwell in his kinetic theory of gases. Others regarded electromagnetic radiation as oscillating back and forth in a cavity, forming systems of standing waves for the individual frequencies. But none of these techniques could produce a theoretical expression that resembled in a sufficiently exact manner the measured distribution of wavelengths in the spectrum.

The answer to the problem finally came from way out in left field. The German physicist Max Planck decided, on the basis of nothing

Blackbodies and Flaming Fires

For most physicists blackbody radiation is an area of their subject to be noted in the textbooks and then neglected. Lawrence Cranberg, a consulting physicist who lives and works in Austin, Texas, thought differently. His response to the energy crisis was to use the blackbody concept to design a more effective, more efficient version of the familiar log fire.

Usually log fires send a staggering 90 percent of their heat up the chimney. Cranberg noticed, however, that he could get more heat into his living room by building up his fire in a certain way. What he had done, he realized, was to create inside the fire a blackbody—a cavity that both absorbs and emits the maximum possible amount of radiant heat.

The physicist's next step was to design this blackbody into the fire. Cranberg therefore figured out the frame shown in the picture. This produces a slot-shaped blackbody cavity between the top and bottom

Texas fireframe. (*Lawrence Cranberg*)

logs. Measurements suggest that fires built on the Texas fireframe can channel as much as 30 percent of their heat into a room; they also appear to have the added advantage that they can be lit with just a few sheets of newspaper and no kindling wood.

more than mathematical curve fitting and inspired guesswork, that the classical theories might fit the facts with one slight adjustment: quantization of the radiation emitted by the blackbody. Planck postulated that instead of emerging continuously, the radiant energy was emitted in small, discrete bundles. The energy of each bundle, he assumed, would be proportional to the frequency v of its radiation. He chose the letter h as his proportionality factor, producing an equation that is almost as famous in physics as $E = mc^2$. Planck's equation was

$$E = hv \tag{21.1}$$

In this way Planck was able to fit his equation to the experimental facts of emitted radiation perfectly, even though he had no real ra-

The Quantum Man

Max Planck was a competent physicist settled into a fairly obscure academic career who made one glorious contribution that sparked the revolution in twentieth century physics and then returned to more pedestrian physics. In 1900, as a 42-yr-old faculty member at the University of Berlin, Planck hit on a simple equation that seemed to account for the mystifying distribution of wavelengths observed in the radiation from blackbodies. The snag was that the approach required energy, which was always regarded as continuous, to be split into discrete packets, or quanta.

The scientific world was slow to credit the theory. Even Planck himself sometimes wondered whether he had not somehow performed nothing more than a feat of mathematical manipulation. But then, in 1905, came Einstein's successful application of the quantum concept to the photoelectric effect and, 8 yr later, Bohr's astonishing intuitive proposal that electronic orbits in atoms were quantized. These advances set the seal on quantum theory, and in 1918, Planck received the Nobel Prize in physics for his contribution to the theory.

By then Planck was 60 yr old, and the rest of his scientific life was inevitably anticlimactic. In 1930, he became president of the Kaiser Wilhelm Society of Berlin, which was soon to become the Max Planck Society. He was forced out of that office by the Nazis, who also executed his son during World War II. But with the onset of peace, Planck was reinstated as president of the society, which today still stands for excellence throughout the world of science.

Max Planck in the center of the front row. (*Wide World*)

Planck had no real rationale for his assumption that energy was indeed quantized.

21.2 THE PHOTOELECTRIC EFFECT

tionale for his assumption that energy was indeed quantized. Subsequent studies showed that h had a value of 6.625×10^{-34} J·s, an extraordinarily tiny value, but as we shall see, an exceptionally important one in modern physics.

In the first several months after his announcement of the quantum inspiration, Planck plowed a lonely furrow. Scientists always tend to be somewhat suspicious of new ideas, particularly those without any

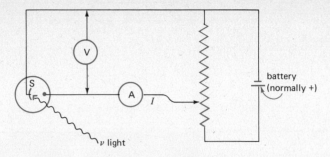

Figure 21.3 Experimental setup for study of the photoelectric effect. Light of frequency ν strikes the metal surface S of the phototube, producing the current I.

obvious fundamental foundation, and many decided to wait for the other shoe to drop, in the form of a successful application of Planck's theory to some other branch of science. Their wait was a short one. In 1905, just 5 yr after Planck's original announcement, the obscure Swiss patent clerk Albert Einstein used the idea of quantization of radiation to explain another experimental finding that had baffled physicists for more than a decade. His explanation of the phenomenon was known as the photoelectric effect.

The photoelectric effect, which was first discovered but not pursued by Heinrich Hertz in 1887, is simply the production of a small electric current when light shines on a metal surface that is part of an electrical circuit. An apparatus often used to investigate the effect is sketched in Fig. 21.3. The metal surface S, which is normally in the shape of a part of a cylinder, acts as one electrode in a vacuum tube; a single piece of wire in the same tube serves as the second electrode. When light is shone on the surface, a small current I passes through the circuit and is measured by the ammeter A, which is actually a sensitive galvanometer. The battery connected to the variable resistor in the circuit is used to study the actual strength of the effect, but the light produces electric current even when the battery is removed.

The use of a galvanometer to measure current was outlined in Chap. 14.

Examination of the current produced by various rays of light in terms of the voltage of the vacuum tube V and the frequency ν of the light reveals seven distinct characteristics for the photoelectric effect:

1. The effect involves the emission of electric charges from the metal surface. These can be identified from their motion in a magnetic field as electrons.
2. The electrons are ejected forcibly by the light, shooting off from the metal surface with kinetic energies $\frac{1}{2}m_e v^2$, where m_e is the rest mass of the electron.
3. The type of metal and the condition of its surface play a large part in the amount of electrical emission in response to the light.
4. There seems to be no delay at all between the impact of the light and the emission of the "photoelectrons."
5. The kinetic energy of the electrons is independent of the intensity of the light.

6. The number of electrons emitted—which is directly measured by the current—is proportional to the intensity of the light.

7. The kinetic energy of the electrons increases with the frequency of the light. However, no emission at all occurs below a certain frequency, which experts refer to as the "threshold frequency." The exact value of this threshold depends on the material of which the photoemitting substance is made and the condition of its surface.

These clear-cut observations clashed at a number of critical points with Maxwell's electromagnetic wave theory. For example, according to the theory, the energy in the wave front striking the metal surface would depend on the intensity of the light; one would, therefore, expect the energy of the emitted photoelectrons to be related to the light intensity also. It was not. Furthermore, one would imagine that a certain amount of energy would have to be built up on the surface before any electrons were emitted, a process which implied a delay between the impact of light and the emission of electrons, particularly in the case of weak intensities of light. But no delay was ever observed. Classical theory failed to predict the threshold frequency effect, and even the energy of the emitted electrons presented a puzzle because wave theory predicted that the energy measured would be contained in wave fronts extending over areas of perhaps as many as 10^8 atoms rather than concentrated in single electrons.

Einstein overcame these difficulties by adopting Planck's quantum ideas, extending the quantization of energy to the electromagnetic waves themselves. He argued that **energy could be absorbed from the electromagnetic field by electrons in the photoelectric process only in quanta of size $h\nu$. These quanta came to be known as photons—particles of light.**

Having postulated the existence of photons, Einstein went on to calculate the energies involved in the photoelectric effect by applying the principle of the conservation of energy. His reasoning was:

If the quantum of energy in the electromagnetic wave that strikes the metal is $h\nu$, then the maximum kinetic energy ($\frac{1}{2}m_e v_{max}^2$) available to any specific electron on the surface of the metal would be this incident energy less the minimum amount of energy required to remove the electron from the metal. This latter quantity is called the work function and is given the symbol ϕ (Greek letter phi). The work-function energy is obviously related to the condition of the metal surface and the type of metal. Furthermore, electrons buried inside the metal would require more energy than the work-function energy to release them from the metal because they might lose energy on the way to the surface.

Applying the principle of conservation of energy to the simplest case of the electron on the metal surface, Einstein wrote the basic photoelectric equation:

$$h\nu - \phi = \tfrac{1}{2}m_e v_{max}^2 \quad \text{or} \quad h\nu - h\nu_0 = \tfrac{1}{2}m_e v_{max}^2 \tag{21.2}$$

Note that some photoelectrons emitted will have less than the maximum kinetic energy given by Eq. (21.2). A more general form of the equation is $h\nu - h\nu_0 - E_p = \frac{1}{2}m_e v^2$, where E_p represents the energy above and beyond the work function that is needed to release more tightly held electrons or energy lost by scattering on the way out.

In the latter equation the work function, the minimum energy required to release an electron from the surface, is written in terms of a quantum of energy $h\nu_0$. Plainly, when the energy of the incoming photon is exactly equal to $h\nu_0$, the threshold energy for emission, then an electron that it contacts is released from the surface, but without any spare kinetic energy. If the frequency of the incoming photon exceeds ν_0, electrons will come away from the surface with a certain kinetic energy. But when the frequency of the light is less than ν_0, no photoemission will occur because the individual photons do not possess enough energy to "knock out" individual electrons.

This treatment solved a number of the questions associated with photoemission. First, it explained why there is a threshold frequency. Second, it showed why the energy of the photoelectrons depended only on the frequency of the light and the nature of the surface (which determines $h\nu_0$) and not on the intensity of the light. It explained the lack of a time delay for photoemission, since a photon of frequency ν greater than ν_0 might knock out an electron immediately upon impact of the light. **The light intensity is a measure of the number of photons. The total energy in a light beam,** which is proportional to light intensity, **is given by the product of the energy of one photon and the total number of photons** incident on the metal surface.

The validity of the equation can be checked using the apparatus shown in Fig. 21.3. In the absence of any direct means of measuring v, the speed of the electrons, experimentalists use some electrical ingenuity to determine the term $\frac{1}{2}m_e v_{max}^2$ in one easy measurement. They simply make the collector electrode negative with respect to the emitting surface by reversing the polarity of the battery in the circuit and increasing the negative voltage on the collecting electrode until the photoinduced current no longer exists. At this point the voltage is known as the "negative stopping voltage," V_s. It is related to the maximum kinetic energy of the photoelectrons by the simple equation

$$eV_s = \frac{1}{2}m_e v_{max}^2$$

Curve **a** in Fig. 21.4 illustrates just what happens in practice in this process. When the voltage of the collecting wire is positive with respect to the emitting surface, a positive current is produced by the photo-

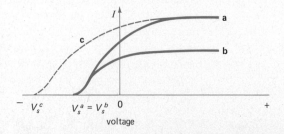

Figure 21.4 Graph of voltage versus current in a phototube. Curves **a** and **b** are for a source of identical frequency but different intensities. Curve **c** represents a higher frequency light with an intensity to give the same current as used in **a**.

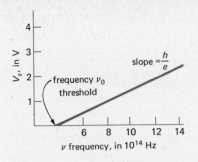

Figure 21.5 Graph of stopping potential versus frequency of the radiation on a phototube.

Millikan had already measured the charge on the electron e by the method described in the Special Topic, Chap. 13.

electric effect. This current is constant above a certain value of positive voltage because the wire electrode collects all freed electrons. However, as the voltage of the collecting wire is first reduced and then made negative, thus repelling the electrons with increasing strength, the amount of current produced by the photoelectric effect gradually falls, until it disappears when the collector has the stopping potential V_s.

The figure also contains a second curve, **b**, measured experimentally with light of the same frequency as **a** but of lower intensity than that used to produce curve **a**. The energy of each photon is exactly the same, $h\nu$, in the two cases. However, the current produced in case **a** is greater because the higher intensity of the light means that more photons reach the emitting surface, thus releasing more electrons overall. The figure also illustrates a third case, **c**, in which the frequency of the light falling on the metal surface is greater than that used in examples **a** and **b**, but the intensity is chosen so that the maximum current is the same as that for curve **a**; since the frequency is greater for case **c**, the amount of energy per photon, and hence the stopping potential, are greater than for curve **a**—that is, more negative.

The next obvious question is: "What is the exact relationship between the stopping voltage and the frequency of light shone on the metal surface?" We can readily find this, at least on a theoretical level, by substituting the stopping potential energy eV_s for the maximum kinetic energy of individual electrons, $\frac{1}{2}m_e v_{\max}^2$, in Eq. (21.2):

$$eV_s = \tfrac{1}{2}m_e v_{\max}^2 = h\nu - h\nu_0$$
$$V_s = \frac{h\nu}{e} - \frac{h\nu_0}{e} \tag{21.3}$$

Equation (21.3) predicts that the stopping voltage will be directly proportional to the frequency of light; as Fig. 21.5 shows, this is exactly the case. In this graph the intercept along the frequency axis at the point where the stopping voltage is zero defines the value of ν_0, the threshold frequency—from which the work function can be derived using the equation $\phi = h\nu_0$—and the slope of the straight line yields the value of h/e, which is an extremely important quantity because it gives physicists the opportunity to measure Planck's constant and compare the value with that obtained from blackbody radiation curves. Robert Millikan actually did this in 1916 and proved that the values of h obtained by the two different techniques were similar enough to establish the validity of the concepts of Planck's constant and the quantization of light in general.

EXAMPLE 21.1

What is the energy of a photon of radiation whose wavelength is 400 nm just at the high-energy end of the visible spectrum, on the boundary of the ultraviolet region?

Solution

The energy of a photon or quantum of radiation is given by the product of the frequency v of the quantum and Planck's constant h. We can readily convert wavelength to frequency using the relationship $c = \lambda v$:

$$E = hv = \frac{hc}{\lambda} = \frac{6.625 \times 10^{-34} \text{ J} \cdot \text{s} \times 3 \times 10^8 \text{ m/s}}{400 \times 10^{-9} \text{ m}}$$

$$= 4.97 \times 10^{-19} \text{ J}$$

EXAMPLE 21.2

A small flea whose mass is 10^{-4} g (10^{-7} kg) falls through a distance H. In the process it obtains a kinetic energy equal to the energy of a single quantum of the most energetic visible light (let $\lambda = 400$ nm for this limit). What is this height H?

Solution

This problem illustrates the size of the quantum of energy associated with visible light. From the previous example we know that the energy of a photon of radiation of wavelength $\lambda = 400$ nm is 4.97×10^{-19} J. As the flea falls through the distance H, its potential energy (mgH) is changed to kinetic energy ($\frac{1}{2}mv^2$) of 4.97×10^{-19} J. We can equate the change in potential energy of the flea to this for the value of hv:

$$4.97 \times 10^{-19} \text{ J} = mgH$$

$$H = \frac{4.97 \times 10^{-19} \text{ J}}{mg} = \frac{4.97 \times 10^{-19} \text{ J}}{10^{-7} \text{ kg} \times 9.8 \text{ m/s}^2} = 5.1 \times 10^{-13} \text{ m}$$

This distance is less than the diameter of an atom and close to the value found for the diameter of the nucleus of an atom—a minute distance.

EXAMPLE 21.3

A typical helium-neon gas laser whose beam of radiation of wavelength $\lambda = 632.8$ nm produces 1 mW of power. How many visible photons of this frequency does it emit per second?

Solution

We can determine the energy in joules of each quantum of laser light if we multiply Planck's constant h by the frequency of the helium-neon laser light. The total energy emitted from the laser is simply the product of the total number of photons and the energy of one photon. We are given the power in the laser beam, which is the energy emitted per second. Hence the number of photons emitted per second is just the power in joules per second divided by the energy of one photon. The energy of a single photon is given by

$$hv = hc/\lambda = 6.625 \times 10^{-34} \text{ J} \times \frac{3 \times 10^8 \text{ m/s}}{6328 \times 10^{-10} \text{ m}}$$

$$= 3.14 \times 10^{-19} \text{ J}$$

$$\frac{\text{Number of photons}}{\text{s}} = \frac{\text{power}}{\text{energy of one photon}}$$

$$= \frac{1 \times 10^{-3} \text{ J/s}}{3.14 \times 10^{-19} \text{ J/photon}} = 3.2 \times 10^{15}$$

EXAMPLE 21.4

When ultraviolet radiation of frequency 1.4×10^{15} Hz falls upon a metal surface, photoelectrons are ejected whose maximum kinetic energy is 2.0 eV. Calculate the energy of the incident photons, the work function of the metal in electronvolts, and the threshold frequency of the metal.

Solution

We determine the energy of a photon of ultraviolet radiation whose frequency is 1.4×10^{15} Hz by multiplying the frequency by Planck's constant h:

$$h\nu = 6.625 \times 10^{-34} \text{ J} \cdot \text{s} \times 1.4 \times 10^{15} \text{ s}^{-1} = 9.28 \times 10^{-19} \text{ J}$$

It is convenient to write such small energies in units of electronvolts, where 1.6×10^{-19} J = 1 eV, and hence

$$h\nu = \frac{9.28 \times 10^{-19} \text{ J}}{1.6 \times 10^{-19} \text{ J/eV}} = 5.8 \text{ eV}$$

This value of $h\nu = 5.8$ eV is the maximum energy incident on an electron. Now, the work function is the minimum energy required to free a photoelectron from the metal. It is the difference between the energy of the incident photon and the maximum kinetic energy that the photoelectrons have after emission from the metal. In terms of the Einstein photoelectric equation, we have

$$h\nu_0 = h\nu - \tfrac{1}{2} m_e v_{\max}^2 = 5.8 \text{ eV} - 2.0 \text{ eV} = 3.8 \text{ eV}$$

The threshold frequency is the frequency ν_0 of the photon whose energy is equivalent to the minimum energy required to free an electron from the metal surface, that is, the work-function energy:

$$h\nu_0 = 3.8 \text{ eV}$$

$$\nu_0 = \frac{3.8 \text{ eV}}{h} = \frac{(3.8 \text{ eV})(1.6 \times 10^{-19} \text{ J/eV})}{6.625 \times 10^{-34} \text{ J} \cdot \text{s}} = 9.1 \times 10^{14} \text{ s}^{-1}$$

Note that we had to change the energy units from electronvolts to joules since Planck's constant h has the units of joule-second. This threshold frequency is in the ultraviolet region of the electromagnetic spectrum.

EXAMPLE 21.5

The work function for potassium is 2.25 eV. If light of wavelength 400 Å falls on potassium, find the stopping potential required to stop the fastest electrons, the kinetic energy of these most energetic electrons, and the speed of the same electrons.

Solution

The stopping potential is the magnitude of the negative potential difference required to stop the fastest electrons from reaching the collector of the phototube. This potential difference V multiplied by the charge on the electron represents the kinetic energy of the most energetic electrons. These are numerically the same in units of volts and electronvolts. Using the Einstein photoelectric equation, we can compute maximum kinetic energy of the electrons:

$$h\nu - h\nu_0 = \tfrac{1}{2}m_e v_{max}^2$$

Recalling that $\nu = c/\lambda$ and $h\nu_0 = 2.25$ eV as stated in the problem, we obtain

$$\tfrac{1}{2}m_e v_{max}^2 = \frac{hc}{\lambda} - 2.25 \text{ eV}$$

$$= \frac{(6.625 \times 10^{-34} \text{ J}\cdot\text{s})(3 \times 10^8 \text{ m/s})}{400 \times 10^{-9} \text{ m}}$$
$$- (2.25 \text{ eV})(1.6 \times 10^{-19} \text{ J/eV})$$

$$= 4.97 \times 10^{-19} \text{ J} - 3.6 \times 10^{-19} \text{ J}$$

$$= 1.37 \times 10^{-19} \text{ J} = \frac{1.37 \times 10^{-19} \text{ J}}{1.6 \times 10^{-19} \text{ J/eV}} = 0.86 \text{ eV}$$

Hence the stopping potential difference is 0.86 V. To compute the speed of the electrons, we solve for the speed from our value of the kinetic energy:

$$\tfrac{1}{2}m_e v_{max}^2 = 1.37 \times 10^{-19} \text{ J}$$

$$v_{max} = \sqrt{\frac{(2)(1.37 \times 10^{-19} \text{ J})}{m}} = \sqrt{\frac{(2)(1.37 \times 10^{-19} \text{ J})}{9.11 \times 10^{-31} \text{ kg}}}$$

$$= \sqrt{0.30 \times 10^{12} \frac{\text{m}^2}{\text{s}^2}}$$

$$= 0.55 \times 10^6 \text{ m/s} = 5.5 \times 10^5 \text{ m/s}$$

21.3 THE DISCOVERY OF THE NUCLEUS

The new quantum theory was just one example of the ferment that was occurring in physics around the turn of the century. Experimental scientists, as well as theoreticians, were participating in amazing discoveries that were quickly to revolutionize our view of the world. Some

Cathode rays were actually the beams of electrons emitted from the cathode (negative electrode) in a gas or vacuum discharge tube and accelerated to the positive anode. The British physicist J. J. Thomson identified these rays as electrons in 1897.

of the most astonishing findings concerned the ultrasmall world of the atom.

During much of the latter half of the nineteenth century, most scientists believed that the world was composed of tiny, round, hard, indivisible atoms that generally combined to form molecules. A number of experiments, among them the Millikan oil-drop experiment, the production of cathode rays, and the photoelectric effect, made it obvious that matters were not quite as simple. Atoms clearly contained negatively charged electrons, which could be removed from the atoms under appropriate circumstances. In addition, in 1895 the German physicist Wilhelm Konrad Röntgen, working at the University of Würzburg, added a much larger complication to the problem of the ultimate nature of matter.

Röntgen discovered that cathode rays which struck a metal target, or even the glass of the vacuum tube in which they were produced, created an entirely new and extremely strange kind of ray. These rays had the astonishing capacity of going through cloth, leather, and even human flesh; only such dense substances as bone and metal seemed able to stop them. In addition to this penetrating power, the new rays were also different from cathode rays insofar as they were not deflected by magnets or electric fields. Because of their incredibly peculiar properties, the rays were dubbed x-rays and immediately put to work in the field of medicine.

Not until 1912 did the nature of x-rays became apparent. Then, a young German crystallographer named Max von Laue discovered, as a result of experimental studies using crystals as diffraction gratings, that the x-rays had wavelengths that were very much shorter than those of visible light. Their astonishing penetrating power resulted from their exceptionally short wavelengths, which were of the order of one-thousandth that of light waves. Thus the x-ray photons are 1000 times more energetic than those of visible light.

Scarcely had Röntgen announced his discovery of x-rays when yet another novel form of radiation hit the physics laboratories. In an effort to discover whether certain kinds of matter might produce x radiation naturally, French physics professor Antoine Henri Becquerel decided to investigate the metal uranium. When his experiments showed that uranium compounds did indeed produce rays—they blackened photographic plates stored near them, as shown in Fig. 21.6—a Polish student of his named Marie Sklodovska Curie started to investigate just how much x radiation the uranium produced. She was amazed to find that some waste eliminated during the purification of a uranium ore (pitchblende, which contains about 80 percent U_3O_8) gave off even more radiation than the uranium compound itself. Within a year, Madame Curie had discovered two new elements in those wastes, which she named polonium and radium, and realized that they produced yet another type of radiation distinct from x-rays. In fact, radioactive substances emitted three types of radiations, as shown in

Figure 21.6 Radioactivity. Left, ordinary photograph of a rock containing uranium compounds. Right, self-photograph of the same rock, taken by means of its radioactivity. (*From J. J. G. McCue, An Introduction to Physical Sciences: The World of Atoms, 2d ed., copyright 1963, The Ronald Press Company, New York*)

The upshot of the discovery of radioactivity and other types of radiation was to destroy forever the concept of the indivisible, billiard-ball atom.

Alpha particles were found to be atoms of the gaseous element helium that had lost their two electrons. Rutherford won the 1908 Nobel Prize in chemistry for this discovery.

Fig. 21.7. This discovery, as we shall see in Chap. 23, was to bear fruit more than 30 yr later, producing new understanding of the fundamentals of matter and eventually leading to the production of the atom bomb.

At the time, however, the upshot of the discovery of this "radioactivity" and other types of radiation was to destroy forever the concept of the indivisible, billiard-ball atom. Atoms plainly had some kind of internal structure, and theoreticians around the world set out to speculate on just what that structure was.

The first theory to emerge was that of Sir John Joseph Thomson, the British scientist who had attained the lofty pinnacle of the professorship of physics at Cambridge University at the age of 27 and who in 1897 had put the final seal on the identification of the electron. Thomson suggested the atom consisted of a large gob of positively charged material through which electrons were scattered like chocolate chips in a cookie. The total negative charge of the electrons, Thomson argued, would be sufficient just to balance the positive charge of the bulk of the atom and thus ensure that the atom was electrically neutral.

Thomson's theory did not long survive the test of experimentation. In 1906, one of Thomson's former students, a New Zealander named Ernest Rutherford, embarked on a series of experiments that involved shooting positively charged products of radioactivity known as alpha (α) particles at very thin sheets of mica. These experiments, which Rutherford started at McGill University in Montreal and continued at

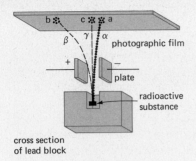

Figure 21.7 Experiment to show that rays from radioactive substances are of three kinds: α, β, and γ.

Manchester University in England, produced evidence of strange deviations, or scattering, to use the technical term. Most of the α particles streamed straight through the layers of mica—or of other substances, such as gold foil—but Rutherford discovered that a few were deflected from their paths by the layers. The tracks of some of these changed by relatively small angles, of the order of 10°; others deviated by a more appreciable 60° or so; and a minute but measurable number of the α particles were, astonishingly, scattered back by the mica along the direction from which they had come, just as if they had been reflected by a mirror.

By 1910, it was clear that Thomson's chocolate chip cookie model (or plum pudding model as it was known in England) of the atom just could not cope with these experiments or more sophisticated scattering studies by Rutherford's students Hans Geiger and Ernest Marsden. "It was then," Rutherford explained later, "that I had the idea of an atom with a minute massive center carrying a charge." Thus was the modern concept of the atom born.

As Rutherford saw it, nearly all the mass of the atom existed inside a very small bundle in the center of the entity, which also contained all the atom's positive charge. Under normal circumstances, α particles aimed in the general direction of any atom would pass straight through the void that surrounded this central "nucleus." Occasionally, however, one of the positively charged particles would come close enough to the nucleus to be repulsed by its positive charge. Rutherford used Coulomb's law of repulsion between two like charges to calculate the paths of α particles fired near the nucleus and came up with an angular distribution for the scattering that agreed remarkably well with the experimental findings of Geiger and Marsden.

Rutherford's model basically assumes that the α particles are scattered by point charges. This is of course a simplifying assumption, and it is actually possible to estimate the upper limit of the size of the "point nuclei" by considering the special case of an α particle that is aimed directly at a nucleus.

Imagine that such an α particle, whose mass is M and charge is $+2e$ (a positive charge whose magnitude is twice that of the electron) starts moving toward a nucleus inside a piece of gold foil, whose charge is $+Ze$, with a speed v. As the particle zeroes in on the nucleus, the Coulomb repulsion slows it down, until it eventually comes to a halt and starts moving back along its original path. If we equate the kinetic energy of the α particle and the electrostatic energy of repulsion between the two positive charges, we can determine how close to the nucleus the α particle stops and turns tail; this distance R_0 gives us at least an upper limit on the size of the nucleus. This equation is

$$\tfrac{1}{2}mv^2 = k\frac{(2e)(Ze)}{R_0}$$

Hence

$$R_0 = \frac{4keZe}{mv^2} = \frac{4kZe^2}{mv^2}$$

The estimate for the value of Z was actually too great, for reasons we shall see in Chap. 23.

Rutherford and his colleagues estimated the value of Z, the number of positive charges on the nuclei in atoms of various different metals, by taking the figure as roughly half the atomic weight of the metals. Using these figures, they concluded that the distance R_0 in metals was about 10^{-14} m—roughly one-thousandth the size of the atoms themselves.

Far from solving the problem of the nature of atoms, these theoretical and experimental discoveries only muddied the general picture. The most important question which arose from Rutherford's pioneering work was that of where the electrons were in the atom. It should be pointed out that if only electrostatic forces act, it is impossible to have a *static, stable* configuration of charges. An answer was to put the electrons in orbit around the positive nucleus, just like planets around a star. But electromagnetic theory flatly stated that any accelerating electric charge (remember that an object moving in a circle must be accelerating) emitted electromagnetic radiation. Thus, the electrons accelerating in orbit would give off electromagnetic radiation (energy). They would fall in toward the nucleus and lose potential energy. Then, there was a problem of the nucleus itself. How, physicists asked, could large numbers of positive charges be squeezed together into tiny spaces, with huge mass densities of about 10^{16} kg/m^3, against the force of electrostatic repulsion? Plainly, all was not well with the theory of atomic structure. It took a combination of astronomical observations, quantum theory, and a Danish genius to put scientists on the right track.

EXAMPLE 21.6

In the Rutherford α-particle scattering experiments, α particles with a speed of 2.09×10^7 m/s struck a thin gold foil. The atomic number of gold—the number of positive charges in a gold atom—was not known exactly at that time. Rutherford estimated that gold had a charge Z on its nucleus of $+100$ electronic charges, a figure that is actually too large. However, assuming this value for Z, compute the distance of closest approach for an α particle that moves directly toward a gold nucleus and hence "reflects" back from the gold foil.

Solution

We must equate the changes in electrostatic potential energy and kinetic energy as the α particle moves from essentially $r = $ infinity, where $E_p = 0$ to the distance of closest approach R_0, where the α particle is stopped and reflected backward. The kinetic energy changes from $\frac{1}{2}mv^2$ at $r = $ infinity to zero at R_0. Hence, we can write

$$\tfrac{1}{2}mv^2 = \frac{(k)(2e)(Ze)}{R_0}$$

$$R_0 = \frac{4kZe^2}{mv^2}$$

We use values of 100 for Z for the gold nucleus as estimated by Rutherford rather than the true value of 79 and substitute 6.64×10^{-27} kg for the mass of the α particle (helium nucleus), 9×10^9 N·m²/C² for k and 2.09×10^7 m/s for v as determined by Rutherford. Hence,

$$R_0 = \frac{4(100)(1.6 \times 10^{-19} \text{ C})^2(9 \times 10^9 \text{ N·m}^2/\text{C}^2)}{(6.64 \times 10^{-27} \text{ kg})(2.09 \times 10^7 \text{ m/s})^2} = 3.18 \times 10^{-14} \text{ m}$$

21.4 THE HYDROGEN SPECTRUM

The astronomical observations that were to be the key to understanding the atom were measurements of the spectrum of atomic hydrogen—that is, hydrogen whose molecules had separated into individual atoms—in the sun. The extremely precise measurements were made in 1861 by the Swedish astronomer Anders Jonas Ångström, whose name is now perpetuated in the ångström unit (Å; 1 Å = 10^{-10} m). The astronomer actually measured just the four spectral lines of atomic hydrogen that appear in visible wavelengths of light; these are the four lines on the extreme left of Fig. 21.8. Close study of the lines indicates that they form a kind of mathematical progression because the spacing between each pair of adjacent lines is a fixed fraction of the spacing between the next pair.

In 1885, Johann Jakob Balmer, who taught mathematics at a girls' high school in Basel, Switzerland, played some mathematical games with the four visible lines and discovered that their wavelengths were given by the formula

$$\lambda \text{ (in meters)} = (3645.6 \times 10^{-10})\left(\frac{n^2}{n^2 - 2^2}\right)$$

In this mathematical series, n can be any integer from 3 upward. Giving n the value of 3 yields the wavelength of what is known as the H_α line. A value of 4 for n corresponds to the next, H_β, line; the value 5 gives the third line in the spectrum, which is referred to as H_γ. The fourth line occurs for $n = 6$ and is indicated as H_δ in Fig. 21.8.

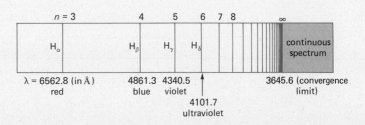

Figure 21.8 Diagram of the Balmer series of atomic hydrogen. (The wavelengths are the values in air.) Higher energy is to the right.

Balmer had absolutely no theoretical justification for his formula, but he was naturally very pleased with the agreement between the formula and Ångström's experimental findings. Indeed, he remarked that the agreement indicated "a striking evidence for the great scientific skill and care with which Ångström must have gone to work." The agreement also gave Balmer confidence to predict lines as yet undiscovered in the spectrum of atomic hydrogen—although he later lost his nerve and withdrew this prediction, with the excuse that the temperatures at which the experiment to determine the lines was performed was unfavorable. Later, Balmer realized that the series of lines would converge to a wavelength of 3645.6×10^{-10} m when n went to infinity. Some of the additional lines did indeed appear when the spectrum of atomic hydrogen was recorded on film that can detect lines in the ultraviolet region, which are invisible to the human eye.

The unpretentious mathematics teacher went one stage further, predicting entire new series of spectral lines whose wavelengths would be given by the formula

$$\lambda = \text{constant} \times \left(\frac{n^2}{n^2 - 3^2}\right)$$

and

$$\lambda = \text{constant} \times \left(\frac{n^2}{n^2 - 4^2}\right)$$

These lines also turned up later, outside the visible region of the spectrum.

Today we normally invert Balmer's formula for his original series, writing it in the form

$$\frac{1}{\lambda} = R\left(\frac{1}{2^2} - \frac{1}{n^2}\right)$$

where $n > 2$ and R is a constant. The formulas for the other series of lines are produced by substituting the squares of other integers for the 2^2 term.

EXAMPLE 21.7

Compute the wavelength of the H_β line in the hydrogen spectrum using Balmer's empirical relationship.

Solution

Balmer's empirical formula is given by

$$\lambda = 3645.6 \times 10^{-10} \left(\frac{n^2}{n^2 - 2^2}\right) \text{ (in meters)}$$

The H_β line shown in Fig. 21.8 is obtained from Balmer's formula when $n = 4$:

$$\lambda = (3645.6 \times 10^{-10}) \left(\frac{4^2}{4^2 - 2^2}\right) m = (3645.6 \times 10^{-10}) \left(\frac{16}{16 - 4}\right) m$$

$$= (3645.6 \times 10^{-10}) \tfrac{4}{3} \, m$$

$$= 4860.8 \, \text{Å}$$

This value is in excellent agreement with the measured value for λ in Fig. 21.8.

21.5 THE ATOM ACCORDING TO BOHR

One of the most critical points about Balmer's mathematical games was that they applied to the hydrogen atom, the simplest type of atomic entity possible, with its single positive charge and single electron. If scientists could solve the structure of this type of atom, then they would undoubtedly have the key to more complex atomic structures. Nevertheless, the hydrogen atom itself seemed, if anything, too complicated for theoreticians in the few years immediately preceding World War I because not only was there the peculiarly ordered series of spectral lines to be accounted for, there remained the intractable problem of where the electron should go to keep the atom stable, as it obviously was in reality.

The Danish physicist Niels Bohr, who went to Cambridge to work with Rutherford just at the time that Rutherford's nuclear model of the atom was announced, solved the problem. In 1913 he published a theory of the hydrogen atom that killed two birds with one stone, putting the electron in a stable position inside the atom and accounting for Balmer's series of spectral lines.

Underlying Bohr's theory was the application of Newtonian mechanics to describe the motion of an electron in a circular orbit around the positively charged atomic nucleus. However, Bohr's major step was to invoke assumptions that placed conditions on the orbits allowed and replaced the classical ideas of the conditions under which radiation is emitted from atoms. The Danish physicist's assumptions can be grouped into the four following statements:

1. The hydrogen atom was like a miniature solar system, with the negatively charged electron moving in a circular orbit around the positively charged nucleus (Fig. 21.9), which in the case of the hydrogen atom was a single, positive particle called a proton.
2. The electron could not move in just any orbit. Bohr restricted its circular motion to paths in which its angular momentum was a simple multiple of Planck's constant divided by 2π. In other words, for an electron of mass m moving at a speed v in a circle of radius r around the proton (whose mass is 1837 times the mass of the electron), the angular momentum was restricted to values given by the equation

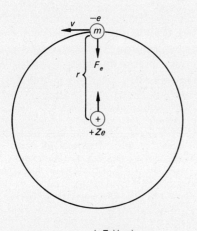

Figure 21.9 Bohr model of the atom. The electron moves in a circular orbit about a positive nucleus of charge $+Ze$.

$$mvr = \frac{nh}{2\pi} \quad \text{where } n = 1, 2, 3, 4 \ldots \tag{21.4}$$

This equation defines a set of orbits in which the electron had specific speed and orbital radius. Thus, where Planck quantized energy and Einstein quantized light, Bohr went one step further and quantized the orbital angular momentum of electrons.

3. According to Bohr's rationale, an orbiting electron would not emit any radiation as long as it remained in one of the "allowed orbits" given by Eq. (21.4). This assumption went directly counter to classical electromagnetic theory, which required that accelerating charges radiate continuously. However, it very neatly solved the problem of the hydrogen atom's stability—even though Bohr, like Planck 10 yr previously when he quantized radiant energy, had no logical reason for making the assumption.

4. Emission and absorption of energy occurred when an electron changed from one allowed orbit to another. Since both the speed and distance from the proton change when the orbit changes, such switches in orbit corresponded to alterations in the electron's energy. Continuing to rely on the quantum concept, Bohr argued that this energy was also quantized:

$$h\nu = E_1 - E_2 \tag{21.5}$$

where E_1 and E_2 are the energies of the electron in two different orbits and ν is the frequency of the radiation. If E_1 is greater than E_2, then the switching electron radiates energy at this frequency; if E_1 is smaller than E_2, the electron absorbs energy at the same frequency.

Given these postulates, we can now, like Bohr, set out to analyze the atom. Let us imagine the general case of a nucleus whose charge is $+Ze$, where Z is the number of protons it contains, and investigate one particular electron that is orbiting the nucleus with speed v in a circle of radius r. The electron and nucleus are attracted to each other by a Coulomb force F, given by

$$F = \frac{(k)(Ze)(-e)}{r^2} = -\frac{kZe^2}{r^2}$$

where k is a constant equal to 9×10^9 N·m²/C².

Now, since the electron manifestly does not fall in toward the proton, this Coulomb attractive force must be identical to the inward, centripetal force, mv^2/r, on the electron. Hence, we can arrive at the following equation:

$$\frac{kZe^2}{r^2} = \frac{mv^2}{r} \tag{21.6}$$

From this equation we can easily obtain the kinetic energy of the electron in its orbit

$$E_k = \frac{1}{2}mv^2 = \frac{1}{2}\frac{kZe^2}{r}$$

The quantity in which we are most interested is the energy emitted or absorbed when an electron moves from one orbit to another. To calculate this figure, we must know the total amount of energy the electron possesses in each orbit, that is, the sum of the kinetic energy and the electrostatic potential energy:

One can neglect gravitational potential energy in this calculation because it is vanishingly small when compared with the other energies, as shown in Example 13.1.

$$E_{\text{tot}} = E_k + E_p$$

$$E_p = -\frac{kZe^2}{r} \qquad E_k = \frac{1}{2}mv^2 = \frac{kZe^2}{2r} \tag{21.7}$$

$$E_{\text{tot}} = \frac{kZe^2}{2r} - \frac{kZe^2}{r} = -\frac{kZe^2}{2r} \tag{21.8}$$

Thus, **the total energy of the electron bound to the positive nucleus in a finite orbit is always negative.** This results from the potential energy of a bound electron always being negative and greater in magnitude than the kinetic energy, which is itself always positive. The maximum energy of the bound electron is obviously zero, a situation that occurs when r is infinite; at this point the electron just frees itself from the proton's influence. Once free, the electron can possess any amount of kinetic energy. The lowest amount of energy that the electron can possess occurs when the radius of its orbits has its smallest value.

Thus far we have not used Bohr's first postulate, which quantizes the angular momentum of the electron. To apply this condition, we must first manipulate the force equation, Eq. (21.6):

$$\frac{mv^2}{r} = \frac{kZe^2}{r^2}$$

Multiplying each side by mr yields

$$m^2v^2 = \frac{kmZe^2}{r}$$

Now, Bohr's quantization condition is given by the equation $mvr = nh/2\pi$, which can be converted to $m^2v^2 = (n^2h^2)/(4\pi^2 r^2)$. Then,

$$m^2v^2 = \frac{kmZe^2}{r} = \frac{n^2h^2}{4\pi^2 r^2}$$

Solving for r, we obtain

$$r = \frac{n^2h^2}{4\pi^2 mZe^2 k} \tag{21.9}$$

All the terms in this equation are constant except n, which can assume only integral values, that is, 1, 2, 3, 4 .... Thus, the radii of the Bohr orbits of the electron vary according to n^2, that is, 1, 4, 9, 16 .... Figure 21.10 shows a series of these allowed orbits.

Figure 21.10 Cause of a series of hydrogen-emission spectral lines. Lines due to jumps of the one electron to the innermost orbit (solid arrows) belong to the Lyman series; those due to jumps from various outer orbits to orbit 2 (broken arrows) are the Balmer series.

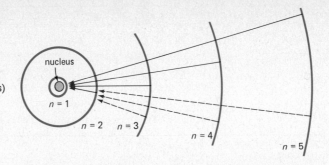

We can use this expression for the radius of the orbits to rewrite our expression for the total energy of the electron [Eq. (21.8)]:

$$E_{\text{tot}} = \frac{-Ze^2k}{2r} = \frac{-2\pi^2 mZ^2 e^4 k^2}{h^2 n^2} = -\frac{\text{constant}}{n^2} \quad (21.10)$$

Since we know the values of m, e, k, and h and can substitute the value of 1 for Z in the case of the hydrogen atom, we can readily obtain the value of the constant in this equation. This constant is usually determined in electronvolts by putting the mass of the electron in kilograms, the charge in coulombs, and Planck's constant h in joule-second, and using the conversion factor 1.6×10^{-19} J = 1 eV. For the hydrogen atom, this gives the expression

$$E_{\text{tot}} = \frac{-13.6 \text{ eV}}{n^2}$$

The energy, then, varies with the inverse of the square of the integer and has its lowest, that is, most negative, value when $n = 1$. This value specifies the most tightly bound electron and corresponds to the smallest orbital radius. In fact, this is the normal situation of the hydrogen atom. Thus, to remove an electron completely from a typical hydrogen atom requires 13.6 eV of energy. Such a removal would make the atom into a hydrogen ion with a single positive charge; we therefore say that 13.6 eV is the ionization energy and 13.6 V is the ionization potential of the hydrogen atom.

21.6 CONFIRMING BOHR'S THEORY

The clincher in our analysis of Bohr's theory of the hydrogen atom comes when we calculate the spectral frequencies of the atom as predicted by the theory and compare them with the measured frequencies. We do this by applying Bohr's fourth postulate: a quantum of energy, $h\nu$, is emitted or absorbed whenever an electron jumps from one orbit, whose energy is E_n, to another, whose energy is E'_n. We often refer to the orbits as energy levels, or energy states. If the electron is to emit energy when it jumps, then E_n must be larger than E'_n; if it is to absorb energy, then it must be smaller. In other words, a jump from an outer orbit to an inner one will result in emission of radiation in the

The Great Dane

Niels Bohr stands second only to Einstein among physicists of the twentieth century. Like Einstein, Bohr turned a complete area of physics upside down, influenced a couple of generations of physicists, and played a critical, if minor, role in the technological pursuit of the atom bomb in World War II.

Bohr received his education in physics—and indulged himself spectacularly in soccer—at the University of Copenhagen. In 1911, replete with his newly won Ph.D., he did what just about every young physics graduate of the time secretly wished to do: he went across the North Sea to Cambridge University to work for Rutherford, who had just proposed the nuclear model of the atom. The young Dane's moment of truth came 2 yr later, when he incorporated Planck's quantum theory into the Rutherford model. Bohr's great leap of faith was the assumption, without any experimental backing, that electrons in the quantized orbits did not emit radiation. However dubious the reasoning behind it, this model accounted for the obvious stability of atoms and agreed remarkably well with the observed spectrum of the hydrogen atom. In 1922, Bohr received the Nobel Prize for physics.

By that time Bohr had returned to the university of his native Copenha-

Niels Bohr. (*Wide World*)

gen. He soon made this university, along with those of Cambridge, Göttingen, and Leiden, a major center for the ferment in physics that lasted through the twenties and thirties. In 1939, as the storm clouds of war were gathering over Europe, the still-active professor took news to the United States of the possibility that uranium might undergo fission. Back in Denmark, Bohr was caught by the German invasion, but in 1943 he escaped, first to Britain and then to the United States, where he worked on the Manhattan project to produce the atom bomb. Before leaving Denmark, Bohr dissolved his golden Nobel prize medal in acid; when he returned at the war's end, he reclaimed it and recast it. He also contributed directly to another Nobel physics prize, through his son Aage, who shared the 1975 award for basic studies on the nature of the nucleus.

---

form of a photon whose energy, $h\nu$, is equal to the energy difference in the two orbits.

The electron in the normal hydrogen atom exists in the lowest possible energy level. If it is to produce spectral lines of emission, therefore, it must somehow be promoted to a higher level by absorption of energy, a process known as excitation. This can be done in an electrical discharge tube. In this type of tube, free electrons are emitted from a hot filament and accelerated by the potential difference (voltage) across the tube. On their way, the electrons hit atoms and knock electrons out of the lowest orbits to higher energy states—or even right out of the atom—to produce ions. This can also be done by light supplying the energy difference. We know from Bohr's fourth postulate that the amount of energy absorbed by any electron which is excited must be a quantum $h\nu$. Once in the higher-energy level, the electron can return to its original low-energy orbit by emitting quanta of energy. If the higher-energy state is represented by n and the state to which the electron drops is n', then, according to the Bohr criterion:

$$h\nu = E_n - E_{n'} = \frac{+2\pi^2 e^4 m Z^2 k^2}{h^2}\left(\frac{1}{n'^2} - \frac{1}{n^2}\right)$$

where $n > n'$. Since $c = \lambda\nu$,

$$\frac{1}{\lambda} = \frac{\nu}{c} = \frac{2\pi^2 e^4 m Z^2 k^2}{h^3 c}\left(\frac{1}{n'^2} - \frac{1}{n^2}\right) = RZ^2\left(\frac{1}{n'^2} - \frac{1}{n^2}\right) \quad (21.11)$$

where $n > n'$. The constant factor R in this equation is known as the Rydberg constant, after the Swedish spectroscopist J. R. Rydberg, and is determined by the values of the fundamental constants in the equation. What is really interesting about the expression, though, is that it is identical with the one derived by Balmer to fit the observed spectrum of hydrogen atoms in the sun. This not only gives meaning to Balmer's mathematical games, in terms of specifying energy levels by the terms n and n', it provides an excellent means of checking the Bohr theory via the actual and calculated values of R. The agreement, worked out in the year 1913, proved to be excellent, providing immediate support for the Bohr theory.

This way of looking at the emission spectrum of hydrogen can be better understood by referring to Fig. 21.11, which shows the various energy states of the hydrogen atom and the transitions between them that produce emission lines. The number n which labels the energy states is known as the principal quantum number, a concept that will become increasingly important as we delve further into atomic theory in the next chapter. The two series of emission lines illustrated in Fig. 21.11 correspond to values of 1 and 2 for n'. The latter is the Balmer series with which we are already familiar; the former, which is entirely in the ultraviolet, is known as the Lyman series, after Theodore Lyman. The lowest line in this energy diagram corresponds to the rock bottom of electron energy, when the particle is in the orbit closest to the nucleus. The highest line is the series convergence limit, at which the electron breaks free of the nucleus. The difference between the two lines is 13.6 eV, the ionization energy of the hydrogen atom. And the lines in between correspond to intermediate orbits. The line immedi-

Figure 22.11 Energy-level diagram for hydrogen.

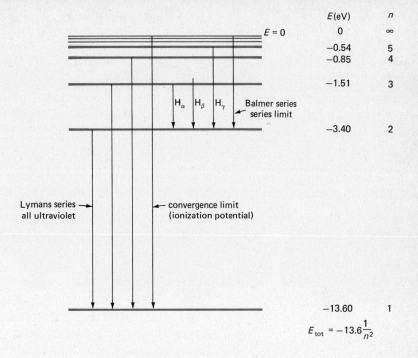

ately above the bottom one, for example, represents an energy of −3.40, or 10.20 eV above the lowest energy level.

Obviously the Balmer series represents transitions from higher energy levels down to the next to the lowest level, and the Lyman series corresponds to drops in energy down to the lowest, and most stable, level of the electron in the hydrogen atom. All the lines shown in Fig. 21.11 and all other transitions between the various energy states have actually been observed by spectroscopists, providing a convincing check on the validity of Bohr's theory.

An important test of energy quantization in the atom was devised by James Franck and Gustav Hertz, two German physicists who in 1914 decided to investigate the process by which the electrons in hydrogen atoms are excited from their normal ground states into excited states from which the emission spectrum results. They used the simple gas discharge tube with extra external circuitry sketched in Fig. 21.12.

In this device electrons from a filament next to the cathode accelerate toward the anode through a gas of hydrogen or perhaps mercury vapor, gaining energy as they make their way toward the positive terminal. When they strike atoms, these free electrons might knock electrons inside the atoms from their ground states into excited states by transferring discrete amounts of energy to them provided they have enough energy. By studying the current in the tube as the voltage was increased, Franck and Hertz were able to monitor these energy changes. The anode in the tube was a metal screen that allowed the

Figure 21.12 Apparatus used for the Frank-Hertz experiment.

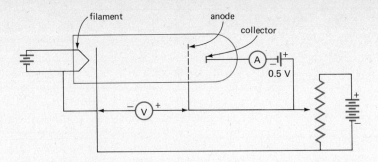

beam of electrons to pass through it onto the collector of the tube which is at a negative potential of 0.5 V with respect to the anode. Therefore the collector tended to repel electrons. Electrons required at least 0.5 eV of energy as they passed the anode if they were to reach the collector and be registered as a current.

As the voltage of the anode was increased above 0.5 V, the current in the collector increased until the voltage across the tube became equivalent to the first excitation potential of the gas in the tube. At this voltage the current dropped because individual electrons had sufficient energy to excite individual atoms. The collisions caused the free electrons to stop and current to drop. The current did not go to zero because not all the accelerated electrons collided with atoms. Further increase in voltage resulted in an increase in collector current until the voltage equaled another of the excitation potentials of the gas or a multiple of the first excitation potential. In the latter situation, a single electron excited two atoms. By looking for drops in the collector current in this way, Franck and Hertz made the separation between energy levels apparent. After the atoms were excited they emitted radiation; because emission follows excitation directly, the appearance of light usually occurs simultaneously with a drop in current.

Another experiment that illustrates the presence of discrete energy levels and another way of exciting atoms is irradiation of atoms with electromagnetic waves. The irradiating light, made up of photons with energy $h\nu$, can excite the atoms in the gas only if this energy is equal to the energy difference between a pair of energy levels of the gas atom. Once excited, the atom may emit "resonance radiation" of the same frequency if the electron returns to the ground state in one jump, or radiation at lower frequency if the electron can make the journey to the ground state by a series of jumps via intermediate levels. In hydrogen, for example, as illustrated in Fig. 21.11, an electron excited to the $n = 4$ from the $n = 1$ energy level would absorb a photon of energy $h\nu = 13.60 - 0.85 = 12.75$ eV. The atom could then return to its ground state by emission of the H_β ($n = 4$ to $n' = 2$) line of the Balmer series and then the first line of the Lyman series ($n = 2$ to $n' = 1$).

One of the first examples of excitation of atoms in a gas by resonance absorption of photons was the observation in 1814 of dark lines (now called absorption spectra) in the spectrum of the sun and later of other stars. The sun is an incandescent source of radiation, but close examination shows that its spectrum is not continuous; instead, it contains some dark lines. These lines, called Fraunhofer lines after the German optician and physicist Joseph Fraunhofer, result from the absorption of photons emitted from the sun by the colder gases that surround the sun or any star. These dark lines match the bright-line emission spectra obtained from the same gases when excited. In fact, the gaseous element helium was first discovered through its dark-line spectrum in the sun, which did not match the emission spectrum of any known element. Absorption spectra can be reproduced in the laboratory by shining white light through the vapor of the element or molecule under investigation and then observing the "holes," or dark lines, in the observed spectrum. The dark lines appear even with subsequent reemission of photons of the same energy, as they are emitted in all directions.

The gaseous element helium was first discovered through its dark-line spectrum in the sun.

The processes stimulating fluorescence and phosphorescence usually result from the absorption of high-energy photons, often ultraviolet photons. In fluorescence, the atoms are first excited by ultraviolet radiation. They then emit visible light as the atom "decays," or deexcitates, to its ground state in more than one jump. In phosphorescence processes, high-energy photons excite the atoms and the electrons remain in the excited states for some length of time. Hence, even after the exciting light has been turned off, the atoms continue to emit, producing materials that "glow in the dark."

EXAMPLE 21.8

From the energy-level diagram given in Fig. 21.11, determine the wavelength of the first line (H_α line). Compare your value with the value of the wavelength of the H_α line measured by Ångström in Fig. 21.8.

Solution

From the energy-level diagram we can obtain the energy corresponding to the photon emitted for this spectral line. The H_α line results from an electronic transition from an excited $n = 3$ to $n = 2$ state, an energy given by the difference between the energies of the two levels $E_n - E_{n'} = -1.51 - (-3.40) = 1.89$ eV. This energy difference $E_n - E_{n'} = 1.89$ eV will equal $h\nu$ for the emitted photon or in terms of wavelength. Recalling $\nu = c/\lambda$, we can write

$$h\nu = \frac{hc}{\lambda} = 1.89 \text{ eV}$$

or

$$\lambda = \frac{hc}{1.89 \text{ eV}} = \frac{(6.625 \times 10^{-34} \text{ J} \cdot \text{s})(3 \times 10^8 \text{ m/s})}{(1.89 \text{ eV})(1.6 \times 10^{-19} \text{ J/eV})} = 657 \times 10^{-7} \text{ m}$$

$$= (657 \times 10^{-7} \text{ m})(10^{10} \text{ Å/m}) = 6570 \text{ Å} \quad \text{or} \quad 657.0 \text{ nm}$$

This line is in the visible region of the spectrum and compares favorably with the value of 6562.8 Å given in Fig. 21.8.

EXAMPLE 21.9

Using the Bohr theory of the atom, determine the values for the energy of the $n = 2$, 3, and 4 levels of hydrogen.

Solution

The Bohr theory of the atom indicates that the total energy of an electron in a particular atomic orbit is given by Eq. (21.10),

$$E_{\text{tot}} = \frac{-2\pi^2 m Z^2 e^4 k^2}{h^2 n^2} = -\frac{\text{constant}}{n^2}$$

The constant can be evaluated, and when expressed in electronvolts its value is given by 13.6 eV. Hence,

$$E_{\text{tot}} = \frac{-13.6 \text{ eV}}{n^2}$$

Therefore the energies of the $n = 2$, 3, and 4 levels are:

Level 2:

$$E_2 = \frac{-13.6}{(2)^2} = -3.40 \text{ eV}$$

Level 3:

$$E_3 = \frac{-13.6}{(3)^2} = -1.51 \text{ eV}$$

Level 4:

$$E_4 = \frac{-13.6}{(4)^2} = -0.85 \text{ eV}$$

EXAMPLE 21.10

Show that all emission spectral lines from excited states in hydrogen to the $n = 3$ level result in wavelengths in the infrared region of the spectrum.

Solution

If we examine the energy-level diagram in Fig. 21.11, it is clear that the highest frequency (most energetic photons) possible from transitions to the $n = 3$ level occur when a free electron at $n = \infty$ ($E_\infty = 0$) drops to $E_3 = -1.51$ eV when $n = 3$. The energy of the photon emitted will be

1.51 eV. All other transitions from excited states above $n = 3$ to the $n = 3$ level will have a lower frequency. If the 1.51-eV photon corresponds to infrared radiation, then all the remaining lines of this spectral series ending in $n = 3$ will also be in the infrared or low-energy region of the spectrum. Converting the 1.51 eV to the wavelength, we have

$$h\nu = \frac{hc}{\lambda}$$

$$\lambda = \frac{hc}{1.51 \text{ eV}} = \frac{(6.625 \times 10^{-34} \text{ J} \cdot \text{s})(3 \times 10^8 \text{ m/s})}{(1.51 \text{ eV})(1.6 \times 10^{-19} \text{ J/eV})}$$

$$= 8.23 \times 10^{-7} \text{ m} \quad \text{or} \quad (8.23 \times 10^{-7} \text{ m})(10^{10} \text{ Å/m}) = 8230 \text{ Å}$$

This wavelength is beyond the red end of the spectrum.

The concept of a planetary atom remains today as a simple and useful picture of the atom's interior, although it is not fully correct. The Bohr model firmly established the ideas of discrete energy levels and the emission and absorption of radiation that occur when the atom's energy changes from one state to another. However, as we shall see in the next chapter, the concept of discrete spatial orbits for the electron can be carried only so far.

The Bohr model of the atom is limited to one-electron atoms because it considers only the interaction of the nucleus and one electron. Hence the theory can be applied only to neutral hydrogen, singly ionized helium, or doubly ionized lithium. The only aspect to change in these special cases is the value of the charge on the nucleus, which is 1 for hydrogen, 2 for helium, and 3 for lithium.

In the years following Bohr's pronouncement of this theory, physicists made attempts to alter the Bohr model to account for multielectron atoms, but with only limited success. All atoms have discrete spectra, but attempts to predict the frequencies of these transitions from a Bohr model, even one modified to account for other interactions, were unsuccessful. The problem of considering exactly the interaction between several electrons and one nucleus is the so-called many-body problem that has not yet been solved exactly, even with the aid of the computational capacity of our large computers. However, as we shall show in the next chapter, physicists can apply simplifications that allow them to determine approximate solutions to the problems of what the electron energy levels are in many-electron atoms. The importance of the Bohr theory and its profound influence upon the development of modern quantum theory and our modern concepts of the atom and nucleus are shown by the fact that physicists still speak in terms of Bohr orbits and the resulting energy states, even though their formalism and concept of the atom are much more complex.

SUMMARY

1. The concept that energy exists in discrete packets, or quanta, was first advanced to explain the relationship between the temperature of a blackbody and the wavelengths of the emitted radiation.
2. This same concept was applied successfully to another phenomenon inexplicable by classical physics: the photoelectric effect.
3. The discoveries of electrons, x-rays, and radioactivity destroyed the concept of the indivisible billiard-ball atom. The chocolate chip cookie model first postulated to replace it was also overturned, by scattering experiments.
4. The scattering experiments suggested an atom that consisted of a massive but small central positive charge. The question "Where did the electrons reside?" remained unanswered.
5. The spectral lines of hydrogen fall in an ordered sequence.
6. Niels Bohr proposed a planetary model of the atom that he quantized by quantization of the angular momentum. Here the electron is restricted to specific orbits from which they do not radiate. This restriction overcame the objection from Maxwell's equations that an orbiting (accelerating) charge would radiate energy. This model was entirely intuitive and had no theoretical evidence to back it. It did, however, account for the experimental observations.
7. The Bohr theory indicates that energy emitted by electrons as they jump from one allowed orbit to another produces the observed spectral lines. It agrees very well with the measured spectrum of hydrogen.

QUESTIONS

1. What are photons, and how do they differ from cathode rays?
2. In what ways do γ- and x-rays differ?
3. What evidence is there that x-rays are electromagnetic waves? That they are photons?
4. If the change in some phenomenon can be made in as small increments or decrements as desired, the phenomenon is said to be variable in a continuous way. If the change can be made only in fixed steps or units, it is variable in a discrete way. (a) Which of the following vary in a continuous, and which in a discrete fashion: (1) flow rate of water, (2) number of pennies per day in a fare box, (3) weight in daily growth of an animal, (4) number of petals on a flower selected at random, (5) speed of a car, (6) first-class postal rates? (b) In the above sense, discuss the differences between the radiation theories of Maxwell and Planck. (c) Planck treated energy the way the early Greek atomic thinkers viewed matter. Elaborate, using these concepts.
5. The study of what phenomenon gave the first experimental verification of the quantum nature of radiation? Who made the observation, and who gave the first satisfactory explanation of the basis of the new theory?
6. Which of the following types of radiation has photons with the least energy, and which with the most energy: (a) radio waves, (b) ultraviolet radiation, (c) red light, (d) x-rays, (e) violet light?
7. If there were no gun capable of shooting projectiles to the height of a given enemy plane, nothing would be gained by increasing the number of guns. (a) Compare this to the observations made of the photoelectric effect with below-threshold frequency light whose brightness was progressively increased. (b) Also consider a similar question for the case where the number of guns that did reach the enemy plane was doubled.
8. On a pure wave theory basis, the total energy of a beam of light depends on both the frequency and the amplitude of the wave. (a) How would the photoelectric effect operate if radiation were a pure wave phenomenon? (b) On what factors does the *total* energy in a beam of light depend in the quantum theory?
9. In the photoelectric effect, (a) on what two quantities does the energy of ejection of an electron depend? (b) How does the number of electrons depend on the intensity of the light if the frequency is held constant?
10. Light falling on a certain metal surface causes electrons to be emitted. What happens to the photoelectric current as the intensity of the light is increased if the frequency of the light is not changed?
11. Einstein's idea of a quantum of light had a definite relation to the wave model. What was it?
12. Why does the photoelectron not have as much energy as the quantum of light that produced the photoelectron?
13. Discuss the dual nature of light. When do we resort to the wave explanation, and when do we use the particle model of light?
14. (a) Tie together in a logical way the following ideas that worried the classical theorists: principle of conservation of energy, lower energy in

orbits of smaller radius r, radiation of electromagnetic energy by orbiting electrons, spiraling of electrons into nucleus. (b) How did Bohr get around these difficulties? When is electromagnetic energy radiated according to the Bohr theory?

15. In terms of the Bohr theory, explain why the hydrogen spectrum contains a number of lines even though the hydrogen atom contains only one electron.

16. (a) What is an excited atom? (b) How is the condition brought about? (c) What happens when the atom returns to its "normal" state?

17. Why could the Bohr theory satisfactorily account for spectral lines from singly ionized helium and doubly ionized lithium but not from neutral helium or lithium?

18. Discuss the Bohr theory of the hydrogen atom. Include the following: (a) State Bohr's assumptions, noting which were contrary to ideas of classical physics. (b) Describe the picture of the atom Bohr developed.

19. The uranium atom has 92 protons and hence 92 electrons. Would you expect, on the basis of the Bohr model, that the size of a uranium atom would be 92 times that of the hydrogen atom?

20. What is the difference between the Thomson model of the atom and Rutherford's model? Explain.

21. Why is it not necessary to use the relativistic energy expression when considering photoelectrons?

22. Explain on the basis of photons the following observed facts concerning the photoelectric effect: (a) photoelectric measurements are (or are not) sensitive to the nature of the photoelectric surface. (b) The intensity of the light does not (or does) affect the number of photoelectrons emitted. (c) The maximum energy of emitted electrons does not (or does) change as a function of the frequency of radiation incident on the photoelectric surface.

23. Describe briefly what measurements were being made when physicists studied blackbody radiation. What new revolutionary idea was employed by Planck to explain blackbody radiation? Why was this such a revolutionary idea?

24. In many areas, such as certain applications in photography, the spectral distribution of the lamp used for illumination is very important. Often lamps are classified by a "color temperature"; e.g., one lamp might have a 1250-K color temperature, another might be an 1850-K lamp. This temperature is related to the blackbody curve that matches the lamp in question. What does the temperature specify, and how do these two lamps differ in spectral output?

PROBLEMS

Photons

1. What is the energy of a light photon that corresponds to a wavelength of 5×10^{-7} m?

2. What is the energy in electronvolts of light photons from a helium-neon laser of wavelength 632 nm?

3. Find the wavelength and frequency of photons whose energies are: 0.1 eV, 1 eV, 10 eV, 100 eV, 1000 eV. Which region of the spectrum do each of these represent?

4. What is the energy of a photon, in electronvolts, in the radio wave region of the spectrum where $\lambda = 200$ m?

5. What is the wavelength of a 1-MeV photon?

6. Why is it that light from a firefly can expose a photographic film, but the 100-MHz electromagnetic radiation from a 50 000-W TV station cannot?

7. What is the minimum voltage that must be applied to an x-ray tube so that the electrons hitting the target in the tube will produce x-ray photons of wavelength 0.1 Å?

8. A red light emits an average wavelength $\lambda = 620$ nm. If the light is a 60-W bulb and 10 percent of the energy emitted is light, how many quanta of red light are emitted each second?

9. The power in a light beam emitted from a helium-neon gas laser is 1 mW. How many photons of $\lambda = 632$ nm are emitted from the laser each second?

10. If the temperature of an object increases from 500 to 1000°C, what is the factor by which the total radiant energy emitted from it changes? Describe how the color of an object would change if its temperature changed in this manner.

Photoelectric Effect

11. A photon of wavelength 400 nm is incident onto a metal with work function 1.3 eV. What is the maximum kinetic energy of the resulting photons?

12. A metal has a work function of 4.5 eV. What is its photoelectric threshold frequency?

13. The threshold frequency for photoemission from a clean silver surface is 1.04×10^{15} Hz. Compute the work function of silver.

14. The photoelectric work function of potassium is 2.0 eV. If light with a wavelength of 400 nm is incident on potassium, find the (a) stopping potential in volts, (b) kinetic energy in electronvolts of the most energetic electrons ejected, (c) largest wavelength radiation that will produce photoelectrons from potassium.

15. Light of wavelength 2×10^{-5} cm falls on an aluminum surface. In aluminum, 4.2 eV are required to remove an electron. What is the kinetic energy of the (*a*) fastest and (*b*) slowest emitted photoelectrons?

16. The work function of a certain phototube is 2.5 eV. Find the energy of electrons emitted by red light of $\lambda = 670$ nm, green light $\lambda = 540$ nm, and violet light $\lambda = 420$ nm.

17. What is the maximum energy, in electronvolts, of the electrons emitted from a surface irradiated by blue light of wavelength 410 nm? Use the fact that photoemission from the metal surface in question ceases when the wavelength of incident light exceeds 600 nm.

18. What is the change in stopping potential for photoelectrons from a surface if the wavelength of light is changed from 4200 to 3800 Å?

19. The stopping potentials for photoelectrons from a certain surface are measured for two wavelengths yielding the following data: when $\lambda = 3660$ Å, $V_s = 1.48$ V, and when $\lambda = 5460$ Å, $V_s = 0.36$ V. From these data, determine the threshold frequency, the threshold wavelength, Planck's constant, and the work function.

20. In a photoelectric experiment it is noted that below a frequency of 5×10^{14} Hz no photoelectrons are emitted. What is the wavelength of the light required in this photoelectric experiment to produce electrons, the fastest of which have a speed of 10^5 m/s?

Rutherford Scattering

21. A proton of kinetic energy 2×10^{-12} J is scattered by a lead nucleus ($Z = 82$). What is the minimum possible distance of closest approach if the lead nucleus is assumed to be a point charge of $+82e$?

22. An α particle of kinetic energy 1.0×10^{-10} J is scattered by a gold nucleus ($Z = 79$). What is the minimum possible distance of closest approach if the gold nucleus is assumed to be a point charge of $+79e$?

23. The α particles from thorium have an energy of 7.33 MeV. How close could such an α particle come to an iron nucleus?

24. A proton beam in a scattering experiment strikes a gold target. In a second experiment, a beam of deuterons with a mass twice that of a proton and a $+1$ electron charge is used. The kinetic energy of the protons and the deuterons is the same in both beams. How will the minimum distance of closest approach to gold nuclei for the two particle beams compare?

Bohr Atom

25. The radius of the first Bohr orbit is 5.3×10^{-9} cm. What is the radius of the fourth Bohr orbit?

26. The energy of an electron in the first Bohr orbit of hydrogen is -13.6 eV. What is the energy of an electron in the third Bohr orbit? Make a simple statement to show how you calculated this energy for the third orbit.

27. What is the minimum frequency photon that can be absorbed by a hydrogen atom in its ground state?

28. In state A, the electronic energy of an atom is -5.87 eV, and in state B its electronic energy is -3.24 eV. What is the frequency of the photon emitted in the transition from B to A?

29. The wavelength with radiation emitted by a Bohr atom when an excited electron in the $n = 4$ energy level (orbit) drops to the $n = 1$ energy level is 973 Å. From this information compute the energy of the electron in the fourth Bohr orbit. The energy (of an electron) in the first Bohr orbit is -13.6 eV. Could the electron make the jump from the $n = 4$ level to the $n = 1$ level in several steps rather than a single jump? Draw an energy-level diagram showing the possibilities for such transitions.

30. Compute the speed of an electron in the first Bohr orbit. Through what potential difference would an electron have to accelerate to obtain that speed?

31. How much energy is required to excite the hydrogen atom from its normal ground state to the $n = 3$ state? How much additional energy is required to ionize the atom once the electron is in the $n = 3$ state?

32. Compute the energy of the photon emitted and the corresponding frequency and wavelength when an excited hydrogen atom decays from the $n = 7$ to $n = 3$ energy state.

33. What is the wavelength of photons emitted from hydrogen atoms when an electron jumps from $n = 4$ to $n = 3$? What kind of radiation is this?

34. What are the largest and shortest wavelengths in the Lyman series of hydrogen that correspond to all transitions from upper excited states to the $n = 1$ level? In which region of the electromagnetic spectrum do these reside?

35. Compute the speed of an electron in a hydrogen atom when it is in the $n = 10$ orbit. Convert this speed to the frequency at which the electron orbits. Then compare this frequency with the frequency of the light emitted when any electron drops from the $n = 11$ to $n = 10$ energy level. What conclusion can you reach? This observation supports the Bohr correspondence principle.

36. Compute the energy of the first Bohr orbit for a singly ionized helium atom in which a single electron remains.

Special Topic
The Solution Without a Problem

Although the basic theoretical work on atomic theory was completed in the first third of this century, applications of that work have continued to multiply since then. One profoundly significant application familiar in daily life is nuclear power. Another, almost as familiar, is the laser. The operation of this source of intense light depends directly on the excitation and deexcitation of atomic electrons.

The fundamental quality of the laser, whose construction was first suggested in 1959 by American physicist Charles Townes, is the natural coherence of its output. The photons that make up a laser beam all have the same frequency, phase, and direction.

The word laser is an acronym for light amplification by stimulated emission of radiation, a phrase that describes the atomic processes responsible for lasing action. The instrument consists of three separate parts: an exciting mechanism, an active medium, and a cavity. In the earliest lasers, which were developed in the early 1960s, photoflash lamps excited the atoms, ruby rods provided the active medium, and mirrors at the two ends of the rods produced the cavity. An analysis of how this ruby laser works indicates the general principles of how lasers as a whole function.

Ruby is actually crystalline aluminum dioxide, mixed with a few chromium atoms. These chromium atoms, which give the mineral its characteristic red color, make up the active medium of the ruby laser. Light fired from a flash lamp excites the chromium atoms by raising their outer electrons to higher energy levels. The electrons immediately drop down to intermediate excited levels, below the highest but above their ground states. They remain in these intermediate states a relatively long time, which makes the whole lasing process possible.

When the electrons finally drop back to their ground states, they emit photons of red light. Initially the light is emitted in all directions. It must still be straightened out to produce the required lasing action. It is here that the cavity comes into play. Any light moving along the length of the rod is reflected back by the mirrors at the ends. The reflected beams of photons stimulate even more emission because many of the chromium atoms are still in their intermediate energy levels. Thus, the light beam along the rod increases in intensity. The mirrors therefore provide a resonant cavity, much like an organ pipe for sound or a string fixed at both ends for visible waves.

Most important, the drop in energy level of an electron in any one atom stimulates the same process in neighboring atoms, which in turn stimulate electrons in yet more atoms to do the same. All this radiation is in phase. The result is a cascade—the stimulated emission of red light that produces coherent radiation.

If the two mirrors did not transmit any light, the stimulated, coherent beam could never get out of the rod. Thus the mirrors are constructed in such a way that usually one is partially transparent. The other is made intentionally "leaky"; it reflects a high proportion, more than 50 percent, but not all the light that impinges on it. As the electrons in the chromium atoms continue to fall to their ground states, the light they emit is channeled out through this partial mirror, in the form of a powerful, coherent beam.

The ruby laser is limited because the lasing action stops once most of the excited chromium atoms have returned to their ground states. The light is therefore emitted in a brief pulse that lasts no longer than a fraction of a second. Other types of laser, including the low-power helium-neon devices commonly used in teaching laboratories, can emit continuous beams of light. The basic principles remain the same, but continuous lasers differ from the pulsed variety in both their method of excitation and active medium.

The medium in the helium-neon laser, for example, is a mixture of the two inert gases, and the excitation process is a continuous electric discharge inside the mixture that keeps a sufficient number of atoms in excited states to allow a continuous beam, at least as long as the discharge continues. The leaky end of the cavity reflects about 97 percent of the light that strikes it. In addition, the process of excitation is a rather complicated one in the helium-neon laser, involving excitation by the electrical discharge of helium atoms,

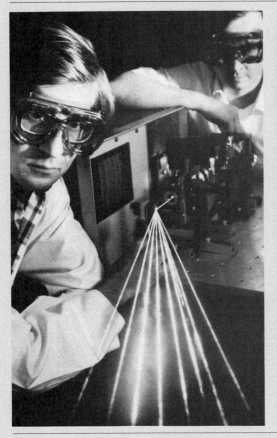
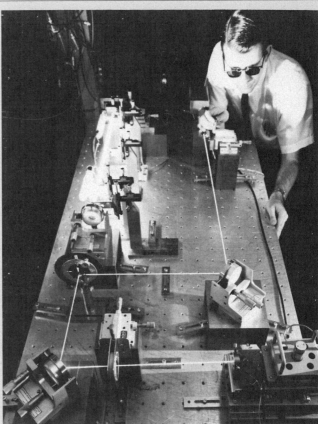

Lasers of all varieties. (*Bell Laboratories*)

followed by transfer of energy to the neon atoms through atomic collisions. This puts the neon atoms into excited states from which they decay only slowly. As they do, they produce three types of light: two at infrared frequencies and one in the visible red region of the spectrum.

Over the years physicists have developed a large variety of pulsed and continuous lasers that emit many different frequencies in the ultraviolet, visible, and infrared regions. Physicists at Bell Telephone Laboratories developed "tunable lasers," whose emission frequencies can be altered over a certain range. Carbon monoxide lasers emit infrared light in such powerful beams that they can burn human skin. And like all other types of laser, the carbon monoxide variety can damage the eyes of anyone careless enough to look directly at the beam.

In the early years of lasers there seemed to be so few applications for the devices that they became known

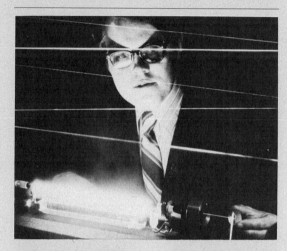

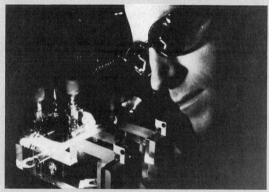

as "the solution without a problem." That situation has changed appreciably. Today lasers are used routinely both inside and outside the laboratory. Ophthalmic surgeons perform delicate retinal operations with lasers. Engineers make tiny welds with precise, powerful beams of laser light. Communications experts are experimenting with laser communications devices in which laser beams carry telephone calls and other signals over distances of a few miles. Weathermen use the "lidar," a laser-based type of radar device. Military scientists have applied lasers to the guidance of "smart bombs," which zero in inexorably on the targets. Holographers could hardly exist without lasers, as Chap. 18 illustrated. And physicists working on nuclear fusion are using ultrapowerful pulsed laser beams in hopes of stimulating a controlled version of the nuclear reaction that produces the awesome power of the sun and the hydrogen bomb.

Moving Into The Atom

- 22.1 THE COMPTON EFFECT
- 22.2 THE SCHIZOPHRENIC NATURE OF LIGHT AND MATTER
- 22.3 THE PROOF OF THE DUALITY
- 22.4 THE QUANTIZED ATOM
- 22.5 THE NATURE OF MATTER WAVES
- 22.6 THE UNCERTAINTY PRINCIPLE

   SHORT SUBJECT: Viewing the Unviewable

- 22.7 QUANTUM NUMBERS

   SHORT SUBJECT: Schrödinger's Cat: Dead or Alive?

- 22.8 PAULI AND THE PERIODIC TABLE
   Summary
   Questions
   Problems

   SPECIAL TOPIC: The Men Who Made a Scientific Revolution

The roaring twenties were years of tremendous vitality, unrivaled prosperity, and outrageous changes in mores. The decade offered much the same experiences to physicists working in the esoteric world of atomic physics. Hardly a month went by without the discovery of some new wrinkle regarding the basic nature of matter that forced theorists to modify their explanations. Hardly a year passed without some Alice in Wonderland finding that turned the world of atomic physics upside down. Taking Bohr's theory of the hydrogen atom as their starting point, physicists spent the decade constructing a theory that involved a strange mixture of waves and particles and set the stage for further probing, in the following decade, into the nucleus itself. The starting point of this particular journey into the unknown was the discovery of a surprising reflection, or scattering, effect that occurred occasionally when x-rays impinged on a solid.

As von Laue showed in 1912, x-rays are actually electromagnetic waves whose frequencies, and therefore energies, are much greater than those of light and ultraviolet radiation. They are produced when extremely excited heavy atoms—atoms containing many electrons in which an inner-orbit electron has been shifted into an orbit far from the nucleus—return to their normal, stable states. These transitions release much more energy than those that lead to light waves because the nuclei have many protons; this increases the energy difference between Bohr orbits.

Because the extremely excited states are unusual under normal circumstances, scientists have to force electrons in atoms into those states. They achieve this by bombarding a metal such as copper or molybdenum inside a vacuum tube with beams of high-energy electrons characterized by potential differences of anywhere between 10 and 50 kV. This bombardment forces the electrons in the heavy metal atoms into very high, unfilled energy levels, or even into what is known as the continuum—a sea of electron states beyond the hold of any individual atomic nucleus. Then, as the electrons return to their normal, lower-energy levels, they give off x-rays.

22.1 THE COMPTON EFFECT

Once produced, x-rays can be scattered by solid materials in two different ways. The more expected type of effect, observed and interpreted around 1912 by the British father-and-son team of William and Lawrence Bragg, involves no change in the wavelength of the x-rays. In Bragg scattering, as it is known, x-rays are actually diffracted by the periodic structure of crystalline solids. The crystal structure actually

The Braggs shared the Nobel Prize for physics in 1915 for this interpretation—the only time in which a father-and-son team has been so honored. Lawrence Bragg was only 25 yr old at the time of the award.

performs as if it were a three-dimensional diffraction grating. In fact, this type of scattering is extensively used to determine the structures of both inorganic and organic crystalline compounds, as we shall see in Chap. 24.

In 1923, the American physicist Arthur Holly Compton, then at Washington University, discovered a different type of x-ray scattering that opened up a variety of theoretical possibilities to physicists. Compton found that, under certain conditions, the wavelength of scattered x-rays was longer than that of the incoming x-rays. He explained this observation, which soon became known as the Compton effect, by proposing that the x-rays could be represented by photons which collided with electrons in the solid.

We can analyze this situation, sketched in Fig. 22.1, by focusing on the interaction between an incoming photon, whose energy is $h\nu$, and an electron in the solid, which we can regard as being effectively at rest. Just as we did in the case of two billiard balls in Chap. 6, we can analyze the collision using the principles of conservation of momentum and conservation of energy.

The assumption that the electron is at rest is justified because the electrons in the solid have kinetic energies of a few electronvolts or less, and the photons have energies of the order of kiloelectronvolts.

We must, however, employ relativistic expressions in our analysis. As Fig. 22.1 shows, the photon collides with the electron, whose rest mass is m_0. After the encounter, the two entities move off at angles α and β with respect to original direction of travel of the incoming photon. To apply the conservation equations, we require the following quantities:

$h\nu$ = initial energy of incoming photon
$m_0 c^2$ = initial energy of electron at rest ($v = 0$), that is, rest mass energy
$h\nu'$ = energy of scattered photon, which is less than $h\nu$
mc^2 = total energy of moving electron after collision

Figure 22.1 Compton scattering. Conservation of momentum and energy in the Compton effect when a photon collides with an electron. The quantity $h\nu/c$ is the magnitude of the photon's momentum.

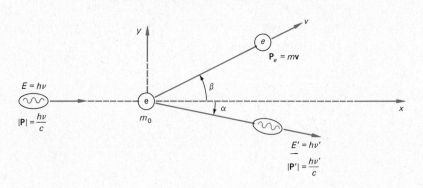

where

$$m = \frac{m_0}{\sqrt{1-(v^2/c^2)}}$$

Using these quantities, we apply the conservation of energy principle, which yields

$$h\nu + m_0 c^2 = h\nu' + mc^2$$

The next step is to prepare two equations—one for each of the two perpendicular axes—that represent the conservation of momentum in the collision. But here we run into a problem: what is the momentum of the incoming photon?

We can solve this problem with a combination of elementary and relativistic physics. In Chap. 6 we defined the linear momentum of any *particle* as the product of its mass and velocity. Thus, in the case of the photon of electromagnetic radiation, x-rays in this case, we have: momentum of photon $p = $ mass of photon $\times c$, where c is the speed of light.

The next question concerns the mass of the photon. What do we mean by "the mass of a photon?" To answer this question we turn to Einstein's special theory of relativity and the relationship $E = mc^2$. Thus, we obtain a mass for the photon of E/c^2. Substituting for E the term $h\nu$ that we already specified as the energy of the incoming photon,

$$m = \frac{h\nu}{c^2} \tag{22.1}$$

our equation for the momentum of the photon becomes

$$p = mc = \left(\frac{h\nu}{c^2}\right)c = \frac{h\nu}{c} = \frac{h}{\lambda} \tag{22.2}$$

Using this equation, we can now apply the law of the conservation of momentum for both the x and y directions. For conservation of momentum in the x direction,

$$\frac{h\nu}{c} = \frac{h\nu'}{c}\cos\alpha + mv\cos\beta$$

where m is the relativistic mass of the scattered electron. For the y component, conservation of momentum yields

$$0 = -\frac{h\nu'}{c}\sin\alpha + mv\sin\beta$$

We can solve the simultaneous equations that express the conservation of energy and momentum in three simple steps:

1. Eliminate the angle β by isolating the terms containing β in both momentum equations, squaring the equations, and adding them. Recall that $\sin^2\beta + \cos^2\beta = 1$.

2. Eliminate v from the equation just obtained by combining the two momentum equations, using the definition for the mass of the electron:

$$m = \frac{m_0}{\sqrt{1-(v^2/c^2)}} \quad \text{or} \quad m^2v^2 = c^2(m^2 - m_0^2) \qquad (22.3)$$

3. Square the energy equation and add it to the result of the sum of the two momentum equations, to yield a final equation:

$$\frac{1}{v'} - \frac{1}{v} = \left(\frac{h}{m_0c^2}\right)(1 - \cos\alpha)$$

or

$$\lambda' - \lambda = \Delta\lambda = \left(\frac{h}{m_0c}\right)(1 - \cos\alpha) \qquad (22.4)$$

Equation (22.4) is an expression for the change in the wavelength of the scattered photon in terms of α, the angle at which the photon is scattered. Figure 22.2 shows the spectra observed when x radiation of wavelength λ is scattered at different angles from a solid. As the equation predicts, radiation of wavelength λ', which is different from λ, is detected at every angle of scattering except zero. This radiation is produced by photons that scatter from unbound electrons in the solid.

The radiation is also accompanied at every angle by some radiation of the original wavelength; this latter radiation is simply explained as involving photons that scatter from electrons still bound to their atoms. Because the electrons are bound, the photons actually interact with the whole atom, whose mass is at least 1837 times that of the individual electrons. Substituting this value for m_0 in Eq. (22.4) yields a value for $\Delta\lambda$ that is so small as to be undetectable.

For the photons that hit free electrons, the maximum change in wavelength occurs when $\cos\alpha$ is -1, which corresponds to a value of 180° for the angle. At this point;

$$\lambda' - \lambda = \left(\frac{h}{m_0c}\right)(1 - \cos 180°) = \frac{2h}{m_0c}$$

$$= \frac{2(6.625 \times 10^{-34} \text{ J} \cdot \text{s})}{(9.1 \times 10^{-31} \text{ kg})(3 \times 10^8 \text{ m/s})}$$

$$= 0.485 \times 10^{-11} \text{ m} = 4.85 \times 10^{-2} \text{ Å} = 4.85 \times 10^{-12} \text{ m}$$

This predicted shift in wavelength agrees remarkably well with the shifts observed in experiments.

Note that the increase in wavelength of the photon corresponds to a decrease in its energy. Furthermore, the theory predicts that $\lambda' - \lambda$ is a constant for a given angle α. Therefore, the smaller the wavelength of the incoming photons, the greater the fractional change in wavelength. As a practical consequence, Compton scattering is applicable only to x-rays or to short-wavelength radiation known as γ radiation. If the

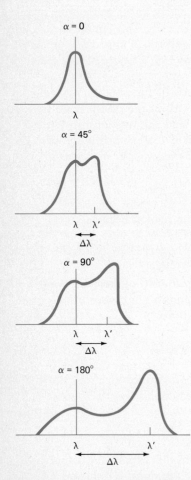

Figure 22.2 Wavelengths of the unchanged scattered x-rays, λ, and the Compton-scattered x-rays, λ', as a function of the scattering angle α of the x-ray.

incoming photons are of lower energy, for example, if they are photons of visible light, then the corresponding wavelengths lying in the region from 4000 to 7000 Å would not be noticeably affected by a change in wavelength of the magnitude predicted, for it is no larger than 0.048 Å, the maximum change in wavelength for 180° Compton scattering. Also note that for longer wavelength radiation, such as visible light, the assumption that the electrons in the target material are free is not valid. In the derivation for the Compton effect, it is assumed that the binding energy of the electrons is small compared to the photon energy.

22.2 THE SCHIZOPHRENIC NATURE OF LIGHT AND MATTER

The Compton effect provides physicists with an excellent check upon relativistic mechanics. Far more importantly, it gives an extraordinary insight into the nature of light, x-rays, and other forms of electromagnetic radiation. Although we treated such radiations as examples of wave motion in Chaps. 20 and 21, Einstein's and Compton's analyses show clearly that electromagnetic radiations also act as if they were photons—particles with specific properties such as momentum and energy.

In effect, **photons can be imagined to have masses that are inertial,** just like the concrete, solid particles with which we are all familiar. But the radiations also show the classic property of waves: diffraction. **Photons,** then, are entities that **have inertial, relativistic masses** given by the equation $m = E/c^2$ **and wavelengths defined in terms of their momenta,** p, **by the expression** $\lambda = h/p$. Defining a wave property such as a wavelength in terms of a particle property like momentum is truly a schizophrenic act, or, to use a scientific euphemism, an integrated piece of physics.

The "apparent" absurdity of this approach was recognized by many scientists. However, in 1924 a 32-yr-old French graduate student born a prince and named Louis de Broglie used the situation to provoke thought rather than mirth. He presented to the Faculty of Sciences at Paris University an astounding theory that drastically altered physicists' concepts of matter.

What de Broglie said was that there was no real reason to accept exclusively the wave theory or the particle theory of radiation. Both theories were correct, he said, and thus radiation must be regarded as a physical characteristic with a dual nature. That was reasonable enough in view of the facts. Then de Broglie added an extension that made the faculty members gasp. If we accept dual nature of radiation, and also concede the concept that nature is inherently symmetrical, reasoned the young physicist, then why should not matter, as well as radiation, have a schizophrenic wave-particle nature? In other words, de Broglie proposed a dual nature for matter: an electron with a particular rest mass should also be possessed of wave properties.

Mathematically, de Broglie took the expression for the wavelength of a photon in terms of its momentum and stated that the same equation would yield the wavelength of any particle, be it an electron,

If we accept the dual nature of radiation, then why should not matter have a schizophrenic wave-particle nature?

an atom, or even a golf ball. This wavelength is defined as $\lambda = h/p$. Using the relativistic definition of momentum yields:

$$\lambda = \frac{h}{p} = \frac{h}{mv} = \frac{h}{\dfrac{m_0 v}{\sqrt{1-(v^2/c^2)}}} \tag{22.5}$$

The larger the mass m, the smaller the wavelength; the higher the speed, the shorter the wavelength.

EXAMPLE 22.1

Calculate and compare the de Broglie wavelengths of a golf ball of mass $= 30$ g moving at 100 mi/h, a 100-eV electron, and a 1-MeV neutron.

Solution

The de Broglie wavelength is given by the ratio of Planck's constant to the momentum of the particle, $\lambda = h/p$. First, we calculate the momentum of these three particles. For the golf ball, the mass is 30 g and the speed is 100 mi/h. Converting 100 mi/h to meters per second, we have

$$v = (100 \text{ mi/h}) \left(\frac{1.61 \times 10^3 \text{ m}}{\text{mi}}\right)\left(\frac{1}{3600 \text{ s/h}}\right) = 44.7 \text{ m/s}$$

The momentum of the golf ball can of course be calculated nonrelativistically as

$$p = mv = (0.030 \text{ kg})(44.7 \text{ m/s}) = 1.34 \text{ kg} \cdot \text{m/s}$$

To calculate the momentum of both the electron and the neutron, we can first calculate their velocity classically using their energy given in the problem. A 100-eV electron is an electron that has accelerated through a potential difference of 100 V. Hence it will have a kinetic energy given by

$$\tfrac{1}{2}mv^2 = eV = (1.6 \times 10^{-19} \text{ C})(100 \text{ V}) = 1.6 \times 10^{-17} \text{ J}$$

The velocity of the electron can be obtained from this equation:

$$v = \sqrt{\frac{2eV}{m}} = \sqrt{\frac{(2)(1.6 \times 10^{-17} \text{ J})}{9.1 \times 10^{-31} \text{ kg}}} = 5.93 \times 10^6 \text{ m/s}$$

This speed is sufficiently short of the speed of light to allow us to use a nonrelativistic calculation. The momentum of the 100-eV electron will be

$$p = mv = (9.1 \times 10^{-31} \text{ kg})(0.593 \times 10^7 \text{ m/s})$$
$$= 5.40 \times 10^{-24} \text{ kg} \cdot \text{m/s}$$

A neutron is a neutral particle of roughly the same mass as the proton. We shall encounter its properties in more detail in Chap. 23.

For the neutron, we repeat the calculation just completed for the electron since the energy of the neutron is given in units of electron-volts—actually, in MeV:

$$\tfrac{1}{2}mv^2 = 1 \text{ MeV} = 10^6 \times 1.6 \times 10^{-19} \text{ J} = 1.6 \times 10^{-13} \text{ J}$$

$$v = \sqrt{\frac{(2)(1.6 \times 10^{-13} \text{ J})}{1.67 \times 10^{-27} \text{ kg}}} = 1.38 \times 10^7 \text{ m/s}$$

The momentum is given by

$$p = mv = (1.67 \times 10^{-27} \text{ kg})(1.38 \times 10^7 \text{ m/s})$$
$$= 2.30 \times 10^{-20} \text{ kg} \cdot \text{m/s}$$

In summary, the momenta for the three "particles" are:

Golf ball at 100 mi/h: $p = 1.34$ kg $\cdot$ m/s
100-eV electron: $p = 5.40 \times 10^{-24}$ kg $\cdot$ m/s
1-MeV neutron: $p = 2.30 \times 10^{-20}$ kg $\cdot$ m/s

The de Broglie wavelengths can now be calculated and compared:

Golf ball at 100 mi/h:

$$\lambda = \frac{h}{p} = \frac{6.625 \times 10^{-34} \text{ J} \cdot \text{s}}{1.341 \text{ kg} \cdot \text{m/s}}$$
$$= 4.94 \times 10^{-34} \text{ m} \quad \text{or} \quad 4.94 \times 10^{-24} \text{ Å}$$

100-eV electron:

$$\lambda = \frac{h}{p} = \frac{6.625 \times 10^{-34} \text{ J} \cdot \text{s}}{5.4 \times 10^{-24} \text{ kg} \cdot \text{m/s}} = 1.23 \times 10^{-10} \text{ m} \quad \text{or} \quad 1.23 \text{ Å}$$

1-MeV neutron:

$$\lambda = \frac{h}{p} = \frac{6.625 \times 10^{-34} \text{ J} \cdot \text{s}}{2.30 \times 10^{-20} \text{ kg} \cdot \text{m/s}}$$
$$= 2.87 \times 10^{-14} \text{ m} \quad \text{or} \quad 2.88 \times 10^{-4} \text{ Å}$$

The wavelengths have been converted to angström units to compare them with other minute distances of the same order. Visible light, for example, comes with wavelengths around 4000 to 7000 Å, whereas x-rays have typical wavelengths in the range 0.1 to 10 Å. Another important distance on this scale is the spacing between atoms in crystals, which generally varies between 1 and 5 Å. Plainly, the 100-eV electrons have wavelengths similar to the atomic spacing in crystals and would thus be expected to undergo diffraction when they encounter crystals, as do x-rays. The wavelength of the 1-MeV neutron, by contrast, is much too small for a beam of such particles to show any diffraction effects on interaction with crystals. However, lower-energy neutrons, which would possess lower momentum, would have longer wavelengths—possibly large enough to undergo diffraction by crystals. Finally, the wavelength of the golf ball is so small—about 10^{-20} times as small as even the atomic nucleus—that we have no hope of being able to observe diffraction effects with it.

There is, in fact, no grating in nature that has a small enough spacing to diffract golf balls, or any other macroscopic objects for that matter. Objects with which we are familiar are simply too massive to show wavelike properties. This may be discouraging in a way for experimentalists who wish to measure everything, but it is reassuring to theoretical physicists because it means that de Broglie's work does not alter the classical description of the movement of macroscopic objects developed by Newton, which works so well in practice. And for the experimenters there is the consolation that plenty of particles, including electrons, protons, neutrons, and even atoms of low energy, do possess de Broglie wavelengths large enough to undergo observable diffraction.

22.3 THE PROOF OF THE DUALITY

Thoughts of actually observing this sort of diffraction were far from the minds of scientists when de Broglie first proposed his remarkable theory. The inspired French physicist put forward the hypothesis without any experimental evidence. It took 3 yr for the first experimental observation of de Broglie's "matter waves" to occur. The men who performed this feat, much to their own amazement because they were completely unaware of de Broglie's work, were the American physicist Clinton J. Davisson and his assistant L. H. Germer of the Bell Telephone Laboratories.

The two men were studying how beams of electrons scattered off crystals. Their technique was to bombard smooth metal surfaces with slowly moving electrons, using the apparatus shown in Fig. 22.3. Electrons with energies of no more than a few electronvolts were emitted from an electron gun and then collimated into a beam that was shot at a target of the metal nickel. The electrons were scattered from the target in various directions. The number sent in any particular direction away from the nickel was measured using a small cup con-

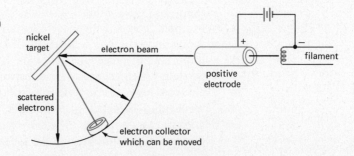

Figure 22.3 Diagram of Davisson-Germer apparatus. A collimated beam of low-energy electrons strikes a nickel target and is scattered and detected by the electron collector.

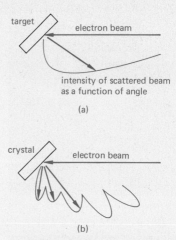

Figure 22.4 Scattering of electrons. The Davisson-Germer experiment. (a) Before heating of nickel strip; (b) after heating and formation of a single crystal of nickel. The curved line in (a) and (b) indicates the distribution of the number of electrons scattered at particular angles.

nected to a galvanometer. The deflection of the galvanometer was proportional to the rate at which electrons were caught by the collector. Since the cup could be moved around in a semicircle, the method could detect the number of electrons deflected in any direction by the nickel.

As is so often the case in significant scientific advances, the experiment required a little bit of luck. The two men started the experiment using as their target a strip of nickel containing many separate crystals [Fig. 22.4(a)]. Midway through the experiment, the vacuum tube containing the strip broke. The experimenters were forced to heat the strip to a high temperature to remove absorbed atmospheric gas from it. This heating, and the slow cooling that followed, caused all the atoms in the nickel target to rearrange themselves into the regular array of a single crystal. In Figs. 22.4(a), (b) the curved lines indicate the distribution of the number of electrons scattered at particular angles with respect to the incoming beam of electrons.

The results before and after the serendipitous breakage of the vacuum tube were astonishingly different. The electrons reflected randomly in all directions from the original polycrystalline nickel, but they were scattered from the single crystal form in the regular pattern shown in Fig. 22.4(b). This pattern was identical in every respect to that one would expect from a diffracted wave—a situation which could occur only if the electrons interacted with many atoms at the same time rather than with one atom alone. The diffracted pattern, in other words, appeared to be the result of collective action by the atoms.

Davisson and Germer found that they could explain the angular peaks of scattered electrons in the "after" situation if they regarded the regular array of atoms in the nickel as a diffraction grating. Using the method of analysis first developed by the two Braggs, which will be detailed in Chap. 24, the two men calculated the wavelength of the "electron waves" in terms of the angles of scattering that produced the pronounced peaks. When they checked the wavelength against de Broglie's prediction (the equation $\lambda = h/mv$, which had finally come to their attention), they found near-perfect agreement. Matter waves, then, moved from theory to fact.

Davisson and Germer were not alone for long as observers of matter waves. Within months of their monumental discovery, George P. Thomson of the University of Aberdeen in Scotland also discovered diffraction effects with a beam of electrons. Thomson, the only son of Cambridge University's J. J. Thomson, who confirmed the existence of the electron, sent a beam of the negatively charged particles through a celluloid film that was no more than 3×10^{-6} cm thick. After passing through the celluloid, the beam passed onto a photographic plate. Around the central spot produced by undeflected electrons, Thomson spotted a series of diffraction rings. The radii of the rings on the photographs were roughly inversely proportional to the speeds (or

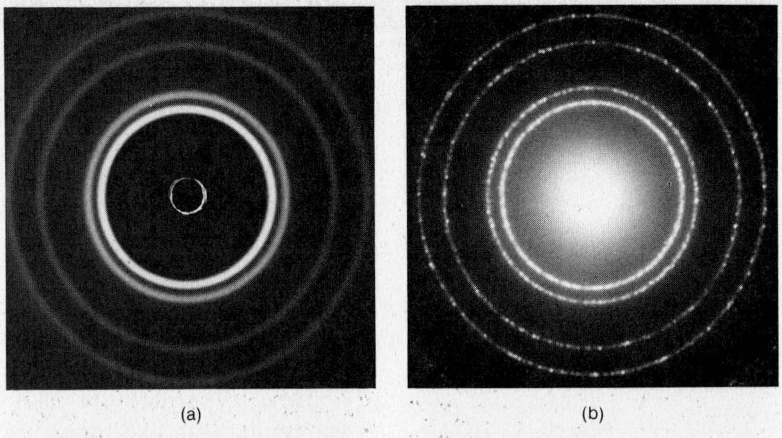

Figure 22.5 A direct comparison of electron and x-ray diffraction. (a) The diffraction pattern obtained by passing x-rays through an aluminum foil. (b) The diffraction pattern obtained by passing a beam of electrons through the same foil. The wavelengths are the same in both cases. (*Education Development Center*)

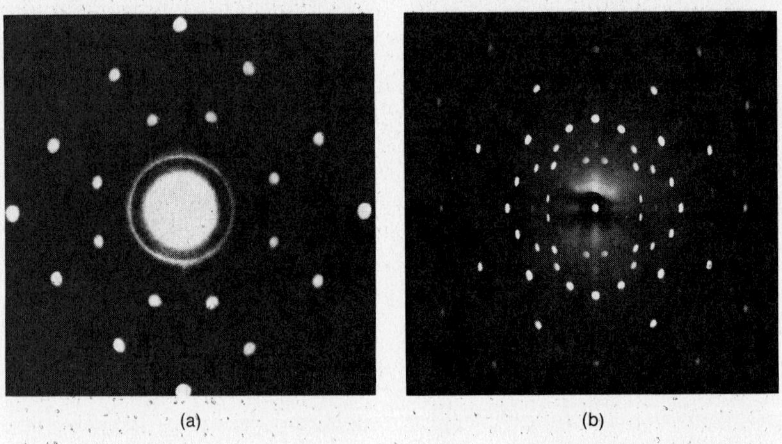

Figure 22.6 Laue diffraction patterns produced when (a) neutrons and (b) photons (x-rays) are incident on a single crystal of NaCl. The difference between these spot patterns and the ring patterns in Fig. 22.5 arises because the aluminum foil used to obtain the ring patterns consisted of many small crystals aligned in all possible directions. A ring pattern is therefore a superposition of a large number of spot patterns with a random distribution of orientations. [(a) E. O. Wollan and C. G. Shull; (b) John Biscoe]

square root of the voltage) of the electrons, thus adding support to the de Broglie equation. Further experiments by Thomson using platinum rather than celluloid foil confirmed the French physicist's picture of the duality of matter even more convincingly. Figure 22.5 shows the similar diffraction patterns obtained by passing x-rays and electrons through polycrystalline aluminum foil.

Since the pioneering experiments of Davisson and Germer and Thomson, electrons, α particles, protons, and neutrons have all shown wavelike behavior in suitably designed experiments. Electrons have actually been diffracted by the edge of a crystal, in just the same way that light is diffracted by a straight edge. Figure 22.6 shows typical diffraction patterns of neutrons and x-rays produced by a single crystal of sodium chloride. This was the type of diffraction actually observed by Davisson and Germer; the bright spots in the pictures correspond to those angles at which the two physicists discovered large numbers of electrons. The men who performed these fundamental studies were suitably rewarded: de Broglie received the Nobel Physics Prize in 1929, and Davisson and Thomson shared the same prize 8 yr later, in 1937.

EXAMPLE 22.2

Davisson and Germer found a strong diffraction peak in the reflected electrons from their nickel crystal at a corresponding angle $\theta = 67.5°$ when the accelerating voltage on the electron gun was 54 V. This was apparently a first-order scattered wave, with $n = 1$ and the lattice spacing $d = 2.15$ Å. The equation relating wavelength and angle of diffraction (to be developed in Chap. 24) is $n\lambda = 2d \cos \theta$, where n is an integer. Compare the de Broglie wavelength of 54-V electrons with the wavelength of the diffracted electron waves.

Solution

The electron wavelength λ depends on v and hence on energy. Using the classical expression for the kinetic energy of an electron in terms of the voltage used to accelerate it, we can write λ directly in terms of the accelerating voltage:

$$\tfrac{1}{2}mv^2 = eV$$

or

$$v = \sqrt{\frac{2eV}{m}}$$

Therefore,

$$\lambda = \frac{h}{mv} = \frac{h}{\sqrt{2emV}} \tag{22.6}$$

The de Broglie wavelength of the 54-V electrons can be calculated directly using this equation

$$\lambda = \frac{h}{\sqrt{2emV}} = \frac{6.625 \times 10^{-34} \text{ J} \cdot \text{s}}{\sqrt{(2)(1.6 \times 10^{-19} \text{ C})(9.1 \times 10^{-31} \text{ kg})(54 \text{ V})}}$$

$$= 1.67 \times 10^{-10} \text{ m} = 1.67 \text{ Å}$$

From the diffraction relation, we obtain a wavelength

$$n\lambda = 2d \cos \theta$$

$$\lambda = \frac{2d \cos \theta}{n} = \frac{(2)(2.15 \text{ Å})(\cos 67.5°)}{(1)} = 1.65 \text{ Å}$$

These two values for the electron wavelength compare very favorably within 1.3 percent.

22.4 THE QUANTIZED ATOM

When Niels Bohr first proposed the quantization of angular momentum and energy levels in the hydrogen atom, he was putting forth nothing more than a shrewd hypothesis that, because it led to agreement with the experimentally observed hydrogen spectrum, quickly gained credibility. More than 15 yr later de Broglie produced a rationale for Bohr's piece of inspiration. He assumed that the hydrogen atom's lone electron behaves as a wave and that the Bohr orbits correspond to those situations in which the electron wave can act as a standing wave.

de Broglie's analogy was the condition for standing waves on a string, as illustrated in Fig. 22.7. Plainly, the standing waves are those for which an integral number of half-wavelengths fit into the length of the string. The French physicist argued that the distance around a Bohr orbit—its circumference—should equal an integral number of wavelengths of the electron wave. Mathematically,

$$n\lambda = 2\pi r$$

or

$$\frac{nh}{mv} = 2\pi r$$

which yields

$$\frac{nh}{2\pi} = mvr$$

This is just the Bohr postulate that quantized angular momentum. The situation is shown graphically in Fig. 22.8 for the Bohr orbits in hydrogen corresponding to $n = 3, 4,$ and 5. No orbits are allowed apart from those corresponding to an integral value of n because the electron wave cannot produce a standing wave for any other orbit.

The man who undertook the basic mathematics that remodeled Bohr's atom in the light of de Broglie's theory of matter waves was Erwin Schrödinger, an Austrian physicist who survived World War I as

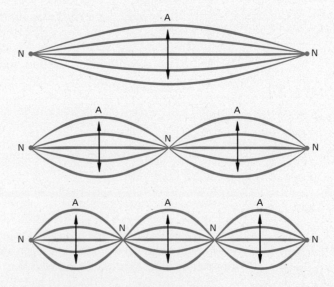

Figure 22.7 Standing waves on a string of finite length firmly held at each end. All quantized phenomena are the result of boundary conditions applied to wave phenomena—similar to standing waves on a string.

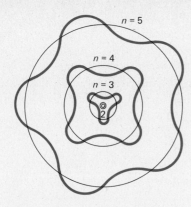

Figure 22.8 de Broglie's concept of matter waves. The electron is restricted to those orbits whose circumference equals an integral number of electron wavelengths.

an artillery officer on the southwestern front and later accepted an appointment at the University of Stuttgart in Germany. In 1925 Schrödinger developed an equation that was, in a sense, the law of motion for matter waves in terms of the masses and potential energies in space of the particles associated with the waves. The potential energies, instead of the forces themselves, were used to describe the forces between the particles. This resulted in an equation very closely resembling the equation for a wave in a material medium.

This expression, now known universally as the Schrödinger equation, plays the same role for material particles that Maxwell's equations play for electromagnetic waves. It allows physicists to obtain in theory—although not always in practice owing to the complicated mathematics involved—the various possible waves associated with any particle of mass m in a particular force field. The solutions to the Schrödinger equation are waves whose amplitudes depend on their spatial positions and the time. And different matter waves can and do interfere and follow the principle of superposition, as do the electromagnetic waves with which we are already familiar.

The Schrödinger equation has any number of solutions, applicable to any energy above zero. In fact, this type of equation was very familiar to scientists and applied mathematicians who had worked on classical boundary value problems. The so-called free solution for a particle of mass m in force-free space provides simply the kinetic energy $E_k = \frac{1}{2}mv^2$ and the wavelength $\lambda = h/p$. For our purposes in examining the atom, one of the most valuable types of solution to the equation is the series of "bound solutions," which correspond to an electron or other particle that is bound in an orbit or confined by nuclear forces to the general area of the nucleus.

A simple example of such a bound solution is a particle that is confined in a box, the box corresponding to a nucleus containing, say, only a proton, or an electron which is bound to the nucleus by the electromagnetic force. We can imagine that the sides of the box are electromagnetic rather than physical—walls of infinitely high potential the electron can never surmount.

Because interference effects reduce the amplitudes of many of the possible waves inside the box, the only possible solutions to the Schrödinger equation are standing waves. These are the de Broglie waves that fit exactly between the boundaries of the box and are identical to the standing waves on a string that is fixed at both ends; thus their amplitude is zero at the wall boundaries. Considering the simplest possible example of a one-dimensional box whose length is l, with walls of infinite potential as shown in Fig. 22.9, it is clear that the longest standing wave occurs when the wavelength is given by the equation $\lambda_1/2 = l$. The next standing wave comes about when $\lambda_2 = l$, the subsequent one when $3\lambda_3/2 = l$, or $\lambda_3 = 2l/3$, and so on. In

Standing waves were dealt with in detail in Chap. 12.

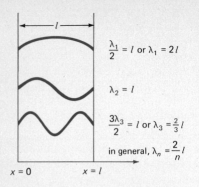

Figure 22.9 A particle in a one-dimensional box of side l. The walls of the box are hard and represent infinitely high potential barriers. The wave amplitude must equal 0 at $x = 0$ or $x = l$ the walls of the box.

general, the nth standing wave occurs at a wavelength $\lambda_n = 2l/n$, where n is an integer. A particle bound in this type of box can possess only the discrete-quantized set of wavelengths given by that equation.

We can compute the energy of an electron in any one of these standing wave states using the de Broglie relation and some classical physics. From de Broglie we know that

$$\lambda = \frac{h}{mv} = \frac{h}{p}$$

The momentum p is therefore given by $p = h/\lambda$. The momentum of an electron in the nth standing wave state will be

$$p_n = \frac{h}{\lambda_n} = \frac{h}{2l/n} = \frac{hn}{2l}$$

To compute the energy we can use the expression

$$E_k = \tfrac{1}{2}mv^2 = \frac{p^2}{2m} = \frac{h^2 n^2}{(2m)(4)(l^2)} = \frac{h^2 n^2}{8ml^2}$$

Clearly, the energy is quantized.

This "wave mechanics" approach of Schrödinger was applied directly to determine the energy of an electron that is bound to the proton in a hydrogen atom. Instead of confining the electron wave to the circumferences of the Bohr orbits, the approach used by de Broglie, Schrödinger confined the electron waves to a potential well, determined by the Coulomb electrostatic force of the proton on the electron. The problem was obviously rather more complex than our one-dimensional box. However, as we shall see, it yielded a set of quantized energy levels that agreed well with those calculated by Niels Bohr in his planetary model of the atom.

22.5 THE NATURE OF MATTER WAVES

Thus far we have treated the wavelike properties such as amplitude of matter waves just as if they had real physical meaning. Is this approach really justified? To find out, we can treat matter waves using exactly the same approach that we used for mechanical or electromagnetic waves in Chap. 17.

In the case of light, the waves consist of oscillating electric and magnetic fields, as we learned in Chap. 17. We can represent the electric field E at a point in space by a cosine function of the sort

$$E = E_0 \cos 2\pi \nu t$$

where E_0 is the amplitude of the wave that oscillates with a frequency ν. The intensity of the light wave is proportional to E^2, as one can determine easily enough by studying the intensity of light in a single- or double-slit experiment. Now, in terms of photons we know that the energy of a beam of light which falls on any particular surface, that is, the intensity of the light, is related directly to the number of photons which hit an area of the surface. Thus the density of photons is proportional to the square of the amplitude: $N \propto E^2$.

We can look at matter waves in just the same way. Instead of the electric field E, we use what we term the **wave function.** This extremely important concept in wave mechanics is denoted by ψ, the Greek psi. We have an understanding of the physical meaning of E. What does ψ represent physically? Why is it analogous to E?

Let us assume that the basic equation for this sine wave function in the case of matter waves takes just the same form as that for more conventional electromagnetic waves. Thus,

$$\psi = \psi_0 \cos 2\pi \nu t$$

By analogy with photons and electromagnetic waves, we can confidently expect the density of electrons—or any other particles with which we are concerned—to be proportional to ψ^2. Thus, we can represent the number of electrons, N_e, falling on a unit of surface area per unit time by the equation

$$N_e = a\psi^2$$

In this equation the factor a is proportional to the number of electrons involved in the system under consideration.

Imagine now that we perform an experiment in which we send electrons with a known speed, and hence a known energy, through a pair of adjacent slits and detect them emerging on the far side of the slits with an electronic collector. The collector will detect each individual electron as a lump of charge. But if we continue to count the emerging electrons for a long time, we shall find their distribution looks exactly like that of light after it passes through the pair of slits in Young's double-slit experiment. This experiment exemplifies the dual nature of matter: the electrons are detected individually as particles, but distributed as a wave, with the intensity ψ^2.

Complex mathematical analysis of this type of experiment shows that a new factor must be assigned to individual particles, the probability of finding them inside a particular region of space. If the wave function of a single particle has a value of ψ_P at a point in space P, then Schrödinger's equation shows that the probability of finding the particle inside a small region of space ΔV which surrounds the point P is equal to $\psi_P^2 \Delta V$.

This line of reasoning indicates the fundamental difference in approach between classical and wave mechanics. In classical mechanics we deal with concrete realities (probabilities of 1). Knowing the position and velocity of a particle at any time and the forces that act on it, we can theoretically compute the particle's position at all future times. In wave mechanics we must be satisfied with probabilities. We

Wave mechanics implies that it is impossible to predict exactly what will happen even in the best of experimental setups.

can no longer, according to wave mechanics, talk about the definite arrival of a wave or particle at any particular point at a given time. Wave mechanics implies that it is impossible to predict exactly what will happen even in the best of experimental setups. We can predict only the probability that different events will happen under different sets of circumstances. The single certainty in wave mechanics is that a particle will be present somewhere in space. This is expressed mathematically by summing over the total volume the probability of finding the particle in each element of volume of space (ΔV_i); that is, $\Sigma \psi_i^2 \Delta V_i = 1$, where Σ = sum over all space.

22.6 THE UNCERTAINTY PRINCIPLE

The introduction of probability into our specifications of the characteristics of matter waves raises a fundamental question: "Can we really measure the location of any particle precisely?" To answer this question we must analyze what we mean by locating particles in wave-mechanical terms.

The important variation here is that of probability, as measured by the factor ψ^2, with position in space. To localize a particle at any particular point in space, the graph of ψ^2 against position must rise to a sharp peak at that point. The sharper the peak, the more precise the location of the particle, as illustrated in Fig. 22.10, which shows the probability distributions of two particles. The most probable position for the particles is the same in both cases, x_0. However, the precision with which the two particles can be located is quite different in the examples. In Fig. 22.10(a) the probability that the particle occupies a large number of different spots around x_0, and even some distance from it, is relatively high. In Fig. 22.10(b), however, the particle almost certainly is to be found in a small pocket of space just around x_0.

How do these distributions come about? Since matter waves are represented by sine (or cosine) functions, the strange distributions shown in Figs. 22.10(a), (b) can be derived only by mixing a number of

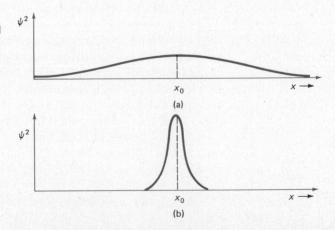

Figure 22.10 Probability of finding a particle in one dimension—a localized wave packet. The area under each curve should be the same.

such waves. The peaked value of ψ^2 in Fig. 22.10(b) requires the superposition of waves of very many wavelengths; interference effects cancel out the intensity at all points except those just around x_0. The broader distribution of Fig. 22.10(a) can be obtained using far fewer individual sine waves of different wavelengths.

Now, the wavelength of any matter wave is related to the momentum, p, by the equation $\lambda = h/p$. Thus, the more wavelengths that are required to put together the intensity distribution ψ^2, the greater the spread in momentum of the waves and hence the particle that those waves specify. There is just one way, in theory, of localizing the matter wave at the point x_0; that is to construct the wave of an infinite number of different waves of different wavelengths. But in doing this theoretical tour de force, one ends up with an infinite spread of momenta. What this all means is that the more clearly specified the position of a particle or matter wave, the greater the uncertainty in its momentum.

This principle, known as the Heisenberg uncertainty principle, was deduced in 1927 by Werner Heisenberg, a German physicist who was then just 25 yr old. The principle states that if you can determine the x component of the momentum of any particle with an uncertainty Δp, then the uncertainty Δx with which you can at the same time determine the x component of the particle's position in space is strictly limited by the equation

$$\Delta x \, \Delta p \geq h \tag{22.7}$$

This is often expressed more precisely as $\Delta x \, \Delta p > \hbar$, where $\hbar = h/2\pi$.

Heisenberg recognized that wave mechanics would collapse if it were possible to measure the momentum and position of a particle with a combined accuracy greater than Planck's modified constant $\hbar$. In essence, his uncertainty principle "protects" wave mechanics. Most importantly, the principle is an inherent property of the physical world, not just an expression of the limitations of instruments and means of measuring positions and momenta. The principle would still be upheld if one had the most theoretically perfect measuring instruments imaginable.

To obtain a better "feel" for the uncertainty principle, let us follow through a single-slit diffraction experiment. Consider a beam of particles incident upon a single slit. The particles act as a wave and are diffracted by the slit before they arrive at the screen. Most of the particles will hit the screen within the central diffraction maximum. As the particles pass through the slit of opening w, they are traveling with a momentum p_0 in the x direction. With a highly collimated beam, the momentum in the y direction before diffraction takes place will be zero, that is, $p_y = 0$. When the particle goes through the slit of width w, its position along the y axis has been determined with uncertainty $\Delta y = w$. We do not know exactly where the particle was in the slit, but we do know that it passed through the slit. Now, since the particles are

> Two years earlier, Heisenberg had derived a complex set of mathematical equations that turned out to be the equivalent of Schrödinger's wave mechanics. The Heisenberg approach was called quantum mechanics. However, the two terms quantum mechanics and wave mechanics are now used interchangeably.

Figure 22.11 Location of particles passing through a narrow slit.

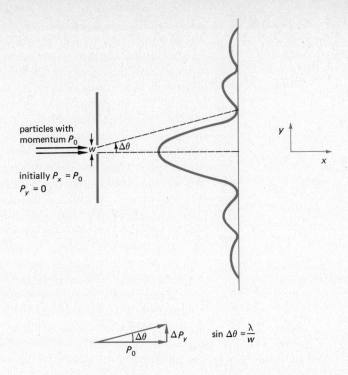

diffracted by the slit, they must have momentum in the y direction, p_y. There is a high probability that particles will not pass straight through the slit. However, most of the particles will arrive at the screen within the central diffraction maximum, which can be specified by the angle $\Delta\theta$ (see Fig. 22.11). Hence we can obtain a rough idea of the spread of momentum in the y direction (Δp_y) by confining our analysis to those electrons arriving in the central diffraction maximum. If we use a little trigonometry, we can determine this spread in Δp_y in the following way.

From Fig. 22.11, the uncertainty in the y component of momentum Δp_y is related to p_0 and $\Delta\theta$ by

$$\tan \Delta\theta = \frac{\Delta p_y}{p_0}$$

For small angles, the tan $\Delta\theta$ is equivalent to the sin $\Delta\theta$. Hence

$$\sin \Delta\theta \sim \frac{\Delta p_y}{p_0}$$

Now, we know from diffraction theory that $\sin \Delta\theta = \lambda/w$. That is the definition of the extent of the central diffraction peak, which is given in terms of the wavelength of the diffracted wave, and the slit opening w. Equating these two values for the sin $\Delta\theta$ yields

$$\frac{\Delta p_y}{p_0} = \frac{\lambda}{w} \quad \text{or} \quad \Delta p_y = p_0 \frac{\lambda}{w}$$

Now the product of the uncertainty in the y coordinate and the y component of momentum is given by

$$\Delta p_y \, \Delta y = \left(p_0 \frac{\lambda}{w}\right) w = p_0 \lambda = p_0 \frac{h}{p_0} = h$$

Here we used the de Broglie relationship to simplify the expression for this product. In our analysis we made some approximations, the most serious one being that the particles are located within the first diffraction maximum. Actually, the probability distribution shown in Fig. 22.11 predicts that a finite number of particles will be diffracted to much larger angles. We used a crude definition of "uncertainty" in position and momentum applicable to this experiment. It shows that in general the product $\Delta p_y \, \Delta y$ will be greater than $\hbar$.

An obvious objection to the uncertainty principle is that one does not notice it in daily life because everyday measurements are so gross by the nature of the sizes of the objects involved and the measuring methods that the effect is completely hidden.

The uncertainty principle is a result of the wave character of matter. We saw that the de Broglie wavelengths of macroscopic objects, like golf balls, are too small to be detected, so it is not surprising that with ordinary macroscopic objects the uncertainty principle has no real consequence. It is only when physicists measure ultrasmall objects, about the size of electrons and other particles, that the uncertainties involved in the uncertainty principle become significant.

The uncertainty principle has a number of more practical consequences in physical measurement. One typical example concerns the value of microscopes as measuring instruments. Because we obtain information on the location of objects by bombarding them with waves, the limit of resolution of any microscope is a wavelength of the radiation employed in the microscope. Our information on the location comes from the pattern produced by the waves that are scattered by the object under investigation, and we can obtain information on the position to only within a wavelength of light. For optical microscopes, the limitation of 1 wavelength of light is roughly 5000 Å. To improve resolution, we must turn to microscopes using ultraviolet light, whose wavelength is shorter than that of visible light. When we reach the resolution limit of ultraviolet light, the next step is the electron microscope for the wavelengths of electrons are typically 1/100 to 1/10 000 times as short as those of the ultraviolet light.

The wavelengths of the electrons used in the electron microscope depend on the energy they acquire from their accelerating voltage. Owing to the de Broglie relationship $\lambda = h/p$, increase of the energy of the electrons increases their momentum and hence reduces their wavelengths. Thus, the more energetic particles can locate detail in objects under the microscope within shorter distances. But in improving the detail this way we reduce the precision with which we can measure the momentum of the particles under study because the more energetic electrons possess more momentum, which is transferred to the particles under the microscope by collision. We are trapped by the wave nature of matter, as expressed by the uncertainty principle.

The principle can also be used to predict a fundamental feature of matter at the absolute zero of temperature. The ideas of kinetic theory outlined in Chap. 11 indicate that the kinetic energies of atoms oscillating about their positions in crystals are proportional to the absolute temperature. Thus, at the absolute zero of temperature, the atoms should be completely still in their fixed lattice positions. But this prediction contradicts the uncertainty principle because it specifies both the position and the momentum of the atoms with complete accuracy at the same time. We must therefore modify our ideas of the kinetic theory to fit in with Heisenberg's principle, allowing the atoms to retain a certain amount of vibratory motion—enough to allow the product of Δx and Δp to equal or exceed Planck's constant—when they reach absolute zero. Physicists term this the "zero point energy," and studies of the motion of atoms within a few thousandths of a degree of absolute zero indicate strongly that this energy is a reality. This finding gives another check on the overall validity of wave, or quantum, mechanics.

Other important applications of the uncertainty principle include a theoretical argument indicating that electrons cannot exist in the atomic nucleus as individual particles—an argument proved in 1932 by the discovery of the neutral particle known as the neutron—and the prediction, also proved, that the individual states of electrons inside the atom can vary slightly in energy. Beyond the laboratory, Heisenberg's uncertainty principle has been applied by philosophers to esoteric arguments about the nature of profound points of logic. It is well to keep in mind that although the uncertainty relation was originally expressed as a limitation on observations, scientists now believe that it has a fundamental validity.

This aspect of the uncertainty principle is regarded as a limitation on experiments in the social sciences. Simply by trying to measure the behavior of people, goes the rationale, one changes that behavior. For example, an observer in a classroom changes the very behavior the observer wants to measure. The limitations on large-scale experiments are of an entirely different magnitude and therefore are hardly related to the uncertainty principle.

EXAMPLE 22.3

The electron in a hydrogen atom is confined within a region of space with a diameter of about 10^{-10} m. Approximately what is the minimum value of the uncertainty in its velocity?

Solution

Let the region of 10^{-10} m represent the uncertainty in the position of the electron. This means that we do not know exactly where it is in the hydrogen atom, according to the wave-mechanical view. Using the uncertainty principle with the equal sign to obtain the minimum value of Δp, we have

Viewing the Unviewable

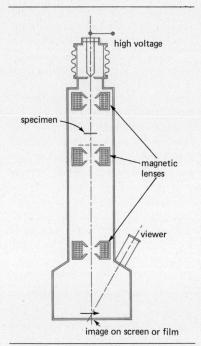

The inside of the electron microscope.

One major practical effect of Heisenberg's uncertainty principle is to limit the size of objects that can be resolved under microscopes which rely on light and radiation of similar wavelengths. Visible light has wavelengths in the range of 4000 to 7000 Å; the principle indicates that even the very best light microscope cannot resolve any object smaller than that. Thus, scientists who wish to view small biological cells, bacteria, viruses, and even large atoms must rely on another, shorter wavelength form of radiation. Taking a leaf from the book of Louis de Broglie, another atomic theorist of the 1920s, scientists have turned to beams of electrons.

In action, the electron microscope operates in much the same way as the optical variety, with appropriate adjustments for electrons' electric charges. The beams are produced using an electron gun operating at an appropriate voltage for the observation at hand. The beams are focused by electric or magnetic "lenses," which alter the speeds and directions of individual electrons in the fashion indicated in Chap. 15. An electron beam focused in this way is shone onto a carbon replica of the object under study (the beam would destroy the original) and forms an image that is projected onto a fluorescent screen or photographic plate.

The original electron microscopes produced greatly magnified but static two-dimensional images. In the mid-1960s, however, researchers developed the scanning electron microscope. The electron beam of this instrument can actually scan the surface of tiny objects, such as insects, bacteria, and blood cells, to produce amazingly clear, detailed, three-dimensional images of the world that is forever closed to our limited vision with light rays.

$$\Delta p = \frac{h}{\Delta x} = \frac{6.625 \times 10^{-34} \text{ J} \cdot \text{s}}{10^{-10} \text{ m}} = 6.625 \times 10^{-24} \text{ kg} \cdot \text{m/s}$$

If $\Delta p = m \, \Delta v$, the uncertainty in speed of the electron will be

$$\Delta v = \frac{\Delta p}{m} = \frac{6.625 \times 10^{-24} \text{ kg} \cdot \text{m/s}}{9.11 \times 10^{-31} \text{ kg}} = 7.27 \times 10^{6} \text{ m/s}$$

22.7 QUANTUM NUMBERS

At this point we are ready to return to what seems to be our regular starting point: the hydrogen atom. How do Schrödinger's equation and

Schrödinger's Cat: Dead or Alive?

Scientists did not quite realize it at the time, but the uncertainty principle put scientific experiments in an entirely new light. Simply by doing an experiment or making an observation, the principle implies, the scientist alters the subject of his or her enquiry. Erwin Schrödinger illustrated the truly fundamental nature of this kind of determinism by a thought experiment involving an atom, a vial of cyanide, and a cat.

Imagine, said Schrödinger, that a cat is locked in a room with a single atom of a radioactive element whose half-life is precisely 1 h and a vial of cyanide that opens only when the atom decays. Until the atom decays, the cat survives happily, but once the atom splits, it causes the vial of cyanide to open, leading immediately to the cat's demise.

What can we say about the cat's state after exactly 1 h has passed? According to quantum mechanics, the feline's state is indeterminate: there is a 50 percent chance that it is alive and a 50 percent chance that it is dead. It is impossible, according to quantum mechanics, to give a simpler answer than alive or dead.

The inquisitive scientist unlocks the door to the room and peers in. What the scientist sees is either a dead cat or a live one—but not a cat in the dead-alive limbo that quantum mechanics forecasts. Plainly, by opening the door the scientist has changed the physical situation of the cat from an indeterminate state—limbo—to a determined one—dead or alive. In making the observation, therefore, the scientist has greatly influenced the experiment. The uncertainty principle states that such influence is the inevitable result of any experiment or observation. Fortunately, few experiments involve single atoms. Most scientific observations deal with quantities of atoms so large that determinate statistical effects predominate, to provide results that make sense despite the uncertainty principle.

The great triumph of wave mechanics was in overcoming the major objection to Bohr's picture of the atom: that the orbiting electrons did not emit radiation as they should according to Maxwell's laws. By wave mechanics, the electron is not an orbiting entity but a particle described by a stationary standing wave. Since it does not accelerate around an orbit, it is not expected to radiate.

the uncertainty principle affect Bohr's rather simple view of the orbital makeup of that fundamental atom? The effect turns out to be not only profound but also extremely useful because it allows physicists and chemists to extend the principles used to understand that simple atom to more complicated mixtures of nucleons and electrons which occur in other atoms.

As a starter, we must note that the idea of probability inherent in the Schrödinger equation is not compatible with fixed orbits for electrons of the type envisioned by Niels Bohr. The electron as viewed in quantum-mechanical terms possesses neither a fixed orbit nor a fixed velocity. Instead, it possesses a wave function ψ at every point in space near the proton such that the factor $\psi^2 \Delta V$ represents the probability of finding the electron at that particular volume, ΔV. It is no longer possible to predict exactly where the electron can be found. We can simply say that its appearance is more likely in regions where ψ^2 is large.

This situation can be represented pictorially by a series of dots whose density corresponds to the value of ψ^2. The density of the dots is sparse in areas where ψ^2 has a small value, corresponding to a low probability that an electron will be present, and high in areas where the value of ψ^2 is relatively large. In the spots where ψ^2 is large, we have a good chance of finding an electron at any particular time.

In the case of the hydrogen atom, Schrödinger's equation yields a series of values for ψ that corresponds to different standing waves, each signifying one of the allowed energy states of the hydrogen atom's lone electron. These **different solutions for ψ are characterized by different values of three characteristic numbers known as quantum numbers.**

The concept of quantum numbers is a critical one in both atomic theory and modern particle physics. Quantum numbers are always integers or half-integers. In fact, the integer n in Bohr's simple concept of the hydrogen atom is itself a quantum number of a rudimentary sort; it signifies which of the possible allowed orbits the electron occupies. The improvement wrought by quantum mechanics is the specification of much more detail about the electron, made possible because the Schrödinger equation specifies three main quantum numbers for each individual wave function; these quantum numbers are represented by the letters n, l, and m_l.

The quantum number n is known as the principal quantum number. It can assume any integral value above zero. This quantum number is very similar to the integer n in Bohr's theory. However, in the Bohr theory n defines the radius of the electron's orbit, but the quantum mechanical n identifies the distance from the proton at which the probability of finding the electron is at its maximum. As one might expect, the value of the radius is just the same in both cases; however, the probability curve derived from Schrödinger's equation does not have a well-defined boundary since we are dealing with "waves" and the uncertainty principle is relevant.

The principal quantum number is also the decisive integer in determining the energy associated with a given cloud of probability. The equation for the energy E is

$$E = \frac{-2\pi^2 m k^2 Z^2 e^4}{h^2 n^2}$$

plus a small contribution from the other quantum numbers.

The first term of this equation is exactly that given by Bohr's theory; the terms produced by the other quantum numbers are all relatively small in comparison—an encouraging fact, because the Bohr theory does explain a great deal of the spectral data obtained for the hydrogen atom. In fact, the situation is rather similar to that involved in the comparison of Newtonian mechanics and the special theory of relativity. The differences produced and predicted by the latter are so small that they become noticeable only under extreme circumstances.

In Bohr's theory, the quantization of angular momentum is associated with the only quantum number, n. In quantum mechanics, by contrast, the quantization of the electron's orbital angular momentum is determined by the second quantum number, l. According to quantum

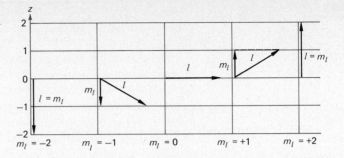

Figure 22.12 The five orientations of the orbital angular momentum relative to the z axis when $l = 2$.

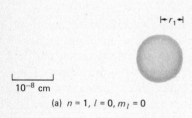

Figure 22.13 Probability clouds for the first two states (the s states, $l = 0$) of the hydrogen atom. The angular momentum is zero.

mechanics, l can possess any value between 0 and $(n - 1)$, where n is the principal quantum number. When n is 1, for example, l can only be 0. When n has the value of 2, l can take the values of 0 or 1. And when n is 3, l can be 0, 1, or 2.

The value of orbital angular momentum specified by this quantum number is roughly $lh/2\pi$ [actually it has a magnitude given by $\sqrt{l(l + 1)}(h/2\pi)$]. This is a vector whose direction is perpendicular to the plane of the orbital motion and along the electron cloud's axis of rotation. Each axis allows two possible directions for the angular momentum, corresponding to circulation of the electron cloud in the clockwise and counterclockwise directions. The orientation of the axis of rotation of the charge cloud is influenced by any magnetic field that happens to exist in the vicinity of the atom. Since some sort of magnetic field is always present, the net field at the general location of the atom sets up a preferred direction in space for the angular momentum of the electron cloud; this direction is normally referred to as the z axis.

The third quantum number, m_l, is known as the orbital magnetic quantum number. Its function is to express the restriction on the direction in which the orbital angular momentum, specified by the l quantum number, can be oriented relative to the z axis. The component of the orbital angular momentum along the z axis is equal to $m_l h/2\pi$. This quantum number can assume $(2l + 1)$ different integral values, ranging from $-l$, through zero, to $+l$. The range of possibilities is illustrated in Fig. 22.12. (The approximate magnitude of the angular momentum is being used.)

In addition to specifying angular momentum, the quantum numbers l and m_l between them affect the shape of the electron cloud around an atomic nucleus, as illustrated in Figs. 22.13 and 22.14, which show the first few electron probability clouds, that is, the clouds of lowest energy, for the hydrogen atom. The minimum energy state, which corresponds to the first Bohr orbit, occurs when $n = 1$, $l = 0$, and $m_l = 0$, as shown in Fig. 22.13(a). The next energy state comes about at a much higher energy, specified primarily by the value of 2 for the principal quantum number n, and is shown in Fig. 22.13(b). This corresponds to the second Bohr orbit. However, because this value of n can be associated with a number of different values for l and m_l, the orbit is not cut and dried.

Figure 22.14 Probability clouds for the p states, the $n = 2$ $l = 1$ energy states of the hydrogen atom.

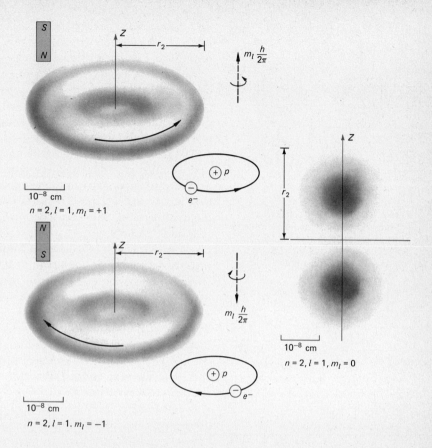

As Fig. 22.14 shows, the probability cloud can take on slightly differing shapes according to the various combinations of values for the minor quantum numbers: l can equal 0 or 1, and m_l can be -1, 0, or $+1$.

One of the most important applications of this type of understanding occurs in modern chemistry; the cloud shapes are used to explain many aspects of the chemical bonding between atoms. The directional nature of many chemical bonds is directly linked to the values of n, l, and m_l and the different shapes of the probability clouds that exist according to various combinations of those quantum numbers.

Before we can apply our system of quantum numbers to chemical phenomena, we must introduce one more quantum number. This refers to the intrinsic angular momentum of the electron or, more simply, the electron's spin. The electron's spin is a fundamental property of the electron, just as mass and electrical charge are. The existence of this quantum number recognizes that the electron whose structure we previously considered a point source of mass and charge actually has a definite, although miniscule, size, which causes it to behave as a tiny magnet. We can visualize the electron as a cloud of charge that is rotating around an axis through its center, just as the earth itself rotates.

The concept of electron spin was advanced in 1925 by Dutch scientists Samuel Goudsmit and H. Ueckla. Goudsmit emigrated to the United States soon after this tour de force.

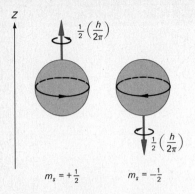

Figure 22.15 The two possible orientations of the intrinsic electron-spin angular momentum with respect to a reference axis (z axis). Usually the reference axis is a magnetic field.

This rotation of charge inevitably produces a magnetic field. The rotation is associated with an intrinsic angular momentum or spin that is as much a characteristic of the electron as its mass and charge. If we specify a spin factor, s, for the angular momentum of the spinning electron that is analogous to the quantum number of l for the orbital momentum of the electron's probability cloud, then the spin angular momentum has a magnitude $s(h/2\pi)$. Careful studies have shown that s possesses only one value—one-half—for the electron.

Just like the orbital angular momentum, the spin angular momentum is influenced by a magnetic field. The magnetic field specifies a z axis, and by analogy with the m_l quantum number, there are $2s + 1$, that is, $(2 \times \frac{1}{2}) + 1$, or 2 possible orientations of the spin angular momentum along this axis. These two possibilities are defined by the spin quantum number m_s, which can take either of the two values $+\frac{1}{2}$ and $-\frac{1}{2}$. These correspond to orientation of the electron's spin along the magnetic field or against it, as illustrated in Fig. 22.15.

In summary, the quantum mechanical relations provide physicists with a set of four quantum numbers, n, l, m_l, and m_s, which in combination influence the energy level of any particular electron in an atom. The principal quantum number, n, specifies the major energy levels of the electron, which are equivalent to those forecast by the simple Bohr theory. The other three numbers produce slight alterations in these energy levels that cause the major spectral lines produced by the electrons to split up into small groups of lines. Fortunately, this splitting is confirmed by careful analysis of atomic spectra. And as a more qualitative check, the concept of probability cloud distributions leads to a variety of different geometric distributions in electron clouds that are fully consistent with the directional aspects of chemical bonding between atoms.

22.8 PAULI AND THE PERIODIC TABLE

A major limitation of Bohr's theory was its limited application; it could treat the one electron atom and nothing else. Quantum theory allows physicists to drop the other shoe and go on from the hydrogen atom to more complicated entities. The fundamental principle to use is that systems in nature tend to the lowest energy state when they are left alone. Thus, the electron in the normal, unexcited hydrogen atom is in the state specified by the quantum numbers $n = 1$, $l = 0$, and $m_l = 0$, corresponding to the first Bohr orbit.

The situation gets more complicated when we start adding other electrons for the cases of helium and other complex atoms. The Schrödinger equations for these atoms must contain factors defining the electrostatic repulsive forces between the electrons and the magnetic forces that exist between the moving, spinning electrons. As a result of these interactions between the many electrons, as well as interactions between the electrons and the positive nuclei, the equations become so complex that they allow only approximate solutions. But these solutions take the form of a series of energy levels specified by the quantum numbers n, l, and m_l, which are related to those predicted for the one-electron hydrogen atom treated by Niels Bohr.

What determines the spread of quantum numbers among the various electrons in a complex atom? More specifically, what is there to prevent all the electrons in a complex atom from occupying the lowest, namely, $n = 1$, $l = 0$, $m_l = 0$, energy level? Theoretical work in the mid-twenties indicated that energy levels have only a certain capacity for electrons; once any particular energy level is filled, extra electrons must go into the next highest level.

The basic restriction that expresses this fact is known as the Pauli exclusion principle, which was formulated in 1925 by the 25-yr-old Austrian physicist, Wolfgang Pauli. The exclusion principle states simply that **no two electrons in the same atom can possess completely identical quantum numbers.** In the hydrogen atom, for example, the lowest energy level is that expressed by the quantum numbers $n = 1$, $l = 0$, and $m_l = 0$; this level is occupied by the single electron. Now, that electron can possess either of two spin states, with a value of $+\frac{1}{2}$ or $-\frac{1}{2}$ for the spin quantum number m_s. Thus, when we move on to the helium atom, with a doubly positive charged nucleus, we can place both the electrons in this lowest energy state, by giving them opposite spins. (Careful studies show that the state represented by $m_s = +\frac{1}{2}$ has slightly more energy than the state $m_s = -\frac{1}{2}$ in magnetic fields.) The probability clouds of both these electrons are spherical; the most probable distance corresponds to the radius predicted by Bohr's theory.

If we carefully look over the specifications for the l and m_l quantum numbers, we find that there are no other combinations which allow the value of n to be 1. Thus the two electrons of the helium atom specified by the sets of quantum numbers $n = 1$, $l = 0$, $m_l = 0$, and $m_s = \pm\frac{1}{2}$ are said to "fill" the $n = 1$ electron shell. Chemically, a filled shell represents extreme stability; helium is an inert gas that reacts with no other elements.

A filled shell represents extreme stability.

Since the $n = 1$ shell is now filled, the third electron of the third element, lithium, must move into the next shell, for which the principal quantum number is 2. This value of n allows values of 0 or 1 for the second quantum number l. Analysis shows that the $l = 0$ state yields a slightly lower energy level than the $l = 1$ situation. And since m_l can only take the value zero when l is zero, the third electron in lithium occupies the lowest possible energy state specified by the combination $n = 2$, $l = 0$, $m_l = 0$, and $m_s = -\frac{1}{2}$ when the atom is in its ground state. The next element, beryllium, has four electrons. The first two go into the $n = 1$ shell, the third into the same state as the third electron of lithium, leaving the state $n = 2$, $l = 0$, $m_l = 0$, and $m_s = +\frac{1}{2}$ for the last electron.

Moving along the periodic table, we come to boron, which has five electrons. With the first four in the same energy states as those of beryllium, we have a range of possibilities for the fifth electron; l must obviously equal 1, because no $l = 0$ slots are available. Therefore, m_l

Table 22.1 Quantum Numbers of the Last Electrons Added to Build Up the First 11 Elements

Element	Atomic Number	n	l	m_1	m_2	
Hydrogen (H)	1	1	0	0	$+\frac{1}{2}$	K shell
Helium (He)	2	1	0	0	$-\frac{1}{2}$	
Lithium (Li)	3	2	0	0	$-\frac{1}{2}$	
Beryllium (Be)	4	2	0	0	$+\frac{1}{2}$	
Boron (B)	5	2	1	-1	$-\frac{1}{2}$	
Carbon (C)	6	2	1	0	$-\frac{1}{2}$	
Nitrogen (N)	7	2	1	$+1$	$-\frac{1}{2}$	L shell
Oxygen (O)	8	2	1	-1	$+\frac{1}{2}$	
Fluorine (F)	9	2	1	0	$+\frac{1}{2}$	
Neon (Ne)	10	2	1	$+1$	$+\frac{1}{2}$	
Sodium (Na)	11	3	0	0	$-\frac{1}{2}$	

can have any of the three values $+1, 0$, and -1, for each of which the spin quantum number can be $\pm\frac{1}{2}$. The state $n = 2, l = 1$ thus opens up six possibilities for placement of the electrons. The configurations actually occupied are shown in Table 22.1. These possibilities end with the element neon, which possesses another completed electron shell, this one the $n = 2$ shell. As we would expect, neon is another inert gas similar to helium.

As a matter of terminology, x-ray physicists and chemists call the $n = 1$ shell the K shell, the $n = 2$ shell the L shell, and subsequent ones M, N, and so on. The l quantum number specifies the shape of the electron probability cloud. Electrons in states for which $l = 0$ have spherical clouds of probability and are known as s-state electrons. When $l = 1$, the electrons are called p-state electrons and possess the nonspherical probability distributions shown in Fig. 22.14.

Addition of further electrons, to compensate for the increased charge on the nucleus as we move up through the periodic table of the elements, gives a series of successive energy levels that corresponds most satisfyingly to the chemical properties of elements expressed by the nature of the periodic table. Sodium, for example, with 11 electrons, is similar to hydrogen and lithium in that it has one electron on the outside in an otherwise empty shell. Because this electron has a somewhat higher energy than those in the inner filled shells, it is quite easily removed from the atom. This explains the chemical observation that hydrogen, lithium, sodium, and other alkali metals in the same column of the periodic table ionize rather easily to form singly charged positive ions. Similar considerations explain the extraordinary amount of order that exists among the elements in the periodic table.

SUMMARY

1. X-rays are scattered by solids in two ways: scattering by atoms, and a process that changes the wavelengths of the incident x-rays. The latter process can be explained if one assumes that x-rays are photons which collide with electrons in the solid.
2. Owing to the inherent symmetry of nature, the dual wave-particle nature of x-rays and all electromagnetic radiation should also apply to matter. Material objects, in other words, possess a wavelike nature.
3. The wavelengths of visible objects are far too small for their wavelike natures to be apparent in everyday life.
4. The wavelike nature of electrons becomes evident when they are reflected from single crystals or diffracted by ultrathin materials.
5. Electrons in orbit around atomic nuclei behave as waves. The Bohr orbits correspond to standing waves.
6. The standing waves of electronic orbits can be understood in terms of quantum mechanics. This type of approach indicates that it is impossible to predict the definite arrival of a wave or particle at any particular point in time. One can only predict the probability of such an event.
7. Heisenberg's uncertainty principle states that the position and momentum of any object can never be precisely determined at the same time. This is an inherent law of nature, not just an experimental limitation.
8. Quantum mechanics indicate that every electron in an atom is characterized by four quantum numbers, which between them describe the electron's probability distribution in the atom.
9. According to the Pauli exclusion principle, no two electrons in the same atom can possess identical quantum numbers.

QUESTIONS

1. (a) What "reasoning" led de Broglie to speculate on the existence of matter waves? (b) Would you say that this is an example of "scientific thinking" or "intuition?" (c) By making what assumptions did he arrive at his formula relating the wavelength λ for a particle of mass m moving with velocity v? (d) Who experimentally verified the existence of these matter waves?
2. In their wavelike aspect, electrons can be made to yield a diffraction pattern much like that in the case of light. What difference in behavior or appearance of the pattern would you expect if the diffraction beam passed between the poles of a magnet before reaching the screen?
3. Review standing waves in Chap.
12. Discuss the orbiting electron as a stationary (standing) wave.
4. How is the Bohr picture of the atom altered by modern ideas of wave mechanics?
5. State the implications of Heisenberg's uncertainty principle in determining the position and momentum of an electron.
6. Distinguish between shells and subshells, energy levels and sublevels. What are their corresponding quantum symbols?
7. Write down all the possible combinations of n, l, m_l, and m_s for the $n = 3$ energy levels.
8. In a 20-electron atom, state the principle that shows why all these electrons do not pile up in the lowest energy level ($n = 1$) according to the "basic principle of nature."
9. What is meant by intrinsic spin? How does the spin of an electron affect the packing of electrons in the atom? Explain.
10. Can you ever determine exactly where a particle is located? Explain.
11. What would be some differences between our world and one in which Planck's constant was $1 \text{ J} \cdot \text{s}$?
12. When electrons or any particles are shot through a narrow slit and captured on the other side, how do you reconcile in terms of the wave-particle duality the distribution or diffractionlike pattern that results?
13. In terms of electron configurations, why are the rare gases so chemically stable?
14. Why do lithium, sodium, potassium, and cesium have very similar chemistry?

PROBLEMS

Compton Scattering

1. In a Compton scattering process, the scattered electron received a speed of 1.5×10^8 m/s from the incident radiation of frequency 1.2×10^{20} Hz. What was the frequency of the scattered radiation?
2. In a Compton scattering experiment, the incident radiation had a frequency 1.0×10^{19} Hz and the scattered radiation had a frequency 0.99×10^{19} Hz. What energy was given to the scattered electron in electronvolts?
3. If the x-rays ($\lambda = 0.1$ Å) are directed on a material, two types of

scattered x-rays are observed. One of the types has a new wavelength that is 0.0413 Å larger at a certain angle than the original 0.1-Å incident x-rays. (*a*) What is the wavelength of the other type of scattered radiation? (*b*) What does one use to determine these wavelengths? (*c*) What is the kinetic energy of the electrons that scatter the indicated radiation?

4. Consider a photon of visible light colliding with an isolated electron. Suppose the photon disappears and its energy goes into changing the kinetic energy of the electron. Show that such a process disobeys the simultaneous conservation of energy and momentum and hence makes the process unlikely. Why is this not a valid objection to the photoelectric effect?

The de Broglie Wavelength—Matter Waves

5. (*a*) What is the de Broglie wavelength of an electron of mass 9.11×10^{-31} kg and a speed of 2×10^8 m/s? (*b*) Compare this wavelength to the electromagnetic spectrum wavelengths and pick out the type of radiation to which it corresponds.

6. What is the de Broglie wavelength of an automobile of mass 1500 kg moving at a speed of 30 m/s (about 60 mi/h)? Comment on the size of this wavelength; could this wavelength be observed?

7. If the de Broglie wavelength of a neutron is 10^{-8} cm, what are its velocity and kinetic energy in electron-volts?

8. Note that the most energetic photoelectrons emitted from a surface have a de Broglie wavelength of 10^{-6} cm. The threshold frequency of 5×10^{14} Hz is required to emit photoelectrons from this material. (*a*) What is the energy of these electrons? (*b*) What is the frequency of the light required to emit these electrons of de Broglie wavelength of 10^{-6} cm?

9. A golf ball has a mass of about 50 g. Assume that it is moving at a speed of 7600 cm/s. What would be its de Broglie wavelength?

10. State Bohr's assumption concerning angular momentum of the electron in the hydrogen atom. From this assumption, compute the de Broglie wavelength for the electron in the hydrogen atom when it is in the first Bohr orbit.

11. An electron is accelerated through a potential difference of 20 000 V in a TV tube. Compute its de Broglie wavelength.

12. Through what voltage must an electron be accelerated to obtain a de Broglie wavelength of 6320 Å, equivalent to the wavelength of red light?

13. If you wished to design a diffraction grating to diffract a rifle bullet of mass 1 g and speed 300 m/s through an angle of 30°, what would be the required separation of the openings?

14. The atoms in a crystal of nickel are about 10^{-10} m apart. To have appreciable diffraction using the crystal of nickel, the wavelength of the waves must be of the same order of magnitude as the atomic separation. Using this length as a wavelength, compute the momentum and energy of electrons with this wavelength.

15. An electron microscope can resolve dimensions that are about 10 times the wavelength of the electron. If the electrons in a particular electron microscope are accelerated to 10 000 eV, what is the minimum structure resolvable with this microscope?

The Uncertainty Principle

16. To determine the position of an electron with an uncertainty of about 10^{-4} cm and at the same time determine its momentum as accurately as possible, which type of radiation would be most suitable: radio waves, microwaves, visible light, x-rays, γ-rays?

17. The x coordinate of a 200-eV proton is known to lie somewhere between $+0.001\,23$ cm and $+0.001\,27$ cm. Approximately what is the minimum uncertainty in the x component of its velocity?

18. A radar pulse lasts for 2.5×10^{-7} s. Approximately what is the uncertainty in the energy of its photons?

19. Batman and Robin's latest foe is the Quantum Kid (son of Heisenberg). The Quantum Kid has somehow constructed a building in which Planck's constant is much larger than usual: h is $6.6 \times 10^{+3}$ J·s. Batman enters a large room in this building and tries to catch the Quantum Kid, who weighs 50 kg. If the Quantum Kid is capable of moving at any speed up to 2 m/s, what is the smallest space into which Batman can localize the Quantum Kid (assuming the principles of modern physics apply)? Would it ever be possible for Batman to know exactly where the Quantum Kid was if the Quantum Kid stopped and stood completely still?

20. If Planck's constant were 10^{-2} J·s, what would be the uncertainty of the speed of a 10-g bouncing ball confined to a region 0.1 m long?

21. The x component of the velocity of an electron lies between 240 and 260 m/s. What is the minimum uncertainty in its x coordinate?

22. A beam of electrons, each with a speed of 10^6 m/s, is incident on a slit 10^{-4} cm wide. What is the width of the central fringe of a diffraction pattern on a screen 1.5 m away?

23. In an experiment, we localize a neutron to 10^{-12} cm. What is the minimum energy it can have?

24. If the uncertainty in the position

of an electron is equal to the diameter of an atom, say, 10^{-10} m, what will be the minimum uncertainty in the momentum and energy of the electron in the atom? Repeat this calculation for an electron confined to the nucleus of dimension 10^{-14} m.

Wave Mechanics and Quantum Numbers

25. If $l = 4$, what are the possible values for m_l? Describe briefly the interpretation of the values of m_l.

26. When $l = 5$, what are the possible values for m_l?

27. A particle of mass m confined to a line with a region L long has allowed energies $E = n^2h^2/8mL^2$. For a $\frac{1}{2}$-g mass confined to a 1-cm region, compute the lowest energy E. Compute the wavelength of the $\frac{1}{2}$-g mass in the lowest state ($n = 1$). Repeat for an electron confined to a line 10^{-8} cm long (the diameter of an atom). Compute the de Broglie wavelength of an electron in the lowest energy state ($n = 1$).

28. The Schrödinger wave picture, when used to describe a particle held in a one-dimensional box of length L, predicts only certain allowable wavelengths given by $\lambda_n = 2L/n$. (a) Give the expression for the allowed momenta. (b) What are the allowed energies?

29. A 1-g marble moves back and forth in a box that is 10 cm long. If the speed of the marble is 10 cm/s, (a) what is the number n of the quantum state of the marble? (b) Why is it possible to vary the energy of the marble in a continuous way even though the quantum description will allow only variations by discrete amounts?

Special Topic
The Men Who Made a Scientific Revolution

The ferment through which the science of physics passed in the 1920s was unique in the scientific world. Great physicists grappled with great problems in physics to produce a revolution of a type never known before or since in human understanding of the universe. Rarely if ever has such a small group of young scientists managed to make such extraordinary progress in any subject in such a short time. Perhaps the best idea of the excitement engendered during those heady years was given by Sam Goudsmit, a professor at the University of Nevada, Las Vegas, who in 1925, as a young graduate student at the University of Leiden in Holland, shared in the discovery of the spin of the electron. "There was so much to be found," he reminisced, "that even the dullards could make important contributions."

That comment was of course a slight exaggeration. There were few dullards among the topnotch physicists of the time because the kind of apprenticeship needed for an academic career tended to filter out all but the most dedicated and dynamic of scientists. In those days, before large government grants for the sciences, university authorities could select the cream of the crop for the scant few physics posts that were available, generally at low salaries. The huge teams that mark modern physics were unthought of in those days, and individuals could alter the foundations of their subject overnight. Werner Heisenberg, for example, hit on the basic concepts of quantum mechanics in the solitary confinement of a late spring visit to the North Sea island of Helgoland, in an effort to fight off a bad attack of hay fever.

The physicists at the frontier of science in the twenties were surprisingly young—apart from a few *eminences grises* such as Niels Bohr, they were generally postdoctoral students in their early to middle twenties—and continually in touch with each other. Grants from the Rockefeller Foundation allowed them to travel quite freely between Cambridge, Copenhagen, Göttingen, Leiden, and other major centers of the revolution to consult and argue with their scientific peers. Niels Bohr in Copenhagen provided the driving force for much of the extraordinary progress of the decade. Whenever Bohr was troubled by a specific problem, Goudsmit recalls, he would summon to Copenhagen the man he judged most likely to solve it. If that researcher failed, Bohr would put in a call for someone else.

Bohr generally put up the guests in his own home, to allow the maximum time for discussions of physics. In his book *Physics and Beyond,* Werner Heisenberg recounted how this type of pressure made Erwin Schrödinger ill: "While Mrs. Bohr nursed him and brought in tea and cake, Niels Bohr kept sitting on the bed talking to Schrödinger."

Face-to-face confrontations of this sort were frequent. At one famous weekend Solvoy conference in 1928, Bohr faced up to Einstein, who could never fully accept quantum mechanics because of its statistical nature. "God does not play dice with the universe," the sage declared. But that weekend, Bohr met every point that Einstein raised.

The era gave rise to a few tragic heroes, most notably Arnold Sommerfeld, professor of theoretical physics at the University of Munich. Throughout the early 1920s Sommerfeld worked on improving Bohr's theory of the hydrogen atom, adding such sophistications as an elliptical rather than circular orbit for the electron. But his labor went for naught with the advent of wave mechanics and the new view of the atom which that particular discipline engendered.

The quantum physicists rarely spent more than one or two waking hours each day away from their subject, but they were not a limited group of men. Bohr regularly went skiing, many of his younger colleagues were keen hikers and bikers, and a number of the group were extremely fond of music. Heisenberg once demanded—and got—a piano in his room before he would accept a visiting professorship at the University of Michigan. Einstein became an authority on Mozart, entertaining himself and others in his later years at Princeton's Institute for Advanced Study by organizing chamber music soirees, at which he played the violin more than competently.

The men also had their personal quirks as physicists. Pauli, for example, the man who formulated the Pauli principle, was also responsible for the "Pauli effect." Whenever this quintessential theoretician visited a laboratory, it seems, experiments in progress went haywire. On one occasion, an experiment by James Franck at the University of Göttingen totally misbehaved, for no apparent reason. Only later did Franck find out that Pauli had been passing through the university town in a train at the very time the experiment went wrong. Pauli himself was extremely proud of his effect.

But, of course, experiments that went right, or at least that produced meaningful data, were the fodder for Pauli and other theoreticians. And all shared the extraordinary sense of wonder at the world that their efforts were revealing. As Heisenberg wrote of his breakthrough in *Physics and Beyond,* "At first, I was deeply alarmed. I had the feeling that, through the surface of atomic phenomena, I was looking at a beautiful interior, and felt almost giddy at the thought that I now had to probe this wealth of mathematical structures nature had so generously spread before me. I was far too excited to sleep...."

23 Inside The Nucleus

23.1 NOW, THE NEUTRON

23.2 NUCLEAR FORCES

23.3 NUCLEAR BINDING ENERGY

23.4 THE STABILITY OF NUCLEI

23.5 RADIOACTIVE DECAY

SHORT SUBJECT: The Mystery of the Missing Neutrinos

23.6 THE DECAY OF RADIOISOTOPES

23.7 NUCLEAR REACTIONS

SHORT SUBJECT: Dating the Past

Fission

SHORT SUBJECT: Inside the Nuclear Reactor

Fusion

23.8 BEYOND THE NUCLEUS
Summary
Questions
Problems

SPECIAL TOPIC: Selected Species from the Subnuclear Zoo

The investigation of the nature of matter by physicists has gone through a number of distinct phases since John Dalton resurrected the atomic theory in 1808, and in each case the advance from one phase to another has involved the breaking down of barriers, in theory if not in fact. Thus, the discoveries of the last few years of the nineteenth century and the first few years of the twentieth represented a thrust by physicists into what was previously thought of as the impenetrable billiard-ball atom. In the middle twenties, theoretical and experimental physicists broke through the barrier of the electronic orbits and realized that the atomic nucleus was also more than a solid billiard-ball-type of entity. Later, physicists figuratively broke through into the nucleus and almost immediately came upon a series of new and unsuspected fundamental particles—the first members of what is occasionally referred to now as the subnuclear "zoo." Indeed, the physicists who now search for fundamental particles and forces with their multimillion dollar, high-energy accelerators are building on the exploits of the quantum physicists who took the first steps in investigating the minuscule pieces of matter at the center of all atoms.

The nature of the nucleus was first glimpsed by Ernest Rutherford, as a result of his scattering experiments in the first decade of this century. His analysis of the scattering of α particles from gold foil indicated that the nucleus had a diameter of roughly 10^{-14} m—about one ten-thousandth of the diameter of the typical atom. If we imagine that an atom is represented by a mighty cathedral, then the object that most closely represents the size of the nucleus inside it is a flea on a tiny church mouse.

Despite its diminutive size, the nucleus contains the predominant amount—more than 99.6 percent—of the mass of the atom that contains it. This implies a huge density for the nucleus. If you consider the volume of the nucleus to be that of a sphere of radius $r = 10^{-14}$ m, then we obtain a volume $V = \frac{4}{3}\pi r^3 = (\frac{4}{3}\pi) \times (10^{-14} \text{ m})^3 = 4.19 \times 10^{-42}$ m^3. For hydrogen, the nucleus contains one proton with a mass essentially equal to an atomic mass unit, 1.66×10^{-27} kg. The density of the hydrogen nucleus will be

$$D = \frac{m}{V} = \frac{1.66 \times 10^{-27} \text{ kg}}{4.19 \times 10^{-42} \text{ m}^3} = 3.96 \times 10^{14} \text{ kg/m}^3$$

This is about 4×10^{11} times the density of water—an all but unimaginable density.

> *If we imagine that an atom is represented by a mighty cathedral, then the object that most closely represents the size of the nucleus inside it is a flea on a tiny church mouse.*

23.1 NOW, THE NEUTRON

Given this extraordinary ratio of mass to volume, we might ask what type of material makes up the atomic nucleus. Rutherford and his colleagues originally believed that the hard centers of all atoms consisted of just protons and electrons, with the simple exception of the hydrogen nucleus, which is nothing more than a single proton. This picture remained fairly convincing up to the twenties, by which time physicists realized that the nature of elements was determined by the net positive charge in their atoms. Scientists assumed that the total numbers of electrons and protons in any atom were equal—to make atoms electrically neutral—and that the electrons were divided between those in orbit and those nestled into the nucleus with all the protons of the atom. The net charge on the nucleus was the difference between the number of protons and electrons inside the nucleus. Thus, as physicists then saw it, the element lithium was believed to possess six protons and three electrons in its nucleus, giving the nucleus a positive charge of 3; three orbiting electrons then made the lithium atom electrically neutral.

There were a few objections to this rather neat theory. For one thing, not all elements seemed to possess atomic masses that were integers of the mass of the hydrogen atom. This was disturbing because the number of protons in the nuclei of different atoms could only take integral values—and since the masses of protons were 1837 times the masses of electrons, the electrons obviously contributed negligible amounts to the atomic masses. How could this apparent existence of fractions of protons in nuclei be explained?

The answer came in 1919, when Professor J. J. Thomson and his assistant Francis W. Aston put into action the mass spectrometer Aston had devised. This device, whose method of operation appeared in Chap. 15, separates beams of different ionized elements from each other. But when the two British physicists sent beams of a single, ionized element—the inert gas neon—through the instrument, they discovered that the beam split into two beams of slightly different atomic masses. Thus was the concept of isotopes born.

Isotopes are different atomic forms of the same element that have the same positive charges on their nuclei but different nuclear masses. Thomson and Aston discovered that neon, for example, has two isotopic forms, with atomic masses 20 and 22 times that of the hydrogen atom; later, a third isotope with a mass 21 times that of hydrogen was discovered. In practical terms, isotopes of the same element are much more difficult to separate from each other than are different elements because their basic chemistry is identical. Isotopes can be parted from each other only by such physical methods that rely on the differences in mass between them as diffusion, centrifugation, and the use of electric and magnetic fields in a mass spectrometer.

Theoretically, isotopes cleared up a few thorny problems. Experts postulated that the nuclei of different isotopes of any particular element contained the same number of excess protons, thus accounting for the fact that their positive charges were identical, but different numbers of the proton-electron pairs that made up the observed atomic masses. Calculations soon showed that the atomic masses of isotopes were very close to integral multiples of the atomic mass of hydrogen.

The simple concept of the nucleus made up of protons and electrons suffered a more damaging blow, however, in the mid-twenties, with the development of the uncertainty principle and the concept of electronic spin. According to the principle, confining electrons in the tiny space occupied by the atomic nucleus would require their momenta to have such a large uncertainty that the electrons could not possibly stay inside the nucleus. The spin problem was just as intractable. Imagine a nitrogen nucleus that, according to the simple picture, consists of 14 protons and 7 electrons. This mixture gives the nucleus the correct nuclear charge, $+7$, and mass, 14. Unfortunately, the numbers just do not produce the correct value for the observed intrinsic angular momentum—the spin—of the nucleus. The value is 1, whereas the spins of all protons and electrons are $\pm\frac{1}{2}$. No combination of these spin values for the 21 particles inside the nucleus according to the simple picture can produce the actual spin of the nitrogen nucleus.

In 1932, the day was saved by an event that was to become increasingly familiar over the next quarter century: the discovery of a new fundamental particle. British physicist James Chadwick of Cambridge University reported that radiation produced when α particles bombarded light elements such as beryllium was actually a new electrically neutral particle, which was dubbed the neutron.

Investigations of the nature of the neutron revealed that, apart from its neutrality, it is remarkably similar to the proton. At about 1838 times the mass of the electron, it is all but identical to the proton in mass, and its spin of $\frac{1}{2}$ is exactly the same as the proton's. One major difference, however, is in survival times. Protons last millions of years in the free state without changing into other particles, but **neutrons last on the average for no more than a few minutes outside the nucleus; they break down into more stable entities.**

Protons and neutrons are called nucleons.

Werner von Heisenberg seized upon the neutron as the solution to the physicists' problems with the structure of the nucleus. Instead of containing protons and electrons, Heisenberg argued, the **nucleus is actually made up of just protons and neutrons.** This picture was rapidly accepted by the physics world, and because of the similarities between protons and neutrons, the two types of particles were given the catch-all name of nucleons. Nitrogen, according to this scheme, has 14 nucleons with spins of $\pm\frac{1}{2}$, a situation that can produce the observed spin of 1 for the nucleus.

The basic structure of the nucleus now becomes relatively clear. **Each nucleus consists of a certain number (Z) of protons, whose value can be any integer from 1 upward, and a certain quantity (N) of neutrons, which can vary upward from zero.** The combination of the two gives the so-called atomic mass number of the nucleus (A). This is an integral number that represents the mass number of the nucleus in terms of the mass of a single nucleon:

$$A = Z + N \tag{23.1}$$

Using the atomic number Z and the mass number A, we can specify a nucleus of a chemical element X by the symbol

$${}^{A}_{Z}X \tag{23.2}$$

An example of the use of this symbolism is ${}^{1}_{1}H$, the symbol for hydrogen, with 1 proton and hence a mass number $A = 1$. Likewise, we describe helium, which has 2 protons and 2 neutrons, as ${}^{4}_{2}He$.

For a given element, Z is fixed; that is, the number of protons and electrons is invariant, and so the use of both the chemical symbol (H, or He) and the Z value as a subscript are redundant. In fact, the subscript is normally omitted, but to assist the reader we shall include it. **The different numbers of neutrons in nuclei give rise to isotopes of elements.** The actual mass of any particular isotope is the approximate product of its mass number A and the quantity known as the atomic mass unit, abbreviated u in the SI system. The mass equivalent of the binding energy, described later in this chapter, must also be discussed.

The value of the atomic mass unit was, until 1959, established by assigning 16.000 to the mass of the lightest and most abundant isotope of oxygen. Now, however, the atomic mass unit is defined as one-twelfth the mass of the atom of the most common isotope of carbon, carbon 12, whose mass number is 12 (6 protons and 6 neutrons). The value of the atomic mass unit is 1.66×10^{-24} g, or 1.66×10^{-27} kg. **The atomic mass of any isotope is now given approximately by the mass number A times the value of 1 u.** Since natural elements are composed of different amounts of the isotopes, the atomic mass listed for an element found in nature is a weighted average of the masses of the isotopes that make up the element in nature.

1 u = 1.66×10^{-27} kg.

For uranium, chemical symbol U, $Z = 92$; a number of isotopes are found in nature, including ${}^{238}_{92}U$, ${}^{235}_{92}U$, and ${}^{234}_{92}U$. The uranium isotope ${}^{238}_{92}U$ contains 92 protons and, since $A = Z + N$, $N = 238 - 92$, or 146, neutrons. The atomic mass is actually 238.051 u, which would equal $(288.051) \times (1.66 \times 10^{-27}$ kg).

Hydrogen possesses three isotopes, known as hydrogen, ${}^{1}_{1}H$; deuterium, ${}^{2}_{1}H$; and tritium, ${}^{3}_{1}H$. Hydrogen contains 1 proton and 0 neutrons; deuterium, 1 proton and 1 neutron; and tritium, 1 proton and 2 neutrons. A nucleus of the hydrogen atom is a proton; the nucleus of a deuterium is called a deuteron; the nucleus of a tritium atom is called a triton.

EXAMPLE 23.1

The element chlorine in nature is made up primarily of two isotopes, with atomic masses of 35 and 37. The relative natural abundance is

$^{35}_{17}$Cl, 75.5 percent, and $^{37}_{17}$Cl, 24.5 percent. What is the atomic mass of chlorine found in nature?

Solution

We need to take the weighted averages of the atomic mass of each isotope. We could look up the experimentally determined atomic masses of these two isotopes, but to a very close approximation the atomic mass in atomic mass units is equivalent to the mass number of the isotope. Hence, the weighted average is essentially equal to

$$(75.5\% \times 35) + (24.5\% \times 37)$$

or

$$(0.755 \times 35) + (0.245 \times 37) = 35.49 \text{ u}$$

Actually, the atomic mass of chlorine determined experimentally is 35.457 u. Had we used the experimental values, found in App. F, for the isotopic masses, we would have obtained this more accurate value.

23.2 NUCLEAR FORCES

One remarkable property of nuclei in general is that they *all* have about the same immense density, 4×10^{11} times that of water, whatever their size and makeup. This situation is quite different from that of macroscopic matter, whose densities vary over a wide range; the densities of solids, for example, vary from less than 1 to over 20. The nuclear densities are uniform because of the nature of the two different forces involved. Whereas atoms in matter are kept intact by the Coulomb force of electrical attraction, **nucleons are kept in close proximity**—separated typically by 1.9×10^{-15} m—**by an entirely different type of force, known as the nuclear binding force.**

Even today, this force, which is often referred to simply as the nuclear force, is something of a scientific puzzle. It has been estimated that scientists have devoted more working hours to studying this force than to investigating any other single scientific question, yet they are still unable to write down a simple—or complex—mathematical equation to describe its nature in the same way that they can mathematically define the law of universal gravitation and the electrostatic Coulomb force. On the other hand, scientists' efforts have yielded a fairly comprehensive picture of the basic characteristics of the nuclear force. These are extremely strange—at least according to the criteria of the electrostatic and gravitational forces with which we are familiar in daily life.

The attractive nuclear force between two protons, two neutrons, or a neutron and a proton varies by no more than 1 percent.

The most astonishing fact about the nuclear force is that its attractive strength between two protons, two neutrons, or a neutron and a proton varies by no more than 1 percent. The force is, in other words, independent of the electric charge on the particles exerting it. Because the force must overcome the natural electrical repulsion between protons, it is clearly stronger than the electrostatic force, at least over the limited

Figure 23.1 The nuclear force. Variation of the nuclear force between two nucleons as a function of their separation from one another.

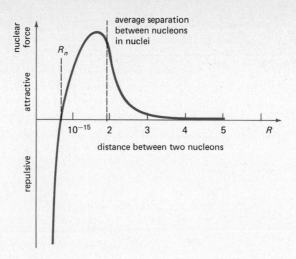

distances involved in the nucleus. Unlike the electrostatic and gravitational force, the nuclear force is essentially a short-range one. At any distance beyond about 10^{-14} m, its value is so small as to be unmeasurable. But as any two nucleons approach each other and pass the 10^{-14}-m distance, the value of the attractive force starts to rise rapidly until it reaches a peak, as shown in Fig. 23.1. At smaller separations the attractive force decreases and eventually, at the point R_n in the figure, changes over to a repulsive force. This particular separation can be regarded as the diameter of a nucleon; in fact, we can view nucleons as billiard balls that cannot be pushed closer together once they are touching without the exertion of an undue amount of force.

Looking at nucleons in this way, one might expect their average separations in nuclei to correspond to the diameter of the nucleon, R_n. In fact, the average separation turns out to be larger than that distance—a direct result of the uncertainty principle. If nucleons are separated by a distance of R_n (see Fig. 23.1), then the uncertainty in their momentum must, according to the principle, be at least h/R_n. However, calculations show that the momenta of nucleons which escape from the nucleus are generally smaller than this value. Nucleons generally settle down with separations of around 1.9×10^{-15} m, which is roughly twice the value of R_n.

This discussion of the characteristics of the nuclear force clearly shows why all nuclei have essentially the same density. The force is just about the same between every pair of nucleons and exists only over a very short range. This is quite different from the electrostatic force that is long range and decreases as $1/R^2$.

The strength of the nuclear force is also influenced by the spin of the nucleons involved.

23.3 NUCLEAR BINDING ENERGY

Although the concept of isotopes accounts for many of the departures of atomic masses from integral multiples of the atomic mass unit, it does not solve the problem entirely. The masses of nuclei are always slightly below the sum of proton and neutron masses; that is, the mass of any nucleus is less than the sum of the masses of the protons and neutrons that make it up. This difference is known as the mass defect. It is this lack of energy (mass) that keeps nuclei from splitting apart.

We can understand this "binding energy" using Einstein's principle of the equivalence of mass and energy. The mass lost when any number of nucleons come together to form a stable nucleus is released in the form of energy, according to the equation $E = mc^2$, so that the nucleons remain bound together. The overall situation is summarized in the following equations:

Mass defect $= Zm_p + (A - Z)m_n -$ mass of nucleus $^A_Z X$
Mass defect $\times c^2 =$ binding energy

As an example, consider the binding energy of the helium nucleus, or α particle, which is made up of 2 protons and 2 neutrons. The masses of the proton, the neutral hydrogen atom, neutrons, and the neutral helium atom $^4_2 \text{He}$ are given in Table 23.1 in terms of atomic mass units, which can be converted to grams using the factor 1.66×10^{-27} kg/u, or to energy units using the conversion 931 MeV/u.

Table 23.1 Atomic Masses of 16 Isotopes

Name	Symbol	Atomic Mass[a]
Neutron	n	1.008 665
Proton	p	1.007 277
Hydrogen	$^1_1 \text{H}$	1.007 825
Deuterium	$^2_1 \text{H}$ or $^2_1 \text{D}$	2.014 102
Tritium	$^3_1 \text{H}$ or $^3_1 \text{T}$	3.016 049
Helium 3	$^3_2 \text{He}$	3.016 029
Helium 4	$^4_2 \text{He}$	4.002 603
Beryllium 8	$^8_4 \text{Be}$	8.005 305
Carbon 12	$^{12}_6 \text{C}$	12.000 000
Nitrogen 14	$^{14}_7 \text{N}$	14.003 074
Oxygen 16	$^{16}_8 \text{O}$	15.994 915
Oxygen 17	$^{17}_8 \text{O}$	16.999 138
Argon 36	$^{36}_{18} \text{Ar}$	35.967 547
Radon 222	$^{222}_{86} \text{Rn}$	222.017 61
Uranium 235	$^{235}_{92} \text{U}$	235.043 9
Uranium 238	$^{238}_{92} \text{U}$	238.050 8

[a] Except for the neutron and proton, the mass given refers to the neutral atom.

Inside the Nucleus

The conversion factor between atomic mass units and mega electronvolts is obtained using the Einstein mass energy equivalence:

$$E = mc^2 = (1.660 \times 10^{-27} \text{ kg/u}) \times (2.998 \times 10^8 \text{ m/s})^2$$
$$\times \left(\frac{1}{1.602 \times 10^{-19} \text{ J/eV}}\right)$$
$$= 9.31 \times 10^8 \text{ eV/u},$$

or

1 u = 1.66×10^{-27} kg is equivalent to 931 MeV

It is often convenient to convert masses in atomic mass units directly to energy in this way.

The mass defect for helium is given by

Mass defect = $2m_p + 2m_n - m_{\text{helium nucleus}}$

Now, the mass of helium given in Table 23.1 is the mass of the neutral helium atom, which contains 2 electrons. To compensate for this, we shall replace the mass of the proton with the mass of a hydrogen atom, thereby using atomic rather than nuclear masses. The error this produces is very slight; it arises from the electrons being bound in helium with a different energy from that in hydrogen. Using the atomic masses in atomic mass units, we can determine the mass defect for the helium atom:

Mass defect = $(2 \times 1.007\,825) + (2 \times 1.008\,665) - 4.002\,603$
$= 2.015\,650 + 2.017\,330 - 4.002\,603$
$= 0.030\,377$ u

This mass converted to energy will be the binding energy:

Binding energy = mass defect $\times c^2 = (0.030\,377 \text{ u})\left(931 \frac{\text{MeV}}{\text{u}}\right)$

$= 28.3$ MeV

If we divide this energy by 4, the number of nucleons in the helium nucleus, we obtain a value for helium's binding energy of about 7 MeV per nucleon.

The concept of binding energy is more than a blackboard exercise. Physicists have confirmed that this type of energy really exists in a procedure known as the photodisintegration of the deuteron, the nucleus of a deuterium atom. The experiment involves the bombardment of deuterium, the isotope of hydrogen whose nucleus contains 1 proton and 1 neutron, by photons whose energies are in the γ-ray region. If the binding energy really exists, then one would expect the deuteron to split apart when struck by a photon with energy greater than the deuteron's binding energy. This effect would manifest itself to experimenters in the form of a threshold frequency of the photon for the disintegration, similar to that in the photoelectric effect. A γ-ray photon with energy greater than the threshold that represents the deuteron's binding energy could cause the nucleus to disintegrate; the excess energy above the threshold would appear as kinetic energy of the proton and neutron. Experiments have shown that this effect does indeed take place, thus proving the validity of the nuclear binding energy and giving physicists the opportunity to work out the binding energies of deuterium and other isotopes by similar techniques.

EXAMPLE 23.2

Compute the binding energy for the deuteron and the threshold frequency of the γ-ray required for photodisintegration of the deuteron.

Solution

Using the masses in Table 23.1, the mass defect is first computed for the deuteron, which consists of 1 proton and 1 neutron:

Mass defect $= 1m_p + 1m_n -$ mass of deuteron

Since the values in Table 23.1 are for deuterium, that is, the atom made of the deuteron plus 1 electron, we should use the mass of the hydrogen atom in place of the mass of the proton to account for the electron. A single electron may not be very massive, but its rest mass energy is 0.50 MeV, and so it cannot be neglected:

$$\text{Mass defect} = m_{^1_1\text{H atom}} + m_n - m_{\text{deuterium}}$$
$$= 1.007\,825 + 1.008\,665 - 2.014\,102$$
$$= 0.002\,388 \text{ u}$$

The binding energy is the mass defect expressed in electronvolts, or

$$\text{Binding energy} = (0.002\,388 \text{ u})\left(931\,\frac{\text{MeV}}{\text{u}}\right) = 2.22 \text{ MeV}$$

To find the threshold frequency, we need only set $h\nu = 2.22$ MeV, or

$$\nu = \frac{2.22 \text{ MeV}}{h} = \frac{(2.22 \times 10^6 \text{ eV})(1.6 \times 10^{-19} \text{ J/eV})}{6.625 \times 10^{-34} \text{ J} \cdot \text{s}}$$
$$= 5.36 \times 10^{20} \text{ Hz}$$

This 2.22-MeV γ-ray is highly energetic and at a very high frequency. Note that at a value of 1 MeV the binding energy per nucleon for the deuteron is much lower than that of the helium nucleus.

23.4 THE STABILITY OF NUCLEI

A plot of the binding energy per nucleon against atomic mass number A, shown in Fig. 23.2, provides an excellent indication of the relative stabilities of different nuclei. As the graph illustrates, the binding energy per nucleon increases sharply at first, when A has relatively small values, reaches a peak at a value of 56 for A, which corresponds to the iron isotope $^{56}_{26}$Fe, and then goes into a steady decline. The decline indicates a decreasing stability that finally manifests itself as chronic instability. No stable nucleus actually has a value for A greater than 215. Nuclei with atomic masses above 215 do occur naturally, but they disintegrate automatically by radioactive decay.

Another method of judging the stability of different nuclei is sketched in Fig. 23.3. Here the number of neutrons is plotted against the number of protons for a variety of different nuclei. In the range of low atomic masses, the most stable nuclei lie along a line that corresponds to equal numbers of protons and neutrons; $^4_2$He, $^{12}_6$C, $^{14}_7$N, $^{16}_8$O, and $^{20}_{10}$Ne are typical examples of the most stable isotopes of particular elements. Once the atomic mass gets above 16, however, the stable

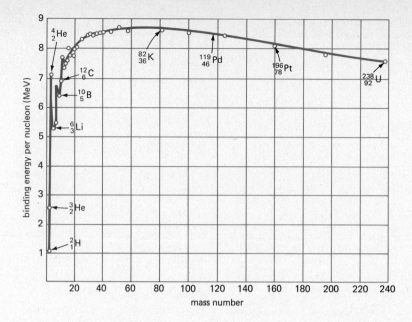

Figure 23.2 The variation of binding energy per nucleon with mass number.

isotopes uniformly possess more neutrons than protons. This reflects the increase in the Coulomb force of repulsion between protons as the number of these positively charged particles increases. The additional neutrons clearly serve to dilute that repulsive force, by providing the extra binding energy necessary to keep the nuclei together. As the value of A continues to increase, we find that proportionally more and more neutrons are needed to keep the nuclei stable, until we come to the atomic mass above which no stable nuclei can exist, no matter how many neutrons are included.

Some other factors enter into physicists' equations that deal with nuclear stability. The nucleus appears to have a structure, according to which nucleons are situated in energy levels described by quantum-mechanical wave equations. Pauli's exclusion principle plays an important part in determining just how these levels are filled, in much the same way that it influences the electronic structure of atoms. Another factor bearing on stability is that large nuclei have much larger surface areas than small ones; as a result, the number of loosely bound nucleons is larger than in smaller nuclei.

23.5 RADIOACTIVE DECAY

Radioactivity was discovered in February 1896 by French physics professor Henri Becquerel, as a direct consequence of Wilhelm Röntgen's finding of x-rays a few months earlier. Becquerel wondered whether some kinds of fluorescent material might give off x-rays spontaneously. He used ultraviolet radiation from sunlight to excite the fluorescence of various crystals, and because physicists knew that x-rays blackened photographic plates, even when the plates were wrapped in

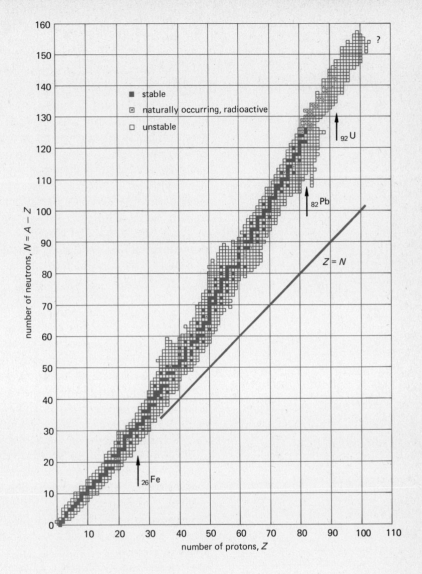

Figure 23.3 Chart of the known nuclei indicating the stability.

Marie Curie became the first person to win a second Nobel Prize, sharing the physics prize in 1903 for her studies on radioactivity and winning the chemistry prize in 1911 for her discovery of the two new elements.

black paper, he placed the crystals on top of photographic plates which had been wrapped to protect them from light. After developing the plates, Becquerel found that the plates under a compound of uranium had indeed been blackened. Unfortunately for Becquerel's hypothesis, the plates darkened anyway, on a cloudy day when the crystals and photographic plates had been stored in a drawer where the crystals could not fluoresce. Apparently the compound gave off some new rays like x-rays, with which it could photograph itself, as shown in Fig. 21.6.

This accidental discovery was not as simple as it first appeared, as shown by Madame Curie (described in Chap. 21). We now know that the radiation from the two elements discovered by Madame Curie, and

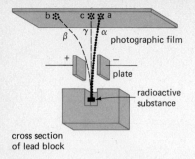

Figure 23.4 Experiment to show that rays from radioactive substances are of three kinds: α, β, and γ.

from other radioactive elements, actually occurs when the unstable atomic nuclei decay, or split apart, thus producing large amounts of energy. As shown in Fig. 23.3, all elements with atomic masses greater than that of bismuth, whose atomic number is 83, are naturally radioactive.

The general result of this decay is ultimate conversion of the unstable isotope to another isotope with a lower atomic mass that has a greater binding energy per nucleon. It does this by emitting rays that can be investigated using the apparatus sketched in Fig. 23.4. The radiation produced by the decay soars up through the hole in the lead block and encounters a strong electric field when it leaves the block. According to the kind of nuclei decaying, this field splits the beam into as many as three different sectors, each of which leaves a black region on the photographic film placed above the apparatus. Thus the radiation consists of up to three independent components.

Studies with a mass spectrometer show that the radiation that falls on region a in Fig. 23.4 consists of helium nuclei with a mass number of 4 and a positive charge of $+2$. This region may be a series of spots, each corresponding to particles of discrete energies. These are the α particles used by Rutherford in the scattering experiments that first gave rise to the concept of an atomic nucleus. Region b in the figure is produced by negatively charged particles since they bend the opposite way in the electric field. They were originally termed beta (β) particles. Their path through the electric field indicates that they are electrons with kinetic energies which vary from zero to some maximum value.

The radiation that impinges on the film at area c in Fig. 23.4 goes right through the electric field without deviation in its path. It is called gamma (γ) radiation because gamma is the third letter in the Greek alphabet, after alpha and beta. Further experiments show that it is actually similar to x radiation, although the frequencies of γ radiation tend to be higher. The big distinction between γ- and x-rays is in the source: nuclear or extranuclear. Transitions between electron energy levels in atoms are responsible for x-rays, and energy changes in the nucleus are responsible for γ radiation. γ-rays are often produced along with α and β decay.

When a nucleus undergoes "α decay," it disintegrates and transforms into a new nucleus with 2 fewer protons and, since the emitted α particle contains 4 nucleons, a mass about 4 u less than that of the original nucleus. We can put this decay in the form of an equation:

$$^A_Z X \rightarrow ^{A-4}_{Z-2} Y + ^4_2 He$$

where X is the original "parent" nucleus and Y is the "daughter" nucleus that results from the α decay. This process is the transmutation of one element to another—the process that was the long-sought objective of the medieval alchemists. It occurs naturally only for nuclei

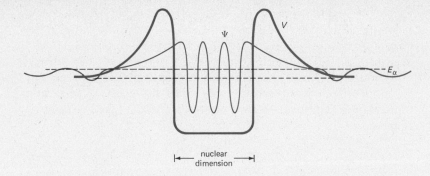

Figure 23.5 A particle, represented by its wave function ψ in a potential well. The energy of the α particle is indicated by E_α.

with large atomic numbers, in which the electrostatic repulsion between the protons is quite large. By releasing the α particle, which is a tightly bound nucleus with a relatively large binding energy per nucleon, the X nucleus in the equation can reduce its own energy. The transformation of the parent nucleus X into the daughter nucleus Y creates a more favorable N/Z ratio.

Mass energy must be conserved in the reaction. In fact, the α particle is emitted with an appreciable kinetic energy, of the order of a few million electronvolts, and the new nucleus also possesses kinetic energy. This energy is compensated for by the loss of mass in the reaction; the sum of the masses of the new nucleus and the α particle is smaller than the mass of the original nucleus.

In physical terms, we can consider the nucleus a potential well—a box that contains the α particle. Looking at it from a quantum-mechanical point of view, this means that the particle is a de Broglie-type wave which is trapped inside a well or box. However, the walls of the well are not infinitely high, and when a wave strikes it, it penetrates into it slightly, rather than being reflected completely. This situation is illustrated in Fig. 23.5. The amplitude of the wave falls rapidly as it penetrates the barrier but does not drop quite to zero. This means that the α particle has a definite, although small, chance of appearing outside the nucleus—as a result not of jumping over the barrier but of somehow tunneling through it. For this reason the effect is called tunneling. This effect is caused by the probabilistic nature of quantum mechanics. The actual probability that a particular α particle will tunnel out of its nucleus is determined by the nature of the nucleus that contains it.

β decay is somewhat more complicated than α decay because it involves the production of an entirely new particle. It occurs, in effect, when a neutron transforms itself into a proton by the creation of an electron, which is emitted, and a particle called the antineutrino, which is also emitted. The reaction is

$$n \rightarrow p^+ + e^- + \bar{\nu}$$

The symbol ν used here should not be confused with the frequency of electromagnetic radiation.

where $\bar{\nu}$ is the antineutrino. Just as in the case of α decay, there is a small change in mass during the reaction. This lost mass emerges in the form of kinetic energies of the products of the reaction. β decay is most likely to take place in a nucleus that contains a large excess of neutrons.

Although the loss of mass is readily explainable, the appearance of the antineutrino is totally unexpected—and at first glance unnecessary. Conservation of electric charge is obeyed by either of the two β-decay equations:

$$n \rightarrow p + e^- \quad \text{or} \quad {}^A_Z X \rightarrow {}^A_{Z+1} Y + e^-$$

Furthermore, the recoil of the daughter nucleus and the electron seems adequate to conserve both energy and momentum in the reaction because the electron speeds away from the reaction with a much higher speed than the far more massive daughter nucleus, as does a cannonball from a cannon.

This reasoning implies, however, that all the electrons emitted from a particular isotope should have just about the same energy. Unfortunately, observations show that this is not the case: the kinetic energies of the emitted electrons range from zero to the amount expected on the basis of mathematical analysis, indicating that some of the anticipated kinetic energy is missing.

This reaction creates another problem when one considers conservation of angular momentum. The neutron, proton, and electron all have spins of $+\frac{1}{2}$ or $-\frac{1}{2}$; no combination of the positive and negative spins in the equation, whereby a neutron splits up into just a proton and an electron, can conserve this spin angular momentum.

In 1930, two years before the neutron was discovered, physicists were worried only about the apparent missing energy. Then, Wolfgang Pauli, who was only 30 yr old, followed up his exclusion principle with yet another revolutionary theoretical concept: he suggested that the energy missing from the β-decay reaction was carried off by a new particle, which he named the neutrino, after the Italian for "little neutral one."

The neutrino has a number of peculiar properties, none more strange than its rest mass, which is precisely zero. This value is necessary to fit the observed facts because some electrons are emitted from β-decay reactions with all the available kinetic energy. To conserve angular momentum in the equation, the particle must also have a spin of $\pm\frac{1}{2}$. As it happened, the properties that Pauli determined for the neutrino did not quite fit the facts, but it soon became apparent that its antiparticle, the antineutrino, solved all the problems of β decay. Particles and their antiparticles with rest mass annihilate each other whenever they meet.

The neutrino itself, rather than its antiparticle, is involved in a form of radioactive decay known as positron emission. This is often regarded as a type of β decay. The reaction is equivalent to the breakup of a proton into a neutron, a positron, which is the positively charged antiparticle of an electron, and a neutrino. In terms of a parent-daughter pair, this reaction is

$${}^A_Z X \rightarrow {}^A_{Z-1} Y + e^+ + \nu$$

One nucleus that emits positrons in this way is the sodium isotope with a mass number of 22, which can be produced in nuclear reactions.

Why did Pauli have to pull an entirely new particle out of the hat? After all, he already had photons available, in the form of γ-rays. However, the spin angular momentum of γ-ray photons is too large to allow conservation of angular momentum, and the photons are also

The Mystery of the Missing Neutrinos

For particle physicists of the 1930s, the concept of the neutrino overcame a number of roadblocks in their drive to understand the nature of matter at its most fundamental level. By contrast, for today's astronomers the neutrino—or more strictly the absence of neutrinos—is the focus of a major mystery: how does the sun really generate its copious amounts of energy?

Theories developed in the 1930s indicated that the sun acts as what we now recognize as a huge hydrogen bomb, producing vast quantities of heat through various cycles of simple nuclear reactions. According to the theory, all the different cycles create neutrinos in the center of the sun. So penetrating are neutrinos that they should be able to pass through the sun's outer layers and reach the vicinity of earth. There, if the theory is correct, a certain proportion of these "solar neutrinos" should be detected by a device that consists basically of thousands of gallons of cleaning fluid, buried deep in the ground to prevent any particles except the penetrative neutrinos from reaching it.

However, a number of years of observation by a detector placed 1 mi under solid rock in the Homestake Gold Mines in Lead, South Dakota, by Dr. Raymond Davis, Jr., of the Brookhaven National Laboratory, revealed a disquieting effect. The device actually detected less than a third of the number of neutrinos predicted by the supposedly well-established theory of solar energy production. Thus astronomers and theoretical physicists are now back at their drawing boards. They know from the evidence of their own eyes that the sun shines. But just how the sun and other stars manage to do so is still a source of wonder and puzzlement.

---

readily observable, using the photoelectric and Compton effects. Since no one had observed photons in β decay, it was clear that they were not involved in the reaction. Neutrinos do not react readily with matter. In fact, not until 1956 were they first observed. The observation by Clyde Cowan and Frederick Reines, two physicists from Los Alamos Scientific Laboratory, in experiments at the nuclear reactor at Savannah River, set the seal on Pauli's suggestion, although it had already been incorporated into the body of theory concerning elementary particles by physicists who did not doubt its existence.

The correct expressions for β decay (Fig. 23.6) are, therefore,

$$n \to p + e^- + \bar{\nu} \quad \text{or} \quad {}^{A}_{Z}X \to {}^{A}_{Z+1}Y + e^- + \bar{\nu}$$

23.6 THE DECAY OF RADIOISOTOPES

The radioactive decay of natural or created radioisotopes can take place by α or β emission. The loss of the α or β particle changes the character of the nucleus, transforming it from an excited state of energy to either a lower excited state or a stable ground state of a different nuclear species. The two reactions often involve the emission of γ-rays

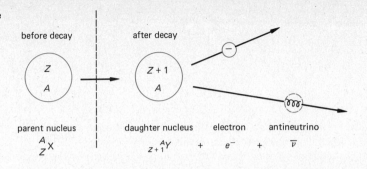

Figure 23.6 β decay. A neutron in the parent nucleus has become a proton in the daughter nucleus.

A nucleus does not decay instantly. No one can say whether a particular nucleus will decay in the next second or in a billion years' time.

as well as the particles. But γ-rays only remove energy from nuclei; their production does not transform a parent nucleus into a new daughter nucleus.

The rate at which unstable nuclei decay is governed by probabilities according to the laws of quantum mechanics. A nucleus does not decay instantly. Instead, its breakup is governed by statistics. No one can say whether a particular nucleus will decay in the next second or in a billion years' time. Physicists can, however, predict the probability that a nucleus will decay within either of those periods. Thus, when they deal with large numbers of radioactive nuclei, scientists can predict the rate of decay with remarkable accuracy.

This probability for decay is expressed by **the half-life.** Denoted by the Greek symbol tau (τ), **the half-life is the period within which every nucleus of a paticular isotope has a 50 percent chance of decaying.** It is termed the half-life because half of all the nuclei in any particular collection actually undergo decay during the period of a half-life. Thus, if we start with 1000 nuclei of an isotope whose half-life is 3 min, we know that only about 500 of the nuclei will remain unchanged after 3 min. After 6 min, twice the half-life, roughly 250 more of the nuclei will have decayed, leaving only 250. And after 3 more minutes, only about 125 nuclei will be left in their original state.

To take the most general case, consider a number of N identical nuclei that undergo α decay. As the example in the previous paragraph indicates, the number of nuclei that actually decay in any interval of time is directly proportional to the number present at the start of the interval. Let us say that the fraction of unchanged nuclei present after 1 s is f, in other words, that Nf nuclei remain after 1 s. Thus, $N(1-f)$ nuclei have disintegrated within the second. Since we start the second with Nf original nuclei, we know that only Nf^2 will remain after 2 s. During the third second, $Nf^2(1-f)$ of these split apart, leaving Nf^3 after 3 s.

The rate at which the nuclei decay, $\Delta N/\Delta t$, is known as the activity of the isotope. It is related to the number of nuclei by the constant of proportionality, λ, which is termed the decay constant and is given by the equation

Figure 23.7 Typical radioactive decay curve.

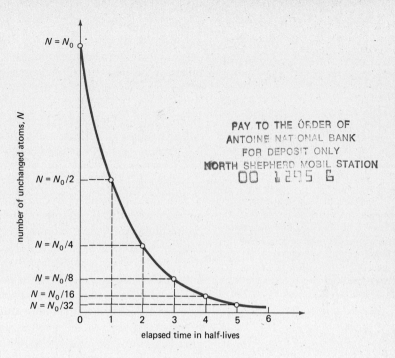

$$-\frac{\Delta N}{\Delta t} = \lambda N \tag{23.3}$$

The negative sign indicates that N is decreasing at time goes on. This equation is simply an exponential decay process, which is shown graphically in Fig. 23.7. This relates the number of nuclei N remaining after time t to the original number N_0 at $t = 0$ by the equation

$$N = N_0 e^{-\lambda t} \tag{23.4}$$

Geiger counters and other instruments designed to detect radiations from radioactive materials are outlined in Chap. 24.

However, since it is difficult to measure N_0 at any time, Eq. (23.3) is the more useful one; the rate of disintegration can be obtained simply and accurately using a Geiger counter or similar device. Mathematical analysis of Eq. (23.4) leads to an expression for the half-life in terms of the decay constant:

$$\tau = \frac{1}{\lambda} \ln 2 = \frac{0.693}{\lambda}$$

Hence, Eq. (23.3) can be rewritten

$$-\frac{\Delta N}{\Delta t} = \frac{0.693}{\tau} N \tag{23.5}$$

The half-lives of isotopes vary from less than one-billionth of a second to more than 1 billion years. The uranium isotope $^{238}_{92}U$, for example, has a half-life of 4.5 billion years, which equals the estimated age of our solar system. Therefore, just half of all the uranium 238 that was present on earth when our planet was born is still here in its original form. The less common isotope $^{235}_{92}U$ has a half-life of only 0.7 billion years; it has, therefore, gone through six half-lives since the birth of the solar system, leaving just $1/2^6$, or 0.0156, of the original amount still on earth. Figure 23.7 indicates that no radioactive isotope present at the creation of the solar system could now exist in detectable amounts on earth unless its half-life is at least 0.1 billion years. Many short-lived isotopes do exist on earth, of course, but these are inevitably created either by humans, or as a result of the decay of longer-lived radioisotopes in what are known as natural radioactive decay schemes.

A natural radioactive decay series is a succession of decays by the emission of α or β particles that leads from one particular parent radioisotope through a number of daughter radioisotopes which are equally or more unstable, to a stable, nonradioactive end product. The three series of this type known in nature all lead to a stable isotope of lead, after starting with the elements uranium, actinium, and thorium. Figure 23.8 shows the uranium series; the nuclear mass numbers of the isotopes are plotted against their atomic numbers. β decay, which adds a positive charge to the nucleus, is represented on the graph by a horizontal move of 1 unit to the right; α decay, which produces a loss of two positive charges and four nucleon (4 u) masses, is represented by a diagonal move downward to the left. One point to note about this figure is that the chain of decay branches at a number of points, which means that the nuclei have a choice, in effect, over the way in which

The half-life of free neutrons, isolated from atomic nuclei, is just 12 min. The half-life of free protons would be infinite since the proton is stable in terms of the age of the universe, although recent theoretical studies suggest that it may have a half-life of about 10^{33} yr.

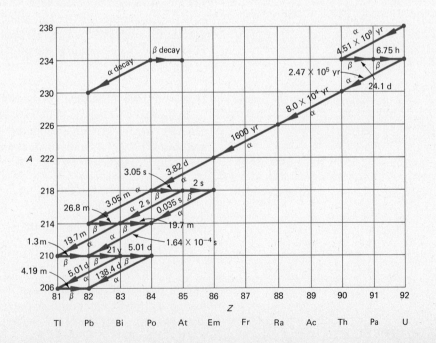

Figure 23.8 Uranium radioactive decay scheme.

they decay. Any individual nucleus can only decay in one particular way, but the branching illustrates once again that radioactive decay is a probabilistic process when applied to collections of nuclei.

EXAMPLE 23.3

The following scheme shows the final stages of the thorium radioactive series that ends in the stable lead isotope $^{208}_{82}\text{Pb}$, specifying an initial and final product and the means of decay. Complete the table by producing the atomic numbers, atomic weights, and symbols for the three intermediate isotopes, and indicate any pairs of isotopes of the same element in the series.

$$^{216}_{84}\text{Po} \;\underset{\substack{\alpha \\ \tau = 0.16\text{ s}}}{\rightarrow}\; ^{212}_{82}\text{Pb} \;\underset{\substack{\beta \\ \tau = 10.6\text{ h}}}{\rightarrow}\; ^{212}_{83}\text{Bi} \;\underset{\substack{\alpha \\ \tau = 60.6\text{ min}}}{\rightarrow}\; ^{208}_{81}\text{Tl} \;\underset{\substack{\beta \\ \tau = 3.1\text{ min}}}{\rightarrow}\; ^{208}_{82}\text{Pb}$$

23.7 NUCLEAR REACTIONS

Scientists need not rely on natural processes to transmute isotopes of elements. They can stimulate nuclear reactions in the laboratory and the field by bombarding nuclei with appropriate particles. The first nuclear reaction was identified by Ernest Rutherford in 1919; he produced oxygen 17 by bombarding nitrogen gas with α particles. The reaction can be written as follows:

$$^{4}_{2}\text{He} + ^{14}_{7}\text{N} \rightarrow ^{17}_{8}\text{O} + ^{1}_{1}\text{H}$$

In this case, just as in that of radioactive decay, the reaction conserves charge and mass number: both sides of the equation contain 18 mass units and 9 positively charged protons. The reaction also conserves energy, as long as we remember the interchangeability of mass and energy. If the sum of the rest mass energy of the nuclei involved in any nuclear reaction is greater than the rest mass energy of the products, then the excess mass is converted into kinetic energy of the products. If, on the other hand, the sum of the rest mass energies of the reacting nuclei is smaller than that of the products, the reaction will require an infusion of kinetic energy from the reacting nuclei in order to take place at all. This is the case in the reaction observed by Rutherford. The α particle must possess several mega electronvolts of energy if it is to spark the nuclear reaction. Of course, the α particle must have sufficient kinetic energy to overcome the Coulomb repulsion of the positive nucleus if it is to penetrate into the nucleus to produce any type of nuclear reaction. However, due to the wave nature of the α particle, lower-energy α particles can tunnel this Coulomb barrier. The lower the energy of the α particle, the less likely this is to occur.

The model for this and many other nuclear reactions can be understood using a concept developed by Niels Bohr, which is known as the compound nucleus. Bohr proposed that the reacting nuclei first

Dating the Past

One of the best-known applications of nuclear reactions beyond the world of physics is carbon 14 dating—the estimation of the age of certain archeological artifacts from the amount of a radioactive isotope that they contain. Invented in the late 1940s by University of Chicago chemist Willard Libby, the method has proved invaluable to scholars investigating the roots of humans and other creatures.

The technique is made possible by cosmic rays, which react with nitrogen atoms in the atmosphere to produce atoms of the radioactive isotope carbon 14. The nuclei of this isotope contain six protons, as does the normal carbon 12, but eight neutrons instead of 6. The isotope splits up by β decay but has the relatively long half-life of 5568 yr.

Carbon 14 is important in dating artifacts because all living creatures take in a certain amount of carbon 14 from the air when they breathe. The proportion of carbon 14 to carbon 12 is, therefore, constant in every living thing. But once the organism dies and stops breathing, its intake of the radioactive isotope stops. The carbon 14 continues to decay, however, and thus the proportion of that isotope to carbon 12 gives a good indication of the amount of time that has elapsed since the organism died. The amount of carbon 14 in such archeological artifacts as bones and wood chippings is determined from their radioactive emission; the total quantity of carbon is determined by simple chemical analysis.

The technique does have some limitations. Because of the tiny activity that remains, it cannot now be applied to artifacts more than about 50 000 yr old. Furthermore, because scientists suspect that the amount of carbon 14 in the atmosphere has varied over the centuries, the method is not entirely reliable by itself. But calibration of carbon 14 ages using more reliable but less general techniques such as tree rings have ironed out many of these difficulties, and physicists have recently developed a technique that counts carbon 14 atoms directly, using accelerators, instead of waiting for them to decay. The latter technique should extend the age of artifacts to be dated and allow dating using miniscule quantities of fossils.

Carbon 14 dating apparatus. (Elizabeth Ralph, University of Pennsylvania Museum)

combine to produce a compound nucleus which contains all the nucleons from the two nuclei. This compound nucleus subsequently decays into two different nuclei. In the case of Rutherford's reaction, the α particle and the nitrogen nucleus form a compound nucleus that contains 18 nucleons and 9 protons. This is the fluorine isotope $^{18}_{9}F$. This compound nucleus, whose half-life of 10^{-9} s is very short by our standards but nevertheless long on the nuclear scale, then emits a proton, to produce the oxygen 17 isotope. The overall reaction is

$$^4_2\text{He} + {}^{14}_{7}\text{N} \rightarrow [{}^{18}_{9}\text{F}] \rightarrow {}^{17}_{8}\text{O} + {}^{1}_{1}\text{H}$$

Other examples of other nuclear reactions are

$$^2_1\text{H} + {}^{26}_{12}\text{Mg} \rightarrow [{}^{28}_{13}\text{Al}] \rightarrow {}^{24}_{11}\text{Na} + {}^{4}_{2}\text{He}$$

$$^1_0 n + {}^{27}_{13}\text{Al} \rightarrow [{}^{28}_{13}\text{Al}] \rightarrow {}^{27}_{12}\text{Mg} + {}^{1}_{1}\text{H}$$

The same isotope is formed as a compound nucleus in these two reactions. However, different products are formed because the compound nuclei are in different states of excitation. The second reaction, in which a neutron is absorbed by aluminum, is a particularly easy way to produce nuclear reactions. Because they have no net charge, neutrons are able to penetrate the positive nucleus and react with it. Thermal neutrons, so-called because they have energies of only a fraction of an electronvolt, easily produce nuclear reactions.

The two different means by which the compound nuclei split up in our examples illustrate another important factor in nuclear reactions: the products of almost all nuclear reactions include a proton or a neutron, or an electron or an α particle.

EXAMPLE 23.4

The reaction between a thermal neutron and a lithium 6 nucleus yields an α particle and a triton. Considering the kinetic energy of the thermal neutron (about 0.025 eV) as insignificant, determine the amount of kinetic energy shared by the product nuclei.

Solution

This reaction can be written as

$$^1_0 n + {}^{6}_{3}\text{Li} \rightarrow {}^{3}_{1}\text{H} + {}^{4}_{2}\text{He}$$

A triton is a nucleus of tritium, the isotope of hydrogen, which contains 1 proton and 2 neutrons.

We consider only the rest mass energies. If the mass of the reactants is larger than that of the products, the reaction will proceed and the excess energy will appear as kinetic energy of the triton and α particle. The masses of the nuclei in atomic mass units are as follows:

$$m_n = 1.008\ 665$$

$$m_{{}^6\text{Li}} = 6.015\ 126$$

$$m_{{}^3\text{H}} = 3.016\ 049$$

$$m_{{}^4_2\text{He}} = 4.002\ 603$$

Summing the masses of the reactants, we obtain $1.008\ 665 + 6.015\ 126 = 7.023\ 791$. The sum of the products is $3.016\ 049 + 4.002\ 603 = 7.018\ 652$. Thus, the reactants are more massive than the products by

$$7.023\ 791 - 7.018\ 652 = 0.005\ 139$$

Since 1 u is equivalent to 931 MeV, this mass difference corresponds to

$$\left(931 \frac{\text{MeV}}{\text{u}}\right)(0.005\ 139\ \text{u}) = 4.78\ \text{MeV}$$

This energy would be shared by the two products in such a way that momentum is also conserved.

Fission   Undoubtedly the most famous, or notorious, type of nuclear reaction is nuclear fission, which occurs when a heavy nucleus absorbs a particle, usually a neutron, and then breaks up into two nearly equal fragments, rather than splitting up by α or β decay. German scientists found that when the isotope of uranium $^{235}_{92}\text{U}$ absorbed a neutron, reactions of the following type resulted:

$$^1_0 n + {}^{235}_{92}\text{U} \rightarrow [{}^{236}_{92}\text{U}] \rightarrow {}^{141}_{56}\text{Ba} + {}^{92}_{36}\text{Kr} + 3\,{}^1_0 n$$

$$^1_0 n + {}^{235}_{92}\text{U} \rightarrow [{}^{236}_{92}\text{U}] \rightarrow {}^{140}_{54}\text{Xe} + {}^{94}_{38}\text{Sr} + 2\,{}^1_0 n$$

The half-life of the $^{236}_{92}\text{U}$ is of the order of 10^{-12} s, making its breakup an instantaneous reaction in our time scale. This nucleus ($^{236}_{92}\text{U}$) can split up in a number of ways besides the two shown here. This occurs because of the probabilistic nature of nuclear processes and that several nuclei of equal stability and binding energy exist in the midregion of the periodic chart. These nuclei are more stable than the heavy $^{235}_{92}\text{U}$ or $^{236}_{92}\text{U}$ nucleus, as seen from the binding energy per nucleon shown in Fig. 23.2.

In the fission process a tremendous amount of energy is generated owing to the loss of mass during the course of the reaction. Since the binding energy of the extra neutron is very large, no kinetic energy is required to set off the reaction of $^{235}_{92}\text{U}$ with a neutron. To see how much energy is produced in the reaction, which chemists call exothermic, we can compute the rest mass difference between reactants and products:

For the reactants:

$$\text{Mass of } {}^1_0 n = 1.008\ 7\ \text{u}$$

$$\text{Mass of } {}^{235}_{92}\text{U} = 235.043\ 9\ \text{u}$$

$$\text{Sum} = 236.052\ 6\ \text{u}$$

For the products:

$$\text{Mass of } {}^{141}_{56}\text{Ba} = 140.913\ 9\ \text{u}$$

$$\text{Mass of } {}^{92}_{36}\text{Kr} = 91.897\ 3\ \text{u}$$

$$\text{Mass of } 3\,{}^1_0 n = 3.026\ 1\ \text{u}$$

$$\text{Sum} = 235.837\ 3\ \text{u}$$

Subtracting the sum of the mass of the products from the sum of the

Figure 23.9 Generalized diagram of a chain reaction. The reaction is self-sustaining if at least one neutron from each fission is captured by a uranium nucleus and another fission is thereby induced. The reaction is explosive, with a tremendous release of energy if, on the average, several neutrons per fission induce additional fissions, thus causing a rapid buildup of fissions. Some of the particles emitted and waves representing energy released are not shown.

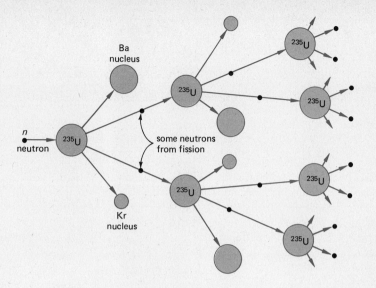

mass of the reactants, we find that 0.215 3 u is lost in the reaction. Converting this to grams, multiplying by c^2, and converting the result to electronvolts, one finds that this mass difference is equivalent to 200 MeV. This is a vast amount of energy when compared with typical chemical energies, such as the 13.6 eV required to ionize hydrogen. The 200 MeV released in this reaction appears mostly as kinetic energy of the products.

What makes this type of reaction exceptional is that it produces one or two more neutrons than the single one that starts it off. If all the neutrons produced in each individual reaction are absorbed by further nuclei, the process will rapidly run out of control, as shown in Fig. 23.9. This is known as a chain reaction. If the reaction is encouraged to continue in this way, the result is the incredibly violent explosion of an atom bomb, replete with its somber mushroom cloud. *If a large mass of uranium 235 were accidentally allowed to undergo an uncontrollable chain reaction, the result would be a disastrous explosion.* But if the number of neutrons that move from one nucleus to another is strictly controlled to ensure that the number of reactions per second remains constant, the reaction forms the basis of a stable nuclear reactor. The heat produced in this type of reactor is used to boil water to steam in just the same way as heat from any other power source. The steam then turns a turbine to generate electric energy. Nuclear engineers control the reaction in their plants by inserting cadmium and boron rods—or rods of other material—that absorb neutrons. Because of the extraordinary precautions introduced in the design of nuclear power plants, there is little danger that a reactor will run out of control. Even if it did run out of control, a reactor would not explode like a bomb. However, critics of nuclear power assert that reactors could, under some circum-

Inside the Nuclear Reactor

Just like coal and oil-fired power plants, nuclear reactors have a single, simple purpose: to produce the heat necessary to convert water into steam that can drive an electric generator. But unlike other forms of a power plant, nuclear reactors involve an inherent risk of radioactive pollution that must be guarded against in every bolt, joint, and weld.

The actual working fuel of the first generation of nuclear reactors—those currently in use—is the uranium isotope U 235. Since this comprises only 0.7 percent of naturally occurring uranium, its proportion must be artificially increased, in the process known as enrichment. Enriched uranium, containing about 3 percent U 235, is formed into long cylindrical fuel rods, which are inserted into the reactor.

Atoms of U 235 split up when struck by neutrons to produce two large fragments, a huge amount of energy, and either two or three more neutrons that can in turn split up more U 235 atoms. Since the reaction occurs most efficiently when the oncoming neutrons are traveling relatively slowly, reactor cores are de-

Nuclear plant. (*U.S. Nuclear Regulatory Commission*)

stances, release dangerous radiation into the atmosphere—as happened on a small scale at Three Mile Island, Pennsylvania, in March 1979—or melt down.

Fusion A glance at the bottom left of the graph of binding energy per nucleon versus mass number in Fig. 23.2 indicates one more nuclear process that might potentially yield large amounts of energy. The binding energy per nucleon increases rapidly from one mass number to another when those numbers are very small. This means that if we fuse two light nuclei to form a fresh nucleus, we shall increase the average binding energy per nucleon, thus releasing an amount of energy equal to the total number of nucleons times the increase in binding energy per nucleon. The two nuclei must first be brought together against the force signed to slow down their neutrons to appropriate speeds. The vehicle for this deceleration is known as the moderator; heavy water (deuterium oxide) and the form of carbon known as graphite are the two most common moderators in use. But even with the moderators in place, the reactor requires a certain amount of U 235 fuel to "go critical" and undertake its chain reaction. This amount, however, is little more than 200 g.

Of course, the technologists who run reactors must be able to shut them down in response to an emergency or the need for regular maintenance. This is achieved using rods of materials such as cadmium and boron that absorb slow neutrons. Insertion of such rods into the reactor core stops the nuclear reaction; removal allows it to restart. When the reactor is operating, its energy is picked up in the form of heat by circulating pressurized water. This water then heats up another stream of water, converting it into the steam that drives a turbine.

This type of reactor, known as the light water reactor, represents just one means of capturing energy from uranium by nuclear reactions. A more futuristic version is a type that uses the common isotope of the metal, U 238, as its fuel. In the course of producing energy, the U 238 is converted into the human-made isotope plutonium 239. Since the plutonium can itself be used as fuel for conventional reactors in place of U 235, the U 238 reactor actually "breeds" fuel while producing heat; it is therefore known as the breeder reactor.

When nuclear reactors first started to produce energy commercially in the mid-1950s, they were hailed as the flagships of scientific progress. But a decade later, critics began to raise disquieting questions about their safety. No one argues that a nuclear reactor will run amok and explode like an atomic bomb; the fuel mix makes that well-nigh impossible. But some scientists fear that nuclear accidents could release large amounts of dangerously radioactive gases into the atmosphere and that terrorists could obtain enough plutonium from reactors to make their own atom bombs. Furthermore, critics of the nuclear industry point out that no one has decided on a satisfactory burial ground for the increasingly large amounts of extremely radioactively "hot" wastes that are produced by reactors. Nuclear proponents argue that these concerns are minor when viewed against the benefits of nuclear power, but in the light of the Three Mile Island accident, the controversy seems likely to continue for many years.

of their natural electrostatic repulsion, a process that requires a great deal of energy. Bringing the nuclei within about 5×10^{-15} m, at which point the attractive nuclear force will pull them closer together and stimulate the fusion reaction between them, requires kinetic energies corresponding to temperatures of at least 10 million degrees. Thus it is hardly surprising that fusion reactions are unusual on the earth.

We do, however, see fusion in action whenever we gaze at the sun or other stars. The gravitational force that pulls matter together in these huge cosmic objects gives nuclei more than enough kinetic energy to reach the conditions necessary for fusion. In the process, it also tears all the electrons from their nuclei, producing what physicists call a plasma of freely roaming electrons and nuclei. Experts still do not fully understand all the details of the fusion reactions in stars, but they know that the most prevalent reaction is the combination of 4 protons to produce 1 helium nucleus in the so-called proton cycle:

$$^1_1H + ^1_1H \rightarrow ^2_1H + e^+ + \nu$$

$$^1_1H + ^2_1H \rightarrow ^3_2He$$

$$^3_2He + ^3_2He \rightarrow ^4_2He + ^1_1H + ^1_1H$$

These reactions are equivalent to

$$4^1_1H \rightarrow ^4_2He + 2e^+ + 2\nu$$

We can readily compute the mass change to energy in this reaction by taking the mass difference between 4 hydrogen atoms and 1 helium atom. We consider the 2 positrons created in the total reaction as energy in the same way we consider the 2 neutrinos as energy. We can do this because the positrons, being antimatter, very quickly annihilate 2 electrons, producing energy in the form of γ-rays.

$$\text{Mass difference} = 4m_H - m_{He} = 4(1.007\,825) - 4.002\,603$$
$$= 0.0287 \text{ u}$$

This mass difference in energy units is equivalent to

$$0.0287 \text{ u} \times 931 \text{ MeV/u} = 26.7 \text{ MeV}$$

This 26.7 MeV is much less than the 200 MeV obtained from the fission of $^{235}_{92}U$. However, the fusion process is inherently about eight times more efficient than fission because only 4 protons produce 26.72 MeV of fusion energy, whereas it takes the 235 nucleons to produce the 200 MeV of fission energy.

Another typical series of thermonuclear fusion reactions in stars is known as the carbon cycle:

$$^{12}_6C + ^1_1H \rightarrow ^{13}_7N \rightarrow ^{13}_6C + e^+ + \nu$$

$$^{13}_6C + ^1_1H \rightarrow ^{14}_7N$$

$$^{14}_7N + ^1_1H \rightarrow ^{15}_8O \rightarrow ^{15}_7N + e^+ + \nu$$

$$^{15}_7N + ^1_1H \rightarrow ^{12}_6C + ^4_2He$$

In any of the stellar thermonuclear processes there must be a balance between the explosive nature of the fusion processes that give up energy and the gravitational attractive force that holds the star together.

Production of thermonuclear fusion on earth, which may start with reactions like the second one of the proton cycle, requires a huge amount of energy to initiate the reaction. At present, the only method of producing this energy is to explode a small atomic—fission—bomb. The fusion process that results from this vast infusion of energy is itself a violent explosion—the hydrogen bomb, which has an explosive power of up to millions of tons of TNT.

Despite more than 25 yr of research, physicists thus far have been unable to tame the reaction and produce the fusion equivalent of fission power stations. The problems of heating a plasma to sufficiently high temperatures and containing it for long enough to allow it to produce a net amount of energy have been overwhelming. However, experts in many nations believe that they will sooner or later manage to produce a peaceful fusion reactor, in which the plasma is produced by enromous electric arcs or powerful laser pulses and confined in strong magnetic fields. Optimistic researchers confidently expect that fusion will be supplying the major proportion of the world's energy by the third decade of the next century. Fusion certainly has inherent advantages over fission as a source of power. Its raw material is the virtually limitless element hydrogen, by contrast with the scarce uranium that provides the fuel for fission, and its major product is the clean, nonradioactive inert gas helium instead of the highly radioactive isotopes that fission reactions create.

23.8 BEYOND THE NUCLEUS

The story of the structure of the nucleus is far from complete. The discoveries of such particles as the neutron and the neutrino were no more than the starting points for a multimillion dollar effort to probe even further into the basic nature of matter, using both natural and human-made collisions between particles. The natural collisions are those that occur when cosmic rays from outer space arrive at great speeds at the top of the earth's atmosphere and strike atoms and nuclei in that wispy region. The human-made encounters take place in particle accelerators, in which various types of atomic particles are accelerated to energies of millions, and even billions, of electronvolts. At present, these types of experiments are lending some credence to a theory first proposed in the early 1960s, which asserts that the truly fundamental particles of matter are entities known as "quarks." Among the unusual properties attributed to quarks is the possession of electric charges that are fractions—either $\frac{1}{3}$ or $\frac{2}{3}$—of the charge on the electron and proton. Some physicists believe that there are at least six types of quark, which combine in twos and threes to produce all the other particles in the subnuclear zoo. But whether this picture, and the overall concept of quarks, will stand up to experimental scrutiny remains to be seen.

SUMMARY

1. The nucleus has a diameter of about one-thousandth that of the atom but contains more than 99.9 percent of the atom's mass.
2. Nuclei consist of positively charged protons and electrically neutral neutrons. Isotopes of the same element have the same number of protons but different numbers of neutrons.
3. The atomic mass number of a nucleus is the total number of protons and neutrons—collectively known as nucleons—in the nucleus.
4. Nucleons are held together in the nucleus by the nuclear force. This is a short-range force whose magnitude is independent of the charge on the nucleons.
5. The masses of nuclei are always lower than the total mass of their nucleons taken individually. This

difference, known as the mass defect, represents the binding energy that keeps the nucleus together.

6. A graph of binding energy per nucleon against atomic mass number shows that nuclei become decreasingly stable as their mass numbers increase above 56. No stable nucleus has a mass number greater than 215.

7. Radioactive decay is the disintegration of an unstable nucleus, producing ultimately another nucleus with lower atomic mass and greater binding energy per nucleon.

8. In α decay, the unstable nucleus emits an α particle. In β decay, the nucleus emits an electron. Both processes can be accompanied by emission of γ radiation.

9. β decay is further accompanied by emission of a new particle, known as the antineutrino, whose rest mass is zero.

10. Unstable nuclei do not decay instantly. Their rate of decay is governed by probabilities according to the laws of quantum mechanics.

11. The decay of each type of nucleus is characterized by its half-life. This is the period in which half of all the nuclei in a collection decay. Half-lives vary from less than a billionth of a second to more than 1 billion years.

12. Nuclear reactions can often be understood using the concept of compound nuclei.

13. Nuclear fission occurs when an isotope of uranium struck by a neutron splits into two large chunks, two or three neutrons, and a great deal of energy. The neutrons produced in this process can, in turn, stimulate further reactions. If the reaction is allowed to continue uncontrolled, it becomes a cascade reaction that produces an explosion of the atomic bomb variety. If the reaction is kept under control, it can produce the heat necessary to drive an electric generator in the form of a nuclear reactor.

14. Nuclear fusion is the reaction by which the sun produces its heat and the hydrogen bomb obtains its power.

QUESTIONS

1. Identify the following: nucleons, isotopes, neutrinos, α particles, β particles, γ-rays, positrons.

2. Describe fission and fusion, distinguishing clearly between them. Which is the more efficient process for energy production, and which is the cleaner process from an environmental standpoint?

3. In which region of the periodic table do elements have the most isotopes?

4. Why is it reasonable to expect, in terms of the forces between nucleons, that nuclei ultimately become too large to be stable?

5. The angular momentum of the deuteron is $1(h/2\pi)$. Is this in agreement with a model of the deuteron consisting of one proton and one neutron?

6. Why do neutrons make much better particles to stimulate nuclear processes than α particles or protons?

7. Describe the characteristics of the neutrino.

8. What are the possible spin angular momenta for the $^6_3$Li nucleus if it contains six protons and three electrons, and if, alternatively, it contains three protons and three neutrons?

9. Atomic masses are generally used to compute mass defect and the energies in nuclear reactions. Does this necessitate a correction?

PROBLEMS

Nuclear Properties

Atomic masses are listed in Table 23.1 and App. F.

1. How many neutrons are there in 1 g of $^{12}_6$C?

2. If nucleons are separated by a distance of 10^{-15} m, calculate an average density of nuclear matter and compare it with the density of atomic matter, say, $\rho = 13.6$ g/cc for mercury.

3. How many (a) neutrons, (b) protons, and (c) nucleons exist in $^{197}_{79}$Au?

4. Which nucleus is made up of three α particles? Is this a particularly stable nucleus? How do you know?

5. Complete the reactions:
(a) $^1_0n + {}^{16}_8O \rightarrow$ ____
(b) $^{10}_5B + {}^4_2He \rightarrow {}^{13}_7N +$ ____
(c) $^{235}_{92}U + {}^1_0n \rightarrow {}^{107}_{43}Tc +$ ____ $+ 5{}^1_0n$
(d) $^{64}_{29}Cu \rightarrow e^+ +$ ____

6. In the following nuclei, (a) how many electrons exist in the neutral atom? (b) How many protons are in the nucleus, and (c) how many neutrons are in the nucleus?

$^{238}_{92}U \qquad ^{57}_{26}Fe \qquad ^{107}_{47}Ag$

7. In the following reaction, what

particle or particles are emitted?

$^{55}_{24}Cr \rightarrow ^{55}_{25}Mn + ?$

8. Complete the following reactions:

(a) $^{210}_{84}Po \rightarrow ^{206}_{82}Pb + $ _____

(b) _____ $+ ^{12}_{6}C \rightarrow ^{13}_{6}C + ^{1}_{1}H$

(c) $^{226}_{88}Ra \rightarrow ^{222}_{86}Rn + $ _____

(d) $^{214}_{82}Pb \rightarrow ^{214}_{83}Bi + $ _____

(e) $^{9}_{4}Be + ^{4}_{2}He \rightarrow ^{12}_{6}C + $ _____

(f) $^{11}_{5}B + ^{4}_{2}He \rightarrow ^{14}_{7}N + $ _____

(g) $^{27}_{13}Al + n \rightarrow ^{27}_{12}Mg + $ _____

9. $^{11}_{6}C$ emits a β^+ particle (positron). Identify the daughter product isotope.

10. Compute the binding energy of a hydrogen atom, using the masses of the atom, the proton, and the electron.

11. Compute the mass defect, binding energy, and binding energy per nucleon for the triton, the nucleus of the tritium atom.

12. Compute the binding energy and the binding energy per nucleon of ^6Li.

13. What is the binding energy per particle for $^{16}_{8}O$ and $^{17}_{8}O$?

14. Will the following processes go by themselves, or do they require input energy?

$^{211}_{83}Bi \rightarrow ^{211}_{84}Po + $ _____

$^{210}_{84}Po \rightarrow ^{206}_{82}Pb + $ _____

Mass of $^{211}_{83}Bi = 210.987\ 29$

Mass of $^{211}_{84}Po = 210.986\ 66$

Mass of $^{210}_{84}Po = 209.982\ 87$

Mass of $^{206}_{82}Pb = 205.974\ 48$

Radioactive Half-Life

15. If an isotope has a half-life of 10 s, what fraction decays in the first second?

16. If a sample initially emits 10^8 α particles per second, how many particles will it emit after three half-lives have passed?

17. A pure radioactive element has a mass of 5×10^{-6} kg; 1 h later only 1×10^{-6} kg of the element remains. What is its half-life?

18. If an isotope has a half-life of 1000 yr, how much remains after (a) 500, (b) 750, and (c) 875 yr have passed?

19. If we begin an experiment with 100 g of pure radioactive $^{55}_{24}Cr$ that undergoes β decay with a half-life of 3.5 min, how many grams of $^{55}_{24}Cr$ will remain after 14 min?

20. A radioactive sample gives 1000 counts per minute, and 5 h later it gives 750 counts per minute. What is the half-life of the decay?

21. If the half-life of radioactive barium, $^{140}_{56}Ba$, is 12 days, how long will it be before 1000 g of radioactive barium are reduced to 250 g?

22. How many decays per second are there in 1 kg of $^{238}_{92}U$? The half-life of $^{238}_{92}U$ is 4.5×10^9 yr.

23. A solution contains an isotope that emits β particles with a half-life of 15.34 days. At that time, a Geiger counter records 490 counts per minute. What will be the count rate from this solution 46.02 days later?

24. Tritium, ^3_1H, is radioactive and decays with a half-life of 12.5 yr. If a sample of water is taken and found to have 10^{-20} tritium atoms per original hydrogen atom, how many tritium atoms will there be per hydrogen atom 50 yr later, assuming that the sample of water is isolated and no new tritium is created in the sample.

25. A wooden artifact found buried shows a radioactive rate of 11.6 counts per minute per gram of carbon present. If living wood is tested, the count is found to be 15.3 counts per minute per gram of carbon. The decaying nucleus is ^{14}C, with a half-life of 5600 yr. How old is the artifact?

Nuclear Reactions

26. Is the isotope $^{40}_{20}Ca$ stable as far as α decay is concerned?

27. What is the energy in MeV of the α particle emitted from $^{226}_{88}Ra$? Assume the nucleus does not recoil.

28. The isotope $^{210}_{84}Po$ emits an 8.95-MeV α particle. Calculate the mass of the nucleus remaining after this decay.

29. Is the decay $^{16}_{8}O \rightarrow ^{12}_{6}C + ^{4}_{2}He$ possible? How much energy (say, in the form of a γ-ray) would be required to initiate this reaction, or will it go by itself?

30. Consider the energy involved in the decay of $^{234}_{92}U$ to $^{218}_{84}Po$ by the emission of four α particles. Is the process whereby $^{234}_{92}U$ emits 2 ^8Be nuclei energetically more favorable than the decay by four α particles? Masses of $^{218}_{84}Po = 218.077\ 02$ and $^{234}_{92}U = 234.114\ 08$.

31. If the maximum energy for β particles emitted from $^{32}_{15}P$ is 1.704 MeV, compute the mass of $^{32}_{15}P$ using the mass of $^{32}_{16}S$ as 31.972 07.

32. A proton strikes a tritium nucleus. What energy must the proton have if this reaction is to take place?

$^{1}_{1}H + ^{3}_{1}T \rightarrow ^{3}_{2}He + ^{1}_{0}n$

33. (a) What is the minimum energy photon required to disintegrate the $^{7}_{3}Li$ isotope into an α particle and a triton? (b) What would be the frequency of such a photon?

34. If the kinetic energy given to the proton and neutron after photodisintegration of a deuteron is 0.85 MeV, what was the frequency of the incident photon?

35. (a) If photodisintegration of a

deuteron is caused by absorption of a photon with wavelength of 5×10^{-11} cm, how much kinetic energy is given to the escaping neutron and proton? (b) What additional information is required to allow you to know how the energy is split between the neutron and proton? What additional conservation law would you have to apply here?

36. A 5-MeV neutron will produce fission in $^{238}_{92}$U, producing two equal mass fission products and three neutrons. Write down the reaction.

37. Consider the fission of ^{235}U, which obeys the following equation:

$$^1_0n + {}^{235}_{92}U \rightarrow {}^{138}_{54}Xe + 8{}^1_0n + \underline{\qquad}$$

Identify the missing product isotope. Compute the approximate energy release in this reaction. Mass of $^{138}_{54}Xe = 137.956$ u. Mass of the other product $= 89.9359$ u.

38. Consider thermonuclear fusion in stars following the carbon cycle. Compute the net energy emitted in this process. You can consider the reaction in total if you add all input particles and all product particles carefully:

$$^{12}_{6}C + {}^1_1H \rightarrow {}^{13}_{6}C + e^+ + \nu$$
$$^{13}_{6}C + {}^1_1H \rightarrow {}^{14}_{7}N$$
$$^{14}_{7}N + {}^1_1H \rightarrow {}^{15}_{7}N + e^+ + \nu$$
$$^{15}_{7}N + {}^1_1H \rightarrow {}^{12}_{6}C + {}^4_2He$$

39. In the fission of $^{235}_{92}$U in a nuclear reactor, about 211 MeV of energy are released in the form of kinetic energy of particles. How many fissions are required to sustain a 100-MW reactor if 80 percent of the reaction energy goes to steam?

Special Topic
Selected Species from the Subnuclear Zoo

In the thirty-odd years since the events outlined in this chapter, both theoretical and experimental particle physics have changed almost beyond recognition. Studies of extremely high-energy particles, produced naturally by cosmic rays in the upper atmosphere or artificially in particle accelerators, have revealed the existence of more than 200 different elementary particles and thrown into confusion the simple idea of the early 1930s that matter ultimately consists of just a few recognizable units such as protons, neutrons, and electrons. This picture was always a little too good to be true, and the concept of the antineutrino showed that real life was, indeed, more complicated. Later came the discoveries or inventions of mesons and the huge variety of other elementary particles that now make up the "subnuclear zoo."

Today physicists recognize that the link between nucleons actually consists of two separate forces, known as the strong and the weak nuclear forces. The strong force is the strongest known in the universe; the weak force falls between the electromagnetic and gravitational forces in strength.

Elementary particles are classified into two groups, hadrons and leptons, according to whether they react with the strong or weak force. Protons and neutrons are examples of hadrons; electrons and positrons are examples of leptons. Physicists believe that all four forces in nature are characterized by a specific type of particle. In the case of the electromagnetic force,

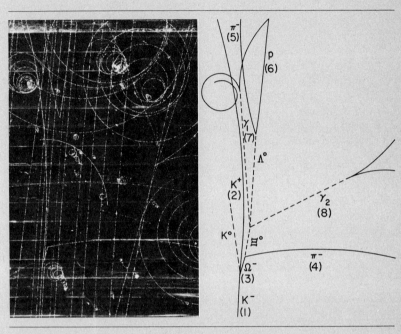

First evidence of the Ω particle. (*Brookhaven National Laboratory*)

this particle is the familiar photon. For the strong force, the mediating particle is known as the meson; a number of different types of mesons are known. The particle associated with gravity has so far defied all efforts to identify it, although one physicist claimed to have some evidence for its existence, as indicated in Chap. 4. Likewise, the particle associated with the weak nuclear force has yet to be discovered, although physicists have enough faith in its existence to have given it a name: the intermediate boson.

What does the extraordinary range of particles in the subnuclear zoo mean to the long-held idea that matter ultimately consists of just a few building blocks? Physicists can provide two different answers to that question. According to one theory, the members of the subnuclear zoo are themselves made up of more fundamental entities; according to another theory, the particles already identified are as fundamental as physicists can possibly obtain.

The first approach has narrowed in recent years to the idea of truly fundamental particles, known as quarks. The concept was first proposed in the early 1960s by Murray Gell Mann of the California Institute of Technology and Israeli physicist Yuval Ne'eman. Gell Mann, a well-

Codiscoverers of the $\Psi = J$ particle. (*Brookhaven National Laboratory; Stanford University*)

rounded personality in cultural terms, called the process the eightfold way, after a Zen Buddhist ideal, and discovered after coining the term "quarks," a phrase from James Joyce's *Ulysses:* "Three quarks for Mr. Mark."

The basic element of the eightfold way is the proposal that all known particles can be gathered together in sets of symmetrical groupings, according to their quantum numbers. The quantum numbers of particles are somewhat more complicated than the four that characterize electrons in atoms; they include figures which indicate the particles' electric charges, masses, and properties describable only in abstruse mathematical terminology, such as the quantum number for "strangeness." The first great triumph of the eightfold way came in 1964, when experimenters at the Brookhaven National Laboratory identified a particle known as the Ω minus, whose properties and very existence had been predicted according to the eightfold way.

Inherent in the mathematics of the theory is the idea that all the known elementary particles consist of just a few quarks. According to the theory, quarks are peculiar in a number of ways, none more so than their apparent possession of electric charges only one- or two-thirds the magnitude of the charge on the electron. If quarks do exist, they are the only particles in creation whose charges are not integral multiples of the electronic charge. Not surprisingly, quarks seem to be very elusive, according to both theory and experiment. The theory indicates that two or more quarks combine in a number of ways to form the familiar—and unfamiliar—particles identified in accelerators and cosmic ray showers. Separating single quarks would, therefore, require large amounts of energy.

A major difficulty with this theory has been the failure of physicists to identify anything resembling a quark, even though they have used more and more powerful accelerators. Some theorists have tried to explain this by figuratively putting quarks behind bars: they suggest that quarks are quintessentially unobservable because they are somehow packaged in impenetrable "bags" and linked to each other by "strings."

A smaller group of theorists takes a more down-to-earth approach to the situation. These physicists argue that the lack of experimental evidence for quarks indicates simply that they do not exist. Instead, these experts believe that there are no particles more fundamental than the ones now known. The basis of this theory, which was first proposed in 1959 by Geoffrey Chew of the Lawrence Berkeley Laboratory and known as either the bootstrap theory or nuclear democracy, is that new particles are actually created artificially. Increasing the energies of particles by accelerating them in atom smashers, Chew argues, provides a means of adding extra mass to them, according to Einstein's mass-energy relationship. Physicists can, therefore, create endless varieties of different particles simply by adjusting the energies of their accelerators. Rather than particles, the bootstrappers say, the real fundamentals in nature are a few physical principles, such as the law of conservation of mass energy and (somewhat incongruously) the Heisenberg uncertainty principle.

The conflict between the proponents of the two theories escalated in late 1974, when teams at Brookhaven and Stanford University independently discovered an entirely new and unanticipated type of particle known as the ψ or J. Those discoveries, and later similar findings in subsequent months, were viewed with glee by particle physicists of every stripe. Team leaders Samuel Ting and Burton Richter received the accolade of the 1976 Nobel Prize in physics. Supporters of quarks believed that the new particles contained quarks marked by a new type of quantum number known as charm. Bootstrappers saw the finds as further examples of entities created artificially. But fresh finds and new developments in theory have increasingly supported the idea of quarks. Order, it seems, is slowly coming to the subnuclear zoo, even if physicists cannot really recognize the fundamental animals in the menagerie.

Radiation

24.1 THE GENERATION OF AN X-RAY BEAM

24.2 RADIATION DETECTORS

SHORT SUBJECT: Measurement of X-Ray Wavelength

24.3 HOW RADIATION AFFECTS MATTER

24.4 PARTICLE RADIATION

24.5 RADIATION DOSIMETRY

24.6 RADIATION IN MEDICINE

SHORT SUBJECT: Nonionizing Radiation: A Long-Term Risk?

Summary
Questions
Problems

SPECIAL TOPIC: The Radiation Around Us

The burning sensation on one's skin after an afternoon in the sun, the scorched rock cavern blasted by a hydrogen bomb test miles beneath the Nevada test site, the simple x-ray photograph of a fractured bone taken in the emergency room of a hospital, and a match-stick model of a complex organic chemical such as insulin all involve radiation. The suntan is produced by ultraviolet light from the sun that manages to penetrate the earth's blanket of ultraviolet-absorbing ozone. The scorching of the rock beneath the test site results from the many forms of penetrating radiation produced in the hydrogen bomb explosion. The x-ray picture and match-stick structure of insulin are created by careful application of controlled amounts of x-rays to an unfortunate accident victim and to crystals of insulin, followed, in the case of the crystals, by extensive computerized analysis of the results.

As these examples show, both artificial and natural radiation play a surprisingly large role in our lives. In this chapter we shall focus on understanding the basis of artificial radiation, particularly insofar as it is applied to the medical world through x- and γ-rays, protons, and neutrons. Since x radiation is easily the most wide-ranging in its applications, we shall concentrate on this form as a means of understanding the nature of helpful and harmful radiation in general.

24.1 THE GENERATION OF AN X-RAY BEAM

Wilhelm Röntgen discovered x-rays in 1895 during studies of low-pressure gas discharges with an evacuated cathode ray tube. The radiation he identified as entirely new was produced when the cathode rays—electrons—struck either the end of the tube or a piece of metal near the end. Much the same type of system is still used today to create x-rays in hospitals and laboratories, as shown in Fig. 24.1. Electrons from a heated filament are accelerated by a large potential difference across an evacuated tube to a target of some appropriate metal, such as tungsten or molybdenum. The impact with the target slows down the

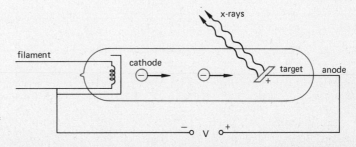

Figure 24.1 X-ray tube. Electrons boiled off from the heated cathode are accelerated by a large potential difference and strike the metal target, which acts as an anode. The decelerating electrons produce a continuous spectrum of x-radiation called bremsstrahlung. In addition, the electrons excite metal atoms in the target, resulting in the emission of characteristic x-rays from the target.

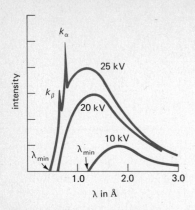

Figure 24.2 X-ray spectrum of molybdenum for three accelerating voltages. On the wavelength axis, λ increases to the right, meaning that energy is decreasing. Note that it requires in excess of 20 kV to remove an electron from the K shell of molybdenum, which is first required before K radiation can be emitted.

electrons. Now, it is a prediction of electromagnetic theory, disobeyed only occasionally in quantum conditions, that accelerating or decelerating charges emit electromagnetic radiation. Recall Bohr's hypothesis that in the case of electrons circling a nucleus, no energy was emitted unless an electron jumped from one orbit to another. However, in systems of such size that quantum effects are not important, classical theory requires radiation of energy from accelerating charges. This is exactly what the electrons do when they decelerate on impact with the metal target. If their energy is large enough—modern x-ray tubes typically utilize accelerating potentials of up to 300 kV—the photons of radiation that they emit have sufficient energy to be termed x-rays. Not all the kinetic energy of the electrons is translated into x radiation, however; a great deal goes into heating up the metal target, which consequently must be cooled with oil or some other fluid during the x-ray production. If it were not cooled, the target would eventually become so hot that it would melt.

It is the strength of the accelerating voltage that determines the ultimate frequency, and hence the energy, of the x-rays produced by an x-ray tube. This is not to say that a certain applied voltage produces x-rays of fixed frequency; any specific voltage actually creates a broad spectrum of x radiation, as shown in Fig. 24.2. The differences produced by different voltages involve the shift of the spectrum toward the short-wavelength (high frequency and high energy) end of the spectrum with increased voltage. Higher energy means that the x-rays have extra penetrating power.

One major characteristic of the x-ray spectra in Fig. 24.2 is that for a given accelerating voltage no radiation emerges with less than a certain wavelength. We can understand the reason for this **cutoff wavelength**, which obviously represents a **maximum amount of energy** in the x-ray spectrum, if we assume that the highest energy x-ray that any particular electron can produce results when the electron gives up all the energy it possesses in the form of a single x-ray photon. This maximum is thus the total kinetic energy of the individual electron, which we can regard as being accelerated from rest through the voltage V that exists across the x-ray tube.

Equating the energy of the photon to this maximum energy of the electron enables us to calculate the maximum energy and minimum wavelength of the x-rays for a particular accelerating voltage V:

$$E_{max} = h\nu_{max} = \tfrac{1}{2}mv^2 = eV \quad \text{or} \quad \frac{hc}{\lambda_{min}} = eV$$

since $c = \lambda\nu$. Hence,

$$\lambda_{min} = \frac{hc}{eV} \tag{24.1}$$

The reason for the wide spread of the spectrum of x-ray wavelengths now becomes obvious. Most of the accelerated electrons do not yield all their kinetic energy in the form of a single photon. Instead, they give up a greater or lesser proportion to the atoms of the target in the form of heat, which increases the temperature of the metal target. The photons of x-ray radiation are then produced by the energy remaining. The overall result is the continuous spread of wavelengths, shown in Fig. 24.2. This continuous spectrum of x radiation is called either the background radiation or, more technically, bremsstrahlung, after the German for "braking radiation."

EXAMPLE 24.1

Determine an expression for the minimum wavelength in angström units of the bremsstrahlung produced by electrons previously accelerated through a potential difference V.

Solution

The maximum kinetic energy available is eV. This energy can be emitted as one photon when an electron undergoes deceleration. Equating $h\nu = eV$ and employing $\lambda = c/\nu$ and the known values of those quantities, we have

$$\lambda = \frac{hc}{eV} = \frac{(6.625 \times 10^{-34} \text{ J} \cdot \text{s})(3 \times 10^8 \text{ m/s})}{1.6 \times 10^{-19} \text{ C} \times V}$$

$$= \frac{1.24 \times 10^{-6} \text{ m}}{V} = \left(\frac{1.24 \times 10^{-6} \text{ m}}{V}\right)\frac{1}{10^{-10} \text{ m/Å}}$$

$$= \frac{12\ 400 \text{ Å}}{V} \tag{24.2}$$

EXAMPLE 24.2

What is the shortest wavelength x-ray that can be emitted in a color TV picture tube when the electrons hit the fluorescent screen if the accelerating voltage is 25 000 V?

Solution

From Example 24.1 we know what the shortest wavelength possible is related to the accelerating voltage by

$$\lambda = \frac{1.24 \times 10^4 \text{ Å}}{V} = \frac{1.24 \times 10^4 \text{ Å}}{2.5 \times 10^4} = 0.496 \text{ Å}$$

Bremsstrahlung does not produce the whole picture as far as the spectrum produced by x-ray tubes is concerned. As Fig. 24.2 shows, some of the x-ray spectra contain noticeable peaks at certain specific wavelengths. The intensities of the x-rays emitted at these wavelengths are pronounced and clearly evident against the background of the bremsstrahlung.

We can explain the emission of these peaks in the x-ray spectrum as the result of what is, in effect, a direct hit by an accelerating electron on an electron in an inner energy level (Bohr orbit) of an atom in the metal target. The impact excites one of the electrons in the atom from a low-energy level into a much higher state of energy. This process leaves a "hole" in the atom's internal electronic structure. An electron in an energy level above the one vacated, including the electron moved upward by the impact, drops down into this hole by emitting energy. If the difference between the original energy level of the falling electron and the state to which it drops is great enough, that energy will take the form of x radiation and superimpose the peaks (K_α and K_β) on the continuous x-ray background in Fig. 24.2.

This process is familiar from our studies of Chap. 21. It is exactly the same procedure as that by which atoms emit visible light; the only difference is the amount of energy involved in the process. The electronic transitions necessary to start off this means of generating x-rays are those from the inner shells of electrons, such as the K, L, and M shells, to the outer valence electron states, or even to the continuum where the electrons effectively free themselves of the influence of individual atoms and diffuse around in a kind of "sea" made up of all the atoms in the metal. Visible light is generated by transitions of outer-shell (valence) electrons.

The origin of the characteristic x-rays is illustrated in Fig. 24.3. Imagine first that an electron is excited from the K ($n = 1$) shell into an unoccupied or unfilled shell anywhere from $n = 4$ to $n = \infty$. A number of x-rays are possible as the atom returns to its normal ground state. For example, an electron in the L ($n = 2$) shell can drop to the hole in

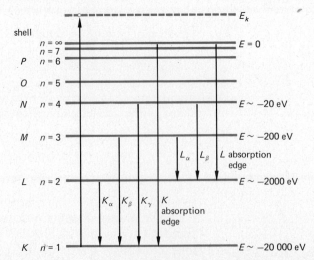

Figure 24.3 Representative x-ray energy-level diagram. An energetic electron collides with a metal atom, exciting an electron from a low-energy level to a vacancy at an upper level or to the continuum. This is indicated by the long, vertical arrow on the extreme left. The excited atom returns to its ground state as electrons fill vacancies at the lower energy levels.

the K ($n = 1$) shell, emitting in the process an x-ray photon called a K_α x-ray. Subsequently, an electron may drop from the M ($n = 3$) shell to the L ($n = 2$) shell, emitting a different energy x-ray photon labeled L_α. Alternatively, the second photon of the cascade of x-rays emitted from a single excited atom might have been the L_β transition from $n = 4$ to $n = 2$. The smaller energy jumps occurring above $n = 4$ are not x-rays but ultraviolet, visible, or infrared (heat) photons, according to the wavelength involved. In the x-ray emission spectrum in Fig. 24.2, the K_α and K_β characteristic lines are shown for molybdenum when the accelerating voltage is above the 20 kV necessary to remove a K electron. A series of K-shell photons are expected, ranging from the K_α ($n = 2$ to $n = 3$) photons up to a maximum energy indicated as the K absorption edge in Fig. 24.3. This edge represents the highest-energy x-ray that could be emitted from an atom when a free electron from the continuum drops into the K ($n = 1$) shell. It is only the binding energy of these inner shell electrons, held by the large electrostatic attraction of the nucleus, that provides the high energy associated with x-rays. Hence heavy atoms such as copper, tungsten, iron, and molybdenum in their metallic state are used as targets in x-ray tubes.

Even before physicists understood the complexities of x-ray production they were able to use observations of x radiation in their basic studies of the chemical elements.

Even before physicists understood the complexities of x-ray production they were able to use observations of x radiation in their basic studies of the chemical elements. In 1913, about when Bohr's model of the hydrogen atom was the talk of the physics world, the young British physicist Henry Gwyn-Jeffreys Moseley discovered that each individual element produced K_α radiation at a different frequency. That observation by itself provided a simple means of identifying elements. But Moseley carried his work further. When he arranged the wavelengths of the K_α lines in the order of the atomic masses of the elements that produced them, he hit upon a striking amount of order—along with some gaps that he correctly attributed to undiscovered elements. In a remarkable piece of perspicacity, Moseley then proposed that "There is in the atom a fundamental quantity, which increases by regular steps as we pass from one element to the next. This quantity can only be the charge on the central positive nucleus."

Moseley did not live to see his forecast proved. He died in 1915 at 27 on the Gallipoli peninsula, a victim of an abortive military campaign of World War I.

Within a few years Moseley's inspired guess was proved absolutely correct. As a result of his work, the metal cobalt, with an atomic mass of 58.94, was given the atomic number—the positive charge on its nucleus—of 27 and placed in the periodic table before nickel, which has a lower atomic mass (58.69) but a larger atomic number (28).

EXAMPLE 24.3

Using Eq. (21.11) for the frequency of the emission lines from hydrogen based on the Bohr theory, determine the frequency and energy in electronvolts of the K_α line for molybdenum ($Z = 42$), a common x-ray target material.

Solution The equation for the frequency of emission as given by Bohr, Eq. (21.11), must be altered since the Bohr model is restricted to a one-electron atom. Considering the K_α line, which is a transition from the $n = 2$ to $n = 1$ energy level, we need only replace Z in Eq. (21.11) by $(Z - 1)$ since the $n = 2$ electrons are attracted to the nucleus effectively by $Z - 1$ positive charges. The remaining electron in the $n = 1$ energy level shields, or screens, the outer electrons from the nucleus. From Eq. (21.11),

$$\nu = \frac{2\pi^2 e^4 m k^2 Z^2}{h^3}\left(\frac{1}{n'^2} - \frac{1}{n^2}\right)$$

For the K_α line in molybdenum, $n' = 1$, $n = 2$, and $Z \to Z - 1 = 42 - 1$

$$\nu = \frac{2\pi^2 e^4 m k^2}{h^3}(42 - 1)^2\left(\frac{1}{1^2} - \frac{1}{2^2}\right)$$

$$\nu = \frac{2\pi^2 e^4 m k^2 (41)^2}{h^3} \cdot \frac{3}{4}$$

$$= \frac{3\pi^2}{2}\frac{(1.6 \times 10^{-19}\text{ C})^4 (9.1 \times 10^{-31}\text{ kg})(9 \times 10^9\text{ N}\cdot\text{m}^2/\text{C}^2)^2}{(6.625 \times 10^{-34}\text{ J}\cdot\text{s})^3}(41)^2$$

$$\nu = 4.13 \times 10^{18}\text{ Hz}$$

This frequency corresponds to an energy in electronvolts of

$$h\nu = (6.625 \times 10^{-34}\text{ J}\cdot\text{s})(4.13 \times 10^{18}\text{ Hz}) \times \frac{1}{1.6 \times 10^{-19}\text{ J/eV}}$$

$$= 17.1 \times 10^3\text{ eV} \quad\text{or}\quad 17.1\text{ keV}$$

The wavelength of this x-ray can be computed from the frequency:

$$\lambda = \frac{c}{\nu} = \frac{3 \times 10^8\text{ m/s}}{4.13 \times 10^{18}\text{ Hz}} = 0.73\text{ Å}$$

These values compare rather favorably with the experimentally observed values of $\lambda = 0.72$ Å and $E = 17.4$ keV.

24.2 RADIATION DETECTORS

Having surveyed the science and technology of x-ray production, our next step is to look at methods of detecting these and other rays. The first methods used by Röntgen to investigate his peculiar discovery were fairly rudimentary but nevertheless effective in a primitive way. Sophisticated versions of these techniques are still used today to investigate the presence of various forms of radiation.

At the beginning of his x-ray studies, Röntgen detected the new rays by looking at the fluorescence they produced in certain chemicals. A number of compounds, among them zinc sulfide and other metal salts, absorb short-wavelength, high-energy x-rays—and also the even shorter wavelength, more energetic γ-rays—and then emit photons in the longer wavelength visible region of the spectrum. In the modern adaptation of this type of detector, a layer of tiny crystals of fluorescent material is placed on a glass plate to produce a fluorescent screen. This screen is an "image converter" that allows the experimenter to "see" x-rays as they strike it.

Röntgen also used photographic plates to detect x-rays, thus pioneering the method used by thousands of hospital radiologists. When x- or γ-rays strike a photographic emulsion, they produce photoelectrons in much the same way that light produces electrons in the photoelectric effect we examined in Chap. 21. These photoelectrons produce a latent image in the emulsion just as if the photoelectrons had come from a source of light. On development, the area of the photographic plate exposed to the x-rays produces a dark image. Today, the number of photoelectrons that each x-ray creates is enhanced by adding atoms of high atomic number to the emulsion of plates used for purposes such as medical x-rays. In addition, an intensifier screen that is essentially a fluorescent material can be placed in contact with the film. This also improves the contrast of the images.

Further experimentation by Röntgen, this time on changes in the conductivity of air when it was irradiated by x-rays—that is, exposed to the radiation—led to yet another method of detecting these peculiar rays. The x-rays ionize the air by occasionally removing an electron from an atom or molecule in it. The amount of ionization, which is directly proportional to the intensity of x radiation, can be measured using the device known as an ionization chamber. This is sketched in Fig. 24.4. The actual design of the construction of an ionization chamber is governed by the type and energy of the radiation that is to be measured with the particular chamber.

The chamber is usually cylindrical in shape, with the cylinder being generally made of metal or occasionally of a plastic that must be coated with graphite or an alternative material to make it electrically conducting. Detection of low-energy x radiation (or charged particle radiation such as α particles) requires a very thin entrance window; this is usually made of nylon, Mylar, or mica sheet as thin as 0.001 mm. The chamber usually contains air at or near atmospheric pressure. The

Figure 24.4 Diagram of an ionization chamber.

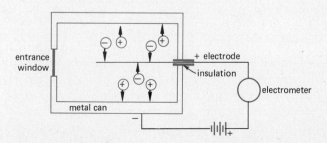

Measurement of X-Ray Wavelengths

To use x-rays in the manner undertaken by Moseley, one must determine the energy or wavelength of the x-ray photons emitted from the target material of the x-ray tube. The data shown in Fig. 24.2 required direct experimental determination of the wavelength of the x-rays emitted when the accelerating voltage for the electron beam of the x-ray tube was altered. How does one measure the wavelength of x-ray electromagnetic waves or x-ray photons? The technique most commonly used involves diffraction and interference of x-rays—a result of the wave nature of this form of electromagnetic radiation. To determine the wavelength of visible electromagnetic waves requires a Young's double slit, or, more commonly, a multiple-slit diffraction grating of the sort described in Chap. 18. In our discussion of the diffraction grating for light waves in Chap. 18 we found that the diffraction of a wave depended upon the ratio of the wavelength of the light to the size of the spacing of the slits or openings through which the light was passing. Since x-ray wavelengths are much shorter than visible light, gratings with smaller spacings were required to produce sufficient diffraction of x-rays. Mechanically ruling gratings suitable for the short x-ray wavelengths is not practical.

In 1912, William L. Bragg devised a technique for diffracting x-rays. Bragg considered how x-rays are scattered by the orderly array of the atoms in a single crystal lattice. Imagine x-rays as waves falling at some angle upon a plane of atoms in a crystal. Each atom scatters x-rays and, according to Huygens' principle, becomes a source of expanding spherical wave fronts. Such wave fronts interfere with wave fronts from other atoms and for the most part combine destructively. However,

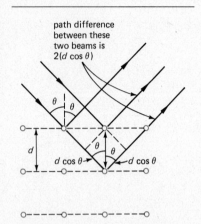

Scattering of x- or γ-rays from atoms in a crystal. Construction for deriving the Bragg diffraction (reflection) relation $n\lambda = 2d \cos \theta$.

if two conditions are met, these scattered wavelets will combine constructively, providing an observable

electrical circuit outside the chamber is used to produce a potential difference between the anode inside the chamber and the metal walls.

When a beam of x-rays ionizes the air, the electrons produced by the effect tend to collect on the anode, and the positive ions move toward the chamber walls. This drift produces a small electric current that can be detected with an ammeter or sensitive electrometer in the circuit. The size of this current gives a good indication of the intensity of the x-rays responsible for ionizing the air in the chamber. These devices have the advantage over fluorescence and photographic detectors of responding quickly and sensitively to x radiation.

The detection device that strikes an instant chord with the general public is the Geiger tube, the Geiger-Müller counter, invented in 1913 by Rutherford's German assistant Hans Geiger. The device, shown in "diffraction pattern." First, all wavelets combine destructively, except those that are scattered at an angle equal to their angle of incidence, as shown in the figure. This condition is exactly the condition for optical reflection—the angle of incidence equals the angle of reflection—which is why Bragg scattering is often called Bragg reflection.

The second condition for constructive interference is that the scattering or reflections from the several layers in the crystal must interfere constructively. That is, the reflected waves from several parallel crystal planes must be in phase with one another. Essentially, the incoming wave-front incident upon a crystal plane at angle θ is reflected from the atoms on the first surface and partially from successive planes in the crystal as the wave penetrates the crystal. The reflected waves interfere as illustrated in the figure. The effective path difference between waves reflected from successive planes separated by lattice spacing d is $2d\cos\theta$ since the waves travel from one plane to the next and upon reflection must travel an equal distance back out of the crystal to the previous crystal layer. For constructive interference in the reflected wave at an angle θ, the path difference between reflected waves from successive planes of atoms must equal an integral number of wavelengths. The Bragg reflection condition becomes

$$n\lambda = 2d\cos\theta \quad (24.3)$$

Note that Eq. (24.3) is often written in terms of the angle the incoming radiation makes with the surface of the crystal, and hence, the sine of that angle replaces $\cos\theta$. By measuring the angles θ for which the x-ray waves produce constructive interference, one can determine the wavelength λ of the electron waves, since the distance d between the planes of atoms in the crystal can be determined, at least for cubic crystals, from the density of the crystal and the knowledge of the atomic masses.

Bragg reflection can take place only in the direction 2θ from the original x-ray direction. To analyze the spectrum of x-ray wavelengths using a crystal, one need only direct a beam of x-rays collimated by passing through a small opening in a piece of lead onto a crystal and, while rotating the crystal, examine the reflected beam at the angle 2θ. Rotating the crystal changes the quantity $2d\cos\theta$ and hence the wavelength of the reflected x-rays at angle 2θ. This technique can be used to measure the wavelength, to analyze the wavelength distribution from an x-ray source, or to select a single wavelength out of a continuous-wavelength source.

Fig. 24.5, is simply a cylindrical metallic gas discharge tube with a wire running down the length of its axis. The wire is maintained at a large positive potential, generally between $+800$ and $+1000$ V, with respect to the walls of the tube. In addition to the wire, the tube contains a low-pressure gas.

Just like the ionization chamber, the Geiger tube relies for its effect on ionization. However, owing to the low pressure of the gas inside the tube, only a few of the x- or γ-rays entering the cylindrical device actually manage to ionize gas atoms. But although the ions that are produced make their way to the tube's outer wall and the electrons proceed to the wire, they collide with other atoms on the way, ionizing those atoms into more electrons and ions. These charged particles in turn start out for the appropriate electrode and slam into yet more

Figure 24.5 The Geiger-Müller tube consists of a thin metal cylinder with a center wire, which acts as an anode. Usually, a thin end window is provided so β rays (electrons) can be detected. γ rays usually will pass through the metal side walls, but electrons (β rays) will not. Unless an especially thin (delicate) window is provided and the source is very close, α particles will not enter the tube.

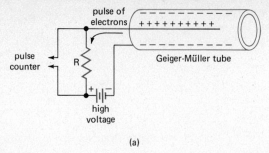

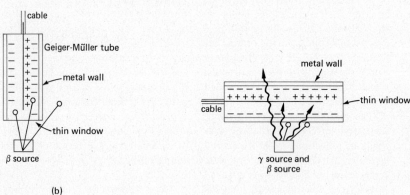

atoms. The overall result is an avalanche of charge through the gas. This pulse of electric discharge can be electrically detected, amplified, and used to activate an audio circuit to produce the characteristic clicking sound, or to drive an electronic counter. By measuring the number of discharges per unit time, the device provides an estimate of the intensity of the x-rays, or whatever other form of radiation is being measured.

The Geiger counter is inexpensive to construct, reliable, and simple to operate. Unfortunately, it has a major shortcoming: after a single discharge of electricity, the device takes an appreciable amount of time to react to the next set of charged particles produced as a result of incoming radiation. This "dead time" or recovery time represents the period during which the ions revert back to neutral atoms after the discharge. Since the time is of the order of milliseconds—thousandths of a second—the effect places a severe limitation on the counting rate of the Geiger instrument.

A preferable alternative, which can detect lower-energy electromagnetic radiation as well as x-rays, is the scintillation counter shown in Fig. 24.6. The radiation strikes a clear crystal, such as sodium iodide, that has been modified by the introduction of a small amount of thallium compounds. As a result of these impurities, the crystal fluoresces when it is exposed to high-energy radiation, producing photons of visible light that enter a photomultiplier tube. Inside the tube, the

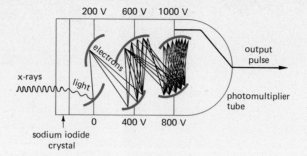

Figure 24.6 Schematic diagram of a scintillation counter.

photons produce electrons by the familiar photoelectric effect; these electrons are accelerated to the next electrode, where they knock out additional electrons in another chain reaction. Eventually the photomultiplier produces an electric pulse that, if the device is calibrated properly, indicates the energy of the incoming radiation. Unlike the Geiger counter, the photomultiplier tube requires a very short dead time. Thus the scintillation counter is extremely useful for measuring large amounts—a large flux—of x radiation and other types of radiant energy.

24.3 HOW RADIATION AFFECTS MATTER

X- and γ-rays react with matter in two basic ways, involving the retention or loss of energy by the radiation. We already discussed examples of the scattering of x- or γ-rays, during which no energy was lost by the radiation. This occurred in the data on Compton scattering, where two scattered wavelengths are observed, one that experiences no change in wavelength or loss in energy and the other (the Compton scattered wavelength) which does experience an energy loss. The former results in the scattering of the x-ray photon by the whole atom and is the type of scattering that occurs at each atom in Bragg reflection. In Bragg reflection, coherent, nonabsorbing scattering of electromagnetic waves by the whole crystal lattice is responsible for this most useful phenomenon. Bragg scattering was used starting about 1915, the year it was discovered, to determine the structures of metals and simple crystalline compounds. Over the past 25 yr the technique has been extremely useful in investigating the atomic structure of such fundamental organic molecules as DNA, proteins, and enzymes.

The other form of interaction between x radiation and matter involves the encounters with individual electrons resulting in Compton scattering or the photoelectric effect or in excitation. This type of interaction is inelastic and involves the transfer of sometimes sizable amounts of energy between the radiation and its target. The predominant reaction of this type is the photoelectric effect, at least in the case of any electromagnetic radiation with energies less than 0.1 MeV. This includes visible, ultraviolet, and a portion of the x-ray region of the electromagnetic spectrum. The atom or metal absorbs an electromagnetic photon through the medium of one of its bound electrons. In excitation by absorption of a photon, the electron is raised to a higher energy level, thus exciting the atom. X-rays of higher energy also produce Compton scattering or lead to the creation of electron-positron pairs, the processes featured in Chaps. 21 and 22.

It is the inelastic scattering processes that account for many of the properties the x-ray pioneers noticed in the new radiation, among them the processes which led to the construction of radiation detectors. But from the start the most profound effect of x radiation was one that involved its relative lack of interaction with any sort of matter—the penetrating power of the radiation.

Röntgen himself discovered that a huge variety of different substances were more or less transparent to x-rays. Black cardboard presented essentially no barrier to the x radiation, nor did normal paper. A fluorescent screen lit up brightly in response to x-rays, even when Röntgen placed a 1000-page book between his discharge tube and the screen. Human flesh also appeared to be largely transparent to the radiation, and Röntgen photographed his hand by x radiation, producing an image in which the bones were clearly distinguishable from the less dense tissue. Thick blocks of board were penetrated equally well, and even a plate of aluminum 15 mm thick did not stop the rays entirely, although it did manage to diminish the intensity of the beam quite substantially.

In later years physicists undertook to measure the penetrating power of x-rays in more quantitative fashion. Their basic objective was to study this penetration as a function of the energy of the radiation in much the same way that one might measure the transmission of visible light through varying thicknesses of colored glass to determine the absorption coefficient of the glass.

The arrangement for a simple absorption measurement for x-rays or other electromagnetic radiation is shown in Fig. 24.7. The x-rays from the source are collimated to obtain a narrow beam of x-rays. A mono-

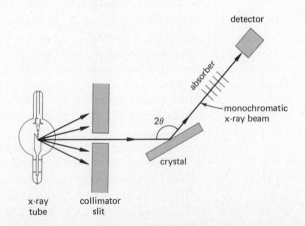

Figure 24.7 Schematic diagram of the apparatus for x-ray absorption experiments. A small opening in a piece of lead serves as a collimator. The x-rays are scattered from a crystal to produce, by means of Bragg reflection, a "monochromatic" beam of x-rays. By rotating the crystal, different wavelength x-rays can be selected.

chromatic (single wavelength) beam is obtained by Bragg reflection from a single crystal. This beam of monoenergetic photons then passes through a varying thickness of absorber into the detector. (To measure the optical absorption in the visible region of the spectrum, the x-ray tube would, of course, be replaced by a light source, and the crystal would be replaced by a prism that would provide a monochromatic light beam.) The result of a typical experiment is shown in Fig. 24.8, which indicates the transmitted intensity plotted as a function of thickness of the absorber. This is an exponential absorption curve very similar to that for radioactive decay. It represents a process whereby a small thickness of absorber ΔX reduces the intensity I by an amount ΔI proportional to both the intensity and the thickness. If I represents the transmitted intensity and I_0 the intensity of the incident beam, this relation can be written

$$\Delta I = (I - I_0) = -\mu I \Delta X \tag{24.4}$$

where μ, the proportionality constant, is called the linear absorption coefficient of the absorbing material. It represents the fraction of energy removed from the beam per unit of length of path in the material. The negative sign in Eq. (24.4) expresses the fact that the transmitted intensity I is less than the initial intensity, I_0. This equation can readily be written in exponential form using calculus, as follows:

$$I = I_0 e^{-\mu X} \tag{24.5}$$

where X represents the total thickness of the absorber. The exponential function describes very well the absorption curve found in Fig. 24.8. The absorption coefficient μ is equal to the fractional decrease in x-ray intensity per unit thickness of absorber. Experiments reveal that the linear absorption coefficient depends not only upon the absorbing material but also on the energy of the x-ray photons. The dependence on energy just reflects the efficiency of the various atomic absorption mechanisms (photoelectric effect, Compton scattering, and electron-positron pair production) in absorbing energy as a function of x-ray

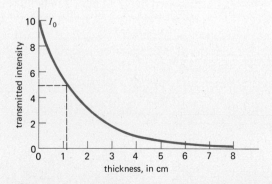

Figure 24.8 The transmitted intensity of ^{60}CO γ radiation absorbed in lead as a function of lead thickness. A layer of 1.2 cm of lead reduces the intensity of γ radiation to 50 percent of I_0, the initial intensity.

energy. It is useful to express the absorption of radiation by a particular material in terms of the thickness of absorber required to reduce the transmitted beam intensity to one-half that of the incident beam. This thickness is named the half-value layer, in analogy with the half-life of radioactive decay processes. The distance that an individual x-ray photon travels in a material before it interacts with an atom or electron cannot be predicted. However, one can predict (and/or determine experimentally) the distance in which a photon has a 50 percent probability of interacting. The half-value layer ($X_{1/2}$) is related to the absorption coefficient, as can be seen from the exponential function that governs the absorption process. If we let $I = I_0/2$ and take the logarithm of the exponential equation, we can obtain the half-value layer in terms of the absorption coefficient:

$$I = I_0 e^{-\mu X}$$

$$\frac{I_0}{2} = I_0 e^{-\mu X_{1/2}}$$

$$\tfrac{1}{2} = e^{-\mu X_{1/2}}$$

$$\ln 2 = \mu X_{1/2}$$

Expressed in logarithms to the base 10,

$$2.30 \log 2 = \mu X_{1/2}$$

$$X_{1/2} = \text{half-value layer} = \frac{0.693}{\mu} \tag{24.6}$$

Table 24.1 gives a few values of the half-value layer for body tissue and lead when exposed to photons of varying energies. The half-value layer clearly increases with the energy of the photons, which means that the absorption coefficient decreases or, alternatively, that the penetrating power of the photon increases with energy, as one would expect. For example, a thickness of 0.131 cm of the body tissue would reduce the beam intensity of a 0.01-MeV x- or γ-ray by one-half. On the other hand, 1-MeV γ-rays would not be reduced in intensity by that proportion until they had penetrated 9.8 cm. The table also indicates that lead, a denser material, is much more effective than tissue at absorbing x- or γ-ray photons.

It is common to rewrite the product μX in the exponential expression as

$$\mu X = \left(\frac{\mu}{\rho}\right) X \rho$$

where ρ is the density of the material. The quantity $\mu_m = \mu/\rho$ is known as the mass absorption coefficient; it is a coefficient that has been normalized to the density of the material.

Table 24.1 Half-Value Layer for γ- or x-rays in Tissue and Lead for Several Energies

Energy (MeV)	Half-Value Layer (cm)	
	Body Tissue	Lead
0.01	0.131	7.6×10^{-4}
0.1	4.05	1.2×10^{-2}
0.5	7.2	0.42
1.0	9.8	0.89
5.0	23.0	1.52

EXAMPLE 24.4

The mass absorption coefficient for aluminum for x-rays with an energy of 40 keV is $\mu_m = 0.06$ m²/kg. Find the linear absorption coefficient for aluminum and the half-value layer. The density of aluminum is 2.7×10^3 kg/m³.

Solution

The mass absorption coefficient and the linear absorption coefficient are related by

$$\mu_m = \frac{\mu}{\rho}$$

where ρ is the density of the material. Therefore,

$$\mu = \rho\mu_m = (2.7 \times 10^3 \text{ kg/m}^3)(0.06 \text{ m}^2/\text{kg}) = 0.16 \times 10^3/\text{m}$$

According to Eq. (24.6), the definition of the half-value,

$$X_{1/2} = \frac{0.693}{0.16 \times 10^3/\text{m}} = 4.3 \times 10^{-3} \text{ m} = 0.43 \text{ cm}$$

24.4 PARTICLE RADIATION

A comparison of the data in Tables 24.1 to 24.3 indicates that x- and γ-rays are much more effective in penetrating matter than are particle radiations of the same energy, such as α particles, electrons, positrons, and ions. On the other hand, neutrons generally have very large ranges in all materials except for those containing large numbers of light atoms, such as hydrogen atoms. All these particles lose energy when they penetrate materials, but they react with matter in noticeably different ways from x- and γ-rays. They therefore deserve a section to themselves.

α particles are obtained from radioactive nuclei that occur naturally or have been created via a nuclear reaction. They are the heaviest of all naturally occurring radioactive emissions and travel through matter

A 5-MeV α particle makes about 166 000 collisions before it comes to rest.

usually in straight lines. When they do interact with atoms, the α particles convert them into ions. The α particles, which are doubly charged helium ions themselves, lose about 30 eV on the average whenever they knock an electron out of an atom. Thus, a 5-MeV α particle makes about 5 MeV/30×10^{-6}, or 166 000, collisions before it comes to rest. This represents a great deal of damage.

The number of collisions is not the only criterion when one considers the amount of damage that any form of radiation is likely to produce. Equally important is the distance that the radiation can travel through any particular material before it comes to a halt. This distance is called the range. Figure 24.9 shows how the range of α particles in air varies with their energy. The value goes from less than 1 cm for particles with 2 MeV of energy to roughly 8 or 9 cm for 9-MeV particles. Obviously, the particles have an inherently short range, for air is an extremely wispy material. Indeed, the penetration of α particles in denser materials such as body tissue or aluminum is no more than a small fraction of a centimeter, as shown in Table 24.2. Thus the particles represent little danger when they are outside the human body. But these radiations can cause considerable internal damage if they are created inside the body, from a radioactive material that has been swallowed. They are then dissipating their entire energy within a small volume of possibly sensitive tissue. The worst case imaginable is the concentration of the α-producing isotope in some organ of the body, such as the liver or brain.

The same general principles that apply to α particles also relate to protons, which, after all, have half the charge and one-fourth the mass of an α particle. β-rays, which are high-energy electrons, travel through materials in a similar manner, ionizing atoms as they move. Because they are much less massive than α particles and protons, however, they are more readily deflected by electrons in the atoms which make up any material that the β-rays penetrate. Thus, β-rays tend to wriggle through materials rather than move in straight lines. And because their mass and charge are both lower than those of α particles, electrons manage to travel somewhat farther than the much slower α particles of the same

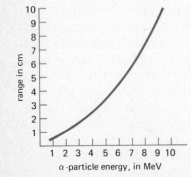

Figure 24.9 Range of α particles in air.

Table 24.2 Range of α Particles in Air, Body Tissue, and Aluminum for Various Energies

Energy (MeV)	Range (cm)		
	Air	Body Tissue	Aluminum
1.0	0.55	0.33×10^{-2}	0.32×10^{-3}
2.0	1.04	0.63×10^{-2}	0.61×10^{-3}
3.0	1.67	1.00×10^{-2}	0.98×10^{-3}
4.0	2.58	1.55×10^{-2}	1.50×10^{-3}
5.0	3.50	2.10×10^{-2}	2.06×10^{-3}

Table 24.3 Range of β Particles in Air, Body, Tissue, and Aluminum for Various Energies

Energy (MeV)	Range (cm)		
	Air	Body Tissue	Aluminum
0.1	12.0	0.0151	0.0043
0.5	150	0.18	0.059
1.0	420	0.50	0.15
2.0	840	1.00	0.34
3.0	1260	1.50	0.56

energy. In fact, the range of β-rays increases rapidly as their energy increases, as illustrated in Table 24.3. The problem of protecting against penetration of electrons is rather more complicated than that of stopping just one type of particle. In addition to ionizing atoms in their paths, high-energy electrons produce a continuous spectrum of x-rays by the bremsstrahlung effect. These x-rays penetrate beyond the range of the electrons that produce them. A thickness of material sufficient to stop the electrons alone is therefore not enough; extra shielding material is essential to stop the x radiation produced in this way.

Neutrons, unlike x- and γ-rays or charged particles, react with matter only when they collide directly with a "particle." They are not charged and do not interact with electrons and protons with the long-range electrostatic force. Even atoms are not dense enough to stop uncharged neutrons. These particles can only collide with nuclei, during which collisions they may knock out protons or the entire nucleus, or even produce a nuclear reaction. Neutrons penetrate very effectively over long ranges and require extremely thick shielding to be stopped. The most efficient method of halting these particles is to "moderate" them, as in nuclear reactors, by forcing them to strike light nuclei. The principles of conservation of energy and momentum suggest that the most effective moderators for neutrons are hydrogen nuclei—protons. Since neutrons and protons have almost identical masses, a neutron that hits a proton is likely to stop in its tracks, whereas the proton moves ahead with the energy that the neutron brought in. Since the proton is positively charged, it is unlikely to travel very far in the material. The process is similar to the billiard-ball collision in Chap. 5. In addition to this type of collision, neutrons of rather low energy are captured by some nuclei such as ^{10}B or ^{6}Li, resulting in a nuclear reaction. This stops the neutron and produces energetic charged nuclei, resulting in large amounts of local damage from the energy release.

24.5 RADIATION DOSIMETRY

Doctors who use radiation in medical treatments and public health officials who are concerned with the protection of the populace from high-energy photons and particle radiations need to know more about the penetration of radiation than the simple ranges. Medical personnel have developed a number of criteria to judge the amount of exposure of their patients to radiation and the likely effects of such exposure.

One criteria is known as the exposure dose, which for x-rays is expressed in röntgens, R. A single röntgen is the quantity of x-rays that produces 1.61×10^{12} pairs of ions in 1 g of air. Actually, **the röntgen is defined as the "quantity of x or γ radiation such that the associated corpuscular emission per 0.001 293 g of air produces, in air, ions carrying 1 electrostatic unit of quantity of charge of either sign."** The mass of 1 cm³ of dry atmospheric air is 0.001 293 g at 0°C and 760 mmHg. Measurements in ionization chambers show that 1 R deposits 87.6 ergs, or 8.76×10^{-6} J, of energy in 1 g of air.

A more critical measure of the medical effects of radiation on people is the so-called radiation absorbed dose (rad), which is measured in terms of the unit known as a rad. **A rad is the amount of radiation that deposits 100 ergs** (10^{-5} J) **of energy on 1 g of tissue, or whatever other substance it impinges on.** This unit applies to all forms of radiation, including particles. It also differs from the exposure unit in that it actually measures the amount of ionization which a photon or other particle of radiation produces. Rads measure only the energy actually absorbed by some material. A γ-ray that passes clear through the human body without affecting any atoms does not contribute to the dose in any way. Because different materials have different absorbing properties, the absorbed dose for a given exposure dose can vary quite markedly. The absorbed dose per röntgen (rad/R) in bone, for example, can be as much as 10 times that in tissue.

Even the absorbed dose of radiation does not give a complete picture of the damage that radiation of different sorts can cause in the human body. For medical experts, the most meaningful unit for measuring radiation damage is that known as the rem, which is an abbreviation for röntgen equivalent man. **One rem is defined as the amount of radiation that causes the same biological effect as 1 rad of 200-kV x-rays.** The value of the rem for any particular form of radiation is obtained by multiplying 1 rad by the quantity known as the relative biological effectiveness (RBE):

$$1 \text{ rem} = 1 \text{ rad} \times \text{RBE}$$

The RBE of any particular type of radiation is the ratio of the dose of x-rays in rads to the dose in rads of the radiation under study that produces exactly the same biologic effect as the x-rays. Thus, the RBE for x-rays of 200 kV is 1 by definition. The RBE for electrons is found by thorough investigation to be similar to that of x- or γ-rays and has a value of 1 RBE. The RBE for neutrons varies from 2 to 10; for protons it varies from 8 and 10; and for α particles it falls in the range 15 to 20. Note that the RBE is variable because it depends on the kind of biological damage, its extent, and the tissue under consideration. α particles are between 15 and 20 times more likely to produce biologic damage than the same absorption dose of x-rays. A dose of 50 rads of α particles is equal to between 15×50 and 20×50, that is, 750 and 1000 rems.

Federal standards currently require that the general population of the country should under no circumstances receive more than 0.50 rem of human-made radiation each year from all sources except medical ones. These sources include nuclear power stations, luminous clocks and watches, and artificial radioactive isotopes. Furthermore, people who are exposed to radiation in their work, such as hospital radiologists and nuclear reactor technicians, must not by law receive more than 5 rem/yr. In the event of catastrophic exposure, experts believe that only half the people who receive 400 rem over a short period of time (a few days) can survive the exposure. Tables 24.4 and 24.5 list the effects of exposure of humans to varying amounts of radiation and the amounts to which we are actually exposed in our daily lives.

Table 24.4 Effects of Doses of Whole-Body γ Radiation

Dose (rad)	Effect
0–15	No observable effect
25–100	Changes in blood
100–200	Moderate blood change; vomiting in 5–50% of cases within 3 h; complete recovery within a few weeks (except for blood-forming system)
200–600	Severe blood changes; vomiting in 50–100% of cases within 3 h; loss of hair within 2 weeks; hemorrhaging and infection; death in 0–80% of cases within 2 months
600–1000	Severe blood changes; vomiting within 1 h; loss of hair; hemorrhaging and infection; death in 80–100% of cases within 2 months

Table 24.5 Estimates of Whole-Body Dose Absorbed by the Average Person in the United States in 1970[a]

Source	Absorbed Dose (rem)
Natural background (sea level)	0.102
Global fallout	0.004
Nuclear power	0.000 003
Diagnostic treatment	0.072
Radio pharmaceuticals	0.001
Occupational sources	0.000 8

[a] From National Academy of Sciences, Report on Radiation Standards, 1972.

24.6 RADIATION IN MEDICINE

The ability of x-rays to produce clear pictures that differentiated between human bone and tissue and between organs of different densities was recognized almost immediately after Röntgen discovered the radiation, and the rays were pressed into medical service well before their nature was understood in even a rudimentary way. Doctors found that the contrasts could be enhanced in some of these situations, and specific organs highlighted in x-ray pictures, if the patient were fed some material with a high absorption coefficient that made its way to a specific organ or region of the body. Barium salts are the best example of this technique. Consumed in ice cream or a liquid suspension, these salts provide high-contrast x-ray photographs of the stomach and intestines.

All barium salts are extremely poisonous to humans. However, those used in x-ray studies, most commonly barium sulfate, are so insoluble in water that not enough of them can possibly dissolve in body fluids to do any harm.

X-rays are not particularly effective at distinguishing tumors from normal tissue, and so the health professionals turn to another technique that relies on radiation. This is the use of so-called radioactive tracers—radioactive elements that can be monitored on their way through the body by the radiation they emit. Studies over the years have shown that certain radioactive pharmaceuticals tend to be selectively absorbed by tumors. Identification of the characteristic radiation from these isotopes as they decay enables radiologists to pinpoint the location of tumors in the body. The number of isotopes that can be used for these studies is quite limited. The isotope should last long enough to allow a search for the tumor, but not so long that it makes the patient radioactive on a long-term basis. A half-life of a matter of minutes is the most reasonable. It is also preferable to choose an isotope that decays by emission of γ radiation alone; such radiation will dissipate a minimum amount of energy into the body to produce ionization effects, and other forms of emission, namely charged particles, could produce considerable internal damage by ionizing atoms in a small region of the body.

Isotopes should last long enough to allow a search for the tumor, but not so long that they make the patient radioactive on a long-term basis.

Body damage from radiation is not necessarily to be avoided. Used carefully, such radiation can actually treat diseases. For example, a radioactive isotope of iodine (^{131}I) is routinely used to treat goiter and other diseases of the thyroid, the gland in the throat. The iodine isotope concentrates in the thyroid. As it decays by β emission, the isotope delivers the energetic β-rays directly to the diseased organ. Other parts of the body at relatively large distances from the thyroid are hardly affected by the radiation.

Other forms of radiation treatment are directed at patients externally. Probably the most important source of this type of treatment is the cobalt isotope ^{60}Co. The isotope changes by β decay to the nickel isotope ^{60}Ni, which in turn emits two penetrating γ-rays with energies of 1.17 MeV and 1.33 MeV. These energies are higher than those obtained from common x-ray machines, and the setup has the advantage of being relatively inexpensive. Furthermore, the half-life of the isotope is so long that it retains its potency over long periods of time.

Nonionizing Radiation: A Long-Term Risk?

Television transmitters, electrical power lines, radar stations, microwave ovens, and CB radios are examples of an unusual and possibly hazardous form of pollution. They all give off radiation that is different from radioactivity in nature but might resemble it insofar as causing harm to humans is concerned. A number of protest groups have demonstrated against what they claim are the health dangers of such "nonionizing radiations"—although most experts refuse to concede that the sources present any imminent risk. "This is not a crisis now," said Janet Healer of the National Telecommunications and Information Administration, a government body. "We need to know more about it, but we are not bathed in seas of dangerous radiation."

Unlike ionizing radiation, the nonionizing form does not cause any obvious changes in cells exposed to it. What it does is heat up tissues. The process is exactly the same as the one that takes place in microwave ovens, but of course at a far less-intensive level of heating. According to the Environmental Protection Agency, more than 98 percent of the United States' population is exposed to less than one-thousandth of the level of nonionizing radiation regarded as the minimum necessary to cause any evidence of damage by heating.

Nevertheless, critics of government scientists point to studies suggesting, among other things, that the number of heart attacks and cancer cases increased in a remote part of Finland after a Soviet radar station went into action nearby; very low levels of nonionizing radiation can cause headaches, irritability, and loss of memory, as witnessed by experiments in eastern Europe; and microwaves can also cause chromosomal changes in animal cells. Testing such claims will undoubtedly take many years. But the United States government and other authorities are looking more critically at the possible hazards of the nonionizing radiation that surrounds us—although in relatively small fluxes.

In action, the cobalt source is kept in a lead container that has a small hole through which the γ radiation can be emitted. The radiologist directs the γ-rays in the direction of the diseased cells, which are usually cancer cells. The radiation starts to kill these cells, but because it is impossible to aim only at the diseased areas, it also destroys healthy cells around them. Thus, the patient can only receive periodic radiation treatments to allow the healthy tissues time to repair between doses. This type of treatment must of necessity represent a careful balance between the amount of radiation necessary to kill the diseased cells and the amount that will not permanently harm the healthy ones.

SUMMARY

1. Beams of x-rays are produced by the impact of electrons on heavy metal targets. The beams consist of continuous spectra of wavelengths above a specific cutoff point and occasional peaks of radiation.

2. X-rays and other forms of radiation can be detected by a variety of devices that rely on fluorescence, ionization, and other physical effects.

3. A major criterion for the effect of radiation on matter is the penetrating power of the radiation. The intensity of a beam of x- or γ-rays in any substance absorbing it varies exponentially with the distance that the beam has traveled through the material.

4. Particles react with matter in a markedly different way from x- and γ-rays. The effect is measured in terms of both the distance that particles travel through any material and the damage which occurs when they collide with atoms.

5. The effects of radiation on the human body are defined in terms of units of measure known as röntgens, rads, and rems.

6. X-rays and certain specific isotopes find application in medical diagnosis. Other isotopes and particular types of radioactivity are applied by physicians to the treatment of disease.

QUESTIONS

1. High-energy x-rays are difficult to detect photographically since they pass through the photographic plate losing little or no energy. How could you improve the image without altering the film?
2. Describe how x-rays are produced experimentally.
3. How are γ-rays produced?
4. In terms of atomic structure, based upon the Bohr model of the atom, explain the mechanism for characteristic x-ray emission.
5. Considering what you have learned about the nucleus and quantum mechanics, where do γ-rays originate?
6. Why are the fluoroscope machines used in shoe stores "to see how well shoes fit" no longer used?
7. When a dentist or the dental technician takes an x-ray of your teeth, they do not stand next to you. Why? Why do they seem to be little concerned about the fact that you are being "irradiated?"
8. Why is it important that pregnant mothers not be exposed to x radiation?
9. What are the basic differences in mechanisms by which neutrons and α particles interact with matter?
10. If you wished to stop neutrons, would you use material made of very light nuclei, medium-weight nuclei, or very massive nuclei like lead? Why?
11. What is bremsstrahlung, and what is its origin?
12. How can measurement of the wavelength and hence energy of the K_α radiation from different atoms yield information on the relative atomic number of these atoms?
13. How does an ionization chamber work?
14. What is the difference between an ionization chamber and a Geiger-Müller detector?
15. Can a Geiger counter be used to detect α, β, and γ radiation equally well? Describe the limitations of the Geiger counter with regard to these radiations.
16. Describe how a scintillation counter works and discuss its advantages over the Geiger-Müller detector.
17. Three processes by which x- or γ-rays interact with matter are the Compton effect, the photoelectric effect, and electron-positron pair production. Distinguish between each of these three processes.
18. Under what conditions would α particles be very dangerous to you?
19. Often we speak of something being radioactively "hot." What is meant by this term, and why do you suppose the word hot was chosen to be used this way?
20. What is meant by "moderating" neutrons?
21. Distinguish between the röntgen, the rem, and the rad.
22. When using radioactive nuclei as tracers in the human body, what are some restrictions with regard to the choice of radioactive isotope?

PROBLEMS

X-Ray Production and Diffraction

1. What is the shortest-wavelength x-ray emitted from an x-ray tube operating at 50 kV?
2. What is the wavelength of the most energetic x-ray produced in an x-ray tube where the accelerating voltage is 300 kV?
3. If an electron beam in a color television tube is accelerated through 30 kV, what will be the wavelength of the highest energy x-rays emitted from the face of the tube when the electrons strike the screen?
4. What wavelength x-rays are diffracted at an angle of 5° with respect to the crystal surface in first order from a crystal with the planes of atoms separated by $d = 2.634$ Å?
5. The spacing between atomic planes in NaCl is 2.820 Å. A monochromatic beam of x-rays has its first-order Bragg reflection peak at 10° with respect to the planes from which the beam is diffracted. What is the wavelength of the radiation, and at what angle will the second-order diffraction pattern occur?
6. A particular cobalt target in an x-ray tube emits the characteristic K series of lines for cobalt and also weak K lines due to two impurities. The wavelengths of the K_α lines are 1.785 Å for cobalt, 2.285 Å for one

impurity, and 1.537 Å for the other impurity. Calculate the atomic number for each element, and identify the impurity elements.

7. Determine the frequency of the K_α line from tungsten.

Half-Value Thickness— Radiation Absorption

8. How many half-value layers are required to reduce an x-ray beam to one-eighth its original value? To one-eightieth its original value?

9. For 0.7-Å x-rays, the linear absorption coefficient of aluminum is 14.3 cm^{-1} and for lead 1593 cm^{-1}. What thickness of each metal is required to reduce the x-rays to one-tenth the incident value?

10. If 4.9 cm of aluminum reduces a monoenergetic γ-ray beam to 40 percent of its incident intensity, what are the half-value thickness and the linear absorption coefficient?

11. Calculate the thickness of lead required to shield a 1-MeV γ-ray source so that no more than 1 percent of the radiation gets through the shield. The absorption coefficient is 0.71 cm^{-1}.

12. X-rays of 0.2-Å wavelength have a mass absorption coefficient of 4.90 cm^2/g for lead. (a) What is the half-value thickness for lead? (b) What thickness is required to reduce the original beam to $\frac{1}{64}$ of its incident value?

13. The half-value thickness for lead is 1.20 cm for ^{60}Co radiation. (a) Compute the reduction in intensity after the radiation has passed through 3.6 cm of lead. (b) Compute the mass absorption coefficient. The specific gravity of lead is 11.5. (c) Write down an expression for the intensity of the radiation as a function of thickness.

14. The mass absorption coefficient for aluminum is 0.61 cm^2/g. Find the (a) linear absorption coefficient of aluminum, and (b) half-value layer and thickness of absorber needed to reduce the intensity of x-rays to $\frac{1}{48}$ of the incident value. The specific gravity of aluminum is 2.7.

Dosimetry

15. A 100-g sample of water receives 500 rad. Calculate the resulting rise in temperature.

16. From the definition of the röntgen, calculate the energy needed to produce a single ion pair in air.

17. What is the mass of 50 mCi (millicuries) of radioactive phosphorus $^{32}_{15}$P with a half-life of 14.3 days? One curie emits 3.70×10^{10} particles per second.

18. How many decays per second come from 1 kg of ^{238}U whose half-life is 4.51×10^9 yr? Express the answer in terms of curies, where 1 Ci is the quantity of radioactive material producing 3.70×10^{10} decays per second.

19. What thickness of lead would be required to shield 1 Ci of ^{60}Co so that the dose rate would be reduced to 0.005 R/h at a distance of 1 m? The half-value thickness for ^{60}Co γ-rays is 1.20 cm of lead. ^{60}Co gives 1.33 R/h at 1 m from a 1-Ci source.

Special Topic
The Radiation Around Us

Shortly after Wilhelm Röntgen discovered x-rays, the American inventor Thomas Alva Edison and his assistant Clarence Dally set out to develop a means of detecting the strange rays using fluoroscopy. The invention worked, but its price was immeasurably high. As a result of the work, Dally developed lesions on his arms, hands, and scalp that became carcinogenic; he died from cancer at a relatively early age.

This fatality was one of many caused directly by scientific investigations into the new world of the atom around the turn of the century. Radiologists who checked the operation of their tubes by examining bones that they held with their left hands in front of a fluorescent screen that they held in their right hands developed malignancies of the left hand with alarming frequency in later life. Women who painted radioactive spots on luminous watches tended to fall victim to lip cancer, for many had the habit of licking their paint brushes, with traces of radium compound on them, before dipping them in the radioactive paint. Certainly, x-rays, radioactivity, and similar forms of radiation opened up by the explorers of the atom do not differ qualitatively from other, more normal types; ultraviolet rays, after all, can cause nonfatal skin cancer and the fatal variety of the disease known as melanoma. However, these more recently discovered forms of radiation do represent a risk of a much greater order of magnitude. No one can afford to relax around machines that produce x-rays and similar types of radiation, for even if a victim does not suffer immediate illness as a result of excessive exposure, he or she can never be sure that the radiation is not producing a long-term, delayed internal effect like that of a time bomb.

The archetypal example of the dangers of radiation are the Japanese children who were in their mothers' wombs and exposed to the searing power of the atom bombs dropped on Hiroshima and Nagasaki. A study of 450 000 such children by the Harvard School of Public Health indicated that they had up to a 30 percent greater risk of developing cancer—and particularly leukemia and cancer of the central nervous system—than children not exposed to the radiation as fetuses. For some children the effects of the bomb were much more immediate; they were born with abnormally small heads and extremely low intelligence quotients.

The enormity of an atom bomb explosion is an exceptional type of exposure, but all indications are that rather more mundane forms of radiation can have serious long-term effects. For a few years in the late 1940s, Michael Reese Hospital in Chicago treated tonsillitis in children with x radiation rather than the more conventional surgery. After the project, which was discontinued in 1949, nothing more was heard of it until 1974. Then, the hospital started to call in patients who had been treated with x-rays for medical checkups. Over the years, a surprisingly large proportion of the patients had developed tumors of the thyroid, the gland in the throat. The "patient recall" represented an effort by the hospital to catch the potential tumors in the early stages, at which point the tumors were most likely to respond to treatment.

Some radiation incidents have had a touch of the bizarre. In early 1976, United States officials protested to the Russians that Soviet eavesdroppers were blanketing the top floors of the United States embassy in Moscow with microwaves that could be physically harming embassy workers and residents. Microwaves are used routinely in radars, long-distance telephone transmissions, and microwave ovens (in which they cook food by causing its atoms to vibrate rapidly and thus heat up by friction), and the effects of exposure to large intensities are well documented. Radar beams can cook human cells as surely as they cook food, and animal studies suggest that high intensities alter the biologic operation of the brain. But medical personnel know very little about the effects of long exposure to low intensities of the radiation. The possibilities range from sleeplessness to sterility in men, and just possibly to alterations in the brain's operation. If nothing else, the incident illustrated the chronic lack of knowledge that still attends the field of radiation medicine.

What sort of radiation risk is run by someone who does not work in a sensitive embassy, a laboratory studying radioactivity, or a hospital radiology unit? No one can avoid some effects of radiation, for the background from cosmic rays, luminous watches, and other sources amounts to about 120 mrems/yr. To that dose one might add about five more millirems per year for people who spend their lives near nuclear power plants (and possibly 5000 mrems/yr for workers inside the plants), a dose of anywhere between 20 and 400 mrems for each chest x-ray, and about 400 to 6400 mrems for a dental x-ray that involves four separate exposures. A color television emits up to 13 mrems/h of operation, and microwave ovens tend to leak a small but measurable amount of radiation. However, the overall exposure to artificial forms of radiation appears to be rather less than that caused by the general, natural background. Nevertheless, doctors, dentists, nuclear power experts, and other specialists in fields that involve radiation exposure are becoming more and more circumspect regarding the possible dangers which their technologies hold in store for the public in coming years.

Epilogue

One major theme of the 24 chapters of this book has been the universal nature of physics. Just like other sciences, physics cannot be treated as a subject in isolation. Physics both nurtures and feeds on other areas of science and technology. Archeological dating using the carbon 14 isotope, elucidation of the structure of DNA and other complex molecules that are fundamental to life, and an understanding of the nature of the chemical bond between atoms are typical examples of the contributions of physicists and the concepts of physics to other branches of research. On the other hand, the development of physics over the years has benefited from inputs from other sciences. Brownian motion, which gave the first concrete indication of the existence of molecules, was discovered by a botanist. The periodic table of the elements, which stimulated such fundamental advances as the Pauli exclusion principle, was the work of a chemist as, indeed, was the initial conception of nuclear fission. And physicists' understanding of the nature of magnetism has been greatly enhanced by the field work of geologists.

Today, the interdisciplinary nature of science in general seems more important than ever. Certainly, many fundamental advances have been made by researchers out of their natural element—scientists who have turned their attention to specialties different from their own and who in doing so have brought fresh approaches and asked different questions. Undoubtedly, overfamiliarity with a subject can be as restricting as the lack of adequately deep knowledge.

Another important theme has been the relationship between science and technology—between development work, which is aimed at specific applications and products on the one hand, and pure research, pursued for no purpose beyond expanding the breadth of our knowledge on the other hand. Just as various fields of science provide the necessary stimuli for each other, so do research and development advance in tandem. The results of fundamental research often suggest new technological means of improving our lot (or occasionally of worsening it, in the form of new weapons), and the search for specific types of technology often stimulates fresh understanding of scientific basics. Fortunately perhaps, no scientist or policymaker has yet discovered how to identify in the basic research laboratory those projects that are guaranteed to produce technological applications. What is clear is that an understanding of the methods and principles of physics equips any scientist with a valuable tool with which to pursue her or his career.

Appendixes

A MATHEMATICS
B CONVERSIONS
C PHYSICAL CONSTANTS
D TABLES OF USEFUL DATE
E NATURAL TRIGONOMETRIC FUNCTIONS
F TABLE OF ELEMENTS, ABUNDANT ISOTOPES

A Mathematics

The mathematical background necessary for your understanding of discussions in most chapters of this book is summarized in this appendix. In addition, specific mathematics reviews have been inserted in specific chapters where the material has a particularly important use.

The mathematical inserts within the chapters include the following:

Powers of 10 Notation, Chap. 1	*page 4*
Vector Addition and Subtraction, Chap. 2	*page 17*
Trigonometry Review, Chap. 2	*page 19*
Slope, Chap. 2	*page 30*
The Dot Product, Chap. 5	*page 129*
Vector Product, Chap. 7	*page 182*
Simultaneous Equations, Chap. 14	*page 439*

ARITHMETIC It is assumed that you are familiar with the four fundamental operations of arithmetic—addition, subtraction, multiplication, and division—and understand the meaning of the common terms used therewith.

The equal sign ($=$) indicates an equality. For example, $2 = 2$.

The inequality sign ($\neq$) indicates an inequality and is read "is not equal to." For example, $3 \neq 2$.

The approximately equal sign is $\cong$. For example, $2.1 \cong 2.0$ is read "2.1 is approximately equal to 2.0."

The sign $>$ is read "is greater than." For example, $3 > 2$.

The sign $<$ is read "is less than." For example, $2 < 3$.

The sign $\geq$ is read "greater than or equal to."

Multiplication is represented by the multiplication sign. The multiplication sign is $\times$, brackets ( ), or $\cdot$. Thus, 3 times $2 = 3 \times 2 = (3)(2) = 3 \cdot 2 = 6$. In multiplication, the answer is called the "product."

When letters are used in place of numerals, the product may be indicated without the multiplication sign. For example, A multiplied by B, instead of being written $A \times B$, may be expressed as AB. Also, 4 times B may be written as $4B$.

In multiplication, 5×0 means that you take 5 zeros and add them; 0×5 means that you take none of the 5s and add them. The product of multiplying anything by zero is zero.

The division sign ($\div$), a horizontal line, or a slant line (/) indicates a fraction.

Fractions A fraction is a number that may be considered as an indicated division. The *numerator* (on top) is divided by the *denominator* (on bottom). Thus, we can write

$$\frac{1}{4}, \frac{3}{8}, \frac{1}{2}, \frac{3}{4}, \frac{11}{8}$$

The *dividend* is the numerator of a fraction, and the *divisor* is the denominator.

An integer, or whole number, may be considered as a fraction with 1 as the denominator. Thus,

$$4 = \frac{4}{1} \qquad 333 = \frac{333}{1}$$

In a *proper* fraction, the numerator is smaller than the denominator, and the value of the fraction is less than 1. Thus,

$\frac{3}{4}$ is less than 1   or $\frac{3}{4} < 1$

In an *improper* fraction, the numerator is larger than the denominator. Thus, $\frac{11}{8}$ is larger than 1, or

$$\frac{11}{8} > 1$$

An improper fraction may be expressed as a mixed number (whole number and fraction), and vice versa. Thus,

$$\frac{11}{8} = 1\frac{3}{8} \qquad 8\frac{1}{2} = \frac{17}{2}$$

In fractions, $\frac{0}{2}$ "asks" how many 2s fit into zero? The answer is, none. Also, $\frac{2}{0}$ "asks" how many zeros fit into 2? The answer is, an infinitely large number. The symbol for an infinitely large number is the one that is used for infinity (∞).

Reciprocals The reciprocal of a fraction is the inverse of the fraction; that is, it is the fraction turned upside down. The reciprocal of $\frac{3}{4}$ is $\frac{4}{3}$; the reciprocal of 8 is the inverse of $\frac{8}{1}$, which is $\frac{1}{8}$.

Division is the inverse of multiplication. Thus, dividing by a fraction is the same as multiplying by the reciprocal of the fraction. For example,

$$36 \div \frac{4}{3} = 36 \times \frac{3}{4} = 27$$

Similarly,

$$36 \times \frac{1}{4} = 36 \div 4 = \frac{36}{4}$$

The value of the fraction is unchanged if both the numerator and denominator are multiplied by the same number (not including zero as a number). For example,

$$\frac{2}{3} = \frac{2}{3} \times \frac{4}{4} = \frac{8}{12}$$

Similarly, the value of a fraction is unchanged if both numerator and denominator are divided by the same number. For example,

$$\tfrac{8}{12} = \frac{\tfrac{8}{4}}{\tfrac{12}{4}} = \tfrac{2}{3}$$

The familiar "canceling" is an application of this rule.

Warning: adding (or subtracting) the same number to the numerator and to the denominator changes the value of the fraction. For example,

$$\frac{2}{3} \neq \frac{2+4}{3+4} = \frac{6}{7} \quad \text{that is} \quad \tfrac{2}{3} \text{ is not equal to } \tfrac{6}{7}$$

In handling "compound fractions," in which the numerator (or the denominator, or both) is a fraction, a rule is "to divide by a fraction, invert, and multiply." Thus,

$$\frac{\tfrac{A}{B}}{\tfrac{C}{D}} = \frac{A}{B} \div \frac{C}{D} = \frac{A}{B} \times \frac{D}{C} = \frac{AD}{BC}$$

For example,

$$\frac{\tfrac{2}{3}}{\tfrac{4}{5}} = \tfrac{2}{3} \div \tfrac{4}{5} = \tfrac{2}{3} \times \tfrac{5}{4} = \tfrac{10}{12}$$

Also, in handling "compound fractions," a rule is: "To divide a fraction by a fraction, divide the product of the extremes by the product of the means." Thus,

$$\frac{\tfrac{A}{B}}{\tfrac{C}{D}} = \frac{A \times D}{B \times C} = \frac{AD}{BC}$$

The "extremes" are A and D; the "means" are B and C.

As an example of the latter rule for handling "compound fractions" (using the same numerical fractions as in the first rule),

$$\frac{\tfrac{2}{3}}{\tfrac{4}{5}} = \frac{2 \times 5}{3 \times 4} = \frac{10}{12}$$

When working with "compound fractions" it is better not to use any slanting lines to indicate division; use only horizontal fraction lines, with the main fraction line longer. This helps avoid confusion.

Any quantity is not changed in value when it is multiplied or divided by 1. Thus,

$$\frac{A}{\tfrac{B}{C}} = \frac{\tfrac{A}{1}}{\tfrac{B}{C}} = \frac{AC}{1B} = \frac{AC}{B}$$

For example,

$$\frac{2}{\frac{3}{4}} = \frac{\frac{2}{1}}{\frac{3}{4}} = \frac{2 \times 4}{1 \times 3} = \frac{2 \times 4}{3} = \frac{8}{3}$$

or

$$2 \div \frac{3}{4} = \frac{2}{1} \times \frac{4}{3} = \frac{2 \times 4}{1 \times 3} = \frac{8}{3}$$

Also

$$\frac{\frac{A}{B}}{C} = \frac{\frac{A}{B}}{\frac{C}{1}} = \frac{A \times 1}{B \times C} = \frac{A}{BC}$$

For example,

$$\frac{\frac{2}{3}}{4} = \frac{\frac{2}{3}}{\frac{4}{1}} = \frac{2 \times 1}{3 \times 4} = \frac{2}{12} = \frac{1}{6}$$

or

$$\frac{2}{3} \div 4 = \frac{2}{3} \div \frac{4}{1} = \frac{2}{3} \times \frac{1}{4} = \frac{2 \times 1}{3 \times 4} = \frac{2}{12} = \frac{1}{6}$$

Percentages Percentage is the fractional part of a number expressed in "parts per hundred." Thus, $\frac{6}{10}$ of a group means, 6 out of 10, or 60 out of 100, or 60 percent. Similarly, 40 percent of 20 means

$$\frac{40}{100} \times 20 = 0.40 \times 20 = 8$$

Decimals Decimals are special fractions in which the denominator is some power of 10. Thus,

$$0.1 = \frac{1}{10} \qquad 0.03 = \frac{3}{100} \quad \text{or} \quad \frac{3}{10^2}$$

$$438.2 = 438 + \frac{2}{10} \qquad 35.48 = 35 + \frac{48}{100}$$

Multiplying a number by 10 is equivalent to moving the decimal point one place to the right; multiplying by 100, two places to the right; by 1000, three places to the right; and so on. For example,

$$4.538 \times 10 = 45.38$$
$$4.538 \times 100 = 453.8$$
$$4.538 \times 1000 = 4538$$

Dividing a number by 10 is equivalent to moving the decimal point one place to the left; dividing by 100, two places to the left; by 1000, three places to the left; and so forth. For example,

$$4538. \div 10 = 453.8$$

$$4538. \div 100 = 45.38$$
$$4538. \div 1000 = 4.538$$

Powers or Exponential Notation

When a number appears as a factor more than one time, the multiplication can be expressed efficiently in *exponential* form, or *power* form. The *base* is the number that appears as the factor, and the *exponent* (or the power) is the number of times the base is multiplied by itself. Thus,

$$A \times A = A^2$$

For example,

$$3 \times 3 = 3^2 = 9$$

Conversely, A^2 ("A squared," or "A to the second power") means $A \times A$.

Similarly, "A cubed" (A^3, or A to the third power) means "A times A times A"; thus,

$$A^3 = A \times A \times A$$

For example,

$$3^3 = 3 \times 3 \times 3 = 27$$

As another example,

$$2 \times 2 \times 2 \times 2 \times 2 = 2^5 = 32$$

where 2 is the base and 5 is the exponent (or power). 2^5 is read "two to the fifth power." Similarly,

$$10 \times 10 \times 10 = 10^3 = 1000$$

Conversely, 5^4 means

$$5 \times 5 \times 5 \times 5 = 625$$

Multiplication of Exponentials

The definition of exponents permits the efficient multiplication of quantities raised to some exponent. Because

$$5^4 = 5 \times 5 \times 5 \times 5 \quad \text{and} \quad 5^3 = 5 \times 5 \times 5$$

it follows that

$$5^4 \times 5^3 = (5 \times 5 \times 5 \times 5) \times (5 \times 5 \times 5) = 5^7, \text{ or } 78\,125$$

The same answer results by adding the exponents:

$$5^4 \times 5^3 = 5^{4+3} = 5^7, \text{ or } 78\,125$$

This leads to the following *multiplication rule*:

To multiply numbers expressed as powers of the same base, add the exponents.

In general,

$$A^m \times A^n = A^{m+n}$$

Division of Exponentials Similarly, this operation may be carried out as follows:

$$\frac{5^5}{5^3} = \frac{5 \times 5 \times 5 \times 5 \times 5}{5 \times 5 \times 5} = 5 \times 5 = 5^2 = 25$$

This leads to the following *division rule:*

To divide numbers expressed as powers of the same base, one should subtract the exponent of the denominator from the exponent of the numerator.

In general,

$$\frac{A^m}{A^n} = A^{m-n}$$

For example,

$$\frac{5^5}{5^3} = 5^{5-3} = 5^2 = 25$$

Is it true that $5^0 = 1$? Because

$$\frac{5^5}{5^5} = 1$$

then

$$\frac{5^5}{5^5} = 5^{5-5} = 5^0 = 1$$

Negative Exponents The concept of exponents has been extended to negative powers. Applying the rule for the division 5^3 divided by $5^5 = 5^{-2}$; as a result,

$$\frac{5^3}{5^5} = 5^{3-5} = 5^{-2}$$

In terms of the original definition of exponents, 5^{-2} is meaningless. However, by carrying out the division in the extended form,

$$\frac{5^3}{5^5} = \frac{5 \times 5 \times 5}{5 \times 5 \times 5 \times 5 \times 5} = \frac{1}{5 \times 5} = \frac{1}{5^2}$$

Therefore,

$$5^{-2} = \frac{1}{5^2}$$

Similarly,

$$10^{-2} = \frac{1}{10^2} = \frac{1}{100}$$

$$\frac{1}{10^3} = 10^{-3} = \frac{1}{1000}$$

and

$$\frac{1}{10^{-6}} = 10^6$$

Any factor may be moved from the numerator to the denominator (or vice versa) by changing the sign of the exponent. Thus, a positive exponent in the denominator becomes negative in the numerator, and vice versa.

Raising to a Power To raise a number already in power form to another power, multiply the exponents. Thus,

$$(2^5)^3 = 2^5 \times 2^5 \times 2^5 = 2^{5 \times 3} = 2^{15}$$

Roots The opposite of raising a quantity to a given power is called "extraction of a given root." The root symbol is $\sqrt{}$ (called the "radical sign"). The desired root is indicated by a numeral as follows, $\sqrt[3]{}$. If no root number is used, the number 2 is understood. Thus, $\sqrt{A}$ is read "the square root of A"; $\sqrt[3]{A}$ is read "the cube root of A."

The expression $\sqrt{25}$ (read "the square root of 25") means to find the number that multiplied by itself will give 25. The number is obviously 5 because $5^2 = 25$. Similarly, the expression $\sqrt[3]{64}$ (read "the cube root of 64") means to find a number that, taken three times and multiplied, will give 64. The number is obviously 4 because $4^3 = 4 \times 4 \times 4 = 64$. Roots may also be indicated by placing a fraction as an exponent; for example,

$$A^{1/2} = \sqrt{A} \quad \text{and} \quad A^{1/3} = \sqrt[3]{A}$$

APPROXIMATIONS Much mathematical reasoning can be carried out with approximate numerical values. The scientist often rounds off numbers, uses rough estimates of numbers, and makes estimates of complicated mathematical expressions. The ease with which the answers are obtained compensates for the loss of exactness. The use of round numbers permits placing the emphasis on the reasoning involved.

Accuracy of Data All scientific data derived from measurement have a limited accuracy. Certain errors are inherent in the method of measurement. The accuracy is indicated by the number of significant figures. Thus, the statement that the equatorial diameter of the earth is 7926.68 mi indicates that it is known accurately through five significant figures (with some uncertainty in the sixth figure). The accuracy of most measurements does not exceed three or four significant figures. For most purposes, however, accuracy of even four significant figures is not necessary, and consequently the numbers are rounded off.

Rounding Off To round off a number, select the significant figure to which the accuracy is desired. Then, (1) if the next digit is less than 5, replace all succeeding digits by zeros and drop the decimals; (2) if the next digit is 5 or greater, increase the significant figure by 1 and replace the succeeding digits by zeros.

Note that by using scientific notation you can better indicate the significance of zeros. For example, a value of 700 is ambiguous since the zeros may or may not be significant. To indicate that they are significant, we could write 7.00×10^2. To indicate that they are not significant, we could use 7×10^2.

The method is illustrated in the following successive steps in rounding off the equatorial diameter of the earth formerly used:

7926.7 to five significant figures
7927   to four significant figures
7930   to three significant figures
7900   to two significant figures
8000   to one significant figure

The latter two numbers might be more clearly written as 7.9×10^3 and 8×10^3.

Approximation Calculations The rounding off of numbers and the methods of notation of numbers in science make possible the estimation of answers quickly. The following two examples illustrate the method.

EXAMPLE A.1 Assume you have to determine the ratio of the diameter of the earth to that of the moon. Given the diameter of the earth = 12 760 km; the diameter of the moon = 3480 km. The desired answer requires the division of 12 760 by 3480. By long division, the answer is 3.67.

The scientist usually proceeds as follows: The diameter of the earth is rounded off to 13×10^3 km, that of the moon, to 3.5×10^3 km and the ratio is quickly seen to be about 4, actually 3.7. The answer is sufficiently accurate for most purposes and permits a meaningful comparison of the relative size of the two bodies.

EXAMPLE A.2 How much time is required for a microwave signal (electromagnetic wave) to go from a radio telescope on the earth to the moon and be reflected back to the earth, given the speed of light and other electromagnetic waves is 186 324 mi/s? The average distance of moon from earth is 239 000 mi. Approximations frequently used are 186 000 and 240 000:

$$\text{Speed of light} = 186\,000 \text{ mi/s}$$
$$= 1.9 \times 10^5 \text{ mi/s}$$
$$\text{Distance of moon} = 240\,000 \text{ mi}$$
$$= 2.4 \times 10^5 \text{ mi}$$
$$\text{Time for round trip} = \frac{2 \times (2.4 \times 10^5)}{1.9 \times 10^5}$$
$$= \frac{4.8 \times 10^5}{1.9 \times 10^5} = 2.5 \text{ s}$$

Order of Magnitude Accuracy to one significant figure is sometimes sufficient for rough estimates. Numbers indicating only order of magnitude are indicated by the symbol $\sim$, which may be read "is the order of." The symbol $\sim$ is also frequently used and read "approximately."

For example, in comparing the diameter of the sun (864 000 mi) to that of the earth ($\sim$8000 mi) to obtain the relative size of sun and earth, a rough approximation is made; thus,

$$\frac{\sim 900\,000 \text{ mi}}{\sim 8000 \text{ mi}} \text{ is } \sim 100$$

The sun is thus approximately 100 times as large as the earth, or the sun is 2 orders of magnitude larger than the earth.

ALGEBRA Algebra is that branch of mathematics that treats the relations and properties of quantities by means of letters and symbols instead of numbers. Algebra is essentially generalized arithmetic. The symbolism facilitates performing the operations and permits placing the emphasis on relations.

Symbolism Any letter can be used as the symbol of any number, known or unknown. Known quantities are usually represented by letters from the beginning of the alphabet (usually a, b, c, d), and unknown quantities, by letters from the end of the alphabet (usually x, y, z). Quantities may be positive ($+$) or negative ($-$).

To suggest the meaning of the symbol, the letter used is often the first letter of the word for the quantity symbolized. For example, p may stand for *p*ressure; t for *t*ime; f for *f*orce. In any discussion, the meanings of the symbols should be stated clearly.

Subscripts The use of subscripts on symbols is acceptable practice. This helps avoid the use of many different symbols in a discussion and helps differentiate among the symbols. D_E may stand for the diameter of the planet earth; D_S, the diameter of the sun. Similarly, the velocity at the end of 1 s may be indicated by v_1; at the end of the second second, by v_2. Also, the original volume, or volume in the first case, may be indicated by V_0 or V_1; the volume in the second case, by V_2.

Equations The equation is the central part of algebra. An equation is a mathematical expression of equality and is established by placing on each side of the "equals" sign quantities that are known or assumed to be equal. The left-hand member is equal to the right-hand member.

Solving Equations Once an equation is established, it may be subject to certain manipulations without destroying the equality beween its left- and right-hand members. Equality is not destroyed by any operation performed on one side of an equation, provided that the same or equivalent operation is performed on the other side.

There are some common manipulations that will not destroy the equation. Such manipulations may be expressed by the following rules and axioms.

1. If equals are added to equals, the sums are equal.
Examples:

If $A = B$, then $A + C = B + C$.
If $X = Y$, then $X + 8 = Y + 8$.
As $10 = 10$, then $10 + 8 = 10 + 8$.

2. If equals are subtracted from equals, the remainders are equal.
Examples:

If $A = B$, then $A - C = B - C$.
If $X = Y$, then $X - 8 = Y - 8$.
As $10 = 10$, then $10 - 8 = 10 - 8$.

3. If equals are multiplied by equals, the products are equal.
Examples:

If $A = B$, then $AC = BC$.
If $X = Y$, then $8X = 8Y$.
As $10 = 10$, then $8 \times 10 = 8 \times 10$.

4. If equals are divided by equals, the quotients are equal.
Examples:

If $A = B$, then $\dfrac{A}{C} = \dfrac{B}{C}$.

If $X = Y$, then $\dfrac{X}{8} = \dfrac{Y}{8}$.

As $10 = 10$, then $\dfrac{10}{2} = \dfrac{10}{2}$.

5. Like powers of equals are equal.
Examples:

If $A = B$, then $A^2 = B^2$.
If $X = Y$, then $X^2 = Y^2$.
As $4 = 4$, then $4^2 = 4^2$, or $16 = 16$.
If $C = 3 \times 10^{10}$, then $C^2 = 3^2 \times (10^{10})^2$, or $C^2 = 9 \times 10^{20}$.

6. Like roots of equals are equal.
Examples (extracting the square roots of both sides):

If $A^2 = 25 \times 10^{10}$
$\sqrt{A^2} = \sqrt{25 \times 10^{10}}$
$A^{2/2} = \sqrt{25} \times \sqrt{10^{10}}$
$A = 5 \times 10^5$
Also, $X_1^2 = 16 X_2^2$
$\sqrt{X_1^2} = \sqrt{16 X_2^2}$
$X_1 = 4 X_2$

In summary, the basic rule for manipulating equations is that an equation remains an equation (an equality) if the same operation is performed on both the left-hand member and the right-hand member. In the foregoing equations, the quantities such as A, B, and C may be complicated quantities that are merely symbolized by those letters.

Solving an equation means getting the unknown quantity by itself on one side of the equality sign. The most common types of equations to be solved in this book are variations of the following:

$$\dfrac{A}{B} = \dfrac{C}{D}$$

We shall now give a very simple mechanical rule for solving this equation, and will then illustrate it and explain why it works.

Rule: Any quantity in the equation

$$\dfrac{A}{B} = \dfrac{C}{D}$$

may be moved, provided it is moves across the equality sign *and* across the fraction line; that is, the equation remains an equation under all the following manipulations:

7. Moving B, leaving 1 in its place,

$$\frac{A}{B} = \frac{C}{D} \qquad \frac{A}{1} = \frac{BC}{D}$$

8. Moving A,

$$\frac{A}{B} = \frac{C}{D} \qquad \frac{1}{B} = \frac{C}{AD}$$

9. Moving C,

$$\frac{A}{B} = \frac{C}{D} \qquad \frac{A}{BC} = \frac{1}{D}$$

10. Moving D,

$$\frac{A}{B} = \frac{C}{D} \qquad \frac{AD}{B} = \frac{C}{1}$$

To understand why the foregoing manipulations are valid, note that

11. Rule 7 is equivalent to multiplying both members by B:

$$\frac{A}{B} = \frac{C}{D} \qquad \frac{AB}{B} = \frac{CB}{D} \qquad \frac{A\cancel{B}}{\cancel{B}} = \frac{CB}{D} \qquad \frac{A}{1} = \frac{CB}{D}$$

Similarly, rule 8 is equivalent to dividing both members by A; rule 9 is equivalent to dividing both members by C; rule 10 is equivalent to multiplying both members by D. You should prove these statements by performing the indicated operations.

If one of these quantities is regarded as unknown, we can solve for it by getting it alone in the *numerator* on one side of the equation:

12. Solving for A,

$$\frac{A}{B} = \frac{C}{D} \qquad \frac{A}{1} = \frac{BC}{D} \qquad A = \frac{BC}{D}$$

13. Solving for B,

$$\frac{A}{B} = \frac{C}{D} \qquad \frac{A}{1} = \frac{BC}{D} \qquad \frac{AD}{1} = \frac{BC}{1} \qquad \frac{AD}{C} = \frac{B}{1} \qquad B = \frac{AD}{C}$$

14. Solving for C,

$$\frac{A}{B} = \frac{C}{D} \qquad \frac{AD}{B} = \frac{C}{1} \qquad C = \frac{AD}{B}$$

15. Solving for D,

$$\frac{A}{B} = \frac{C}{D} \qquad \frac{AD}{B} = \frac{C}{1} \qquad \frac{D}{B} = \frac{C}{A} \qquad \frac{D}{1} = \frac{BC}{A} \qquad D = \frac{BC}{A}$$

You should learn to solve any of these types of equations (12 to 15) very quickly in your head by inspection, writing down only the answer.

Some other operations commonly employed in rearranging the terms of an equation to facilitate solution are in reality simplified applications of axioms 1 to 4.

16. Transposing a term from one side to the equation to the other and changing its sign.
Example: If

$$X + A = B + C$$

and we wish to obtain an expression for X in terms of A, B, and C, we may *transpose* the A to the other side of the equal sign and change its sign, obtaining

$$X = B + C - A$$

This is equivalent to the axiom, "If equals are subtracted from equals, the remainders are equal."

Likewise, transposing a negative term from one side to the other and changing its sign to positive is equivalent to "adding equals to equals."

17. *Cross multiplying,* or multiplying each numerator by the denominator of the opposite side of the equation.
Example: If

$$\frac{A}{B} = \frac{C}{D} \qquad \text{then} \qquad AD = CB$$

This is the same as multiplying each side of the original equation by BD; thus,

$$\frac{A}{\not{B}} \times \not{B}D = \frac{C}{\not{D}} \times B\not{D} \qquad \text{or} \qquad AD = CB$$

By cancellation, there remains $AD = CB$. Therefore, cross multiplying is the same as "equals multiplied by equals."

Quadratic Equations Most equations in this text involve linear relationships in which the unknown quantity appears raised only to the first power. The word "linear" originates from such relationships being represented as straight lines on graphs. In a few instances in this text we find that the unknown appears raised to the second power. When this happens we have a quadratic equation. Many quadratic equations can be solved by factoring the equation, but in all cases the quadratic formula may be used to obtain the solutions in a straightforward manner. For example, a quadratic equation with coefficients a, b, and c can be solved as indicated:

$$ax^2 + bx + c = 0$$

$$x = \frac{-b \pm \sqrt{b^2 - 4ac}}{2a}$$

The two solutions occur due to the plus and minus signs. Often only one solution fits a physical problem, so the result obtained from the equation must be examined in light of the realities of the problem.

EXAMPLE A.3

$$3x^2 = 16x - 5$$

The first step in applying the formula is to put all terms on one side of the equation. Then, identify a, b, and c:

$$3x^2 - 16x + 5 = 0$$

$$a = 3 \quad b = -16 \quad c = 5$$

$$x = \frac{-(-16) \pm \sqrt{(-16)^2 - (4)(3)(5)}}{(2)(3)}$$

$$x = \frac{16 \pm \sqrt{256 - 60}}{6} = \frac{16 \pm \sqrt{196}}{6} = \frac{16 \pm 14}{6}$$

$$x = \tfrac{2}{6} \text{ and } \tfrac{30}{6} \quad \text{or} \quad x = \tfrac{1}{3} \text{ and } 5$$

There is no information here that would enable us to select the appropriate answer with physical meaning. Sometimes it is obvious that one solution has no meaning. For example, if we were solving for time and one answer were negative, it would have no physical meaning since negative time is not usually meaningful. We usually set up an equation based at the start of some physical process so that the beginning is represented by $t = 0$.

PROPORTIONALITY Proportionality and the related concepts of ratio and proportion are considered in this section.

Ratio A ratio is a fraction, one number divided by another. Thus, the ratio of 6 to 4 is 6/4. The ratio of 6 to 4, for example, tells how many times 6 is as large as 4. The ratio of 6 to 4 may be written as 6:4; of 4 to 6 as 4:6.

Proportion A proportion is a statement of the equality of two ratios. The equality can be stated as an equation. For example,

$$\tfrac{6}{4} = \tfrac{3}{2}$$

The proportion may also be written as $6:4::3:2$. It is read "six is to four as three is to two."

Proportionality Much quantitative thinking and experimentation in science is concerned with establishing a relation between one quantity and another quanity. Equality of ratios may be used. The concept of proportionality is essential.

Often one quantity depends on another quantity. For example, the length of the circumference of a circle is related to the length of the diameter. Simple relations can be expressed in simple mathematical forms. In many cases the relation between two quantities is not a simple one. We should attempt to determine if the relation is "direct" or "indirect" (inverse).

Direct Proportionality If one quantity increases as a second quantity increases, the two quantities are said to be "directly" proportional to each other. Also, if one quantity decreases as another quantity decreases, the two quantities are "directly" proportional to each other. (The "direct" proportion may be to the first power, to the second power—that is, squared—or to some other power.)

A good example of a direct proportionality is the relation of the length of the circumference C of a circle to the length of the diameter D. Because the circumference increases as the diameter increases (or the circumference decreases as the diameter decreases), the circumference of a circle is directly proportional to the diameter. This direct proportionality is expressed with symbols as follows:

$$C \propto D$$

which is read "C is proportional to D."

In changing a proportionality to an equation, a constant (some number, perhaps accompanied with units) must be introduced. The proportionality $C \propto D$ is transformed into an equation as follows:

$$C = k \times D$$

where k is the proportionality constant. The value of k in this example has been given the name π, which is approximately $\tfrac{22}{7}$. Thus, $C = \tfrac{22}{7} \times D$.

A quantity may be directly proportional to the square of another quantity. That the area A of a circle is directly proportional to the square of the radius, r^2, can be read from the equation $A = \pi r^2$. Other relationships can be interpreted in a somewhat similar manner.

Two concepts of proportionality can be expressed as an equality of two ratios. As an example, if

$$A = kB \quad \text{and} \quad C = k'D$$

then

$$\frac{A}{C} = k'' \frac{B}{D} \quad \text{where} \quad k'' = \frac{k}{k'} \quad \text{and} \quad \frac{A}{C} \propto \frac{B}{D}$$

Indirect (Inverse) Proportionality

If one quantity increases as a second quantity decreases (or vice versa), the two quantities are indirectly (inversely) related to each other.

For example, if the temperature of an enclosed volume of gas is held constant, what is the relation between the volume V of that gas and its pressure P? The results of experimentation indicate that if the volume is halved, the pressure is doubled; if the volume is decreased to one-third the original volume, the pressure is tripled. Thus pressure is inversely proportional to the volume:

$$P \propto \frac{1}{V}$$

Mathematically, it can be seen in the foregoing proportionality that a smaller volume V would result in an increased pressure P, and a larger volume would result in a decreased pressure. A smaller number V divided into 1 gives a larger number P, and a larger number V divided into 1 gives a smaller number P.

This inverse proportionality between pressure and volume can be expressed as an equality of two ratios. As an example, if $P_1 = k\left(\frac{1}{V_1}\right)$ and $P_2 = k\left(\frac{1}{V_2}\right)$, then dividing one equation by the other and canceling the k's we have

$$\frac{P_1}{P_2} = \frac{V_2}{V_1}$$

The equality is verified by numbers recorded as experimental data. (When the volume is halved, the pressure is doubled—an inverse relation.)

An inverse proportion may be to the first power, to the second power, or to some other power. An excellent example is the following: The intensity of illumination l from a light source is inversely proportional to the square of the distance d from the source:

$$l \propto \frac{1}{d^2} \quad \text{or} \quad l = \frac{k}{d^2}$$

The intensity of illumination on a surface is only one-fourth as much if the distance from the surface to the source of light is doubled. Tripling the distance decreases the intensity to one-ninth. This type of proportionality is known as an "inverse square" proportionality, and this generalization is called the "inverse square law."

Interpreting Equations

Equations express the relation between the quantities involved. Consider the equation $B = k(Ll^3/bd^3)$, where B is the bending, due to load l, of a bar of length L, width b, and depth d; k is a constant depending on the material of the bar. This equation gives us the following relations:

Bending varies directly as load: $B \propto L$
Bending varies directly as cube of length: $B \propto l^3$
Bending varies inversely as width: $B \propto 1/b$
Bending varies inversely as cube of depth: $B \propto 1/d^3$

Properly interpreting the relations given by equations enables us to solve problems involving the quantities of the equation.

The idea of a constant of proportionality was used in Chap. 4 with the determination of G in the equation for Newton's law of universal gravitation. The constant permitted proceeding from a proportionality to an equation.

Chapter 2 provides another illustration of the proportionality constant. The distance s a car travels at uniform speed is proportional to the time t:

$$s \propto t$$

The ratio of the measured distance in a given direction to the measured time is the speed v; therefore,

$$\frac{s}{t} = v$$

The proportionality constant is v, the uniform speed:

$$s = vt$$

Graphs Data, or quantities given or determined by experimentation, are plotted as points on a graph grid, and a curve is drawn for these points. The curve is used in the determination of values between the values of the known data. This procedure is shown in Example A.4. An understanding of the procedure used there should make it easier for you to understand and use graphs.

EXAMPLE A.4

Plot the values of F and S on a graph grid and draw a curve for the points plotted. What is the value of F in N (newtons) when $S = 8$ m? The corresponding values of F and S are

$F(N)$	S (m)
120	20
475	10
1860	5
2900	4

Solution

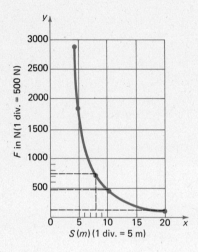

Figure A.1

To plot quantities on a graph, an x axis, which is horizontal, and a y axis, which is vertical, are used. The intersection of these mutually perpendicular axes is the origin 0. Lines drawn parallel to these axes and equidistant from one another form a grid, as shown in Fig. A.1. One of the quantities is then plotted along the y axis and the other along the x axis. A proper choice of the scale to be used is important; the idea is to make each division represent a convenient number and to use nearly all the available space.

Values for F are to be plotted along the y axis. Because the largest value of F is 2900 N and six divisions are available, a convenient scale would be 1 division = 500. Do not let 1 division = $\frac{2900}{6}$ = 438.3 N because this would be very inconvenient. A convenient scale for S along the x axis would be 1 division = 4 m or 1 division = 5 m. To plot the point $F = 120$ N, $S = 20$ m, estimate the position of $F = 120$ N along the vertical axis; the point must be horizontally opposite this and vertically up from $S = 20$. This locates the point; a small dot has been drawn to indicate this point. Similarly, over from $F = 475$ and up from $S = 10$ locate the second point; a dot is used to indicate this point. The other two points are located in the same manner, and a smooth curve is then drawn through the points. Do not connect adjacent points by straight lines; draw the smooth curve that the points as a whole seem to indicate.

The question was, "What is the value of F when $S = 8$ m?" In Fig. A.1 the broken lines from $S = 8$ to the curve and from the curve to the vertical axis indicate that for $S = 8$ m, $F = 720$ N.

B Conversions

LENGTH
1 in. = 2.54 cm
1 m = 39.37 in
1 ft = 0.3048 m
12 in. = 1 ft
3 ft = 1 yd
1 yd = 0.9144 m
1 km = 0.621 mi
1 mi = 1.61 km
1 mi = 5280 ft
1 Å = 10^{-10} m
1 μm = μ = 10^{-6} m = 10^4 Å
1 light-year = 9.46×10^{12} km

AREA
1 m^2 = 10^4 cm^2 = 10.76 ft^2
1 ft^2 = 0.0929 m^2 = 144 in^2
1 in^2 = 6.452 cm^2

VOLUME
1 m^3 = 10^6 cm^3 = 6.102×10^4 $in.^3$
1 ft^3 = 1728 in^3 = 2.83×10^{-2} m^3
1 L = 1000 cm^3 = 1.0576 qt = 0.0353 ft^3
1 ft^3 = 7.481 gal = 28.32 L

MASS
1000 kg = 1 t (metric ton)
1000 g = 1 kg
1 slug = 14.57 kg
1 u = 1.66×10^{-27} kg

FORCE
1 N = 10^5 dyn = 0.225 lb
1 lb = 4.45 N
1 ton = 2000 lb

PRESSURE
1 bar = 10^5 N/m^2 = 14.50 lb/$in.^2$
1 atm = 760 mmHg = 76.0 cmHg = 14.7 lb/$in.^2$ = 1.013×10^5 N/m^2
   (1.013 bar = 1013 mbar)

ENERGY 1 J = 0.7376 ft · lb = 10^7 ergs
1 cal = 4.184 J
1 Btu = 252 cal = 778 ft · lb
1 eV = 1.6×10^{-19} J
931.5 MeV is equivalent to 1 u

POWER 1 hp = 550 ft · lb/s = 0.746 kW
1 W = 1 J/s = 0.738 ft · b/s

MAGNETIC INDUCTION B 1 T = 10^4 G

C Physical Constants

Name	Symbol	Value
Absolute zero	0 K	$-273.15\,°C$
Acceleration due to gravity at earth's surface	g	$9.81\ m/s^2$ or $32.2\ ft/s^2$
Avogadro's number	N_0	6.02×10^{23}
Boltzmann's constant	k	$1.381 \times 10^{-23}\ J/K$
Electric constant	k	$8.988 \times 10^9\ N \cdot m^2/C^2$
Electronic charge	e	$1.602 \times 10^{-19}\ C$
Gas constant per mole	R	$8.314\ J/K$ or $0.0821\ L \cdot atm/mol \cdot K$ or $1.987\ cal/mol \cdot K$
Gravitational constant	G	$6.673 \times 10^{-11}\ N \cdot m^2/kg^2$
Joules equivalent	J	$4.184\ J/cal$ or $4184\ J/kcal$
Magnetic constant	k'	$10^{-7}\ N/A^2$
Planck's constant	h	$6.625 \times 10^{-34}\ J \cdot s$
Rest mass of electron	m_e	$9.110 \times 10^{-31}\ kg$
Rest mass of proton	m_p	$1.6726 \times 10^{-27}\ kg$
Rest mass of neutron	m_n	$1.6749 \times 10^{-27}\ kg$
Speed of light in vacuo	c	$2.9979 \times 10^8\ m/s$

Solar System Data

Mass of earth	$5.98 \times 10^{24}\ kg$
Mass of sun	$1.99 \times 10^{30}\ kg$
Mass of moon	$7.34 \times 10^{22}\ kg$
Equitorial radius of earth	3963 mi, or 6380 km
Radius of sun	$6.97 \times 10^8\ m$
Radius of moon	$1.74 \times 10^3\ km$
Mean distance of earth to moon	$3.84 \times 10^5\ km$
Mean distance of earth to sun	$150 \times 10^6\ km$, or $93 \times 10^6\ mi$, or 1 AU
Length of year	365 days, or $3.16 \times 10^7\ s$

D Tables of Useful Data

Table	Title	Page
2.1	Units of Length	9
3.1	Metric Units of Mass	63
3.2	Coefficients of Friction	78
	Useful Planetary Data, Chap. 4	118
8.1	Specific Gravity	217
8.2	Elastic Moduli	221
8.3	Surface Tension	228
10.1	Coefficients of Thermal Expansion at 0°C	272
10.2	Specific Heat Capacities	279
11.1	Saturated Vapor Pressure of Water at Temperatures from 0 to 150°C	312
14.1	Resistivity (ρ) and Temperature Coefficients of Resistance (α) at 0°C	422
18.1	Index of Refraction for Yellow Sodium Light	558
23.1	Atomic Masses of Some Isotopes	741

E Natural Trigonometric Functions

Angle		Sine	Cosine	Tangent	Angle		Sine	Cosine	Tangent
Degree	Radian				Degree	Radian			
0°	.000	0.000	1.000	0.000					
1°	.017	.018	1.000	.018	46°	0.803	0.719	0.695	1.036
2°	.035	.035	0.999	.035	47°	.820	.731	.682	1.072
3°	.052	.052	.999	.052	48°	.838	.743	.669	1.111
4°	.070	.070	.998	.070	49°	.855	.755	.656	1.150
5°	.087	.087	.996	.088	50°	.873	.766	.643	1.192
6°	.105	.105	.995	.105	51°	.890	.777	.629	1.235
7°	.122	.122	.993	.123	52°	.908	.788	.616	1.280
8°	.140	.139	.990	.141	53°	.925	.799	.602	1.327
9°	.157	.156	.988	.158	54°	.942	.809	.588	1.376
10°	.175	.174	.985	.176	55°	.960	.819	.574	1.428
11°	.192	.191	.982	.194	56°	.977	.829	.559	1.483
12°	.209	.208	.978	.213	57°	.995	.839	.545	1.540
13°	.227	.225	.974	.231	58°	1.012	.848	.530	1.600
14°	.244	.242	.970	.249	59°	1.030	.857	.515	1.664
15°	.262	.259	.966	.268	60°	1.047	.866	.500	1.732
16°	.279	.276	.961	.287	61°	1.065	.875	.485	1.804
17°	.297	.292	.956	.306	62°	1.082	.883	.470	1.881
18°	.314	.309	.951	.325	63°	1.100	.891	.454	1.963
19°	.332	.326	.946	.344	64°	1.117	.899	.438	2.050
20°	.349	.342	.940	.364	65°	1.134	.906	.423	2.145
21°	.367	.358	.934	.384	66°	1.152	.914	.407	2.246
22°	.384	.375	.927	.404	67°	1.169	.921	.391	2.356
23°	.401	.391	.921	.425	68°	1.187	.927	.375	2.475
24°	.419	.407	.914	.445	69°	1.204	.934	.358	2.605
25°	.436	.423	.906	.466	70°	1.222	.940	.342	2.747

E Natural Trigonometric Functions (*cont'd.*)

Angle		Sine	Cosine	Tangent	Angle		Sine	Cosine	Tangent
Degree	Radian				Degree	Radian			
26°	.454	.438	.899	.488	71°	1.239	.946	.326	2.904
27°	.471	.454	.891	.510	72°	1.257	.951	.309	3.078
28°	.489	.470	.883	.532	73°	1.274	.956	.292	3.271
29°	.506	.485	.875	.554	74°	1.292	.961	.276	3.487
30°	.524	.500	.866	.577	75°	1.309	.966	.259	3.732
31°	.541	.515	.857	.601	76°	1.326	.970	.242	4.011
32°	.559	.530	.848	.625	77°	1.344	.974	.225	4.331
33°	.576	.545	.839	.649	78°	1.361	.978	.208	4.705
34°	.593	.559	.829	.675	79°	1.379	.982	.191	5.145
35°	.611	.574	.819	.700	80°	1.396	.985	.174	5.671
36°	.628	.588	.809	.727	81°	1.414	.988	.156	6.314
37°	.646	.602	.799	.754	82°	1.431	.990	.139	7.115
38°	.663	.616	.788	.781	83°	1.449	.993	.122	8.144
39°	.681	.629	.777	.810	84°	1.466	.995	.105	9.514
40°	.698	.643	.766	.839	85°	1.484	.996	.087	11.43
41°	.716	.656	.755	.869	86°	1.501	.998	.070	14.30
42°	.733	.669	.743	.900	87°	1.518	.999	.052	19.08
43°	.751	.682	.731	.933	88°	1.536	.999	.035	28.64
44°	.768	.695	.719	.966	89°	1.553	1.000	.018	57.29
45°	.785	.707	.707	1.000	90°	1.571	1.000	.000	∞

F Table of Elements, Abundant Isotopes

Atomic Number Z	Symbol	Average Atomic Mass[a]	Element	Mass Number A	Relative Abundance (%)	Mass of Isotope[b]
1	H	1.00797	Hydrogen	1	99.985	1.007 825
			Deuterium	2	0.015	2.014 102
			Tritium	3		3.016 049
2	He	4.002	Helium	3	0.00015	3.016 029
				4	100—	4.002 603
3	Li	6.939	Lithium	6	7.52	6.015 126
				7	92.48	7.016 005
4	Be	9.012	Beryllium	9	100—	9.012 186
5	B	10.811	Boron	10	18.7	10.012 939
				11	81.3	11.009 305
6	C	12.011	Carbon	12	98.892	12.000 000
				13	1.108	13.003 354
7	N	14.006	Nitrogen	14	99.635	14.003 074
				15	0.365	15.000 108
8	O	15.999	Oxygen	16	99.759	15.994 915
9	F	18.998	Fluorine	19	100	18.998 405
10	Ne	20.183	Neon	20	90.92	19.992 440
				22	8.82	21.991 384
11	Na	22.990	Sodium	23	100—	22.989 773
12	Mg	24.312	Magnesium	24	78.60	23.985 045
13	Al	26.981	Aluminum	27	100	26.981 535
14	Si	28.086	Silicon	28	92.27	27.976 927
15	P	30.974	Phosphorus	31	100	30.973 763
16	S	32.064	Sulfur	32	95.018	31.972 074
17	Cl	35.453	Chlorine	35	75.4	34.968 854
				37	24.6	36.965 896
18	Ar	39.948	Argon	40	996	39.962 384
19	K	39.102	Potassium	39	93.08	38.963 714
20	Ca	40.08	Calcium	40	96.97	39.962 589
21	Sc	44.956	Scandium	45	100	44.955 919
22	Ti	47.90	Titanium	48	73.45	47.947 948
23	V	50.942	Vanadium	51	99.76	50.943 978
24	Cr	51.996	Chromium	52	83.76	51.940 514
25	Mn	54.938	Manganese	55	100	54.938 054
26	Fe	55.847	Iron	56	91.68	55.934 932
27	Co	58.933	Cobalt	59	100	57.933 19
28	Ni	58.71	Nickel	58	67.7	57.935 34
				60	26.23	59.930 32

F Table of Elements, Abundant Isotopes (*cont'd.*)

Atomic Number Z	Symbol	Average Atomic Mass[a]	Element	Mass Number A	Relative Abundance (%)	Mass of Isotope[b]
29	Cu	63.54	Copper	63	69.1	62.929 59
30	Zn	65.37	Zinc	64	48.89	63.929 14
31	Ga	69.72	Gallium	69	60.2	68.925 68
32	Ge	72.59	Germanium	74	36.74	73.921 15
33	As	74.9216	Arsenic	75	100	74.921 58
34	Se	78.96	Selenium	80	49.82	79.916 51
35	Br	79.909	Bromine	79	50.52	78.918 35
36	Kr	83.30	Krypton	84	56.90	83.911 50
37	Rb	85.47	Rubidium	85	72.15	84.911 71
38	Sr	87.62	Strontium	88	82.56	87.905 61
39	Y	88.905	Yttrium	89	100	88.905 43
40	Zr	91.22	Zirconium	90	51.46	89.904 32
41	Nb	92.906	Niobium	93	100	92.906 02
42	Mo	95.94	Molybdenum	98	23.75	97.905 51
43	Tc	(99)	Technetium	98		97.907 30
44	Ru	101.07	Ruthenium	102	31.3	101.903 72
45	Rh	102.905	Rhodium	103	100	102.904 80
46	Pd	106.4	Palladium	106	27.2	105.903 20
47	Ag	107.870	Silver	107	51.35	106.904 97
48	Cd	112.40	Cadmium	114	28.8	113.903 57
49	In	114.82	Indium	115	95.7	114.904 07
50	Sn	118.69	Tin	120	32.97	119.902 13
51	Sb	121.75	Antimony	121	57.25	120.903 75
52	Te	127.60	Tellurium	130	34.49	129.906 70
53	I	126.9044	Iodine	127	100	126.904 35
54	Xe	131.30	Xenon	132	26.89	131.904 16
55	Cs	132.905	Cesium	133	100	132.905 09
56	Ba	137.34	Barium	138	71.66	137.905 01
57	La	138.91	Lanthanum	139	99.911	138.906 06
58	Ce	140.12	Cerium	140	88.48	139.905 28
59	Pr	140.907	Praseodymium	141	100	140.907 39
60	Nd	144.24	Neodymium	144	23.85	143.909 98
61	Pm	(147)	Promethium	145		144.912 31
62	Sm	150.35	Samarium	152	26.63	151.919 49
63	Eu	151.96	Europium	153	52.23	152.920 86
64	Gd	157.25	Gadolinium	158	24.87	157.924 10
65	Tb	158.924	Terbium	159	100	158.924 95
66	Dy	162.50	Dysprosium	164	28.18	163.928 83

F Table of Elements, Abundant Isotopes (cont'd.)

Atomic Number Z	Symbol	Average Atomic Mass[a]	Element	Mass Number A	Relative Abundance (%)	Mass of Isotope[b]
67	Ho	164.930	Holmium	165	100	164.930 30
68	Er	167.26	Erbium	166	33.41	165.930 40
69	Tm	168.934	Thulium	169	100	168.934 85
70	Yb	173.04	Ytterbium	174	31.84	173.939 02
71	Lu	174.97	Lutetium	175	97.40	174.940 89
72	Hf	178.49	Hafnium	180	35.44	179.946 81
73	Ta	180.948	Tantalum	181	100	180.947 98
74	W	183.85	Tungsten	184	30.6	183.950 99
75	Re	186.2	Thenium	187	62.93	186.955 96
76	Os	190.2	Osmium	192	41.0	191.961 41
77	Ir	192.2	Iridium	193	61.5	192.963 28
78	Pt	195.09	Platinum	195	33.7	194.964 82
79	Au	196.967	Gold	197	100	196.966 55
80	Hg	200.59	Mercury	202	29.80	201.970 63
81	Tl	204.37	Thallium	205	70.50	204.974 46
82	Pb	207.19	Lead	208	52.3	207.976 64
83	Bi	208.980	Bismuth	209	100	208.980 42
84	Po	(210)	Polonium	210		209.982 87
85	At	(210)	Astatine	211		210.987 50
86	Rn	(222)	Radon	222		222.017 61
87	Fr	(223)	Francium	221		221.014 18
88	Ra	(226)	Radium	226		226.025 36
89	Ac	(227)	Actinium	225		225.023 14
90	Th	(232)	Thorium	232	100	232.038 21
91	Pa	(231)	Protactinium	231		231.035 94
92	U	(238)	Uranium	233		233.039 50
				235	0.715	235.043 93
				238	99.28	238.050 76
93	Np	(237)	Neptunium	239		239.052 94
94	Pu	(242)	Plutonium	239		239.052 16
95	Am	(243)	Americium	243		243.061 38
96	Cm	(248)	Curium	245		245.065 84
97	Bk	(249)	Berkelium	248		248.070 305
98	Cf	(249)	Californium	249		249.074 70
99	Es	(254)	Einsteinium	254		254.088 11
100	Fm	(257)	Fermium	252		252.082 65
101	Md	(258)	Mendelevium	255		255.090 57
102	No	(259)	Nobelium	254		254
103	Lw	(260)	Lawrencium	257		257

[a] Values in parentheses are the mass number of the most stable, long-lived of the known isotopes.
[b] Masses of neutral atoms based on $^{12}_{6}C = 12$ u.

Answers to Odd-Numbered Problems

Chapter 2

1. 1.98 m, 198 cm **3.** (a) 0.173 m, (b) 73 ft/s, (c) 80 km/h, (d) 2.99×10^8 m/s **5.** 200 ft², 2.88×10^4 in.², 18.6 m²
7. (a) 2640 m, (b) 339 m, SW, (c) NE **9.** 40 km; 28.8 km at 33.7° E of N **11.** $v_h = 301$ km/h, $v_v = 109$ km/h
13. $F_H = 132$ lb, $F_V = 176$ lb **15.** 138 lb at 55.3° from the 80-lb vector **17.** (a) 11.2 m, 63.4° from the 5-m vector (N of E),
(b) 19.7 m, 10° N of E, (c) 8.66 m, 60° N or W, (d) 6.17 m/s, 20° S of W, (e) 3.0 mi, 31° S of W **19.** 2.5×10^{-2} s/frame
21. 3.57 cm/s **23.** 12.5 m **25.** 333 m/s **27.** (a) 15 km/h, (b) +15 km/h, (c) −15 km/h **29.** 632 km/h at 71.6° N of W
31. 33.5 m/s at 63.4° forward of perpendicular to the car **33.** 68.2° from the vertical **35.** (a) 77.5° downstream from
perpendicular to the bank, (b) 0.25 h, (c) 2.25 km **37.** (a) 5 km/h, 0, −6.5 km/h, (b) last few minutes,
(c) 6.67 km/h, 4.17 km/h **39.** The same **41.** (a) 0.63 m/s², (b) 20 m, (c) 1.25 m/s **43.** (a) 25 m/s, (b) 62.5 m, (c) 90 m
45. 2 s, 18 m **47.** −5 m/s², 40 m, 5 m/s **49.** 3640 km/h² or 0.28 m/s² **51.** 4.85 h **53.** 1.62 m **55.** Yes, 4 m
57. (a) 8.7 s, (b) 75 ft, (c) $v_{car} = 52$ ft/s, $v_{truck} = 35$ m/s **59.** 53.8 ft **61.** (a) 122.5 m, (b) 49 m/s, (c) $v = 24.5$ m/s,
(d) 7.4 percent of speed of sound **63.** (a) 10 m, (b) 1.43 s, (c) 2.86 s, (d) −14 m/s **65.** (a) 45.4 m, (b) 4.08 s,
(c) 29.8 m/s **67.** 29.6 m/s **69.** 11 m **71.** 23.6 m below the balloon, 25.6 m/s relative to the earth, 21.6 m/s relative to the
balloon **73.** six times farther **75.** (a) 24.5 ft/s at 82° from horizontal, (b) 3.1 ft **77.** 63.4 ft/s

Chapter 3

1. 10 N down **3.** 0.32 N **5.** 1600 N **7.** 0.2 m/s² **9.** 880 N **11.** 9.4 m/s² **13.** 0.5 slug, 500 ft
15. 2.98 m/s² to right **17.** $g \sin \theta$ **19.** 180 lb **21.** 45.2 lb **23.** 135 lb at 72° from horizontal to the right
25. $T = 163$ N, $F = 220$ N **27.** 501 N at 30°, 614 N at 45° **29.** 6 N **31.** (a) 1.33 m/s²,
(b) 3.67 N **33.** (a) 0.5 N, (b) 12 kg **35.** (a) 2.5 m/s², (b) 500 N **37.** 51.5 N **39.** (a) 2.45 m/s², (b) 0.61 m/s²
41. $T_1 = 156$ N, $T_2 = 52$ N **43.** 1.23 m/s², $25.7 \times$ MN (M in kg) **45.** 7.67 s **47.** (b) 0.25, (c) yes, v = constant and $a = 0$,
(d) yes, $a = 0$.

Chapter 4

1. 2.45 m/s², 1.09 m/s² **3.** 1.41 earth radii **5.** 8 m/s² **7.** (a) $2F$, (b) $\frac{1}{2}F$, (c) $2F$, (d) $\frac{1}{2}F$, (e) $\frac{1}{4}F$, (f) $4F$, (g) $\frac{1}{9}F$, (h) $4F$,
(i) $16F$, (j) $8F$, (k) $\frac{1}{8}F$ **9.** $F_s = 169 F_m$ **11.** 0.20 of 1 revolution, or 7.4×10^9 km **13.** 2.16 m/s², 0.22 g
15. 0.5 g downward **17.** 5.00 ft/s² **19.** 1.69×10^{-2} m/s² **21.** (a) 1.1×10^{-3} m/s, (b) 1.1×10^{-4} m/s² inward
23. 0.45 r/s **25.** (a) 20.3 ft/s, (b) 0, (c) 1.96 ft/s² **27.** 2.0 m/s **29.** 22.1 m/s (0.035 r/s) **31.** 294 ft/s², 147 lb
33. 22.4 m/s **35.** 3.1 m/s **37.** 12 h, 2.02×10^7 m **39.** At top, $T = m\left(\dfrac{v^2_{top}}{r} - g\right)$, at bottom, $T = m\left(\dfrac{v^2_{bottom}}{r} + g\right)$, no
41. 0.74 m/s, 2.48 N **43.** 15.3°

Chapter 5

1. 1500 ft·lb **3.** (a) 12 N·m, (b) 12 J, (c) 12×10^7 ergs, (d) 3.0 N **5.** 0 **7.** More work to push. (a) 1.13×10^4 J,
(b) 2.1×10^4 J **9.** (a) 529 J, (b) 529 J **11.** (a) −60 J friction, 181 J pulling, (b) 121 J, (c) 9.0 m/s **13.** -9.1×10^4 ft·lb
15. (a) 66.4 J, (b) 66.4 J **17.** (a) 68.6 (m) J, (b) 68.6 (m) J, (c) 68.6 (m) J, (d) 68.6 (m) J, (e) 11.7 m/s, (m) = mass in kg
19. 4.5×10^5 J **21.** (a) 2×10^{-6} J, (b) 4.5×10^{-6} J, 8×10^{-6} J, 32×10^{-6} J, (c) 16 **23.** 10 J **25.** 6.7×10^6 m
27. 3.75×10^5 km **29.** 44.3 m/s, 159 km/h, 99 mi/h **31.** (a) 170 ft·lb, (b) 33 ft/s **33.** 7 m/s **35.** 6.25 m
37. (a) 9×10^4 J at both times, (b) 1.15 km, (c) changed to heat, energy is conserved, mechanical energy is not conserved
39. (a) 12×10^4 ft·lb, (b) 0, (c) 16 ft below starting point, (d) 5.6×10^4 ft·lb, (e) 6.4×10^4 ft·lb, (f) 0, (g) 12×10^4 ft·lb,
(h) 8.4×10^4 ft·lb **41.** (a) 47.4 J, (b) 9.7 m/s, (c) 6.7 m/s, 22.4 J, (d) 2.3 m **43.** (a) 19.8 m/s, (b) 14 m/s, (c) 50 m

Chapter 6

1. 21.9 kg·m/s down the alley   **3.** 1.5 kg·m/s NE   **5.** 2.66×10^{-23} kg·m/s direction not specified
7. 5.0×10^{-23} kg·m/s   **9.** 4.04 kg·m/s 68.2° from the vertical toward the pitcher   **11.** 54.2 N upward   **13.** (a) 200 N, (b) he will be accelerated opposite the direction of the water's motion   **15.** 10 ft/s opposite to the projectile
17. 0.38 m/s in opposite direction   **19.** (a) 3.0 kg·m/s, (b) 3.75×10^{-2} m/s   **21.** 6 min   **23.** 510 cm/s   **25.** 0.62 m/s in direction the more massive car was moving, yes   **27.** 0.5 kg at 143.1° from velocity of 2-kg piece and 126.9° from velocity of the 1-kg piece   **29.** (a) 23.2 J, (b) 3.05 kg·m/s, (c) 3.05 N·s, (d) 232 N   **31.** (a) In direction the smaller car had been moving, (b) 0.33 m/s in direction of the 1000-kg car, (c) kinetic energy is lost (5330 J lost)   **33.** (a) 281 m/s, (b) 0.7 m/s, (c) 7.88×10^3 N   **35.** (a) $v_B = 19.8$ m/s, $v_C = 14.0$ m/s, $v_D = 19.8$ m/s, (b) 348 J, (c) it hits the wall, 10 m, it could have traveled 17.8 m, (d) 0, (e) 0   **37.** (a) v_0, (b) the neutron stops, (c) the neutron would rebound with $(9/11)v_0$
39. (a) $(2/5)v_i$, (b) 16/25

Chapter 7

1. (a) -160 N·m, (b) -101.4 N·m, (c) -60 N·m, (d) -138.6 N·m (out of page)   **3.** -0.16 N·m, -0.32 N·m
5. 55.9 cm   **7.** 22.7 in. from head, 16.8 in. from ground   **9.** 0.065×10^{-8} cm below O atom and midway between H atoms   **11.** 4.5 ft from center   **13.** $T_1 = 1.42 \times 10^3$ N, $T_2 = 1.32 \times 10^3$ N   **15.** 970 N   **17.** 90.5 N horizontal, 120 N at 40.9° from horizontal   **19.** (b) 1273 lb, (c) 900 lb horizontal to the left, 200 lb up   **21.** At top $F_H = 12.50$ lb, $F_V = 3.35$ lb up; at base $F_H = 12.50$ lb, $F_V = 46.64$ lb up   **23.** (a) 41.9 rad/s, (b) 126 rad, (c) 628 cm   **25.** 32 rad, 16 rad/s
27. (a) 0.3 r/s² or 0.6π rad/s, (b) 13.8 r   **29.** (a) 0.21 rad/s², (b) 12.6 rad/s, (c) 11.3×10^3 rad, (d) 679 m, (e) 151 cm/s
31. (a) 9.16 m/s, (b) twice the speed of the center along the road, (c) 0.055 m/s²   **33.** 2.36×10^{-2} slug·ft²
35. (a) 0.173 rad/s, (b) 2.81×10^{-4} N·m, 3.53×10^{-3} N, (c) 9.7×10^{-3} J, (d) 34.6 rad   **37.** 4.8×10^6 kg·m²/s   **39.** 2 r/s
41. 2.82×10^{34} kg·m²/s, nothing, remains constant   **43.** (a) 2.35 rad/s, (b) 0.986 rad/s   **45.** 2.86 J   **47.** $v_{end} = 8.2$ m/s, $v_{center} = 4.1$ m/s   **49.** Ball   **51.** 0.014 kg·m²

Chapter 8

1. 1.05   **3.** 0.81 kg, 13.6 kg   **5.** 0.9 N   **7.** 55.5 N/cm² or 5.55×10^5 N/m²   **9.** 1.2×10^{-2}, compressive strain
11. 61.3 ft/s²   **13.** 2.5 cm   **15.** 1.77×10^3 lb   **17.** (a) 5.98×10^6 N/m², (b) 6.64×10^{-4}, (c) 2.46×10^{-2} cm
19. (a) Bottom, (b) 1.0×10^9 N/m²   **21.** 49 N/m²   **23.** 1.95×10^3 N/cm²   **25.** 7.14×10^{-5} **27.** 2×10^6 N/m²
29. 24 cm   **31.** 6 m

Chapter 9

1. (a) 706 N, (b) 3920 N/m², (c) 1960 N/m², (d) 470 N   **3.** 33.9 ft   **5.** 860 mmHg   **7.** 19.9×10^4 N/m²   **9.** 1.51×10^5 N
11. 123 N   **13.** 1.21×10^5 N/m²   **15.** 4.62×10^3 N or 1.04×10^3 lb   **17.** (a) 9900 lb, (b) the hoist will not rise
19. 0.74 N   **21.** 1.32 kg   **23.** 0.917   **25.** 0.025 m³, 245 N   **27.** 1.6 m   **29.** 11.4   **31.** (a) 50 m/s, (b) 0.75×10^6 N/m²
33. 1.98 m/s   **35.** 3.37 mmHg $P_{lower\ inside}$ **37.** (a) 36 m/s, (b) $P_{upper} = 1$ atm, (c) $P_{lower} = 7.29$ atm, $\Delta P = 6.37 \times 10^5$ N/m²   **39.** 2.66 mm/s   **41.** 64 N/m²   **43.** 4.6 m/s

Chapter 10

1. 93.2°F   **3.** $-40°$ **5.** 19 999 727°C   **7.** 15.041 cm   **9.** 92.046 m   **11.** 48.058 cm²   **13.** 0.52 cm³   **15.** 12.4 gal
17. $-74.8°$C   **19.** (a) 7.6×10^{-3}, (b) 6.9×10^8 N/m²   **21.** 49 cal/g   **23.** 29.1°C   **25.** 0.40 cal/g·°C   **27.** 10.6°C
29. 52.8 kcal   **31.** (a) 31.6 h, (b) 1.7×10^3 h   **33.** 200 m   **35.** About 60 times more through the window
37. (a) 4.67×10^5 Btu/h, (b) 401 lb   **39.** 2.1×10^4 Btu   **41.** 0.47 kg

Chapter 11

1. 6.8 mol   **3.** 0.0446 mol   **5.** 10.8°C   **7.** 2.64 L   **9.** 3.26 atm   **11.** 50.5 atm   **13.** (a) $2P$, (b) $P/2$, (c) P, (d) $4P$
15. 1.33 L   **17.** 358   **19.** 2.42×10^4, 2.45×10^{19} at 1 atm   **21.** 3.25   **23.** (a) 27 lb/in.², (b) 19 lb/in.²
25. 2.5×10^4 m/s   **27.** 10^3 N   **29.** 1.75 kg/m³   **31.** 333 K   **33.** $-10.5°$C   **35.** (a) 1.64×10^5 K, (b) 7.38×10^3 K
37. 200°C   **39.** $P_{H_2} = 1$ atm, $P_{N_2} = 2.1$ atm, $P_{tot} = 3.1$ atm   **41.** $P_{O_2} = 0.79$ atm, $P_{He} = 3.15$ atm, $\rho = 1.56 \times 10^{-3}$ g/cm³
43. 59 percent   **45.** 85 percent   **47.** 5×10^3 J   **49.** 271°C   **51.** 817 J   **53.** 900 cal   **55.** 382 cal   **57.** 62.5 percent

Chapter 12

1. 5 N/cm **3.** 0.7 m **5.** $(\frac{1}{3})$ N/cm **7.** 0.75 kg **9.** 6.28 s **11.** 10 cm **13.** 0.248 m **15.** 21.7 lb
17. (a) 0.94 s, (b) 0.45 m **21.** (a) 75.8 m/s² at endpoints, (b) 3.02 m/s at mid-point, (c), (d) 0 at $t = 4$ s and $t = (\frac{1}{8})$ s
23. (a) 10 cm, (b) 2 Hz, (c) 0.5 s, (d) 31.6 N/m, (e) 0 **25.** (a) 8.84×10^3 cm/s², (b) 2.53×10^3 cm/s², (c) 3.37 m/s **27.** 2 m/s
29. 0.91 m **31.** 69 ft, 5.5×10^{-2} ft **33.** 1 min **35.** 5.06×10^3 m/s **37.** 1.65×10^{10} N/m² **39.** 632 m/s **41.** 40 m/s
43. 67.1 Hz (3d harmonic) **45.** 20 Hz, 40 Hz, 60 Hz **47.** 430 Hz **49.** 3 m, 348 m/s **51.** 37.5 cm **53.** 7
55. 0.76 m, 482 Hz **57.** 311 Hz **59.** 525 Hz **61.** 96.5 dB **63.** 1.27×10^{-2} W/m², 101 dB

Chapter 13

1. 6.25×10^{18} electrons **3.** For 121 lb ($m = 55$ kg), $Q = 1.22 \times 10^{-4}$ C **5.** (a) opposite direction, attractive force, magnitude unchanged, (b) F doubles, (c) 4F, (d) F/4, (e) (16/9)F, (f) unchanged, (g) it moves away, F decreases **7.** 2.94 m
9. (a) $F_g/F_e = 2.40 \times 10^{-43}$, (b) the two orbit about a center of mass midway between them **11.** 5.76×10^2 N
13. (a) 320 N to right, (b) 383 N to left **15.** 7.20×10^2 m/s² **17.** 0.54×10^{-7} C **19.** 0.48×10^{-7} C **21.** (a) $(-5$ m, 0),
(b) 0, (c) -1.8×10^4 V, (d) -0.45 J **23.** (a) 3.71 N/C at 14° below the x axis toward the -20×10^{-9}-C charge,
(b) 1.48×10^{-5} N in a direction opposite the electric field, (c) 9 V, (d) 1.8×10^{-5} J **25.** (a) 4 cm, (b) 3.52×10^7 N/C, (c) 0,
(d) accelerate toward the negative charge, (e) -6.75×10^5 V **27.** 16 J, external **29.** (a) 1.25×10^{29} eV, 2×10^{10} J,
(b) 5.98×10^4 kg **31.** -23.0×10^{-19} J or 14.4 eV **33.** (a) 1.82×10^{-12} J, (b) 4.67×10^7 m/s
37. 3×10^6 V/m downward **39.** (a) 1.1×10^6 V/m toward the left, (b) 4.4 N **41.** (a) 180 N attraction,
(b) 2.7×10^8 V/m toward the -4-μC charge, (c) 1.8×10^6 V **43.** 800 V/m **45.** 35.4 V/m **47.** 10^6 V/m **49.** 39.2 μC
51. (a) 1.2×10^{-4} J, (b) 1.2×10^{-4} J, (c) 60 V, (d) -1.2×10^{-4} J of work since the electric field will do the work

Chapter 14

1. (a) 4 A, (b) 1.5×10^{21} electrons **3.** 1.55×10^{-4} m/s **5.** 2.53×10^{-2} Ω, yes **7.** 12 Ω in series, 1.33 Ω in parallel
9. 2.48 Ω **11.** $4 \times 10^{-3}/°$C **13.** 12 Ω **15.** -0.285 Ω, yes, the coefficient α is not constant **17.** 8.92×10^{-3} cm²
19. 0.55 A **21.** 30 kΩ **23.** 10 in parallel **25.** (a) 0.87 Ω, (b) 5.22 V **27.** (a) 242 Ω, (b) 4.55×10^{-2} A, (c) 11.0 V,
(d) 110 V, (e) 0 V **29.** 0.91 A **31.** 420 W, 24 W **33.** 20 A, 2.4 kW **35.** (a) The wires will overheat—start a fire,
(b) yes, with 250 W to spare, (c) yes, with 1058 W to spare **37.** (a) 1500 W, (b) 50 V, (c) increases by 45 Ω **39.** 0.92 A from battery and in the single 6-Ω resistor, 0.46 A in each 6-Ω resistor at the bottom, 0.31 A in 12-Ω resistor at the top, 0.61 A in the 6-Ω resistor near the top **41.** $I_\epsilon = I_r = 11.83$ A, $I_2 = 9.03$ A, $I_4 = 2.78$ A, $I_5 = 1.39$ A **43.** $I_3 = 0.266$ A to the left
45. $I_6 = 0.271$ A, up; $I_{4,5} = 0.036$ A, toward right; $I_3 = 0.235$ A, toward the right **47.** 3×10^6 Ω, 50 μA **49.** Add a 0.1-MΩ in series assuming the resistance of the movement is negligible **51.** (a) Essentially 5 MΩ in series $(5 \times 10^6 - 10)$,
(b) 1.3×10^{-5} Ω in parallel **53.** 30 Ω **55.** 1 V if the output circuit remains open **57.** 5.77 Ω

Chapter 15

1. 2.4 N, toward the east **3.** 4.8×10^{-16} N **5.** (b) $F = qvB = B\sqrt{\dfrac{2q^3V}{m}}$, (c) $\dfrac{1}{B}\sqrt{\dfrac{2mV}{q}}$, (d) R, V, B **7.** 5.62 m
9. 5.6×10^7 m/s, 9×10^3 eV **11.** $R_{He} = 2R_H$ **13.** 0.018 m **15.** 7.5×10^{-6} T downward **17.** Attraction
19. $M = evr/2$, $M = eL/2m$ **21.** 1.33×10^{-5} T circular around current **23.** 3.28×10^{-6} T about 1/20 of earth's field
25. 2×10^{-6} T **27.** (a) 0, (b) 1.0×10^{-5} T toward observer if right-hand wire has an upwardly directed current **29.** 1.02 A
31. 1.37×10^{-3} T out of paper **33.** 2.92 N upward at 30° from plane at right angles to wire **35.** 0.75 N outward
37. 4.9 A **39.** 1.92×10^{-4} N/m attraction **41.** $\theta = \tan^{-1} DIB/Mg$ **43.** 1.6×10^{-2} N·m **45.** (a) 1.96×10^{-2} N·m,
(b) 2.26×10^{-2} N·m plane of loop is parallel to B **47.** 0.6 T

Chapter 16

1. (a) 0, (b) 1.44×10^{-3} Wb, (c) 1.02×10^{-3} Wb **3.** 1.2×10^{-5} A counterclockwise, if you look along the direction of B
5. 0.05 V **7.** (a) Looking in the direction of B the induced current is clockwise (b) 9×10^{-3} V, (c) 9×10^{-2} A
9. 2.25×10^{-3} V **11.** (a) 0.75 V, (b) west, (c) 0 **13.** 1.5 m/s **15.** (a) Into the pages, (b) clockwise **17.** 4.17 T
19. 44, 0.44 A **21.** 433 rad/s, 69 Hz **23.** 0.18 V **25.** 1 A, 0.27 A **27.** 216 V, 550 A

Chapter 17

1. 2×10^{-3} C   **3.** (a) 89 pF, (b) 8.9×10^{-9} C   **5.** (a) 40 V, (b) 60 μF   **7.** (a) 2.4×10^{-3} C, (b) 2.4×10^{-3} C, (c) 24 μF, (d) $\frac{1}{2}$
9. $C = 1.6\,\mu$F, 38.4 μC, 19.2 V   **11.** 290 Ω **13.** 833 Ω, 200 V   **15.** 6.67×10^{-5} s, 0.75 A   **17.** 1.89×10^3 Ω **19.** 8-Ω resistance, 24-mH inductance   **21.** 2.5 μF   **23.** 10.1 to 1.13 pF   **25.** 0.25 A   **27.** 2.18 A   **29.** (a) 54 W, (b) 0.1 H
31. (a) 75.4 Ω, (b) 885 Ω, (c) 816 Ω, (d) 0.135 A, (e) 0.123, (f) $-83°$ **33.** 3.7 mA   **35.** (FM):3.4 m $-$ 2.8 m, (AM):545 m $-$ 188 m   **37.** 5.45×10^2 nm, 0.545 μm   **39.** 3×10^{18} Hz   **41.** Microwaves

Chapter 18

1. (a) 3×10^8 m, (b) 7.5   **3.** 2.95×10^8 m/s   **5.** 2.21×10^8 m/s   **7.** 2.12×10^8 m/s   **9.** 8×10^{-7} m   **11.** 10^{-3} m
13. (a) 0.63°, (b) 11.0 mm   **15.** 6×10^{-7} m   **17.** First   **19.** (a) 17.5° to 31.7° or 14.2° (first), 36.9° to 90° or 53.1° (second), but only part of second order visible, (b) no overlap
21. (a) 3 mm, (b) 2.26 mm   **23.** $w \approx d$ **25.** 1.04×10^{-7} m   **27.** $(m + \frac{1}{2})(1.55 \times 10^{-7})$ m
29. 143 nm with air on both sides of the film   **31.** 3.41×10^{-3} mm

Chapter 19

1. 19.5°   **3.** 62.1°   **5.** 63.5°   **7.** (a) 2.31×10^8 m/s, (b) 50.3°, (c) 0.58 in.   **9.** 3.33 ft   **11.** 51.2°
13. (a) $\phi = 48.9°, \theta_r = 41.1°$, (b) $\phi = 29.0°, \theta_r = 61.0°$ **15.** (a) Refracted at 53°, reflected at 45°, (b) 62.5°, (c) 2.0×10^8 m/s, (d) 7.5 cm   **17.** Virtual image, $q = -60$ cm, erect image, $M = 1$ **19.** (a) $q = -0.22$ cm, $y' = 0.11$ cm, (b) $q = -60$ cm, $y' = 25$ mm, (c) $q = 65$ cm, $y' = -25$ mm   **21.** 0.3125 m or 0.1125 m
23. 11.1 cm, $M = -0.38$, 28.9 cm, $M = -2.62$ **25.** -240 cm   **27.** $p = 17.1$ mm, $M = -14.6$ **29.** 1.47 to 1.67 cm
31. (a) 10.3 cm, (b) inverted, (c) -39 **33.** 43.5 m   **35.** -3 cm, virtual, erect, $M = 0.75$
37. 25 cm in front of first lens, virtual   **39.** 20   **41.** (a) virtual, inverted, 276 cm from eyepiece, (b) $M = -48$, (c) 0.52
43. (a) 27.4 cm, (b) 0.056, (c) 12   **45.** (a) 24 cm, image size $= 0.6$ cm, (b) 31 cm to the left of diverging lens
47. $-60, +100$ cm, $+27.3$ cm   **49.** (a) 30 cm in front of mirror, real, inverted, (b) 20 cm, (c) 40 cm
51. 3.5 cm to left of lens   **53.** $q = -2.5$ cm, virtual, erect   **55.** 75 cm   **57.** 62.5

Chapter 20

1. 3.07×10^3 m/s   **3.** (a) 200 ft tall $\times$ 268 ft long, (b) 0.67 s, (c) 0.50 by you, 0.67 by woman on the street
5. Reduced by 0.02 cm^3 **7.** 0.998 c **9.** 528.9 m/s   **11.** 0   **13.** (a) 0.97 c, (b) 4×10^{-4} s   **15.** 0.996 c
17. Unchanged, increased by 23.9 kg   **19.** 9.39×10^8 eV   **21.** (a) 0.18 MeV, (b) 1.23×10^{-30} kg, (c) 0.67 c
23. 3.16×10^{-5} g/s

Chapter 21

1. 2.48 eV   **3.** 2.42×10^{13} Hz, 1.24×10^{-5} m, infrared; 2.42×10^{14} Hz, 1.24×10^{-6} m infrared; 2.42×10^{15} Hz, 1.24×10^{-7} m, ultraviolet; 2.42×10^{16} Hz, 1.24×10^{-8} m, ultraviolet; 2.42×10^{17} Hz, 1.24×10^{-9} m, x-rays
5. 1.24×10^{-12} m or 1.24×10^{-2} Å   **7.** 1.24×10^5 V   **9.** 3.2×10^{15} **11.** 1.81 eV   **13.** 4.31 eV   **15.** (a) 2.01 eV, (b) 0
17. 0.96 eV   **19.** 4.63×10^{14} Hz, 0.648×10^{-6} m, 6.63×10^{-34} J $\cdot$ s, 1.92 eV   **21.** 9.45×10^{-15} m   **23.** 1.02×10^{-14} m
25. 8.5×10^{-8} cm   **27.** 2.47×10^{15} Hz   **29.** -0.83 eV; yes   **31.** 12.09 eV, 1.51 eV   **33.** 1.88×10^{-6} m infrared
35. 219×10^3 m/s, 6.6×10^{12} Hz, 5.7×10^{12} Hz

Chapter 22

1. 1.01×10^{20} Hz   **3.** (a) 0.1 Å, (b) crystal diffraction, (c) 36.3-keV Compton electrons   **5.** 2.71×10^{-2} Å, (b) γ-rays
7. 3.96×10^3 m/s, 8.18×10^{-2} eV   **9.** 1.7×10^{-34} m   **11.** 8.68×10^{-2} Å   **13.** 4.42×10^{-33} m   **15.** 1.23 Å   **17.** 16 cm/s
19. 10 m, no   **21.** 5.8×10^{-6} m   **23.** 2.1×10^5 eV
25. 4, 3, 2, 1, 0, -1, -2, -3, -4, component of angular momentum in assigned direction   **27.** (a) 1.1×10^{-60} J, (b) 2×10^{-2} m, (c) 6.02×10^{-18} J, 2Å   **29.** (a) 3×10^{26}, (b) n is so large

Chapter 23

1. 3.0×10^{23} **3.** (a) 118, (b) 79, (c) 197 **5.** (a) $^{17}_{8}\text{O}$, (b) ^{1}n, (c) $^{124}_{49}\text{In}$, (d) $^{64}_{28}\text{Ni} + \nu$ **7.** $e^- + \bar{\nu}$ **9.** $^{11}_{5}\text{B}$
11. 9.106×10^{-3} u, 8.48 MeV, 2.83 MeV/nucleon **13.** 7.97 MeV/nucleon for ^{16}O, 7.75 MeV/nucleon for ^{17}O **15.** 0.067
17. 0.43 h **19.** 6.25 g **21.** 24 days **23.** 61.3 counts/min **25.** 2240 yr **27.** 4.79 MeV **29.** 7.16 MeV
31. 31.973 90 u **33.** (a) 2.47 MeV, (b) 6.0×10^{20} Hz **35.** (a) 0.26 MeV between the neutron and proton, (b) direction of breakup (almost equal split), conservation of momentum **37.** $^{90}_{38}\text{Sr}$, 85 MeV **39.** 3.7×10^{18} fissions/s

Chapter 24

1. 0.25 Å **3.** 0.41 Å **5.** 5.55 Å, will not occur **7.** 1.3×10^{19} Hz **9.** 0.161 cm, 1.45×10^{-3} cm **11.** 6.5 cm
13. (a) $(\frac{1}{8})I_0$, (b) 5.0×10^{-2} cm^2/g, (c) $I = I_0 e^{-58x}$ (x in meter) **15.** $1.2 \times 10^{-3}\,°\text{C}$ **17.** 0.18×10^{-6} g **19.** 9.67 cm

Index

Aberrations, lenses, 615–616
Absolute temperature scale (Kelvin), 266, 275, 298
Absolute zero, 267, 275, 298, 323, 428, 720, 816
Absorption, of radiation, 780
Acceleration, 29–38
  angular, 188
  apparent weight, 104, 111
  centripetal, 102, 188, 340
  circular motion, 99
  due to gravity, 32–34
  instantaneous, 29
  simple harmonic motion, 337
  uniform, 35
Accelerator:
  Cockcroft-Walton, 404
  electromagnetic, 467
  electrostatic, 404
Addition, vector, 17
Adiabatic processes, 317, 344
Agonic line, 455
Air track, 159
Algebra review, 804
Alpha (α) particle, 746
Alternating current (ac), 500–503, 517–536
Ammeter, 440–442, 476
Ampère, André Marie, 418
Ampere (A), unit of current, 390, 418, 473
Amplifier, 538
Amplitude, wave, 336, 337, 363
Amplitude modulation, 363
Angle:
  critical, 592
  of deviation, 595
  of incidence, 557, 590
  of minimum deviation, 596
  of reflection, 557, 590
  of refraction, 557, 590
  trigonometric functions, 818
Ångström, Anders Jonas, 681
Ångström, hydrogen spectrum, 681
Ångström (Å) unit, 560, 678
Angular acceleration, 188
Angular momentum (*See* Momentum, angular)
Angular motion, 187
  equations for, 191
Angular variables, 187
Angular velocity, 188

Annihilation, 657
  antiparticle, 657
Answers to problems, 823–827
Antimatter, 657
Antineutrino, 747
Antinode, 352, 359
Aorta, 262
Aphelion, 92
Apollo space flight, 32, 53, 88, 545
Apparent weight, 104, 111
Aquaplaning, 79
Archimedes' principle, 243–249
Area, 9, 814
  Kepler's law of, 93
Arecibo radio telescope, 550
Aristotle, 8, 210
Arithmetic review, 796–813
Artery, 262
Astigmatism, 627
Aston, Francis W., 736
  mass spectrometer, 462, 736
Astronomical telescope, 613
Atmosphere, 237–243
  pressure, 239
Atom, 210, 384
  model of, 683–693, 726–728
Atomic clock, 14, 648
Atomic mass unit (u), 300, 738, 742
Atomic number, 737, 820–822
Atomic weight, 300, 820–822
Atwood machine, 72
Aurora borealis, 466
Avogadro, Amadeo, 300
  hypothesis, 300
Avogadro's number, 300, 307, 816

Back electromotive force, 507, 517
Ballistic pendulum, 164–165
Balmer, Johann Jakob, 681
Balmer series, 681–683, 688
Bands, electronic, 792
Bar, 240
Bardeen, Cooper, Schrieffer (BCS) theory of superconductivity, 428
Bardeen, John, 428, 536
Barometer, 238–240
Battery, electric, 419–421
Beats, 362
Becquerel, Antoine Henri, 677, 744
Bel, 364
Bernoulli, Daniel, 252

Bernoulli's principle, 252–256
Beta (β) decay, 747–749
Beta (β) particles, 746
Bimetallic strip, 272
Binoculars, prism, 593
Binomial expansion, 633
Biot and Savart, law of, 470
Black hole, 122, 662
Blackbody radiation, 666
Bohr, Niels, 683, 687, 732
  atomic theory, 384
  hydroden atom, 683–693
Boltzmann, Ludwig, 308, 667
  Stefan-Boltzmann law, 667
Boltzmann's constant, 308, 816
Bone, as building material, 224, 232
Boyle, Robert, 296, 301
Boyle's law, 296–297
Bragg diffraction, 776
Bragg, Lawrence, 701, 776
Bragg, William, L., 701, 776
Brahe, Tycho, 91
Brans, Carl, 662
Bremsstrahlung, 771
British engineering system of units, 65
British system of weights and measures, 9, 11, 814
British thermal unit (BTU), 280
Broglie, Louis de, 705
  atom model, 713
  dual nature of matter, 705–711
Brownian motion, 211
Bulk modulus, 223, 343
Bouyant force, 247–249

Calorie (cal), 277
Capacitance, 519–525
Capacitive reactance, 521
Capacitors, 519–525
  in parallel, 523
  parallel-plate, 406, 521
  in series, 522
Capillary action, 228, 262
Carbon 14 dating, 754
Carnot cycle, 320–324
Carnot, Sadi, 320
Cartesian coordinates, 20
Cathode ray tube (CRT), 512–513
Cathode rays, 677, 769
Cavendish, Henry, 97

Index

Cells, electric, 419–421
  in parallel, 431
  in series, 431
Celsius, Anders, 268
Celsius temperature scale, 266–270
Center of gravity, 175–177
  locating, 181–187
Center of mass, 175
Centigrade (Celsius) temperature scale, 266–270
Centimeter (cm), 9
Centrifugal force, 114
Centripetal acceleration, 102, 188, 340
Centripetal force, 103
  electric charges, 461
  satellites, 103
Chadwick, James, 737
Chain reaction, 757
Charge, electric:
  carrier, 385, 419–421
  conservation of, 436
  on the electron, 427
  fundamental, 386
  by induction, 388
  moving, magnetic force on, 455–463
Charles, Jacques, 298
Charles' law, 298
Choke, 518
Chromatic aberration, 615–616
Circular motion, 101
Circulatory system, 262–263
Cockcroft, John, 404
Cockcroft-Walton accelerator, 404
Coefficients:
  of expansion, 270–272
  of kinetic friction, 78
  of resistance, 422
  of static friction, 78
  of thermal expansion (table), 272
Coherence, 562, 581
Collisions, 161–166, 305
Commutator, 477
Component, 19
Compound microscope, 611
Compressibility, 223
Compton, Arthur Holly, 702
Compton effect, 701–705
Compton scattering, 780
Concorde, 346
Conduction:
  of electricity, 385, 418–421
  of heat, 286–290
Conductivity, thermal, 288–289
  table of, 288
Conductor, electrical, 383–386
Conservation:
  of angular momentum, 195
  of charge, 436
  of energy, 124, 133–143, 161–173, 317, 436
  of momentum, 155, 158–167
Convection, 286, 290
Conversions, unit, 814–815
Copernicus, Nicholas, 8, 91
Cosine, 19

Cosmology, 121–122
Coulomb, Charles Augustin, 389
Coulomb force, 389
Coulomb's law of electrostatics, 389–392, 539
Critical angle, 592
Cross product, 182, 458
Curie, Marie Sklodovska, 677, 745
Curie temperature, 453
Current, 418
  alternating (ac), 500–503, 517–536
  direct (dc), 418–427, 517
  electric, 418–427
  induced electric, 488–508
  root-mean-square (rms), 526
Current loop, 473–475
Cyclotron, 467

Dating, carbon 14, 754
Davisson, Clinton J., 705
  Davisson-Germer experiment, 708–711
Decibel (dB), 364
Declination, magnetic, 455
Density, 213–216
  of nuclear matter, 735
  optical, 566
Deuterium, 738, 742
Deuteron, 738, 742
Dewpoint, 314
Dicke, Robert, 662
Dielectric, 519, 524
Diffraction, 554
  with lenses, 612
  of particles, 709–711
  single-slit, 572–574
Diffraction grating, 575
Diffusion, 211
Diode, 533
Direct current (dc), 418–427, 517
Dispersion, 558, 595
Displacement:
  angular, 187
  linear, 15, 17
  wave motion, 333, 336, 339
Distance, 9, 54
Domains, magnetic, 454
Doppler effect, 367–374, 649
Dose, gamma (γ) rays, 787
Dosimetry, 785
Dot (scalar) product, 129
Drift velocity, 421
Dry cell, 419–421
Dyne (dyn), 63

Ear, 379–380
Earth:
  atmosphere of, 244
  gravitational field of, 114
  magnetic field of, 454–455, 467, 486
  tides, 108–111
Eddy currents, 498, 506–507
Efficiency of engines, 320

Einstein, Albert, 2, 6, 541, 629, 635, 638–639, 732
  gravitation, 90, 121
  photoelectric effect, 670
  photograph of, 5, 638
  special theory of relativity, 635–659
Elastic limit, 219
Elastic moduli (table), 221
Electric current, 418–427
Electric field, 393–398, 407
  changing, 519
  lines, 393–394, 407–409
Electric oscillation, 530–532
Electric potential, 398–410
Electric shock, 440
Electrical circuit symbols, 431
Electrodes, 420
Electromagnetic induction, 488–508
Electromagnetic waves, 332, 538–546
Electromagnets, 474
Electromotive force (emf), 424, 488
  induced, 489–508
Electron:
  charge on, 390, 415, 816
  intrinsic spin, 475, 726
  mass of, 816
  wave nature, 706
Electron microscope, 54, 721
Electronvolt (eV), 399
Electroscope, 387–389
Electrostatic force, 389
Electrostatics, 381–416
  accelerator, 404
  copying by, 387
  potential, 398–410
Elementary particles, 765
Elements, table of, 820–822
Ellipse, 92
Energy, 124–144
  conservation of, 124, 133–143, 161–173, 317, 436
  electrical, 398, 427
  fission, 756
  of flywheel, 198
  fusion, 758
  geothermal, 151
  gravitational potential, 139
  internal, 317
  kinetic, 128–130, 191
    angular, 191
    linear, 128
    relativistic, 636–640
  levels in atom, 686
  mechanical, 124–150, 319
  nuclear binding, 740–744
  potential, 130–143, 400
  production, 149–152
  quantum of radiation, 668
  relativistic, 655–658
  satellite, 141
  solar, 151
  thermal, 265, 285–286, 317–319, 665
  thermodynamics, first law of, 289, 317
  units of, 126–127, 815
  zero-point, 720

Entropy, 319, 322, 454
Equilibrium:
 rotational, 179
 translational, 74
Equipotential, 407–410
Erg, 126
Ether, 540
Evaporation, 282, 312
Excitation potential, 690
Exclusion principle, Pauli's, 727–728, 747
Expansion, thermal, 270–277
 coefficients of (table), 272
Exponential decay, 518, 520, 751, 781
Eye, 626

Fahrenheit, G. D., 268
Fahrenheit, temperature scale, 266
Farad (F), 520
Faraday, Michael, 490, 474, 503
 electric field, 393
 law, 490, 492, 539
Farsightedness, 627
Ferromagnetism, 453, 454
Field:
 electric, 393–398, 407
  changing, 519
  lines, 393–394, 407–409
 gravitational, 114
 magnetic, 451–476
 See also Magnetic field
Field theory, 6
Filter, frequency, 518, 535
Fission, 756
Fizeau, Armand Hippolyte Louis, 541, 587
Fluids, 235–258
 motion of, 252–258
 static, 249–251
Fluorescence, 691
Flux, magnetic, 491, 503
Flywheel, energy of, 198
Foci, 598
Foot, 12
Foot-pound (ft · lb), 127
Force, 56–82
 buoyant, 247–249
 centrifugal, 114
 centripetal, 103
  electric charges, 461
  satellites, 103
 Coulomb, 389
 electromotive, 424, 488
  induced, 489–508
 electrostatic, 389
 frictional, 77–82
 gravitational, 94
 magnetic, 457–466
 See also Magnetic force
 Newton's laws, 57–77
 nuclear, 739–744
 See also Nuclear force
 units of, 63, 814
 weak, 6
Foucault, Jean, 207, 588
Fourier analysis, 366

Frames of reference, 24, 630
 inertial, 630, 636
Franck, James, 689, 732
 Franck-Hertz experiment, 689
Franklin, Benjamin, 382
 electric charge, 387
Fraunhofer, Joseph, 573, 691
 dark line spectra, 691
 diffraction, 573
Free fall, 32–38
Frequency, 189, 335
 angular, 189
 in Doppler effect, 367
 of filter, 518, 535
 fundamental, 354
 harmonic, 355
 and pitch, 364
Friction, 77–82
 coefficients of (table), 78
 in fluids (viscosity), 256–259
 static, 77
Frictional force, 77–82
Fusion, nuclear, 758–761
 latent heat of, 281

G (constant of universal gravitation), 95, 98, 816
g (free fall acceleration), 33, 98, 816
Gabor, Dennis, 581, 583
Galileo, 2, 6, 8, 10, 210
 free fall, 33
 picture of, 2, 10
 speed of light, 587
Galvanometer, 440–445, 476
 d'Arsonval, 476
Gamma (γ) rays, 746
 electromagnetic spectrum, 543
Gas constant, 275, 302
Gas laws, ideal, 301–309, 344
Gases, 211, 235, 296–316
Gauge pressure, 241, 297
Gauss, Johann Karl Friedrich, 459
Gauss (G), unit of magnetic induction, 459
Geiger, Hans, 679
Geiger-Müller counter, 777–778
Gell Mann, Murray, 5
Generator:
 ac, 497–502
 dc, 502
 electrostatic, 404
Geothermal energy, 151
Germer, L. H., 708
Gilbert, William, 382
Goddard, Robert H., 162
Goudsmit, Samuel, 725, 732
 electron spin, 725
Graphs, 812
Gravitation:
 field, 114
 law of universal, 94–99
 satellites, 107
 and solar system, 102–107
 tides, 108
 waves, 115, 662

Gravitational force, 94
Gravitational potential energy, 139
Gravity, 34, 62, 64, 98, 102–116
 center of, 175–177
 and relativity, 662
 specific (*See* Specific gravity)
Gyroscope, 197–200, 207–208

Hafele, Joseph, 648
Half-life, 750–753
Half-value layer, 782
Harmonics, 354–362
Heat:
 as form of energy, 266, 285, 317
 latent, 281–283
 mechanical equivalent of, 283–285, 317
 specific, 279–281
  table of, 279
 transfer of, 286–291
Heat death, 322
Heisenberg uncertainty principle, 716–721
Heisenberg, Werner, 717, 732, 737
Hemosonde, 370
Henry, Joseph, 518
Henry (H), unit of induction, 518
Hertz, Gustav, 689
 Franck–Hertz experiment, 689
Hertz, Heinrich, 338, 544–546, 670
 photoelectric effect, 670
Hertz (Hz), unit of frequency, 189, 335, 544
Holes, 534
Holography, 580–583
Hooke, Robert, 219, 296
Hooke's law, 219, 336–346
Horsepower (hp), 143, 283
Humidity, 314–316
Huygens, Cristiaan, 553
Huygens' principle, 553–556
Hydraulic press, 250
Hydroelectric power, 152
Hydrogen atom, model of, 683–686
Hydrogen, energy levels of, 689
Hydrogen spectrum, 681
Hydrostatics, 243, 246–251
Hygrometer, 315

Ideal gas laws, 301–309, 344
Image formation, 599–607
 analytical, 602–607
 graphical, 599–602, 618–621
Impedance, 528
Impulse, 155, 171
Inclination, magnetic, 455
Incompressible, 235
Induced electric current, 488–508
Induced electromotive force, 489–508
Inductance, 517–519
Induction:
 charging by, 388
 electromagnetic, 488–508
 magnetic, 458–459, 469–471
 in transformer, 502–508
Inductive reactance, 518

Inertia, 57, 61
  rotational, 191
Inertial mass, 60–61, 159
Inertial reference frames, 630, 636
Infrared radiation, 543, 666
Insulator:
  electrical, 383–387, 521
  thermal, 288
Integration, 156
Intensity:
  electromagnetic waves, 714
  matter waves, 715
  sound waves, 363
Interference:
  air wedge, 569, 571
  beats, 362
  of light, 561–577
  matter waves, 712
  mechanical waves, 351–363
  Michelson interferometer, 569–570, 632, 634
  of reflected light, 566–572
  sound waves, 353–363
Internal energy, 317
Ionization chamber, 775
Ionization potential, 686
Ionosphere, 244
Ions, 419
Isotherm, 310–311
Isothermal processes, 310–311, 320
Isotopes, 736, 788
  atomic mass (tables of), 741, 820–822

Jacks, automobile, 136
Joule, James, 284
  mechanical equivalent of heat, 284
Joule (J), unit of energy, 127, 284, 427
Joule's law, 428
Junction, p-n, 534
Junction transistor, p-n-p, 536

Keating, Richard, 648
Kelvin, Lord, 269, 322
Kelvin scale, 267, 270, 298, 308, 323
  absolute zero, 267, 275, 298, 323, 428, 720, 816
Kepler, Johannes, 91
Kepler's three laws, 92–94, 105
Kilocalorie (kcal), 277
Kilogram (kg), 14, 63
Kilowatt (kW), 143
Kinematics, 7–47
Kinetic energy (See Energy, kinetic)
Kinetic theory, 211, 285–286, 305–309
Kirchhoff, Gustav, 435
Kirchhoff's laws, 435–439

Laser, 697
Latent heat, 281–283
Land, Edwin H., 578
Length, 9, 814
  relativistic contraction, 644–646
Lens:
  aberrations, 615–616
  power, 608

Lens equation, 603
Lens-makers' equation, 607–609
Lenses, 596–616
  aberrations of, 615–616
  combinations, 609–615
  concave, 598
  converging, 596
  convex, 596
  diverging, 598
  focal length of, 598–602
  magnifying power, 613
  resolution of, 612
Lenz, Heinrich F. E., 492
Lenz's law, 492
Lever, 136, 178–181
  arm of, 178–181
Levitation, magnetic, 498
Lift, hydraulic, 250
Light, 553–627
  corpuscular theory of, 553
  diffraction of, 554, 572–577
  dispersion of, 558
  duality of, 553–554
  electromagnetic spectrum of, 541–545
  interference of, 561–577
  pipe, 593, 597
  polarization of, 577–580
  reflection of, 553, 559, 572, 590–595, 616–621
  refraction of, 553, 555–561, 590–616
  speed of, 53, 538–544, 587–588, 636, 816
  wave theory of, 554
Light-year, 54
Lightning, 408–410
Linear expansion, 270
Lines of force, 407
Liquid crystals, 214–215
Liquids, 211–213, 235–258
  expansion of, 273–277
  in motion, 252–258
  viscosity of, 256–258
Lissajous figures, 513–515
Lodestone, 452
Longitudinal wave, 337
Lorentz, Hendrik, 631
  contraction, 645
Lorentz transformation, 643
Loudness, 363–365
Lyman series, 688

Machines, simple, 136–137
Mach number, 651
Magdeburg hemispheres, 237
Magnetic field, 451–476
  from bar magnet, 451
  of current-carrying wire, 468–471
  near current loop, 471–472
  of earth, 454–455, 467, 486
Magnetic force:
  on charges, 455–463
  on current-carrying wires, 463, 471–480
Magnets:
  permanent, 452–455

Magnets, (cont'd.):
  superconducting, 498
Magnification:
  angular, 613
  lens, 601, 603
  mirrors, 600
Manometers, 241
Marconi, Guglielmo, 546
Marsden, Ernest, 679
Mass, 61–64
  atomic unit, 738–814
  center of, 175
  of earth and planets, 118, 816
  of electron, 816
  energy, 655–658
  inertial, 60–61, 159
  relativistic, 655–658
  rest, 656
  spectrometer, 460, 736
  units of, 63
Mass defect, 741
Mathematics review, 796–813
Matter:
  conversion to energy, 655
  dual nature of, 705–712
  nuclear, 753
  states of, 211
  wave nature of, 705, 714–716
Maxwell, James Clerk, 309, 538
  equations, 538–540, 630, 671
  speed distribution, 309–312
Mechanical energy, 124–150, 319
Meter (m), unit of length, 9
Meters, 479
Metric system of measure, 9, 11, 14
MeV, 405, 724
Michelson, Albert A., 541, 569
  interferometer, 569–570, 632, 634
  search for ether, 631–635
Microscope, 601
  compound, 611
  electron, 54, 721
Microwaves, 544
Millibar (mbar), 240
Millikan, Robert A., 415
  oil-drop experiment, 415
  photoelectric effect, 673
Mirrors, 616–621
  focal length of , 618–621
  plane, 616–617
  spherical, 618–621
Modulus, elastic, 219–226
  table of, 221
  wave speed, 343
Mole, 300
Molecules, 211, 300
Moment of a force (torque), 179
Moment of inertia, 191–192
Moment, magnetic, 478
Momentum, 154–167
  angular, 182, 195–197, 712, 724–726
    conservation of, 195
  conservation of, 155, 158–167
  impulse-momentum theorem, 155

Momentum, (*cont'd.*):
   intrinsic angular, 725
   of photon, 703
   orbital angular, 724
   relativistic, 655
Monopole, magnetic, 456
Morley, E. W., 541, 570, 631
Moseley, Henry Gwyn-Jeffreys, 773
Motion, 7–47
   angular, 187–200
      equations for, 191
   Brownian, 211
   circular, 101
   planetary, 102–107
   projectile, 38–47
   relative, 24, 630
   relativistic, 635–655
   simple harmonic, 336–341
   uniformly accelerated, 35
Motor, dc, 477–480
Muon, 650
Musical sounds, 363–367
   intensity, 363
   loudness, 363
   pitch, 365
   quality, 366

Nearsightedness, 627
Nervous system, 450
Net force, 60
Neuron, 450
Neutrino, 747–749
Neutron, 737, 747–749
   decay of, 748–749
Newton, Sir Isaac, 2, 57, 90, 96, 154
   biographic sketch, 96
   law of gravitation, 94–99
   optics, 553
   picture of, 5
   unit of force (N), 63
Newton's laws, 57–77
   first, 57, 630
   second, 60–82
   third, 66–69
Newton's rings, 569
Node, 352, 359
Nonlinear devices, 533
Nonreflective coatings, 568
Normal force, 78
Northern lights, 466
Nuclear binding energy, 740–744
Nuclear force, 739–744
   decay, 743–753
   fission, 150, 756
   reactions, 753–761
Nuclear fusion, 758–761
   latent heat of, 281
Nuclear power, 150–152
Nuclear power reactor, 757–759
Nucleon, 737
Nucleus, 736–767
   compound, 754
   density of, 735
   diameter, 680, 735
   discovery of, 676

Nucleus, (*cont'd.*):
   stability, 743
   transmutation, 744

Octave, 366
Oersted, Hans Christian, 468, 476
Ohm, Georg Simon, 422
   unit of resistance (Ω), 422
Ohm's law, 422, 528
Oil-drop experiment (Millikan), 415
Optical activity, 579
Optical density, 566
Optical flat, 569
Optics, fiber, 597
Oscillation, 340
   electric, 530–532
   natural frequency, 356, 531
Oscilloscope, 512
Overtones, 354–362, 366

Pair production, 657, 780
Paramagnetism, 453
Pascal, Blaise, 249
   unit of pressure (Pa), 240
Pascal's principle, 249
Pauli, Wolfgang, 6, 727, 732
Pauli's exclusion principle, 727–728, 747
Pendulum, 334, 338
   period, 338
Perihelion, 92
Period, 93, 101, 189, 335
   of pendulum, 338
   simple harmonic motion, 340
Periodic table, 726–728, 820–822
Periodic waves, 346–349
Permittivity, 390
Phase shift, 527–530
Phase velocity, 349
Phase, wave, 348–350
Photoelectric effect, 669–676, 779
Photon, 668, 670
   in Compton effect, 702
Pion, 649
Pirellitires, 79
Pitch, 363, 365, 367
Planck, Max, 669
Planck's constant, 668, 669, 683, 718
Plane mirrors, 616–617
Planetary data, 118, 816
Planetary motion, 102–107
p-n junction, 534
p-n-p junction transistor, 536
Poiseulle, Jean Léonard M., 258
Poiseulle's law, 258
Polarization of light, 577–580
   electrical, 385
Poles, magnetic, 453, 456, 486
Positron, 657, 780
Potential difference (electrical), 398–410, 419
Potential energy:
   electrical, 398–410
   gravitational, 130
Potential excitation, 690
Potential gradient, 405–406

Potential ionization, 686
Potentiometer, 444
Power, 143
   electrical, 428, 525–529
   hydroelectric, 152
   of lens, 608
   nuclear, 150–152
   radiation, 666
   sound waves, 363
Power factor, 529
Powers of 10 notation, 4
Precession, 199
Pressure, 216–218, 235–252
   atmospheric, 237–243, 244–246
   diastolic, 243
   gauge, 241, 297
   hydrostatic, 243–252
   partial, 313–316, 328
   saturated vapor, 310, 312–316
   systolic, 243
   vapor, 310, 312–314
Prism binoculars, 593
Probability and matter waves, 716–728
Projectile motion, 38–47
Ptolemy, 8, 90
   theory of the solar system, 98
Pulley, 137
Pulsars, 551
Pythagoras, 6

Quadratic equations, 808
Quantum:
   of angular momentum, 684, 712, 724
   of electrical charge, 392, 416
   of electromagnetic energy, 668, 670
Quantum mechanics, 717
Quantum numbers, 688, 721–728
Quantum theory of radiation, 665
Quark, 389, 761, 765

Rad, 786
Radar waves, 543
Radian, 187
Radiation, 769–792
   absorption of, 780
   detectors, 774–776
   dose, 785–789
   heat, 286, 666
   infrared, 543, 666
   in medicine, 788
   quantum aspects, 665
   ultraviolet, 543
Radio telescope, 550
Radio waves, 545
Radioactivity, 678, 743–753
   dating, 754
   decay, 744
   half-life, 750–753
Radioisotopes, 750, 788
Rainbow, 600
Range:
   of α particles, 784
   of β particles, 785
   in projectile motion, 39, 49
Rarefraction, 336

Ray, light, 558, 590
Reactance:
  capacitive, 521
  inductive, 518
Real image, 601
Rectifier, 533, 535
Red shift, 373, 649
Reference circle, 339
Reflection:
  diffuse, 591
  of light, 553, 559, 590–595, 616–621
  of mechanical waves, 349–352
  plane mirrors, 616–617
  specular, 591
  spherical mirrors, 618–621
  total internal, 592–595
Refraction, 553, 555–561, 590–616
  index (table), 558
  lenses, 596–616
Refrigerator, 311
Relative humidity, 314–316
Relativity, 629–662
  Galilean, 629, 637–643
  general theory, 659, 662
  and length contraction, 644
  and mass increase, 655
  space and time, 643
  special theory of, 635–659
  and time dilation, 642
  velocity transformations, 652–655
rem, 786
Resistance:
  electrical, 421–439
  internal, 424, 430–431
  measurement of, 441–444
  in parallel, 433
  in series, 432
  temperature coefficient of, 422, 426
Resistivity, 422
  table, 422
Resolution of lenses, 612
Resonance, 352–362
  cavity for laser, 697
  electrical, 528–532
  mass, 656
  , 626
  hand rule, 459, 469
  and screw rule, 182, 458–459
  Olaf, 587
  Wilhelm Konrad, 744, 769
  , 786
  769, 775
  square (rms) current, 526
  quare speed, 306, 312
  uare voltage, 526
  mics, 190–200
  179
  212, 265, 283
  valent of heat, 284
  678
  , 678–680

Satellite energy, 141
Satellites, 107
Scalars, 15, 22, 235
Schrödinger, Erwin, 712, 722, 732
Schrödinger's wave equation, 712–716
Schwinger, Julian, 6
Scintillation counter, 779
Scott, David, 32–33
Seasons, causes of, 93
Second (s), unit of time, 13
Semiconductor, 385, 418, 422, 534
Semiconductor diode, 534
Shear modulus, 221, 343
Shock, electric, 440
Shock wave, 333, 374
Shunt resistor, 441
Significant figures, 13, 803
Simultaneous equations, 439
Sine, definition, 19
Slip rings, 499
Slope, 30
Slug, unit of mass, 63
Snell, Willebrod, 558
Snell's law, 558, 592
Solar energy, 151
Solar system, data, 118, 816
Solenoid, 474
Solids, 211
Sommerfeld, Arnold, 732
Sonic boom, 346, 373–374
Sound, quality of, 363, 366
Sound waves, 333, 343, 352–374
  Doppler effect, 367
  intensity, 363
  musical, 363
  See also Musical sounds
  speed, 333, 343–346
Space time, 637
Specific gravity, 213–216, 245–246
  table, 219
Specific heat (*See* Heat, specific)
Spectra:
  absorption, 691
  continuous, 666
  discrete, 681, 686
  emission, 688
  of hydrogen, 681
Spectroscope:
  diffraction grating, 574–577
  prism, 595
Spectrum, electromagnetic, 541–545
Speed, 22
  average, 22, 27, 29
  of electromagnetic waves, 538–544
  instantaneous, 27
  of light, 53, 538–544, 587–588, 636, 816
  of molecules, 309–312
  rms, 302, 312
  of sound, 333, 343–346
Spherical aberration, 615
Spin-intrinsic magnetic moment, 475, 725
Standard temperature and pressure, 301, 304
Standing waves, 352–361, 712–714

States of matter, 211
Static friction, 77
Stefan, Josef, 667
  Stefan-Boltzmann law, 667
Stokes, Sir George Gabriel, 258
Stokes' law, 258, 415
Strain, 219–226, 271
Stratosphere, 244
Streamline flow, 252
Stress, 219–226
  shear, 221
Sublimation, 281
Submarine, 249
Superconductivity, 428
Supergravity, 6
Superposition, principle of, 350, 561
Surface tension, 226–229
  table, 228
Synchrotron, 467
Systolic pressure, 243

Tables, reference, 816–822
Tachyons, 640
Tacoma Narrows Bridge, 331, 354
Tangent, 19
Tank circuit, 530
Temperature, definition, 285
Temperature scales, 266–270
  absolute, 267
  Celsius (centigrade), 266–270
  Fahrenheit, 266–270
  Kelvin, 266–270, 298, 308
  origin, 268
Tension, 69
Terminal velocity, 254, 415
Terminal voltage, 430
Tesla, Nikola, 459)
Tesla (T), unit of magnetic induction, 459
Texas fire frame, 668
Thermal conductivity, 288–289
Thermal energy, 265, 285–286, 317–319, 665
Thermal expansion, 270–277
Thermocouple, 278
Thermodynamics, 316–324
  first law, 289, 317
  second law, 311, 319
  third law, 323
Thermometers, 275, 278
  gas, 278
  liquid in glass, 275, 278
  resistance, 426–427
Thermostat, 272
Thin films, reflection from, 566–569
Thompson, Benjamin, 212, 265, 283
Thomson, George P., 709
  diffraction of electrons, 709
Thomson, Sir John Joseph, 415, 677, 736
  model of atom, 678
Threshold frequency photoelectric effect, 670–676
Tides, 108–111
Timbre, 366
Time, 13, 637, 647
  dilation, 647–652

Torque, 177–187, 478
Torsion balance, 97
Townes, Charles, 697
Transformer, 502–508
Transistor, 536–538
  amplifier, 538
Transmission lines, high-tension, 506
Transverse waves, 333–336, 540
  electromagnetic, 540
  light, 540
  on string, 333–336
Trigonometry:
  functions, 818–819
  review, 19
Tritium, 738, 755
Triton, 738, 755
Troposphere, 237
Tyus, Wyomia, 52

Ultrasound, 370
Ultraviolet radiation, 543
Uncertainty principle, 716–721
Uniformly accelerated motion (*See* Motion, uniformly accelerated)
Unit conversions, 814–815
Universal gas constant, 302
Uranium decay series, 752

Van Allen belts, 466
Van de Graaff, 404
Van de Graaff accelerator, 404
Vapor pressure, 310, 312–314
  of water, 312
Vaporization, latent heat of, 281
Variables, angular, 187
Vectors, 15–23
  addition and subtraction, 17
  scalar (dot) product, 129
  vector or cross product, 182–183, 458

Velocity, 23, 100
  angular, 188
  drift, 421
  relative, 23–27
  relativistic transformations, 652–655
  terminal, 254, 415
Vibration, 331
  air columns, 333, 358
  of string, 332, 334, 346, 353
Virtual image, 601
Viscosity, 252, 256–259, 415
Volt (V), 399
Volta, Alessandro, 419
Voltage, terminal, 430
Voltmeter, 440–443, 476
Volume, 9
  expansion, 272
von Guericke, Otto, 237
von Laue, Max, 677
  x-ray crystallography, 677, 701

Walton, Ernest, 404
  Cockcroft-Walton accelerator, 404
Water:
  thermal expansion, 276
  waves, 335
Watt, James, 143
  electrical power, 428
Watt (W), unit of power, 143, 365
Waves:
  electromagnetic, 332, 538–546
  matter, 714–716
  optics, 554–583
  sound, 363
Wave front, 367, 558, 590
Wave speed:
  in fluid, 343–346
  in solid, 343
  on string, 341–343

Wavelength, 335, 347
  de Broglie, 705
  of electromagnetic waves, 540
Wave motion, 331–380, 554–583
  longitudinal, 333, 357–361
  periodic, 346–349
  transverse, 333–336, 540
Weber, Joseph, 115
  gravity waves, 115
Weber (Wb), unit of magnetic flux, 491
Weight, 62–65, 104
  apparent, 104, 111
  atomic, 300, 820–822
  gram-equivalent, 300
  molecular, 300, 309
Weightlessness, 111–114
Wheatstone bridge, 443
Wheatstone, Sir Charles, 443
Wien, Wilhelm, 667
Wien displacement law, 667
Work, 125–128, 265, 317–321, 399
Work function, 671

Xerox, 387
X-ray spectrum, 770
X-rays, 538, 543, 769–775
  Compton scattering, 780
  diffraction of, 701, 776
  part of electromagnetic spectrum, 538–544
  production of, 769–774

Young, Thomas, 219, 336, 561
  double-slit interference, 561–566
  modulus, 219, 336, 343

Zero, absolute, 267, 275, 298, 323, 428, 720, 816
Zero-point energy, 720